Microlepidoptera of Europe
Volume 6

Peter Huemer & Ole Karsholt

Gelechiidae II

(Gelechiinae: Gnorimoschemini)

Microlepidoptera of Europe

Volume 6

EDITED BY
P. HUEMER, O. KARSHOLT & M. NUSS

Peter Huemer & Ole Karsholt

Gelechiidae II

(Gelechiinae: Gnorimoschemini)

Apollo Books
Stenstrup
2010

Editors: Peter Huemer, Tiroler Landesmuseen, Innsbruck, Ole Karsholt, Zoological Museum, Natural History Museum of Denmark, Copenhagen and Matthias Nuss, Senckenberg Museum of Zoology, Dresden

Technical editor: Peder Skou, Stenstrup, Denmark

Colour plates: Geert Brovad, Copenhagen, Denmark

Genitalia plates: Stefan Heim & Peter Huemer, Innsbruck, Austria

This publication should be cited as:
Huemer, P. & Karsholt, O., 2010. Gelechiidae II (Gelechiinae: Gnorimoschemini). – *In* P. Huemer, O. Karsholt & M. Nuss (eds): *Microlepidoptera of Europe* 6: 1-586. Apollo Books, Stenstrup.

ISBN 978-87-88757-87-3
ISSN 1395-9506

Authors' addresses:
Peter Huemer, Tiroler Landesmuseen Betriebsgesellschaft m.b.H., Naturwissenschaften, Feldstr. 11a, A-6020 Innsbruck, Austria
p.huemer@tiroler-landesmuseen.at

Ole Karsholt, Zoological Museum, Natural History Museum of Denmark, Universitetsparken 15, DK-2100 Copenhagen, Denmark
okarsholt@snm.ku.dk

Date of publication: 01.11.2010

Microlepidoptera of Europe

Volume 6

EDITED BY
P. HUEMER, O. KARSHOLT & M. NUSS

Peter Huemer & Ole Karsholt

Gelechiidae II

(Gelechiinae: Gnorimoschemini)

Apollo Books
Stenstrup
2010

Editors: Peter Huemer, Tiroler Landesmuseen, Innsbruck, Ole Karsholt, Zoological Museum, Natural History Museum of Denmark, Copenhagen and Matthias Nuss, Senckenberg Museum of Zoology, Dresden

Technical editor: Peder Skou, Stenstrup, Denmark

Colour plates: Geert Brovad, Copenhagen, Denmark

Genitalia plates: Stefan Heim & Peter Huemer, Innsbruck, Austria

This publication should be cited as:
Huemer, P. & Karsholt, O., 2010. Gelechiidae II (Gelechiinae: Gnorimoschemini). – *In* P. Huemer, O. Karsholt & M. Nuss (eds): *Microlepidoptera of Europe* 6: 1-586. Apollo Books, Stenstrup.

ISBN 978-87-88757-87-3
ISSN 1395-9506

Authors' addresses:
Peter Huemer, Tiroler Landesmuseen Betriebsgesellschaft m.b.H., Naturwissenschaften, Feldstr. 11a, A-6020 Innsbruck, Austria
p.huemer@tiroler-landesmuseen.at

Ole Karsholt, Zoological Museum, Natural History Museum of Denmark, Universitetsparken 15, DK-2100 Copenhagen, Denmark
okarsholt@snm.ku.dk

Date of publication: 01.11.2010

Contents

Abstract

The Gnorimoschemini (Gelechiidae: Gelechiinae) of Europe are reviewed. A total of 211 species is recognized, including two species not yet found in Europe with certainty and two others recently introduced. Descriptions and diagnoses are accompanied by colour figures of the adults, usually depicting variation, and black and white photographs of male and female genitalia characters. Notes on distribution and bionomics are added for every species.

15 new species are described: *Gnorimoschema nupponeni* sp. n. (Russia, Ukraine), *Scrobipalpa skulei* sp. n. (Slovenia), *Scrobipalpa dorsolutea* sp. n. (Russia), *Scrobipalpa corleyi* sp. n. (Portugal), *Scrobipalpa oleksiyella* sp. n. (Russia), *Scrobipalpa tokari* sp. n. (Croatia), *Scrobipalpa heimi* sp. n. (Russia), *Scrobipalpa jariorum* sp. n. (Bulgaria), *Scrobipalpa abstrusa* sp. n. (Russia), *Scrobipalpa hendrikseni* sp. n. (Greece), *Scrobipalpa deutschi* sp. n. (Russia, Algeria), *Scrobipalpa mixta* sp. n. (Macedonia, ?Albania, Ukraine), *Scrobipalpa postulatella* sp. n. (Spain), *Scrobipalpa halimioniella* sp. n. (France), *Caryocolum baischi* sp. n. (Greece).

The following 42 species group names are synonymized and two taxa are revised as subspecies: *Lita porcella* Heinemann, 1870 syn. n. of *Scrobipalpa acuminatella* (Sircom, 1850); *Scrobipalpa sibirica* Bidzilya, 2009 syn. n. of *Scrobipalpa occulta* (Povolný, 2002); *Scrobipalpa deleta* Povolný, 1981 syn. n. of *Scrobipalpa clintoni* Povolný, 1968; *Euscrobipalpa bezengensis* Povolný, 2001 syn. n. of *Scrobipalpa caucasica* (Povolný, 2001); *Scrobipalpa submagnificella* Povolný, 1977 syn. n. of *Scrobipalpa heretica* Povolný, 1973; *Scrobipalpa glaserorum* Povolný, 1977 syn. n. of *Scrobipalpa bradleyi* Povolný, 1971; *Scrobipalpa fraterna* Povolný, 1969 syn. n. of *Scrobipalpa selectella* (Caradja, 1920); *Scrobipalpa milleri* Povolný, 1977 syn. n. of *Scrobipalpa niveifacies* Povolný, 1977; *Gnorimoschema hyoscyamivora* Gerasimov, 1940 and *Gnorimoschema pseudobsoletellum* Povolný & Gregor, 1955 syn. n. of *Scrobipalpa indignella* (Staudinger, 1879); *Lita atriplicella* var. *infumatella* Fuchs, 1901 syn. n. of *Scrobipalpa atriplicella* (Fischer von Röslerstamm, 1841); *Scrobipalpa hartigi* Povolný, 1977 syn. n. of *Scrobipalpa suaedella* (Richardson, 1893); *Scrobipalpa rezniki* Piskunov, 1990 and *Scrobipalpa rezniki turkmenica* Piskunov, 1990 syn. n. of *Scrobipalpa dagmaris* Povolný, 1987; *Scrobipalpa artemisiella mongolensis* Povolný, 1969 syn. n. of *Scrobipalpa artemisiella* (Treitschke, 1833); *Scrobipalpa japonica* Povolný, 1977 syn. n. of *Scrobipalpa rebeli* (Preissecker, 1914); *Gelechia eremaula* Meyrick, 1891, *Gelechia lyciella* Walsingham, 1900, *Lita desertella* Rebel, 1901, *Phthorimaea bertramella* Lucas, 1940, *Teleia leroyella* Lucas, 1950, *Gnorimoschema reisseri* Povolný & Gregor, 1955 and *Gnorimoschema philolycii* Hering, 1957 syn. n. of *Scrobipalpa portosanctana* (Stainton, 1859); *Scrobipalpa trebujenae* Povolný, 1977 syn. n. of *Scrobipalpa salinella* (Zeller, 1847); *Lita caliacrae* Caradja, 1932 syn. n. of *Scrobipalpa salicorniae* (Hering, 1889); *Phthorimaea argentinae* (Povolný, 1989) syn. n. of *Phthorimaea operculella* (Zeller, 1873); *Opacopsis buvati* Povolný, 1992 syn. n. of *Ephysteris deserticolella* (Staudinger, 1871); *Doryphora? praticolella* Christoph, 1872 as ssp. , syn. n. and *Opacopsis gallica* Povolný, 1992 syn. n. of *Ephysteris insulella* (Heinemann, 1870); *Ephysteris treskensis* Povolný, 1964, *Ephysteris hispanica* Povolný, 1981 and *Ephysteris foulonsensis* Povolný, 1981 syn. n. of *Ephysteris diminutella* (Zeller, 1847); *Ephysteris olympica* ssp. *monticola* Povolný, 1981 syn. n. of *Ephysteris olympica* Povolný, 1968; *Phthorimaea microcasis* Meyrick, 1929, *Hedma abzacella* Dumont,

1932, *Teleia hyoscyami* Amsel, 1935, *Megalocypha polioptera* Janse, 1960 and *Ephysteris aellographa* Janse, 1960 syn. n. of *Microlechia chretieni* Turati, 1924; *Lita roseella* Zetterstedt, 1839 syn. n. of *Cosmardia moritzella* (Treitschke, 1835); *Lita crepusculella* Teich, 1889 syn. n. of *Caryocolum viscariella* (Stainton, 1855); *Caryocolum emarginatum* Huemer, 1988 syn. n. of *Caryocolum stramentella* Rebel, 1935; *Phthorimaea horoscopa* Meyrick, 1926 as ssp., syn.n. of *Caryocolum blandella* (Douglas, 1852).

The following genus group names are synonymized: *Hedma* Dumont, 1932 syn. n. and *Megalocypha* Janse, 1960 syn. n. of *Microlechia* Turati, 1924.

The following new combinations are introduced: *Scrobipalpa karadaghi* (Povolný, 2001) comb. n.; *Scrobipalpa acuta* (Povolný, 2001) comb. n.; *Scrobipalpa alterna* (Falkovitsh & Bidzilya, 2006) comb. n.; *Scrobipalpa grossoides* (Povolný, 2001) comb. n.; *Microlechia rhamnifoliae* (Amsel & Hering, 1931) comb. n.; *Microlechia karsholti* (Nupponen, 2010) comb. n.

The following taxa are recalled from synonymy: *Gnorimoschema nilsi* Huemer, 1996 sp. rev.; *Gnorimoschema hoefneri* (Rebel, 1909) sp. rev.; *Scrobipalpa divisella* (Rebel, 1936) sp. rev.; *Scrobipalpa salicorniae* (Hering, 1889) sp. rev.; *Scrobipalpa spergulariella* (Chrétien, 1910) sp. rev.; *Scrobipalpa thymelaeae* (Amsel, 1939) sp. rev.; *Scrobipalpa halymella* (Millière, 1864) sp. rev.; *Klimeschiopsis terroris* (Hartig, 1938) sp. rev.; *Microlechia* Turati, 1924 gen. rev.

The species status of the following taxa is revised: *Gnorimoschema alaskense* Povolný, 1967 stat. rev.; *Scrobipalpa stabilis* Povolný, 1977 stat. n.; *Ephysteris inustella gredosensis* (Rebel, 1935) stat. rev.; *Ephysteris insulella praticolella* (Christoph, 1872) stat. n.; *Lutilabria lutilabrella olympica* Huemer, 1993 stat. n.; *Caryocolum blandella horoscopa* (Meyrick, 1926) stat. n.

Finally the synonymy of *Gnorimoschema antiquum* Povolný, 1966 syn. rev. and the generic combinations of *Scrobipalpa lagodes* (Meyrick, 1926) comb. rev. and *Scrobipalpa ergasima* (Meyrick, 1916) comb. rev. are revised.

Lectotypes are designated for the following taxa: *Phthorimaea calaritanella* Amsel, 1952; *Gelechia portosanctana* Stainton, 1859; *Gelechia lyciella* Walsingham, 1900; *Gelechia salinella* Zeller, 1847; *Teleia hyoscyamella* Rebel, 1912; *Oecophora moritzella* Treitschke, 1835; *Lita roseella* Zetterstedt, 1839.

Introduction

General remarks

Gelechiidae is one of the most grossly neglected families of micromoths both in Europe and on a global scale. Only a few specialists are dealing with this group, despite its enormous diversity with more than 4700 described and thousands more unnamed species. Particularly the lack of concise identification literature was and still is a major hindrance for identification of species. More than 10 years ago we therefore decided to publish altogether four volumes dealing with the European fauna of Gelechiidae. The first volume devoted to the tribes Teleiodini and Gelechiini appeared at the end of the last century (Huemer & Karsholt, 1999) and shortly after we started with the preparation of a book on Gnorimoschemini. Though well aware of numerous taxonomic problems at that time we could not imagine the prodigious effort which was finally necessary to finish this work.

Today's knowledge of Gnorimoschemini is largely based on the enormous taxonomic work which was undertaken by the late Prof. Dalibor Povolný. Starting in the late forties of the last century he introduced the tribe Gnorimoschemini into literature in 1964. In more than four decades of scientific work on the group Povolný formally decribed several hundred taxa and published numerous papers including a monograph of the Palaearctic species (except *Caryocolum*) (Povolný, 2002a). In spite of his efforts identification of Gnorimoschemini remained difficult, not least because of the overall similarity of several species and the grossly lacking differential diagnoses. To complicate matters further, several species have been described from unique specimens, sometimes even in bad condition, and most former authors used traditional genitalia preparation instead of the more appropriate unrolling technique (see Genitalia preparation). In consequence important characters were frequently invisible in the accompanying drawings or photographs of earlier work. Last but not least the variation of important taxonomic characters was and still is insufficiently described for several taxa due to lack of appropriate material. Molecular data which could help to resolve many of the aforementioned problems are still largely lacking and eagerly awaited for future studies.

Our work was seriously delayed by the inaccessibility of many of the type specimens accumulated in Povolný's private collection. Only after his death in 2004 this material became available for the scientific community and it will be returned to the legal owners in future by the staff of the MZMB, Brno, where the collection is preserved presently.

Furthermore we had to consider important and very rich material collected during expeditions to the southern Ural Mountains by colleagues from Finland. Several species turned up being first records for the European fauna or new to science. Though we have not been able to identify all taxa to species level so far a good portion is described herein.

The difficulties in identification of Gnorimoschemini are reflected in the dichotomous keys to genera where some genera turn up several times. Taking into account these difficulties and well aware that amateur lepidopterists in particular only exceptionally use keys, we decided not to produce a key to species level. However, important diagnostic characters are discussed in some detail within descriptive species sections. These sections strictly follow the format and style of former volumes of Microlepidoptera of Europe, with the only exception that we have decided to accept original spellings of species names (Sommerer, 2002).

A concise phylogenetic analysis of Gnorimoschemini at species level is still lacking and only a few genera such as *Caryocolum* have been revised on a global scale (Huemer, 1988). As a consequence the arrangement of species in several genera was hitherto organized alphabetically (see Povolný, 2002a). In this book the sequence of species is merely based on morphological similarities. An integrative taxonomical approach combining morphological, ecological and molecular data will be one of the major tasks for the future.

Collecting methods

Identification of Gnorimoschemini during field work is a difficult task and only advisable with long experience. Therefore voucher specimens are frequently necessary to avoid misidentifications.

Collecting methods for recording Gnorimoschemini are manifold and mainly depend on the habits of single species. As in several other groups the most promising results are obtained with artificial light where numerous species may be attracted. Particularly short-wave light is recommendable and a large majority of taxa may be found with this method. Comparatively few species are at least to some extant diurnal. These taxa are on the wing during sunshine or more frequently at dawn or dusk and can be collected by sweeping over the vegetation with a net. Several hidden species, e.g. those with flightless females, can be flushed from the vegetation with a bee-smoker. However, often the most effective method of collecting Gnorimoschemini is breeding of larvae to immatures, which also provides valuable information about the bionomics of species. Several otherwise rarely observed species, e.g. of the genera *Scrobipalpa* and *Caryocolum*, are easy to breed.

Material should be brought home alive in small glass tubes and set freshly in the laboratory. Alternatively the material can be killed immediately with cyanide or ethyl acetate and set after a couple of hours. Softening of material should be avoided since freshly dried specimens are most suitable for molecular sequencing. Nuclear genes are best preserved when fresh material is transferred into absolute alcohol and kept in a refrigerator.

Genitalia preparation

Preparation of genitalia is essential for identification of the majority of Gnorimoschemini. We used the standard preparation technique described for microlepidoptera (Robinson, 1976) with the adaptations introduced for Gelechiidae and described as 'unrolling technique' (Pitkin, 1986; Huemer, 1987). In this technique the male genitalia are unrolled by cutting the membranous connection between the vinculum and the tegumen on the right-hand side, if necessary the female genitalia are unrolled by longitudinally cutting the membranous medial zone of tergite VIII. For a long time genitalia of Gnorimoschemini have been dissected more or less following the standard preparation technique without unrolling the male and/or female genitalia. However, in many cases important structures such as the vinculum, valva and sacculus are insufficiently visible and difficult to compare. We therefore decided to unroll the male genitalia whereas females were only unrolled in exceptional cases when important characters such as processes of segment VIII became better visible.

The genitalia preparation followed following procedure:

1) breaking off the abdomen

2) maceration of the abdomen in 10% potassium hydroxide (KOH) solution in a double-boiler for about 5-10 minutes
3) washing in clean water
4) staining with 1% mercurochrome suspension in water (sclerotized structures) and/or chlorazol black (membranous structures) (optional)
5) dissecting and cleaning in clean water
6) dehydrating in absolute (isopropyl) alcohol
7) embedding in Euparal

In certain cases, when our material was solely based on older and not yet unrolled slides we carefully tried to remount and unroll most of these slides. Only few delicate exceptions are figured as standardized slides in ventrodorsal view.

Notes on the illustrations

The specimens are figured in 3 times magnification. Priority is given to show variation of the involved species over figuring perfect specimens. The label data are given in a standardized form as: Sex / country / province / locality / altitude / date / collector(s) / genitalia slide number / collection. Collectors are listed alphabetically according to surname. Abbreviations for finding and breeding of larva(e) are cited literally ("la.", 'ex. la', 'e.l.' etc.) as there are no standards for using these terms among European microlepidopterists.

The large majority of genitalia figures is based on slides dissected by the authors or the late H. Hendriksen. Older slides with former conventional preparation techniques proved unsatisfactory and were only used in exceptional cases. The genitalia photographs were taken on an Olympus BH2 microscope by S. Heim or sometimes P. Huemer at variable magnification. Male genitalia and phallus are figured at the same scale for each individual species, but may differ between species. In a few exceptions the phallus is slightly enlarged. Female genitalia with segment VIII and signum are figured at the same scale but if necessary with interspecifically different magnification. Structures of the abdominal segments, particularly the sternite and tergite VIII of the male, are not figured due to the low value for species identification.

Acknowledgments

During the preparation of the present volume we have received help and support from a lot of people. We are especially indebted to Klaus Sattler, BMNH, London, U.K. and Oleksiy Bidzilya, ZMKU, Kiev, Ukraine for answering numerous questions, to the latter and Kari Nupponen, Espoo, Finland for repeatedly sending material for our study, and to Niels Peder Kristensen, ZMUC, Copenhagen, Denmark for his continuous support. Access to the ZMUC collections was kindly supported by COBICE (Copenhagen Biosystematics Centre) (P.H.).

The authors would like to thank the following colleagues for the loan and/or donation of material and/or for various other support: Antonio M. F. Aguiar, Camacha, Madeira, Portugal; Marcos Báez, La Laguna. Tenerife, Spain; Joaquin Baixeras Almela, Valencia, Spain; Vasiliy Anikin, Saratov, Russia; Giorgio Baldizzone, Asti, Italy; Petr Baňař, MZMB, Brno, Czech Republic; Salvatore Bella, Catania, Italy; Günter Baisch, Biberach, Germany; Karel Cerny, Innsbruck, Austria; Martin Corley, Faringdon, U.K.; Anthon Cox, Mol, The Netherlands; Helmut Deutsch, Lavant, Austria; Jordi Domingo Calabuig, Valencia, Spain; Günter Ebert, LNK, Karlsruhe, Germany; Gustav Elsner, Prague,

Czech Republic; Michael Fibiger, Sorø, Denmark; Sabine Gaal-Haszler, NHMW, Vienna, Austria; Christian Gibeaux, Avon, France; Cees Gielis, Lexmond, The Netherlands; Lancelot A. Gozmány (†), HNHM, Budapest, Hungary; Frans Groenen, Luyksgestel, The Netherlands; Theo Grünewald, Landshut, Germany; Heinz Habeler, Graz, Austria; Dietger Hausenblas, Stuttgart, Germany; Axel Hausmann, ZSM, Munich, Germany; Robert J. Heckford, Plympton, U.K.; Marcel Hellers, Luxemburg; Carsten Hviid, Virum, Denmark; Povilas Ivinskis, Vilnius, Lithuania; Maurice G. M. Jansen, Wageningen, The Netherlands; Gert Jeppesen, Idestrup, Denmark; Roland Johansson, Växjö, Sweden; Jari Junnilainen, Vantaa, Finland; Lauri Kaila, ZMUH, Helsinki, Finland; Jari-Pekka Kaitila, Vantaa, Finland; Zoltán Kovács, Miercurea Ciuc, Romania; Vasiliy Kravchenko, TAU, Tel Aviv, Israel; András Kun, HNHM, Budapest, Hungary; Keld Gregersen, Sorø, Denmark; Patrice Leraut, MNHN, Paris, France; Houhun Li, Nankai University, Tianjin, China; Maria Luisa Lobo Lima, Sal, Cape Verde Islands; Martin Lödl, NHMW, Vienna, Austria; Gérard Luquet, MNHN, Paris, France; Alexandr Lvovsky, ZIAP, St. Petersburg, Russia; Toni Mayr, Feldkirch, Austria; Wolfram Mey, ZMHU, Berlin, Germany; Joel Minet, MNHN, Paris, France; Carlo Morandini, MFSN, Udine, Italy; Lucio Morin, Monfalcone, Italy; Marko Mutanen, Zoological Museum, Oulu, Finland; Jaques Nel, La Ciotat, France; Erik J. van Nieukerken, RMNH, Leiden, The Netherlands; Timo Nupponen, Espoo, Finland; Mark Parson, Wareham, U.K; Linda Pitkin, BMNH, London, U.K.; Dalibor Povolný (†), Brno, Czech Republic; Jerry Powell, EME, University of California, Berkeley, U.S.A; Susanne Randolf, NHMW, Vienna, Austria; Jozef Razowski, ISEZ, Kraków, Poland; Emili Requena, Igualada, Spain; Ignác Richter, Prievidza, Slovakia; Gaden Robinson (†), BMNH, London, U.K.; Nikos Roditakis, Plant Protection Institute, Heraklion, Greece; Walter Ruckdeschel, Übersee, Germany; Paul Sammut, Rabat, Malta; Luigi Sannino, Scafati, Italy; Victor Sartos, Universitat Autònoma de Barcelona, Spain; Nikolay Savenkov, LDM, Riga, Latvia; Jürg Schmid, Illanz, Switzerland; Willibald Schmitz, Bergisch Gladbach, Germany; Andreas Segerer, ZSM, Munich, Germany; Sergey Sinev, ZIAP, St. Petersburg, Russia; Peder Skou, Stenstrup, Denmark; Bjarne Skule, Veksø, Denmark; Frantisek Slamka, Bratislava, Slovakia; Alexey Solodovnikov, ZMUC, Copenhagen, Denmark; Marja van der Straten, Plant Protection Service, Wageningen, The Netherlands; Reinhard Sutter, Bitterfeld, Germany; Jukka Tabel, Hartola, Finland; Gerhard Tarmann, TLMF, Innsbruck, Austria; Zdenko Tokár, Šal'a, Slovakia; Robert Trusch, LNK, Karlsruhe, Germany; Kevin Tuck, BMNH, London, U.K.; Christian Wieser, LMKÄ, Klagenfurt, Austria; Josef Wimmer, Steyr, Austria; Hugo van der Wolf, Nuenen, The Netherlands.

Henning Hendriksen (†), ZMUC, Copenhagen, kindly assisted with making genitalia slides. Photographs of the adults were taken by Geert Brovad, ZMUC, Copenhagen, those of labial palpi by Peter Buchner, Schwarzau am Steinfeld, Austria, and those of most of the genitalia by Stefan Heim, TLMF, Innsbruck, and we are most grateful to these colleagues. Last but not least we want to thank Martin Corley, Faringdon, U.K. and Matthias Nuss, Dresden, Germany for careful checking of the manuscript.

Abbreviations of museums, institutions and private collections

ANIK	coll. Vasiliy Anikin, Saratov, Russia.
BALD	coll. Giorgio Baldizzone, Asti, Italy.
BELL	coll. Salvatore Bella, Catania, Sicily, Italy.
BMNH	The Natural History Museum, London, U.K.
BNMC	Bündner Naturmuseum, Chur, Switzerland.

CORL	coll. Martin Corley, Faringdon, U.K.
ELSN	coll. Gustav Elsner, Prague, Czech Republic.
EME	Essig Museum of Entomology, University of California, Berkeley CA, USA.
ETHZ	Eidgenössische Technische Hochschule, Zürich, Switzerland.
GRÜN	coll. Theo Grünewald, Landshut, Germany.
HNHM	Hungarian Natural History Museum, Budapest, Hungary.
ISEZ	Instytut Systematyki i Ewolucji Zwierzat, PAN, Krakow, Poland.
JUNN	coll. Jari Junnilainen, Vantaa, Finland.
KAIT	coll. Jari-Pekka Kaitila, Vantaa, Finland.
LDM	Latvian Museum of Natural History, Riga, Latvia.
LMKÄ	Landesmuseum Rudolfinum Kärnten, Klagenfurt, Austria.
LNK	Landessammlungen für Naturkunde, Karlsruhe, Germany.
MFSN	Museo Friulano di Storia Naturale, Udine, Italy.
MNHN	Muséum National d'Histoire Naturelle, Paris, France.
MZMB	Moravské Zemské Muzeum, Brno, Czech Republic.
NHMW	Naturhistorisches Museum, Vienna, Austria.
NHRS	Naturhistoriska Riksmuseet, Stockholm, Sweden.
NUPP	coll. Kari & Timo Nupponen, Espoo, Finland.
RMNH	National Museum of Natural History Naturalis, Leiden, The Netherlands.
SDEI	Senckenberg Deutsches Entomologisches Institut, Müncheberg, Germany.
TAU	Tel Aviv University, Tel Aviv, Israel.
TLMF	Tiroler Landesmuseum Ferdinandeum, Innsbruck, Austria.
TOKA	coll. Zdenko Tokár, Šal'a, Slovakia.
ZIAP	Zoological Institute, Russian Academy of Sciences, St. Petersburg, Russia.
ZMHU	Zoologisches Museum der Humboldt-Universität, Berlin, Germany.
ZMKU	Zoological Museum, Kiev, Ukraine.
ZMUC	Zoological Museum, Natural History Museum of Denmark, Copenhagen, Denmark.
ZMUH	Zoological Museum, University of Helsinki, Helsinki, Finland.
ZSM	Zoologische Staatssammlung, Munich, Germany.

Classification of Gelechiidae

The higher classification of Gelechioidea has made considerable advances during the last decades but is still a matter of ongoing discussion (Bucheli, 2009; Bucheli & Wenzel, 2005; Heikkilä & Kaila, 2010; Hodges, 1978, 1998; Minet, 1990; Kaila, 2004; Kuznetzov & Stekolnikov, 1984; Sinev, 1992). The family concept of Gelechiidae is less controversial and most authors follow the widely accepted synthesis presented by Hodges (1998). Accordingly Gelechiidae are defined by the following characters: gnathos with pair of articulated and symmetric lateral sclerites with an articulated mesal hook (synapomorphy); absence of CuP on the forewing (parallelism); and three polymorphic characters of the hindwing. The monophyly of Gelechiidae is supported by a morphology based phylogenetic analysis by Kaila (2004) who found one unambiguous synapomorphy for Gelechiidae, the presence of a row of narrow scales ventrally on the forewing vein R of females, a character shared by the Gnorimoschemini. A first attempt at a combined molecular and morphological analysis is based on a small sample of taxa (Bucheli & Wenzel, 2005) but suggests that morphological characters that unite taxa may not be homologous.

The suprageneric classification of Gelechiidae is much disputed and even the number of subfamilies remains unstable. Hodges (1986) divided Gelechiidae into the three subfamilies Gelechiinae, Dichomeridinae and Pexicopiinae, differing by characters of the adult abdominal sternum II. Later he added the south-east Asian Physoptilinae as the sister-group of the aforementioned subfamilies (Hodges, 1998). Kuznetzov & Stekolnikov (1984) initially distinguished only two subfamilies whereas in a later publication they recognized six (Kuznetzov & Stekolnikov, 2001). Further concepts have been introduced, e. g. by Leraut (1993) and Sinev (1992, 1993). Recently Ponomarenko (2005, 2006, 2009) presented a classification based on a comparative morphological analysis particularly of the male genitalia and its musculature system. She distinguished five subfamilies: Physoptilinae, Anomologinae, Gelechiinae, Anacampsinae and Dichomeridinae. However, none of these concepts has received thorough acceptance from the scientific community so far. Hodges (1998) in the most widely accepted review characterized the Gelechiinae by abdominal sternum II with a pair of venulae and a pair of apodemes and the antenna usually without pecten (present as a single slender scale in *Bryotropha* and in some species of *Metzneria, Monochroa* and *Aristotelia* and a full pecten in the *Apatetris*-group). However, Ponomarenko (2006, 2009) showed that several genera associated with the Gelechiinae by Hodges (*op. cit.*) in fact agree in the second sternite with most genera of Dichomeridinae. She concluded that the peculiarities of adult abdominal sternum are diverse within the family and thus unsuitable for establishing taxa at the subfamily rank. Despite these shortcomings the monophyly of Gelechiinae is still (according to Ponomarenko, 2006) based on two autapomorphies: valva with mediobasal process and male genital segment with glands, which are however secondarily reduced in several groups.

The subdivision of Gelechiinae into tribes is a further matter of ongoing discussion. Karsholt & Riedl (1996) distinguished seven tribes in Europe. According to these authors Gelechiinae include the tribes Apatetrini, Anomologini, Litini (formerly Teleiodini), Gelechiini, Gnorimoschemini, Anacampsini, and Chelariini. This subdivision was also followed by Lee *et al.* (2009) for the North America fauna. Other authors such as Povolný (2002b) add Metzneriini as an eighth tribe. On the contrary Ponomarenko (2006) only separated three gelechiinae tribes, Gnorimoschemini, Gelechiini and Litini, based on various apomorphic character states of the genitalia and their muscles, whereas all other tribes were moved to different subfamilies. The systematic position of several suprageneric taxa still remains disputed (Bidzilya & Karsholt, 2008) and needs further analysis.

Morphology and systematics of Gnorimoschemini

Gnorimoschemini was introduced by Povolný (1964) who basically defined the tribe by genitalia characters. Despite his repeated, detailed and extensive descriptions of adults and preimaginal stages, he never supported the Gnorimoschemini by synapomorphies – maybe with the exception of the reduction of muscle m7 in the male genitalia (Povolný, 2002a, referring to Kuznetzov & Stekolnikov, 2001). Identification of the group is merely based on a combinations of characters (Povolný, 2002b). Indeed Gnorimoschemini is weakly separated from Gelechiini with which it shares the conspicuous dilation of the lateral parts of the vinculum as synapomorphy but it is characterized as morphologically homogenous and more specialized than Gelechiini. Later the monophyly of Gnorimoschemini has been supported based on a hook-like signum and a lateral zone of microtrichia near the ostial area (Huemer & Karsholt, 1999; Ponomarenko, 2006).

Diagnosis (following the terminology introduced by Povolný (various papers), Huemer (1988), Jansen (1999) and Kristensen (2003)). Moth (Text-figs. 1-3) with a high degree of overall similarity; small to medium-sized, wingspan ranging from 6 to 24 mm; labial palpus recurved, second segment usually with brush of short scales on outer surface; forewing usually with one plical and two discal spots. Male genitalia (Text-figs. 4-6) with uncus usually broadly hood-like; gnathos with hook-shaped to spatulate mesal sclerite, or reduced, culcitula usually well-developed; teguminal pedunculi large; valva (cucullus *sensu* Ponomarenko) simple, mostly digitate to rod-shaped; sacculus (parabasal process *sensu* Povolný) short to long, frequently with pointed apex; vinculum frequently supporting ridge from anterior margin to medial part, posterior margin with one to two pairs of small projections to distinct vincular process (sacculus process *sensu* Povolný); saccus short to long, frequently rod-like to tongue-shaped; phallus (aedeagus *sensu auctorum*) short to long, rarely with small spines, sub-apical hooklet usually present, coecum inflated. Female genitalia (Text-figs. 7-10) with sclerotized subgenital plates, partially covered with microtrichia, foamy sculpture or smooth, rarely with process-like sclerites; ventromedial part of sternite VIII with or without paired depressions, areas of foamy sculpture and/or microtrichia present or reduced; subostial sclerites present or absent; antrum funnel-shaped to tubular or largely reduced, colliculum shortly ring-shaped to reduced; posterior part of ductus bursae with or without lateral sclerites; corpus bursae normally well defined; signum usually with basal plate and a strong distal signum-hook, rarely reduced. Povolný (1991) furthermore published an extensive synthesis on preimaginal characters which is not repeated here because it lacks differential diagnoses and a phylogenetic analysis.

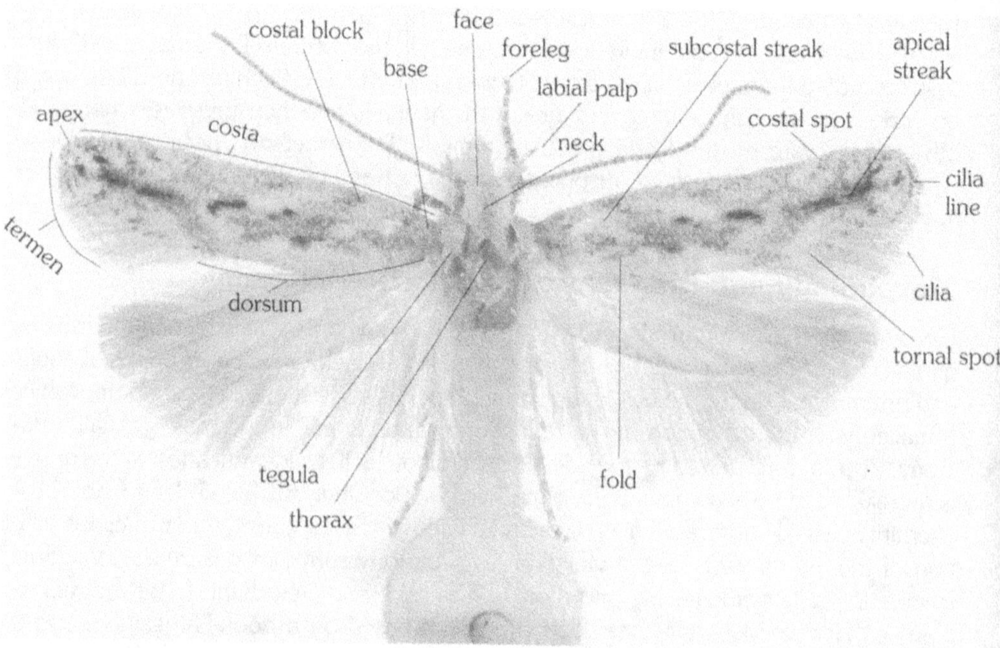

Text-fig. 1. Wing markings of Gnorimoschemini.

The generic classification of Gnorimoschemini was for a long time merely based on overall similarities and intuition. In 1988, Povolný & Šustek analysed 169 genitalia characters of 44 genera with methods of numerical taxonomy. Based on this analysis they propose a phylogeny which includes several informal groups without taxonomic status: symmetrischemoid, gnorimoschemoid, scrobipalpoid, scrobipalpuloid, ephysteroid, hedmoid and caryocoloid Gnorimoschemini (see also Povolný, 1991a). How-

Text-fig. 2. Head with labial palpi of *Scrobipalpa acuminatella* (Sircom).

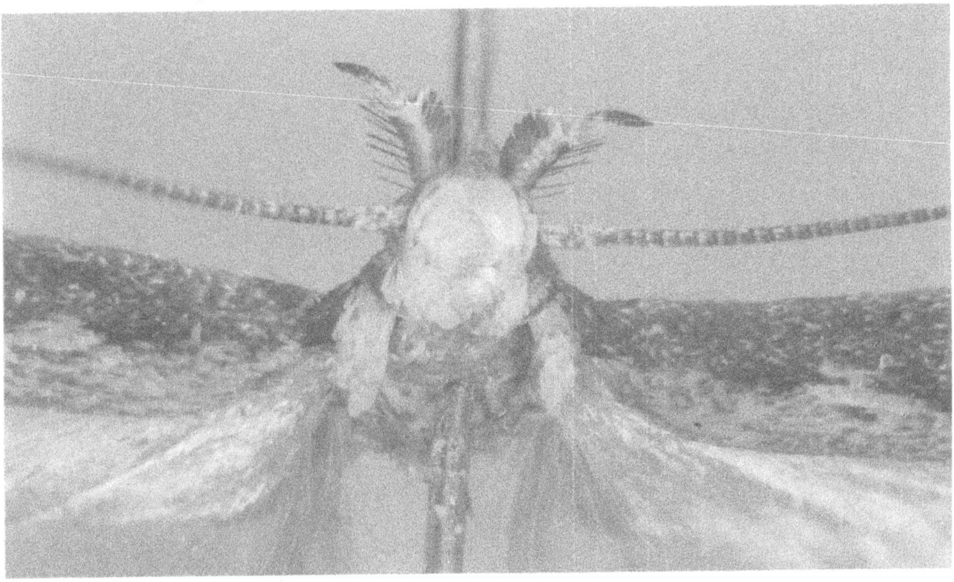

Text-fig. 3. Head with labial palpi of *Pogochaetia solitaria* Staudinger.

ever, these groups remain weakly defined by morphology and their status is uncertain.

Gnorimoschemini are diverse with altogether about 900 described species attributed to 44 genera and still many undescribed taxa (Povolný, 2002b). Most lineages are known from the Neotropics. The Nearctic region is particularly important with alto-

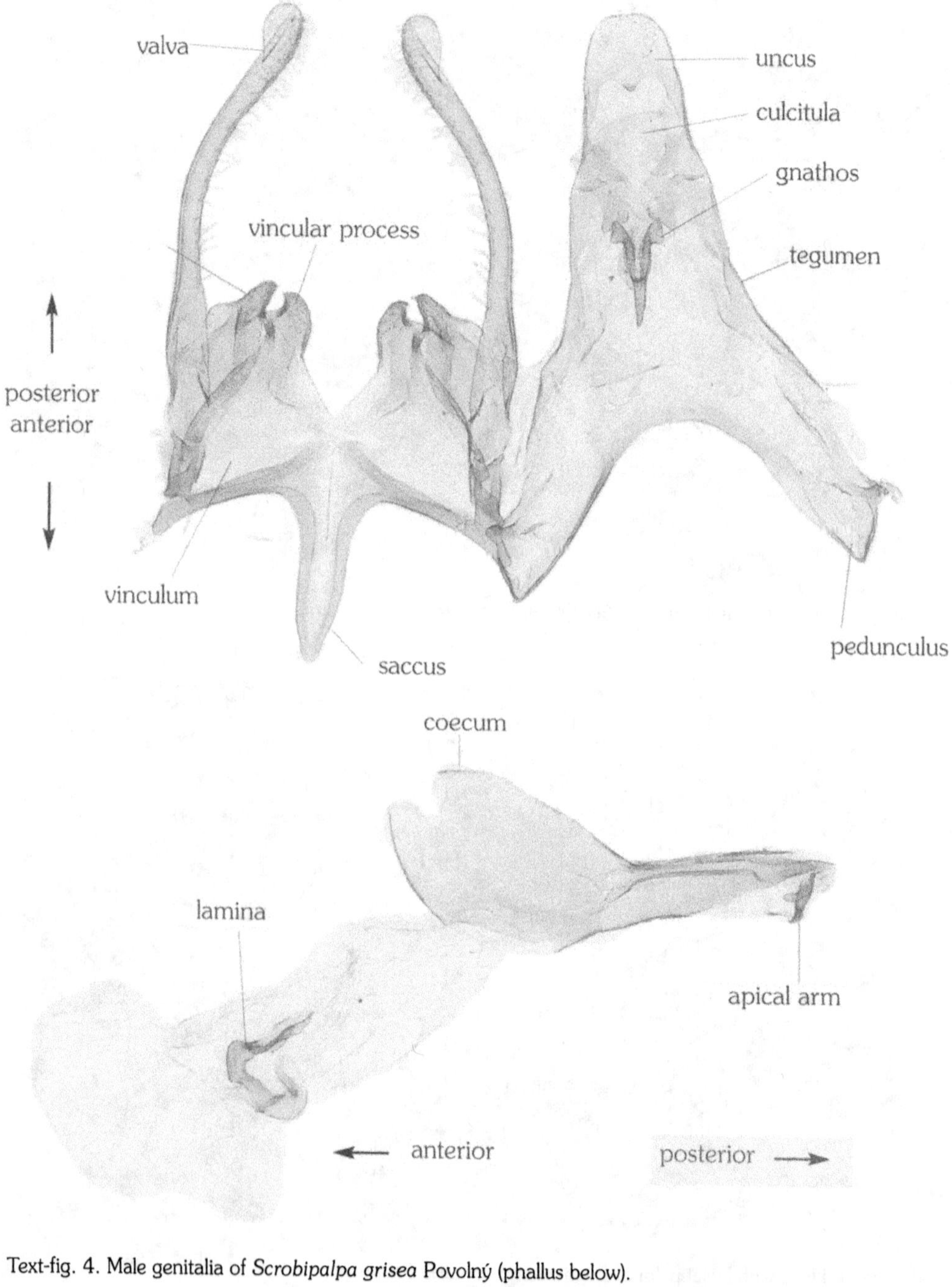

Text-fig. 4. Male genitalia of *Scrobipalpa grisea* Povolný (phallus below).

gether 180 species, among them 95 species of *Gnorimoschema* (Lee *et al.*, 2009). Highest species diversity is known in the Palearctic region with more than 300 described species of which more than 200 classified into 20 genera are known from the European continent.

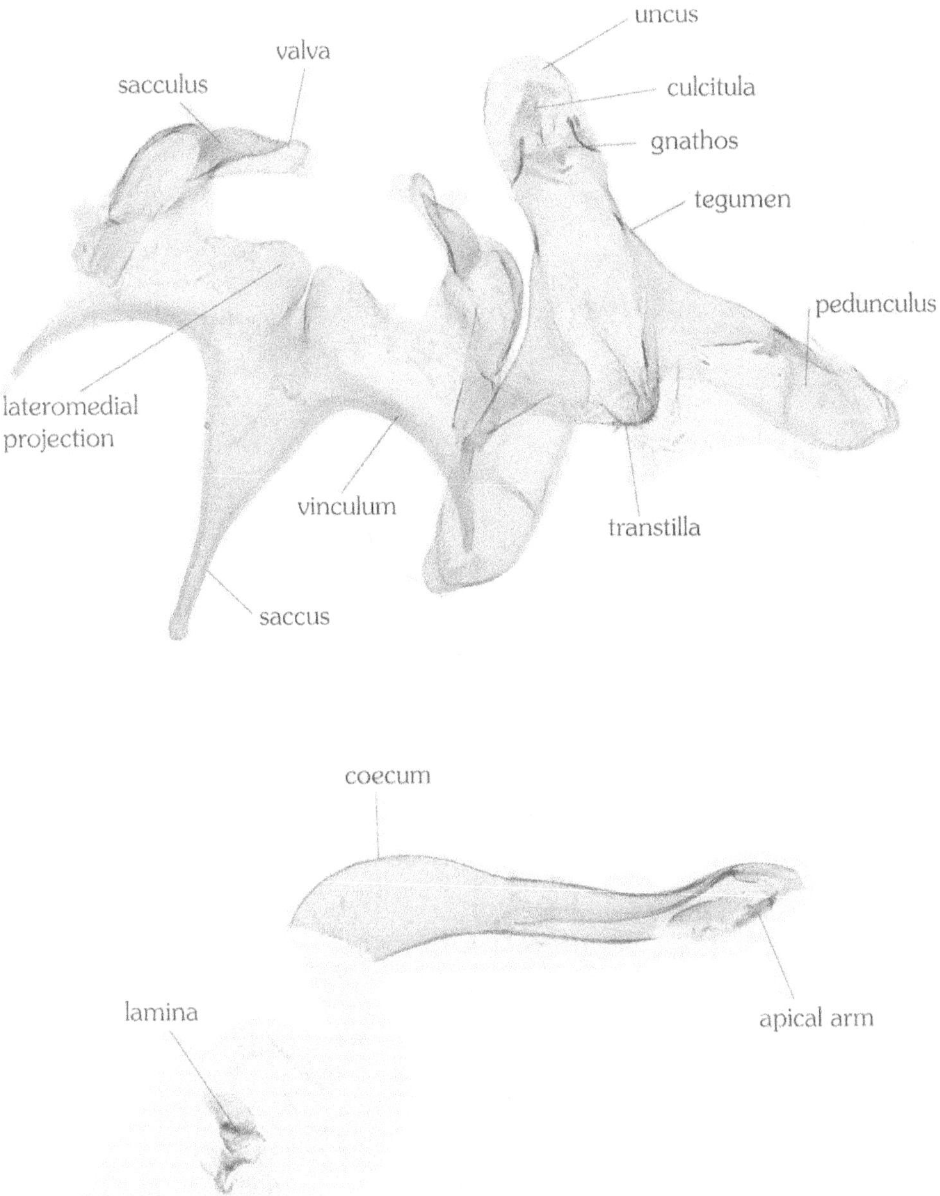

Text-fig. 5. Male genitalia of *Caryoculum proxima* (Haworth) (phallus below).

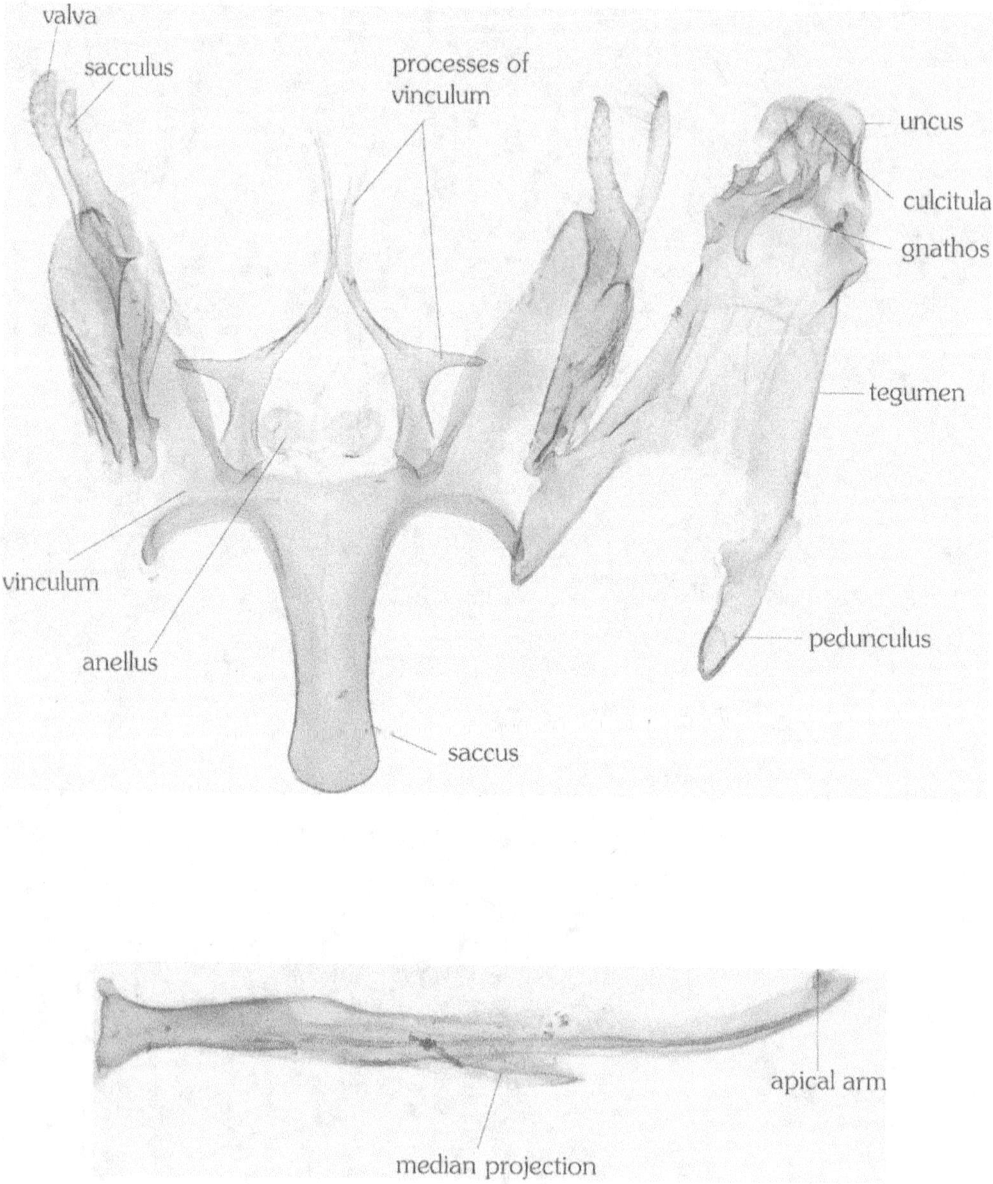

Text-fig. 6. Male genitalia of Sattleria melaleucella (Constant) (phallus below).

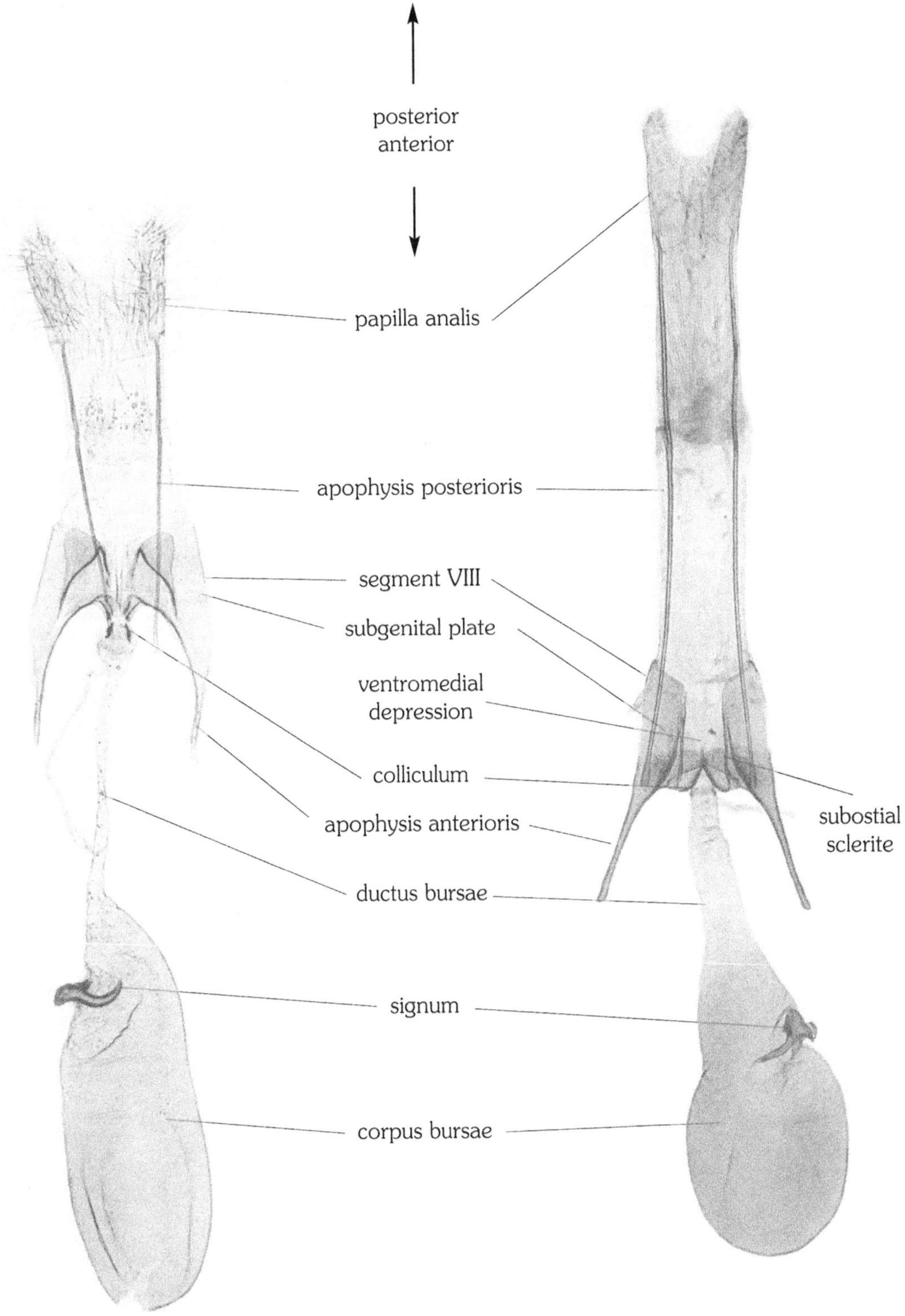

Text-fig. 7. Female genitalia of *Gnorimoschema hoefneri* (Rebel).
Text-fig. 8. Female genitalia of *Scrobipalpa lutea* Povolný.

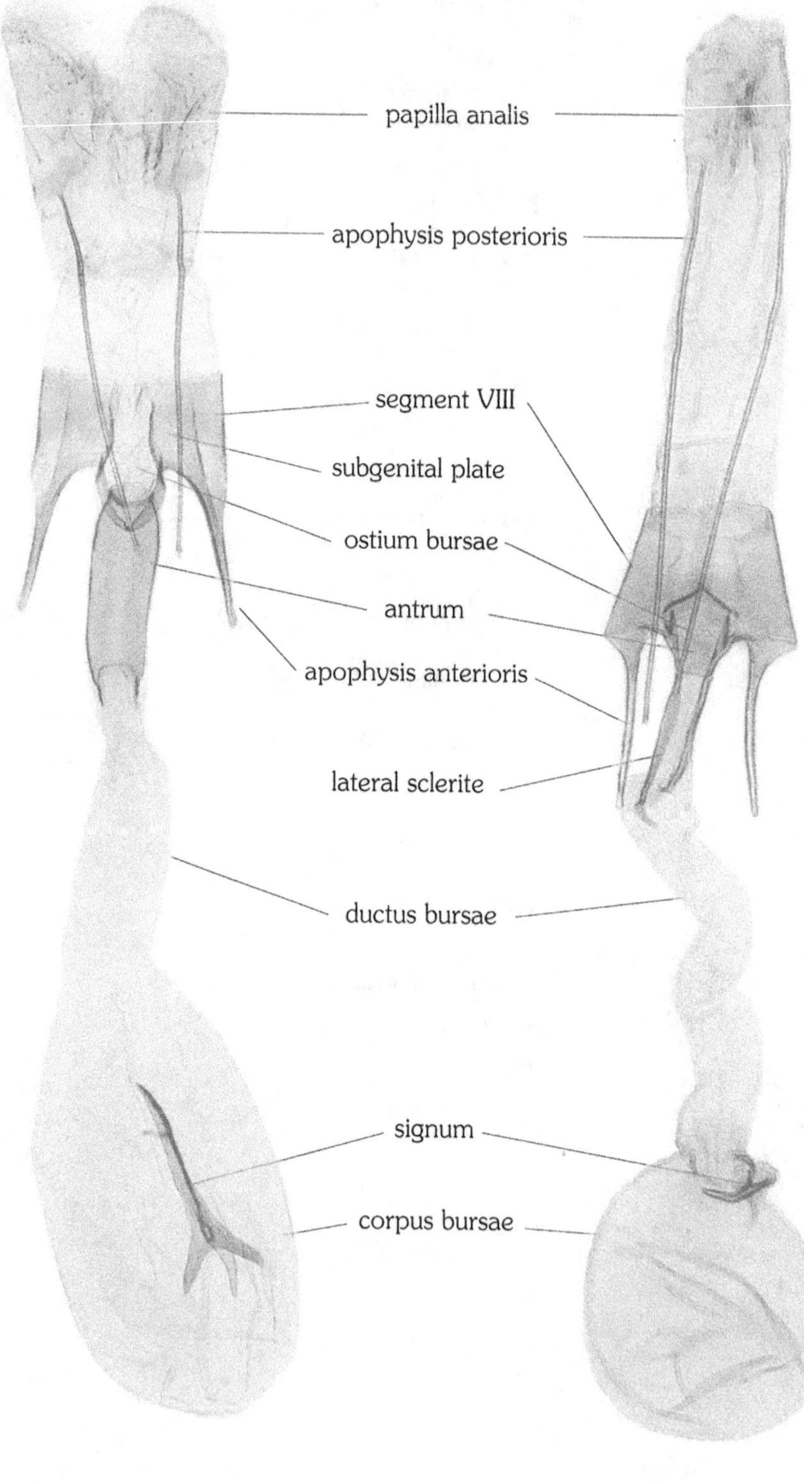

Text-fig. 9. Female genitalia of *Ephysteris diminutella* (Zeller).
Text-fig. 10. Female genitalia of *Caryocolum leucomelanella* (Zeller).

Bionomics of Gnorimoschemini

The morphology of the eggs, larvae and pupae of the Gnorimoschemini falls outside the concepts of this series. We can refer to more or less comprehensive treatments on these by, e.g., Bland *et al.* (2002), Jansen (1999), Nel (1999a) and Stehr (1987); for the pupae especially Patočka (1989) and Patočka & Turčani (2005). We are aware that larvae and also pupae of closely related species may have good diagnostic characters (Nel, 1999a; Patočka, *op. cit.*). However, it is very important that descriptions of larvae and pupae are correctly associated with the bred adults and correctly identified, and the aim of this book is to facilitate the identification of the adults. Larval feeding of gnorimoschemine species of any genus is, as in most other Gelechiidae, restricted to single genera or families of plants (Huemer, 1988: 443; Povolný, 2002a: 21). The predominant host-plant families of the Gnorimoschimini are Asteraceae, Caryophyllaceae, Chenopodiaceae, Poaceae, Solanaceae and Zygophyllaceae (table 1).

Besides their normal host-plants some species are at least occasionally able to feed on more or less distantly related plant species. Examples of such exceptional host-plants is one larva of *Gnorimoschema bodillum* feeding on *Myrica gale* (Myricaceae) among hundreds of larvae feeding on *Salix repens* (Salicaceae), or the reports of larvae of *Scrobipalpa salicorniae* (normally feeding on *Salicornia* and other Chenopodiaceae) from *Spergularia* (Caryophyllaceae) and *Aster* (Asteraceae).

Table 1. Host-plant families for European genera of Gnorimoschemini; families in brackets are less often used.

Genus	Host-plant families
Gnorimoschema	Asteraceae, Salicaceae, (Equisetaceae), (Myricaceae)
Scrobipalpopsis	Asteraceae
Scrobipalpa	Asteraceae, Chenopodiaceae, Solanaceae, (Brassicaceae), (Caryophyllaceae), (Juncaginaceae), (Lamiaceae), (Plantaginaceae), (Plumbaginaceae), (Polygonaceae), (Thymelaeaceae)
Turcopalpa	Unknown
Scrobipalpula	Asteraceae
Phthorimaea	Solanaceae
Tuta	Solanaceae
Keiferia	Solanaceae
Ephysteris	Poaceae
Ochrodia	Zygophyllaceae
Vladimirea	Zygophyllaceae
Microlechia	Solanaceae, (Rhamnaceae)
Cosmardia	Caryophyllaceae
Lutilabria	Unknown
Klimeschiopsis	Unknown (Caryophyllaceae?)
Caryocolum	Caryophyllaceae
Tila	Caryophyllaceae
Pogochaetia	Caryophyllaceae
Agonochaetia	Unknown
Sattleria	Caryophyllaceae, (Saxifragaceae)

Host plant records in the literature should be taken with care. Various mistakes are caused by misidentifications of larvae or moths (e.g., records of *Scrobipalpa portosanctana* from Tamaricaceae and Rhamnaceae) or misidentifications of host-plants, but also the changing nomenclature of both plants and moths has resulted in errors. A curious example of misidentifications of host-plants deals with three species belonging to different genera: 1) *Scrobipalpa artemisiella*, named after *Artemisia* (Asteraceae), although its larva lives on *Thymus* (Lamiaceae). Its bionomics was confused with that of 2) *Scrobipalpula psilella*, having a larva which lives on *Artemisia*. Most literature records on the bionomics of 3) *Gnorimoschema streliciella* state it to be associated with *Thymus*, although its larva lives on *Antennaria* (Asteraceae), the larva feeding on *Thymus* being that of *S. artemisiella*.

Larvae of European Gnorimoschemini have diverse life strategies, feeding within spun leaves, shoots, flowers and seed capsules. Some are leaf-miners, others feed within stems and fruits, and some are gall makers, or feed in the tuber of, e.g. potatoes. Most species are restricted to a single life strategy, but some can, e.g., start as leaf-miner and later live within spun leaves. Larvae of a few species are able to feed on/in all parts of the host-plants. Many species, especially of *Caryocolum,* have only a single generation each year, whereas others have two or more generations. In north Europe most species have only a single generation, and the larva of *Scrobipalpopsis petasitis* and probably some *Sattleria* spp. need two years to complete the life cycle. The individual species may pass the winter as egg, larva or pupa. *Caryocolum junctella*, *Cosmardia moritzella, Scrobipalpa hyoscyamella* and (occasionally) *S. costella* hibernate in the adult stage.

The host-plants of Gnorimoschemini are almost exclusively low-growing plants, and the habitats of most species are consequently more or less open land. They occur from coastal to alpine habitats at altitudes from 0 to 3500 m. Many species of *Scrobipalpa* are restricted to salt marshes or inland saline habitats, whereas other species (especially of *Caryocolum*) live on calcareous slopes or similar habitats with a specialized flora. Their connection with particular host-plants and habitats makes species of the Gnorimoschemini suitable indicators for conservation management.

Other species of Gnorimoschemini are considered as pests of agricultural crops. This is especially the case with the two introduced species *Phthorimaea operculella* and *Tuta absoluta*, both originating from South America. *P. operculella* (Potato Tuber Moth) was introduced to Europe more than 100 years ago. It is now well established in warmer areas, and there is a huge literature on the bionomics, harm and control of this species. *T. absoluta* (Tomato Leafminer) was only introduced into Europe in 2007, and it has already invaded large parts of the Mediterranean area and also been found in a number of places in Central Europe. It is a serious threat to tomato growing, and at the moment probably the economically most important invasive Lepidoptera species in Europe. Other Gnorimoschemini pests introduced from outside Europe include *Scrobipalpa aptatella* (on tobacco) and *Keiferia lycopersicella* (on tomato). Both species are only doubtfully recorded from Europe and are apparently not able to maintain populations so far. A further invasive pest species is *Tecia solanivora* (Povolný, 1973) (Central American Potato Tuberworm). It was introduced to the Canary Islands in 1998 and has since been an important pest on potatoes (Báez, *in litt.*; Povolný, 2004; Vives Moreno, 2003: 97, 107). It has not yet been found in continental Europe. Links to figures of larvae, adults and the genitalia are given by EPPO (2009). Native European pest species include *Scrobipalpa ocellatella* (on beets) (Gregor & Povolný, 1954; Povolný, 1955, 1991b) and *S. ergasima* (on eggplant) (Povolný, 1979b: 145-147).

Key to the European genera of Gnorimoschemini

The keys are frequently based on symplesiomorphic characters due to the lack of synapomorphies. Only European species are considered.

Male genitalia

1. Gnathos with hook-like to spatulate or sub-oval mesal sclerite 2
– Gnathos without or with small plate-like mesal sclerite 18
2 Uncus broad, apicomedially pointed; sacculus hook-shaped (Figs. 1-12) .... *Gnorimoschema* (p. 37)
– Uncus and sacculus of different shape 3
3 Gnathos-hook broadly spatulate (Figs. 119-123) .... *Scrobipalpula* (p. 194)
– Gnathos of different shape 4
4 Vinculum with pair of needle-shaped, frequently forked processes .... (Figs. 202-211) .... *Sattleria* (p. 297)
– Vinculum without needle-shaped processes 5
5 Uncus slender, pointed, about length of tegumen (Fig. 126) ..... *Keiferia* (p. 206)
– Uncus broadly rounded, much shorter than tegumen 6
6 Sacculus largely reduced 7
– Sacculus well developed 10
7 Valva with hump-like projection at basal third, distal part covered with long setae (Fig. 125) .... *Tuta* (p. 204)
– Valva without hump-like projection near base, distal part covered with short setae 8
8 Gnathos with sub-oval mesal sclerite (Fig. 124) .... *Phthorimaea* (p. 202)
– Gnathos with mesal hook 9
9 Vinculum large; saccus broadly sub-rectangular (Fig. 117) .... *Scrobipalpa ergasima* (p. 191)
– Vinculum of moderate size; saccus slender (Figs. 14-15) .... *Scrobipalpa* (*partim*) (p. 56)
10 Sacculus long and needle-shaped (Figs. 137-139) .... *Microlechia* (p. 225)
– Sacculus of different shape 11
11 Sacculus widely separated from valva 12
– Sacculus near to valva 13
12 Vincular process well developed, strongly projected posteriad (Fig. 13) .... *Scrobipalpopsis* (p. 54)
– Vincular process largely reduced (Fig. 116) .... *Scrobipalpa peterseni* (p. 190)
13 Saccus broadly sub-rectangular (Fig. 118) .... *Turcopalpa* (p. 192)
– Saccus slender, moderately to distinctly pointed 14
14 Tegumen slender; sacculus slender digitate 15
– Tegumen moderately broad; sacculus of different shape 16
15 Valva broad (Fig. 135) .... *Ochrodia* (p. 221)
– Valva slender (Figs. 127-134) .... *Ephysteris* (p. 208)
16 Sacculus about as long as valva, knife-shaped (Fig. 140) ...... *Cosmardia* (p. 229)
– Sacculus distinctly shorter than valva 17
17 Vinculum without anteriomediad, sclerotized transverse sclerites; saccus longer than vinculum (Fig. 136) .... *Vladimirea* (p. 223)
– Vinculum with anteriomediad, sclerotized transverse sclerites, saccus

shorter than vinculum (Figs. 16-115) .... *Scrobipalpa (partim)* (p. 56)
18 Anellus with needle-shaped sclerites at least of about length of valva 19
– Anellus without or with indistinct additional sclerites 21
19 Apex of valva sub-triangular (Fig. 197) .... *Tila* (p. 290)
– Apex of valva of other form, rounded or pointed 20
20 Valva digitate, without medial projection (Fig. 198) .... *Pogochaetia* (p. 292)
– Valva distomedially widened or with hump-like projection
(Figs. 199-201) .... *Agonochaetia* (p. 294)
21 Sacculus broadly digitate, curved; gnathos with small mesal sclerite
(Figs. 141-143) .... *Lutilabria* (p. 231)
– Sacculus of different shape; mesal sclerite of gnathos reduced 22
22 Valva and sacculus of similar length and shape
(Figs. 144-146) .... *Klimeschiopsis* (p. 235)
– Valva and sacculus of different length and shape
(Figs. 147-196) .... *Caryocolum* (p. 238)

Female genitalia

1 Apophyses anterioris basally adjoined 2
– Apophyses anterioris basally well separated 3
2 Apophysis anterioris straight (Fig. 126) .... *Keiferia* (p. 206)
– Apophysis anterioris strongly bent (Fig. 135) .... *Ochrodia* (p.221)
3 Signum with small to moderate basal plate, distal hook well developed,
rarely small rounded plate without hook, or reduced 4
– Signum with large subtriangular to arrow-shaped basal plate and
anteriomedial thorn 20
4 Signum without hook 5
– Signum hook well developed 8
5 Sternite VIII with longish lateral flaps or drop-like dorsal processes;
signum small to moderately sized, rounded plate (Figs. 193, 196)
.... *Caryocolum (partim)* (p. 238)
– Sternite VIII with pair of drop-like lateral processes; signum a longish plate
or reduced 6
6 Antrum tubular, about length of apophysis anterioris; signum reduced
(Figs. 202-211) .... *Sattleria* (p. 297)
– Antrum funnel-shaped, at most half length of apophysis anterioris;
signum plate-like 7
7 Segment VIII subrectangular; antrum short (Fig. 197) .... *Tila* (p. 290)
– Segment VIII anteriolaterally projected; antrum about half as long as
segment VIII (Fig. 198) .... *Pogochaetia* (p. 292)
8 Sternite VIII with foamy sculpture 9
– Sternite VIII without foamy sculpture 12
9 Apophyses anterioris outwardly directed; antrum distinctly longer than
segment VIII (Fig. 125) .... *Tuta* (p. 204)
– Apophyses anterioris directed anteriad; antrum shorter than segment VIII 10
10 Sternite VIII with transverse zone of foamy sculpture medially
(Fig. 124) .... *Phthorimaea* (p. 202)
– Sternite VIII with various patches or field of foamy sculpture 11
11 Signum with long and slender base; antrum broadly funnel-shaped, at least

half as long as apophysis anterioris (Figs. 119-123)........ *Scrobipalpula* (p. 194)
– Signum of different shape, antrum reduced (Figs. 23-26, 29-117) .... *Scrobipalpa (partim)* (p. 56)
12 Antrum large, sub-oval (Fig. 118).. *Turcopalpa* (p. 192)
– Antrum absent to moderately sized, rarely large and funnel-shaped to cylindrical 13
13 Sternite VIII usually with flap-like sclerotizations; ductus bursae narrow; signum hook stout, at acute angle to basal plate (Figs. 1-12) .... *Gnorimoschema* (p. 37)
– Sternite VIII rarely with flap-like sclerotizations, otherwise of different shape; ductus bursae moderately broad; signum hook slender to stout, usually at about right angle to basal plate 14
14 Anterior margin of sternite VIII usually with lateromedial humps; antrum tubular, of moderate size to large; signum hook evenly curved, large and slender (Figs. 137-139) .... *Microlechia* (p. 225)
– Anterior margin of sternite VIII without lateromedial humps; antrum absent or of moderate size and funnel-shaped; signum hook variable 15
15 Apophysis anterioris medially dilated (Fig. 13) .... *Scrobipalpopsis* (p. 54)
– Apophysis anterioris rod-like.. 16
16 Ductus bursae with ring-shaped colliculum (Figs. 14-22, 27-28) .... *Scrobipalpa (partim)* (p. 56)
– Ductus bursae without ring-shaped colliculum... 17
17 Apophysis anterioris short, at most half length of segment VIII (Fig. 140) .... *Cosmardia* (p. 229)
– Apophysis anterioris long, at least length of segment VIII 18
18 Tergite VIII much longer than sternite VIII with oblique sclerotized anterior margin (Figs. 141-143)... *Lutilabria* (p. 231)
– Tergite and sternite VIII of about same length, without oblique sclerotized anterior margin 19
19 Antrum broadly funnel-shaped, short; signum hook long and weakly curved (Figs. 144-146).. *Klimeschiopsis* (p. 235)
– Antrum of various shape, rarely broad and short funnel with additional sclerites; signum hook moderate, curved (Figs. 147-192, 194-195) .... *Caryocolum (partim)* (p. 238)
20 Antrum broadly funnel-shaped; signum with longish, subtriangular basal plate (Figs. 199-201) .... *Agonochaetia* (p. 294)
– Antrum slender tubular to largely reduced; signum arrow-shaped 21
21 Sternite VIII medially with various sclerotized ridges and/or membranous zones; antrum short to long, tubular (Figs. 127-134)........... *Ephysteris* (p. 208)
– Sternite VIII smooth, without medial membrane or sclerotizations; antrum absent (Fig. 136) .... *Vladimirea* (p. 223)

Check-list of European Gnorimoschemini

GELECHIIDAE Stainton, 1854
GELECHIINAE Stainton, 1854

Gnorimoschemini Povolný, 1964

Gnorimoschema Busck, 1900
Lerupsia Riedl, 1965
Neoschema Povolný, 1967
1. soffneri (Riedl, 1965)
    antiquum Povolný, 1966 syn. rev.
    montanum Povolný, 1966 (subspecies)
2. herbichii (Nowicki, 1864)
    pusillella (Rebel, 1893)
    tengstroemiella (Joannis, 1910)
    pazsiczkyi (Rebel, 1913)
    parentesella (Toll, 1936)
    tengstroemi (Hackman, 1946)
    mongoliae Povolný, 1973 (subspecies)
    kamchaticum Povolný, 1977 (subspecies)
3. bodillum Karsholt & Nielsen, 1974
4. nupponeni sp. n.
5. robustella (Staudinger, 1871)
    syrphetopa (Meryick, 1926)
6. steueri Povolný, 1975
7. epithymella (Staudinger, 1859)
    brunneomaculella (Hackman, 1946) (subspecies)
    boerneri (Amsel, 1952) (subspecies)
    kirgisicum Povolný, 1994 (subspecies)
8. nordlandicolella (Strand, 1902)
    cyceonodes (Meyrick, 1924)
    eucausta (Meyrick, 1929)
    fennicella (Hackman, 1946)
9. nilsi Huemer, 1996 sp. rev.
10. valesiella (Staudinger, 1877)
    diabolicella (Hering, 1924)
    charcoti (Meyrick, 1934)
    hackmani (Schantz, 1952)
11. streliciella (Herrich-Schäffer, 1854)
    cinctipunctella (Erschoff, 1877) (subspecies)
    mongolorum Povolný, 1969 (subspecies)
12. hoefneri (Rebel, 1909) sp. rev.

Scrobipalpopsis Povolný, 1967
13. petasitis (Pfaffenzeller, 1867)
    petasitella (Staudinger, 1867)
    petasitae (Heinemann, 1870) (emendation)

Scrobipalpa Janse, 1951
Ilseopsis Povolný, 1965
Euscrobipalpa Povolný, 1967
Ergasiola Povolný, 1967
14. aptatella (Walker, 1864)
    heliopa (Lower, 1900)
15. kasyi Povolný, 1968
16. notata (Povolný, 2001)
17. acuminatella (Sircom, 1850)
    pulliginella (Sircom, 1850)
    cirsiella (Stainton, 1851)
    porcella (Heinemann, 1870) syn. n.
    ingloriella (Heinemann, 1870)
    gracilella (Stainton, 1871)
18. skulei sp. n.
19. hungariae (Staudinger, 1871)
20. adaptata (Povolný, 2001)
21. brahmiella (Heyden, 1862)
22. vasconiella (Rössler, 1877)
    drahomirae Povolný, 1966
23. dorsolutea sp. n.
24. amseli Povolný, 1966
25. hyssopi Nel, 2003
26. montanella (Chrétien, 1910)
27. corleyi sp. n.
28. chrysanthemella (Hofmann, 1867)
    opificella (Mann, 1878)
29. proclivella (Fuchs, 1886)
    rancidella auctt., (*nec* Herrich-Schäffer, 1854) (misident.)
30. frugifera Povolný, 1969
    hypothetica Povolný, 1973
31. oleksiyella sp. n.
32. smithi Povolný & Bradley, 1964
33. occulta (Povolný, 2002)
    sibirica Bidzilya, 2009 syn. n.
34. grisea Povolný, 1969
    uralensis Povolný, 1973 n. inv.
35. usingeri Povolný, 1969
36. clintoni Povolný, 1968
    linella Piskunov, 1975
    deleta Povolný, 1981 syn. n.
37. reiprichi Povolný, 1984
38. obsoletella (Fischer von Röslerstamm, 1841)
    miscitatella (Clarke, 1932)
    bipunctella (Hartig, 1941)
    calaritanella (Amsel, 1952)
    hospes Povolný, 1964
39. feralella (Zeller, 1872)
    rebeliella (Hauder, 1917)

40. halonella (Herrich-Schäffer, 1854)
41. perinii (Klimesch, 1951)
42. phagnalella (Constant, 1895)
    staehelinella (Caradja, 1920) n. inv.
43. tokari sp. n.
44. karadaghi (Povolný, 2001) comb. n.
45. heimi sp. n.
46. acuta (Povolný, 2001) comb. n.
47. soffneri Povolný, 1964
48. jariorum sp. n.
49. murinella (Duponchel, 1843)
    culminicolella (Staudinger, 1871)
    excelsa (Frey, 1880)
50. wiltshirei Povolný, 1966
    obrteliana Povolný, 1971 (subspecies)
51. caucasica (Povolný, 2001)
    bezengensis Povolný, 2001 syn. n.
52. pauperella (Heinemann, 1870)
    klimeschi Povolný, 1967
53. spumata (Povolný, 2001)
54. arenbergeri Povolný, 1973
55. nana Povolný, 1973
    caroxyli (Falkovitsh & Bidzilya, 2006) (subspecies)
56. heretica Povolný, 1973
    submagnificella Povolný, 1977 syn. n.
57. bigoti Povolný, 1973
    tunesica Povolný, 1979 (subspecies)
58. dorsoflava (Povolný, 1966)
59. magnificella Povolný, 1967
60. abstrusa sp. n.
61. superstes Povolný, 1977
62. plesiopicta Povolný, 1969
63. bradleyi Povolný, 1971
    glaserorum Povolný, 1977 syn. n.
    meyricki auctt., (*nec* Povolný, 1971) (misident.)
64. selectella (Caradja, 1920)
    fraterna Povolný, 1969 syn. n.
65. alterna (Falkovitsh & Bidzilya, 2006) comb. n.
66. lutea Povolný, 1977
67. niveifacies Povolný, 1977
    milleri Povolný, 1977 syn. n.
68. indignella (Staudinger, 1879)
    hyoscyamivora (Gerasimov, 1940) syn. n.
    pseudobsoletellum (Povolný & Gregor, 1955) syn. n.
    grossa Povolný, 1966
69. punctata (Povolný, 1996)
70. lagodes (Meyrick, 1926) comb. rev.
71. deluccae Povolný, 1966
72. atriplicella (Fischer von Röslerstamm, 1841)

atrella (Thunberg, 1788) (homonym)
detersella (Clemens, 1860) (homonym)
infumatella (Fuchs, 1901) syn. n.
brackenridgiella (Busck, 1903)
chenopodiella (Busck, 1916)
arogantella Povolný, 1967
altajica Povolný, 1969

73. suaedella (Richardson, 1893)
flavidorsella (Amsel, 1952)
hartigi Povolný, 1977 syn. n.
74. solitaria Povolný, 1969
75. dagmaris Povolný, 1987
rezniki Piskunov, 1990 syn. n.
turkmenica Piskunov, 1990 syn. n.
76. suasella (Constant, 1895)
77. hendrikseni sp. n.
78. traganella (Chrétien, 1915)
79. bazae Povolný, 1977
80. artemisiella (Treitschke, 1833)
ancillella (Bruand d'Uzelle, 1851)
paniculatella (Novickij, 1924)
mongolensis Povolný, 1969 syn. n.
oreocyrniella (Petry, 1904) (subspecies)
syriaca Povolný, 1967 (subspecies)
81. stangei (Hering, 1889)
saltenella (Meess, 1910)
82. suaedivorella (Chrétien, 1915)
detersipunctella (Toll, 1947)
83. bryophiloides Povolný, 1966
84. algeriensis Povolný & Bradley, 1964
85. deutschi sp. n.
86. disjectella (Staudinger, 1859)
87. mixta sp. n.
88. rebeli (Preissecker, 1914)
fuscella (Klimesch, 1938)
japonica Povolný, 1977 syn. n.
89. gallicella (Constant, 1885)
90. ustulatella (Staudinger, 1871)
91. postulatella sp. n.
92. nitentella (Fuchs, 1902)
seminella (Pierce & Metcalfe, 1935)
93. costella (Humphreys & Westwood, 1845)
costimaculella (Bruand d'Uzelle, 1859)
94. hyoscyamella (Stainton, 1869)
95. portosanctana (Stainton, 1859)
eremaula (Meyrick, 1891) syn. n.
lyciella (Walsingham, 1900) syn. n.
desertella (Rebel, 1901) syn. n.
bertramella (Lucas, 1940) syn. n.

leroyella (Lucas, 1950) syn. n.
reisseri (Povolný & Gregor, 1955) syn. n.
philolycii (Hering, 1957) syn. n.
gallincolella auctt., (*nec* Mann, 1872) (misident.)

96. vicaria (Meyrick, 1921)
tineiformis Povolný, 1967

97. ocellatella (Boyd, 1858)
ocellatella (Stainton, 1859) (homonym)
submissella (Stainton, 1859)
horticolella (Rössler, 1866)
clarella (Caradja, 1920)
obscurior (Rebel, 1927)
orientale (Gregor & Povolný, 1954)
portosanctana auctt., (*nec* Stainton, 1859) (misident.)

98. pulchra Povolný, 1967

99. gecko (Walsingham, 1911)

100. hannemanni Povolný, 1966
furva Povolný, 1969 (subspecies)
gamanthi (Falkovitsh & Bidzilya, 2006) (subspecies)

101. erichi Povolný, 1964

102. divisella (Rebel, 1936) sp. rev.

103. voltinella (Chrétien, 1898)

104. suaedicola (Mabille, 1906)
suaedicola (Amsel, 1939) (homonym)
mabillei Povolný, 1971

105. monochromella (Constant, 1895)

106. samadensis (Pfaffenzeller, 1870)
plantaginella (Stainton, 1883)
brunhildae (Schawerda, 1921)
zimmermanni (Zimmermann, 1923) n. inv.
mariae (Zimmermann, 1926)
testacella (Rebel, 1935)
echo (Meyrick, 1937)

107. salinella (Zeller, 1847)
omachella auctt., (*nec* Oberthür, 1888) (misident.)
zernyella (Rebel, 1918)
corsicanum (Gregor & Povolný, 1954)
ignotum (Gregor & Povolný, 1954)
trebujenae Povolný, 1977 syn. n.

108. spergulariella (Chrétien, 1910) sp. rev.

109. salicorniae (Hering, 1889) sp. rev.
caliacrae (Caradja, 1932) syn. n.

110. halimioniella sp. n.

111. thymelaeae (Amsel, 1939) sp. rev.

112. halymella (Millière, 1864) sp. rev.

113. camphorosmella Nel, 1999

114. stabilis Povolný, 1977 stat. n.

115. instabilella (Douglas, 1846)
lagunella (Chrétien, 1910)

strobilacella (Caradja, 1920) n. inv.
salsolella (Amsel, 1935)
halymiphaga (Amsel, 1952)
116. peterseni (Povolný, 1965)
117. ergasima (Meyrick, 1916) comb. rev.
hyoscyamella (Rebel, 1912) (homonym)
mignatella (Caradja, 1920) n. inv.
intestina (Meyrick, 1921)
mirabile (Gregor & Povolný, 1955)
pervada (Clarke, 1962)

Turcopalpa Povolný, 1973
118. glaseri Povolný, 1973

Scrobipalpula Povolný, 1964
119. psilella (Herrich-Schäffer, 1854)
nocturnella (Staudinger, 1859)
pallidella (Heinemann, 1870)
killiasii (Frey, 1880)
astericolellum (Hering, 1957) n. inv.
asiatica Povolný, 1968 (subspecies)
120. ramosella (Müller-Rutz, 1934)
121. seniorum Povolný, 2000
ptarmicae (Hering, 1957) n. inv.
compositella (Povolný, 1964) n. inv.
122. diffluella (Frey, 1870)
cacuminum (Frey, 1870)
diffluella (Heinemann, 1870)
bellidiastri (Klimesch, 1951)
uniflorellum (Hering, 1957) n. inv.
123. tussilaginis (Stainton, 1867)
tussilaginella (Heinemann, 1870)
retusella (Rebel, 1891)

Phthorimaea Meyrick, 1902
124. operculella (Zeller, 1873)
terrella (Walker, 1864) n. obl.
tabacella (Ragonot, 1879)
sedata (Butler, 1880)
solanella (Boisduval, 1874)
argentinae Povolný, 1989 syn. n.
piscipellis auctt., (*nec* Zeller, 1873) (misident.)
epicentra auctt., (*nec* Meyrick, 1909) (misident.)

Tuta Kieffer & Jørgensen, 1910
125. absoluta (Meyrick, 1917)

Keiferia Busck, 1939
126. lycopersicella (Walsingham, 1897)

Ephysteris Meyrick, 1908
Microcraspedus Janse, 1958
Opacopsis Povolný, 1964
Echinoglossa Clarke, 1965
127. promptella (Staudinger, 1859)
despectella (Walker, 1863)
petiginella (Mann, 1867)
parvula (Staudinger, 1879)
cacomicra (Walsingham, 1908)
chersaea Meyrick, 1908
oschophora (Meyrick, 1910)
crystallista (Meyrick, 1911)
dispensata (Meyrick, 1921)
fanatica (Meyrick, 1921)
xanthorhabda (Gozmány, 1951)
australiae Povolný, 1977
128. tenuisaccus Nupponen, 2010
129. deserticolella (Staudinger, 1871)
albocapitella (Rebel, 1928)
buvati (Povolný, 1992) syn. n.
130. insulella (Heinemann, 1870)
insularis (Staudinger, 1871)
praticolella (Christoph, 1872) (subspecies) comb. n., syn. n., stat. n.
gallica (Povolný, 1992) syn. n.
131. diminutella (Zeller, 1847)
lunaki (Hartig, 1941)
treskensis Povolný, 1964 syn. n.
hispanica Povolný, 1981 syn. n.
foulonsensis Povolný, 1981 syn. n.
132. inustella (Zeller, 1847)
delminiella (Rebel, 1904)
gredosensis (Rebel, 1935) (subspecies)
133. olympica Povolný, 1968
monticola Povolný, 1981 syn. n.
134. iberica Povolný, 1977

Ochrodia Povolný, 1966
135. subdiminutella (Stainton, 1867)
jamaicensis (Walsingham, 1897)
bucolica (Meyrick, 1904)
zygophyllella (Rebel, 1912)
ericnista (Meyrick, 1914)
ferritincta (Turner, 1919) (subspecies)
ochrodeta (Meyrick, 1923)
extorris (Meyrick, 1923)
crocoleuca (Meyrick, 1923)
unitella (Turati, 1930)
tribulivora (Dumont, 1931)
pulverea (Janse, 1950)

turgida (Janse, 1951)
pentamacula Janse, 1958
infallax (Gozmány, 1960)
tractatum (Gozmány, 1960)

Vladimirea Povolný, 1967
Distinxia Povolný, 1967
136. glebicolorella (Erschoff, 1874)
submaculata Povolný, 1967

Microlechia Turati, 1924 gen. rev.
Hedma Dumont, 1932 syn. n.
Megalocypha Janse, 1960 syn. n.
137. rhamnifoliae (Amsel & Hering, 1931) comb. n.
rhamnifoliae (Amsel, 1935)
138. chretieni Turati, 1924
microcasis (Meyrick, 1929) syn. n.
micradelpha (Walsingham, 1900) (homonym)
hyoscyamella (Amsel & Hering, 1931) (homonym)
abzacella (Dumont, 1932) syn. n.
hyoscyami (Amsel, 1935) syn. n.
polioptera (Janse, 1960) syn. n.
aellographa (Janse, 1960) syn. n.
139. karsholti (Nupponen, 2010) comb. n.

Cosmardia Povolný, 1965
140. moritzella (Treitschke, 1835)
morizella (Geyer, 1836)
roseella (Zetterstedt, 1839) syn. n.

Lutilabria Povolný, 1965
141. lutilabrella (Mann, 1857)
olympica Huemer, 1993 (subspecies) stat. n.
142. volgensis Anikin & Piskunov, 1996
143. prolata Junnilainen & Nupponen, 2010

Klimeschiopsis Povolný, 1967
144. kiningerella (Duponchel, 1843)
atralbella (Palm, 1947)
145. discontinuella (Rebel, 1899)
146. terroris (Hartig, 1938) sp. rev.

Caryocolum Gregor & Povolný, 1954
147. fischerella (Treitschke, 1833)
148. tischeriella (Zeller, 1839)
149. alsinella (Zeller, 1868)
albifrontella (Heinemann, 1870)
tristella (Heinemann, 1870)
semidecandriella (Tutt, 1887)

semidecandrella (Threlfall & Stainton, 1887)
150. viscariella (Stainton, 1855)
crepusculella (Teich, 1889) syn. n.
151. albifaciella (Heinemann, 1870)
behenella (Constant, 1890)
152. vicinella (Douglas, 1851)
inflatella (Chrétien, 1901)
albescens (Bankes, 1909) n. inv. (misident.)
suffusa (Bankes, 1909) n. inv. (misident.)
153. bosalella (Rebel, 1936)
154. amaurella (Hering, 1924)
viscariae (Schütze, 1926)
155. oculatella (Thomann, 1930)
ochraceella (Thomann, 1929) (homonym)
156. leucofasciatum Huemer, 1989
157. petryi (Hofmann, 1899)
rougemonti (Rebel, 1907)
repentella (Chrétien, 1908)
benanderi (Hering, 1933)
158. baischi sp. n.
159. repentis Huemer & Luquet, 1992
repentella auctt., (*nec* Chrétien, 1908) (misident.)
160. siculum Bella, 2008
161. inflativorella (Klimesch, 1938)
xuthella (Rebel, 1941)
census Gozmány, 1954)
162. saginella (Zeller, 1868)
coussonella (Chrétien, 1908)
163. cauligenella (Schmid, 1863)
164. trauniella (Zeller, 1868)
165. peregrinella (Herrich-Schäffer, 1854)
melantypella (Mann, 1877)
166. delphinatella (Constant, 1890)
fiorii (Klimesch, 1953)
167. provinciella (Stainton, 1869)
168. mucronatella (Chrétien, 1900)
poschiavensis (Rebel, 1936)
169. leucomelanella (Zeller, 1839)
gypsophilae (Stainton, 1869)
170. mazeli Huemer & Nel, 2005
171. leucothoracellum (Klimesch, 1953)
172. schleichi (Christoph, 1872)
syriacum Povolný, 1977
dianthella (Chrétien, 1925) (subspecies)
hackeri Derra, 1985
improvisella (Rebel, 1936) (subspecies)
arenariella (Benander, 1937) (subspecies)
173. marmorea (Haworth, 1828)
manniella (Zeller, 1839)

marmorella (Doubleday, 1859) (emendation)
marmoreum auctt.
pulchra (Wollaston, 1858) (subspecies)
pulchrum auctt.

174. pullatella (Tengström, 1848)
pulla (Tengström, 1848)
subtractella (Walker, 1864)
livoniella (Teich, 1898)
agricolaris (Meyrick, 1933)

175. stramentella (Rebel, 1935)
emarginatum Huemer, 1988 syn. n.

176. hispanicum Huemer, 1988

177. confluens Huemer, 1988

178. gallagenellum Huemer, 1989

179. fraternella (Douglas, 1851)
intermediella (Hodgkinson, 1897)

180. klosi (Rebel, 1917)

181. interalbicella (Herrich-Schäffer, 1854)
quadrella (Fabricius, 1794) (homonym)

182. laceratella (Zeller, 1868)
thurneri (Pinker, 1953)

183. blandella (Douglas, 1852)
maculea (Haworth, 1828), (*nec* Fabricius, 1794) (emendation, misident.)
horoscopa (Meyrick, 1926) (subspecies) syn. n., stat. n.

184. blandelloides Karsholt, 1981

185. jaspidella (Chrétien, 1908)

186. proxima (Haworth, 1828)
maculiferella (Douglas, 1851)
maculivicinella (Bruand d'Uzelle, 1859)
horticolla (Peyerimhoff, 1871)
proximum auctt.

187. blandulella (Tutt, 1887)

188. arenbergeri Huemer, 1989

189. tricolorella (Haworth, 1812)
contigua (Haworth, 1828)
acernella (Herrich-Schäffer, 1854)

190. fibigerium Huemer, 1988

191. junctella (Douglas, 1851)
aganocarpa (Meyrick, 1935)

192. cassella (Walker, 1864)
melanotephrella (Erschoff, 1877)
albifasciella (Toll, 1936)
subvicinella (Hackman, 1946)
falellum Piskunov, 1975

193. moehringiae (Klimesch, 1954)

194. petrophila (Preissecker, 1914)
kemnerella (Palm, 1947)
petrophilum auctt.

195. huebneri (Haworth, 1828)

hubnerella (Doubleday, 1866) (emendation)
knaggsiella (Stainton, 1866)
196. kroesmanniella (Herrich-Schäffer, 1854)
huebneri auctt., (*nec* Haworth, 1828) (misident.)

Tila Povolný, 1965
197. capsophilella (Chrétien, 1900)

Pogochaetia Staudinger, 1879
Pogonochaetia Rye, 1881 (emendation)
Chaetopogon Rye, 1881 (emendation)
198. solitaria Staudinger, 1879
ocymoidella (Walsingham, 1900) (subspecies)
cabreretsi Povolný, 1981

Agonochaetia Povolný, 1965
Sautereopsis Povolný, 1965
199. terrestrella (Zeller, 1872)
muestairella (Müller-Rutz, 1922)
200. intermedia Sattler, 1968
201. quartana Povolný, 1990

Sattleria Povolný, 1965
202. arcuata Pitkin & Sattler, 1991
203. pyrenaica (Petry, 1904)
204. angustispina Pitkin & Sattler, 1991
205. izoardi Huemer & Sattler, 1992
206. breviramus Pitkin & Sattler, 1991
207. melaleucella (Constant, 1865)
mariae (Frey, 1867) n. inv.
fusca (Burmann, 1954)
208. triglavica Povolný, 1987
basistrigella Huemer, 1997 (subspecies)
basistrigella (Müller-Rutz, 1934) n. inv.
209. styriaca Pitkin & Sattler, 1991
210. marguareisi Huemer & Sattler, 1992
211. dzieduszyckii (Nowicki, 1864)
tatrica (Gregor & Povolný, 1955)

Systematic treatment of the genera and species of Gnorimoschemini in Europe

Gnorimoschema Busck, 1900

Gnorimoschema Busck, 1900: 227.
Type species: *Gelechia gallaesolidaginis* Riley, 1869.
Lerupsia Riedl, 1965: 461.
Type species: *Lerupsia soffneri* Riedl, 1965.
Neoschema Povolný, 1967b: 81.
Type species: *Gnorimoschema* (*Neoschema*) *klotsi* Povolný, 1967.

DIAGNOSIS.– Adult. Wingspan 8-20 mm. Labial palp upturned; segment 2 roughly scaled below forming two ridges with a groove in between, colour variable, but lighter on upper and inner surface; segment 3 shorter than segment 2, smoothly scaled, often with one or two black rings. Antenna greyish black, more or less distinctly lighter ringed. Forewing more or less slender, mostly lighter or darker greyish or brownish from dark lighter-based scales and mottled with orange or brown, especially in fold, subcostal stripe and around black spots; light costal and tornal spots; termen with many black scales; cilia mostly grey with darker cilia line(s). Hindwing greyish.

MALE GENITALIA.– Uncus short and broad, medioapically pointed; gnathos-hook short to long and slender to broad; tegumen anteriorly strongly widened; valva extending to about apex of uncus, curved in basal half, apically and slightly constricted medially, apically scarcely to strongly inflated; sacculus short to long, hook-shaped; vinculum with small to large emargination, vincular process reduced, hump-like; saccus short, broadly subtriangular to subrectangular; phallus moderately short to long, almost straight to weakly curved, distal part slender, with small dentate subapical ridge, coecum distinctly inflated.

FEMALE GENITALIA.– Apophysis posterioris moderately short; segment VIII subquadrate, with membranous tergite and sclerotized sternite; sternum VIII medially membranous with numerous microtrichia, subgenital plates sclerotized and smooth or with additional sclerites; apophysis anterioris rod-like, slightly shorter to distinctly longer than segment VIII, inner margin occasionally with sclerotized rim prolonged towards ostium bursae; antrum membranous to short funnel; ductus bursae with ring-shaped to cylindrical colliculum; ductus bursae narrow; corpus bursae well defined, pyriform to globular; signum close to entrance of ductus bursae into corpus bursae, with moderately small base and strong more or less curved hook.

DISTRIBUTION.– Holarctic; with about 95 species in the Nearctic region, around 20 species in the Palaearctic region and three species occurring in the Holarctic region (Povolný, 2002a: 19; Lee *et al.*, 2009: 23-25).

BIONOMICS.– Insufficiently known and only described for few Palaearctic taxa. Nearctic species are frequently endophagous and some taxa cause galls. Palaearctic species are as far as known leaf-miners on Asteraceae, Salicaceae, Myricaceae and Lamiaceae.

REMARKS.– Some species with a vast range tend to geographical variation which is reflected in different wingspan and forewing colour of disjunctive populations. However, variation in genital characters is low and many of the described taxa are treated only as subspecies (Povolný, 2002a).

1. *Gnorimoschema soffneri* (Riedl, 1965)

Lerupsia soffneri Riedl, 1965: 461, figs. 61-62, 80.
Gnorimoschema antiquum Povolný, 1966c: 400, figs. 5-6, 22-24, 41, **syn. rev.**
Gnorimoschema antiquum montanum Povolný, 1966c: 402, fig. 6. Subspecies.

DIAGNOSIS.– Adult. Wingspan 10-12 mm. Labial palp white; segment 2 mottled with some black; segment 3 with two black rings. Antenna black, ringed with whitish grey. Head white, greyish towards neck; thorax and tegula as forewing. Forewing with white-based black scales partly overlaid by light orange-brown, especially along dorsum, in fold, subcostal stripe, around black spots and in apical part of wing; costal and tornal spots indistinct, cream-coloured. Hindwing grey.

Variation. A moderately variable species. The dark scales on the forewing can be grey or black, and they can be more or less overlaid with light orange-brown. Sometimes the black spots are hidden behind orange-brown scales. A specimen from Corsica is unusually dark, having also a black patch in the middle of the forewings about 1/3 from base. In specimens from Turkey the antennae are distinctly ringed black and white.

Similar species. *G. soffneri* is similar to *G. herbichii* but distinguished by its relatively small size and the variegated light grey and brown forewings with large orange-brown markings around the spots. *G. herbichii* is on average larger and has larger black markings in the middle of the forewing.

MALE GENITALIA.– Uncus broadly subrectangular, medioapically projected, moderately pointed; gnathos-hook short; valva extending almost to apex of uncus, strongly curved and slightly constricted medially, distal part scarcely inflated; sacculus moderately long, hook-shaped, gap to vincular process small; posterior margin of vinculum strongly projected posteriolaterally, with largely reduced vincular process, medially with small U-shaped emargination and with weak hump-like projection mediolaterally; saccus broadly subrectangular, short, extending slightly below apex of pedunculus; phallus long, stout, slightly curved medially, distal part slender, with small dentated subapical ridge, coecum inflated, about half length of phallus. Very similar to *G. herbichii*, but differs in the broader uncus and sacculus, the smaller medial humps of the vinculum and some characters of the phallus.

FEMALE GENITALIA.– Apophysis posterioris about two times length of segment VIII; sternite VIII subrectangular, broader than long; subgenital plate smooth, with folded and rounded posterior edge, lateromedial edge delimiting membranous medial zone which is covered with microtrichia; anterior margin of sternite VIII strongly sclerotized, transverse, without medial opening; apophysis anterioris slightly shorter than segment VIII, broad at base, distal two-thirds rod-like; antrum broadly funnel-shaped; colliculum short; ductus bursae with distinct transition to pyriform corpus bursae; signum with small basal plate and stout hook with distinct apical curvature. Differs from the overall similar *G. herbichii* in the reduced extent of microtrichia, the straight anterior margin of sternite VIII and the longer and basally broadened apophysis anterioris.

DISTRIBUTION.– The nominotypical race is restricted to southern Europe, from Spain to

Bulgaria; also in Turkey and Iraq. *G. soffneri montanum* occurs in the mountains of the Middle East (Iran, Afghanistan).

New records. France (Corsica): 1 ♀, 15 km SSE Bastia at l'Etang de Biguglia, 10 m, 20.vi.2005, leg. Skule, gen. slide 6401 Hendriksen (ZMUC). Greece (Crete): 2 ♂, Lassithi distr., Makrigialos, 20.v.1993, leg. Johansson (ZMUC). Confirmed record. Italy (Sicily): 2 ♂, 2 ♀, Nebrodi, San Fratello, Torrente Ingànno, 300 m, 10.-11.ix.2002, leg. Karsholt *et al.*, gen. slide 3866 Hendriksen (ZMUC).

BIONOMICS.– Host-plant and early stages unknown. The adults have been observed in a single generation from early June to late July. A specimen from Iraq collected in late October indicates bivoltinism in this area. Material of *G. antiquum montanum* was collected from early April to August. European populations are restricted to steppe habitats from sea-level to about 1400 m. In Afghanistan the species has been observed up to elevations of 3000 m (Povolný, 1992a: 227). The moth is attracted to light.

REMARKS.– *Lerupsia soffneri* was described from one male collected by Soffner at Nessebar in Bulgaria (Riedl, 1965: 461). The original description is illustrated with a photograph of the moth and drawings of its genitalia and the abdominal base. As can be seen from the picture the hindwings of the holotype were rolled up lengthwise so that the arched termen was hidden, and that misled Riedl to place *Lerupsia* in the Momphidae. The genitalia of the holotype are figured in a lateral position but clearly present characters of *Gnorimoschema*, with which *Lerupsia* was formally synonymized by Sattler (1968: 118). Based on the original description Karsholt & Nielsen (1974: 91) established *L. soffneri* as a senior synonym of *Gnorimoschema antiquum* (Povolný, 1966). This synonymy was not accepted by Povolný (1982: 122-125) who depicted the genitalia in a dorso-ventral position, allegedly from the same genitalia slide (prép. gen. 111/R) that Riedl had pictured in a lateral position.

The male genitalia of *G. herbichii* and *G. soffneri* are very similar, and the main diagnostic character (the size of the medial humps of the vinculum and the incision between them) is difficult to interpret in Povolný's figure, because of damage to this structure, and it is completely hidden in Riedl's original drawing. A safe identification can therefore not be made from these figures alone. Regrettably Povolný does not figure or discuss the external appearence of the holotype at all.

The holotype of *L. soffneri* was stated to be in the collection of J. Soffner in Stassfurt, Germany. After his death Soffner's collection went into the collection of the SDEI, but without this specimen. Povolný (1982: 122, 125) stated that he had studied (at least the genitalia slide of) the holotype. At our request Petr Baňař searched for the holotype of *L. soffneri* in the Povolný collection in the MZMB and found a specimen labelled as holotype, but he could not find the accompanying genitalia slide. We were able to examine the specimen which indeed bears all the original labels of the holotype. However, it is not the specimen figured by Riedl!

The photograph of the holotype in the original description shows a rather small species (wingspan stated to be 9 mm). However, the specimen in the MZMB with unrolled hindwings has a different wing pattern and, especially, a larger white subbasal dot on the right forewing (probably damage caused by setting the specimen). This dot is missing in the photograph of the holotype (Riedl: 1965: fig. 80). It is also a larger and more slender-winged specimen which fits to *G. herbichii*, whereas the original photograph of the holotype largely agrees with *G. antiquum*. The synonymy of *G. antiquum* with *L. soffneri* is moreover supported by the fact that Soffner collected both taxa in the same locality, Nessebar in Bulgaria (Karsholt & Nielsen, 1974: 461).

We conclude that the specimen in the MZMB, although bearing Riedl's original labels, is a fake. Moreover, as the genitalia slide of the holotype is currently lost, we cannot prove that Povolný made his drawing in dorso-ventral position from this laterally embedded slide (he did so in a few other cases) rather than from another specimen. To summarize, the original description of *L. soffneri* points towards *G. antiquum*, and in the absence of evidence to the contrary we consider *G. soffneri* as a valid species with *G. antiquum* as its junior subjective synonym – as previously established by Karsholt & Nielsen (1974: 91) as the first revisors.

G. soffneri montanum from Afghanistan and Iran differs from the nominotypical subspecies in the unicolorous yellowish loamy to ochreous forewings without distinct markings (Povolný, 2002a: 22). However, intermediate forms are known from north-western Iran (Povolný, 1992a: 227).

2. *Gnorimoschema herbichii* (Nowicki, 1864)

Gelechia herbichii Nowicki, 1864: 17, pl. 1, fig. 6.
Lita pusillella Rebel, 1893: 47.
Gelechia (*Lita*) *tengstroemiella* Joannis, 1910: 296.
Lita pazsiczkyi Rebel, 1913: 173, fig.
Lita parentesella Toll, 1936: 407, pl. 49, fig. 18.
Phthorimaea tengstroemi Hackman, 1946: 61, figs. 2, 5.
Gnorimoschema herbichi [sic] *mongoliae* Povolný, 1973a: 19, figs. 4, 14, 22. Subspecies.
Gnorimoschema herbichi [sic] *kamchaticum* Povolný, 1977d: 218, fig. 14. Subspecies.

DIAGNOSIS.– Adult. Wingspan 10-15 mm. Labial palp white; segment 2 more or less mottled with blackish brown; segment 3 with two broad, black rings. Antenna black, ringed with light grey. Head cream-coloured; neck and tegula light greyish brown; thorax blackish grey. Forewing greyish from white-based black scales and overlaid by whitish, light brown and orange-brown; black patches at base, at 1/4 and 3/4 along costa, in middle of wing (distinct) and posterior to tornus; spots in fold and at 1/3 and 2/3 in middle of wing surrounded by orange-brown; tegumen and apical area with many black scales. Hindwing whitish grey with darker margins.

Variation. *G. herbichii* exhibits both individual and geographical or ecological variation. Specimens from sand dune areas in north-western Europe have segment 2 of the labial palps almost pure white, more distinctly ringed antennae and more white on the forewings, e.g. white costal and tornal spots forming a postmedial fascia. Specimens from north-eastern Europe have rather variegated forewings with many orange-brown scales, and the black spots can be covered with orange-brown. Specimens from western Europe are darker with a more blackish ground colour, and they have a black thorax.

Similar species. *G. herbichii* is characterized by its slender, variegated forewings with a conspicuous black patch about 1/3 from base. Due to its variability it can be confused with other species of *Gnorimoschema*. Dark specimens may resemble *G. streliciella*, which can be separated by its more blackish forewings and its dark grey hindwings. *G. bodillum* q.v.; *G. soffneri* q.v.

MALE GENITALIA.– Uncus moderately narrow, subrectangular, medioapically projected, pointed; gnathos-hook short; valva extending almost to apex of uncus, strongly curved and slightly constricted medially, distal part weakly inflated; sacculus long, slender, hook-shaped, with small gap to vincular process; posterior margin of vinculum strongly projected posteriolaterally, with largely reduced vincular process, medially with small U-shaped emargination and with distinct hump-like projection mediolaterally; saccus broadly subrectangular, short, extending slightly below apex of pedunculus; phallus long, stout, slightly curved medially, distal part slender, with small sub-apical ledge, coecum inflated, less than half length of phallus. Differs from the otherwise very similar *G. soffneri* q.v. in the more slender uncus and sacculi, the medial humps of the vinculum and the subapical ledge of the phallus.

FEMALE GENITALIA.– Apophysis posterioris about two times length of segment VIII; sternite VIII subrectangular, broader than long; subgenital plate smooth, with folded and weakly rounded posterior edge, lateromedial edge delimiting membranous medial zone which is densely covered by microtrichia; anterior margin of sternite VIII strongly sclerotized, transverse, medially sinuous without opening; apophysis anterioris distinctly shorter than segment VIII, rod-like throughout; antrum broadly funnel-shaped; colliculum short; ductus bursae with distinct transition to pyriform corpus bursae; signum with small basal plate and moderately stout hook with weak apical curvature. Similar to *G. soffneri* q.v. but differs in the more extensive microtrichia, the sinusoid transverse anterior margin of sternite VIII and the shorter apophysis anterioris slender throughout.

DISTRIBUTION.– Scattered records from Spain and France to Central Europe, and from Scandinavia to eastern Europe including Russia. Outside Europe recorded from Iraq, Mongolia, Kamchatka and Canada (Povolný, 1966c: 400, 2002a: 24) and also from Siberia (Bidzilya, 2009: 4).

BIONOMICS.– The larva has been recorded from *Equisetum palustre* L. (Equisetaceae) (Kaitila, 1996: 93), which has proved to be a misidentification of *E. arvense* L. (M. Mutanen, *in litt.*). Mature larvae were observed in spun galleries beneath the host-plant in the sand in early July. At the same time pupae and adults were found indicating the major larval period in June (Kaitila, 1996: 102). In Denmark the larva has been found feeding on *Salix repens* L. (Salicaceae) in coastal sand dunes, living in a similar way to *G. bodillum* (Gregersen & Kjeldgaard, 2009: 236-240). In northern Europe the adult occurs from early May to early August, in the Mediterranean region from early April to mid-September, exceptionally until late October (Iraq), probably in two generations. The moth can be disturbed from the host-plants during sunny afternoons and can be attracted to light. The species seems to prefer sandy soil but the habitat is insufficiently documented.

REMARKS.– Povolný (2002a: 23-24) separates three subspecies: *G. herbichii herbichii* from Europe, with large wingspan and light greyish forewing colour with reddish brown spots; *G. herbichii mongoliae* from Transbaikal and Mongolia, with smaller wingspan and distinctly darker grey-brown forewings; *G. herbichii kamtchaticum* from Kamchatka and possibly also Canada (Ontario) with dark forewings and distinct dots. However, there seems to be some variation within these subspecies, e.g. European specimens become darker towards the north of the range (Kaitila, 1996: 103).

A male syntype of *Lita parentesella* from ISEZ, Kraków, labelled 'Bydgorrer 24.VI.1930' (slide 1679 ♂ Toll) was examined by us and agrees with *G. herbichii*.

The repeated usage of *G. herbichi* (e.g. Povolný, 2002a: 23-24) is a misspelling.

3. *Gnorimoschema bodillum* Karsholt & Nielsen, 1974

Gnorimoschema bodillum Karsholt & Nielsen, 1974: 91, figs. 1-9.

DIAGNOSIS.– Adult. Wingspan, male 8-12 mm; female 10-14 mm. Labial palp white; apical half of segment 3 black. Antenna black, ringed with whitish grey. Head, thorax and tegula as forewing. Forewing light grey-brown, mottled with yellowish and white, especially on costal half; costa white with grey-black patches near base and at 1/4 and 3/4; base grey-brown; black spots in fold and at 2/5 and 3/5 in middle of wing surrounded by orange-brown scales; subcostal stripe light yellow-brown; a black patch posterior of tornus; termen variegated black and white. Hindwing grey, lighter towards base.

Variation. There is some variation from lighter to darker in the grey-brown scales of the forewing. Whereas the spots in fold are normally black the ones in the middle of the forewing are often without black scales, thus being orange-brown. Worn specimens look darker than fresh ones. Females are generally larger than males.

Similar species. *G. bodillum* is similar to *G. herbichii* but distinguished by its rather small size, the pure white segment 2 of the labial palps and by its light grey-brown forewings with rather indistinct spots. Whitish specimens of *G. herbichii* have slightly more slender forewings with many black scales, distinct spots and a light postmedial fascia.

MALE GENITALIA.– Uncus broadly subrectangular, medioapically projected, moderately pointed; gnathos-hook broad, long; valva extending almost to apex of uncus, strongly curved and slightly constricted medially, apically distinctly inflated; sacculus moderately strong, hook-shaped, with small gap to vincular process; posterior margin of vinculum strongly projected posteriolaterally, with moderately distinct broadly shouldered and pointed vincular process, medially with broad U-shaped emargination and additonal deep medial incision, without mediolateral projections; saccus subtriangular, apically broadly rounded, short, extending below apex of pedunculus; phallus long, straight, distal part slender, with small apical hook, coecum strongly inflated, about two fifths length of phallus. The emargination of the posterior margin of the vinculum is unique within the European fauna.

FEMALE GENITALIA.– Apophysis posterioris about two times length of segment VIII; sternite VIII subrectangular with anteriorly projected lateral parts; subgenital plate smooth, with folded posterior edge, sclerotized parts medially separated by narrow membranous zone which is covered with microtrichia; anterior margin of sternite VIII broadly concave, moderately sclerotized, without small medial opening; apophysis anterioris about two-thirds length of segment VIII, rod-like; antrum small, funnel-shaped; colliculum longish, posterior part cylindrical; ductus bursae with abrupt transition to pyriform corpus bursae; signum at right side of corpus bursae, basal plate small, signum-hook straight with moderately curved apex. The smooth and unstructured subgenital plate is characteristic for this species.

DISTRIBUTION.– Only known from coastal areas of the North Sea in Denmark and Germany. Recently also recorded from Siberia (Taimyr Peninsula) (Bidzilya, 2005: 14), although the figured specimen is darker than northern European samples and also genitalia show slight differences.

BIONOMICS.– The larva has been recorded from *Salix repens* L. (Salicaceae) and exceptionally *Myrica gale* L. (Myricaceae) (Karsholt & Nielsen, 1974: 95) and is exclusively restricted to sand dunes. It produces sandy tubes below the surface and constructs a system of galleries. The larva feeds from a tube on the lowest leaves of plants that have been sanded over but which are not completely covered. Pupation takes place in a separate tube close to the sand surface. The adult occurs from June to August, often together with various larval stages. Overlapping generations or a certain dispersion of generations seems possible (Karsholt & Nielsen, 1974: 96). The imago has not been observed flying freely but it runs or jumps on the sand surface for short distances during day time.

REMARKS.– *G. bodillum* is stated to be closely related to *G. vastificum* Braun, 1929, from North America (Powell & Povolný, 2001: 6), a species also restricted to sand dunes. We could examine specimens of *G. vastificum* from California and found several differences in both adults and, especially, the male genitalia. The similar biology with feeding on *Salix* spp. leaves from silken tubes hidden in sand is found also in other more or less distantly related *Gnorimoschema* species like *G. herbichii* and *G. vibei* (Wolff, 1954) from Greenland (Karsholt, *et al.*, in press).

4. *Gnorimoschema nupponeni* sp. n.

DIAGNOSIS.– Adult. Wingspan 12-14 mm. Labial palp creme-white; segments 2 and 3 each with two black rings, most distinct in segment 3. Antenna black, indistinctly ringed with grey. Head grey; thorax and tegula blackish grey. Forewing blackish grey from white-based black scales and mottled with orange, especially in fold, subcostal stripe and (partly) surrounding the black spots; a black streak from middle of wing to apex; light costal and tornal spots. Hindwing light grey.

Variation. The type series is rather constant. One female has lighter labial palps, head and forewings, the latter with a brown longitudinal streak.

Similar species. *G. nupponeni* sp. n. is characterized by its blackish grey forewings with a moderate number of contrasting orange scales, and by a black streak through its apical part. It resembles *G. steueri* which has more greyish forewings with less distinct black spots and no apical streak.

MALE GENITALIA.– Uncus broad and short, medioapically strongly projected to acute tip; gnathos-hook long and broad, massive; valva shorter than uncus, strongly bent and slightly constricted medially, apically strongly inflated; sacculus sickle-shaped, short and stout; vincular process reduced to weak hump, posterior margin of vinculum with small, shallow U-shaped emargination and V-shaped medial incision; saccus subrectangular, moderately long, extending to about apex of pedunculus; phallus moderately short, straight, distal part slender, with small lateral dentation near apex, coecum strongly inflated, globular, about two-fifths length of phallus. Differs from similar species such as *G. robustella* and *G. steueri* e.g. in the shorter sacculus and the different shape of the phallus.

FEMALE GENITALIA.– Apophysis posterioris at most up to three times length of segment VIII; sternite VIII subrectangular, broader than long; subgenital plate smooth, lateromedial edge delimiting a narrow and membranous medial zone, weakly folded and narrow sclerotized band extending from inner edge of subgenital plate to base of apophysis anterioris; anterior margin of sternite VIII weakly concave, sclerotized, with small medial opening; apophysis anterioris slightly longer than segment VIII, broader at base, distal three-quarters slender, rod-like; antrum without specific sclerotization; colliculum short, ring-shaped; ductus bursae about 1.5 times length of apophysis anterioris, distinctly separated from sub-oval corpus bursae; signum a large and extremely stout hook with moderate curvature, basal plate small. Differs from otherwise similar species such as *G. robustella* and *G. steueri* q.v. particularly in the position and shape of the sclerotized fold which is unique in the genus.

DISTRIBUTION.– Only known from the type-localities in Russia (southern Ural Mountains) and the Ukraine (Crimea).

BIONOMICS.– Early stages and host-plant unknown, although the latter is possibly *Aster oleifolius* (Lam.) Wagenitz (= *Linosyris villosa* (L.) DC.) (Asteraceae) because the moth was found in numbers by sweeping on this plant (Nupponen, *in litt.*). The moth has been collected by day and most frequently at about midnight at light. The adults have been collected from late August to late September.

MATERIAL EXAMINED.– Holotype. ♂, RUSSIA S-Ural Orenburg oblast Donskoje village 6km W mount Verbljushka 27.viii.2000 K. Nupponen leg.' 'GU 09/1280 ♂ P. Huemer' (NUPP).

Paratypes. Russia: 5 ♂, 1 ♀, same data as holotype but gen. slide 6426 Hendriksen (NUPP); 3 ♂, 28.viii.2000, gen. slide 3160 Hendriksen (NUPP; ZMUC); 4 ♂, same data as holotype but 9.ix.2002 (NUPP); 1 ♂, same data as holotype but 10.ix.2002 (NUPP); 3 ♂, same data as holotype but 5.ix.2009 (NUPP); 6 ♂, same data as holotype but 6.ix.2009 (NUPP); 1 ♂, S-Ural, Orenburg oblast, near Burannoe village, 29.viii.2000, leg. Nupponen, gen. slide 4997 Hendriksen (NUPP); 1 ♂, 12 km SE Kuvandyk, 2.ix.2000, leg. Nupponen (NUPP); 97 ♂, 1 ♀, S-Ural, Orenburg district, 40 km WSW Orsk, spring near Guberlja River, 7.ix.2009, K. Nupponen leg. (NUPP); 54 ♂, 3 ♀, ditto, but 8.ix.2009 (NUPP, TLFM, ZMUC); 17 ♂, 1 ♀, S-Ural, Orenburg district, 40 km W Orsk, near Guberlja village, 9.ix.2009, leg. K. Nupponen (NUPP); 1 ♂, Orenburg district, Kumak river, 220 m, 10.ix.2009, leg. Nupponen (NUPP). Ukraine: 1 ♂, Crimea, Karadagh, 26.ix.1988, leg. Budashkin, gen. slide 6329 Hendriksen (ZMUH); 1 ♂, same data but 27.-28.ix.2004, gen. slide 220/08 ♂ Bidzilya (ZMKU).

ETYMOLOGY.– The name is dedicated to our colleague Kari Nupponen who significantly contributed to our work with most valuable material from Russia.

5. *Gnorimoschema robustella* (Staudinger, 1871)

Gelechia robustella Staudinger, 1871: 312.
Phthorimaea syrphetopa Meyrick, 1926: 278.

DIAGNOSIS.– Adult. Wingspan 16-17 mm. Labial palp white; segment 2 weakly mottled with brownish on outer surface; segment 3 almost blackish brown. Antenna grey, indistinctly lighter ringed. Face white; neck, thorax and tegula brown. Forewing brown, mottled with blackish brown, orange-brown and white, the latter especially at base, at 1/6 on costa, between three black spots in fold and as a faint, angulated fascia from costal to tornal spot; black spots at 2/5 and 3/5 in middle of wing surrounded by orange-brown; termen with black spots. Hindwing light grey.

Variation. The above description is based on few, probably faded specimens. Fresh specimens are possibly more greyish.

Similar species. *G. robustella* is characterized by its large size, almost white segment 2 of the labial palps and by its rather variegated forewings.

MALE GENITALIA.– Uncus broad, medioapically projected with distinct tip; gnathos-hook long, stout; valva stout, extending to about tip of uncus, evenly curved, without distinct apical inflation; sacculus sickle-shaped, evenly curved and strongly hooked; vincular process broadly hump-shaped, posterior margin of vinculum with broad and V-shaped medial incision; saccus subtriangular, short, extending distinctly below apex of pedunculus; phallus straight, distal part moderately long and slender, with small apical hook, coecum distinctly inflated.

FEMALE GENITALIA.– Apophysis posterioris about three times length of sternite VIII; sternite VIII short, distinctly broader than long; subgenital plate with slightly oblique, transverse and strongly sclerotized fold, medially with short hump, laterally extending almost to lateral margin of sternite, anterior margin sinusoid; area between subgenital plates small, membranous, with few microtrichia; apophysis anterioris about length of segment VIII, with indistinct and gradual transition to sternite VIII, broad at base, distal half rod-like; colliculum irregularly sclerotized, small; ductus bursae long, about 7 times length of segment VIII, narrow, distinct transition to large and pyriform corpus bursae; signum near entrance at right side of corpus bursae, basal plate small, signum-hook stout, hardly curved.

DISTRIBUTION.– Known from south-western Russia (Krasnoarmeysk and the southern Ural Mountains) and Kazakhstan (Uralsk, Lake Indersky).

BIONOMICS.– Host-plant and early stages unknown. According to the few available data the adults have been collected in June on saline steppes.

6. *Gnorimoschema steueri* Povolný, 1975

Gnorimoschema steueri Povolný, 1975: 190, figs. 1-3, 6-9.

DIAGNOSIS.– Adult. Wingspan 12-15 mm. Labial palp cream-white; segment 2 mottled with grey and black; segment 3 with two ill-defined black rings. Antenna black, indistinctly lighter ringed. Head grey with light face; thorax and tegula dark grey, the latter with lighter tip. Forewing grey from white-based black scales and mottled with some orange-brown, especially along fold, in subcostal stripe, around black spots (rather sparsely) and at 3/4 from base. Hindwing whitish grey.

Variation. The examined specimens show only slight variation from lighter to darker grey. In some specimens the medial part of the forewings from base to apex is darker than the costal and dorsal parts. Females are slightly smaller than males. Worn specimens look lighter than fresh ones.

Similar species. *G. steueri* is characterized by its indistinct black spots, which are weakly suffused by orange-brown scales. *G. epithymella* has a black patch from 1/4 of costa to the black spots on the fold, and broadly orange-brown surrounded spots. *G. nupponeni* sp. n. q.v.

MALE GENITALIA.– Uncus broad and short, medioapically strongly projected to acute tip; gnathos-hook long and broad, massive; valva shorter than uncus, strongly bent and slightly constricted medially, distal part significantly inflated; sacculus sickle-shaped, long and stout, evenly curved; vincular process largely reduced to indistinct hump, posterior margin of vinculum with small, shallow U-shaped emargination and V-shaped medial incision; saccus broadly rounded to subrectangular, short, extending below apex of pedunculus; phallus moderately short, straight, distal part slender, with lateral dentation and small apical hook, coecum strongly inflated, about half length of phallus.

FEMALE GENITALIA.– Apophysis posterioris almost four times length of segment VIII; sternite VIII subrectangular, slightly broader than long; subgenital plate smooth, latero-medial edge delimiting a narrow and smooth membranous zone, sinusoid and broadly folded sclerotization from anterior margin to central part of sternite with straight longitudinal inner margin; anterior margin of sternite VIII weakly concave, sclerotized, with small medial opening; apophysis anterioris slightly longer than segment VIII, weakly broader at base; antrum a short ring; colliculum irregularly ring-shaped with knob-like sclerotization; ductus bursae about two times length of apophysis anterioris, distinctly separated from sub-oval corpus bursae; signum a large and extraordinarily stout hook with moderate curvature, basal plate small. The shape of the sclerotization of sternite VIII and the signum make this species unmistakable.

DISTRIBUTION.– Only known from widely separated localities in Central Europe, between France and Slovakia. An isolated record from Transbaikal (Ponomarenko, 2008: 96) needs verification.

BIONOMICS.– The larva has been recorded from *Aster amellus* L. (Asteraceae) (Vávra, 1999: 157). Last instar larvae were observed feeding in large whitish blotch mines in leaf blades. They changed mines within one plant several times and were able to pass to another host-plant if the affected plant was too small or destroyed. The mines communicated with the petiole and the root-neck by a narrow gallery. 1 to 2 cm long resting tubes made from silk and grains were constructed at the base of the host-plant close to the soil surface. Pupation took place in late July below the soil surface in cocoons covered with sand and grains (Vávra, *op. cit.*).The adults have been collected in August and early September at light. Favourite habitats are south-facing limestone hills at low altitudes.

7. *Gnorimoschema epithymella* (Staudinger, 1859)

Gelechia epithymella Staudinger, 1859: 242.
Phthorimaea brunneomaculella Hackman, 1946: 60, figs. 3, 6. Subspecies.
Phthorimaea boerneri Amsel, 1952: 123, fig. 29. Subspecies.
Gnorimoschema epithymellum kirgisicum Povolný, 1994a: 61, figs. 3, 8. Subspecies.

DIAGNOSIS.– Adult. Wingspan 12-14mm. Labial palp white; segment 2 mottled with black; segment 3 with two black rings. Antenna black, ringed with light grey. Face cream-coloured; neck grey; thorax and tegula dark grey. Forewing covered with lighter-based blackish grey scales and mottled with black and orange-brown, the latter especially in fold, short subcostal stripe and broadly around black spots; an oblique, black patch from 1/4 of costa towards middle of wing; light costal and tornal spots indistinct. Hindwing light grey.

Variation. *G. epithymella* exhibits a pronounced geographical variation. The above description is based on a specimen of the nominotypical subspecies from Spain. Specimens from north-eastern Europe (spp. *brunneomaculella* – not figured) have the black scales on the forewings clearly white-based and the brownish scales more scattered over the wings. Such specimens look more or less brown, and with the black spots sometimes being covered with orange-brown. Ssp. *boerneri* from Sardinia is much lighter, having more whitish labial palps and whitish grey to light grey forewings with black base, a distinct black patch from 1/4 of costa to the black spots on the fold, and much black in the apical part. Specimens within the above mentioned populations show some individual variation. Worn specimens of ssp. *boerneri* become lighter, whereas specimens of ssp. *epithymella* and ssp. *brunneomaculella* become darker. Populations from the Alps which are attributed to the nominotypical subspecies are characterized by the greater extent of orange-brown colour on the forewing.

Similar species. Due to its geographical variability *G. epithymella* can be mistaken for a number of other species, especially of the genus *Scrobipalpa*, but males can be separated by the more slender uncus and the strongly inflated tip of the valva. *G. epithymella brunneomaculella* is very similar to some forms of the slightly larger *G. nordlandicolella*; most specimens of the latter have lighter orange-brown scales on the forewing and more prominent spots. *G. steueri* q.v.

Male genitalia.– Uncus moderately long, slender, medioapically projected to weak tip; gnathos-hook small, stout; valva extending to about middle of uncus, distinctly bent and slightly constricted medially, apex extraordinarily inflated; sacculus a short, stout, evenly curved hook; vincular process largely reduced to indistinct hump, posterior margin of vinculum with broad and shallow U-shaped emargination and small V-shaped medial incision; saccus subrectangular, short, extending distinctly below apex of pedunculus; phallus short, straight, distal part slender, with small apical hook, coecum distinctly inflated. Differs from other species e.g. in the strong inflation of the valva, the hook-like sacculus and the subrectangular saccus.

Female genitalia.– Apophysis posterioris about three times length of segment VIII; sternite VIII subrectangular, broader than long; subgenital plate smooth, with folded posterior edge, lateromedial edge delimiting a narrow and smooth membranous medial zone; oblique and longitudinal narrow band ridge-like sclerotization in medial part of subgenital plate; anterior margin of sternite VIII concave, strongly sclerotized, with rather small medial opening; apophysis anterioris slightly longer than segment VIII, broadened at base, distal two thirds rod-like; antrum membranous; colliculum a short and smooth ring; ductus bursae distinctly separated from pyriform corpus bursae; signum a large and moderately stout hook with distinct apical curvature, basal plate small. Differs from *G. nordlandicolella* q.v. in the more narrow band ridge of the subgenital plate, the shape of the colliculum and the comparatively slender signum-hook.

Distribution.– Scattered and isolated records from Spain and France to Central Europe (exclusively the Alps), Sardinia, and from Scandinavia to the Baltic states and Byelorussia. Also found on the Kola Peninsula of north-western Russia (Kozlov & Kullberg, 2006: 126). Outside Europe recorded from Algeria, and Kyrgyzstan (Povolný, 2002a: 23) and recently from Siberia (Bidzilya, 2005: 15).

Bionomics.– The larva has been recorded from *Solidago virgaurea* L. in Finland (Kaitila, 1996: 102) and from *Aster alpinus* L. (Asteraceae) in Austria (Huemer, pers. obs.), the latter erroneously published as *A. amellus* L. (Kaitila, *op. cit.*, Povolný, 2002a: 23). On *S. virgaurea* the larva feeds from May to June inside the lower stem, leaf stalks or lower leaves, spinning a silken gallery beside the root. On

A. alpinus it has been observed mining leaves. The adults occur from July to August in Central and northern Europe, and in April and May in Spain. They are attracted to artificial light. Preferred habitats are steppe rock formations and grassland up to an altitude of about 1200 m in the Alps and 2500 to 2900 m in Kyrgyzstan.

REMARKS.– Populations of *G. epithymella* are disjunctive and sometimes differ in their external appearance. Some have been named and are occasionally treated as subspecies (Povolný, 2002a: 23): *G. e. epithymella* from North Africa, western and Central Europe, with variegated forewings; *G. e. boerneri* from Sardinia, with occasionally lighter dorsum and lack of brownish scales of the forewings; *G. e. brunneomaculella* from north-eastern Europe, with dark brown forewings and some reddish suffusion; *G. e. kirgisicum* from Kyrgyzstan, with strong reddish brown suffusion of the forewings. However, no differences could be found in genitalia characters between these populations.

8. Gnorimoschema nordlandicolella (Strand, 1902)

Gelechia (Lita) nordlandicolella Strand, 1902: 21.
Phthorimaea cyceonodes Meyrick, 1926: 278.
Phthorimaea eucausta Meyrick, 1929: 492.
Phthorimaea fennicella Hackman, 1946: 60, figs. 1, 4.

DIAGNOSIS.– Adult. Wingspan 14-17 mm. Labial palp white, mottled with blackish brown on lower and outer surface of segment 2 and with black on segment 3. Antenna black, ringed with light grey. Head light grey with lighter face; thorax and tegula as forewing. Forewing light greyish brown from white-based black scales and mottled with light orange-brown, especially in fold, subcostal stripe and around spots; and with white-based black scales along veins; costal and tornal spots white; termen with many black scales. Hindwing grey.

Variation. The amount of lighter or darker orange-brown scales of the forewing is variable. Especially in specimens with many such scales the black spots and the costal and tornal spots are indistinct. Specimens from the mountains of central Turkey are darker from more blackish scales, having more distinct spots and indistinct costal and tornal spots.

Similar species. *G. nordlandicolella* is a rather characterless *Gnorimoschema*. It may be recognized by its relatively large size, the light greyish brown forewings and the white costal and tornal spots. *G. nilsi* q.v., *G. epithymella* q.v.

MALE GENITALIA.– Uncus broadly subrectangular, medioapically projected, moderately pointed; gnathos-hook short and slender; valva extending to about middle of uncus, basal half strongly bent, distal part straight with moderately inflated apex; sacculus moderately long and strongly curved, crescent-shaped, with distinct gap towards posterior margin of vinculum; posterior margin of vinculum without separated vincular process, posteriolaterally projected, medially scarcely emarginated, slightly incised in middle and with hump-like projection in lateral part; saccus subtriangular, broad at base, evenly tapered to slightly rounded or truncate apex, extending to about apex of pedunculus; phallus short, distal part slender with small apical hook, coecum strongly inflated. Similar to *G. nilsi* q.v. but differs in several genitalia characters, particularly the smaller gnathos-hook and the less inflated valva.

FEMALE GENITALIA.– Apophysis posterioris about three times length of segment VIII; sternite VIII subrectangular, broader than long; subgenital plate smooth, with folded posterior edge, lateromedial edge delimiting a narrow and smooth membranous medial zone; oblique and longitudinal broad band ridge-like sclerotization in medial part of subgenital plate; anterior margin of sternite VIII concave, strongly sclerotized, with small medial opening; apophysis anterioris longer than segment VIII, broadened at base, distal two-thirds rod-like; antrum membranous; colliculum a moderately long ring with irregular lateral sclerites; ductus bursae distinctly separated from pyriform corpus bursae; signum a large and stout hook with distinct apical curvature, basal plate small. Differs from *G. epithymella* q.v. particularly in the broader band and slightly serrate ridge of the subgenital plate and the massive signum-hook.

DISTRIBUTION.– Locally recorded from Norway, Sweden, Finland and Estonia (Jürivete, 2007: 13). Furthermore known from Russia (Kola Peninsula, Novgorod, Irkutskaja oblast), Turkmenistan, Kyrgyzstan and an unpublished record from Turkey. Records from Canada and the U.S.A. (Texas, Arizona) need further examination (see Remarks).

New record. Turkey: 1 ♀, Kayseri, 5km NW Erciyes Daği, 2000 m, 4.vii.1987, leg. Fibiger, gen. slide 4399 ♀ Hendriksen; 1 ♂, ditto, but 22.vii.1986 (ZMUC).

BIONOMICS.– Host-plant and early stages unconfirmed. The adults have been observed from early June to mid-August, preferably at light. The habitat is insufficiently documented but from the available evidence *G. nordlandicolella* seems to be restricted to sandy and steppe areas. In northern Europe the species mainly occurs at low altitudes but in Central Asia it has been observed in xeromontane habitats from 2200 m up to about 2900 m. Adults can be swept in the evening. In a locality in Finland moths were collected in numbers on and around *Erigeron acer* L. (Asteraceae), growing in bare sand. Supposed larvae, later found on this plant, did not mine the leaves but rather gnawed them from the underside (larva in sand), but as the rearing was unsuccessful its life history remains to be confirmed. The moth has been observed at light, although almost all specimens have been collected by net. It is active at dusk but is often observed during daytime as well. It flies low down as other related species do (Mutanen, *in litt.*).

REMARKS.– The taxonomic status of populations from North America (Arizona) remains unresolved since small genitalia differences (Huemer, 1996; Povolný, 1998, 2002a) can be observed which contradict proposed conspecificity of Palaearctic and Nearctic populations.

9. *Gnorimoschema nilsi* Huemer, 1996 **sp. rev.**

Gnorimoschema nilsi Huemer, 1996: 78, figs. 1, 3, 5-6, 11-12, 17-18, 21-22.

DIAGNOSIS.– Adult. Wingspan 14-17 mm. Labial palp white; segment 2 mottled with blackish brown on lower and outer surface; segment 3 with two broad black rings. Antenna black, ringed with light grey. Head grey; thorax light grey; tegula grey with white tip. Forewing light grey from white-based black and grey scales and mottled with white and light orange-brown, the latter especially in fold, around black spots and in apical area along veins; darker areas at base and before indistinct white costal and tornal spots. Hindwing grey.

Variation. The amount of orange-brown scales of the forewing is variable and occasionally the costal part is largely covered with such scales.

Similar species. *G. nilsi* is very similar *G. nordlandicolella*. It differs by the darker head, by the lighter, greyish forewings and the more indistinct white costal and tornal spots.

MALE GENITALIA.– Uncus broadly subrectangular, medioapically projected, distinctly pointed; gnathos-hook long and broad, massive; valva extending to about middle part of uncus, basal half strongly bent, distal part straight with distinctly inflated apex; sacculus very short and strongly curved, crescent-shaped, with indistinct gap towards posterior margin of vinculum; posterior margin of vinculum without separated vincular process, posteriolaterally projected, medial part with deep V-shaped emargination and slightly hump-like projection in lateral part; saccus subtriangular, broad at base, evenly tapered to slightly rounded apex, extending to about apex of pedunculus; phallus short, distal part slender with small apical hook, coecum strongly inflated. Differs from the closely related *G. nordlandicolella* q.v. in several characters such as the broader and longer gnathos-hook, the more strongly inflated apex of the valva, the small gap between sacculus and vinculum and the deep emargination of the vinculum.

FEMALE GENITALIA.– Unknown.

DISTRIBUTION.– Only known from the type locality in Austria (East Tyrol: National Park Hohe Tauern) and from a few localities in France (Hautes Alpes) (Nel, 2003c: 398) and an unpublished record from Italy.

New record. Italy: 1 ♂, Cuneo, V. Varaita, Colle dell'Agnello, 2750 m, 12.vii.1998, leg. Baldizzone, gen. slide 6400 Hendriksen (BALD).

BIONOMICS.– Host-plant and early stages unknown. The adults occur from mid-July to mid-August. They fly freely in the early morning just after sunrise. Only few specimens have been observed during dusk and at light but several adults have been attracted to pheromones. Despite searching no females have been found yet and the female sex may be brachypterous. The habitats are steep, south to south-west exposed meadows within the Seslerio-Semperviretum association, partially with scattered vegetation on sandy soil at high altitudes of about 2500 to 2750 m.

REMARKS.– The taxonomic state of *G. nilsi* as a separate species is disputed and Povolný (1998, 2002a) considers it merely as a synonym of *G. nordlandicolella*. However, several constant differences in the male genitalia (see above and Huemer, 1996) and mtDNA clearly indicate two species.

10. Gnorimoschema valesiella (Staudinger, 1877)

Lita valesiella Staudinger, 1877: 205.
Lita diabolicella Hering, 1924: 81.
Phthorimaea charcoti Meyrick, 1934: 59.
Phthorimaea hackmani Schantz, 1952: 19.

DIAGNOSIS.– Adult. Wingspan 16-18 mm. Labial palp white, mottled with black, especially on segment 3. Antenna black, indistinctly lighter ringed. Face cream-white; neck, thorax and tegula greyish black. Forewing greyish black mottled with white-based black scales and some orange, the latter in fold and around distinct black spots; faint white costal and tornal spots. Hindwing grey.

Variation. A rather constant species. The population from Iceland differs in being slightly larger and by having more scattered white scales especially in the apical part of the forewing, and by the distinct white costal and tornal spots, which can be fused into a fascia. Specimens from Sierra de Gredos in Spain are distinctly lighter coloured compared with specimens from north and Central Europe.

Similar species. *G. valesiella* is characterized by its large size, and by the rather plain greyish black forewings with suffusion of orange-brown scales only in the fold and around the black spots. In the slightly smaller *G. streliciella* the forewings are more variegated from more white and orange-brown scales.

MALE GENITALIA.– Uncus subrectangular, short, medioapically projected, with pointed apex; gnathos-hook long and broad, massive; tegumen moderately stout; valva extending to about middle of uncus, strongly constricted and bent medially, distal part with distinctly inflated apex; sacculus long and slender, S-shaped; vincular process largely reduced to small setose lobe, posterior margin of vinculum with moderately shallow and narrow V-shaped emargination; saccus sub-ovate, with broadly rounded apex, extending below apex of pedunculus; phallus short, distal part slender with small apical hook, coecum distinctly separated, strongly inflated and globular. The male genitalia show some individual variation and are very similar overall to *G. streliciella* q.v. and *G. hoefneri* q.v., mainly differing in the more slender sacculus.

FEMALE GENITALIA.– Apophysis posterioris about 2.5 times length of segment VIII; sternite VIII subrectangular, broader than long; subgenital plate smooth, posterior edge with broad and strongly sclerotized fold, outwardly separated by deep incision, lateromedial edge broadly folded from posterior to anterior edge of sternite; area between subgenital plates small, membranous, with few microtrichia; anterior margin of sternite VIII concave, strongly sclerotized, with small medial opening; apophysis anterioris longer than segment VIII, distinctly widened in basal half, distal half rodlike; antrum short; colliculum short, ring-shaped with irregular lateral sclerotization; ductus bursae almost two times length of apophysis anterioris, distinctly separated from pyriform corpus bursae; signum a large and stout hook with distinct curvature, basal plate small. Easily recognizable by the peculiar sclerotization near the anterior margin of sternite VIII.

DISTRIBUTION.– Arctic-alpine distribution; local records from the Alps, Caucasus (Povolný, 1966c: 396), Sierra de Gredos and Pyrenees, northern Europe to Iceland and Greenland; recently also found in Transbaikal (Povolný, 2001a: 206).

BIONOMICS.– The larva has been recorded from *Solidago virgaurea* L. (Asteraceae) where it mines the lower leaves from a spun gallery beneath the leaf under the ground surface (Kaitila, 1996: 103). The species has also been reared from *Erigeron acer* L. (Asteraceae) in Finland (Mutanen, *in litt.*). The adult occurs from June to the middle of August. The preferred habitats are dry or sandy slopes and alpine scree with sparse vegetation. Vertical distribution depends on the latitude, and ranges from about 600 m in the northern part of its range to nearly 3000 m in the Alps.

REMARKS.– Specimens from Spain are distinctly lighter in forewing colour, but identical with populations from other areas in genitalia structures. Povolný (1967b: 75, 1998: 338) recorded *G. valesiella* from Alaska, as a subspecies *G. valesiella alaskense* Povolný, 1967. However, based on differences in the genitalia, *Gnorimoschema alaskense* Povolný (**stat. rev.**) is here considered as distinct from *G. valesiella*.

11. Gnorimoschema streliciella (Herrich-Schäffer, 1854)

[no genus] *streliciella* Herrich-Schäffer, 1853: pl. 67, fig. 495.
Gelechia streliciella Herrich-Schäffer, 1854: 171.
Gelechia cinctipunctella Erschoff, 1877: 344. Subspecies.
Gnorimoschema mongolorum Povolný, 1969a: 4, pls 1-5, figs. 1-10, pl. 32, fig. 31. Subspecies.

DIAGNOSIS.– Adult. Wingspan 14-15 mm. Labial palp white; segment 2 mottled with black on lower and outer surface; segment 3 with two broad black rings. Antenna black, indistinctly lighter ringed. Face white; neck, thorax and tegula as forewing. Forewing with white-tipped black scales and mottled with white and orange-brown; black spots large, surrounded by orange-brown; an oblique whitish fascia from costa at 1/5; white scales between and around black spots; white costal and tornal spots forming a fascia; termen with white and black scales. Hindwing dark grey. Upperside of abdomen black.

Variation. There is some variation in the amount of white and orange-brown scales of the forewings. Occasionally there is an orange-brown patch near the base of the forewings. *G. streliciella* was in the past (e.g. Povolný, 2002a: 25-26) considered as a geographically variable species. Some of the former subspecies are here regarded as distinct species (see also remarks).

Similar species. *G. streliciella* is characterized by its black forewings with large orange-brown surrounded spots. *G. herbichii* q.v.; *G. hoefneri* q.v.; *G. valesiella* q.v.

MALE GENITALIA.– Uncus moderately long, slender, subrectangular, with pointed apex; gnathos-hook comparatively short; tegumen moderately slender; valva extending to about distal part of uncus, strongly curved medially, distal part weakly inflated; sacculus long and slender, strongly bent, hook-like; vincular process largely reduced, setose, posterior margin of vinculum straight, with moderately shallow V-shaped emargination; saccus subtriangular, with moderately broad apex, long, extending slightly below apex of pedunculus; phallus moderately short, distal part slender, coecum strongly inflated. Differs from *G. valesiella* q.v. by characters such as the broader and less curved sacculus and the longer saccus and from *G. hoefneri* q.v. by the smaller gnathos, the more slender sacculus and the straight posterior margin of the vinculum.

FEMALE GENITALIA.– Apophysis posterioris about three times length of segment VIII; sternite VIII subrectangular, about same length as breadth; subgenital plate smooth, posterior edge with subtriangular and strongly sclerotized fold, outwardly demarcated by longitudinal ridge which extends to about middle of sternite; lateromedial edge of subgenital plate folded; area between subgenital plates moderately small, membranous, with some microtrichia; anterior margin of sternite VIII concave, strongly sclerotized, with small medial opening; apophysis anterioris slightly longer than segment VIII, widened in basal half, distal half rod-like; antrum short; colliculum moderately long, ring-shaped; ductus bursae about two times length of apophysis anterioris, distinctly separated from oval corpus bursae; signum moderately large and stout hook with distinct curvature, basal plate small. Differs from *G. hoefneri* q.v. by the longer than broad segment VIII and the shorter longitudinal ridge of sternite VIII.

DISTRIBUTION.– Known from Scotland, Scandinavia, northern Germany, parts of the Baltic states, Poland and the southern Ural mountains; further populations are recorded from Central Asia, Siberia to China and from the Far East (Bidzilya, 2009: 5; Povolný, 2002a: 25-26). The species disappeared from its localities in The Netherlands, Germany, Poland, Denmark and Great Britain already in the first half of the 20th century. Records from Belgium, France and Sweden could not be confirmed from voucher specimens and may be based on misidentified material. Records from the Alps refer to *G. hoefneri*.

Bionomics.– The larvae feed on *Antennaria dioica* L. (Gaertn.) (Asteraceae). The life history is similar to that of *G. valesiella*, except for the host-plant. The larva lives under the ground surface in a silken gallery mining leaves from the underside. The mine resembles that of *Scrobipalpa murinella*. It seems possible that the ground should not be too mossy (as is often the case in places with *Antennaria*), but there must be sand as the larvae make sandy tubes on the ground (Mutanen *et al.*, 2003: 10; Mutanen, *in litt.*). References to *Thymus* (Lamiaceae) as a possible host-plant for *G. streliciella* (e.g., Povolný, 1967d: 201; 2002a: 25) are erroneous. Vertical distribution: in Europe at low altitudes, in Mongolia found from 600 to about 2400 m.

Remarks.– Povolný (2002a: 25-26) separates four subspecies: *G. s. streliciella* from northern Europe and the northern part of Central Europe, with variegated dark and reddish brown forewings; *G. s. mongolorum* from Central Asia and recent records from the Urals (Junnilainen *et al.*, 2010: 46), with lighter greyish brown forewings; *G. s. cinctipunctellum* from the Amur region, with very large wingspan, greyish colour of the forewings and largely reduced markings. However, the latter is only known from a single male and has also been considered as a synonym of *G. streliciella* (Ponomarenko, 2008: 328). *G. s. hoefneri* is here treated as separate species due to differences in genitalia structures.

12. *Gnorimoschema hoefneri* (Rebel, 1909) **sp. rev.**

Gelechia (*Lita*) *hoefneri* Rebel, 1909: 331.

Diagnosis.– Adult. Wingspan 17-20 mm. Labial palp white; segment 2 mottled with black on lower and outer surface; segment 3 with two black rings. Antenna black, indistinctly lighter ringed. Face white; neck and tegula light grey; thorax blackish grey, posteriorly white. Forewing blackish grey from white-tipped black scales and mottled with greyish brown and orange-brown, especially in fold, along dorsum and around black spots; costal and tornal spots white, separate. Hindwing light grey. Upperside of abdomen cream-grey.

Variation. The few examined specimens are rather constant.

Similar species. *G. hoefneri* is characterized by its large size and its blackish grey forewings with black and orange markings. It resembles *G. streliciella*, which is smaller and has a distinct postmedial fascia, darker hindwings and upperside of abdomen black.

Male genitalia.– Uncus short, broad, subrectangular, with projected apex; gnathos-hook long, stout; tegumen broad, stout; valva extending to about distal part of uncus, strongly curved medially, distal part inflated; sacculus long and slender, stronly bent, hook-like; vincular process largely reduced, setose, posterior margin of vinculum obliquely emarginated, with moderately shallow V-shaped medial incision; saccus subtriangular, stout, with broad apex, extending slightly below apex of pedunculus; phallus moderately short, distal part slender, coecum strongly inflated. Differs from *G. valesiella* in the broader and less curved sacculus and the longer saccus and from *G. streliciella* in the stouter sacculus and vinculum, the larger gnathos and the posterior margin of the vinculum.

FEMALE GENITALIA.– Apophysis posterioris about three times length of segment VIII; sternite VIII subrectangular, broader than long; subgenital plate smooth, posterior edge with subtriangular and strongly sclerotized fold, outwardly demarcated by longitudinal ridge which extends to anterior edge of sternite; lateromedial edge of subgenital plate folded; area between subgenital plates small, membranous, with some microtrichia; anterior margin of sternite VIII concave, strongly sclerotized, with small medial opening; apophysis anterioris longer than segment VIII, widened in basal two-thirds, distal part rod-like; antrum short; colliculum moderately long, ring-shaped; ductus bursae about two times length of apophysis anterioris, distinctly separated from oval corpus bursae; signum a moderately large and stout hook with distinct curvature, basal plate small. Similar to *G. streliciella* but differs in the broader than long segment VIII and the distinctly longer longitudinal ridge of sternite VIII.

DISTRIBUTION.– Endemic to the south-eastern Alps (Italy, Austria, Slovenia).

BIONOMICS.– Host-plant and early stages unknown. The adult occurs from late June to early August. It is attracted to light and flies freely in the late evening. Preferred habitats are alpine scree on calcareous soil at altitudes from about 1000 m to 2400 m.

REMARKS.– The taxonomic state of *G. hoefneri* is disputed. Povolný (2002a: 25) treated it as a subspecies of *G. streliciella*. However, differences in external appearance as well as in genitalia structures are distinct and first molecular data give clear evidence to two species. Hence, we consider both taxa specifically distinct.

Scrobipalpopsis Povolný, 1967

Scrobipalpopsis Povolný, 1967b: 110.
Type species: *Gelechia petasitis* Pfaffenzeller, 1867.

DIAGNOSIS.– Adult. See species description.

MALE GENITALIA.– Uncus with convex posterior margin, arched, dorsolateral part with microsetae; gnathos-hook longish, pointed; valva shorther than or extending nearly to apex of uncus, apically weakly inflated; sacculus basally near vincular process; saccus long, broader at base, apically pointed; phallus slender, almost straight to weakly curved, apex with small subterminal sclerite, coecum moderately inflated.

FEMALE GENITALIA.– Segment VIII longer than broad; sternum VIII medially membranous, subgenital plates broadly convex, sclerotized and smooth without additional sclerites; apophysis anterioris rod-like, medially with or without hump-like expansion, about length of segment VIII, inner margin occasionally with sclerotized rim prolonged towards ostium bursae; antrum tubular; corpus bursae well defined, pyriform to globular; signum a small to large plate with a strong hook.

DISTRIBUTION.– A single species occurring in the Palaearctic region. Five species recorded from the Nearctic region. Records from the Neotropics (Povolný, 2002a: 19) probably refer to the genus *Tecia*.

Bionomics.– As far as known the larvae are leaf-miners of various Asteraceae. Acording to Povolný (2002a: 27) the only neotropical species probably feeds in potato tubers. However, no reference is given to the identity of this species and in fact the genus *Scrobipalpopsis* is not mentioned from the Neotropics in an earlier review paper of Neotropical Gnorimoschemini (Povolný, 1994b).

Remarks.– *Scrobipalpopsis* was incorrectly treated as a synonym of *Ptycerata* Ely, 1910 by Hodges (1983: 22). Later the genus was synonymized with the genus *Tecia* Kieffer & Jørgensen, 1910 from South America (Hodges & Becker, 1990: 76) but this synonymy was rejected by Povolný (1993: 93) based on differences in genitalia, e.g. the long segment VIII and the lack of an antrum in *Scrobipalpopsis*, and the distinct porrect labial palpus of *Tecia*.

13. *Scrobipalpopsis petasitis* (Pfaffenzeller, 1867)

Gelechia petasitis Pfaffenzeller, 1867: 79.
Gelechia petasitella Staudinger, 1867: 211.
Gelechia petasitae Heinemann, 1870: 215; unjustified emendation of *Gelechia petasitis* Pfaffenzeller, 1867.

Diagnosis.– Adult. Wingspan 17-21 mm. Labial palp white mottled with some grey-brown scales on lower and outer surface. Antenna black, more or less distinctly lighter ringed. Head, thorax and tegula light grey; face and tip of tegula lighter. Forewing light brownish grey; one distinct black spot in fold and two at 2/5 and 3/5 in middle of wing; termen with distinct black spots. Hindwing broad, grey.

Variation. The colour of this species varies from whitish grey-brown to rather dark grey. Females are generally lighter than males.

Similar species. *S. petasitis* can be recognized by its large size, the broad wings and the uniformly coloured forewings with three distinct black spots and a row of distinct terminal dots.

Male genitalia.– Coremata absent; uncus broadly convex, lateral part folded with some dorsolateral microsetae; culcitula large; gnathos-hook long and stout, broad at base; valva distinctly shorter than uncus, rather broad and stout; sacculus widely separated from valva, basally close to vincular process, knife-shaped with pointed apex; vincular process posteriorly projected, about length of sacculus, apically slightly curved outwards to moderate tip; posterior margin of vinculum with deep, V-shaped medial emargination; saccus long and stout, broader at base, apically pointed; phallus moderately slender, straight, apex with small subterminal hooklet and additional sclerite, coecum inflated.

Female genitalia.– Apophysis posterioris moderately long; dorsoposterior margin of segment VIII sclerotized, band-shaped, turned into largely sclerotized ventromedial part ; sternite VIII distinctly longer than broad, lateral margins parallel, anteriolateral margins concave; subgenital plate smooth with broadly convex inner margin; apophysis anterioris about length of segment VIII, moderately broad at base with distinct medial dilation, distal part rod-like; colliculum moderately sclerotized, tubular; ductus bursae short, with smooth transition to pyriform corpus bursae; signum large, horn-shaped, with rather small base and extremely large hook.

DISTRIBUTION.– Arctic-alpine species, recorded from the Alps, northern Europe and an unpublished record from northern Siberia (Taimyr Peninsula) (Savenkov, pers. comm.). The suspected occurrence in the eastern U.S.A. needs confirmation (Povolný, 2002a: 27).

New record. Russia: 1 ♂, N. Siberia, Taimyr, 25 km S Kayak village, valley of Kotuy river, leg. Savenkov (LDM).

BIONOMICS.– The larva has been recorded from *Petasites paradoxus* (Retz.) Baumg. (Asteraceae) where it feeds as a leaf-miner. The large leaf-mine starts from the tip towards the base and up to four larvae can live within it. The fully grown larva overwinters in the mine or between spun parts of the leaves and pupates outside the hibernaculum in early spring (Hofmann, 1867: 202). In North Europe the host-plant is *Petasites frigidus* (L.) Fr. Svensson (1991: 68-69) figured the leaf-mine and reported between 5 and 10 larvae from each mine. In North Europe the species has a two-year life cycle. In the first year the larva lives gregariously, in the second solitarily (Mutanen, *in litt.*).The adult occurs from mid-May to late July, depending on the elevation of its habitat. It is attracted to light but has also been observed flying around the host-plant in the early evening. Preferred habitats in central Europe are riverine systems and alpine scree on calcareous soil up to 1800 m in the Alps.

REMARKS.– The species was almost simultaneously recognized by Staudinger, Hofmann and Pfaffenzeller. Due to an editorial error the first description was attributed to the last author (Dohrn, 1867: 210).

Scrobipalpa Janse, 1951

Scrobipalpa Janse, 1951: 199.
  Type species: *Gelechia heliopa* Lower, 1900.
Ilseopsis Povolný, 1965: 481.
  Type species: *Ilseopsis peterseni* Povolný, 1965.
Euscrobipalpa Povolný, 1967a: 212.
  Type species: *Scrobipalpa grossa* Povolný, 1966.
Ergasiola Povolný, 1967a: 232.
  Type species: *Phthorimaea ergasima* Meyrick, 1916.

DIAGNOSIS.– Adult. Wingspan 7-21 mm. Labial palp upturned; segment 2 roughly scaled below forming two ridges with a groove in between, colour variable, but lighter on upper and inner surface; segment 3 of about same length as segment 2, smooth scaled, often with one or two black rings. Antenna plain coloured or more or less distinctly ringed with light and dark. Forewing rather slender, mostly lighter or darker greyish or brownish, from dark lighter-based scales, rarely black or white; one black spot in fold and two at 2/5 and 3/5 in middle of wing, often surrounded by orange or brownish scales; dorsum sometimes lighter; a black longitudinal stripe in some species; fold and sub-costal stripe often yellowish; apical area normally darker than rest of the wing; cilia mostly grey with darker cilia line(s). Hindwing lighter or darker grey with greyish cilia. A couple of species have pink scales on labial palps, forewings and legs. *Scrobipalpa peterseni* has broadly lanceolate and pointed, slightly brachypterous forewings, and the head possesses an irregular frontal process.

MALE GENITALIA.– Sternite VIII broad, posteriorly rounded, frequently with pair of coremata; tergite VIII broadly tongue-shaped; uncus broad, rounded off apically, narrower than tegumen; gnathos with distinct, short to longish hook and small to large culcitula; tegumen widened from uncus to pedunculi, shallowly to deeply emarginate anteriorly; pedunculi distinct; valva slender, digitiform to clavate, apically weakly swollen, longish, reaching below or beyond uncus, frequently curved ventrad (in slide inwards); sacculus short, variably shaped, frequently with apical tooth; vinculum broad, posterior margin deeply to weakly emarginate, lateromedially with pair of variable processes; saccus short to longish, slender to broad, distal part usually tapered, apex frequently pointed, lateral and posterior margins of saccus strongly sclerotized with a gap posteriomedially; phallus stout and rather short, rarely slender and long, apex with sclerotized arm, coecum strongly inflated.

Female genitalia.– Apophysis posterioris comparatively long; segment VIII subrectangular, wide and longish, mediodorsally membranous; sternite VIII with or without more or less extended foamy structure and or microtrichia; subgenital plate heavily sclerotized, normally separated from ventromedial depression by longitudinal edge, more or less distinct longitudinal depression present; pairs of longitudinal ventromedial depressions, usually membranous with foamy sculpture and/or microtrichia, medially fused, anteriorly separated and frequently extending beyond anterior edge of sternite; apophysis anterioris rod-like, slightly shorter to distinctly longer than segment VIII, inner margin occasionally with sclerotized rim prolonged towards ostium bursae; intrasegmental membrane frequently with subostial sclerite, extending from base of apophysis anterioris towards ostium; antrum membranous; ductus bursae with ring-shaped colliculum; corpus bursae usually well defined, frequently covered with microtrichia; signum a strong hook of various shape, with or without basal spines, with moderately small base.

Distribution.– With 104 species *Scrobipalpa* is the most species rich genus of Gelechiidae in Europe. Worldwide there are around 400 species-group names for the nearly 290 recognized species. Most species (about 87%) occur in the Palaearctic region, the remaining species being scattered over the Afrotropical region (about 8%), the Nearctic region (about 3%), the Oriental region (less than 1%) and Australia (less than 2%). There are undoubtedly still many undescribed species to be discovered, especially outside Europe, but even for the European fauna we know of several probably undescribed species which could not be described in this book due to insufficient material.

Bionomics.– The host-plants include a wide range of different families although a considerable number of species feeds on Chenopodiaceae, Asteraceae and Solanaceae. Numerous species are halophytic or occur in eremic habitats and in mountainous areas.

Remarks.– Povolný (1989a) treated *Scrobipalpa*, *Euscrobipalpa* and *Ergasiola* as separate genera, and *Ilseopsis* as a subgenus of *Euscrobipalpa*, though the former was described two years earlier and should therefore have become the valid generic name. More recently *Ilseopsis* was resurrected to generic rank (Povolný, 2002a: 81). However, Sattler (1988) already discussed the synonymy of *Ilseopsis* with *Scrobipalpa* based on the genitalia and external characters, which fall within the range of one genus; this also applies to *Euscrobipalpa* which consequently was considered as a junior synonym of *Scrobipalpa* (Ponomarenko & Park, 2007: 59). The four taxa *Scrobipalpa*, *Euscrobipalpa*, *Ilseopsis* and *Ergasiola* are here treated as congeneric due to their similarity in genitalia structures which does not justify a generic subdivision.

Piskunov (1995: 250) synonymized *Homotima* Diakonoff, 1954 (from New Guinea) with *Ilseopsis* Povolný, based on the figure of the genitalia of the type species *H. purpurata* Diakonoff, 1954. Whereas that figure depicts a species of the Gnorimoschemini (probably of the genus *Kiwaia* Philpott) this is not the case with the moth, and genitalia and moth of two different species have been mixed. We here follow the opinion of K. Sattler (*in litt.*) that the genus name *Homotima* should be based on the moth and not on the wrong figure of its genitalia.

The genitalia of many species are very similar and diagnostic differences are sometimes difficult to observe. It is in general strongly recommended to dissect specimens, particularly from the Mediterranean region.

The terminology of genital structures is in accordance with that already used for other genera of gnorimoschemines such as *Caryocolum* and *Sattleria*. It is mainly based on the presumed homology of characters and differs partially from the terminology used by Povolný in numerous papers. Particularly the parabasal process (of the valva) is called sacculus in this paper. The sclerotized apical arm of the phallus is called subterminal spine by Povolný and cornutus by Jansen (1999).

14. *Scrobipalpa aptatella* (Walker, 1864)

Gelechia aptatella Walker, 1864a: 636.
Gelechia heliopa Lower, 1900: 417.

DIAGNOSIS.– Adult. Wingspan 12-16 mm. Labial palp cream-coloured; segment 2 mottled with red-brown and black and with a black ring at base; segment 3 about half the length of segment 2, with a black ring at base and near tip. Antenna black, ringed with light grey. Neck and tegula as forewing; face and thorax lighter. Forewing brown, mottled with lighter and darker scales; spots rather large, indistinct, dark brown; faint black spots at base, 1/5 and 3/3 at costa. Hindwing dark grey; cilia grey with yellowish base.

Variation. The few examined specimens are rather constant.

Similar species. *S. aptatella* is unique among European *Scrobipalpa* in having segment 3 only half the length of segment 2. It is moreover characterized by the brownish forewings with indistinct markings. *P. operculella* q.v.

MALE GENITALIA.– Uncus comparatively broad and long, subrectangular; valva extending to about tip of uncus or slightly exceeding uncus, slightly curved apically with weakly inflated apex; sacculus very short, rounded; vincular process tooth-like, pointed, distinctly longer than sacculus; posterior margin of vinculum with deep V-shaped emargination; saccus comparatively long and stout, broad, extending far beyond apex of pedunculus; phallus long and slender, basally weakly inflated. Differs from the otherwise similar *S. kasyi* particularly in the less curved valva, the longer vincular process and the much longer saccus.

FEMALE GENITALIA.– Segment VIII short, anteriorly broadened; subgenital plate smooth, inwardly edged by indistinct longitudinal fold; ventromedial depression narrow, membranous, covered with numerous microtrichia; antrum broadly tubular, extending to about middle of apophysis anterioris; apophysis anterioris rod-like, broad-based, about length of segment VIII; colliculum short; ductus bursae moderately broad, distally with abrupt transition to corpus bursae; corpus bursae oval; signum on the right side of corpus bursae, basal plate moderately small; signum-hook large and stout, strongly curved at almost right angle in middle, without additional spines. Very similar to *S. kasyi* but differs in the medially sclerotized antrum, the shorter and stouter apophysis anterioris and the larger signum.

DISTRIBUTION.– Listed from France, Italy and former Yugoslavia by Povolný (1996a: 113) but also figured in a map for Greece by the same author (Povolný, 1991b: 286). However, Povolný (*in litt.*) has informed us that all these records were not based on examined material, but that he had once examined specimens of *S. aptatella* 'reared from mines on tobacco' from Bulgarian Macedonia. The status of *S. aptatella* in Europe needs confirmation. Originally distributed in eastern Australia this species was later introduced into large parts of the oriental and afrotropical regions (Povolný, 2002a: 84).

BIONOMICS.– The larva of *S. aptatella*, well known as the Tobacco Stemborer, is a pest of *Nicotiana tabacum* L., but it has also been recorded from several other Solanaceae: *Nicotiana rustica* L., *Datura* L., *Solanum dubium* Fresen. (Robinson *et al.*, 2007), *Solanum melongena* L. and *Solanum tuberosum* L. (Povolný, 2002a: 83-84). It is a miner of leaves and stems.

REMARKS.– Povolný (2002a: 83) erroneously applied the junior synonym *S. heliopa* as the valid name, merely based on the wide usage of the latter name in agricultural papers (see also Povolný, 1966a: 133).

15. Scrobipalpa kasyi Povolný, 1968

Scrobipalpa kasyi Povolný, 1966b: 403, figs. 23-24, 9.

DIAGNOSIS.– Adult. Wingspan 11-13 mm. Labial palp cream-coloured, segment 2 mottled with black on outer and lower surface; segment 3 with two black rings. Antenna distinctly ringed black and light. Head cream-coloured, mottled with grey towards neck; thorax and tegula as forewing. Forewing blackish brown with orange longitudinal stripes along subcosta and in fold; black spots surrounded by orange scales; a yellow fascia from costal to tornal spot; termen with black spots.

Variation. The examined specimens show only slight variation. When becoming worn specimens lose the black scales and look lighter than fresh ones. Povolný (2002a: pl. 11, fig. 1) figures a very dark specimen.

Similar species. *S. kasyi* is characterized by its blackish brown and orange forewings with a yellow postmedial fascia. *S. vasconiella* and *S. brahmiella* have more distinct blackish and orange forewings without a yellow postmedian fascia.

MALE GENITALIA.– Uncus broad at base, strongly narrowing towards rounded apex; culcitula well developed, gnathos-hook strong and long; valva shorter than uncus, strongly curved in basal third, apically without inflation; sacculus a largely reduced hump, about even width from base to rounded apex; vincular process digitate, slender, almost extending to apex of sacculus, apex with short tip directed outwards, separated from sacculus by wide space; posterior margin of vinculum with very deep and broadly V-shaped medial emargination, hardly separated from vincular process;

saccus narrowing in medial part, apex knob-like, moderately long, slightly exceeding apex of pedunculus; phallus longish, slender, with weakly inflated coecum. *S. aptatella* q.v.

FEMALE GENITALIA.– Segment VIII short, anteriorly broadened; subgenital plate smooth, inwardly edged by indistinct longitudinal fold, posteriorly with membranous notch; ventromedial depression narrow, membranous, covered with numerous microtrichia; antrum broadly tubular, extending almost to middle of apophysis anterioris, medially with small longitudinal membrane; apophysis anterioris moderately slender, rod-like, base gradually broadened, slightly longer than segment VIII; colliculum short; ductus bursae evenly widened with indistinct transition to corpus bursae; corpus bursae oval; signum on the left side of corpus bursae, basal plate moderately small; signum-hook large and stout, strongly curved at an almost right angle near middle, without additional spines. *S. aptatella* q.v.

DISTRIBUTION.– Restricted to the south-eastern Europe (Macedonia, Greece, unpublished record from Bulgaria).

New record. Bulgaria: 3 ♂, Kresna, 31.v.-2.vi.2002, leg. Kaitila (KAIT, ZMUC).

BIONOMICS.– The larva feeds on *Alyssum montanum* L. (Brassicaceae) (Povolný, 1966b: 403) but its behaviour has not been described in detail yet. The adults have been collected from late May to mid-July.

16. Scrobipalpa notata (Povolný, 2001)

Euscrobipalpa notata Povolný, 2001a: 193, figs. 16, 39.

DIAGNOSIS.– Diagnosis. Wingspan 9-10 mm. Segment 2 of labial palp cream-white, outer and lower surface mottled with black; segment 3 black with a white ring. Antenna black, indistinctly ringed with light grey. Head cream-white; neck, thorax and tegula as forewing. Forewing covered with light-based black scales; black spots distinct, followed by yellowish scales; yellowish stripes in fold and as a subcostal stripe. Hindwing light grey.

Variation. Specimens can be almost black, variegated with yellow or orange, or dark greyish due to more exposed light-based scales. Worn specimens look more grey. This is the case with the holotype.

Similar species. *S. notata* is characterized by the dark forewing colour, the distinct black spots surrounded by yellowish and by the orange-yellow stripes on the forewing. It may resemble *S. artemisiella*, which is of similar size, but that species has orange scales also at the dorsum of the forewing. *S. occulta* q.v.

MALE GENITALIA.– Uncus slender, gradually narrowing to truncate posterior margin; gnathos-hook distinct, long; tegumen deeply emarginate anteriorly; valva moderately slender, extending almost to apex of uncus, weakly curved, with slightly pointed apex; sacculus moderately broad and short, slightly widened from base to apical part, apex with broadly rounded outer margin and with distinct inwardly directed tooth at inner margin; vincular process largely reduced to small hump; posterior margin of vinculum with broad and shallow emargination; saccus long and moderately broad, strongly tapered towards apex, anteriorly extending far beyond apex of pedunculus; phallus stout, coecum weakly inflated, with small sclerotized arm apically. The reduced vincular process makes this species unmistakable.

FEMALE GENITALIA.– Segment VIII distinctly broader than long; anterior margin of sternite VIII strongly sclerotized, transverse rim with small medial opening; subgenital plate with evenly convex posterior and inner margin, smooth and without foamy sculpture or microtrichia; ventromedial depressions subtriangular, with distinct lateral edges, rounded apex extending below anterior margin of sternite VIII, medially well separated by anteriorly widened V-shaped gap, covered with few microtrichia proximally; apophysis anterioris rod-like, almost without transition to sternite VIII, about twice length of segment VIII; colliculum short; ductus bursae moderately narrow, with abrupt transition to small, spherical corpus bursae; signum on the right side of corpus bursae; signum-hook large, without curvation, basal plate large.

DISTRIBUTION.– Only known from the type locality in the Ukraine (Crimea) and from Russia (southern Ural Mountains) (Junnilainen *et al.*, 2010: 49).

BIONOMICS.– Host-plant and early stages unknown. The holotype was collected in late July, additional specimens from Russia from early to late June at light.

REMARKS.– *S. notata* was described from a single male with distorted genitalia, missing the phallus; furthermore the first thoracic segment and the hindwings have been broken and subsequently glued to the rest of the specimen. Apart from the holotype we have been able to examine a small series of both sexes from the southern Ural mountains.

17. *Scrobipalpa acuminatella* (Sircom, 1850)

Gelechia acuminatella Sircom, 1850: lxxii.
Gelechia pulliginella Sircom, 1850: lxxii.
Gelechia cirsiella Stainton, 1851: 4.
Lita porcella Heinemann, 1870: 253, **syn. n.**
Lita ingloriella Heinemann, 1870: 259.
Gelechia gracilella Stainton, 1871: 97.

DIAGNOSIS.– Adult. Wingspan 10-16 mm. Labial palp fuscous, more or less mottled with cream, especially on inner surface. Antenna black, slightly lighter ringed. Head light grey, lighter at frons; thorax and tegula as forewing. Forewing ground colour grey from fuscous and light-based scales, more or less covered with light brown, especially between veins; black spots rather indistinct. Hindwing grey.

Variation. A variable species. The colour of the labial palps varies from almost yellow with only little black to dark fuscous only slightly mottled with cream. In some specimens there are more or less distinct light rings in both segments at middle and at tip. The forewings vary from light yellow-brown to blackish grey and may be either nearly plain coloured or variegated. The black spots may be distinct or invisible. Females tend to be slightly smaller with slightly more pointed forewings than males.

Similar species. Due to its variability *S. acuminatella* is a difficult species to characterize. Most specimens look nearly plain grey-brown to the naked eye, and they miss the characteristics of related species like *S. arenbergeri* and *S. skulei* sp. n. (orange streaks) or *S. halonella* (black terminal dots). *S. amseli* q.v.; *S. pauperella* q.v.; *S. hungariae* q.v.; *S. clintoni* q.v.; *S. proclivella* q.v.; *S. atriplicella* q.v.

MALE GENITALIA.– Coremata absent; uncus sub-oval, apex broadly rounded; valva extending slightly below or to about tip of uncus, slightly curved with weakly inflated apex; sacculus short, outer margin convex, apex with distoventral tooth; vincular process small, sickle-shaped, extending below tip of sacculus posteriorly; posterior margin of vinculum transverse, without distinct hump, medially with broad V-shaped emargination; saccus stout, broadly subtriangular, anteriorly exceeding apex of pedunculus; phallus short, stout, basally inflated, ductus ejaculatorius with distinctly sclerotized lamina. Particularly the characteristic posterior margin of the vinculum make this species unmistakable.

FEMALE GENITALIA.– Segment VIII short, anteriorly broadened, anterior edge of sternite VIII with broad and strongly sclerotized rim and small medial opening; subgenital plate smooth, laterally straight, posteriorly with rounded edge, inwardly edged by short fold; ventromedial depression membranous in posterior part, weakly sclerotized plate anteriorly, covered with numerous microtrichia; apophysis anterioris rod-like, abruptly broad-based, about length of segment VIII; colliculum short; entrance of ductus bursae narrow, distally widening with gradual transition to pyriform corpus bursae; signum on the right side of corpus bursae, basal plate small, signum-hook strongly curved at a right angle at about two-thirds, without additional spines.

DISTRIBUTION.– Widely distributed in most of the European countries, only absent in the northernmost part of Scandinavia and not recorded from Bulgaria and Crete. Also known from western and Central Asia (incl. Afghanistan), Siberia and eastern China (Bidzilya, 2009: 5).

New records. Andorra: 1 ♂, by Pto. de Envalira, 2300 m, 1.viii.1988, leg. M. Fibiger, gen. slide 3557 Hendriksen (ZMUC). Croatia: 3 ♂, Velebit Mts., 1 km W Brusane, 27.vi.2003, leg. Skule & Hviid, gen. slide 4762 Hendriksen (ZMUC).

BIONOMICS.– The larva has been found on a wide range of species belonging to the family Asteraceae: *Cirsium palustre* (L.) Scop., *C. oleraceum* (L.) Scop., *C. heterophyllum* (L.) Hill (Schütze, 1931: 201), *C. arvense* (L.) Scop., *Tussilago farfara* L., *Carlina vulgaris* L. (Jansen, 1999: 44), *Cirsium vulgare* (Savi) Ten. (Corley *et al.*, 2006), *Carduus nutans* L. (Schütze, 1931: 201), probably also *Cirsium erisithales* (Jacq.) Scop. and *Carduus defloratus* L (Sattler, 1987c: 443). Records of *Onopordum* L., *Cnicus benedictus* L., *Artemisia campestris* L., *Serratula tinctoria* L. *Centaurea scabiosa* L., *Centaurea* spp. (Sattler, *op. cit.*) and *Tanacetum vulgare* L. (Povolný 2002a: 28) probably refer to other species of *Scrobipalpa* and need verification. The larva occurs in two generations from June to October (Schütze, 1931: 201), according to Povolný (2002a: 28) the species is suspected to be polyvoltine. Full-grown larvae were found in July and in the second half of September in The Netherlands (Jansen, 1999: 44). However, in the Alps and the northern parts of Great Britain (Sattler, 1987c: 444) the species is univoltine. The larva causes a large, branched upperside leaf-mine, which is changed from time to time, preferably in the lower leaves. It frequently hides below the midrib of the host-plant. Pupation takes place in a cocoon inside the mine and the pupa hibernates. The adult occurs from April to June and again from July to late September, exceptionally even in early November (Jansen, 1999: 44). It can be found by sweeping stands of the host-plants during the day and it is attracted to light. *S. acuminatella* occurs in a wide range of different habitats from coastal dunes to alpine meadows. Vertical distribution from sea-level to about 2200 m.

REMARKS.– *Lita porcella* was described from an unspecified number of specimens collected in the surroundings of Regensburg (Germany) (Heinemann, 1870: 253). A couple from the Herrich-Schäffer collection (BMNH) is labelled '*porcella* ? HS Slg.' but there is no clear evidence that this material should be considered as syntypic though it fits the original description. Nonetheless a dissection of the female genitalia proved the conspecifity with *S. acuminatella* and we therefore synonymize *Lita porcella* with the latter taxon.

Lita acuminatella f. *monticolella* Rebel, 1932, erroneously considered as synonym of *S. acuminatella* by Povolný (2002a: 28), is a species of *Mirificarma* (Pitkin, 1984: 37).

18. *Scrobipalpa skulei* sp. n.

DIAGNOSIS.– Adult. Wingspan 12-15 mm. Segment 2 of labial palp fuscous, with cream-coloured inner surface and light rings before middle and at tip; segment 3 black with distinct white rings at middle and at tip. Antenna black, indistinctly lighter ringed. Head light grey; thorax and tegula as forewing. Forewing orange-brown, densely covered with black, light-based scales, especially in costal half; longitudinal streaks of orange scales near costa and in fold; black spots surrounded with orange; cilia dark grey beyond black cilia line. Hindwing grey.

Variation. The few known specimens vary in size.

Similar species. *S. arenbergeri* is on average smaller, has less orange on the forewings and darker hindwings. *S. halonella* has black dots along the termen.

MALE GENITALIA.– Coremata absent; uncus sub-oval, apex broadly rounded; tegumen with broadly sinus-shaped emargination anteriorly; valva slightly shorter than or extending to about tip of uncus, almost straight with inflated apex; sacculus short, outer margin convex, apex with distoventral tooth; vincular process small, sickle-shaped, extending below tip of sacculus posteriorly, gradually passing into broadly V-shaped medial emargination of posterior margin of vinculum; saccus stout, slender, anteriorly slightly exceeding apex of pedunculus; phallus short, stout, basally inflated, apically with slender hook. Similar to *S. hungariae* q.v. but differs e.g. in the rounded emargination of the tegumen and the smaller and narrower valva, sacculus and saccus.

FEMALE GENITALIA.– Segment VIII short, distinctly broader than long, anteriorly widened; anterior edge of sternite VIII with broad and strongly sclerotized rim and small medial opening; subgenital plate laterally smooth, straight, posteriorly with rounded edge, inwardly edged by short fold and sclerotized longitudinal edge; ventromedial depression without foamy sculpture or microtrichia, smooth and sub-oval lateral zone, anterior part with short humps and smooth sclerotization in between; apophysis anterioris about 1.5 times length of segment VIII, abruptly broad-based; colliculum short; ductus bursae longish, distally widening with abrupt transition to pyriform corpus bursae; signum on the left side of corpus bursae, basal plate small, signum-hook stout, strongly curved at apical third, basally with additional tooth. Differs from the similar *S. acuminatella* e.g. in the short and broad, completely smooth sternite VIII and the shape of the signum.

DISTRIBUTION.– Slovenia.

BIONOMICS.– Host-plant and early stages unknown. The adults have been collected in a single generation from late June to early July at light. The localities range from altitudes of about 450 to 900 m.

MATERIAL EXAMINED.– Holotype. ♂, 'Slovenia, SW part, 950m 11 km above Kozina Stavnik Mts. 950 m 45°32'N 13°58'E 30.vi.2003, B. Sklue & C. Hviid' 'gen. Præparat No. 4763 ♂ H. Hendriksen' (ZMUC).

Paratypes. Slovenia. 1 ♀, same data as holotype but gen. slide 4764 Hendriksen (ZMUC); 1 ♂, 5 km S Kozina, at Presnica, 450 m, 29.vi.2003, leg. Skule & Hviid, gen. slide 4755 Hendriksen (TLMF).

ETYMOLOGY.– The new species is dedicated to Bjarne Skule who collected most valuable material of Gelechiidae during the last decades.

19. *Scrobipalpa hungariae* (Staudinger, 1871)

Gelechia hungariae Staudinger, 1871: 300.

DIAGNOSIS.– Adult. Wingspan 18-20 mm. Labial palp covered with dark brown, white-tipped scales. Antenna dark brown. Face light grey-brown; neck and tegula with dark brown, white-tipped scales; thorax dark brown. Forewing rather broad, dark brown, mottled with many white-tipped scales, especially in apical part. Black spots in fold and in middle of wing hardly visible. Hindwing grey.

Variation. Only two specimens have been examined.

Similar species. *S. acuminatella* and other plain brownish species. *S. hungariae* can be recognized by being larger with broader wings, and by the indistinct black spots.

MALE GENITALIA.– Coremata absent; uncus compact, basally broad, apex narrower, rounded; culcitula large, gnathos-hook strong, long; tegumen with subrectangular emargination anteriorly; valva slightly extending to about apex of uncus, weakly curved, with inflated apex; sacculus short, about width of valva, apically flattened, indistinct apicoventral tooth; vincular process weakly developed, small, distinctly shorter than sacculus, gradually passing into broadly V-shaped medial emargination of posterior margin of vinculum; saccus stout, rounded apex, distinctly exceeding apex of pedunculus; phallus comparatively short and stout, coecum strongly inflated. The compact uncus and the shape of the sacculus and the emargination of the vinculum are characteristic for *S. hungariae*.

FEMALE GENITALIA.– Apophysis posterioris very short, about two times length of apophysis anterioris; segment VIII short, about three times broader than long; anterior margin of sternite VIII with broad and strongly sclerotized rim and small medial opening; subgenital plate posteriorly rounded, inner margin folded, smooth and without foamy sculpture, extended field of microtrichia anterior of inner margin; ventromedial depression subrectangular, sclerites with distinct anterior and lateral demarcation at right angle, broader than subgenital plate, nearly fused in middle, covered with numerous microtrichia; apophysis anterioris rod-like, gradually broad-based, about 1.5 times length of segment VIII; colliculum short; ductus bursae moderately short, gradually widened with indistinct transition to corpus bursae; corpus bursae almost globular; signum on the right side of corpus bursae, basal plate large, signum-hook exceptionally large and evenly curved, without additional spines. Differs from the extremely similar *S. adaptata* in the extended field of microtrichia of sternite VIII, the smaller corpus bursae and the larger signum. However, only a single female has been examined.

DISTRIBUTION.– Only known from few localities in eastern Austria, Hungary, Croatia and the Ukraine. Records from the Italian and Swiss Alps (Povolný, 1968b: 18; 2002a: 47) are extremely doubtful and most likely based on mislabelling (Sattler, 1969: 395).

BIONOMICS.– Host-plant and early stages unknown. The adult occurs in two generations from late March to early June and from July to August. Preferred habitats are swamps and wet meadows (Sattler, 1987c: 445). The species is attracted to light.

REMARKS.– The female 'paralectotype' figured by Povolný (1968b: pl. 15, fig. 75) depicts another species of *Scrobipalpa* and is probably based on a mislabelling or a mixture of slides (Sattler, 1969: 394).

20. Scrobipalpa adaptata (Povolný, 2001)

Euscrobipalpa adaptata Povolný, 2001a: 182, figs. 2, 30.

DIAGNOSIS.– Adult. Wingspan 10-11 mm. Labial palp fuscous, mottled with cream-white on lower and inner surface. Antenna blackish grey, only slightly lighter ringed. Head and tegula grey mottled with light grey; thorax blackish grey. Forewing blackish grey mottled with lighter grey, black and yellowish; black spots small, surrounded by yellow-brown; indistinct yellow-brown stripes in fold and along subcosta; termen with black scales. Hindwing light grey.

Variation. Apart from the rather worn holotype only a few specimens were examined.

Similar species. *S. adaptata* is characterized by its rather uniform blackish grey forewings with small black spots surrounded by yellow-brown. Based on the available material a detailed comparison with similar species cannot be made.

MALE GENITALIA.– Uncus long, tapered towards rounded apex; gnathos-hook strong, long; tegumen with broadly rounded emargination anteriorly; valva shorter than uncus, distal part curved, slightly inflated; sacculus moderately long and broad, almost two times width of valva near base, distally narrowing, dorsal margin slightly convex, inner margin slightly cut off near apex, apical tooth indistinct; vincular process weakly developed, small, distinctly shorter than sacculus, gradually passing into moderately narrow and shallow V-shaped medial emargination of posterior margin of vinculum; saccus slender, rod-like, distinctly exceeding apex of pedunculus; phallus comparatively short and slender, apical arm distinct and slightly curved, coecum strongly inflated. Similar to *S. hungariae* and several other species but the emargination of the tegumen, the shape of the valva and sacculus, the smaller medial emargination of the vinculum and the slender saccus are characteristic for *S. adaptata*. Furthermore the male genitalia are distinctly smaller than those of *S. hungariae*, due to the smaller size of the adults.

FEMALE GENITALIA.– Apophysis posterioris very short, about two times length of apophysis anterioris; segment VIII short, about three times broader than long; anterior margin of sternite VIII with broad and strongly sclerotized rim and small medial opening; subgenital plate posteriorly rounded, inner margin folded, smooth and without foamy sculpture, small field of microtrichia along inner margin; ventromedial depression strongly demarcated, subrectangular sclerites with distinct anterior and lateral demarcation at right angle, broader than subgenital plate, nearly fused in middle, covered with numerous microtrichia; apophysis anterioris rod-like, gradually broad-based, about two times length of segment VIII; colliculum short; ductus bursae short, gradually widened with indistinct transition to corpus bursae; corpus bursae pyriform; signum on the right side of corpus bursae, basal plate large, signum-hook large and evenly curved, without additional spines. *S. hungariae* q.v., *S. brahmiella* q.v., *S. dorsolutea* sp. n. q.v.

DISTRIBUTION.– *S. adaptata* was described from a single male collected in the southern Ukraine. Furthermore known from few localities in eastern Austria (Povolný, 1967a: 215, pl. 3, fig. 17, as *S. brahmiella*) and unpublished records from Slovakia and the Czech Republic. Recently published from the southern Ural Mountains (Russia) (Junnilainen *et al.*, 2010: 46).

New records. Slovakia: 1 ♂, Dev. Kobyla, 2.v.1993, leg. Liska, GU 02/1166 ♂ Huemer (ELSN). Czech Republic: 1 ♀, Moravia, Boretice, Hor. Ochozy, 10.viii.1996, leg. Elsner, GU 02/1167 ♀ Huemer (ELSN).

BIONOMICS.– Host-plant and early stages unknown. The few specimens known to date have been collected in May and again from late July to August, thus indicating bivoltinism.

REMARKS.– The female of *S. adaptata* was misidentified and its genitalia figured under the name *S. hungariae* by Elsner *et al.* (1999: pl. 65, fig. 197) and Povolný (2002a: pl. 68, fig. 674). We have been able to examine the holotype, deposited in ZMKU.

21. Scrobipalpa brahmiella (Heyden, 1862)

Gelechia brahmiella Heyden, 1862: 175.

DIAGNOSIS.– Adult. Wingspan 12 mm. Labial palps bone white to cream-coloured; segment 2 mottled with black on lower and outer surface; segment 3 with two black rings. Antenna ringed black and white. Face cream-coloured; neck grey; thorax and tegula as forewing; forewing black and orange; costal and apical area black; black patches in middle of wing at 1/3 and 1/2; black spots indistinct. Hindwing grey.

Variation. The few specimens examined show only slight variation.

Similar species. *S. brahmiella* is characterized by its orange forewings with black markings. *S. vasconiella* closely resembles *S. brahmiella*, and it is uncertain if the two species can be reliably separated without examination of the genitalia. *S. brahmiella* has less black scales on the forewing and a less distinct or missing light fascia at 3/4. *S. kasyi* q.v.

MALE GENITALIA.– Uncus comparatively slender, slightly tapered, apically flattened or slightly emarginated; gnathos-hook long and slender; valva slightly shorter than uncus, with inflated apex; sacculus short, outer margin almost straight, apically rounded without distinct tooth; vincular process slightly broader than sacculus basally, subtriangular, extending below tip of sacculus posteriorly, continuous with very broad V-shaped medial emargination of posterior margin of vinculum; saccus stout, short, pointed, anteriorly slightly exceeding apex of pedunculus; phallus comparatively long and stout, coecum inflated. Differs from the similar *S. vasconiella* in the broader vincular process and the shape of the sacculus.

FEMALE GENITALIA.– Segment VIII broader than long; anterior margin of sternite VIII strongly sclerotized, transverse rim with small medial opening; subgenital plate with oblique and folded inner margin, smooth and without foamy sculpture or microtrichia; ventromedial depressions extending below anterior margin of sternite VIII, subrectangular, with distinct anterior and oblique lateral edges, nearly fused in middle, covered with numerous microtrichia; apophysis anterioris rod-like, almost without transition to sternite VIII, about 1.5 times length of segment VIII; colliculum small; ductus bursae gradually widened, with abrupt transition to spherical corpus bursae; signum on the right side of corpus bursae, basal plate moderately large with additional tooth, signum-hook very long, weakly curved, without additional spines. Differs from *S. vasconiella* in the larger ventromedial field of microtrichia and from *S. adaptata* in the smaller ventromedial depression and the shape of the signum.

DISTRIBUTION.– Only known from scattered localities in Central and eastern Europe, ranging from Belgium and France through parts of southern Germany to Slovakia and the Ukraine. Recently recorded from Russia (southern Ural Mountains) (Junnilainen *et al.*, 2010: 47), and we also examined material from the Volgograd area. A record from eastern Austria (Povolný, 1967a: 215, pl. 3, fig. 17) is erroneous and refers to *S. adaptata*.

New record. Greece: 1 ♂, Makedonia / Thessalia, Olympos, c 1000 m, 21.-26.v.1990, leg. Karsholt, gen. slide 2435 Hendriksen (ZMUC).

BIONOMICS.– The larva has been recorded from *Jurinea cyanoides* (L.) Reichenbach (Heyden, 1862: 175), *J. humilis pyrenaica* (Gren & Godron) Nyman (Millière, 1876: 330) and *J. mollis* (L.) Reichenbach (Asteraceae) (Povolný, 2002a: 36). It mines the leaves of the host-plant, feeding from the tip. The infested leaf becomes swollen and withered. Frass is deposited in the apical part of the leaf which is also used for hibernation of the larva. The fully grown larva leaves the mine in spring and pupates on the ground surface (Heyden, 1862: 175). The larval stage was observed in two generations from June to August and from October to May (Sattler, 1987c: 446; Schütze, 1931: 200). The adult occurs in May and from late August to September (Sattler, *op. cit.*), according to Schütze (*op. cit.*) from June to September. It is rarely observed at light. The habitats are xeric meadows at low altitudes.

22. *Scrobipalpa vasconiella* (Rössler, 1877)

Lita vasconiella Rössler, 1877: 377.
Scrobipalpa drahomirae Povolný, 1966b: 400, figs. 11-15, 12.

DIAGNOSIS.– Wingspan 9-15 mm. Labial palp cream-coloured mottled with black, especially on lower and outer surface; apical half of segment 3 black. Antenna black, more or less distinctly lighter ringed. Face cream-coloured; neck grey; thorax and tegula as forewing. Forewing greyish black, especially along costa, dorsum and in apical half, with orange markings, especially in middle of wing; black spots surrounded by orange; postmedial fascia angulated, orange. Hindwing grey.

Variation. The extent of orange scales on the forewing is variable, and the forewing may look almost orange or black with some orange spots. *S. vasconiella* is also rather variable in size.

Similar species. The clear orange patches in the greyish black forewing are only found in a couple of other *Scrobipalpa* species: *S. brahmiella* q.v. and *S. kasyi* q.v.

MALE GENITALIA.– Uncus comparatively slender, slightly tapered towards flattened apex; valva slightly shorter than uncus, with inflated apex; sacculus moderately short, outer margin convex, with distinct apical tip turned inwards; vincular process narrower than sacculus, subtriangular, extending below tip of sacculus posteriorly, continuous with very broad and deep V-shaped medial emargination of posterior margin of vinculum; saccus stout, tapered towards apex, short, slightly exceeding apex of pedunculus; phallus moderately short, stout, coecum inflated. Rather similar to *S. brahmiella* but differs e.g. in the pointed sacculus.

FEMALE GENITALIA.– Segment VIII moderately short, broader than long; anterior margin of segment VIII strongly sclerotized, transverse rim with small medial opening; subgenital plate with oblique and folded inner margin, smooth and without foamy sculpture or microtrichia; ventromedial depression strongly demarcated, subrectangular sclerites with distinct anterior and strongly oblique lateral demarcation, sclerites nearly fused in middle with few microtrichia; apophysis anterioris rod-like, distinctly broad-based, about 1.5 times length of segment VIII; colliculum small; ductus bursae gradually widened, with abrupt transition to spherical corpus bursae; signum on the right side of corpus bursae, basal plate moderately large, signum-hook very long, weakly curved, with few additional spines. Differs from *S. brahmiella* and *S. adaptata* e.g. in the largely reduced ventromedial field of microtrichia and the more oblique lateral edge of the ventromedial depression.

DISTRIBUTION.– Scattered localities in the Mediterranean, from the Canary Islands and Madeira to Spain and Portugal, France, Sardinia and from Italy to Greece and south-western Russia. Also occurring in Turkey and North Africa. Furthermore recorded from from the Middle East (Iran, Afghanistan) (Povolný, 1977b: 196).

New record. Italy (Sicily): 1 ♀, Nebrodi, San Fratello, Torrente Ingànno, 300 m, 10.-11.ix.2002, leg. Karsholt *et al.*, gen. slide 3888 Hendriksen (ZMUC).

BIONOMICS.– Host-plant and early stages unknown. The adult has been observed from April to July and in a second generation from September to October in steppe habitats (Povolný, 2002a: 76).

REMARKS.– The type-material of *Lita vasconiella* was examined by Povolný (1987: 82).

23. Scrobipalpa dorsolutea sp. n.

DIAGNOSIS.– Adult. Wingspan 11 mm. Labial palp cream-white; segment 2 overlaid with light grey-brown on outer surface; segment 3 mottled with some grey-brown. Antenna black, indistinctly lighter ringed. Head cream-white; central part of thorax cream-white, lateral parts and tegula grey-brown. Costal and central part of forewing light grey-brown, dorsal part light yellow; sub-costal stripe long, yellow; a distinct small black spot in fold; other spots indistinct; termen with black scales. Hindwing light grey.

Variation. The few specimens from the type series show minor variation in the amount of cream-white and grey-brown on the labial palps, and in the amount of yellow on the forewings.

Similar species. *S. dorsolutea* is characterized by bi-coloured thorax and forewings. It is very similar to *S. divisella*, which has a distinct black spot at 3/5 on the forewings, and to *S. voltinella*, which normally has a black patch in the apical area.

Male genitalia.– Coremata absent; uncus moderately slender, slightly tapered towards evenly round apex; valva distinctly shorter than uncus, evenly curved, with truncate apex; sacculus about half length of valva, of same width from base to apex, evenly curved with convex outer margin, without apical tip; vincular process broader and much shorter than sacculus, subtriangular with outwardly turned apical tooth, inner margin continuous with very broad and moderately shallow V-shaped posteriomedial emargination of vinculum; saccus stout, slightly tapered towards truncate apex, exceeding apex of pedunculus; phallus moderately short, stout, coecum inflated. The characteristic shape of the valva and sacculus are unique within *Scrobipalpa* whereas other characters such as the anterior margin of the vinculum are shared with several other species.

Female genitalia.– Some characters of the single female paratype are difficult to state since the slide is medially distorted. Apophysis posterioris short, about three times length of apophysis anterioris; segment VIII moderately short, about two times broader than long; anterior margin of sternite VIII with broad and strongly sclerotized rim and small medial opening; subgenital plate posteriorly rounded, inner margin folded, longitudinal depression with small field of foamy sculpture; ventromedial depression strongly demarcated by lateral fold, covered with small-sized foamy sculpture; apophysis anterioris rod-like, slightly longer than segment VIII; colliculum short; ductus bursae short, gradually widened with indistinct transition to corpus bursae; corpus bursae pyriform; signum on the right side of corpus bursae, basal plate large, signum-hook large and weakly curved, without additional spines. Similar to *S. adaptata* particularly in the large signum but differs in the longer apophysis posterioris and the longer segment VIII with some foamy sculpture.

Distribution.– Only known from the southern Ural Mountains in Russia.

Bionomics.– Host-plant and early stages unknown. The type-material was collected in May.

Material examined.– Holotype. ♂, RUSSIA S-Ural Orenburg district Donskoje village 6km W mount Verbljushka 23.5.2004 K. Nupponen leg.' 'GU 09/1288 ♂ P. Huemer' (NUPP).

Paratypes. 1 ♂, same data as holotype (NUPP); 1 ♀, same data as holotype, but 13.v.1999, gen. slide Hendriksen, 6428 (NUPP); 1 ♂, same data as holotype, but 24.v.2004, gen. slide 6384 Hendriksen (ZMUC); 1 ♂, S-Ural, Orenburg district, 20 km S Pokrovka village, Schibendy valley, 28.v.2004, leg. Nupponen, genitalia in tube (NUPP).

Etymology.– The name refers to the yellow colour of the forewing dorsum.

24. *Scrobipalpa amseli* Povolný, 1966

Scrobipalpa amseli Povolný, 1966b: 394, figs. 1-3, 13.

Diagnosis.– Adult. Wingspan 11-15 mm. Labial palp cream-coloured mottled with fuscous, especially on inner and upper surfaces; tip light. Antenna black, ringed with light grey. Head with cream-coloured face and grey neck; thorax and tegula as forewing. Forewing grey from black, white-based scales and with many scattered light orange-brown scales; black spots indistinct, faintly surrounded with light orange-brown; termen with indistinct black spots. Hindwing grey.

Variation. The specimens examined vary in the distinctness of the black spots. In one specimen there is a faint, bent light fascia at 3/4 of the forewing.

Similar species. Very similar to *S. acuminatella* and *S. pauperella*. *S. amseli* looks lighter grey and less brown compared to these species; *S. hyssopi* q.v.

MALE GENITALIA.– Uncus slender, tapered towards rounded apex; valva extending to about tip of uncus, slightly curved, distally scarcely inflated; sacculus stout, broad, distodorsal part broadly rounded, inner margin almost straight, with inwardly pointed apex; vincular process shorter than valva, broad at base, strongly tapered towards shouldered and outwardly pointed apex, inner margin conterminous to moderately shallow, V-shaped medial emargination of posterior margin of vinculum; saccus slender, slightly exceeding apex of pedunculus; phallus moderately short, with distinct apical sclerite, coecum strongly inflated.

FEMALE GENITALIA.– Apophysis posterioris exceptionally short and stout, only about two times length of apophysis anterioris; segment VIII short, distinctly broader than long, anteriolateral edge with concave emargination; anterior margin of sternite VIII with short sclerotized rim, medial zone with broad membranous opening; subgenital plate subrectangular, with lontitudinal proximal depression covered with numerous microtrichia, anteriorly with short longitudinal rim on inner margin, merged with inner margin of apophysis anterioris; ventromedial depressions adjoined, broadly subrectangular, covered with microtrichia, anterior margin below anterior edge of sternite VIII; apophysis anterioris broad-based, slightly longer than segment VIII; colliculum short; ductus bursae broad with indistinct transition to pyriform corpus bursae; signum on the right side of corpus bursae, basal plate small with additional tooth; signum-hook slender, strongly curved at a right angle in middle.

DISTRIBUTION.– Only known from south-eastern France (Alps) and from an unpublished record from Spain.

New record. Spain: 1 ♂, Teruel, Albarracin, 1200 m, 25.-26.vi.1992, leg. Skou & Skule, gen. slide 5995 Hendriksen (ZMUC).

BIONOMICS.– Host-plant and early stages unknown. The adult has been collected in xeric habitats from June to July at an altitude from about 600 to 1200 m (Povolný & Luquet, 1983: 63).

REMARKS.– The area of microtrichia of the female subgenital plate was not figured by Povolný (2002a: pl. 75, fig. 804) though it is distinct in all slides we have examined.

25. *Scrobipalpa hyssopi* Nel, 2003

Scrobipalpa hyssopi Nel, 2003b: 49, figs. 9-10.

DIAGNOSIS.– Adult. Wingspan 11 mm. Labial palp cream-white; segment 2 mottled with black on lower and outer surface; segment 3 with a broad, black ring before tip. Antenna black, ringed with light grey. Head and tegula grey; thorax as forewing. Forewing with white-based and white-tipped scales, densely mottled with light brown; black scales along costa and on veins in middle and distal part of the wing; no visible spots. Hindwing light grey.

Variation. The description is based on the holotype.

Similar species. *S. hyssopi* is characterized by its light brown forewings with white-based and tipped scales along the veins in the costal and distal parts, and by its missing spots. It resembles other species of *Scrobipalpa*, especially *S. amseli*, *S. pauperella* and *S. proclivella*, but these species have normally more or less distinct black spots.

MALE GENITALIA.– Uncus comparatively slender, apically tapered, rounded; gnathos-hook strong; valva slightly exceeding apex of uncus, nearly straight, without visible apical inflation; sacculus short, sickle-shaped, strongly curved inwards and tapered towards pointed apex; vincular process shorter than sacculus, broadly rounded projection, tapered towards apex; posterior margin of vinculum with comparatively deep V-shaped medial emargination; saccus moderately broad at base, distal part slender, shorter than vinculum, extending to about apex of pedunculus; phallus comparatively short and slender, coecum distinctly inflated. The shape of the sacculus and the vincular process are characteristic for this species.

FEMALE GENITALIA.– Apophysis posterioris exceptionally short and stout, only about two times length of apophysis anterioris; segment VIII short, distinctly broader than long; subgenital plate subrectangular, smooth, with proximal depression, inner margin a broadly sclerotized longitudinal fold, extending from posterior margin to base of apophysis anterioris, anterior margin of sternite VIII with short longitudinal rim; ventromedial depressions adjoined, broadly subrectangular, sclerotized, smooth; apophysis anterioris rod-like, about two times length of segment VIII; ductus bursae slender with gradual transition to pyriform corpus bursae; signum with moderately small basal plate, additional tooth on inner side; signum-hook slender, strongly curved in basal half.

DISTRIBUTION.– Only known from two specimens which were both collected in southern France (Hautes-Alpes).

BIONOMICS.– The larva has been found in June and early July feeding on spun leaves of *Hyssopus officinalis* L. (Lamiaceae). A single male emerged in mid-July (Nel, 2003b: 49), and a further female was collected in late June (Nel, 2006: 53). The habitat is a relict steppe biotope with stands of *Juniperus thurifera* L. (Cupressaceae) at an altitude of about 1400 m.

REMARKS.– The species is insufficiently characterized and additional material is required. We have been able to examine the holotype and the only known female, both deposited in TLMF. The pin of the holotype has a piece of *Hyssopus officinalis* L. with the exuvia glued to a piece of paper. The genitalia description of the female is probably incomplete since some parts of the anteriomedial edge seem to have been lost during preparation.

26. *Scrobipalpa montanella* (Chrétien, 1910)

Lita montanella Chrétien, 1910: 272.

DIAGNOSIS.– Adult. Wingspan 12-13 mm. Labial palp white mottled with much black on lower and outer surface of segments 2 and segment 3. Antenna black, ringed with grey; thorax and tegula dark grey. Forewing blackish grey, mottled with some white and yellow-brown scales; spots indistinct, black, not surrounded by yellow-brown; faint yellow-brown streaks on fold and sub-costa; black dots on termen. Hindwing grey.

Variation. The few examined specimens are rather constant. The black spots can vary in size, and occasionally they are followed by a few yellow-brown scales. Worn specimens look lighter than fresh ones.

Similar species. *S. montanella* is characterized by its greyish black forewings with few indistinct markings. It resembles several other greyish black *Scrobipalpa* species, e.g. *S. proclivella*, *S. chrysanthemella* and dark specimens of *S. acuminatella*, and dissection of the genitalia will often be necessary for a reliable identification.

MALE GENITALIA.– Uncus broad, slightly narrowing towards broadly rounded apex; valva stout, shorter than uncus, basally slightly inflated, apex rounded; gnathos-hook longish; sacculus longish and broad, about width of valva, with almost parallel margins, apical part rounded with indistinct inwardly directed tip; vincular process wider and distinctly shorter than sacculus, slightly crescent-shaped with an indistinct outwardly directed tooth, gap between sacculus and vincular process about width of sacculus; posterior margin of vinculum with narrow and shallow U-shaped emargination; saccus broad at base, strongly tapered to pointed apex, extending beyond apex of pedunculus; phallus stout, with distinctly inflated coecum, small sclerotized arm apically.

FEMALE GENITALIA.– Segment VIII short, about two times broader than long; anterior margin of sternite VIII transverse, strongly sclerotized rim, medially with small membranous opening; subgenital plate with convex and folded inner margin and anteriomedial depression, smooth throughout; ventromedial depression smooth, subrectangular, medially merged, anterior margin extending below anterior edge of sternite VIII, folded with medial incision; apophysis anterioris rod-like, stout, broad-based, slightly longer than segment VIII; colliculum short; ductus bursae broad, with indistinct transition to globular corpus bursae; signum-hook stout, strongly curved near middle, basal plate small.

DISTRIBUTION.– Only known from isolated localities in southern France and from an unpublished record from Sicily.

New record. Italy (Sicily): 1 ♂, 1 ♀, Etna, Mt. Arso, 11 km N Paterno, 1050 m, 2.-4.v.1987, leg. Fibiger, gen slide Huemer GU 02/1146, Karsholt 4803 (ZMUC).

BIONOMICS.– The larva has been found on *Anthemis cretica* L. ('as *Anthemis montana* L.') (Asteraceae) (Lhomme, 1935-1949: 621). However, beyond this information the biology is hardly known.

REMARKS.– *Lita montanella* was described from an unspecified number of specimens bred from *Anthemis cretica* L. (Asteraceae) in the area of Foulon (Herault, France) (Chrétien, 1910: 272). The original description states *Anthyllis montana* L. (Fabaceae) as the host-plant but this mistake was later corrected by Lhomme (1935-1949: 621). Povolný (1983a: 173) selected a female specimen, bred on the 9.v.1904, as the lectotype. The identity of this species was disputed and he concluded that it could eventually be an extreme variation of *S. murinella*. The latter species was also collected at the type locality, but without indication that these specimens were part of the type series. Despite our efforts we have not been able to examine the lectotype which possibly was never returned to the MNHN (Minet, *in litt.*). However, the female genitalia, figured by Povolný (1983a: 167, fig. 18), fully correspond with a paralectotype in the MNHN examined by us and with material from Sicily.

27. *Scrobipalpa corleyi* sp. n.

DIAGNOSIS.– Adult. Wingspan 11-13 mm. Labial palp cream-coloured, mottled with some blackish brown on lower and outer surface of segment 2 and especially towards tip of segment 3. Antenna black, ringed with light grey. Face cream-coloured; neck light grey; thorax and tegula as forewing. Forewing light yellow-brown from white-based black scales mottled with yellow, especially along fold, between tornus and black spot at 3/5 and in apical part; black spots in fold and at 2/5 and 3/5, the latter double, surrounded with some light yellow; apical area somewhat darker; termen variegated black and yellowish. Hindwing light grey.

Variation. The examined specimens show only slight variation.

Similar species. *S. corleyi* sp. n. is characterized by the light yellow-brown forewings with many, scattered, white-based black scales and distinct black spots. It resembles *S. niveifacies*, which has a black ring on segments 2 and 3 of the labial palps.

MALE GENITALIA.– Coremata absent; uncus broad, apex evenly rounded; gnathos-hook moderately short; valva shorter than uncus, with distinctly inflated apex; sacculus short and slender, outer margin weakly curved near apex with small tooth directed inwards; vincular process almost reaching length of sacculus, about 1.5 times breadth of sacculus basally, strongly tapered towards outwardly pointed apex, inner margin convex, gradually passing into very broad V-shaped medial emargination of posterior margin of vinculum; saccus moderately slender, tapered towards pointed apex, anteriorly slightly exceeding apex of pedunculus; phallus stout, coecum moderately inflated, subapical hook well developed. The broad vincular process together with the shape of the uncus are characteristic and separate *S. corleyi* sp. n. from related species such as *S. chrysanthemella* q.v.

FEMALE GENITALIA.– Segment VIII slightly broader than long, with parallel outer edges; anterior margin of sternite VIII with oblique sclerotized ridge and broad medial opening; subgenital plate smooth and without foamy sculpture or microtrichia, longitudinal depression in proximal part, scarcely folded posterior and inner edges at right angle, inner and outer margin oblique and almost parallel; ventromedial depression membranous, trapezoidal, narrower than subgenital plate, anteriorly widened with slightly rounded lateral projections and weak medial emargination, nearly fused in middle and covered with numerous microtrichia in this part; subostial sclerites well developed; apophysis anterioris rod-like, abruptly broad-based, about 1.5 times length of segment VIII; colliculum short; ductus bursae narrow, with abrupt transition to oval corpus bursae; signum with moderately large basal plate, signum-hook moderately long, slender, strongly curved medially, without additional spines. Very similar to *S. chrysanthemella* but differs in the narrower ventromedial depression of sternite VIII and the oblique anterior margin.

DISTRIBUTION.– Only known from Portugal.

BIONOMICS.– The larva has been found on *Phalacrocarpum oppositifolium* (Brot.) Willk. (Asteraceae). Corley (*in litt.*) gives a detailed description of the life-history. Accordingly the larva feeds in a terminal shoot of the plant, mining the leaves from the base and usually causing some damage to the stem apex. The mined leaves are spun together in a withered bunch on one side of the shoot immediately below the shoot-tip. No frass was observed in the mines, so it appears that the mine is open at the basal end of the leaf. More than one leaf is mined by each larva. The larva was found in September and pupated in a tough cocoon amongst debris in October. Adults emerged in captivity in May and early June of the following year. In the field moths were observed from late May to mid-July, flying at dusk and at light, occasionally also by day, probably as a result of slight disturbance. The moth is found associated with the foodplant which grows on ledges on cliffs and among boulders in the open or under trees. During July no signs of larval damage were observed and probably the species is univoltine. Vertical distribution from about 700 to 1600 m.

MATERIAL EXAMINED.– Holotype. ♂ 'Portugal BB Serra da Estréla Torre, 1600-1950 m 10.-11.vii.1986 O. Karsholt' 'GU 01/1093 ♂ P. Huemer' (ZMUC).

Paratypes. Portugal. 1 ♂, 1 ♀, same data as holotype, but gen. slides 1165 Hendriksen, GU 01/1094 ♀ P. Huemer; 1 ♀, Quinta da Volta, Trinta, Beira Alta, 750m, 27.v.2004, leg. M.F.V. Corley, gen slide ♀ 2240; 1 ♂, Serra da Estrela, 1500 m. 25.v.2004, leg. M.F.V. Corley, gen. slide ♂ 2176; 1 ♂, 2 ♀, ditto, but 7.v.2005 ex l. ix.04 (*Phalacrocarpum oppositifolium*), leg. M.F.V. Corley; 2 ♂, ditto, but 15.v.2005 ex l.; 1 ♂, 1 ♀, ditto, but 16.v.2005 ex l., gen. slide GEL 1108 ♂ Huemer; 1 ♂, ditto, but 18.v.2005 ex l.; 1 ♀, ditto, but 20.v.2005 ex l.; 1 ♂, ditto, but 27.v.2005 ex l. ix.04; 1 ♀, ditto, but 9.vi.2005 ex l.; 1 ♀, ditto, but 9.vi.2005 ex l., gen. slide GEL 1109 ♀ Huemer (CORL; TLMF).

ETYMOLOGY.– This species is dedicated to our colleague Martin Corley who bred the species and added a lot to our work.

28. Scrobipalpa chrysanthemella (E. Hofmann, 1867)

Gelechia chrysanthemella Hofmann, 1867: 202.
Lita opificella Mann, 1878: 499.

DIAGNOSIS.– Adult. Wingspan 13-14 mm. Labial palp cream-coloured mottled with blackish grey on lower and outer surface. Antenna black, indistinctly lighter ringed. Face cream-coloured; neck, thorax and tegula as forewing. Forewing covered with black, whitish-based and whitish-tipped scales, giving it a dark greyish appearance; one black spot in fold and two black spots in middle of wing at 2/5 and 3/5; a faint, light, angulated fascia at 4/5; termen with black scales. Hindwing grey.

Variation. The species is fairly constant. The light post-median fascia may be absent.

Similar species. *S. chrysanthemella* is characterized by its dark greyish forewings with indistinct markings.

MALE GENITALIA.– Uncus moderately slender, slightly tapered, apex flattened and emarginated; valva shorter than uncus, with weakly inflated apex; sacculus short, outer margin curved, apically rounded with distinct tooth directed inwards; vincular process reaching length of sacculus, about 1.5 times breadth of sacculus basally, strongly tapered towards pointed apex, outer margin concave, inner margin convex, gradually passing into very broad V-shaped medial emargination of posterior margin of vinculum; saccus moderately slender, tapered towards apex, anteriorly extending to or slightly exceeding apex of pedunculus; phallus stout, coecum weakly inflated. Differs from similar species such as *S. proclivella* and *S. corleyi* sp. n. in the shape of the sacculus and the vincular process and particularly the characteristic uncus.

FEMALE GENITALIA.– Segment VIII broader than long, anteriorly widened; anterior margin of sternite VIII with broad and strongly sclerotized transverse ridge and small medial opening; subgenital plate smooth and without foamy sculpture or microtrichia, longitudinal depression in proximal part, distinctly folded posterior and inner edges at right angle, inner and outer margin oblique and parallel; ventromedial depression covered with numerous microtrichia, broadly trapezoidal, broader than subgenital plate, anteriorly widened with slightly rounded lateral projections and weak medial emargination, nearly fused in middle; subostial sclerites well developed; apophysis anterioris rod-like, abruptly broad-based, about 1.5 times length of segment VIII; colliculum short; ductus bursae narrow, with abrupt transition to oval corpus bursae; signum with small basal plate, signum hook moderately long, slender, evenly curved, without additional spines. Differs from the closely related *S. corleyi* sp. n. in the broader ventromedial depression and the transverse anterior margin of sternite VIII.

DISTRIBUTION.– Locally recorded from eastern France through parts of Central Europe to the Ukraine; recently also recorded from Hungary (Fazekas, 2002: 309). Turkey (Povolný, 2002a: 48). Records from Lithuania (Ivinskis 2004: 113) refer to *S. proclivella*, and a record from Denmark (Povolný, 2002a: 48) is wrong.

BIONOMICS.– The larva has been recorded from *Leucanthemum vulgare* Lam. on several occasions. Further records refer to *Leucanthemum atratum* (Jacq.) DC. (Klimesch, 1958: 270), *Tanacetum corymbosum* (L.) Schultz (Meess, 1910: 367) and *Artemisia absinthium* L. (Asteraceae) (Povolný, 2002a: 48). However, the last host-plant may be attributed to *S. proclivella*. From August to October the larva causes a large, light coloured and pellucid blotch-like mine in the tip of a basal leaf. Several short galleries are directed towards the centre of the leaf. The tenanted leafmine is slightly puckered, later it becomes desiccated. The mines are changed from time to time. Pupation takes place in the mine or between leaf-litter (Hofmann, 1867: 204; Schütze, 1931: 187). The adult occurs from April to May. *S. chrysanthemella* is predominantly found in shady and moist places within forests at low altitudes and in the montane zone.

29. Scrobipalpa proclivella (Fuchs, 1886)

Lita proclivella Fuchs, 1886: 68.
Gelechia rancidella auctt., (*nec* Herrich-Schäffer, 1854); misidentification.

DIAGNOSIS.– Adult. Wingspan 8-13 mm. Labial palp bone-white; segment 2 mottled with fuscous on outer and lower surface; segment 3 with two indistinct, blackish rings. Antenna black, ringed with grey. Head, thorax and tegula with grey, light-tipped scales. Forewing blackish grey with longitudinal stripes of light brown scales; one black spot in fold and two black spots in middle of wing at 2/5 and 3/5. Hindwing grey.

Variation. Specimens with many light brown scales between the veins look light brown, whereas specimens with few such scales look dark grey. The black spots can be more or less distinct, and in some specimens they are hardly visible. Rarely the apical area is lighter than the rest of the forewing. Segment 3 of the labial palp can have two broad black rings, or it can be more or less blackish. Specimens from the southern Ural Mountains are smaller (wingspan 8-10 mm) than specimens from other parts of Europe (Junnilainen *et al.*, 2010: 49).

Similar species. Very similar to *S. acuminatella*, *S. pauperella* and *S. amseli* and not always separable from these without examination of the genitalia. *S. proclivella* is on average smaller and looks more greyish, due to many light-based grey scales on the forewing. None of the three above mentioned species have two black rings on segment 3 of the labial palps. *S. hyssopi* q.v.

MALE GENITALIA.– Coremata absent; uncus broad, apex weakly emarginate; valva shorter than uncus, with weakly inflated apex; sacculus short, distal part narrowing to distinct, inwardly directed apical tooth; vincular process of equal length to sacculus, about two times breadth of sacculus basally, separated from sacculus by a small gap, weakly tapered towards pointed apex, apical part with convex inner margin; vincular process continuous with very broad and deep V-shaped medial emargination of posterior margin of vinculum; saccus moderately slender, tapered towards apex, anteriorly slightly exceeding apex of pedunculus; phallus stout, coecum broadly inflated. The shape of the uncus, the sacculus and the vincular process are characteristic.

FEMALE GENITALIA.– Segment VIII broader than long; subgenital plate with small depression posteriad of apophysis anterioris, covered with weak foamy sculpture, inner and outer margin oblique and parallel-sided, anterior part of outer margin excised, inner margin folded; ventromedial depression membranous, anteriorly turned outwards, with large and broadly rounded projections, anteriorly separated by broad and rounded sinus and extending distinctly beyond anterior margin of sternite VIII, posteriorly fused with sclerotized area, largely covered with mictrotrichia except for apical part of projections; apophysis anterioris rod-like, with gradually broadened base, about length of segment VIII; digitate subostial sclerites; colliculum short; ductus bursae broad, with abrupt transition to rather small and oval corpus bursae; signum on the right side of anterior part of ductus bursae, small additional spines on inner side of basal plate; signum-hook stout and rather short, weakly curved in middle.

DISTRIBUTION.– Locally recorded from Spain to France and through Central Europe, parts of south-eastern Europe to the Ukraine and European Russia (Ponomarenko, 2008: 97). Asian Russia (Transbaikal) (Bidzilya *et al.*, 1998: 47). Recently also found in Finland (Mutanen *et al.*, 2008).

BIONOMICS.– The larva has been recorded from *Artemisia absinthium* L., *Tanacetum vulgare* L., *Leucanthemella serotina* (L.) Tzvelev (Sattler, 1987c: 446) and *Achillea clavennae* L. (Asteraceae) (Klimesch, 1951b: 108). In Latvia and in northern Italy a further host-plant is *Artemisia vulgaris* L. (Klimesch, 1938b: 45; Savenkov, pers. comm.). The larva initially feeds within a longitudinally folded leaf-mine, similar to species of *Phyllonorycter* Hübner, 1822 (Gracillariidae), later between two or more spun leaves which are partially mined. The larva occurs in two generations, from May to June and September to October. Pupation takes place on the soil or rarely in the mine. Probably the pupal stage hibernates. The adult occurs in two generations from April to May and in July (Fuchs, 1886: 68; Klimesch, 1951b: 108; Sattler, 1987c: 447). At higher altitudes of the Alps, the species is probably univoltine with adults flying in July only (Huemer, pers. obs.). Habitats vary from a canal-verge and an industrial site with sandy soil in the Netherlands (Jansen, 1999: 70) to alpine pastures in Austria (Huemer, pers. obs.). Vertical distribution from lowlands to about 1600 m in the Alps.

REMARKS.– The erroneous synonymization of *Gelechia rancidella* Herrich-Schäffer, 1854, with *S. proclivella* (Povolný, 1967a: 224) was corrected by Sattler (1978: 59) who transferred the former taxon to the genus *Athrips* Billberg, 1820 (Gelechiidae).

30. *Scrobipalpa frugifera* Povolný, 1969

Scrobipalpa frugifera Povolný, 1969a: 8, figs. 30-31.
Scrobipalpa hypothetica Povoný, 1973a: 20, figs. 3, pl. 6, fig. 21.

DIAGNOSIS.– Adult. Wingspan 12 mm. Labial palp cream-white mottled with blackish brown on lower and outer surface of segment 2 and especially on segment 3. Antenna blackish brown, indistinctly lighter ringed. Head, thorax and tegula as forewing; face lighter. Forewing light grey-brown from black and brown, white-based scales and mottled with orange; black spots in fold separated by orange; black double-spots in middle of wing at 2/5 and 3/5 surrounded by some orange-brown scales; indistinct whitish costal and tornal spots; some black scales along dorsum and in apical part. Hindwing light grey.

Variation. The above description is base on one greasy specimen. The colour figures by Povolný (2002a: pl. 8, fig. 8, pl. 10, fig. 12) seem to be based on darker specimens.

Similar species. *S. frugifera* is characterized by its greyish brown forewings with black double-spots.

MALE GENITALIA.– Uncus subrectangular, long and slender; gnathos-hook long; valva slender, weakly curved, with scarcely inflated apex, extremely long and extending far beyond tip of uncus; sacculus moderately long and broad, outer margin convex, apex inwardly pointed; vincular process narrow, distally pointed, slightly shorter than sacculus; posterior margin of vinculum with deep U-shaped medial emargination; saccus long and moderately stout, broad at base, distally tapered, extending beyond apex of pedunculus; phallus short and stout, with distinct apical hook, coecum strongly inflated. Differs from *S. oleksiyella* sp. n. in the much broader sacculus and the comparatively small process of the vinculum. The above description is based on a traditional slide and figures from the original description.

FEMALE GENITALIA.– Segment VIII slightly longer than broad; subgenital plate largely covered with microtrichia, foamy sculpture absent, slightly oblique and straight inner edge distinctly sclerotized, longitudinal depression weakly developed; ventromedial depression widened anteriorly, covered with microtrichia posteriorly, anterior part wrinkled and smooth, distinct longitudinal ridge parallel to inner edge of subgenital plate, depressions anteriorly separated by U-shaped emargination, posteriorly merged; apophysis anterioris about length of segment VIII, abruptly broad-based; subostial sclerites broad, inwardly curved; colliculum very short; ductus bursae slender, with abrupt transition to clearly defined oval corpus bursae; signum on the right side of corpus bursae; signum-hook large, weakly curved, several strong spines on inner curve from base to apical third, basal plate small.

DISTRIBUTION.– Mongolia and widely distributed throughout Siberia (Bidzilya, 2009: 5; Ponomarenko, 2008: 96). The only European record is based on a single female from Kalmykia (Russia) (Anikin & Piskunov, 2003: 70).

BIONOMICS.– Host-plant and early stages unknown. The adults have been collected from mid-May to mid-August.

REMARKS.– *S. frugifera* was described from five males collected at four different localities in Mongolia (Povolný, 1969a: 9). The holotype from Bulgan aimak was never figured, though Povolný (2002a: pl. 34, fig. 308) later identified a specimen from

South Gobi aimak as such. We have been able to study a paratype from HNHM. *S. hypothetica* was described from three females collected in Mongolia, the holotype comes from a locality only 9 km away from the type locality of *S. frugifera*. Bidzilya (2009: 5) syonynmized both taxa based on sympatrically collected males and females.

31. *Scrobipalpa oleksiyella* sp. n.

DIAGNOSIS.– Adult. Wingspan 11-12 mm. Segment 2 of labial palp cream-white mottled with light brown; segment 3 almost black with an indistinct ring in middle. Antenna black, ringed with light grey. Head cream-yellow with grey neck; thorax and tegula as forewing. Forewing grey from white-tipped black scales and mottled with whitish yellow and brown; black patches near base and at 1/4 on costa and before white costal spot; black spots surrounded by orange-brown in fold and at 2/5 and 3/5 in middle of wing, the latter oblong, angulated; distinct white costal and tornal spots fused; apical area black. Hindwing light grey.

Variation. The forewings can be more or less covered with whitish yellow scales, and such specimens look more variegated, or whitish grey with black apex.

Similar species. *Scrobipalpa oleksiyella* sp. n. is characterized by its grey forewings with contrasting black, whitish yellow and orange-brown markings, matching several *Caryocolum* species rather than a *Scrobipalpa*. It is superficially similar to *S. reiprichi*, but that species has segment 2 of the labial palp blackish.

MALE GENITALIA.– Coremata absent; uncus subrectangular, long and slender; culcitula moderately large, gnathos hook short, stout; tegumen with deep emargination anteriorly; valva slender, weakly curved, with scarcely inflated apex, extremely long and extending far beyond tip of uncus; sacculus short, outer margin weakly convex, distally tapered towards pointed apex; vincular process large, about two times width of and slightly longer than sacculus, inner margin strongly curved to rounded apex, outer margin short and concave; posterior margin of vinculum with deep U-shaped medial emargination; saccus long and stout, broad at base, moderately tapered distal part, extending distinctly beyond apex of pedunculus; phallus short and stout, with distinct apical hook, coecum strongly inflated. Differs from *S. frugifera* in the narrower sacculus and the large process of the vinculum and from *S. occulta* in the longer tongue-shaped eighth sternite and tergite, the distinctly longer valva and the more slender vincular process and the shape of the saccus.

FEMALE GENITALIA.– Segment VIII slightly broader than long; anterior edge of sternite VIII with short transverse ridge; subgenital plate moderately narrow with inner and outer edges almost lontigudinal and parallel-sided, inner and posterior edge folded, medially small zone of microtrichia, subgenital plate otherwise smooth; ventromedial depressions broad, weakly sclerotized, basally fused, broadly digitate anterior parts extending to anterior margin of segment VIII, edged with folded margins, laterally adjoined to subgenital plate, anteriorly separated by narrow and deep emargination; subostial sclerites sword-shaped, transverse, adjoined; apophysis anterioris rod-like, with gradual transition to anterior edge of subostial plate, slightly longer than segment VIII; colliculum short; ductus bursae long and narrow, with abrupt transition to clearly defined corpus bursae; corpus bursae globular; signum on the right side near entrance of corpus bursae; signum-hook longish, almost straight, with small spines basally, basal plate moderately large, subtriangular. Indistinguishable from *S. occulta* (see Remarks).

DISTRIBUTION.– Only known from the type locality in Russia (southern Ural Mountains), and possibly also from Kyrgyzstan.

BIONOMICS.– Host-plant and early stages unknown. The adults have been collected from early June to late August.

MATERIAL EXAMINED.– Holotype. ♂ 'RUSSIA S-Ural Orenburg's district Pokrovka village 20 km S Schibendy valley, 4.vi.1998, T. & K. Nupponen leg.' 'gen. Præparat No. 3988 ♂ H. Hendriksen' (NUPP).

Paratypes. 21 ♂, 12 ♀, same locality as holotype, but 3.-7.vi.1998, J. Junnilainen leg, gen. slides GU 02/1121 ♂, GU 02/1122 ♀ Huemer (JUNN, TLMF, ZMUC); 6 ♂, 1 ♀, 4.vi.1998, gen. slide 3156 Hendriksen, 15 ♂, 11 ♀, 5.vi.1998, 14 ♂, 14 ♀, 7.vi.1998, 4 ♀, 17.vii.1998, gen. slide 3155 Hendriksen, 1 ♂, 21.vi.1999, 2 ♂, 23.vi.1999, T. & K. Nupponen leg.; 1 ♂, 31.vii.2000, T. Nupponen leg.; 1 ♂, 11.vii.2001, 1 ♂, 1 ♀, 29.vi.2003, 1 ♂, 2.vii.2003, 1 ♂, 28.v.2004, 7 ♂, 2 ♀, 2.viii.2005, 4 ♂, 3 ♀, 3.viii.2005, 1 ♂, 2 ♀, 4.viii.2005, gen. slide 5450 Hendriksen, 1 ♂, 20.viii.2006, 6 ♂, 21.viii.2006, 7 ♂, 22.viii.2006, K. Nupponen leg. (NUPP, ZMUC).

ETYMOLOGY.– Dedicated to our colleague Oleksiy Bidzilya in recognition of his important contribution to Palaearctic Gnorimoschemini.

REMARKS.– The female genitalia of this new species seem indistinguishable from *S. occulta* as the allegedly different shape of the basal plate of the signum depends on the position. However, the male genitalia and the adults differ distinctly.

In our opinion the female genitalia figured by Povolný (1996b: 32, fig. 11; 2002a: pl. 72, fig. 735) as the female of *S. heretica* from Kyrgyzstan belong to *S. oleksiyella* n. sp. (see remarks under *S. heretica*).

32. Scrobipalpa smithi Povolný & Bradley, 1964

Scrobipalpa smithi Povolný & Bradley, 1964: 555, figs. 4-6.

DIAGNOSIS.– Adult. Wingspan 11-12 mm. Segment 2 of labial palp bone-white mottled with blackish brown on lower and outer surface; segment 3 almost black with a white ring. Antenna black, ringed with light grey. Face cream-white; neck, thorax and tegula dark grey. Forewing variegated grey, black, orange-brown and yellow; base, an ill-defined costal block and apical area mostly black; black spots surrounded by orange-brown; fold and subcostal stripe orange, and costal and tornal spots yellowish orange. Hindwing grey.

Variation. There is some variation in the amount of yellowish and orange-brown scales on the forewing. In light specimens the base and an area between costal and tornal spots are orange-brown.

Similar species. *S. smithi* is characterized by its variegated forewings. It is similar to *S. suaedicola*, but that species has the dorsum of the forewings uniformly orange-brown. *S. deluccae* is very similar, although its costal and tornal spots are less distinct and whitish.

MALE GENITALIA.– Uncus moderately slender, apex rounded; gnathos-hook short; valva moderately slender, weakly curved, slightly shorter than uncus, apically slightly inflated; sacculus comparatively short and moderately slender, distodorsal part rounded with inwardly directed tip, inner margin concave; vincular process broadly thumb-like, about 1.5 times width but shorter than sacculus, outer margin concave with acute tip in distal part; posterior margin of vinculum with moderately deep U-shaped emargination; saccus slender, gradually tapered to pointed apex, comparatively long, extending far beyond apex of pedunculus; phallus slender, with strongly inflated coecum, distinct though short sclerotized apical arm.

FEMALE GENITALIA.– Segment VIII slightly broader than long; subgenital plate covered with numerous microtrichia except for smooth lateral and anterior margin, inner edge distinct, longitudinal; ventromedial depression largely covered with microtrichia, sclerotized anterior part smooth with convex inner margin and smooth transition to anterior edge of subgenital plate, V-shaped anteriomedial emargination between depressions; subostial sclerite distinct, slender sword-like, medially adjoined; apophysis anterioris rod-like, broad-based, abruptly separated from anteriorly projected subgenital plate, about length of segment VIII; colliculum short; ductus bursae long and narrow; corpus bursae abruptly separated from ductus bursae, globular; signum on the right side of corpus bursae; signum-hook slender, weakly curved, additional spines at basal third, basal plate moderately small.

DISTRIBUTION.– Locally distributed in Spain and the Ukraine (A. Bidzilya, pers. com.). Also recorded from North Africa, Turkey, the Near East (Syria, Lebanon), Mongolia (Povolný, 2002a: 70) and China (Bidzilya & Li, 2010). A record from Germany (Thüringen) (Steuer, 2002) is based on misidentification (Gerstberger, 2003: 138). Published records from France, Hungary (Szabòky *et al.*, 2006: 71) and Austria are incorrect (see Remarks).

BIONOMICS.– Host-plant and early stages unknown. The adults have been collected from May to August and in early October at light. According to Povolný (2002a: 70) the species is probably associated with xerothermic grassland at altitudes up to about 3400 m (Mongolia).

REMARKS.– Povolný's figure of the female genitalia (2002a: pl. 61, fig. 568) is based on a specimen from France, Aude, Villedeigne (Burmann *leg.*), gen. slide 'Povolný Brm 1635'. We have been able to examine this specimen and slide which in fact belong to *S. atriplicella*. Furthermore two female specimens from Hungary (Budapest, Ujhelyi, 9.-10.iv.1927 (Burmann coll. *in* LNK), genitalia slide 'Povolný Brm 3847') and from Austria (Elsner *et al.*, 1999: pl. 67, fig. 212, pl. 17, fig. 212a) belong to *S. atriplicella*.

33. Scrobipalpa occulta (Povolný, 2002)

Euscrobipalpa occulta Povolný, 2002a: 59, pl. 6, fig. 22, pl. 69, fig. 689.
Scrobipalpa sibirica Bidzilya, 2009: 9, figs. 6-7, 14, 19, **syn. n.**

DIAGNOSIS.– Adult. Wingspan 11-12 mm. Segment 2 of labial palp blackish brown, whitish on upper and inner surfaces; segment 3 black with a white ring. Antenna black, indistinctly lighter ringed. Face cream-coloured; neck, thorax and tegula almost black. Forewing with black white-based scales and mottled with orange and yellow-white, especially in fold, around black spots and as an interrupted fascia from costal to tornal spot; a black spot at 3/5 in middle of wing extended towards tornus. Hindwing grey, lighter in central part.

Variation. The black spot at 3/5 in the middle of the forewing can be double or fused into one elongate spot. Specimens from Siberia have many light brownish scales on the forewing.

Similar sprecies. *S. occulta* is characterized by its almost black forewings mottled with orange and yellow-white. It is similar to *S. notata*, which has more greyish black forewings with less orange and hardly any yellow-white scales.

MALE GENITALIA.– Coremata absent; uncus subrectangular; culcitula moderately large, gnathos hook short, stout; tegumen with deep emargination anteriorly; valva slender, weakly curved, with slightly inflated apical part, extending to about tip of uncus; sacculus short, outer margin weakly convex, distally tapered towards pointed apex; vincular process exceptionally large, about 2.5 times width of and longer than sacculus, inner margin strongly curved to rounded apex, outer margin short and concave; posterior margin of vinculum with deep U-shaped medial emargination; saccus slender, long, extending distinctly beyond apex of pedunculus; phallus rather short, with distinct apical hook, coecum strongly inflated. Somewhat similar to *S. oleksiyella* sp. n. but differs in the much shorter valva and the more slender saccus.

FEMALE GENITALIA.– As in *S. oleksiyella* sp. n. q.v.

DISTRIBUTION.– Only known from the type locality in Turkey, Asian Russia (Altai Mountains), and from hitherto unpublished material from south-western Russia and the southern Ural Mountains.

New record. European Russia: 1 ♀, Krasnoarmeysk [=Sarepta], vi.1860, leg. H. Christoph, gen. slide 26589, (BMNH); 2 ♂, S Ural Cheliabinsk district, Arkaim reserve near Amurskii village, 18.vi.1996, leg. Nupponen, Kaitila, Junnilainen & Ahola, gen. slides 2933 Hendriksen, GU 01/1104 ♂ Huemer (JUNN, ZMUC); 1 ♀, S Ural Cheliabinsk district, Miass, Ilmen State Res., 28.vi.1996 leg. Nupponen, Kaitila, Junnilainen & Ahola, gen. slide GU 09/1314 ♀ Huemer (NUPP).

BIONOMICS.– Host-plant and early stages unknown. The holotype of *S. occulta* was collected in mid-September, further specimens from mid-June to mid-August at altitudes up to 2200 m in the Altai.

REMARKS.– *S. occulta* was described from a single female collected in Turkey by M. and W. Glaser.

S. sibirica was described from a series of altogether 16 males and 5 females collected in several localities in the Altai Mountains (Russia) (Bidzilya, 2009: 9). The taxon agrees with *S. occulta* in female genitalia. The alleged differences of the adults are in our opinion not convincing and figures of holotype and a paratype (Bidzilya, *op. cit.*, figs. 6-7) vary considerably. The holotype of *S. occulta* has the same ground colour as *S. sibirica* and the reduced forewing pattern fits very well to the holotype of the latter. Putative differences in the shape of the signum are based on different preparation technique (lateral/dorsoventral view). Furthermore, the microtrichia on the subgenital plate, not figured by Povolný (2002a: pl. 69, fig. 689), vary in their extent. Taking into consideration the hitherto unknown, vast distribution we have no doubt that *S. sibirica* is conspecific with *S. occulta* and synonymize both taxa.

34. Scrobipalpa grisea Povolný, 1969

Scrobipalpa grisea Povolný, 1969a: 10, figs. 42, 44, 46, 11.
Scrobipalpa grisea f. *uralensis* Povolný, 1973a: 22, 11, 16, fig. 7, pl. 3, fig. 8; unavailable, infrasubspecific.

DIAGNOSIS.– Adult. Wingspan 16-18 mm. Labial palp cream white, mottled with blackish brown, especially on segment 3. Antenna black, ringed with grey. Head white, mottled with grey-brown on neck; thorax and tegula as forewing. Forewing grey from white-tipped black scales, more or less mottled with light brown; black spots in fold and (double) at 2/5 and 3/5 in middle of wing, surrounded by brownish scales; indistinct light costal and tornal spots separated by a dark streak; termen with black scales. Hindwing light grey.

Variation. A variable species. The forewings can be almost plain grey or light brownish, and they can be variegated with black patches. The black spots can be more or less distinct. The form *uralensis* Povolný, 1973 refers to such variegated specimens. Specimens of *S. grisea* have a tendency to become greasy.

Similar species. Despite its rather variable forewing colour and pattern *S. grisea* can be separated from most similar species by its rather large size. The even larger *S. plesiopicta* has more black and less brown on the forewings. *S. indignella* q.v.

MALE GENITALIA.– Uncus long, moderately slender, scarcely tapered, with rounded apex; gnathos-hook short; tegumen long, anterior margin with moderately deep emargination; valva slender, extending to about tip of uncus, weakly curved, with broadly inflated apex; sacculus small, subtriangular, basally about width of valva, distal part tapered towards inwardly pointed apex; vinculum projected posteriad; vincular process small, shorter than sacculus, basally about 1.5 times width of sacculus, tapered towards crescent-shaped and outwardly pointed apex, gap between sacculus and vincular process small; posterior margin of vinculum with narrow, moderately deep V-shaped medial emargination; saccus moderately slender at base, distal part slightly tapered, strongly exceeding apex of pedunculus; phallus moderately slender, with small apical hook, coecum strongly inflated.

FEMALE GENITALIA.– Segment VIII slightly broader than long; subgenital plate medially covered with numerous microtrichia, inner edge distinct, straight, oblique, small anteriomedial depression weakly wrinkled; ventromedial depression covered with microtrichia except for wrinkled and sclerotized anterior part, convex inner margin and smooth transition to anterior edge of subgenital plate, V-shaped anteriomedial emargination between depressions; subostial sclerite distinct, slender sword-like, medially adjoined; apophysis anterioris rod-like, broad-based, abruptly separated from anteriorly projected subgenital plate, about length of segment VIII; colliculum short; ductus bursae long and moderately narrow; corpus bursae abruptly separated from ductus bursae, globular; signum on the right side of corpus bursae; signum-hook stout, weakly curved, additional spines along inner curvature of basal two thirds, basal plate moderately small.

DISTRIBUTION.– In Europe restricted to the southern Ural Mountains of Russia (Junnilainen *et al.*, 2010: 47). Also known from Siberia to Central Asia, Mongolia and from Korea (Bidzilya, 2009: 5).

BIONOMICS.– Host-plant and early stages unknown. The adults have been collected in early June (Ural mts.) and from June to July (Mongolia) at altitudes between 300 and 3000 m. The species is attracted to light.

REMARKS.– Even though Povolný (1973a: 11) writes that *S. grisea* f. *uralensis* may possibly represent a subspecies from the Urals, the name is deemed infrasubspecific because it was described as a form (ICZN, 1999: Art. 43.6.3). Our study of a long series of *S. grisea* from the Urals has shown that the form *uralensis* (being larger with more distinct markings) falls within the variation of the species.

35. *Scrobipalpa usingeri* Povolný, 1969

Scrobipalpa usingeri Povolný, 1969a: 19, pls. 22-23, figs. 81-82, 85-86, pl. 31, fig. 24.

DIAGNOSIS.– Adult. Wingspan 11 mm. Labial palp cream-coloured; segments 2 and 3 each with two black rings. Antenna black, ringed with whitish grey. Head with slight projection at frons, cream coloured, greyish towards neck; thorax and tegula as forewing. Forewing light brownish grey from white-based black and light brown scales; fold and fused costal and tornal spots whitish yellow; black spots rather indistinct; apical area with many black scales. Hindwing whitish grey, with light grey cilia.

Variation. Only two worn and probably faded type specimens were examined.

Similar species. The condition of the examined specimens does not allow a differential diagnosis. A frontal projection is, besides *S. usingeri*, only known in *S. peterseni*, where it is larger.

MALE GENITALIA.– Uncus slender, slightly tapered towards flattened or slightly excised apex; valva distinctly shorter than uncus, with inflated apex; sacculus long, slender, outer margin distally convex, distinct apical tip turned inwards; vincular process much broader and distinctly shorter than sacculus, subtriangular, with outwardly directed apical tooth, inner margin continuous with very broad and deep almost U-shaped medial emargination of posterior margin of vinculum; saccus moderately slender, tapered towards apex, exceeding apex of pedunculus; phallus moderately short, stout, coecum inflated.

FEMALE GENITALIA.– Segment VIII slightly longer than broad; anterior margin of sternite VIII sclerotized, obliquely rounded; subgenital plate with longitudinal depression, covered with small-sized foamy sculpture almost throughout, inner edge long and straight; ventromedial depressions covered with small-sized foamy sculpture sculpture and posteriorly some microtrichia, lateral parts sclerotized and fused with sternite VIII to a large extent, anteriomedially separated by deep and narrow emargination, posteriomedially merged; apophysis anterioris with gradual transition to anterior edge of sternite VIII, slightly longer than segment VIII; colliculum short; ductus bursae with gradual transition to corpus bursae; signum on the right side of corpus bursae; signum-hook moderately stout, weakly curved, basal half with several additional spines.

DISTRIBUTION.– Russia (Volgograd area) (Povolný, 1980: 250). We could examine the specimen upon which Povolný based this record, but not its genitalia. It has more strongly marked forewings compared with other specimens of *S. usingeri* examined by us, and its identity should be confirmed from additional material from Russia. Described from Mongolia, further records from Palestine and Saudi Arabia.

BIONOMICS.– Host-plant and early stages unknown. The adults have been observed from February to July up to altitudes of about 1600 m (Povolný, 2002a: 75). They are attracted to light.

REMARKS.– We have been able to examine the holotype, deposited in HNHM.

36. *Scrobipalpa clintoni* Povolný, 1968

Scrobipalpa clintoni Povolný, 1968a: 113, figs. 1-5.
Scrobipalpa linella Piskunov, 1975: 868, figs. 15a, b.
Scrobipalpa deleta Povolný, 1981: 386, figs. 19, 57, **syn. n.**

DIAGNOSIS.– Adult. Wingspan 10-15 mm. Labial palp cream-coloured mottled with black, especially on lower and outer surface. Antenna blackish brown, weakly lighter ringed. Head light yellow-brown; thorax and tegula as forewing. Forewing light reddish brown mottled with black, especially along veins; a black spot in fold and another in middle of wing at 3/5; termen with black scales. Hindwing grey.

Variation. The amount of black scales on the forewing is variable. Specimens can be almost plain light reddish brown or can have the veins black; occasionally larger areas between the veins are also black. The black spots are often indistinct. The species varies also in size, probably depending on the quality of food.

Similar species. Whereas specimens with black veins are rather characteristic, light plain-coloured specimens may be mistaken for, e.g. *S. acuminatella*. However, *S. clintoni* lacks light brown or yellowish brown scales as found in that species. *S. acuta* q.v.

MALE GENITALIA.– Coremata absent; uncus moderately slender, gradually narrrowing towards rounded or slightly flattened apex, with two distinct setae apically; gnathos-hook stout; valva shorter than uncus, weakly curved, with inflated apex; sacculus moderately short and broad, distal part of outer margin rounded, apex with distinct tooth directed inwards, gap between sacculus and vincular process slightly narrower than sacculus; vincular process shorter than sacculus, basally wider than sacculus, strongly tapered towards outwardly pointed apex; posterior margin of vinculum with extremely deep V-shaped emargination; saccus slender, evenly tapered towards pointed apex, long, anteriorly considerably projecting beyond apex of pedunculus; phallus moderately slender, coecum inflated.

FEMALE GENITALIA.– Segment VIII with about same length and breadth; anterior margin of sternite VIII membranous; subgenital plate with weakly rounded posterior and slightly oblique and straight, folded inner margin, covered with numerous microtrichia, particularly along longitudinal depression; ventromedial depression subrectangular, covered with numerous microtrichia throughout, strongly demarcated laterally and anteriorly, anterior margin proximally with rounded projection, depressions anteriomedially separated by short V-shaped incision, nearly fused in middle; subostial sclerites indistinct; apophysis anterioris rod-like, broad-based, about length of segment VIII; colliculum short; ductus bursae about same width throughout, with abrupt transition to oval corpus bursae; signum on the right side of posterior third of corpus bursae, large basal plate with several additional spines, signum-hook large and stout, weakly curved.

DISTRIBUTION.– Coastal areas of north-western and northern Europe. Also in Latvia and Byelorussia (holotype of *S. linella*). Recently recorded from the Ukraine (Bidzilya & Budashkin, 2004: 64) and from Russia (southern Ural Mountains) (Junnilainen *et al.*, 2010: 47). The two specimens of *S. remota* from the Ukraine published by Povolný (2001a: 205) are misidentifications and in fact fully agree with *S. clintoni*. The record from France (holotype of *S. deleta*) remains unconfirmed.

BIONOMICS.– The host-plant is *Rumex crispus* L. (Polygonaceae) (Povolný, 1968: 114). In Latvia *S. clintoni* has been reared from stems of *Rumex confertus* Willd. (Savenkov & Šulcs, 2004: 55). The larva feeds in the stem of the host-plant; it ejects pale brown frass from holes situated mostly at the nodes; all parts of the stem can be tenanted. There are two generations with larvae in June-July and in the autumn. Pupation takes place in the stem, where a small hole betrays its presence (Bradford, 1979: 126; Karsholt, pers. obs.). The adult has been recorded from May to June, and from middle of July into August. Habitats are stony seashores like shingle beaches and dry coastal commons with the host-plant. In Latvia along rivers. Whereas the adult may be difficult to detect *S. clintoni* can more easily be found in the larval or pupal stage.

REMARKS.– *Scrobipalpa deleta* was described from a single male specimen (Povolný, 1981: 387). The type locality of this species is uncertain. The holotype label 'V. d'Ar. 5.6.[18]98' possibly refers to Villar-d'Arene in the French Hautes-Alpes, but some doubts remain (Povolný & Luquet, 1983: 64). It is also surprising that no further records from this well explored area became available during the last 100 years. We examined the holotype (but not its genitalia) when still on loan to the late D. Povolný. It corresponds well with other specimens of *S. clintoni*. Its male genitalia figured in Povolný (2002a: pl. 21, fig. 190) and also the diagnosis of the specific characters mentioned by Povolný (1981: 387-388) fully support this synonymy.

37. *Scrobipalpa reiprichi* Povolný, 1984

Scrobipalpa (Euscrobipalpa) reiprichi Povolný, 1984: 453, figs. 1-3.

DIAGNOSIS.– Adult. Wingspan 10-11 mm. Labial palp cream-white; segment 2 mottled with black; segment 3 with two ill-defined, black rings. Antenna black, ringed with light grey. Face cream-coloured; neck, thorax and tegula as forewing. Forewing grey brown from brownish and dark grey, white-tipped scales and mottled with black and orange; black areas at base, as an outwardly directed block from 1/4 on costa reaching two-thirds to dorsum; another such block from before white postmedial fascia towards tornus; a black spot in fold surrounded by orange; apical area with many black and some orange scales. Hindwing grey.

Variation. The above description is based on a few specimens from Norway which show only slight variation, especially in the amount of blackish scales on the forewing. The black spots can be more or less covered with the black costal blocks. Worn specimens become lighter. Specimens from the eastern European population seem, according to the figures by Elsner *et al.* (1999: pl. 18, fig. 218) and Povolný (2002: pl. 8, fig. 7) to be darker than the Norwegian ones. A specimen from northern Greece is even darker, having the labial palps and the forewings almost covered with blackish grey-brown scales so that the black costal bloches become less distinct. This specimen, however, has the black spots distinct and surrounded by orange. In the genitalia it differs especially in the more slender saccus and uncus. Due to lack of additional material we tentatively refer it to *S. reiprichi*.

Similar species. *S. reiprichi* can be recognized by the two large costal blocks on the forewing, a character which is almost unique among European *Scrobipalpa* and only shared with *S. oleksiyella* sp. n. q.v. A similar wing pattern is furthermore found in some species of *Caryocolum*, and *S. reiprichi* can easily be mistaken for a species of that genus, especially when still alive, sitting with folded wings, e.g. on rocks.

MALE GENITALIA.– Uncus comparatively long and slender, slightly tapered, rounded apex, distal margin flattened or slightly emarginated; gnathos-hook long and slender; valva slightly shorter than uncus, with inflated apex; sacculus moderately short, outer margin evenly curved convex, inner margin slightly concave, apex with distinct inwardly directed tooth; vincular process digitate, projected posteriad, narrower than sacculus basally, extending slightly below tip of sacculus posteriorly, separated from sacculus by narrow gap; vincular process continuous with very broad and deep V-shaped medial emargination of posterior margin of vinculum; saccus stout, long and moderately broad with weakly tapered apex, anteriorly extending far beyond apex of pedunculus; phallus long and moderately slender, coecum inflated. Similar to some other species of *Scrobipalpa* such as *S. ferruginosa* (Povolný, 2001) but can be distinguished easily by external characters.

FEMALE GENITALIA.– Segment VIII subrectangular, about same length and breadth, anterior edge of sternite VIII membranous, with small medial opening; subgenital plate narrow, laterally straight, longitudinal inner edge folded, parallel to outer margin, anterior margin obliquely turned towards apophysis anterioris, narrow zone of microtrichia along longitudinal depression of subgenital plate; ventromedial depression subrectangular, about half breadth of segment VIII, extending slightly below anterior margin of segment VIII, covered with microtrichia except for anteriomedial part; apophysis anterioris rod-like, abruptly broad-based, about length of segment VIII; colliculum short; ductus bursae long, distally widening with abrupt transition to oval corpus bursae; corpus bursae covered with microtrichia; signum on the right side of corpus bursae; signum-hook stout, weakly curved, with small additional spines in basal half, basal plate stout.

DISTRIBUTION.– Only known from few localities in Slovakia, Norway (Karsholt *et al.*, 1986), and Hungary (Povolný, 2002a: 65). A male specimen from Greece (Mount Olympus, 1700 m) shows some small differences in the colour of the forewings and in genitalia but is probably conspecific with *S. reiprichi*. Recently also recorded from China (Bidzilya & Li, 2010) and the Russian Far East (Sakhalin) (Bidzilya, 2009: 11).

BIONOMICS.– Host-plant and early stages unknown. The adults have been collected in a single generation from May to July. The species occurs on rocky slopes with xeric steppe vegetation.

38. *Scrobipalpa obsoletella* (Fischer von Röslerstamm, 1841)

Lita obsoletella Fischer von Röslerstamm, 1841: 225, pl. 79, figs. a-k.
Gnorimoschema miscitatella Clarke, 1932: 66, pl. 2, fig. 1, pl. 3, fig. 1.
Phthorimaea bipunctella Hartig, 1941: 158, pl. 6, fig. 7.
Phthorimaea calaritanella Amsel, 1952: 128, fig. 33.
Scrobipalpa obsoletella ssp. *hospes* Povolný, 1964: 354, pl. 11, fig. [70], fig. 11.

DIAGNOSIS.– Adult. Wingspan 11-15 mm. Labial palp cream-white; segment 3 mottled with black on lower and outer surface; segment 2 with two, often confluent rings. Antenna black, ringed with light grey. Face cream-white; neck, thorax and tegula as forewing. Forewing covered with grey, light-based scales and mottled with black and light brown, the latter rather evenly dispersed over the wing and especially around the three black spots, in fold and in middle of wing at 2/5 and 3/5; apical area with several black scales, especially along termen. Hindwing light grey.

Variation. A rather variable species. The forewing can be more or less covered with light brown scales; when these are missing it looks buff grey. The black spots are normally distinct, but can be absent. The one at 3/5 is sometimes angulated and can rarely be fused with a dark tornal spot. In some specimens there is an ill-defined, postmedial fascia.

Similar species. Because *S. obsoletella* is such a common and widespread species it can be confused with a number of other *Scrobipalpa* species. It can normally be recognized by its light grey or light brown forewings with normally three black spots with only few or no surrounding orange-brown scales. *S. nitentella* q.v.; *S. indignella* q.v.; *S. abstrusa* sp. n. q.v., *Turcopalpa glaseri* q.v.

MALE GENITALIA.– Coremata absent; uncus moderately slender, apically rounded; gnathos-hook short; valva extending slightly beyond apex of uncus, weakly curved, with inflated apex; sacculus moderately short and slender, almost even width from base to apex, small apical tooth directed inwards, narrow gap between sacculus and vincular process; vincular process broadly sickle-shaped, almost length of sacculus, apically with outwardly directed tip; vincular process continuous with deep and broadly V-shaped emargination of posterior margin of vinculum; saccus slender, tapered towards apex, anteriorly projecting distinctly beyond apex of pedunculus; phallus long, moderately slender, coecum weakly inflated.

FEMALE GENITALIA.– Segment VIII slightly broader than long; anterior edge of sternite VIII membranous; subgenital plate moderately narrow with inner and outer edges almost longitudinal and parallel-sided, inner edge folded, anteriorly small zone of granulated and wrinkled structure extending into base of apophysis anterioris, subgenital plate otherwise smooth; ventromedial depressions broad, smooth and largely membranous, basally fused, digitate anterior part extending beyond anterior margin of segment VIII, laterally adjoined to subgenital plate, anteriorly separated by broad emargination; subostial sclerite sword-shaped, curved posteriad, adjoined; apophysis anterioris rod-like, with gradual transition to anterior edge of subostial plate, slightly longer than segment VIII; colliculum short, moderately broad; ductus bursae long and narrow, with abrupt transition to clearly defined corpus bursae; corpus bursae irregularly oval; signum on the right side near entrance of corpus bursae; signum-hook large and stout, weakly curved, with numerous strong spines on concave side at basal two-thirds.

DISTRIBUTION.– Widely distributed in Europe, ranging from Portugal to the southern Ural Mountains and from Scandinavia to the Mediterranean, though mainly ocurring along the coasts. Outside Europe recorded from Turkey and Iran to Asian Russia (Transbaikal) and Mongolia. A single specimen from South Africa is probably introduced there. Also recorded from North America, though the identity of a male figured by Povolný (2002a: pl. 4, fig. 33) is dubious.

New records. Greece (Crete): 1 ♂, 8 km W Heraklion, 5.xi.1991, leg. Fibiger, gen. slide 1154 Hendriksen (ZMUC). Malta: 1 ♂, Mellieha Bay, 25 m, 5-11.iv.2004, leg. Skule, gen. slide 4971 Hendriksen (ZMUC).

BIONOMICS.– The larva has been recorded especially from *Atriplex* spp., e.g. *A.*

glabriuscula Edmondston (Langmaid, 1995: 277), *A. littoralis* L. (Kaitila, 1996: 94), *A. halimus* L. (Povolný, 2002a: 59), *A. tatarica* L. (Falkovitsh & Bidzilya, 2006: 68) and also *Chenopodium* spp. (Chenopodiaceae) (Schütze, 1931: 87), both on ruderal and halophytic species. It usually feeds in the stem and ejects the frass through a small hole at the base of a stem or near a lateral branch (Schütze, *op. cit.*). Frass is spun together by silk threads. The larval stage has been observed gregariously with up to 36 larvae in a single stem (Stüning, 1988). Also recorded from irregular blotch-mines in the leaves (Hering, 1957) and between spun fruits (Langmaid, *op. cit.*). The larva occurs in two overlapping generations, from June to September (Jansen, 1999: 61). However, in some areas such as Finland it seems to be univoltine (Kaitila, 1996: 94). Pupation takes place in the stem close to a hole which is covered with the epidermis, or rarely between spun leaves or on the ground surface (Jansen, 1999: 61; Schütze, *op. cit.*). Hibernation takes place in the pupal stage. The adult occurs in May and from July to August. However, in the North the species is univoltine with adults flying in June and July only (Kaitila, *op. cit.*). Habitats vary from salt-marshes and brackish inland sites to steppes and deserts as well as human settlements.

REMARKS.– Povolný (1964: 352; 1966a: 134-136) erroneously synonymized *P. calaritanella* with *S. instabilella*. Later on, after having studied the 'holotypus' (even though no holotype is stated in the original description) of *P. calaritanella* he corrected this mistake (Povolný, 1967a: 222). *Phthorimaea calaritanella* was described from a male and a female type ('Tipi') collected around Cagliari in Sardinia, 17.vii and 2.viii.1936, respectively. The type series moreover included 11 syntypes from the same area from 16.vii and 9.viii.1936. Among the 5 specimens located by us in the LNK, a female is labelled 'Typus, leg. H. Amsel' and 'Cagliari, Sardegna, 9.8.1936, H. G. Amsel'. Despite its 'Typus'-label it should hence be one of the syntypes of *P. calaritanella*. There are moreover two males labelled 'Paratypus, leg. H. Amsel', and 'Cagliari, Poetto, 17.7.1936, H. G. Amsel'. One of them has an additional label 'G. U. 54'. This is, however, not among the slide numbers for the type-material of *P. calaritanella* listed in the original description (though it is a male of *S. obsoletella*). In order to stabilize this rather confusing situation the other male, which according to its label was collected at the place and date given for the male 'Tipi' is here published as the lectotype. It is labelled: 'Paratypus leg. H. G. Amsel [and on reverse side:] *Phthorimaea caralitanella* [sic!], Cagliari, Poetto, 17.7.1936, H. G. Amsel, ex coll. H. G. Amsel, Gen. prep. Nr. 5045 O. Karsholt, Lectotype *Phthorimaea calaritanella*, O. Karsholt design., 2001'.

S. obsoletella hospes was described from a single male from South Africa. However, later the species was stated to be introduced (Povolný, 2002a: 59) and in consequence the subspecies state is erroneous.

39. Scrobipalpa feralella (Zeller, 1872)

Gelechia feralella Zeller, 1872: 110.
Gelechia rebeliella Hauder, 1917: (29).

DIAGNOSIS.– Adult. Male. Wingspan 13-16 mm. Labial palp cream-white mottled with fuscous, especially on outer surface of segment 2, less on inner surface of segment 2 and on segment 3. Antenna black. Head, thorax and tegula as forewing. Forewing black with a somewhat lighter shade from light-based scales; costal and tornal spots cream-white, almost confluent. Hindwing grey. Female. Differing from male in:

wingspan 10-12 mm. Labial palps less mottled with fuscous. Antenna slightly lighter ringed. Forewing slightly broader with two broad, clear, white fasciae at 1/3 and 2/3, the former not reaching dorsum. In dorsal side of abdomen the posterior margin of each segment with white scales.

Variation. Individual variation within this species is slight.

Similar species. *S. feralella* is unique in *Scrobipalpa* for its striking sexual dichroism, and neither the male nor the female should be mistaken for other species in the genus.

MALE GENITALIA.– Uncus long, slender, distally tapered, apex rounded; gnathos-hook distinct; valva extending slightly below tip of uncus, moderately inflated apex; sacculus moderately long and slender, even width from base almost to apex, apically rounded with small inwardly turned tooth; vincular process shorter than sacculus, basally about same width, separated from sacculus by narrow gap, subtriangular, with distinct apical tip; vincular process continuous with broad and moderately shallow V-shaped medial emargination of posterior margin of vinculum; saccus broad at base, distal part stout, evenly tapered towards rounded apex, extending far beyond apex of pedunculus; phallus long, with weakly inflated coecum.

FEMALE GENITALIA.– Segment VIII short, distinctly broader than long; anterior margin of sternite VIII strongly sclerotized, medially merged to very short and broad antrum; subgenital plate of sub-quadrate shape, posterior and inner margin straight and at right angle, broadly folded, lontitudinal depression almost from posterior to anterior margin of subgenital plate, covered with small-sized foamy sculpture; ventromedial depressions adjoined, broadly subrectangular, posterior part covered with microtrichia, anteriomedial indistinctly demarcated from sternite VIII, anteriolateral edged by rounded fold, extending below anterior edge of sternite VIII; apophysis anterioris rod-like, slightly shorter than segment VIII; colliculum short; ductus bursae extremely broad with indistinct transition to small, pyriform corpus bursae; signum on the right side of corpus bursae; signum-hook slender, weakly curved, basal plate small and without additional spines.

DISTRIBUTION.– Only known from few localities in Austria (Nordtirol, Oberösterreich, Niederösterreich, Steiermark), Switzerland (Graubünden) and northern Italy (Südtirol).

BIONOMICS.– Host-plant and early stages unknown. The adults have been collected from late May to late August, exceptionally as early as March (Schmid, 2006), most likely in two generations. They occur both on calcareous montane scree-formations and on xeric grassland. Moths are attracted to light.

REMARKS.– The synonymy of *S. rebeliella* with *S. feralella* is based on strong evidence. Both taxa were known from one sex only, the former from the males, the latter from the females. The largely corresponding distributional area and the overall similarity of the adults, differing only in extent and number of white bands of the forewing, furthermore make the validity of *S. rebeliella* as a separate species suspicious. Recently J. Wimmer (Steyr) and P. H. collected male *S. rebeliella* and female *S. feralella* sympatrically and on several occasions on the same night at light (Austria: Nationalpark Kalkalpen (Oberösterreich), Nordkette (Tirol); Italy: Dolomites (South Tyrol)). In consequence both taxa have been considered conspecific, representing only an example of sexual dichroism (Huemer & Karsholt, 2008).

40. *Scrobipalpa halonella* (Herrich-Schäffer, 1854)

Gelechia halonella Herrich-Schäffer, 1854: 178.

DIAGNOSIS.– Adult. Wingspan 12-13 mm. Labial palp cream-coloured mottled with blackish brown on lower and outer surface of segment 2 and especially on segment 3. Antenna dark brown. Head, thorax and tegula as forewing. Forewing covered with light based grey-brown scales and mottled with black and light orange, the latter especially in fold, as a faint subcostal streak and in tornal area; black spots in middle of wing indistinct; termen with diffuse black dots. Hindwing grey.

Variation. The few examined specimens show only slight variation.

Similar species. *S. halonella* is hardly separable from other species with greyish brown forewings like *S. acuminatella*, *S. amseli* and *S. pauperella* without examination of the genitalia. The grey-brown face and the black spots in the termen seem to be quite characteristic for *S. halonella*.

MALE GENITALIA.– Uncus moderately slender, tapered towards rounded apex; gnathoshook strong, short; anterior margin of tegumen with broad and deep emargination; valva slightly excceding uncus, slender, distally curved, with weakly inflated apex; sacculus long, basally about width of valva, distal part tapered towards inwardly pointed apex; vincular process small, about half width and length of sacculus, tapered towards outwardly pointed apex, gap between sacculus and vincular process about width of sacculus; posterior margin of vinculum with moderately deep V-shaped medial emargination; saccus moderately broad at base, distal part slender, exceeding apex of pedunculus; phallus stout, with distinct apical hook, coecum moderately inflated.

FEMALE GENITALIA.– Segment VIII short, broader than long; subgenital plate laterally straight, posteriorly with rounded edge, anterior margin with broad and strongly sclerotized rim and broad membranous opening medially, longitudinal depression with foamy sculpture from base of apophysis anterioris towards middle of sternite, laterally with few microtrichia; ventromedial depressions small, subovate with rounded inner margin, anteriorly diverging, covered with small-sized foamy sculpture and microtrichia, separated by smooth membrane; apophysis anterioris slender rod-like, abruptly broad-based, about two times length of segment VIII; colliculum short; ductus bursae short, gradually widened with indistinct transition to corpus bursae; corpus bursae pyriform; signum on the right side of corpus bursae, basal plate rather small with small additional spines, signum-hook moderately curved in middle.

DISTRIBUTION.– Only known from scattered localities in France, eastern Austria, Hungary and northern Italy; recently also found in Slovakia (Tokár *et al.*, in press) and in Russia (southern Ural Mountains) (Junnilainen *et al.*, 2010: 48); a record from Poland (Elsner *et al.*, 1999: 41) needs confirmation; records from southern Germany are based on misidentifications of three different species of gelechiids (Hausenblas, 2006: 11-12).

BIONOMICS.– The larva is a leaf-miner on *Centaurea scabiosa* L. (Sattler, 1987c: 452), *Artemisia campestris* L. and *A. absinthium* L. (Asteraceae) (Elsner *et al.*, 1999: 41). The mine is similar to that made by *S. acuminatella.* Its galleries are lined with silk and filled with excrements, so the larva is well hidden. It makes several mines, occasionally changing the leaf. Pupation takes place outside the mine and the pupa hibernates. *S. halonella* is most probably univoltine with a single generation of larvae from late May to July and adults in April-May the following year (Sattler, *op. cit.*). The species is restricted to steppe habitats.

41. ***Scrobipalpa perinii*** (Klimesch, 1951)

Phthorimaea perinii Klimesch, 1951a: 31, figs. 7-9.

DIAGNOSIS.– Adult. Wingspan 10-12 mm. Labial palp white mottled with some mid-brown at outer surface of segment 2 and on tip of segment 3. Antenna black. Head and thorax white; tegula black. Forewing with yellowish white costa and dorsum, and a broad, black longitudinal streak from base to apex, surrounded by some orange-brown scales; no visible spots; apical area with a mixture of light brown and white-based black scales. Hindwing light grey.

Variation. The examined specimens show only slight variation.

Similar species. *S. perinii* differs from other European *Scrobipalpa* by the contrasting, black and yellowish white forewing pattern and the likewise contrasting white thorax and black tegulae.

MALE GENITALIA.– Uncus moderately slender, apically tapered and evenly rounded; gnathos-hook strong; valva slightly shorter than uncus, curved, with distinctly inflated apex; sacculus long and moderately broad, about three times as long as wide, sickle-shaped, curved inwards and tapered towards pointed apex, widely separated from vincular process; vincular process distinctly shorter than sacculus, strongly projected posteriorly, tapered towards outwardly pointed apex; posterior margin of vinculum with deep V-shaped medial emargination; saccus broad, distal part tapered and stout, slightly exceeding apex of pedunculus; phallus comparatively short and slender, coecum strongly inflated. The shape of the sacculus and the vinculum is characteristic for *S. perinii*.

FEMALE GENITALIA.– Segment VIII short, about two times broader than long; anterior margin of sternite VIII with transverse sclerotized rim, medially a small membranous opening; subgenital plate with oblique inner edge, narrow longitudinal depression extending from posterior third into base of apophysis anterioris, covered with foamy sculpture and posteriorly with some microtrichia; ventromedial depression narrow, basal two-thirds merged, covered with microtrichia, anterior parts digitate with folded lateral and anterior margins, extending almost to anterior edge of sternite VIII, separated by broadly V-shaped emargination; apophysis anterioris stout, rod-like, gradually broad-based, about length of segment VIII; colliculum short; ductus bursae moderately broad with abrupt transition to oval corpus bursae; signum on the right side near entrance of corpus bursae, basal plate small without additional tooth; signum-hook moderately stout, strongly curved at a right angle near middle.

DISTRIBUTION.– Locally recorded from south-eastern Europe (northern Italy to Macedonia, and unpublished record from Greece), recently also found in southern France (Nel, 2001: 102).

New record. Greece: 1 ♂, Olymp, southern slope, Leptokaria-Karia, ca. 12 km W Leptokaria, 920 m, 17.v.1988, leg. P. Huemer (TLMF).

BIONOMICS.– The larva has been found on *Centaurea scabiosa* L. and *Carduus* L. (Asteraceae) (Klimesch, 1951a: 31). It causes a large, blotched and branched leaf-mine, visible from both sides. Frass is ejected from the mine or deposited on the midrib. The larva frequently hides below the midrib of the host-plant. Pupation takes place in a coccon outside the mine. In its natural habitat the adult has been observed from mid-May to June and in early October (Klimesch, 1968: 123), indicating a possible bivoltinism. In the laboratory material from larvae collected in July and August emerged in August and again in April of the following year (Klimesch, 1951a: 31); other bred material dates from June. The species has been collected in xeric steppe habitats at moderately low altitudes up to about 1200 m.

42. ***Scrobipalpa phagnalella*** (Constant, 1895)

Lita phagnalella Constant, 1895: 52.
Lita staehelinella Caradja, 1920: 101; unavailable, nomen nudum.

DIAGNOSIS.– Adult. Wingspan 10-13 mm. Labial palp cream-coloured, mottled with black on lower and outer surface of segment 2 and especially segment 3. Antenna black, ringed with grey. Face cream-coloured; neck grey; thorax and tegula as forewing. Forewing greyish brown from white-based dark grey and light brown scales; a small black spot in fold and two large, elongate spots at 2/5 and 3/5 in middle of wing, all surrounded by yellowish brown; a black streak between the black spot at 3/5 and a black spot in apex. Hindwing grey.

Variation. The amount of light brown and dark grey scales of the forewing is variable, resulting in lighter and darker specimens. The black spots and the black streak in the apical part of the wing are variable in size, form and distinctness. Some specimens have the central part of the forewing between the spots blackish grey.

Similar species. *S. phagnalella* is characterized by the greyish brown forewings with elongate black spots and an elongate black streak towards apex. It resembles *S. disjectella*, but differs in the more uniformly coloured greyish brown forewings and by having the black spots and apical streak less pronounced.

MALE GENITALIA.– Uncus moderately broad and short, apically evenly rounded; gnathos-hook with characteristic tongue-shaped part; valva extending to about tip of uncus, stout, weakly curved, with inflated apex; sacculus moderately short and slender, narrowing from base to apical part, apex with rounded outer margin and with small and inwardly directed tooth at inner margin, gap between sacculus and vincular process about 1.5 times width of sacculus; vincular process projected posteriorly, shorter than sacculus, tapered towards outwardly directed apex; posterior margin of vinculum with moderately shallow V-shaped medial emargination; saccus broad and stout, distal part slightly tapered with rounded apex, projecting beyond apex of pedunculus; phallus comparatively short and slender, coecum strongly inflated. The shape of the gnathos is unique within *Scrobipalpa*.

FEMALE GENITALIA.– Segment VIII short, about 3 times broader than long; anterior margin of sternite VIII strongly sclerotized, transverse rim with small medial opening; subgenital plate densely covered with microspines to a large extent, spiny zone anteriorly extending into base of apophysis; ventromedial depressions extending below anterior margin of sternite VIII, longitudinal, with distinct oblique lateral edges except for posterior margin, separated in middle, covered with numerous microspines throughout, membrane ventrad of depressions spiny; apophysis anterioris stout, with abrupt transition to sternite VIII, about 2.5 times length of segment VIII; colliculum short; ductus bursae broad, gradually widened, with distinct transition to spherical corpus bursae; signum on the right side at entrance of corpus bursae, signum-hook short and stout, medial part strongly arched, basal plate small, without additional spines.

DISTRIBUTION.– Only known from few isolated localities in southern France, Spain and from hitherto unpublished records from Greece (Crete) and Morocco. According to Povolný (2002a: 62) also recorded from Algeria and doubtfully from Corse, and according to Karsholt (2009) from Portugal and Sardinia. Recently also recorded from the Volga-Don region in Russia (Ponomarenko, 2008: 97), which requires verification.

New records. Greece (Crete): 1 ♂, Agia Galini, Nom. Rethymon, ca. 150 m, 21.iii.1995, leg. Ruckdeschel, gen. slide GEL 1072 ♂ Huemer (TLMF). Morocco: 1 ♂, 1 ♀, 25 km S Essaouira, Sidi Kaouki, 100 m, 28.iii.2005, leg. Karsholt, gen. slides 5067, 5068 Hendriksen (ZMUC).

BIONOMICS.– The larva has been recorded from *Phagnalon saxatile* (L.) Cass. and *P. rupestre* (L.) DC. (Asteraceae) and feeding on the leaves until late autumn (Constant, 1895: 53; Povolný, 2002a: 62). The parenchyma is completely eaten and the larva is hidden among some silk. Infested leaves become twisted and inflated to appear like a shoot. The adults emerged in September and October but on Crete the moth was collected in late March, indicating either a second generation or hibernation of the adult.

REMARKS.– According to Povolný (1967a: 229; 2002a: 72) *Lita staehelinella* Caradja, which is in fact a nomen nudum, belongs to *S. suasella*.

43. Scrobipalpa tokari sp. n.

DIAGNOSIS.– Adult. Wingspan 11 mm. Labial palp cream-coloured; segment 2 mottled with blackish brown; segment 3 with two blackish rings. Antenna ringed light and dark. Face cream-coloured; neck, thorax and tegula with grey light-tipped and light-based scales. Forewing with blackish brown and orange scales; an orange streak in the fold; black spots rather indistinct. Hindwings light grey.

Variation. The type series shows only slight variation.

Similar species. *S. tokari* sp. n. resembles *S. skulei* sp. n. but is smaller, has less variegated forewings and different genitalia.

MALE GENITALIA.– Uncus broad at base, strongly narrowing towards rounded apex; culcitula well developed, gnathos-hook strong and long; valva exceeding tip of uncus, weakly curved, apically inflated; sacculus comparatively short and broad, about even width from base to almost apex, dorsoapically strongly rounded with distinct tooth turned inwards; vincular process distinctly shorter than sacculus, basally broader than sacculus, distally strongly tapered, with outwardly turned tip, separated from sacculus by small gap; posterior margin of vinculum with deep V-shaped medial emargination, scarcely separated from vincular process; saccus moderately narrow, evenly tapered towards apex, short, exceeding apex of pedunculus; phallus stout, basal half with distinctly inflated coecum. Differs from similar species such as *S. karadaghi*, *S. heimi* sp. n. and *S. acuta* particularly in the curvation of the sacculus and the shorter and broader vincular process.

FEMALE GENITALIA.– Unknown.

DISTRIBUTION.– Croatia.

BIONOMICS.– Host-plant and early stages unknown. The few adults known to date were collected in April and September at light. The localities are situated at altitudes between 50 and 300 m and have typical Mediterranean grassy and bushy vegetation with scattered trees (Tokár, *in litt*).

MATERIAL EXAMINED.– Holotype. ♂, 'CROATIA [3 km NE] Zaostrog [300 m] 6.09.2002, leg. Z. Tokár' 'GU 09 1289 ♂, P. Huemer' (TOKA).

Paratypes. 4 ♂, same data but, gen. slides 4180, 6275 Hendriksen, gen. slides in tube Tokár 7961, 7965 (TOKA, ZMUC); 1 ♂, 5 km SE of Pirovac, Šibenek, 50 m, 25.-28.iv.2001, leg. Srnka (TOKA).

ETYMOLOGY.– The species is dedicated to Zdenko Tokár who collected the type series and made important contributions to European lepidopterology.

44. *Scrobipalpa karadaghi* (Povolný, 2001) comb. n.

Euscrobipalpa karadaghi Povolný, 2001a: 192, figs. 14, 37.

DIAGNOSIS.– Adult. Wingspan 10 mm. Labial palp cream-white, slightly mottled with blackish brown. Antenna distinctly ringed black and cream-white. Head, thorax and tegula as forewing. Forewing cream-coloured, mottled with black, brown and some orange scales; black spots indistinct, followed by orange scales, especially in fold; costal and tornal spots of light scales. Hindwing light grey.

Variation. The above description is based on the holotype. Two specimens from the southern Ural Mountains are slightly larger (wingspan 11-12 mm) and have distinctly darker labial palps, head, thorax, tegulae and forewings.

Similar species. *S. karadaghi* is characterized by its light forewings covered with black, brown and orange scales (especially in the fold) and by its indistinct markings. *S. heimi* sp. n. q.v.

MALE GENITALIA.– Uncus broad at base, strongly narrowing towards rounded apex; culcitula well developed, gnathos-hook strong and long; valva slightly exceeding tip of uncus, weakly curved, apically inflated; sacculus comparatively long and broad, evenly tapered towards slightly pointed apex; vincular process distinctly shorter and narrower than sacculus, sickle-shaped, distally tapered, with outwardly turned tip, separated from sacculus by wide space; posterior margin of vinculum with comparatively deep V-shaped medial emargination, scarcely separated from vincular process; saccus narrow, evenly tapered towards apex, short, scarcely exceeding apex of pedunculus; phallus longish, stout, with distinctly inflated coecum. Differs from related species such as *S. tokari* sp. n. q.v.; *S. heimi* sp. n. q.v. and *S. acuta* q.v. particularly in the long and comparatively slender sacculus.

FEMALE GENITALIA.– Unknown.

DISTRIBUTION.– Only known from the type locality in the Ukraine (Crimea), and unpublished records from Russia (southern Ural Mountains).

New record. European Russia: 2 ♂, S Ural, Orenburg oblast, 20 km S Pokrovka village, Schibendy valley, 28.v. & 1.vi.2004 leg. K. Nupponen, gen. slides 5467♂, 5469 ♂ Hendriksen (NUPP, ZMUC).

BIONOMICS.– Host-plant and early stages unknown. The holotype was collected in early August, additional specimens in late May to early June.

REMARKS.– *S. karadaghi* was described from a single male only. The colour figure of the moth by Povolný (2002: pl. 7, fig. 8) is too dark.

45. *Scrobipalpa heimi* sp. n.

DIAGNOSIS.– Adult. Wingspan 9-10 mm. Labial palp cream-white, mottled with blackish brown. Antenna black, ringed with grey. Head cream-coloured, mottled with light brown in neck; thorax and tegula as forewing. Forewing with white-based light brown and grey brown-scales, mottled with some black along costal margin and in apical area, and some yellow scales, especially in fold; blackish spots at 2/5 and 3/5 in middle of wing indistinct. Hindwing light grey.

Variation. The examined specimens vary from light grey-brown to light yellowish brown, females being slightly lighter than males. In strongly marked specimens grey-black spots are present in the fold and at 2/5 and 3/5 in middle of the forewing, but one to all of them can be missing. Only some specimens have yellow scales along the fold.

Similar species. *Scrobipalpa heimi* sp. n. is characterized by its small size and by its nearly uniformly coloured light greyish brown forewings with indistinct spots, which are not surrounded by brownish scales. It is very similar to *S. karadaghi*, which has darker forewings, more distinctly ringed antennae and costal and tornal spots of light scales. *S. acuta* is larger, with darker, brownish grey head and forewings.

MALE GENITALIA.– Uncus broad at base, strongly narrowing towards rounded apex; culcitula well developed, gnathos-hook strong and long; valva slightly exceeding tip of uncus, scarcely curved, apically moderately pointed; sacculus comparatively long and broad, about even width from base almost to apex, dorsoapically rounded with distinct tooth turned inwards; vincular process small and distinctly shorter and narrower than sacculus, sickle-shaped, distally tapered, with outwardly turned tip, separated from sacculus by wide space; posterior margin of vinculum with comparatively deep V-shaped medial emargination, scarcely separated from vincular process; saccus narrow, evenly tapered towards apex, short, scarcely exceeding apex of peduncululus; phallus longish, stout, with distinctly inflated coecum. Very similar to *S. karadaghi* and *S. acuta* but differs in the pointed rather than inflated apex of the valva and the broader and more distinctly pointed sacculus. *S. tokari* sp. n. q.v.

FEMALE GENITALIA.– Segment VIII short, almost two times broader than long, anterior margin of sternite VIII without sclerotized rim, medially with narrow incision; subgenital plate largely covered with distinct foamy sculpture in anteriomedial part, extending slightly into basal part of apophysis anterioris on inner side, lateromedial part with some microtrichia; ventromedial depression covered with distinct foamy sculpture, anteriorly weakly convex with folded margin demarcating anterior edge of segment VIII, posteriomedially merged and covered with few microtrichia; apophysis anterioris rod-like, with gradual transition to anterior edge of sternite VIII, slightly longer than segment VIII; colliculum short and small; ductus bursae rather short, gradually broadened, with indistinct transition to pyriform corpus bursae; signum on the right side of corpus bursae, basal plate small without additional tooth; signum-hook relatively short, slender, strongly curved at an almost right angle in middle. Differs from *S. acuta* particularly in the larger field of foamy sculpture with adjoined zone of microtrichia. *S. soffneri* q.v.

DISTRIBUTION.– Only known from Russia (southern Ural Mountains).

BIONOMICS.– Host-plant and early stages unknown. The type series was collected in late May at light.

MATERIAL EXAMINED.– Holotype. ♂. 'RUSSIA S-Ural Cheliabinsk district near Moskovo village 26.v.1998 T. & K. Nupponen leg.' 'gen. Præparat No. 3177♂ H. Hendriksen' (NUPP).

Paratypes. Russia. 12 ♂, 2 ♀, same data as holotype, but gen. slides 3178 Hendriksen, GU 03/1197 Huemer (NUPP, TLMF, ZMUC); 2 ♂, 1 ♀, same data as holotype, but leg. Junnilainen and gen. slides 2965 Hendriksen, GU 01/1109 ♀, GU 01/1110 ♂ Huemer (JUNN, ZMUC); 1 ♂, S-Ural, Orenburg district, 20 km S Pokrovka village, Schibendy valley, 26.v.1998, leg. T. & K. Nupponen; 1 ♂, same data but 4.vi.1998; 2 ♂, same data but 17.vii.1998 (NUPP).

ETYMOLOGY.– This new species is dedicated to Stefan Heim who supported our work with most of the genitalia photographs.

46. Scrobipalpa acuta (Povolný, 2001) **comb. n.**

Euscrobipalpa acuta Povolný, 2001a: 182, figs. 1, 29.

DIAGNOSIS.– Adult. Wingspan 11-13 mm. Labial palp cream-white, mottled with fuscous, especially on segment 2. Antenna dark grey, only slightly lighter ringed. Head grey; thorax and tegula dark grey. Forewing greyish brown from many brownish and blackish light based scales and mottled with light orange-brown; rather indistinct black spots in fold and at 2/5 and 3/5 in middle of wing; apical part of wing with more black scales. Hindwing grey.

Variation. The few more or less worn specimens examined show slight variation. The black spots at 3/5 in the middle of the wings can be double.

Similar species. *S. acuta* is characterized by its nearly plain brownish grey forewings with only rather indistinct black spots, the apical one sometimes being double. It resembles unicolorous specimens of *S. clintoni*, but in that species the veins are black at least at the apex. *S. heimi* sp. n. q.v.

MALE GENITALIA.– Uncus long, slender, tapered towards rounded apex; gnathos-hook longish; valva extending slightly below tip of uncus, inflated apex; sacculus long and slender, tapered from base to apex, outer (distodorsal) margin rounded with acute tip, inner (distoventral) margin straight; vincular process long, sickle-shaped, extending below tip of sacculus posteriorly, conterminous to deep V-shaped medial emargination of posterior margin of vinculum; saccus slender, extending slightly beyond apex of pedunculus. The slender uncus, saccus and vincular process are characteristic for *S. acuta*.

FEMALE GENITALIA.– Segment VIII short, broader than long, anterior margin of sternite VIII without sclerotized rim, medially with narrow incision; subgenital plate with distinct foamy sculpture extending slightly into basal part of apophysis anterioris on inner side, posterior edge broadly rounded on inner side; ventromedial depression with distinct foamy sculpture, subrectangular with broad lateral folds, medially merged, anteriorly demarcating anterior edge of segment VIII; apophysis anterioris rod-like, with gradual transition to anterior edge of sternite VIII, slightly longer than segment VIII; colliculum short and small; ductus bursae rather short, gradually broadened, with indistinct transition to pyriform corpus bursae; signum on the right side of corpus bursae, basal plate small without additional tooth; signum-hook relatively short, slender, strongly curved at an almost right angle in middle. *S. heimi* sp. n. q.v.

DISTRIBUTION.– Only known from the Ukraine and unpublished records from Russia (southern Ural Mountains).

New record. European Russia: 1 ♂, S Ural, Orenburg oblast, 6 km W Donskoje village, mount Verbljushka 13.v.1999, Nupponen leg., gen. slide 3167 Hendriksen (NUPP).

BIONOMICS.– Host-plant and early stages unknown. The holotype was collected in late August, further specimens in May, indicating bivoltinism.

REMARKS.– *S. acuta* was described from a single male only. In the holotype the right forewing is missing, and the right hindwing is kept in a gelatin capsule. Furthermore the phallus of the holotype was lost during preparation (Povolný, 2001a: 182). The colour figure of the moth by Povolný (2002: pl. 7, fig. 9) is too dark.

47. *Scrobipalpa soffneri* Povolný, 1964

Scrobipalpa soffneri Povolný, 1964: 357, pl. 4, fig. 24, pl. 5, fig. 33, fig. 4.

DIAGNOSIS.– Adult. Wingspan 11-14 mm. Labial palp pure white, mottled with light brown on outer surface. Antenna black. Head, thorax and tegula pure white. Forewing white mottled with light brown and black, especially in apical half; black spots at 1/3 and 3/5 in middle of wing; costa blackish brown; termen and cilia with many black scales. Hindwing grey with lighter cilia. Underside of forewing dark grey.

Variation. The few examined specimens varies in the amount of light brown and black scales on the forewing, and in the distinctness of the black spots. The colour of the hindwings varies from light to dark grey. Specimens from Bulgaria have the forewings covered with much more black and brown scales in the apical part. This may, however, be a matter of fresh and worn specimens.

Similar species. *S. soffneri* is characterized by its black antenna, the pure white head, thorax and tegulae, and the white to light brown forewings with no black spot in the fold.

MALE GENITALIA.– Coremata absent; tergite VIII with concave anterior margin; uncus slender, apically tapered and rounded; gnathos-hook long, slender; valva stout, slightly curved, with weakly inflated apex, extending to about tip of uncus; sacculus moderately long and slender, gradually tapered towards pointed apex; vincular process reaching almost length of sacculus, basally about 1.5 times width of sacculus, distally strongly tapered with shouldered apical part, apex broadly rounded on inner margin, with outwardly directed tooth, gap between vincular process and sacculus moderately narrow; posterior margin of vinculum with V-shaped medial emargination; saccus short, rod-like, apex rounded, extending beyond apex of pedunculus; phallus short, with distinct apical hook, coecum strongly inflated. Very similar to *S. jariorum* sp. n. q.v. but differs e.g. in the shape of the sacculus.

FEMALE GENITALIA.– Segment VIII short, almost two times broader than long, anterior margin of sternite VIII without sclerotized rim, medially with narrow incision; subgenital plate largely covered with distinct foamy sculpture in anteriomedial part, extending slightly into basal part of apophysis anterioris on inner side, lateromedial part with some microtrichia; ventromedial depression covered with distinct foamy sculpture, anteriorly weakly convex with folded margin demarcating anterior edge of segment VIII, posteriomedially merged and covered with few microtrichia; apophysis anterioris rod-like, stout, with gradual transition to anterior edge of sternite VIII, slightly longer than segment VIII; colliculum short and small; ductus bursae rather short, gradually broadened, with indistinct transition to pyriform corpus bursae; signum on the right side of corpus bursae, basal plate small without additional tooth; signum-hook relatively short, slender, strongly curved at an almost right angle in middle. Practically indistinguishable from *S. heimi* sp. n. q.v. with minute differences in the larger size, which is however related to the size of the adults, and the more stout apophysis anterioris.

DISTRIBUTION.– Only known from isolated localities in Bulgaria, and hitherto unpublished records from the southern Ural Mountains and Volgograd district (Russia) and Turkmenistan.

New records. Russia: 1 ♂, South Ural, Orenburg district, near Burannoe village, 12.v.2005, leg. K. Nupponen, gen. slide Huemer 08/1266 (NUPP). 1 ♂,Volgograd district, 20 km SW Frolovo village, 17.v.2005, leg. K. Nupponen, gen. slide Nupponen 1/11.x.2005 (NUPP). Turkmenistan: 3 ♂, Repetekas, 29.iv.1988, leg. Ivinskis, gen. slide Hendriksen 2198 (ZMUC).

BIONOMICS.– Host-plant and early stages unknown. The adults have been collected from late April to May and in July. According to Povolný (2002a: 70) the species is associated with sandy coasts but it also occurs in inland habitats.

REMARKS.– One paratype in ZSM labelled as a paratype of *S. soffneri* belongs to *Ephysteris insulella praticolella* (Heinemann).

48. *Scrobipalpa jariorum* sp. n.

DIAGNOSIS.– Adult. Wingspan 13-15 mm. Labial palp pure white; each segment with two faint, black rings. Antenna ringed black and white. Head white, mottled with some brown in neck; thorax and tegula as forewing. Forewing white with scattered white-based black scales and some yellow-brown; black spots weak, surrounded by yellow-brown; a black streak from black spot at 3/5 towards apex; termen and cilia with black scales. Hindwing light grey.

Variation. The examined specimens show slight variation with regard to the amount of white-tipped black scales on the forewing. The black streak in the apex of the forewings can be more or less distinct.

Similar species. *S. jariorum* sp. n. is characterized by its pure white labial palps with two faint, black rings, and by the white forewings with scattered black scales.

MALE GENITALIA.– Coremata absent; tergite VIII with almost straight anterior margin; uncus slender, apically rounded; gnathos-hook moderately long, slender; valva stout, almost straight, with weakly inflated distal half, extending to about tip of uncus; sacculus moderately short and slender, outer margin convex, gradually tapered towards pointed apex; vincular process stout, slightly shorter than sacculus, basally slightly broader than sacculus, distally strongly tapered with shouldered apical part, apex broadly rounded on inner margin, with outwardly directed tooth, gap between vincular process and sacculus moderately narrow; posterior margin of vinculum with V-shaped medial emargination; saccus short, rod-like, apex rounded, extending beyond apex of pedunculus; phallus rather short, with distinct apical hook, coecum strongly inflated. Similar to *S. soffneri* but differs in the straight anterior margin of tergite VIII, the shape of the valva and the more stout sacculus and vincular process.

FEMALE GENITALIA.– Unknown.

DISTRIBUTION.– Only known from Bulgaria.

BIONOMICS.– Host-plant and early stages unknown. The type locality are south and east exposed limestone slopes close to the Black Sea coast with rich vegetation. Adults have been collected from late May to late June at light and by net in the evening and also in the early morning, mainly flying around *Ephedra* sp. (Ephedraceae) and *Artemisia maritima* L. (Asteraceae), the latter being a possible host-plant of the larva (Junnilainen & Kaitila, *in litt.*; Karsholt, pers. obs.).

MATERIAL EXAMINED.– Holotype. ♂ 'Bulgaria (NE) 5 km E Balich Tuzlata 10-100 m 29.-30.v.2010 O. Karsholt' (ZMUC).

Paratypes. Bulgaria, 3 ♂, same data as holotype (TLMF, ZMUC); 3 ♂, same locality, but 26.v.2002, leg. Kaitila, gen. prep. Huemer GU 08 1272 ♂ (KAIT, ZMUC); 15 ♂, same locality but 26.v.2002, leg. Junnilainen, gen. slide Nupponen 1/06.xii.2005; 1 ♂, same data but 27.-28.vi.2001; 2 ♂, same data but 28.v.2002 (JUNN, NUPP).

ETYMOLOGY.– This new species is dedicated to Jari Junnilainen and Jari-Pekka Kaitila who collected important material of Gnorimoschemini.

49. *Scrobipalpa murinella* (Duponchel, 1843)

Lita murinella Duponchel, 1843: 458, pl. 85, fig. 7.
Gelechia culminicolella Staudinger, 1871: 306.
Gelechia excelsa Frey, 1880: 363.

DIAGNOSIS.– Adult. Wingspan 9-11 mm. Labial palp white mottled with brown, especially on outer and upper surfaces; segment 3 darker than segment 2. Antenna blackish brown. Head, thorax and tegula with grey-brown, light-tipped scales; face lighter. Forewing grey-brown mottled with some black and white; one black spot in fold and two in middle of the wing at 2/5 and 3/5; apical area with many black, white-tipped scales. Hindwing grey.

Variation. In dark specimens the black spots become inconspicuous. Specimens from the high Alps have the forewings more or less covered with whitish-grey scales. Females are slightly smaller than males and have the apex of the forewing more pointed.

Similar species. *S. murinella* is recognized by its small size and rather plain grey-brown forewings. *S. artemisiella* is on average larger and has the medial part of thorax yellow and many orange or light brown scales on the forewing. *S. montanella* q.v.

MALE GENITALIA.– Coremata absent; uncus slender, apically rounded; gnathos-hook long; valva extending to about apex of uncus, weakly curved, with slightly inflated apex; sacculus moderately broad and long, tapered towards inwardly directed apical tooth, gap between sacculus and vincular process about width of sacculus; vincular process broadly projected, subtriangular, distally tapered, much shorter than sacculus; posterior margin of vinculum with broad and shallow V-shaped emargination; saccus short and slender, tapered towards apex, anteriorly slightly projecting beyond apex of pedunculus; phallus moderately slender, coecum strongly inflated.

FEMALE GENITALIA.– Segment VIII slightly broader than long; anterior margin of sternite VIII oblique, sclerotized rim, medially with small membranous opening; subgenital plate with longitudinal inner edge, lateromedial and posterior part smooth, anterior depression with large zone of distinct foamy sculpture from about middle of subgenital plate to base of apophysis anterioris; ventromedial depressions broadly digitate, anteriomedially separated by narrow emargination, apices turned outwards and extending to anterior margin of segment VIII, membranous basal half fused and covered with microtrichia, anterior half covered with distinct foamy sculpture; apophysis anterioris slightly shorter than segment VIII, rod-like; colliculum short; ductus bursae moderately broad, gradually widened towards corpus bursae; corpus bursae clearly demarcated from ductus bursae, oval; signum on the right side of corpus bursae; signum-hook short, moderately curved, basally with few additional spines, basal plate small.

DISTRIBUTION.– Boreomontane species, occurring in the Alps, Carpathians and from Ireland and Great Britain through northern Europe; also recorded from the Ukraine (Budashkin & Bidzilya, 2005: 24), European Russia and Siberia (Ponomarenko, 2008: 96). Records from the Czech Republic (Povolný, 1996a: 114), Hungary (Sattler (1987c: 453) and from some localities at lower altitudes in Central Europe (e.g. Kasy, 1983: 19) are dubious or erroneous and need confirmation. A record from southern Germany (Bavaria: Regensburg) is correct and based on the lectotype (Sattler, *op. cit.*).

BIONOMICS.– The larva has been bred from *Antennaria dioica* L. (Gaertn.) (Asteraceae) with certainty, whereas all other published host-plants such as *Leontopodium* R. Br. ex Cass., *Gnaphalium* L. and *Anthemis* L. are unconfirmed or erroneous. In fact some of these records refer to *Scrobipalpula psilella* (Sattler, 1987c: 453). The larva mines the leaves and stems but it also feeds between spun leaves, flowers and seeds-heads. Mined leaves are completely hollowed out towards the tip. The larva is univoltine from late June to September. Pupation takes place among plant debris from August onwards. The pupa hibernates until the following spring (Klimesch, 1943: 153-156). Benander (1929: 140-141) described the larva and its biology in Sweden in details. The adult is on the wing from April to June, depending on the altitude of its habitat. Preferred habitats are alpine scree and rock formations and dry grassland with stands of the host-plant up to altitudes of about 3000 m in the Alps.

REMARKS.– *Gelechia cacuminum* Frey, 1870, considered as conspecific with *S. murinella* by some authors (e.g. Povolný, 1967a: 221) was recently synonymized with *Scrobipalpula diffluella* (Frey, 1870) (Huemer & Karsholt, 1998: 59). However, records of *S. cacuminum* in the sense of Klimesch (1943), Burmann (1951) and Huemer & Tarmann (1993) refer to *S. murinella*.

Gelechia culminicolella was described from an unspecified number of specimens collected from mid-July to early August at high altitudes in Switzerland (Valais). A female syntype examined by us is conspecific with *S. murinella* (Huemer & Karsholt, 1998: 59).

50. *Scrobipalpa wiltshirei* Povolný, 1966

Scrobipalpa wiltshirei Povolný, 1966b: 406, figs. 30, 14.
Scrobipalpa obrteliana Povolný, 1971a: 33, figs. 9, 31-32, 50. Subspecies.

DIAGNOSIS.– Adult. Wingspan 10-11 mm. Labial palp cream-white mottled with some black at lower and outer surface of segment 2 and at base and near tip of segment 3. Antenna black, ringed with cream-grey. Face cream-white; neck, thorax and tegula light brown. Forewing light brown, mottled with darker brown and black; black spots in fold and at 2/5 and 3/5 in middle of wing; black spots at costa at base and 1/5; an indistinct, light postmedial fascia; apical area with many black scales. Hindwing light grey.

Variation. The amount of (white-based) black scales on the forewing is variable. In some specimens the black spots are distinct, in others they are covered with black scales.

Similar species. *S. wiltshirei* is characterized by the brown colour and by having the labial palps light with distinct black rings. The form with brown scales covering the black spots resembles *S. salinella*.

MALE GENITALIA.– Uncus slender, mitre-shaped; gnathos-hook very long, slender; anterior margin of tegumen with moderately deep emargination; valva slender, distally curved, with inflated apical part, extending far beyond tip of uncus; sacculus long and slender, outer margin weakly convex, distally tapered towards pointed apex; vincular process largely reduced to small outwardly directed tooth, about one-third length of sacculus, gap between vincular process and sacculus almost two times width of sacculus; posterior margin of vinculum with shallow V-shaped medial emargination; saccus short, basally broad, distally narrowing to pointed apex, extending to about apex of pedunculus; phallus rather short, with small apical hook, coecum strongly inflated. The shape of the uncus and the long valva are characteristic characters.

FEMALE GENITALIA.– Segment VIII short, about two times broader than long, anterior margin of sternite VIII with transverse sclerotized rim, medially with narrow incision; subgenital plate with strongly convex inner edge, anteriomedially with large zone of foamy sculpture extending to basal part of apophysis anterioris; ventromedial depressions broad, extending slightly beyond anterior margin of sternite VIII, largely covered with foamy sculpture, posterior quarter merged and covered with some microtrichia, anterior margin smooth with V-shaped medial incision; apophysis anterioris rod-like, with abrupt transition to anterior edge of sternite VIII, slightly longer than segment VIII; colliculum short and small; ductus bursae broad, gradually widened with weak turning and abrupt transition to clearly defined corpus bursae; corpus bursae large, oval; signum on the right side of corpus bursae, with distinct basal plate; signum-hook large and stout, strongly curved, with numerous spines on concave side up to about middle of hook.

DISTRIBUTION.– In Europe restricted to the Iberian Peninsula; furthermore recorded from the Canary Islands, North Africa, and from Cyprus to the Middle East.

BIONOMICS.– Host-plant and early stages unknown. The adults have been collected from February to October and they are attracted to light. *S. wiltshirei* has been observed in saline habitats up to altitudes of about 1100 m.

REMARKS.– *S. wiltshirei* Povolný, 1971, currently *S. walsinghami* Povolný, 1971, from North Africa is a primary homonym of *S. wiltshirei* Povolný, 1966 (Povolný, 1971c).

The nominotypical subspecies occurs from North Africa to the Middle East and European populations are regarded as subspecies *S. wiltshirei obrteliana*, mainly defined by their larger size and the more brownish colour of the forewing.

51. *Scrobipalpa caucasica* (Povolný, 2001)

Euscrobipalpa caucasica Povolný, 2001a: 182, fig. 6.
Euscrobipalpa bezengensis Povolný, 2001a: 183, figs. 3, 31, **syn. n.**

DIAGNOSIS.– Adult. Wingspan 12-14 mm. Labial palp white; segment 2 mottled with greyish brown; segment 3 mottled with black. Antenna black, ringed with light grey. Face light; neck and tegula light grey; thorax blackish grey. Forewing dark grey from white-based black scales and mottled with light brown, especially in fold and along veins in apical area; black spots rather indistinct. Hindwing grey.

Variation. Only the worn holotypes of *S. caucasica* and *S. bezengensis* were examined.

Similar species. The condition of the holotypes does not allow a differential diagnosis from other species.

MALE GENITALIA.– Uncus slender, apex slightly rounded; valva exceeding tip of uncus, apically strongly broadened with rounded tip; sacculus long and slender, distodorsal part rounded with acute tip, inner margin straight; vincular process long, sickle-shaped, pointed, extending below tip of sacculus posteriorly, conterminous to comparatively deep V-shaped medial emargination of posterior margin of vinculum; saccus slender, gradually tapered to pointed apex, distinctly exceeding apex of peduncculus; phallus short, stout, with distinctly inflated coecum.

FEMALE GENITALIA.– Unknown.

DISTRIBUTION.– Only known from the type locality Bezengi in the northern Caucasus Mountains (Russia).

BIONOMICS.– Host-plant and early stages unknown. The holotypes were collected in early June.

REMARKS.– *S. bezengensis* was collected in the same locality and on the same date as the holotype of *S. caucasica*. Both male holotypes are, apart from being worn, very similar and share a rather long segment 2 of the labial palps, blackish grey thorax and dark grey forewings, only slightly mottled with light brown. They differ only slightly in the broader forewings of *S. bezengensis*, with a darker black area in the middle whereas the genitalia are identical. We therefore consider both taxa conspecific.

Povolný (2002a, pl. 7, fig. 12) published a reconstructed picture of the holotype of *S. bezengensis*.

52. *Scrobipalpa pauperella* (Heinemann, 1870)

Lita pauperella Heinemann, 1870: 256.
Scrobipalpa (*Euscrobipalpa*) *klimeschi* Povolný, 1967a: 220, pl. 6, figs. 37-41, pl. 16, fig. 6.

DIAGNOSIS.– Adult. Wingspan 12-15 mm. Labial palp fuscous, more or less mottled with cream-white especially on inner surface; no distinct light rings. Antenna black, slightly lighter ringed. Head grey, lighter at frons; thorax and tegula as forewing. Forewing ground colour dark grey from black lighter-based scales and more or less covered with orange-brown, especially between veins; black spots surrounded by orange-brown. Termen with indistinct black spots. Hindwing grey.

Variation. A rather variable species. The colour of the labial palps may be dominated by cream, white or fuscous. The forewings vary from nearly fuscous to more or less orange-brown, depending on the amount of orange scales. Females tend to be slightly smaller and with slightly more pointed forewings.

Similar species. Very similar to some forms of *S. acuminatella*. *S. pauperella* has slightly more slender forewings, a darker ground colour of these, and the black spots are more distinctly surrounded with orange-brown. *S. pauperella* can also be confused with similar species like *S. arenbergeri* q.v.; *S. halonella* q.v.; *S. amseli* q.v.; *S. proclivella* q.v.; *S. hyssopi* q.v.

MALE GENITALIA.– Coremata absent; uncus slender, gradually narrrowing towards rounded apex; gnathos-hook distinct; valva extending to about or slightly beyond tip of uncus, almost straight, slender with inflated apex; sacculus long and slender, dorsodistally rounded, apex with inwardly directed tooth, gap between sacculus and vincular process at most 1.5 width of sacculus; vincular process distinctly shorter and basally wider than sacculus, distal part abruptly tapered with outwardly pointed apex; posterior margin of vinculum with moderately deep V-shaped emargination; saccus moderately broad, evenly tapered towards apex, anteriorly projecting beyond apex of pedunculus; phallus stout, coecum weakly inflated, short sclerotized arm apically.

FEMALE GENITALIA.– Segment VIII short, broader than long; anterior margin of sternite VIII with transverse, sclerotized rim, medially without membranous opening; subgenital plate with broadly rounded posterior edge, longitudinal depression with large field of foamy sculpture anteriorly, posteriorly with microtrichia, foamy sculpture extending into basal part of apophysis anterioris; ventromedial depressions merged, anterior third covered with foamy sculpture, posterior two-thirds membranous, covered with microtrichia, lateral edge broadly folded, anterior part broadly rounded and reaching far beyond anterior edge of segment VIII, apically separated by small V-shaped emargination; apophysis anterioris rod-like with gradual transition to anterior edge of sternite VIII, slightly shorter than segment VIII; colliculum short; ductus bursae moderately short and broad; corpus bursae well distinguished from ductus bursae, oval; signum on the right side of corpus bursae, basal plate small with few additional spines; signum-hook moderately long, slender, strongly curved at an almost right angle in middle. Similar to *S. arenbergeri* but differs in the more extended and fine textured foamy sculpture, the broader ventromedial depressions and the presence of additional spines of the signum. *S. spumata* q.v.

DISTRIBUTION.– Insufficiently known due to misidentifications. Verified records are known from England, southern Scandinavia, France to Slovakia and Hungary, Russia and the Ukraine (Bidzilya & Budashkin, 2009: 17) and hitherto unpublished records from northern Spain. According to Povolný (2002a: 52) also occurring in Afghanistan and Asian Russia (Transbaikal). Recently recorded from China (Bidzilya & Li, 2010).

New records. Spain: 1 ♂, Vizc, Errigoiti, Olibarra, 9.viii.1997, leg. Wolschrijn, gen. slide 2780 Hendriksen (ZMUC). Luxemburg: 1 ♂, Oberanven, Arneschi, 11.v.2001, leg. Hellers, gen. slide Hendriksen (ZMUC).

BIONOMICS.– The larva has been recorded from *Centaurea scabiosa* L., *Petasites albus* (L.) Gaertner and possibly *Serratula tinctoria* L. (Asteraceae) (Sattler, 1987c: 449). In Finland reared from *Cirsium palustre* (L.) Scop. and *C. helenioides* (L.) Hill (Asteraceae) (Mutanen, *in litt.*). According to Povolný (1983: 16) the larva mines the leaves and the species occurs in three generations. However, due to mixture with other species (see Sattler, *op. cit.*) the larval habits are insufficiently documented. Verified data exist of larvae feeding in stems of *Centaurea scabiosa* during July. In Finland *S. pauperella* has been bred from larvae mining leaves of *Centaurea phrygia* L. in July and August (Tabell, *in litt.*). Observations of adults from April to early June and again in July-August indicate bivoltinism. The habitats include various types of meadows and pastures at low altitudes.

REMARKS.– The identity of *Lita pauperella* was stated by Sattler (1989: 9) based on the examination of a single syntype, unfortunately without abdomen. Therefore this taxon is regarded as nomen dubium by some authors (Elsner *et al.*, 1999: 42). However, we believe that it serves stability best to use the oldest name, *S. pauperella*, for this species.

Sattler (1989: 7) argued that the holotype of *S. klimeschi* differs in the male genitalia from other specimens of *S. pauperella*, and he considered them to be different species. Among specimens collected from Görliz near Bilbao in northern Spain by J. Wolschrijn we dissected males corresponding both with the holotype of *S. klimeschi* and with *S. pauperella*, and we hence consider them as conspecific.

53. Scrobipalpa spumata (Povolný, 2001)

Euscrobipalpa spumata Povolný, 2001a: 197, figs. 19, 41.

DIAGNOSIS.– Adult. Wingspan 10-12 mm. Labial palp white; segment 2 mottled with blackish grey; segment 3 with two black rings. Antenna ringed black and light grey. Head light grey; neck darker; thorax and tegula greyish brown. Forewing light brown mottled with white-based black scales, especially on costa and along veins in apical half. No visible black spots. Hindwing grey.

Variation. There is some variation in the mixture of light brown and white-based black scales on the forewing, resulting in specimens looking either more light brown or grey.

Similar species. *S. spumata* is characterized by the almost light brown forewings with white-based black scales along the veins and by having no visible spots. Similar specimens are found in populations of the larger *S. samadensis*. *S. lutea* is slightly smaller, and has more or less visible spots.

MALE GENITALIA.– Coremata absent; uncus subrectangular, longish; gnathos-hook distinct; valva extending slightly beyond tip of uncus, weakly curved with slightly inflated apex; sacculus moderately long and slender, outer margin strongly convex, apex with inwardly directed tooth, gap between sacculus and vincular process smaller than width of sacculus; vincular process slightly shorter and basally wider than sacculus, distal part gradually tapered with outwardly pointed apex; posterior margin of vinculum with moderately deep V-shaped emargination; saccus moderately broad, evenly tapered towards apex, anteriorly projecting beyond apex of pedunculus; phallus stout, coecum weakly inflated, short sclerotized arm apically. The characteristic shape of the uncus is of particular diagnostic value.

FEMALE GENITALIA.– Segment VIII short, broader than long; anterior margin of sternite VIII without sclerotized rim; subgenital plate with broadly rounded posterior edge, longitudinal depression with large field of distinct foamy sculpture anteriorly, posteriorly with microtrichia, foamy sculpture extending into basal part of apophysis anterioris; ventromedial depressions merged, anterior third covered with foamy sculpture, posterior two-thirds membranous, covered with microtrichia, lateral edge broadly folded and smooth, anterior part thumb-shaped and reaching far beyond anterior edge of segment VIII, apically separated by small V-shaped emargination; apophysis anterioris rod-like with gradual transition to anterior edge of sternite VIII, slightly longer than segment VIII; colliculum short; ductus bursae moderately short and broad; corpus bursae well distinguished from ductus bursae, oval; signum on the right side of corpus bursae, basal plate small; signum-hook moderately long, slender, strongly curved at an almost right angle near middle. Differs from the otherwise similar *S. pauperella* in the larger sized foamy sculpture of the subgenital plate, the absence of a sclerotized anterior edge of sternite VIII and the longer apophysis anterioris.

DISTRIBUTION.– Only known from Russia (southern Ural Mountains) (new record), and from the type locality in the Tuva region in Siberia.

New record. Russia: 1 ♂, South Ural, Bashkiria, Sakamara river, Jantyshevo village, 21.vi.1996, leg. Nupponen *et al.*, gen. slide GU 01/1084 ♂ Huemer (JUNN); 1 ♀, South Ural, Orenburg district, 6 km W Donskoje village, Mount Verbljushka, 30.v.-2.vi.1998, leg. Junnilainen, gen. slide GU 01/1085 ♀ Huemer (JUNN).

BIONOMICS.– Host-plant and early stages unknown. The adults have been collected from late May to late June at light.

REMARKS.– *S. spumata* was described from a single female collected in Tuva region (Russia). We have been able to examine the holotype from ZMKU. The female genitalia are squeezed and the ventromedial depression is strongly distorted and adjoined, particularly in distal half. However, despite the apparent differences between the original line drawing and our photographs the genitalia structures fully agree in every detail when comparing the slides.

54. *Scrobipalpa arenbergeri* Povolný, 1973

Scrobipalpa arenbergeri Povolný, 1973c: 282, figs. 1-2, 4-5, 9.

DIAGNOSIS.– Adult. Wingspan 11 mm. Labial palp cream-coloured; segment 2 mottled with greyish black on lower and outer surface; segment 3 with two black rings. Antenna black, ringed with light grey. Head and tegula with grey, light-tipped scales; face lighter; thorax greyish black. Forewing covered with light-based and tipped black scales and mottled with orange in fold, as a subcostal stripe, around and between black spots in middle of wing and as faint costal and tornal spots. Hindwing dark grey.

Variation. The few examined specimens show only slight variation.

Similar species. *S. arenbergeri* is characterized by its relatively small size and its blackish grey forewings with many orange scales. It is very similar to *S. pauperella*, but the orange scales on the forewings of that species are more scattered and less clear. Segment 3 of the labial palps of *S. pauperella* lacks black rings. *S. skulei* sp. n. q.v.

MALE GENITALIA.– Uncus broad, apically rounded; valva extending slightly below or at most to apex of uncus, weakly curved with inflated apex; sacculus comparatively long, tapering towards pointed apex; vincular process wider and distinctly shorter than sacculus, slightly curved outwards with distinct tip, inner margin conterminous to comparatively shallow U-shaped medial emargination of posterior margin of vinculum; saccus broad, gradually tapered towards rounded apex, comparatively long, distinctly exceeding apex of pedunculus; phallus stout, with weakly inflated coecum. Very similar to *S. pauperella* but differs e.g. in the tapered sacculus and the broader and longer saccus.

FEMALE GENITALIA.– Segment VIII short, broader than long; anterior margin of sternite VIII with long sclerotized rim, medially with narrow membranous opening, posterior edge broadly rounded on inner side; anteriolateral and medial depression with distinct foamy sculpture, particularly well developed in anterioproximal depression of sternite VIII, extending into basal part of apophysis anterioris; ventromedial depressions adjoined, anteriorly rounded and reaching far beyond anterior edge of segment VIII, with V-shaped medial gap anteriorly; apophysis anterioris rod-like with gradual transition to anterior edge of sternite VIII, slightly shorter than segment VIII; colliculum short; ductus bursae rather short and broad, indistinct transition to pyriform corpus bursae; signum on the right side of corpus bursae, basal plate small without additional tooth; signum-hook moderately long, slender, strongly curved at an almost right angle in middle. Differs from the overall similar *S. pauperella* e.g. in the more honey-comb like lateromedial foamy sculpture, the narrower ventromedial depressions and the absence of additional spines of the signum.

DISTRIBUTION.– Known from a few localities in Austria, Hungary, Slovakia, Croatia and the Ukraine; further unpublished records from Italy and the southern Ural Mountains (Russia); outside Europe known from Asian Russia (Transbaikal) (Bidzilya *et al.*, 1998: 47). The occurrence in Hungary was considered dubious by Szabóky *et al.* (2006: 78), but later confirmed Szabóky *et al.* (2009: 6-7).

New records. Italy: 1 ♂, Friuli-Venezia Giulia, Prealpi Giulie, Alta Valle del Torre, loc. Musi, 650 m, 19.vii.2000, leg. Morin, gen. slide GU 08/1269 ♂ Huemer (MFSN); 4 ♂, 1♀, Friuli-Venezia Giulia, 6 km NW Vivaro, 250 m, 26.iv.-3.v.2008; 1 ♂, 1 ♀, ditto, but 28.vi.2008, leg. Skou & Nilsson, gen. slides 6543 Hendriksen, 5117 Karsholt (ZMUC). European Russia: 1 ♂, S Ural, Orenburg district, near Burranoe village, 18.viii.2006, K. Nupponen leg., gen. slide GU 09/1290 ♂ Huemer (NUPP).

BIONOMICS.– Host-plant and early stages unknown. The adults have been collected from late April to early May and from July to early September in halophytic or wetland habitats at light. Sattler (1987c: 451) suspected two generations and a trophic linkage to Asteraceae. We could examine one specimen which, according to label data, was bred from *Centaurea scabiosa* L. (Asteraceae).

REMARKS.– Povolný (1983b: 14) synonymized *S. arenbergeri* with *S. pauperella* (= *S. klimeschi*) based on individual variation of genitalia slides of both taxa, allegedly falling into the same species range. However, Sattler (1987c: 450) convincingly proved the identity of *S. arenbergeri*, a point of view which was later followed by different authors including Povolný himself (Elsner *et al.*, 1999: 43; Povolný, 2002a: 30).

55. *Scrobipalpa nana* Povolný, 1973

Scrobipalpa nana Povolný, 1973b: 108, figs. 2, 13-14, 28.
Euscrobipalpa nana caroxyli Falkovitsh & Bidzilya, 2006: 77. Subspecies.

DIAGNOSIS.– Adult. Wingspan 8-9 mm. Labial palp cream-coloured; segments 2 and 3 each with two black rings. Antenna black, ringed with light grey. Head cream-coloured, mottled with grey, especially in neck; thorax and tegula as forewing. Forewing with white-based black scales and mottled with light yellow and orange brown; base, costal and central part of wing black; lighter brownish areas along dorsum, in fold and as a subcostal stripe; black spots distinct, surrounded by orange-brown; costal and tornal spots light yellow, separated by brown scales; an orange-brown patch from black spot at 3/5 to before tornal spot. Hindwing light grey.

Variation. The few examined specimens vary in the number of light yellow and orange-brown scales on the forewing.

Similar species. *S. nana* is characterized by its small size, the distinct black spots and the yellowish costal and tornal spots separated by brown scales. *S. heretica* q.v.

MALE GENITALIA.– Uncus comparatively broad, apex with concave emargination and pointed lateral corners; gnathos-hook strong; anterior margin of tegumen with moderately deep emargination; valva slightly shorter than uncus, slender, distally curved, with distinctly inflated apex; sacculus long, broader than valva, slightly sickle-shaped, curved inwards and tapered towards pointed apex; vincular process strongly projected posteriad, slightly shorter than sacculus, broad, tapered towards outwardly pointed apex, gap between sacculus and vincular process about width of sacculus; posterior margin of vinculum with moderately deep V-shaped medial emargination; saccus broad at base, distal part slender, slightly exceeding apex of pedunculus; phallus moderately slender, coecum weakly inflated.

FEMALE GENITALIA.– Segment VIII slightly broader than long; subgenital plate with longitudinal inner edge, depression with large zone of distinct foamy sculpture from posterior third to base of apophysis anterioris, edged with few microtrichia; ventromedial depressions broadly digitate, extending slightly beyond anterior margin of segment VIII, membranous basal half fused and covered with microtrichia, sclerotized anterior and lateral part covered with distinct foamy sculpture, anteriomedially separated by V-shaped emargination; apophysis anterioris about length of segment VIII, rod-like; colliculum small; ductus bursae evenly widened, with indistinct transition to large, pyriform corpus bursae; signum-hook short, strongly curved near middle, basally with few additional spines, basal plate small. *S. heretica* q.v.

DISTRIBUTION.– In Europe only known from Russia (southern Ural Mountains). Outside Europe recorded from Turkey (Anatolia) to northern Iran and Kazakhstan (subspecies *caroxyli*).

BIONOMICS.– Host-plant and early stages of the nominotypical subspecies are unknown. *S. nana caroxyli* was bred from spun leaves of *Caroxylon orientale* (S. G. Gmelin) Tzvelev (Chenopodiaceae) (Falkovitsh & Bidzilya, 2006: 77). The adults have been collected from the second half of May to August at light.

REMARKS.– *Scrobipalpa nana caroxyli* was described from a single female collected in Kazakhstan (Falkovitsh & Bidzilya, 2006: 77). According to the original description it differs from the nominotypical subspecies by less contrasting forewing markings, the longer and more filiform apophyses anteriores and the longer signum.

56. *Scrobipalpa heretica* Povolný, 1973

Scrobipalpa heretica Povolný, 1973b: 107, figs. 5, 22, 27.
Scrobipalpa submagnificella Povolný, 1977e: 330, figs. 9, 25, **syn. n.**

DIAGNOSIS.– Adult. Wingspan 10 mm. Labial palp white; segment 2 mottled with black on lower and outer surface; segment 3 with two black rings. Antenna ringed with black and white. Face cream-coloured; neck grey; thorax and tegula as forewing. Forewing dark grey from white-based and tipped black scales and mottled with brown, especially along dorsum, as a subcostal stripe, around black spots in fold and at 2/5 and 3/5 in middle of wing and between light costal and tornal spot. Hindwing grey.

Variation. The above diagnosis is based on the single European specimen from Spain. Specimens from the the eastern part of the distribution area are somewhat lighter, having the face white, the neck light grey, the brown areas on the forewings lighter and the costal and tornal spots more distinctly white.

Similar species. *S. heretica* is characterized by its small size, and by having the forewings separated into a brownish dorsal part and a greyish costal part. This is not the case in the similar, but slightly smaller *S. nana*; *S. suaedicola* is slightly larger and has a thin black streak separating costal and tornal spots.

MALE GENITALIA.– Uncus broadly subrectangular, apically truncate with strongly concave posterior margin; gnathos-hook small; anterior margin of tegumen with moderately deep and broad emargination; valva almost straight, apical third weakly inflated, extending to about apex of uncus; sacculus long and moderately broad, outer margin curved, inner margin straight, gradually tapered towards inwardly tipped apex; vincular process slender, distinctly shorter than sacculus, tapered to outwardly pointed apex, separated from sacculus by moderate gap; posterior margin of vinculum with shallow V-shaped medial emargination; saccus stout, moderately broad, evenly tapered towards apex, extending to about apex of pedunculus; phallus moderately short, with long apical hook, coecum moderately inflated.

FEMALE GENITALIA.– Segment VIII about same length and breadth; subgenital plate with longitudinal inner edge, depression with small zone of distinct foamy sculpture from anterior third to base of apophysis anterioris; ventromedial depressions broadly digitate, extending far beyond anterior margin of segment VIII, membranous basal half fused and covered with microtrichia, sclerotized anterior parts separated by deep V-shaped emargination and covered with foamy sculpture; apophysis anterioris about length of segment VIII, moderately broad at base, evenly crossing over into rod-like distal part; colliculum small; ductus bursae evenly widened, with indistinct transition to large, pyriform corpus bursae; signum-hook short, strongly curved near middle, basally with few additional spines, basal plate small. Differs from the very similar *S. nana* in the smaller zone of foamy sculpture and the longer and apically projected ventromedial depressions.

DISTRIBUTION.– In Europe only known from southern Russia (Piskunov & Anikin, 2001: 154; Junnilainen *et al.*, 2010: 48), an unpublished record from southern Spain. More widely distributed in Palaearctic Asia, ranging from Turkey to Iran and Kazakhstan. Records from Kyrgyzstan (Povolný, 1996b: 32) are partially based on misidentification (see below).

New record. Spain: 1 ♂, Teruel, Albarracín, 11-24.vii.1987, leg. Wolschrijn, gen. slide 4066 ♂ Hendriksen (ZMUC).

BIONOMICS.– Host-plant and early stages unknown. The adults have been collected from June to August, at altitudes from 750 to 2750 m (Povolný, 2002a: 46, 57).

REMARKS.– *S. heretica* was described from 4 males and a single female, *S. nana* from a single male and two females. The male holotypes of both species were collected syntopically on the same night by the late W. Glaser. Both species are closely related and the differences between them were stated to be found in the smaller size and lighter wings of *S. nana* and furthermore in some genitalia structures. Povolný (1973b: 108) particularly mentioned the short and slender valva as specific character of *S. nana* but later some individual variation of male genitalia was figured (Povolný, 2002a: pl. 28, figs. 249-250, figs. 255-256). The small differences in genitalia seem to support the existence of two species. However, the identity of the female sex remains somewhat disputed. The female of *S. heretica* from the original description fully agrees with *S. nana* and Povolný himself later attributed it to the latter species (Povolný, 1996b: 32). More recently this specimen was figured as *S. nana* (Povolný, 2002a: pl. 72, fig. 741) without giving any hints to its original identity as a paratype of *S. heretica*. Povolný (1996b: 32, fig. 11) furthermore redescribed a quite different female from Kyrgyzstan as *S. heretica*, based on the sympatric occurrence with *S. heretica* males. However, this female belongs to *S. oleksiyella* sp. n. In our opinion females of *S. heretica* and *S. nana*, figured in the original description and coming from the type locality, are to be treated as correctly identified.

S. submagnificella was described from a single and worn male which was collected in Afghanistan (Povolný, 1977e: 330-331). In the original diagnosis the species was compared with *S. magnificella*. However, size, wing markings and colour and particularly the genitalia of *S. submagnificella*, again figured by Povolný (2002a: pl. 26, fig. 247), completely fall within the range of variation of *S. heretica*. We therefore formally synonymize it with the latter.

57. *Scrobipalpa bigoti* Povolný, 1973

Scrobipalpa bigoti Povolný, 1973c: 283, figs. 3, 6-8.
Scrobipalpa bigoti tunesica Povolný, 1979a: 115, figs. 1, 5, 7-8. Subspecies.

DIAGNOSIS.– Adult. Wingspan 7-10 mm. Labial palp white; segments 2 and 3 each with a black ring at base and near tip. Antenna distinctly ringed black and whitish grey. Head, thorax and tegula cream-white mottled with light brown and grey scales; face lighter. Forewing rather short and broad, covered with white-tipped grey, black and orange scales; distinct black spots in fold and at 2/5 and 3/5 in middle of wing surrounded by orange; black spots on costa near base and at 1/4 and 3/4; a light, bent fascia at 3/4; terminal line black, followed by up to four faint, black and grey cilia lines; an orange patch from black spot at 3/5 to tornus. Hindwing whitish grey.

Variation. Females are on average larger than males. The scales of the forewing vary from light grey to black (white-tipped), giving it a lighter or darker appearance.

Similar species. *S. bigoti* is characterized by its small size, its rather short and broad forewings and by the orange patch from tornus to the black spot at 3/5.

MALE GENITALIA.– Uncus broad, apex flattened; gnathos-hook short, small; valva extending to about tip of uncus, without apical inflation; sacculus long and slender, about even width from base to apex, dorsoapically rounded with distinct tooth turned inwards; vincular process slightly shorter and basally broader than sacculus, weakly pointed at apex, without tip; inner margin of vincular process gradually passing into V-shaped medial emargination of posterior margin of vinculum; saccus narrow, slightly tapered distal part with knob-like apex, extending at most to, or slightly beyond, apex of pedunculus; phallus rather short, with weakly inflated coecum, apex pointed. The genitalia show some variation, especially in the width of the sacculus. *S. bigoti* is hardly distinguishable from the North African *S. mahunkai* Povolný, 1979, in the male genitalia (see Remarks). It differs from the also similar *S. cultrata* Povolný, 1971, a species described from Algeria, in the more narrow sacculus and vincular process.

FEMALE GENITALIA.– Segment VIII broader than long, anteriolateral edge with concave emargination; subgenital plate with long and nearly straight inner edge, almost right angle to posterior edge, distinct foamy sculpture from middle of subgenital plate extending into base of apophysis anterioris; ventromedial depressions with well developed foamy sculpture throughout, broadly digitate, reaching far beyond anterior margin, basally fused, distal parts separated by broadly V-shaped emargination and curved outwards; apophysis anterioris about length of segment VIII, moderately broad at base, evenly crossing over into slender and pointed distal part; colliculum small; ductus bursae short and broad, indistinct transition to scarcely separated, small and globular corpus bursae; signum with small basal plate without additional tooth; signum-hook longish and slender, strongly curved at an almost right angle before middle.

DISTRIBUTION.– Known from a few localities in Spain, southern France and from unpublished records from Italy (Sicily) and Greece. Also recorded from Cyprus and Tunisia.

New records. Italy (Sicily): 7 ♂, 3 ♀, 5 km W Pachino, Pantano Longarini, sea lev., 12.ix.2002, leg. Karsholt *et al.*, gen. slide 3786 Hendriksen. Greece: 1 ♂, Kounopelli by Patras, 0 m, 25.ix.2005, leg. & coll. Schmitz, gen. slide 5917 Hendriksen.

BIONOMICS.– Host-plant and early stages unknown. The adults have been collected from April to June and again in September in halophytic habitats (Povolný & Luquet, 1983: 64). They are attracted to light.

REMARKS.– *S. bigoti tunesica* was described from Tunisia. According to Povolný (1979a: 116; 2002a: 35) it differs from the nominotypical race in the larger range of variation which includes completely pale individuals. We have been able to examine specimens from this area which indeed show a remarkable variability both externally and in genitalia.

S. mahunkai Povolný, 1979 was described from a series of males and females from Tunisia. It is extremely similar to *S. bigoti* and may be conspecific. Particularly the males are practically indistinguishable whereas the females seem to be different. However, the type series of *S. mahunkai* could be based on a mixture of males and females belonging to different species. Also individuals from different European populations show some variation in the genitalia, and further revisional studies are desirable to resolve this problem satisfactorily.

There are some further related, taxonomically unrevised species in the deserts of North Africa and Arabia, but all specimens examined by us from Europe seem to belong to *S. bigoti*.

58. *Scrobipalpa dorsoflava* (Povolný, 1996)

Ilseopsis (Euscrobipalpa) dorsoflava Povolný, 1996b: 22, figs. 3, 21.

Diagnosis.– Adult. Wingspan 10-11 mm. Labial palp cream-white; each segment with two black rings. Antenna black, ringed with light grey. Head cream-coloured mottled with grey towards neck; thorax and tegula as forewing. Forewing with white-based and white-tipped black scales mottled with light brown and grey; dorsum including fold light grey-brown from base towards middle of wing; a faint light brown subcostal stripe; black spots at 2/5 and 3/5 distinct, surrounded by light brown; costal and (especially) tornal spot light grey-brown, separated by a black stripe towards apex. Hindwing grey.

Variation. A rather variable species. The light area along dorsum can be more or (in blackish specimens) less distinct. In one specimen the light areas are clear brown. The black spots at 3/5 are sometimes oblong or drop-shaped.

Similar species. *S. dorsoflava* resembles some forms of *S. artemisiella* but has more marbled forewings. The forewings of *S. voltinella* and *S. divisella* are more clearly separated into a dark costal and a lighter dorsal area.

Male genitalia.– Uncus long, moderately broad, distally rounded with weak medial emargination at apex; gnathos-hook short; valva slender, evenly curved, slightly shorter than uncus, apically distinctly inflated; sacculus moderately long and slender, dorsoapically rounded with weakly developed inwardly directed acute tip, ventral margin almost straight; vincular process long and moderately slender, apex with distinct outwardly directed tooth, separated from sacculus by narrow gap; vinculum comparatively long, posterior margin with moderately shallow U-shaped medial emargination; saccus subtriangular, strongly tapered towards pointed apex, extending slightly beyond apex of pedunculus; phallus short, stout, with distinct apical hook, coecum inflated. *S. abstrusa* sp. n. q.v.

Female genitalia.– Segment VIII about same length and breadth; sternite VIII with indistinct, membranous anterior margin; subgenital plate covered with numerous microtrichia posteriorly and increasingly with foamy sculpture in anterior part, particularly along depression near longitudinal inner edge; ventromedial depression broadly subrectangular, distinct digitate processes reaching far beyond anterior margin of sternite VIII, anteriorly separated by V-shaped emargination, posteriomedially adjoined, depression largely covered with microtrichia, anteriorly increasingly with foamy sculpture; apophysis anterioris rod-like, abruptly separated, slightly shorter than segment VIII; colliculum moderately short; ductus bursae broad; corpus bursae well distinguished from ductus bursae, oval; signum on the right side of ductus bursae near entrance to corpus bursae; signum-hook moderately short, slender, weakly curved in middle, short additional tooth near base, basal plate small.

DISTRIBUTION.– Only known from Russia (southern Ural Mountains), Siberia and Kyrgyzstan.

BIONOMICS.– Host-plant and early stages unknown. The adults have been collected from mid-June to mid-August at light.

REMARKS.– *S. dorsoflava* was described from a single male collected by Kaila & Mikkola on 13.viii.1990 near lake Issyk Kul (Kyrgyzstan) (Povolný, 1996b: 22). We have been able to study a small series of additional specimens from Russia (southern Ural Mountains). According to Povolný (1996b: fig. 3) the valva is much shorter than the uncus. However, this 'character' is merely a problem of preparation technique and seems to underly some artificially caused variation. The species seems closely related to *S. kyrana* Povolný, 2001 which was described from a single female from eastern Siberia. Apparent differences between both taxa may be primarily based on different preparation technique of the holotype slide of *S. kyrana* and both taxa may even be conspecific. Already doubted records of *S. kyrana* from Altai (Bidzilya, 2009: 6) refer to *S. dorsoflava*.

59. *Scrobipalpa magnificella* Povolný, 1967

Scrobipalpa (Euscrobipalpa) magnificella Povolný, 1967a: 221, pl. 8, figs. 42-45, pl. 16, fig. 6.

DIAGNOSIS.– Adult. Wingspan 11-12 mm. Labial palp white; base and tip of segment 2 and 3 each with a black ring. Antenna black, rather indistinctly ringed with light grey. Head cream-coloured mottled with light grey-brown; thorax and tegula as forewing. Forewing grey from white-based black scales and mottled with light brown and white; a distinct black spot in fold and more indistinct ones at 2/5 and 3/5 in middle of wing, all surrounded by light brown and cream-white; costal and tornal spots white, more or less fused; a light brown patch before tornal spot. Hindwing white with faint grey margins and light grey cilia. Underside of forewing with costal and dorsal two-thirds and apical third white.

Variation. The forewings can be lighter and darker grey, mottled with more or less light brown and cream-white. The black spots at 2/5 and 3/5 in the middle of the forewing can be distinct or hardly visible.

Similar species. *S. magnificella* is characterized by the white labial palps having base and tip of segment 2 and 3 black, by the grey forewings with a distinct black spot in fold and clear white, fused costal and tornal spots, and by the white hindwings.

MALE GENITALIA.– Uncus broadly subrectangular, apically truncate; gnathos-hook subtriangular, stout; anterior margin of tegumen with deep and moderately narrow emargination; valva stout, slightly curved, distal part sword-like, extending beyond apex of uncus; sacculus long and slender, about three times as long as wide, weakly curved inwards, gradually tapered towards pointed apex; vincular process shorter and slightly broader than sacculus, distally tapered with shouldered inner margin and outwardly pointed apex, separated from sacculus by moderate gap; posterior margin of vinculum with shallow U-shaped medial emargination; saccus stout, broad, apex with transverse sclerite, extending to about apex of pedunculus; phallus moderately short, with long apical hook, coecum weakly inflated. Differs from the similar *S. heretica* e.g. by the broad apex of the saccus and the narrower sacculus.

FEMALE GENITALIA.– Segment VIII slightly broader than long; subgenital plate with longitudinal outer and oblique inner edge, medial part covered with some microtrichia, distinct field of foamy sculpture extending to base of apophysis anterioris; ventromedial depressions broadly digitate, largely sclerotized, narrow posterior part with some microtrichia, broadened anterior part covered with foamy sculpture, extending slightly beyond anterior edge of segment VIII, anteriomedially separated by V-shaped emargination, posteriomedially adjoined; apophysis anterioris abruptly broad-based, about length of segment VIII; colliculum small; ductus bursae broad, with indistinct transition to ill-defined corpus bursae; corpus bursae oval; signum on the right side near entrance of corpus bursae; signum-hook small, strongly curved near middle, without additional spines, basal plate small.

DISTRIBUTION.– Recently found in the Ukraine (Bidzilya, 1997: 123) and Russia (southern Ural Mountains) (Junnilainen *et al.*, 2010: 48). Outside Europe recorded from Syria, Iran, Mongolia (Povolný, 2002a: 54) and China (Bidzilya & Li, 2010).

BIONOMICS.– Host-plant and early stages unknown. The adults have been recorded in May and June up to altitudes of about 1600 m.

60. *Scrobipalpa abstrusa* sp. n.

DIAGNOSIS.– Adult. Wingspan 12-13 mm. Labial palp white mottled with brownish black on lower and outer surface of segment 2 and on base and tip of segment 3. Antenna black, ringed with grey. Head, thorax and tegula grey. Forewing grey from white-based black scales and mottled with light brown, especially in fold, subcostal stripe and in apical half; black spots indistinct, surrounded by some light brown; apical area with many black scales. Hindwing grey.

Variation. The examined specimens show only slight variation.

Similar species. *Scrobipalpa abstrusa* sp. n. is a rather characterless species. The grey forewings with indistinct black spots resemble some forms of *S. obsoletella*, but that species has more light yellowish brown labial palps and forewings (*S. abstrusa* being more clear grey). With our present knowledge *Scrobipalpa abstrusa* sp. n. can hardly be identified without examination of the genitalia.

MALE GENITALIA.– Coremata absent; uncus long, evenly tapered to truncate apex, apically with weak emargination; gnathos-hook stout; valva slender, distal part weakly curved, slightly shorter than uncus, apically distinctly inflated; sacculus stout, moderately long and slender, dorsoapically rounded with weakly developed inwardly directed tooth, ventral margin almost straight; vincular process moderately long and slender, with acute and outwardly directed apex, separated from sacculus by narrow gap; vinculum comparatively long, posterior margin with moderately shallow almost U-shaped medial emargination; saccus subtriangular, abruptly tapered towards pointed apex, extending beyond apex of pedunculus; phallus short, stout, with distinct apical hook, coecum inflated. Differs from *S. marmorella* Povolný, 1969, and from *S. dorsoflava* in the tapered uncus and from the latter also in the comparatively shorter and stout sacculus and the vincular process.

FEMALE GENITALIA.– Segment VIII slightly broader than long; anterior edge of sternite VIII obliquely sclerotized with moderate medial opening; subgenital plate smooth, without microtrichia and foamy sculpture, posteriomedial edge broadly rounded, inner edge oblique and anteriorly with even transition to apophysis anterioris; ventromedial depression widened anteriorly, covered with microtrichia except for slightly wrinkled foremost part, anteriorly separated by U-shaped emargination, digitate apex reaching slightly beyond anterior edge of segment VIII; apophysis anterioris moderately stout, with base gradually widened, rod-like, about length of segment VIII; colliculum short; ductus bursae long, slender; corpus bursae well distinguished from ductus bursae, globular; signum on the right side of corpus bursae; signum-hook moderately slender, strongly curved medially, basal plate large with two pairs of distinct spines. Differs from the similar *S. marmorella* particularly in the shape of the signum and the lack of subostial sclerites.

DISTRIBUTION.– Only known from Russia (southern Ural Mountains and the Volgograd district).

BIONOMICS.– Host-plant and early stages unknown. The adults have been collected from May to late June.

MATERIAL EXAMINED.– Holotype. ♂ 'RUSSIA S Ural Cheliabinsk district, Arkaim reserve near Amurskii village, 14.vi.1996, leg. Nupponen, Kaitila, Junnilainen & Ahola' 'gen. Præparat No. 2886 ♂ Hendriksen' (NUPP).

Paratypes. Russia: 3 ♂, 1 ♀, same data as holotype, but gen. slide 2938, 2970, 2975, 5016 Hendriksen (NUPP, ZMUC); 1 ♂ Cheliabinsk district, Troizkii reserve near Berlin village, 30.vi.1997, leg. K. Nupponen & J. Junnilainen (JUNN, NUPP); 1 ♂, Cheliabinsk district, near Moskovo village, 26.v.1998, leg. Junnilainen, gen. slide 2973 Hendriksen (JUNN); 1 ♂ Cheliabinsk district, 20 km S Mednogorsk near Kidriasovo village, 28.-29.v.1998, leg. Junnilainen, gen. slide 3061 Hendriksen (JUNN); 1 ♂, Orenburg district, 8 km E Novoiletzk, 9.vi.1998, leg. Nupponen, gen. slide 3166 Hendriksen (ZMUC); 1 ♂, Cheliabinsk district, Arkaim reserve, near Amurskii, 15.vi.1999, gen. slide 3170 Hendriksen; 1 ♂, 18.v.2004, leg. Nupponen (NUPP); 23 ♂, Orenburg district, 20 km Pokrovka village, S Schibendy valley, 10. & 11.v.2005, leg. Nupponen (NUPP, TLMF, ZMUC).

Material excluded from type series. Russia: 2 ♂, Volgograd district, near Olhovka, 18.v.2005, leg. Nupponen (NUPP).

ETYMOLOGY.– The name refers to the latin adjective *abstrusus* (= hidden).

61. Scrobipalpa superstes Povolný, 1977

Scrobipalpa superstes Povolný, 1977b: 189, figs. 10-14, 31.

DIAGNOSIS.– Adult. Wingspan 8-12 mm. Labial palp cream-coloured; segments 2 and 3 each with two ill-defined black rings. Antenna black, ringed with light grey. Face cream-coloured; neck light grey-brown; thorax and tegula greyish. Forewing greyish with a mixture of white-based light brown and black scales, the latter especially at base, along costa and in apical area; black spots at base and 1/5 of costa, in fold and at 2/5 and 3/5 in middle of wing; an indistinct, white, postmedial fascia. Hindwing light grey.

Variation. *S. superstes* occurs in a form with greyish and another with light reddish brown forewings. Intermediate forms are relatively rare. There is slight variation in the amount of black and light brown scales on the forewing. The black spots can be almost absent. Sometimes there are a few yellowish scales in the fold.

Similar species. *S. superstes* is a rather characterless species, having either greyish or reddish brown forewings with black suffusion at base, along costa and in apical area. It is easily confused with several other *Scrobipalpa* species from which it is best separated by characters of the genitalia.

MALE GENITALIA.– Uncus moderately slender, gradually narrrowing towards rounded apex; gnathos-hook stout; valva extending to about tip of uncus, almost straight, slender with inflated apex; sacculus moderately short and slender, distal part of outer margin rounded, apex without distinct tooth, gap between sacculus and vincular process slightly narrower than sacculus; vincular process about length of sacculus, basally wider than sacculus, strongly tapered towards outwardly pointed apex; posterior margin of vinculum with deep V-shaped emargination; saccus broad, stout, evenly tapered towards apex, long, anteriorly projecting above apex of pedunculus; phallus stout, coecum inflated, distinct sclerotized arm apically. The male genitalia are indistinguishable from *S. remota*.

FEMALE GENITALIA.– Segment VIII distinctly broader than long; subgenital plate with longitudinal and strongly folded inner edge, lateral and posterior margin smooth, longitudinal depression covered with granulated sculpture and microspines; ventromedial depression broadly subrectangular, sclerotized and covered with microspines and microtrichia, largely fused, anterior margin oblique with V-shaped incision, anteriolateral margin rounded and overlapping with subgenital plate; subostial sclerite distinct, sword-like; apophysis anterioris rod-like, evenly broad-based, slightly longer than segment VIII; colliculum short; ductus bursae broad, abrupt transition to small, oval corpus bursae; signum at right side near entrance of of corpus bursae; signum-hook longish and slender, evenly curved, several additional spines in basal half of curve, basal plate moderately large; signum on the right side of corpus bursae, basal plate small without additional tooth; signum-hook relatively short, slender, strongly curved at an almost right angle in middle.

DISTRIBUTION.– Portugal, southern Spain, and unpublished records from southern France and Italy (Sardinia, Sicily).

New records. France: 1 ♂, Provence, Domaine de Maure Vieille, 21.vi.2001, leg. Hendriksen, gen. slide 3034 Hendriksen (ZMUC). Italy (Sardinia): 1 ♂, Cabras, 21.vii.1981, leg. Baldizzone, gen. slide 3242 Hendriksen (BALD). Italy (Sicily): 5 ♂, 1 ♀, 5 km W Pachino, Pantano Longarini, sea lev., 12.ix.2002, leg. Karsholt *et al.*, gen. slides 3789, 3790, 3791 Hendriksen (ZMUC).

BIONOMICS.– Host-plant and early stages unknown. The adults have been observed in May and June and again in September and October (Povolný, 2002a: 73) at low altitudes. They are attracted to light.

REMARKS.– Male specimens from western Europe with reddish brown forewings have been recorded as *S. remota* (see discussion under 'Species excluded from the European fauna').

62. ***Scrobipalpa plesiopicta*** Povolný, 1969

Scrobipalpa plesiopicta Povolný, 1969a: 17, fig. 76, pl. 31, fig. 21.

DIAGNOSIS.– Adult. Wingspan 17-20 mm. Labial palp bone-white, mottled with blackish brown on lower and outer surface of segment 2 and, especially, on segment 3. Antenna black, ringed with light grey. Head cream-coloured, mottled with light brown on neck; thorax and tegula grey, the latter with light tip. Forewing with light-based grey, light brown and black scales; black patches especially near base in middle of wing between 1/4 and 1/3, along costa and in apical part of the wing; light brown scales along fold, as a subcostal stripe, around black spots and before black apical area. Hindwing light grey.

Variation. The examined specimens show slight variation. The black patch between 1/4 and 1/3 can be covered with light yellow scales (in fresh specimens) or obsolete (in worn specimens).

Similar species. *S. plesiopicta* can be separated from similar species, e.g. *S. obsoletella*, *S. indignella* and *S. grisea* by its large size.

MALE GENITALIA.– Coremata present; uncus broad, apex slightly rounded; culcitula large, gnathos-hook distinct, long; valva moderately stout, weakly curved, scarcely shorter than uncus, apically inflated; sacculus comparatively short and slender, with parallel dorsal and ventral margins, distodorsal part rounded with weak inwardly directed tip; vincular process broadly hump-like, distinctly wider and slightly longer than sacculus, apex with outwardly directed tip; posterior margin of vinculum with moderately shallow V-shaped medial emargination; saccus broad and short, subtriangular, gradually tapered to pointed apex, extending slightly beyond apex of pedunculus; phallus broad, stout, with distinctly inflated coecum, distinct and long sclerotized arm apically. Similar to some species e.g. of the *S. salinella*-group but differs in the much larger overall size of the genitalia and the particularly small sacculus.

FEMALE GENITALIA.– Segment VIII longer than broad; subgenital plate largely covered with microtrichia and/or small-sized foamy sculpture, inner margin oblique; ventromedial depressions digitate, medially separated by small incision, apically covered with small-sized foamy sculpture; subostial sclerite distinct, sword-like; apophysis anterioris short, posterior part broad with indistinct transition to anterior edge of sternite VIII, rod-like distal part about half length of segment VIII; colliculum short and broad; ductus bursae with abrupt transition to clearly demarcated corpus bursae; corpus bursae oval; signum on the right side of corpus bursae, in transition to ductus bursae, of moderate size, basal plate small, signum-hook weakly curved with few additional spines in basal half on concave side.

DISTRIBUTION.– Recorded from the southern Ural Mountains of Russia (Cheliabinsk district, Orenburg district) (Junnilainen *et al.*, 2010: 49), Iran and Mongolia.

BIONOMICS.– Host-plant and early stages unknown. The few adults known to date were collected from mid-May to late July at light in semi-deserts.

REMARKS.– *S. plesiopicta* was described from a single male collected in Mongolia (Povolný, 1969: 17). Interestingly the forewing length is given as 9.8 mm in the original description but only 8.4 mm more recently (Povolný, 2002a: 63).

The above description of the female genitalia is based on the only known female specimen figured by Lvovsky & Piskunov (1989: 543, fig. 28) and later redrawn by Povolný (2002a: pl. 63, fig. 599).

63. *Scrobipalpa bradleyi* Povolný, 1971

Scrobipalpa bradleyi Povolný, 1971a: 24, figs. 2-3, 22-23, 45.
Scrobipalpa glaserorum Povolný, 1977b: 185, figs. 6-9, 27, **syn. n.**
Scrobipalpa meyricki auctt., (*nec* Povolný, 1971); misidentification.

DIAGNOSIS.– Adult. Wingspan 13-15 mm. Labial palp white mottled with some black at lower and outer surface of segment 2 and on tip of segment 3. Antenna black, ringed with lighter grey. Face white; neck, thorax and tegula as forewing. Forewing light grey, from grey white-based scales and mottled with some black and orange; black spots at base and at 1/5 at costa; a patch of slightly raised black scales at dorsum near base; a prominent black spot in fold; a black spot at 2/5 in middle of wing and an angulated black spot at 3/5; these three spots mixed with and surrounded by orange; a white, angulated, postmedial fascia; apical area with many black and orange scales. Hindwing grey.

Variation. *S. bradleyi* is variable in the amount of black scales on the forewing, and some specimens look blackish grey. Also the distinctness of the markings on the forewing shows some variation.

Similar species. *S. bradleyi* is characterized by its rather broad, light grey forewings with black and orange spots and a white postmedial fascia. *S. selectella* has more variegated, less plain light grey forewings, which are often mottled with light brown, grey and black scales. *S. bryophiloides* q.v.

MALE GENITALIA.– Uncus moderately broad, apically rounded; gnathos-hook long; valva shorter than or extending to apex of uncus, weakly curved, with hardly inflated apex; sacculus short and moderately slender, tapered towards apical tooth, gap between sacculus and vincular process about width of sacculus; vincular process tooth-like, slightly shorter than sacculus, basally about same width, tapered towards pointed apex; posterior margin of vinculum with broad and shallow V-shaped emargination; saccus subtriangular, evenly tapered towards apex, anteriorly slightly projecting beyond apex of pedunculus; phallus moderately slender, coecum weakly inflated. The genitalia are similar to a number of species such as *S. meyricki* from North Africa and further taxa of the *S. salinella*-group, from the latter mainly differing in the broader uncus and the more pointed vincular process.

FEMALE GENITALIA.– Segment VIII broader than long, with parallel outer edges; anterior edge of sternite VIII heavily sclerotized without membranous medial zone, transverse transition to apophysis anterioris almost at right angle; subgenital plate with slightly curved inner edge, oblique and longitudinal depression covered with small-sized foamy sculpture and few microtrichia laterally, extending to base of apophysis anterioris; ventromedial zone with longitudinal, weakly sclerotized and broadly digitate depressions, extending almost to anterior edge, lateral edge folded, depressions covered with small-sized foamy sculpture and microtrichia; apophysis anterioris slightly longer than segment VIII, rod-like throughout with moderately abrupt transition to segment VIII; colliculum short; ductus bursae broad, with gradual transition to moderately separated oval corpus bursae; signum on the right side of corpus bursae, basal plate small, signum-hook longish, evenly curved, without additional tooth. Differs from the otherwise similar *S. meyricki* in the larger sternite VIII with almost transverse rather than angular anterior margin.

DISTRIBUTION.– Southern Spain and unpublished records from Italy (Sicily) and France; outside Europe recorded from North Africa (Algeria, Tunisia).

New records. France: 1 ♂, 1 ♀, Camargue, St. Maries de la Mer, 22.viii.1983, leg. Grünewald, gen. slides GU 01/984 ♀, GU 01/985 ♂ Huemer (GRÜN). Italy (Sicily): 4 ♂, 26 ♀, 5 km W Pachino, Pantano Longarini, sea lev., 12.ix.2002, leg. Karsholt *et al.*, gen. slide 3782, 3783, 3786 Hendriksen (ZMUC).

BIONOMICS.– The species was bred from *Arthrocnemum fruticosum* (L.) Moq. (Chenopodiaceae) and adults emerged from late March to mid-April (Povolný, 1971a: 26). Furthermore the adults have been collected from mid-August to early October at light, indicating two generations. *S. bradleyi* prefers halophytic habitats, usually along the coasts.

REMARKS.– *S. bradleyi* was described from 3 males and 15 females, collected in Algeria and Tunisia (Povolný, 1971a: 24, 26). We have examined two paratypes and long additional series from various localities in North Africa, Spain and Sicily.

S. glaserorum was described from 3 males and 7 females, collected in southern Spain (Povolný, 1977b: 185-188). We have examined the male holotype and 4 paratypes which fully correspond with *S. bradleyi* resulting in a new synonymy. One female paratype is conspecific with *S. salinella*.

Povolný (1981: 388-389) erroneously synonymized *S. glaserorum* with *S. meyricki* Povolný, 1971 from North Africa which is darker and more narrow winged, with lighter hindwings, but with only small differences in genitalia. The holotype and several additional specimens of *S. meyricki* have been examined by us. The figures of *S. meyricki* in Povolný (2002a) are a mixture: the adult from North Africa (pl. 5, fig. 16) depicts *S. meyricki*, whereas the male genitalia (pl. 17, fig. 154-155) and the female genitalia (pl. 67, figs. 656-657) refer to *S. bradleyi*.

64. Scrobipalpa selectella (Caradja, 1920)

Gelechia selectella Caradja, 1920: 99.
Scrobipalpa fraterna Povolný, 1969a: 8, figs. 29, 7, **syn. n.**

DIAGNOSIS.– Adult. Wingspan 12-14 mm. Labial palp cream-white; segment 2 mottled with black at base and tip of lower and outer surface; segment 3 with two black rings. Antenna black, ringed with whitish grey. Head cream-coloured, darker towards neck; thorax and tegula as forewing. Forewing covered with light brown, grey, black (all white-based) and white scales; black spots at base and at 1/5 at costa; black spots in fold and at 2/5 and 3/5 in middle of wing surrounded by orange-brown; a white postmedial fascia; apical area with many black scales. Hindwing light grey.

Variation. A rather variable species. Most specimens have the forewings light brown, grey and white. This is sometimes more or less overlaid with black (white-based) scales. In such specimens the postmedial fascia and some white spots in termen are strongly contrasting. The black spot at 3/5 can be angulated.

Similar species. Very similar to *S. bryophiloides* and not always separable from that species without examination of the genitalia. *S. bryophiloides* often has the forewings more or less covered with light brown scales, whereas *S. selectella* has either more white on the forewings or, in darker specimens, they are covered with black, white-based scales. *S. bradleyi* q.v.; *S. solitaria* q.v.

MALE GENITALIA.– Coremata absent; uncus large, broadly subrectangular, apically flattened to slightly emarginated; gnathos-hook strong; valva stout, extending slightly below or at most to apex of uncus, weakly curved with inflated apex; sacculus short, even width from base almost to apex, apically rounded with small inwardly turned tooth; vincular process basally broad, distally narrowing, nearly reaching tip of sacculus, separated from sacculus by moderately broad gap, slightly curved outwards to a distinct tip, inner margin running straight into comparatively shallow V-shaped medial emargination of posterior margin of vinculum; saccus broad at base, strongly tapered distally, extending to about apex of pedunculus or slightly exceeding it; phallus stout, coecum distinctly inflated. Extremely similar to *S. bradleyi* and *S. meyricki* with only small differences e.g. in the shape of the sacculus and the longer vincular process.

FEMALE GENITALIA.– Segment VIII slightly broader than long; anterior margin of sternite VIII sclerotized with small medial opening; subgenital plate with broadly folded and slightly curved inner edge, depression with large foamy sculpture in anteriomedial third, medial and lateromedial part covered with microtrichia; ventromedial depression longitudinal, about width of subgenital plate, with digitate projections reaching slightly beyond anterior margin of sternite VIII, adjoined and covered with microtrichia for the better part, anterior parts with foamy sculpture, separated by V-shaped emargination; subostial sclerite well developed, curved inwards; apophysis anterioris slightly longer thant segment VIII, rod-like; colliculum short; ductus bursae gradually widened, indistinct transition to large, oval corpus bursae; signum on the right side near entrance of corpus bursae; signum-hook short, slender, strongly curved at an almost right angle in middle, basally with minute additional spines.

DISTRIBUTION.– Southern Russia. Recently found in the Ukraine (Bidzilya & Budashkin, 1998: 9) and an unpublished record from Greece. Also known from Kazakhstan, Turkey, Mongolia (Povolný, 1973a: 10), China (Bidzilya & Li, 2010) and Tunisia (Povolný, 1981: 391).

New record. Greece: 1 ♀, Makedonia, Strimondelta, 5 m, 19.-20.ix.2000, leg. Wieser, GEL 1114 ♀ Huemer (TLMF).

BIONOMICS.– The larva has been bred from *Halostachys belangeriana* (Moq.) Botsch. and *Halocnemum strobilaceum* (Pall.) Bieb. (Chenopodiaceae) (Falkovitsh & Bidzilya, 2006: 74). It lives between 3 to 4 twigs which are spun together. The adults have been collected in February and April and from June to September (Povolný, 2002a: 42) indicating bivoltinism.

REMARKS.– *Gelechia selectella* was described from 14 syntypes of both sexes from Uralsk (Kazakhstan) (Caradja, 1920: 99). Popescu-Gorj (1992: 173) published a male with genitalia slide O. Karsholt nr. 4081 as the lectotype. The genitalia of the lectotype agree with those of *S. fraterna* which becomes a synonym. Popescu-Gorj also published an 'allolectotype' which is, however, conspecific with *S. bryophiloides*.

65. Scrobipalpa alterna (Falkovitsh & Bidzilya, 2006) **comb. n.**

Euscrobipalpa alterna Falkovitsh & Bidzilya, 2006: 88, figs. 2, 17.

DIAGNOSIS.– Adult. Wingspan 10-11 mm. Labial palp white; segment 2 mottled with blackish brown; segment 3 with a blackish ring at base and tip. Antenna black, ringed with light grey. Face white; neck light grey; thorax and tegula grey. Forewing grey from white-based black scales; some yellow-brown scales in fold, as a subcostal stripe and especially around the distinct black spots; black patches at 1/4 and 2/4; some light brown scales before blackish apical area. Hindwing light grey.

Variation. The black patches at 1/4 and 2/4 can be distinct, single, double or triple or almost missing. In light specimens the spots are almost covered with light brown scales and the apical area is hardly darker than the rest of the wing.

Similar species. *S. alterna* is characterized by its light grey appearance, and by having the black spots distinctly surrounded first with light brown and then with whitish yellow. *S. alterna* is similar to *S. candicans* Povolný, 1996 (type material examined). The latter has more distinctly ringed white and black labial palps, and more distinct black spots and terminal dots on the forewing.

MALE GENITALIA.– Uncus moderately broad, apically flattened; gnathos-hook long; valva extending to apex of uncus, weakly curved, with inflated apex; sacculus short and moderately broad, almost even width from base to apex, small apical tooth directed inwards, gap between sacculus and vincular process narrower than sacculus; vincular process about length of sacculus, basally same width or slightly broader, distally tapered with outwardly pointed apex; inner edge of vincular process continuous with V-shaped emargination of posterior margin of vinculum; saccus slender, tapered towards rounded apex, anteriorly projected beyond apex of pedunculus; phallus stout, with distinct apical hook, coecum moderately inflated. The male genitalia are indistinguishable from those of *S. lutea*.

FEMALE GENITALIA.– Segment VIII about same length and width; anterior edge of sternite VIII membranous; subgenital plate with inner and outer edges straight and parallel-sided, longitudinal depression with field of distinct foamy sculpture in anterior half extending into base of apophysis anterioris, inner edge of depression folded, smooth and anteriorly widened area proximad of fold; ventromedial depression smooth, subrectangular, with longitudinal lateral fold, apical third digitate, with small field of foamy sculpture, medially separated by V-shaped gap, extending beyond anterior margin of segment VIII; apophysis anterioris rod-like, with gradual transition to anterior edge of sternite VIII, slightly longer than segment VIII; subostial sclerite broadly digitate and inwardly curved; colliculum short; ductus bursae short, gradually widened; corpus bursae moderately demarcated from ductus bursae, oval; signum with moderately large basal field of about 20 additional spines on inner side; signum-hook moderately stout, curved in middle. Differs from the similar *S. lutea* e.g. in the large field of additional spines of the signum.

DISTRIBUTION.– Only known from Russia (southern Ural Mountains) (new record), and from the type locality in south-eastern Kazakhstan.

New record. Russia: 1 ♀, South Ural, Orenburg district, 6 km W Donskoje village, mount Verbljushka, 13.v.1999, leg. K. Nupponen, gen. slide Hendriksen 3175 (NUPP).

BIONOMICS.– Part of the type series was bred from *Krascheninnikovia ceratoides* (L.)

Gueldenst. (Chenopodiaceae) (Falkovitsh & Bidzilya, 2006: 71). The adult was collected in late May and bred material dates from early June and late September (Falkovitsh & Bidzilya, 2006: 85).

REMARKS.– *S. alterna* was described from two males and three females collected in Kazakhstan (Falkovitsh & Bidzilya, 2006: 85). The original figures of genitalia are sketchy and insufficient for identification and unfortunately we were unable to examine type-material. However, according to Bidzilya (*in litt.*) material from the southern Ural Mountains is close to *S. alterna* and we therefore figure it under this species. The population from this area may later prove to belong to a distinct species.

66. *Scrobipalpa lutea* Povolný, 1977

Scrobipalpa lutea Povolný, 1977c: 148, figs. 6, 12.

DIAGNOSIS.– Adult. Wingspan 9-11 mm. Labial palp white; each segment with two brownish black rings. Antenna distinctly ringed black and white. Head white, mottled with light brown on neck; thorax and tegula as forewing. Forewing with yellow-brown and white-tipped black scales; dark areas along costa and dorsum, at 1/4 and in middle of wing and in apex, indistinct black spots densely surrounded by yellow-brown; base, fold, subcostal stripe and tornal area yellow-brown. Hindwing light brown.

Variation. Specimens can be more or less densely covered with yellow-brown. The black spots are sometimes totally covered with yellow-brown. Worn specimens look light yellow, and in such specimens the labial palps are almost white.

Similar species. *S. lutea* is characterized by the white, black-ringed labial palps, and the more or less densely yellow-brown scaling of the forewing. *S. spumata* q.v.

MALE GENITALIA.– As in *S. alterna* q.v.

FEMALE GENITALIA.– Segment VIII slightly broader than long; anterior edge of sternite VIII membranous; subgenital plate with inner and outer edges straight and almost parallel-sided, longitudinal depression with field of distinct foamy sculpture anteriorly, smooth and anteriorly widened area proximad of fold; ventromedial depression smooth, subrectangular, with longitudinal lateral fold, apical third digitate, with small field of foamy sculpture, medially separated by V-shaped gap, extending beyond anterior margin of segment VIII; apophysis anterioris rod-like, with gradual transition to anterior edge of sternite VIII, slightly longer than segment VIII; subostial sclerite broadly digitate and inwardly curved; colliculum short; ductus bursae short, gradually widened; corpus bursae weakly demarcated from ductus bursae, pyriform to oval; signum on the right side of corpus bursae, small additional spines on inner side of basal plate; signum-hook in transition to ductus bursae, slender and longish, weakly to moderately curved in middle. Differs from the similar *S. alterna* particularly in the signum with few additional spines.

DISTRIBUTION.– Only known from Russia (southern Ural Mountains: Cheliabinsk district and Orenburg district) (Junnilainen *et al.*, 2010: 48) and from the type locality in Turkey.

BIONOMICS.– Host-plant and early stages unknown. The adults have been collected between mid-June and late July. They are attracted to light.

REMARKS.– Hitherto only known from the female holotype, *S. lutea* was recently collected in numbers, including both sexes, in the southern Ural mountains by Finnish lepidopterists.

67. *Scrobipalpa niveifacies* Povolný, 1977

Scrobipalpa niveifacies Povolný, 1977b: 189, figs. 15, 32.
Scrobipalpa milleri Povolný, 1977c: 144, figs. 2, 9, **syn. n.**

DIAGNOSIS.– Adult. Wingspan 9-14 mm. Labial palp cream-coloured; segments 2 and 3 each with a black ring at base and near tip. Antenna ringed with black and light grey. Head and tegula cream-coloured, mottled with some grey; thorax as forewing. Forewing light grey mottled with yellowish and black; two black spots in fold and two pairs of spots at 2/5 and 3/5; black spots at costa near base at 1/4, half and 3/4; yellow stripes in fold, as a subcostal stripe, around black spots and as a bent fascia at 3/4. Hindwing light grey.

Variation. A rather variable species in size and colour. Females are on average smaller and lighter than males. Depending of the mixture of scales specimens can look light greyish, light brownish or light yellowish.

Similar species. *S. niveifacies* is characterized by its light labial palps with segments 2 and 3 having black rings, and by the light, variegated forewings with two pairs of black double-spots at 2/5 and 3/5.

MALE GENITALIA.– Uncus long, distally slightly tapered, apex rounded; culcitula well developed, gnathos-hook long; valva extending to about tip of uncus, hardly inflated apex, weakly curved; sacculus moderately short and slender, almost even width from base to apex, apically with rounded hump and with indistinct inwardly turned tooth; vincular process about length of or slightly exceeding sacculus, basally about two times width of sacculus, apically narrowing with broadly rounded crescent-shaped apex, outer margin with minute tooth and gradually declining margin towards to sacculus; posterior margin of vinculum projected, with moderately deep and narrow V-shaped medial emargination, scarcely separated from vincular process; saccus broad at base, evenly tapered towards apex, short, extending to about apex of pedunculus; phallus rather short, with broadly inflated coecum. Particularly the shape of the sacculus is characteristic for this species.

FEMALE GENITALIA.– Segment VIII slightly longer than broad; subgenital plate with evenly curved inner border, smooth without microtrichia or foamy sculpture, anteriomedial part slightly wrinkled; ventromedial depressions smooth, longitudinal, membranous and without microtrichia, largely fused with rounded anterior edges extending to margin of segment VIII and short medial incision; apophysis anterioris with gradual transition to anterior edge of sternite VIII, about length of segment VIII, basally slightly wrinkled; subostial sclerites moderately sclerotized; colliculum comparatively long; ductus bursae short with gradual transition to corpus bursae; corpus bursae oval, covered with microtrichia; signum on the right side of corpus bursae; signum-hook long and stout, slightly curved, without additional spines.

DISTRIBUTION.– Southern Spain (Almeria, Granada), Portugal.

BIONOMICS.– The larva has been recorded in the second half of September in slight spinnings on the very short branches along the stem of *Atriplex halimus* L. (Chenopodiaceae) in an area with saltmarshes and cliffs of sandstone (Corley, *in litt.*).

The adults have been collected at light in April, June and from August to early November, indicating at least two generations.

REMARKS.– *S. niveifacies* was described from a single female collected on 7.x.1975 by Glaser near Baza (Granada, Spain). The worn holotype and its genitalia were figured in the original description (Povolný, 1977b: 189, figs. 15, 32). *S. milleri* on the contrary was described from a single male, again collected in Spain, near Garrucha (Almeria) between 4.-6.x.1974 by Amsel & Rösler (Povolný, 1977c: 144, figs 2, 9). Though recognizing the external similarity of both taxa Povolný (1977c: 146) stated that they are hardly closely related. However, our examination of rich material of both sexes from various localities in Spain clearly proves the conspecificity of both. According to the reprint dealing with *S. niveifacies*, this paper was published on 29.v.1977, whereas the publication date of *S. milleri* is stated as 15.vi.1977 on the cover page. We therefore syonymize *S. milleri* with *S. niveifacies*.

68. Scrobipalpa indignella (Staudinger, 1879)

Bryotropha indignella Staudinger, 1879: 308.
Gnorimoschema hyoscyamivora Gerasimov, 1940: 77, figs. 68-69, **syn. n.**
Gnorimoschema pseudobsoletellum Povolný & Gregor, 1955: 83, fig. 3, **syn. n.**
Scrobipalpa grossa Povolný, 1966b: 400, figs. 17-21, 8.

DIAGNOSIS.– Adult. Wingspan 12-15 mm. Labial palp cream-coloured; each segment with a black, basally slender and distally broader ring. Antenna black, ringed with grey. Head cream-coloured, neck darker; thorax and tegula as forewing. Forewing grey from white-based black scales, and mottled with light brown; distinct black spots surrounded by light brownish scales in fold and in middle of wing at 2/5 and 3/5; fold and subcostal stripe light brown; costal and tornal spots light, forming an indistinct, angulated fascia; black scales also along costa and in termen. Hindwing grey.

Variation. The forewings can be almost uniformly covered with light brown scales and having the black spots indistinct. In more greyish specimens the black spots are normally more distinct. The single spot is sometimes divided into two. In light specimens the labial palps and the head are almost pure cream-coloured. Specimens have a tendency to become greasy.

Similar species. *S. indignella* is characterized by its dark greyish brown forewings with black patches in basal half and many light brown scales. It resembles *S. obsoletella*, but is on average slightly larger and with the forewings normally light greyish brown and not buff grey as in *S. obsoletella*. It is smaller than *S. grisea* and the otherwise very similar *S. plesiopicta*.

MALE GENITALIA.– Uncus moderately slender, apex evenly rounded; gnathos-hook long; valva stout, shorter than uncus, curved, with hardly inflated apex; sacculus moderately short and broad, outer margin curved, apically rounded with distinct tooth directed inwards; vincular process nearly reaching length of sacculus, about width of sacculus basally, strongly tapered towards pointed apex, continuous with broad and deep V-shaped medial emargination of posterior margin of vinculum; saccus stout, broadly rod-like, distal two-thirds with almost even width, apex rounded, anteriorly considerably projecting beyond apex of pedunculus; phallus long and stout, coecum moderately inflated.

FEMALE GENITALIA.– Segment VIII slightly broader than long; anterior edge of sternite

VIII sclerotized with small medial opening; subgenital plate with parallel and straight, slightly oblique outer and inner edges, proximal part more heavily sclerotized than distal part, longitudinal depression with field of small-sized foamy sculpture; ventromedial depressions broadly digitate with folded apical edge, reaching slightly beyond anterior edge of segment VIII, posteriomedially adjoining, anteriorly separated by short V-shaped incision, covered with microtrichia to a large extent; apophysis anterioris rod-like, with gradual transition to anterior edge of sternite VIII, about length of segment VIII; colliculum short; ductus bursae gradually widened towards corpus bursae; corpus bursae clearly demarcated from ductus bursae, oval, covered with microtrichia; signum on the left side of corpus bursae; signum-hook massive, weakly curved medially, basal third with row of short spines, basal plate moderately small.

DISTRIBUTION.– South-western Russia to the southern Ural Mountains; Ukraine. Widely distributed in the Middle East, including Afghanistan (Povolný, 1969c), also recorded from Turkmenistan (Povolný, 2001a: 204) and China (Bidzilya & Li, 2010).

BIONOMICS.– The larvae were found in June feeding on spun leaves and fruits of *Hyoscyamus* L. (Solanaceae). They pupated in mid-June and adults emerged in late June (Gerasimov, 1940: 78). Furthermore earlier larval findings from May and October were mentioned in the same paper. Outside Europe the adults have been collected from March to September, indicating several generations. The species is attracted to light.

REMARKS.– *Bryotropha indignella* was described from 6 specimens collected by Staudinger between June and July 1875 in the area of Amasya (Turkey) (Staudinger, 1879: 308). Its identity remained obscure until the type-material was revised by Sattler (1986: 233-234, pl. 15, figs 1-6), who also figured paralectotypes of both sexes but erroneously synonymized *S. indignella* with *S. obsoletella*. Povolný (1986b: 372) resurrected *S. indignella* as a valid species and furthermore synonymized *Scrobipalpa grossa*, a species he had described earlier based on rich material from Turkey to Afghanistan (Povolný, 1966b: 400-402), with the former.

Gnorimoschema hyoscyamivora was described from an unspecified number of specimens collected in the surroundings of Orenburg in the southern Ural Mountains (Russia) (Gerasimov, 1940: 77). Later this species was considered as junior synonym of *S. hyoscyamella* by Povolný (2002a: 47) who at the same time assumed that Gerasimov's description was based on a mixed series of *S. hyoscyamella* and *S. costella*. However, the rather detailed original description, unfortunately without figure of the adult, did not fit any of these species. The genitalia were figured laterally and altogether quite difficult to interpret, but the extraordinarily large female signum in particular clearly indicated conspecificity with *S. indignella*. This interpretation was further supported by the diagnosis of the adult and by rich material of this species collected by Finnish lepidopterists in the area of Orenburg. Finally Bizdilya (*in litt.*) found syntypes of *S. hyoscyamivora* in the collection of ZIAP and parts of this material was later examined by us with the help of S. Sinev. The dissection of a male syntype proved the alleged conspecificity of *S. hyoscyamivora* and *S. indignella*.

Gnorimoschema pseudobsoletellum, described from a single female collected by Amsel between 15. and 27.iii.1933 on the southern border of the Dead Sea (Povolný & Gregor, 1955: 83-84, fig. 3), is treated as a further synonym of *S. indignella*. The original description gives a rough sketch of the genitalia, but a differential diagnosis to *S. indignella* (and *S. grossa*) was never published. Recently published figures are misleading as they indicate the absence of a foamy sculpture at the base of the apophysis anterioris, and a smaller signum without spines for *S. indignella* compared with *S. pseudobsoletellum* (Povolný, 2002a: pl. 61, figs. 553 and 554-556 resp.). However, these alleged differences are only based on different drawing techniques/observations. *S. indignella* always has the foamy sculpture and also the ventromedial field of microtrichia, characters which were simply not drawn in figs. 553 and 556. The size of the signum varies considerably, even within the same population, as already figured by Povolný (1966b: 401, figs. 20-21). The figures of the male genitalia are similarly difficult to interpret (Povolný, 2002a: pl. 5, figs. 41-42 and 44-45). *S. indignella* seems to have broader sacculi than *S. pseudobsoletellum*. However, Povolný (2002a: pl. 5, fig. 41) clearly shows that these alleged differences are simply based on preparation technique, as the left and the right sacculus are strongly different due to various pressure on the cover slip. It is interesting that the two illustrations depicting *S. pseudobsoletellum* appear to suggest two species due to differences in the saccus. Possibly the specimen depicted in fig. 44 on the same plate, with a very short and slender saccus, is a misidentification of *S. grossoides* (Povolný, 2001) **comb. n.**, or the latter could even prove to be another synonym of *S. indignella*. Summarizing we have no doubt that *S. pseudobsoletellum* is conspecific with *S. indignella* and we therefore formally synonymize it with the latter.

69. Scrobipalpa punctata (Povolný, 1996)

Ilseopsis (Euscrobipalpa) punctata Povolný, 1996b: 26, figs. 6, 25.

DIAGNOSIS.– Adult. Wingspan 11-12 mm. Labial palp white; segments 2 and 3 each with two black rings. Antenna black, ringed with light grey. Head white, mottled with grey, especially towards neck; thorax and tegula as forewing. Forewing grey from black light-based scales and mottled with light brown; black spots at 2/5 and 3/5 double, the latter followed by a black spot or streak; fold with 1-2 black spots and a spot or streak at 1/2; all these black spots followed by light brown scales; fold and subcostal stripe light brown; termen with black scales. Hindwing light grey.

Variation. The amount of brown scales on the forewing is variable, from very few to many; in the latter case specimens look light brown. The black spots can be more or less distinct.

Similar species. *S. punctata* is characterized by its grey or brownish grey forewings with additional black spots, especially in the fold.

MALE GENITALIA.– Uncus long, slender, distally tapered, apex rounded; gnathos-hook comparatively long, stout; valva extending to about tip of uncus, curved, with weakly inflated apex; sacculus moderately short and slender, almost even width throughout, apically rounded with small inwardly turned tooth; vincular process slightly shorter than sacculus, basally broader than sacculus, strongly tapered towards crescent-shaped and outwardly directed apical point, separated from sacculus by narrow gap; posterior margin of vinculum with comparatively broad and moderately deep V-shaped medial emargination; saccus long and slender, rod-like, extending slightly beyond apex of pedunculus; phallus rather short, with weakly inflated coecum.

FEMALE GENITALIA.– Segment VIII about same length and width; subgenital plate moderately membranous, smooth, band-like convex sclerotization along inner margin, extending from posterior margin to base of apophysis anterioris; ventromedial depressions narrow, longitudinal, laterally folded, medially separated by membrane throughout, extending to about margin of segment VIII with small medial incision, depressions covered with microtrichia; apophysis long and stout, with gradual transition to sternite VIII, about twice length of segment VIII; colliculum short; ductus bursae gradually widened with weak transition to rather small and oval corpus bursae; signum on the right side at entrance of corpus bursae; signum-hook stout and rather short, weakly curved in middle, small basal plate with few additional spines.

DISTRIBUTION.– Only known from Russia (southern Ural Mountains) (Junnilainen *et al.*, 2010: 50) and Kyrgyzstan.

BIONOMICS.– Host-plant and early stages unknown. The holotype was collected in late July at an altitude of 2500 m, additional specimens in the last third of June. The species occurs on steppe slopes.

REMARKS.– *S. punctata* was hitherto only known from the female holotype, collected in Kyrgyzstan (Povolný, 1996b: 26). We have been able to study a small series of additional specimens of both sexes from Russia.

70. Scrobipalpa lagodes (Meyrick, 1926) **comb. rev.**

Phthorimaea lagodes Meyrick, 1926: 278.

DIAGNOSIS.– Adult. Wingspan 12 mm. Labial palpus mixed cream-white and loamy-brown; segment 3 with two blackish brown rings. Head, thorax and tegula loamy brown; face lighter. Antenna dark brown, indistinctly lighter ringed, loam-brown towards base. Forewing almost unicolorous loamy brown, darker tipped scales in costal and subapical area; without spots; cilia cream-brown with some darker tipped scales. Hindwing greyish brown, with greyish brown cilia.

Variation. The few examined specimens show only slight variation. One specimen has indistinct darker spots.

Similar species. *S. lagodes* is characterized by its almost unicolorous loamy brown forewings. It may resemble unicolorous forms of *S. samadensis*, but that species has darker labial palps, and the veins in the apical part of the forewing are decidedly lighter than their surroundings. *S. spumata* also has unicolorous forewings, but their brown colour is lighter yellowish, not loamy brown.

MALE GENITALIA.– Shape of uncus invisible in lectotype slide, according to Povolný (1966a: 130, fig. 10) broadly subrectangular; valva exceeding tip of uncus, slightly curved, apically with weakly inflated apex; sacculus short and slender, untipped; vincular process broader, distally weakly curved outwards, with an indistinct tip, apex almost level with tip of sacculus; incision of vinculum invisible in lectotype slide, according to Povolný (1966a: 130, fig. 10) with deep V-shaped emargination; saccus comparatively long, extending far beyond apex of pedunculus; phallus comparatively broad and short, basally inflated, distinct apical arm developed.

FEMALE GENITALIA.– Segment broader than long; anterior margin of sternite VIII oblique; subgenital plate slender, with longitudinally folded inner margin, covered with numerous microtrichia; ventromedial depression much broader than subgenital plate, subrectangular, covered with numerous microtrichia except for membranous anteriomedial part, anterior margin convex, slightly exceeding anterior edge of sternite VIII; apophysis anterioris rod-like, about length of segment VIII; colliculum broad and short; ductus bursae broad, about same width throughout, with abrupt transition to oval corpus bursae; signum on the left side at entrance of corpus bursae, small basal plate with several additional spines, signum-hook slender, moderately long and weakly curved.

DISTRIBUTION.– Only known from northern Kazakhstan; probably *S. lagodes* also occurs in the adjacent parts of Russia.

BIONOMICS.– The larva has been recorded from *Atriplex cana* C. A. Meyer (Falkovitsh & Bidzilya, 2006: 68) where it feeds on the upper leaves. The two type-specimens were collected in June; additional material from Kazakhstan emerged in mid-July.

REMARKS.– *Phthorimaea lagodes* was described from two specimens collected by Bartel at Lake Indersky ('European' part of Kazakhstan) in June 1907. A male specimen, erroneously labelled as female by Meyrick, was selected as the lectotype by Clarke (1969: 163, pl. 78, figs. 4, 4a, 4b) and the adult and its genitalia figured in lateral position. The second male type-specimen lacks the abdomen and two wings. Povolný (1966a: 130, fig. 90; 2002a: pl. 3, fig. 16, pl. 10, fig. 90) figured the lectotype in colour and also the genitalia of this specimen. Interestingly he gives an accurate figure of the genitalia in dorsoventral position, though the slide is still embedded laterally!

71. *Scrobipalpa deluccae* Povolný, 1966

Scrobipalpa deluccae Povolný, 1966b: 398, figs. 9-10, 16, 10.

DIAGNOSIS.– Adult. Wingspan 11 mm. Labial palp white; segment 2 mottled with brownish black; segment 3 with two black rings. Antenna black, indistinctly lighter ringed. Face white; neck and tegula light grey; thorax as forewing. Forewing slender, greyish black from light-based black scales and mottled with orange-brown, especially along dorsum; black spots in fold surrounded by yellow-brown, and at 2/5 and 3/5 in middle of wing surrounded by some orange-brown scales; indistinct light costal and tornal spots; apical area with many light-tipped black scales. Hindwing light grey.

Variation. Only one specimen was examined.

Similar species. *S. deluccae* is characterized by its slender greyish black forewings mottled with orange brown in the dorsal part. It resembles several other *Scrobipalpa* species, e.g. *S. smithi* q.v., and a safe identification will frequently require examination of the genitalia.

MALE GENITALIA.– Uncus long, moderately broad, apex with distinct concave emargination; gnathos-hook short; valva slender, distinctly exceeding tip of uncus, apically weakly pointed, without inflation, posterior part curved; sacculus comparatively broad and short, turned inwards, about even width from base almost to apex, dorsoapically rounded with weakly developed tooth turned inwards; vincular process short and small, digitate, separated from sacculus by narrow gap; vinculum comparatively long, posterior margin with moderately shallow V-shaped medial emargination; saccus broadly subtriangular, evenly tapered towards pointed apex, short, not extending as far as apex of pedunculus; phallus short, stout, with inflated coecum. The male genitalia are unmistakable among European species of *Scrobipalpa*.

FEMALE GENITALIA.– Segment VIII about same length and breadth, anteriolateral edges slightly concave; subgenital plate with predominantly straight, anteriorly convex inner border, oblique at base of apophysis anterioris, anteriomedial zone covered with numerous microspines; ventromedial depressions digitate, covered with microtrichia except for posterior part, anteriorly separated by deep V-shaped emargination, posteriomedially merged; apophysis anterioris with abrupt transition to anterior edge of sternite VIII, slightly longer than segment VIII; subostial sclerites distinct, inwardly curved and approaching in middle; colliculum small and indistinct; ductus bursae with abrupt transition to small and clearly defined, globular corpus bursae; signum on the left side at entrance of corpus bursae; signum-hook large and stout, strongly curved near middle, with minute additional tooth at base, basal plate small.

DISTRIBUTION.– Only known with certainty from the Near East (Lebanon, Syria).

BIONOMICS.– Host-plant and early stages unknown. The few adults known to date have been collected from mid-April to mid-June up to altitudes of about 1900 m.

REMARKS.– *Scrobipalpa deluccae* was described from five specimens from Lebanon, two specimens from Syria and one specimen from Malta. The genitalia of the paratype from Malta was figured in the original description (Povolný, 1966b: 399, fig. 10). We located the moth, labelled 'Malta Gharghur, 12.4.60' and '[genitalia slide] Povolný Lc. 1145' in the LNK. It is a worn specimen belonging to the family Symmocidae. No other specimens of *S. deluccae* has been found in the otherwise relatively well known Gelechiidae fauna of Malta (Karsholt in prep.), and we find it unlikely that this species occurring in the mountains of Lebanon and Syria should also occur there. Most likely a mistake has happened and either the abdomen or the labels have been exchanged.

S. dellucae (Povolný, 1996a: 114) is a multiple misspelling.

72. *Scrobipalpa atriplicella* (Fischer von Röslerstamm, 1841)

Lita atriplicella Fischer von Röslerstamm, 1841: 223, pl. 78, figs. a, e, h.
Tinea atrella Thunberg, 1788: 78; homonym of *Tinea atrella* [Denis & Schiffermüller], 1775.
Gelechia? detersella Clemens, 1860: 162; homonym of *Gelechia detersella* Zeller, 1847.
Lita atriplicella var. *infumatella* Fuchs, 1901: 383, **syn. n.**
Gelechia brackenridgiella Busck, 1903: 894.
Gnorimoschema chenopodiella Busck, 1916: 148.
Scrobipalpa (*Euscrobipalpa*) *arogantella* Povolný, 1967a: 213, pl. 2, figs. 7-9, pl. 16, fig. 1.
Scrobipalpa altajica Povolný, 1969a: 5, figs. 15-17, 2.

DIAGNOSIS.– Adult. Wingspan 11-17 mm. Labial palp bone-white mottled with fuscous, mostly on segment 3. Antenna almost unicolorous black. Head, thorax and tegula as forewing; face lighter. Forewing long and slender covered with white-based blackish scales and mottled with orange-brown; black orange-surrounded spots normally indistinct; costal and tornal spots light, indistinct, separated by a black streak surrounded by orange; apical area blackish. Hindwing grey. Underside of abdomen cream-coloured.

Variation. A rather variable species. In some specimens the forewings have many orange-brown scales, giving them a variegated appearance. There is a tendency towards melanism, especially in populations from northern Europe, and some specimens can be almost black.

Similar species. *S. atriplicella* is characterized by its dark and rather long and slender forewings, which are more or less mottled with orange-brown scales. The cream-coloured underside of the abdomen is a useful character in separating *S. atriplicella* from, e.g., *S. acuminatella* when sorting un-spread material from light traps. *S. ocellatella* q.v.; *Tuta absoluta* q.v.

MALE GENITALIA.– Coremata absent; uncus long, slender, apex with concave emargination; gnathos-hook long; valva slender, distinctly exceeding tip of uncus, without distinct apical inflation, posterior part curved; sacculus comparatively slender, about even width from base to almost apex, dorsoapically rounded with distinct tooth turned inwards; vincular process about length of sacculus, sickle-shaped, strongly tapered distally, apically much smaller than sacculus with outwardly turned tip, separated from sacculus by wide space; inner margin of vincular process runs directly into comparatively shallow V-shaped medial emargination of posterior margin of vinculum; saccus comparatively broad at base, evenly tapered towards apex, extending slightly beyond apex of pedunculus; phallus rather short, with weakly inflated coecum. The male genitalia are best recognized by the long valvae and the indented uncus.

FEMALE GENITALIA.– Segment VIII with about equal length and width; extensive, with small-sized foamy sculpture in anterior and medial part, numerous microtrichia in posterior half, posterior quarter smooth; subgenital plate with long and weakly curved inner border, extensively covered with small-sized foamy sculpture anteriorly and with microtrichia medially, posterior third smooth; ventromedial depressions covered with microtrichia posteriorly and small-sized foamy sculpture anteriorly, digitate, reaching far beyond anterior edge of segment VIII, posteriomedially contiguous, anteriorly separated by deep V-shaped incision; subostial sclerite distinct, sword-like; apophysis anterioris rod-like, with gradual and even transition to anterior edge of sternite VIII, longer than segment VIII, posterior third inwardly covered with small-sized foamy sculpture; colliculum very short; ductus bursae with abrupt transition to clearly demarcated corpus bursae; corpus bursae spherical to oval; signum on the left side of corpus bursae; signum-hook in transition to ductus bursae, slender and weakly arched, basal third with few additional spines on concave side.

DISTRIBUTION.– Widely distributed over a vast part of the Palaearctic region. Numerous records are based on human impact, and the species is frequently found in pastures, e.g. in Central Asia (Povolný, 2002a: 32). *S. atriplicella* was also introduced to North America (Povolný, 1967b: 214).

BIONOMICS.– The larva has been found in two generations from April to late June and from September to October on *Atriplex prostrata* Boucher ex. DC., *A. patula* L., *A. tatarica* L., *Chenopodium album* L., *C. hybridum* L., *C. ficifolium* Smith, several varieties of *Beta vulgaris* L. (Chenopodiaceae) (Elsner *et al.*, 1999: 43; Jansen, 1999: 49), *Halimione portulacoides* (L.) Aellen (Karsholt, pers. obs.) and probably also on other species of Chenopodiaceae. However, records from *Atriplex laciniata* L. (Fischer von Röslerstamm, 1841: 224) and *Cirsium vulgare* (Savi) Ten. (Asteraceae) (Elsner *et al.*, 1999: 43) are probably due to misidentification. In Denmark the larva has caused damage to the introduced South American crop *Chenopodium quinoa* Willd. (Sigsgaard *et al.*, 2008: 1). The eggs are deposited singly on the inflorescence of floral buds or in small groups on the underside of young leaves (Jansen, 1999: 49). The larva initially mines a leaf, later it feeds between spun central leaves or in late summer and autumn on flowers and seeds. Pupation takes place between dry leaves which are spun together, or inside the stem. The pupal stage hibernates. The adult occurs in May and June and again from late June to mid-August. It prefers dry, open grassland and wasteland. *S. atriplicella* is common in cultivated areas.

Adults are sometimes found in September and October, probably representing a partial third generation. Vertical distribution: from lowland localities up to 4000 m in Central Asia.

REMARKS.– *Lita atriplicella* var. *infumatella* was described from a single specimen collected in Norway (Fuchs, 1901: 383) and was regarded as local form of northern Europe. We have been able to examine the male holotype deposited in the HNMH. It is a dark specimen without abdomen; it perfectly matches the generally darker and less variegated populations from Scandinavia.

Lita atriplicella var. *clarella* is not a synonym of *S. atriplicella* (Povolný, 2002a: 32) but of *S. ocellatella* q.v.

73. Scrobipalpa suaedella (Richardson, 1893)

Lita suaedella Richardson, 1893: 241.
Phthorimaea flavidorsella Amsel, 1952: 124, fig. 31.
Scrobipalpa hartigi Povolný, 1977a: 170, fig. 1, 5, **syn. n.**

DIAGNOSIS.– Adult. Wingspan 13-16 mm. Labial palp cream-coloured; segment 2 mottled with black scales; segment 3 with two black rings. Antenna black. Face cream-coloured; head light grey; thorax medially light yellow brown, laterally black; tegula dark grey. Forewing with a mixture of light grey, black and light brown scales, the latter especially as a subcostal stripe, in fold and along dorsum; base black; black spots in subcostal stripe, in fold and at 2/5 and 3/5 in middle of wing, surrounded by light orange-brown scales. Hindwing grey, lighter towards base.

Variation. A variable species. The above description is based on nominotypical specimens from Great Britain, which are normally rather plain coloured with more or less distinct light dorsal area and only variegated by black spots surrounded by orange-brown scales. Forms with a light dorsal area are more common in South Europe. The black spots can be divided. Specimens from Spain are more colourful than specimens from Great Britain. The labial palps and the forewings of such specimens have more black scales, and the light parts of the forewings can have a pinkish tinge. Such specimens occur both with and without light dorsum, and the dorsal part can be light orange-brown. Occasionally there is a black stripe in the apical area. Hindwings of the form occurring in Spain are more whitish than those of the British form.

Similar species. Due to its variability *S. suaedella* can be confused with a number of other *Scrobipalpa* species. It can be recognized by its relatively large size, the black antenna, the light middle of thorax, the (normally) light dorsum of the forewings, the black base and the distinct black spots. *S. stangei* also has the middle of thorax light, but it has more grey than grey-brown forewings and rarely a light dorsum. *S. ocellatella* q.v.

MALE GENITALIA.– Coremata present; uncus long, slender, distally tapered, apex distinctly notched; gnathos-hook short; valva extending slightly below or to about tip of uncus, with weakly inflated apex, slightly curved; sacculus moderately long and slender, even width from base to almost apex, apically rounded with small inwardly turned tooth; vincular process shorter than sacculus, separated from sacculus by narrow space, strongly tapered from base to tip, apex rounded with weakly crescent-shaped part turned outwards; posterior margin of vinculum with comparatively shallow almost U-shaped medial emargination; saccus broad at base, evenly tapered towards apex, extending slightly beyond apex of pedunculus; phallus rather short, with weakly inflated coecum. The male genitalia are similar to those of other species of *Scrobipalpa* such as *S. artemisiella* or *S. stangei* but differ in the notched posterior margin of the uncus.

FEMALE GENITALIA.– As in *S. artemisiella* q.v. and *S. stangei* q.v.

DISTRIBUTION.– From Portugal and Spain through parts of France and Great Britain to northern Germany; also in Corsica and Sardinia; further isolated records from Cyprus, Libya and Turkey. A record from Central Europe (Germany, Thüringen) (Steuer, 2002) is based on misidentification (Gerstberger, 2003: 138).

New record. France (Corsica): 1 ♂, Calvi, 10 m, 18.-19.viii.1998, leg. Karsholt (ZMUC).

BIONOMICS.– The larva has been recorded from *Suaeda vera* Forssk. ex J. Gmelin and *S. maritima* (L.) Dumort. (Chenopodiaceae). It feeds in silken galleries on leaves and blossoms, preferably at the tips of the shoots, occasionally on two shoots which are spun together (Richardson, 1893: 241). Infested shoots become withered and turn brown. According to Povolný (2002a: 71) the larva is mining. It has been found from May to early June, on *S. maritima* also in August (Bland *et al.*, 2002: 170), indicating hibernation in the larval stage. Pupation takes place in a tough cocoon on the ground surface. In England the adult is on the wing in June and July but Povolný (2002a: 71) states a flight period from May to October, probably depending on the locality. Imagines are attracted to light. *S. suaedella* is restricted to halophytic coastal and inland habitats with a high level of salinity.

REMARKS.– *Phthorimaea flavidorsella* was described from one male and one female (both called types in the original description) and 2 female 'paratypes' collected in Sardinia 1936 (Amsel, 1952: 127). Only the two last mentioned syntypical females could be located in the Amsel collection in LNK. They are in good condition and belong to the contrastingly coloured (Mediterranean) form of *S. suaedella*. This is in accordance with the figure of male genitalia in the original description. Povolný (1964: 354) treated *P. flavidorsella* as a subspecies of *S. suaedella*, but later on (1967a: 229) his opinion on its status became less clear.

S. hartigi was described from a single male collected in Libya, Bengasi, 31.iv.1922 by G. C. Krüger. The figure of the male genitalia and a colour aquarelle of the adult (Povolný, 1977a; 2002a) corresponds with *S. suaedella* and we therefore synonymize it with the latter. This is supported by two females of *S. suaedella* in ZMUC caught at the type locality of *S. hartigi* on 20.iv. and 1.v.1922, respectively.

74. *Scrobipalpa solitaria* Povolný, 1969

Scrobipalpa solitaria Povolný, 1969a: 21, fig. 103.

DIAGNOSIS.– Adult. Wingspan 10-11 mm. Labial palp cream-coloured, mottled with blackish brown, especially on segment 3. Antenna ringed black and white. Face cream-coloured; neck, thorax and tegula grey. Forewing light brownish grey from white-based brown and black scales; black areas at base, along costa, dorsum and termen; spots large, brown; whitish areas before and between spots; an oblique white fascia at 4/5. Hindwing grey.

Variation. The few examined specimens show some variation from lighter to darker brownish grey forewings, with a variable amount of white scales in the middle of the wing.

Similar species. *S. solitaria* resembles *S. selectella* and *S. bryophiloides*, but it is smaller and has darker forewings.

MALE GENITALIA.– Coremata absent; uncus comparatively broad, apex with weak emargination; gnathos-hook slender, short; valva extending almost to apex of uncus, slender, distally curved, with weakly inflated apex; sacculus short, basally slightly broader than valva, distal part curved inwards and tapered towards pointed apex, with characteristic fold; vincular process of same length as sacculus, broad, tapered towards outwardly pointed apex, gap between sacculus and vincular process narrow; posterior margin of vinculum with broad and deep U-shaped medial emargination; saccus broad at base, distal part slender, exceeding apex of pedunculus; phallus stout, coecum weakly inflated, subapical sclerite distinctive.

FEMALE GENITALIA.– Segment VIII about same length and breadth; subgenital plate posteriorly and laterally smooth, long subovate depression covered with large field of distinct foamy sculpture, laterally and inwards demarcated by weakly curved edge, lateroproximally with oblique edge; ventromedial depression largely covered with distinct foamy sculpture, basally fused, distal parts digitate, adjoined to subgenital plate and weakly projecting, separated by small V-shaped emargination anteriomedially; apophysis anterioris rod-like, evenly broad-based, slightly shorter than segment VIII; colliculum moderately short; ductus bursae rather long, of about same breadth throughout; corpus bursae well distinguished from ductus bursae, globular; signum on the right side of corpus bursae, basal plate moderately small with additional spines on inner side; signum-hook slender, weakly curved.

DISTRIBUTION.– Only known from Russia (southern Ural Mountains) (new record), the Ukraine (Bidzilya & Budashkin, 2009: 17) and from the type locality in Mongolia.

New record. Russia: 1 ♂, 1 ♀, South Ural, Orenburg district, 6 km W Donskoje village, mount Verbljushka, 18.-19.vi.1999, leg. K. & T. Nupponen, gen. slide 09/1282 P. Huemer (NUPP).

BIONOMICS.– Host-plant and early stages unknown. The few adults known to date have been collected in mid-June and in July.

REMARKS.– *S. solitaria* is extremely similar to *S. chetitica* Povolný, 1974 from Turkey in male and female genitalia and it is virtually impossible to separate them from genitalia characters. However, the adults of *S. chetitica* differ in the distinct brown markings of the forewing (Povolný, 2002a: pl. 2, fig. 21). Therefore and in the absence of sufficient material we hesitate to synonymize both taxa for the time being. Despite this shortcoming the only European record of *S. chetitica*, which is also documented by a colour figure (Bidzilya & Budashkin, 2009: 17, fig. 5), clearly refers to typical *S. solitaria*. We have been able to study one of the two published specimens from ZMKU.

75. Scrobipalpa dagmaris Povolný, 1987

Scrobipalpa (Euscrobipalpa) dagmaris Povolný, 1987: 80, figs. 1-2.
Scrobipalpa rezniki Piskunov, 1990: 296, figs. 8-11, **syn. n.**
Scrobipalpa rezniki turkmenica Piskunov, 1990: 298, figs. 12-15, **syn. n.**

DIAGNOSIS.– Adult. Wingspan 13 mm. Labial palp white; segment 2 mottled with grey and black; segment 3 with two black rings. Antenna black, ringed with whitish grey. Head cream-white mottled with some grey in neck; thorax and tegula as forewing. Forewing variegated with white-based black and orange-brown scales, the latter especially in fold, subcostal stripe, around black spots and as a larger patch at 4/5 of wing; margins of forewing black; indistinct, light costal and tornal spots; hindwing white with grey margins.

Variation. Only the holotype was examined.

Similar species. *S. dagmaris* is characterized by the black margins of the forewings, the larger orange-brown patch at 4/5 and by the white hindwings with grey margins.

MALE GENITALIA.– Coremata present; uncus moderately slender, slightly tapered, apex flattened and weakly emarginate; gnathos-hook short; valva shorter than uncus, evenly curved, with scarcely inflated apical part; sacculus long and slender, about three times as long as wide, apex inwardly pointed, narrow gap between sacculus and vincular process; vincular process about length of sacculus, slender digitate, curved inwards with outwardly pointed apex, continuous with very broad and deep U-shaped medial emargination of posterior margin of vinculum; saccus broad at base, distal part strongly tapering to pointed apex, considerably exceeding apex of pedunculus; phallus stout, coecum distinctly inflated.

FEMALE GENITALIA.– Segment VIII slightly broader than long; sternite VIII with indistinct, membranous anterior margin; subgenital plate largely covered with numerous microtrichia, longitudinal area of foamy sculpture in anterioproximal part of depression; ventromedial depression broadly subrectangular, posterior part membanrous, covered with some microtrichia, anterior part with distinct digitate processes reaching far beyond anterior margin of sternite VIII, proximally covered by large-scaled foamy sculpture, medially separated by broad U-shaped emargination; apophysis anterioris rod-like, abruptly broad-based, about length of segment VIII; subostial sclerites distinct, posteriorly curved and approaching in middle; colliculum short; ductus bursae broad; corpus bursae well distinguished from ductus bursae, small, oval; signum near entrance on the right side of corpus bursae; signum-hook moderately large, slender, strongly curved in apical half, basal plate large without additional spines.

DISTRIBUTION.– Doubtful record from northern Italy. Outside Europe known from Uzbekistan, Turkmenistan and Mongolia.

BIONOMICS.– The larva has been bred from *Haloxylon persicum* Bunge (Chenopodiaceae) (Falkovitsh & Bidzilya, 2006: 78). The holotype of *S. dagmaris* was collected in early August. Further material was collected from May to July at light.

REMARKS.– *S. dagmaris* was described from a single male collected in the surroundings of Gargano near Lago di Garda (Italy) by the heteropterist E. Heiss. However, the locality of the holotype remains doubtful since the species was never re-collected by experienced lepidopterists. Possibly the specimen was mislabelled at a later stage. The holotype has some glue at the base of the hindwings, probably because they, together with thoracic segment III, have been glued to the rest of the specimen.

S. rezniki was described from altogether 5 males collected in Uzbekistan and Mongolia. The figure of the holotype from Uzbekistan leaves no doubt about the conspecificity with *S. dagmaris*. This proposed synonymy was already recognized by A. Bidzilya (*in litt.*).

S. rezniki turcmenica was described from the male holotype and two female paratypes. The main differences from the nominotypical subspecies were stated to be the larger wingspan of the adults and the larger size of the genitalia. These weak character states were already regarded as infrasubspecific by Povolný (2002a: 66).

76. Scrobipalpa suasella (Constant, 1895)

Lita suasella Constant, 1895: 53.

DIAGNOSIS.– Adult. Wingspan 9-10 mm. Labial palp white; segment 2 mottled with greyish black on lower and outer surface; segment 3 with two black rings. Antenna black, indistinctly lighter ringed. Face cream-coloured; neck grey; thorax and tegula as forewing. Forewing dark grey from white-based black scales and mottled with light brown; black spots in fold and at 2/5 and 3/5 in middle of wing surrounded by light brown; black scales from the latter spot towards tornus; fold and subcostal stripe light brown; apical area with many black scales. Hindwing grey.

Variation. There is some variation, especially in the amount of black scales on the forewing, resulting in specimens looking lighter and darker grey. In some specimens there is a weak, light, postmedial fascia, which sometimes is separated by a small, black streak; segment 3 of the labial palps can be nearly unicolorous black.

Similar species. *S. suasella* is characterized by its small size and dark grey, more or less variegated forewings. *S. hendrikseni* sp. n. q.v.

MALE GENITALIA.– Uncus broad, apex truncate with distinct emargination; gnathos-hook distinct; valva slender, weakly curved, slightly shorter than uncus, apically slightly pointed; sacculus moderately short and broad, distodorsal part rounded with weak and inwardly directed tip, inner margin almost straight; vincular process slender, digitate, shorter than sacculus; posterior margin of vinculum with extremely broad and moderately deep, nearly U-shaped medial emargination, continuous with vincular process; saccus broad and moderately short, subtriangular, gradually tapered towards apex, extending far beyond apex of pedunculus; phallus broad, stout, with heavily sclerotized ventral wall and distinctly inflated coecum, distinct sclerotized arm apically. Differs from the closely related *S. hendrikseni* sp. n. q.v. in the characteristic shape of the vincular process, and the broad and comparatively shallow medial emargination of the vinculum.

FEMALE GENITALIA.– Segment VIII slightly broader than long; subgenital plate largely covered with foamy sculpture except for smooth posterior third and lateral margin, inner edge longitudinal, abruptly turned to oblique anterior margin; ventromedial depressions well sclerotized, long and digitate, distinctly edged, reaching from posterior margin beyond anterior margin of segment VIII, widely separated except for base, posterior third covered with microtrichia; apophysis anterioris rod-like, with gradual transition to sternite VIII, slightly longer than segment VIII; colliculum short; ductus bursae long and narrow; corpus bursae well distinguished from ductus bursae, small, spherical; signum on the right side of corpus bursae, small additional spines on inner side of basal plate; signum-hook in transition to ductus bursae, slender and longish, scarcely curved. Very similar to *S. hendrikseni* sp. n. but differs e.g. in the smaller and well separated ventromedial depressions with less extent of microtrichia and the more slender signum.

DISTRIBUTION.– Southern France, Portugal and Spain (Requena, 2009: 46).

BIONOMICS.– In southern France the larva mines the leaves of *Staehelina dubia* L. (Asteraceae). The characteristic blotch-mine covers large parts of the leaf and frass is distributed throughout the mine. Occasionally leaves and petioles are spun together by a slight webbing. The larva occurs in July and August and after hibernation again from late February to March (Constant, 1895: 53; Nel, 1999b: 353; Povolný, 2002a: 72). The adult has been observed from July to early September in dune habitats (France), in May, September and October in scrubby habitats on limestone (Portugal) and in mountainous oak-forest mixed with steppe grassland (Spain).

77. *Scrobipalpa hendrikseni* sp. n.

DIAGNOSIS.– Adult. Wingspan 10-13 mm. Labial palp bone-white; segment 2 mottled with grey and black on lower and outer surface; segment 3 with two black rings. Antenna black, ringed with whitish grey. Face whitish grey; neck and tegula with grey, white-based scales; thorax blackish grey. Forewing grey from grey and black white-based scales and mottled with some orange-brown scales in fold and scattered on the wing; black spots rather indistinct; termen with black scales. Hindwing grey.

Variation. The type series shows only slight variation.

Similar species. *S. hendrikseni* sp. n. is characterized by its greyish forewings with scattered orange-brown scales and rather indistinct black spots. The similar *S. suasella* is smaller (wingspan 9-10 mm) and has slightly darker forewings with more brownish scales.

MALE GENITALIA.– Uncus broad, apex truncate with distinct emargination; gnathos-hook distinct; valva slender, weakly curved, extending to apex of uncus, apically pointed; sacculus moderately short and broad, distodorsal part rounded with weak inwardly directed tip, inner margin almost straight; vincular process moderately long and slender, digitate, shorter than sacculus, inner margin slightly convex; posterior margin of vinculum with broad and deep U-shaped medial emargination, continuous with vincular process; saccus broad and moderately short, subtriangular, gradually tapered towards apex, extending far beyond apex of pedunculus; phallus broad, stout, with heavily sclerotized ventral wall and distinctly inflated coecum, distinct sclerotized arm apically. The larger and curved ventromedial processes and the narrower and deeper emargination of the vinculum are characteristic and differentiate *S. hendrikseni* sp. n. from *S. suasella* q.v.

FEMALE GENITALIA.– Segment VIII slightly broader than long; longitudinal depression of subgenital plate covered with foamy sculpture and laterally with some microtrichia, posterior third and lateral area smooth, inner edge longitudinal, abruptly turned to oblique anterior margin; ventromedial depressions well sclerotized, long and broadly digitate, distinctly edged, reaching from posterior margin far beyond anterior margin of segment VIII, anterior half widely separated by V-shaped gap except for base, posterior half medially merged, posterior two-thirds covered with microtrichia; apophysis anterioris rod-like, with gradual transition to sternite VIII, about length of segment VIII; colliculum short; ductus bursae long and moderately broad; corpus bursae well distinguished from ductus bursae, large, oval; signum horn-shaped, with large, rounded basal plate, additional spines absent; signum-hook in transition to ductus bursae, long and stout, scarcely curved. The shape of the ventromedial depressions and the larger signum-hook are characteristic and differentiate *S. hendrikseni* sp. n. from *S. suasella* q.v.

DISTRIBUTION.– Only known from the type locality in Greece (Prov. Ioannina).

BIONOMICS.– Host-plant and early stages unknown. The type series was collected in late August at light.

MATERIAL EXAMINED.– Holotype ♀. 'Greece prov. Ioánnina, 4 km W Agia Paraskevi, 750 m, 26.viii.2008, Peder Skou leg.' 'GU 09/1286 ♀ Huemer' (ZMUC).

Paratypes. Greece. 4 ♂, 1 ♀, same data as holotype, but gen. slides 1165 Hendriksen, GU 01/1094 ♀, GU 09/1285 ♂ (TLMF, ZMUC).

ETYMOLOGY.– The name is dedicated to the late Henning Hendriksen who supported our work with countless genitalia slides.

78. *Scrobipalpa traganella* (Chrétien, 1915)

Lita traganella Chrétien, 1915: 321.

DIAGNOSIS.– Adult. Wingspan 10-12 mm. Labial palp white; segments 2 and 3 each with two black rings. Antenna black, ringed with whitish grey. Head cream-white, neck mottled with grey-tipped scales; thorax and tegula as forewings. Forewing light grey from black light-based scales and yellowish brown, the latter especially as a thin subcostal stripe, in fold, along dorsum and as a weak postmedial fascia, the latter often separated by a thin black streak; four black spots on costa; 1-2 black spots in fold and two in middle of wing at 2/5 and 3/5 surrounded or overlaid with orange-brown; a black spot in apex. Hindwing light grey.

Variation. A variable species. The amount of white-based black scales of the forewing varies, resulting in a lighter to darker appearance. Sometimes these black scales are replaced by white-based brown scales, or they may have the white part slightly pinkish. The black spots are often more or less covered with orange-brown scales. The black streak in the middle of the postmedial fascia can be absent.

Similar species. *S. traganella* can be recognized by its relatively slender forewings with distinct orange-brown patches surrounding or covering the black spots, and (when present) by the thin black streak in the middle of the postmedial fascia.

MALE GENITALIA.– Uncus broadly subrectangular, apically rounded; gnathos-hook broad, short; anterior margin of tegumen deeply emarginated; valva stout, almost straight, with weakly inflated distal part, extending to about tip of uncus; sacculus moderately long and slender, slightly tapered towards apex, distal part with weakly convex outer margin, inner margin almost straight, apically rounded with small inwardly turned tooth; vincular process slightly shorter than sacculus, basally about width of sacculus, distally strongly tapered, apex rounded with weakly crescent-shaped part turned outwards, gap between vincular process and sacculus about width of sacculus; posterior margin of vinculum with deep and almost U-shaped medial emargination; saccus very short, rod-like, slender, apex knob-like, extending to about apex of pedunculus; phallus rather short, with moderately inflated coecum.

FEMALE GENITALIA.– Segment VIII broader than long; subgenital plate, with weakly convex outer margin, inner edge nearly straight, obliquely directed outwards, longitudinal depression posterior to apophysis anterioris with patch of foamy sculpture anteriorly, otherwise smooth; ventromedial depressions covered with few microtrichia posteriorly, anterior half with foamy sculpture, digitate, reaching to anterior edge of segment VIII, posteriomedially contiguous, anteriorly separated by deep and broad V-shaped incision; apophysis anterioris rod-like, distal part distinctly tapered, shorter than segment VIII; subostial sclerite distinct, sword-like; colliculum short; ductus bursae with moderately abrupt transition to oval corpus bursae; ductus and corpus bursae covered with microtrichia; signum on the left side of corpus bursae; signum-hook large and slender, moderately arched, posterior margin of basal plate with small spines.

DISTRIBUTION.– Southern France (Povolný & Luquet, 1983: 68-70), Portugal and Spain, Canary Islands and unpublished records from the Cape Verde Islands, North Africa to the Middle East and Pakistan (Povolný, 1979a: 113).

New records. Malta: 1 ♀, Mellieha Bay, 25 m, 5.-11.iv.2004, leg. Skule, gen. slide 4973 Hendriksen (ZMUC). Cape Verde Islands: 1 ♀, Sal, Santa Maria, sea level, 2.i.2003, leg. Karsholt, gen. slide 5384 Hendriksen (ZMUC). Morocco: 8 ♂, 9 ♀, 8 km S Sidi Ifni, 50 m, 20.iii.2005, leg. Karsholt, gen slides 5057, 5059, 5060, 5062, 5065 Hendriksen (ZMUC).

BIONOMICS.– The larva has been observed from March to May, mining the twigs of *Bassia muricatus* Moquin, *Salsola vermiculata microphylla* Moquin and *Traganum nudatum* Delille (Chenopodiaceae) (Chrétien, 1915: 322). According to Klimesch (1984: 155) it feeds on various Chenopodiaceae. The adults have been recorded from April to June, and in October; on the Canary and Cape Verde Islands also during the winter months. *S. traganella* is restricted to halophytic habitats.

79. *Scrobipalpa bazae* Povolný, 1977

Scrobipalpa bazae Povolný, 1977b: 184, figs. 1-2, 26.

DIAGNOSIS.– Adult. Wingspan 8-14 mm. Labial palp warm yellowish; segments 2 and 3 each with two black rings. Antenna ringed black and yellowish grey. Head, thorax and tegula as forewing. Forewing yellow, mottled with orange-brown and black; a larger oblique streak from 1/5 at costa to 3/5 in middle of wing; a black streak near apex; apex and termen black. Hindwing grey.

Variation. The lighter or darker orange-brown areas are more or less covering the yellow ground colour. Worn specimens become light yellowish. The species is variable in size.

Similar species. *S. bazae* is characterized by its yellow and orange-brown colour with obliquely longitudinal black markings.

MALE GENITALIA.– Uncus long, slender, apex flattened or with concave emargination; culcitula well developed, gnathos-hook distinct; valva slightly exceeding uncus, moderately curved, with weak apical inflation; sacculus long and slender, about even width from base to apex, dorsoapically rounded with distinct tooth turned inwards; vincular process slightly shorter and distinctly broader than sacculus throughout, crescent-shaped, with outwardly turned tip, separated from sacculus by wide space; inner margin of vincular process gradually passing into comparatively shallow V-shaped medial emargination of posterior margin of vinculum; saccus comparatively broad at base, strongly tapered towards apex, extending to about or slightly beyond apex of pedunculus; phallus rather short, with weakly inflated coecum. *S. bazae* can be recognized by the valva which overtops the uncus and by the particularly slender sacculus in combination with other characters.

FEMALE GENITALIA.– Segment VIII slightly broader than long; anterior margin of sternite

VIII with short sclerotized rim; subgenital plate with strongly folded inner edge, additional lateromedial fold delimiting longitudinal depression of foamy sculpture, extending into base of apophysis anterioris, ventromedial depressions long and digitate, reaching to anterior edge of segment VIII, medially fused, anteriorly rounded, covered with small-sized foamy sculpture; apophysis anterioris comparatively short, about two thirds length of segment VIII, broad at base, distal part rod-like; colliculum short; ductus bursae gradually widened posteriorly, with indistinct transition to pyriform corpus bursae; signum on the right side of posterior part of ductus bursae, basal plate small with indistinct additional tooth; signum-hook short and stout, with weak apical curvature.

DISTRIBUTION.– With certainty only known from the Canary Islands and Spain, also recorded from France (Leraut, 1997: 122). Klimesch, (1984: 155) erroneously gives several additional records ranging from Madeira and North Africa to Turkey, South Russia and Afghanistan.

BIONOMICS.– Host-plant and early stages unknown. The adult occurs from September to November and is attracted to light. On the Canary Islands *S. bazae* is restricted to dunes (Klimesch, 1984: 155).

REMARKS.– Specimens from the Canary Islands differ from nominotypical populations from southern Spain in their forewing markings (Klimesch, 1984: 156) and Povolný (2002a: 34) considers these populations as a supposed separate subspecies.

The female sex of *S. bazae* was hitherto undescribed.

80. Scrobipalpa artemisiella (Treitschke, 1833)

Lita artemisiella Treitschke, 1833: 97.
Anacampsis ancillella Bruand d'Uzelle, 1851: 75.
Lita oreocyrniella Petry, 1904a: 249. Subspecies.
Lita paniculatella Novickij, 1924: 141, figs. 1, 4, 7, 10, 15, 18.
Scrobipalpa artemisiella syriaca Povolný, 1967a: 214. Subspecies.
Scrobipalpa artemisiella mongolensis Povolný, 1969a: 6, figs. 18-19, 3, **syn. n.**

DIAGNOSIS.– Adult. Wingspan 8-13 mm. Segment 2 of labial palp whitish mottled with black on lower and outer surface; segment 3 cream-white with two broad, black rings. Antenna greyish black. Head with light grey frons and dark grey neck; central part of thorax yellow, lateral part dark brown; tegula grey-brown. Forewing blackish brown, more or less mottled with light brown or orange, especially as longitudinal stripes in fold, at subcosta, in middle of wing along 2-3 black spots and along dorsum; a black streak in apex. Hindwing grey, lighter at base.

Variation. *S. artemisiella* shows a remarkable individual and geographical variability and several taxa have been given subspecific rank: *S. artemisiella artemisiella*, *S. artemisiella oreocyrniella*, *S. artemisiella syriaca* and *S. artemisiella mongolensis* (Povolný, 2002a: 31-32). Nominotypical specimens are rather dark and larger than those from the Mediterranean which become increasingly creamy to whitish along the dorsum, head and thorax and also smaller in wingspan. Populations from Corsica and Sardinia, considered as separate subspecies *S. artemisiella oreocyrniella*, are particularly distinct with almost pure whitish dorsum, head and thorax. Specimens from Greece are also small and lighter than those from Central and northern Europe but less distinct than the Tyrrhenian material and with more orange-brown suffusion. They are matching *S. artemisiella syriaca*, which otherwise occurs in the Middle East, ranging from Syria to northern Iran (Povolný, 1967a: 114, Povolný, 2002a: 32). *S. artemisiella mongolensis*, described from Mongolia (Povolný, 1969a: 6), depicts rather dark specimens with no convincing differentiation from nominotypical populations.

Similar species. Due to its variability *S. artemisiella* can be confused with a number of other species, but that is not the case with ssp. *oreocyrniella*. Specimens of *S. artemisiella* can often be recognized by the following characters which are, however, rarely present in one specimen: a relatively small size; a lighter central part of the thorax; a lighter dorsum and a black streak in the apex of the forewing. *S. murinella* q.v.; *S. stangei* q.v.; *S. voltinella* q.v.; *S. notata* q.v.; *S. dorsoflava* q.v.

MALE GENITALIA.– Coremata present; uncus long, slender, distally tapered, apex rounded or slightly concave; gnathos-hook short; valva extending slightly below tip of uncus, weakly inflated apex, posterior part weakly curved; sacculus long and slender, even width from base almost to apex, apically rounded with small inwardly turned tooth; vincular process shorter than sacculus, separated from sacculus by narrow space, strongly tapered from base to tip, apex rounded with weakly crescent-shaped part turned outwards; posterior margin of vinculum with comparatively shallow V-shaped medial emargination, scarcely separated from vincular process; saccus broad at base, evenly tapered towards apex, extending slightly beyond apex of pedunculus; phallus rather short, with weakly inflated coecum. The male genitalia are indistinguishable from those of *S. stangei*.

FEMALE GENITALIA.– Segment VIII longer than broad; anterior margin of sternite VIII sclerotized; subgenital plate covered with numerous microtrichia and anteriorly small-sized foamy sculpture, proximal part smooth, oblique and slightly curved inner edge, smooth subtriangular zone anterioproximad of edge; ventromedial depressions reaching slightly beyond anterior edge of segment VIII, medially merged, covered with microtrichia to a large part, anteriorly smooth and sclerotized with rounded apex and separated by small medial incision; apophysis anterioris rod-like, with gradual transition to anterior edge of sternite VIII, slightly longer than segment VIII; colliculum short; ductus bursae gradually widened towards corpus bursae; corpus bursae clearly demarcated from ductus bursae, pyriform; signum on the right side of corpus bursae, indistinct additional tooth near basal plate; signum-hook rather short, strongly curved at an almost right angle in middle. Indistinguishable from *S. suaedella* and *S. stangei* and also similar to *S. suaedivorella* q.v. and *S. deutschi* sp. n. q.v.

DISTRIBUTION.– Widely distributed throughout Europe. In Asia known from Syria to Mongolia and recently recorded from China (Bidzilya & Li, 2010). However, during the last decades declining in some areas such as the Netherlands (Jansen, 1999: 48). Records from North America (Povolný, 1998: 339; Powell & Povolný, 2001: 19) are based on misidentification (see below).

New record. Andorra: 2 ♂, by Pto. de Envalira, 2300 m, 1.viii.1988, leg. M. Fibiger, gen. slide 3757 Hendriksen (ZMUC).

BIONOMICS.– The larva has been recorded from *Thymus pulegioides* L. (Jansen, 1999: 48), *T. praecox* Opiz (Bradford, 1979: 126) and *T. serpyllum* L. (Lamiaceae) (Schütze, 1931: 167). Records from *Centaurea paniculata* L., *Achillea clavennae* L. and *Artemisia campestris* L. (Asteraceae) (Povolný, 1990b: 194) remain unconfirmed (Sattler, 1987c: 441). Povolný (*op. cit.*) considers *Artemisia* L. (Asteraceae) as a further host-plant, but this remains unconfirmed (for records on *Artemisia* from North America see below). European records of *Artemisia* L. as a host-plant for *S. artemisiella* almost certainly refer to *Scrobipalpula psilella* (Herrich-Schäffer) (Huemer & Karsholt, 1998: 50). Initially the larva makes a short brownish leaf-mine. Later it feeds between spun terminal shoots and stems (Jansen, 1999: 48). The larva feeds in a single generation in May and June, possibly hibernating in an early stage. Pupation takes place between spun leaves. The adult has been recorded from June to August. It flies freely at dusk and is attracted to light. Preferred habitats are dry grassland, sandy and gravelly spots and similar open landscape with dry and hot microclimate. Vertical distribution: from lowland localities up to 3000 m in Kyrgyzstan (Povolný, 1996b: 30).

REMARKS.– Leraut (1997: 122) erroneously used the name *S. artemisiella ancillella* instead of *S. artemisiella oreocyrniella*. However, *Anacampsis ancillella* was described from mainland France (Jounge, Doubs) and was stated to have more reddish forewings and a more whitish head than normal *S. artemisiella* (Bruand d'Uzelle, 1851: 75). Considering the type locality and the description this name cannot be used for *S. artemisiella oreocyrniella*.

S. artemisiella was recorded from North America by Povolný (1998: 339) and Powell & Povolný (2001: 19). We were able to examine a female published as *S. artemisiella* from Canada in ZMUH (Povolný, *op. cit.*). It is misidentified and probably belongs in the *S. salinella*-group. We moreover examined specimens of *Scrobipalpa* reared by J. A. Powell from *Artemisia douglasiana* Besser (Asteraceae) from three sites in California (in coll. EME). They were recorded as *S. artemisiella* by Powell & Povolný (2001: 22). These specimens, although somewhat variable in habitus, differ from *S. artemisiella* in being larger, and in having more uniformly coloured forewings without a lighter dorsum. In the male genitalia the Californian specimens have differently shaped vincular processes, a truncate (not rounded) uncus tip and a more stout saccus. The female genitalia resemble those of *S. artemisiella*, but differ in having less microtrichia on the subgenital plate and, especially, on the ventromedial depression which is almost smooth. Although the Californian specimens, based on the structure of the female genitalia, may be related to *S. artemisiella*, they represent in our opinion a different species, and they have been introduced into literature as an unnamed new species by Povolný himself in an earlier paper (Povolný, 1990b: 195).

81. *Scrobipalpa stangei* (Hering, 1889)

Gelechia stangei Hering, 1889: 299.
Lita saltenella Meess, 1910: 366.

DIAGNOSIS.– Adult. Wingspan 13-16 mm. Labial palp almost blackish grey. Antenna black. Head dark grey; central part of thorax brownish; lateral part of thorax and tegula as forewing. Forewing grey brown, mottled with light brown and orange as longitudinal stripes along dorsum, in fold, at subcosta and along black spots in middle of the wing; veins often black; a black streak from distal black spot to apex. Hindwing grey.

Variation. A rather variable species. The forewing colour can vary from light brownish to almost black, the black spots can be more or less distinct, often the veins are black and sometimes the dorsum is lighter. A female from an inland locality in Poland has the palps, head, hindlegs and dorsum of the forewings yellow.

Similar species. *S. stangei* resembles *S. artemisiella*, but it is larger and specimens should hence not be confused. *S. samadensis* has the central part of thorax concolorous with the rest of the thorax, and the veins of the forewings have black, white-based scales (black in *S. stangei*). *S. suaedella* q.v.; *S. mixta* sp. n. q.v.

MALE GENITALIA.– As in *S. artemisiella* q.v., only differing by the generally larger size which is related to the size of the adults.

FEMALE GENITALIA.– As in *S. artemisiella* q.v. and *S. suaedella* q.v.; also very similar to *S. suaedivorella* q.v.

DISTRIBUTION.– Only recorded from scattered localities ranging from northern Germany, Denmark and southern Scandinavia to Poland and Latvia; also recorded from Austria and Hungary (Fazekas, 2002: 309). From Great Britain only known from ancient material collected in 1882 on the Isle of Wight (Sattler, 1987b: 53). Records from Romania (Kovács & Kovács, 2000: 31) require confirmation, as two of the mentioned specimens examined by us turned out to be *S. salicorniae*. Recently also published from Central Asia (Altai Mountains) (Bidzilya, 2009: 5).

BIONOMICS.– The larva has been recorded from *Triglochin maritima* L. and *T. palustris* L. (Juncaginaceae). It feeds from autumn to June, probably mining the leaves and hibernating in the rootstock (Bland *et al.*, 2002: 181). In spring the larva produces a broad and translucent leaf-mine but also feeds in the shoots, stem or roots. The larva mines flower stalks and leaves; in the attacked leaves a part of the leaf droops and forms an acute angle with the rest of the leaf; frass is deposited inside the angle (Larsen, 1927: 92-93, as *saltenella*); finally the larva descends into the rootstock (Benander, 1965: 16). Pupation takes place outside the larval dwelling (Hering, 1957: 1070) or in the larval mine (Bland *et al.*, *op. cit.*). The adults have been observed from mid-June to early August flying freely at dusk but also attracted to light. *S. stangei* is restricted to halophytic coastal and inland habitats with a moderately high level of salinity.

REMARKS.– The nomenclatorial problems concerning the name *Lita saltenella* have been discussed in detail by Karsholt (1995: 148).

82. *Scrobipalpa suaedivorella* (Chrétien, 1915)

Lita suaedivorella Chrétien, 1915: 320.
Lita detersipunctella Toll, 1947: 114, pl. 4, fig. 23, pl. 6, fig. 41.

DIAGNOSIS.– Adult. Wingspan 15 mm. Labial palp light ochreous; segments 2 and 3 with rings of black scales. Antenna black, ringed with light scales. Face light ochreous; neck, thorax and tegula pinkish red. Forewing light pinkish red mottled with black, especially from base to 3/5 in middle of wing, covering the black, light brown surrounded spots in fold and at 2/5 and 3/5 (the latter divided into two); black spots at base and at 1/5 at costa. Hindwing whitish grey.

Variation. The description is based on the only western European specimen. A specimen from Morocco is less reddish, more yellowish and light brown, with less black. Povolný (2002a: pl. 5, fig. 15) figures a specimen from Algeria, which looks similar, although the yellowish and light brown colour on the forewing is given as light chocolate brown in the text (Povolný, loc. cit., p. 72). A specimen from south-western Russia has a dark brown head and thorax and the light brown tornal part of the forewing is distinctly demarcated from the largely dark brown dorsal part by a dark brown stripe, extending from middle almost to apex of forewing . Furthermore about 5 dark brown spots at the basal line of the cilia are present in the tornal part.

Similar species. *S. suaedivorella* is characterized by its large size, its pinkish red colour and the black markings of the forewing.

MALE GENITALIA.– Uncus moderately broad, apex evenly rounded; gnathos-hook very long; valva slender, weakly curved, extending slightly below tip of uncus, with distinctly inflated apex; sacculus moderately broad and short, almost even width from base to apex, with parallel inner and outer margins, apically convex with distinct inwardly directed tooth medially; vincular process distinctly projected posteriad, almost length of sacculus, separated from sacculus by moderately narrow gap, digitate and strongly tapered from base to tip, apex rounded with weakly crescent-shaped part turned outwards; posterior margin of vinculum with moderately broad and deep V-shaped medial emargination, almost continuous with vincular process; saccus moderately broad at base, evenly tapered towards stout apex, extending to about apex of pedunculus; phallus moderately short and slender, with inflated coecum and distinct apical hook. The male genitalia are similar to species of the *S. instabilella*-complex but differ e.g. by the shape of the phallus and the broader uncus.

FEMALE GENITALIA.– Segment VIII broader than long; subgenital plate covered with numerous microtrichia, anteriorly with small-sized foamy sculpture, oblique and slightly curved inner edge, longitudinal depression with small-sized foamy sculpture, lateral and anteriomedial part between edge and depressions smooth; ventromedial depressions digitate, membranous and covered with microtrichia throughout, medially merged, extending slightly beyond anterior edge of segment VIII, anteriomediallly separated by small incision; apophysis anterioris rod-like, slightly longer than segment VIII, with gradual transition to sclerotized, transverse anterior edge of sternite VIII; colliculum short; ductus bursae gradually widened towards corpus bursae; corpus bursae clearly demarcated from ductus bursae, pyriform; signum on the right side of corpus bursae, indistinct additional tooth near basal plate; signum-hook rather short, strongly curved at an almost right angle in middle. Similar to *S. artemisiella* q.v., *S. suaedella* q.v. and *S. stangei* q.v. with differences in the ventromedial depressions which are covered with microtrichia throughout and not smooth and distinctly sclerotized at the apex.

DISTRIBUTION.– Unpublished records from Spain and from south-western Russia. More widely distributed in North Africa (Tunisia and Algeria), and furthermore recorded from Turkey and Iran (Povolný, 2002a: 72).

New records. Spain: 1 ♀, prov. Zaragoza, 3 km E Cerveruela, at Rio de Huerva, 800 m, 7.vii.2002, leg. Skule, gen. slide 4011 Hendriksen (ZMUC). European Russia: 1 ♂, prov. Volgograd, distr. Pallasovska, lake Elton, loc. Tschernavka, 3.-8.v.2004, leg. Rutjan, gen slide GU 06/1244 ♂ Huemer (ZMKU).

BIONOMICS.– The larva mines in the leaves of various species of *Suaeda* L. (Chenopodiaceae) which are spun together along the twigs (Chrétien, 1915: 321); Falkovitsh & Bidzilya (2006: 75) published bred material from *S. drepanophylla* Litvinov, *S. microphylla* Pallas and *S. physophora* Pallas. The larva has been observed in November and December and the adults emerged between May and July of the following year. According to Povolný (2002a: 72) the moth is on the wing from February to May at lower altitudes and from mid-June to July in mountain areas. *S. suaedivorella* is restricted to halophytic coastal and inland habitats up to 800 m in Europe and 2000 m in Iran.

REMARKS.– *Lita detersipunctella* was described from a single female collected in Iran (Toll, 1947: 114). The synonymy of this taxon with *S. suaedivorella* follows Povolný (1972: 197) who studied the holotype and its genitalia slide.

83. Scrobipalpa bryophiloides Povolný, 1966

Scrobipalpa bryophiloides Povolný, 1966b: 397, figs. 6-7, 6.

DIAGNOSIS.– Adult. Wingspan 12-14 mm. Labial palp cream-white mottled with blackish at base and tip of lower and outer surface of segment 2; segment 3 with two black rings. Antenna black, ringed with whitish grey. Face cream-coloured, neck, thorax and tegula as forewing. Forewing covered with light brown and dark grey white-based scales and mottled with black and white; black spots in fold and at 2/5 and 3/5 in middle of wing surrounded by orange; black scales near base, along costa and in apical area; a white postmedial fascia. Hindwing light grey.

Variation. Some specimens have rather plain light brown forewings, whereas these in other specimens are more variegated light grey, brownish and white. The black spot at 3/5 can be angulated. Sometimes the neck is also cream-coloured.

Similar species. *S. bradleyi* has greyish (not light brown) forewings, and the black spots are more distinct. *S. selectella* q.v.; *S. solitaria* q.v.

MALE GENITALIA.– Uncus remarkably broad, subrectangular, even width towards slightly rounded to almost flattened apex; culcitula well developed, gnathos-hook strong; valva extending to about apex of uncus, weakly curved, with apical inflation; sacculus short and broad, indistinctly tapered towards rounded apex, distinct apical tooth turned inwards; vincular process slightly shorter than and similar in width to sacculus, apically broadly rounded, separated from sacculus by narrow space; inner margin of vincular process gradually passing into very broad and deep medial emargination of posterior margin of vinculum; saccus broad at base, weakly tapered towards rounded apex, distinctly exceeding apex of pedunculus; phallus stout, with weakly inflated coecum. The broad uncus and the rounded vincular process are quite characteristic for *S. bryophiloides*.

FEMALE GENITALIA.– Segment VIII broader than long; sternite VIII with transverse, weakly curved anterior margin; subgenital plate laterally straight, inner edge slightly curved and oblique, posteriolateral part covered with microtrichia, longitudinal depression with foamy sculpture extending to base of apophysis anterioris; ventromedial depressions subtriangular, lateral part with small-sized foamy sculpture and microtrichia, medially membranous; subostial sclerite well developed, transverse; apophysis anterioris moderately stout, abruptly broad-based, shorter than segment VIII; colliculum short; ductus bursae longish, gradually widened with indistinct transition to large, suboval corpus bursae; signum on the right side of corpus bursae, basal plate small, signum-hook small, strongly curved at right angle in middle.

DISTRIBUTION.– Restricted to eastern Europe (Finland, Latvia, Russia, Ukraine) and hitherto unpublished record from Greece. Also known from Turkey to the Middle East, Central Asia and Mongolia (Povolný, 1973a: 9). Recently recorded from China (Bidzilya & Li, 2010).

New record. Greece: 2 ♀, Kos, Tigaki, 16. & 22.ix.1992, leg. Johansson, gen. slides 2929, 3076 Hendriksen (ZMUC).

Bionomics.– The larva has been bred from *Suaeda confusa* Iljin (Chenopodiaceae) (Falkovitsh & Bidzilya, 2006: 75). The adults have been collected from April to September, mainly at light. Preferred habitats are halophytic places such as inland salt lakes.

Remarks.– *Scrobipalpa bryophiloides* was described from one male and four females. A female from Krasnodovsk was originally selected as the holotype. However, the figure of the female genitalia shown in the original description (Povolný, 1966: 396, fig. 7) is drawn from a misidentified paratype which later was attributed to *S. fraterna* (currently *S. selectella*) (Povolný, 1973b: 111). The female holotype, deposited in ZMHU, was erroneously labelled as paratype whereas the only male, expressly stated as paratype in the original description, was labelled as holotype (Bidzilya, *in litt.*).

84. *Scrobipalpa algeriensis* Povolný & Bradley, 1964

Scrobipalpa algeriensis Povolný & Bradley, 1964: 553, figs. 1-3.

Diagnosis.– Adult. Wingspan 12 mm. Labial palp cream-white; segments 2 and 3 each with two black rings. Antenna ringed black and light grey. Head cream-white, neck mottled with grey-tipped scales; thorax and tegula as forewing. Forewing covered with grey, white-based scales and mottled with some black and light yellowish scales; a distinct black spot on costa at 1/5; black spots in fold and at 2/5 and 3/5 in middle of wing only faintly surrounded with yellowish brown; termen with many black scales. Hindwing light grey, whitish towards base.

Variation. The few specimens available show only slight variation with regard to the amount of yellowish scales on the forewing. Povolný (2002a: pl. 3, fig. 7) figures the holotype of *S. algeriensis* as having darker, more brownish and more variegated forewings.

Similar species. *S. algeriensis* is characterized by its rather plain grey forewings with three black spots and a distinct black spot on costa at 1/5.

Male genitalia.– Uncus comparatively narrow, apex with deep concave emargination medially, lateral corners rounded; valva extending slightly below or at most to tip of uncus, almost straight, with slightly pointed tip; sacculus comparatively short and broad, with accessory hump on inner surface, apically with acute tip, directed inwards; vincular process narrower and distinctly shorter than sacculus, distal part constricted, slightly curved outwards with indistinct tip; posterior margin of vinculum with comparatively shallow V-shaped medial emargination, scarcely separated from vincular process; saccus broad, gradually tapered towards rounded apex, comparatively long, distinctly exceeding apex of pedunculus; phallus stout, with weakly inflated coecum. The uncus and sacculus are unique within European species of *Scrobipalpa*. The closely related *S. burmanni* Povolný, 1971, occurs in Tunisia.

Female genitalia.– Segment VIII longer than broad, anteriorly widened; subgenital plate largely covered with numerous microtrichia except for smooth anterior margin, long subovate depression covered with microtrichia and distinctly demarcated by convex inner and outer edge; ventromedial depressions weakly separated from subgenital plate, with short lateral fold, largely covered with microtrichia, anteriorly smooth and broadly rounded, with V-shaped anteriomedial emargination in between; apophysis anterioris rod-like, abruptly broad-based, slightly shorter than segment VIII; colliculum short; ductus bursae moderately narrow; corpus bursae well distinguished from ductus bursae, globular; signum on the right side of corpus bursae, basal plate moderately small with few additional spines; signum-hook slender, strongly curved at an almost right angle in middle.

DISTRIBUTION.– Scattered records from southern Spain. Described from Algeria and also known from Tunisia (Povolný, 2002a: 30). A record from the Czech Republic (Povolný, 1996a: 114) is dubious.

New record. 1 ♂, Morocco, Tinerhir area, 1400-2000 m, 12.iv.1989, leg. Karsholt, gen. slide 2906 Hendriksen (ZMUC).

BIONOMICS.– Host-plant and early stages unknown. The adults have been collected in April and in late September and early October at light. Povolný (1977b: 196) suspects the species to be halophytic, though it occurs in mountain regions.

85. *Scrobipalpa deutschi* sp. n.

DIAGNOSIS.– Adult. Wingspan. 14-15 mm. Labial palp cream-white mottled with grey-brown on lower and outer surface of segment 2 and (especially) on segment 3. Antenna almost black. Head, thorax and tegula grey; face lighter. Forewing light greyish brown from light-based and tipped black and brown scales, and mottled with yellow-white and brown; distinct black spots surrounded by yellow-brown in fold and at 2/5 and 3/5 in middle of wing; subcostal stripe yellowish brown; a distinct black streak from black spots at 3/5 to apex; termen with black spots. Hindwing grey.

Variation. Two specimens from Algeria differ from those from the Urals by (apart from being faded) having a light area between medial black streak and costa, and by the less distinct spots.

Similar species. *S. deutschi* sp. n. resembles *S. disjectella* by its longitudinal pattern. It has more uniformly coloured forewings with less light areas and distinct terminal spots. *S. deutschi* sp. n. lacks the yellow-white subdorsal stripe found in *S. rebeli*.

MALE GENITALIA.– Few coremata present; uncus broadly subrectangular, apically truncate; gnathos-hook short, stout; valva stout, weakly curved, apical part strongly inflated, extending to tip of uncus; sacculus short and broadly subtriangular, outer margin slightly convex, apex without distinct tooth; vincular process subtriangular, slightly longer and more narrow than sacculus, basally nearly fused with sacculus; posterior margin of vinculum with moderately broad and deep U-shaped medial emargination; saccus short and stout, broad at base, strongly tapered towards apex, extending slightly beyond apex of pedunculus; phallus short, stout, with distinct apical hook, coecum weakly inflated. Differs from related species such as *S. disjectella* and *S. mixta* sp. n. particularly in the characteristic shape of the sacculus and the vincular process.

FEMALE GENITALIA.– Segment VIII about same length and breadth, lateral edges straight, anterior edge sclerotized; subgenital plate with longitudinal depression, anteriorly covered with small-sized foamy sculpture, medial part with microtrichia, inner edge indistinct; ventromedial depression broad, largely covered with microtrichia, anterior part of depressions smooth and strongly sclerotized with convex anterior margin extending far beyond anterior margin of segment VIII, anteriomedially a narrow V-shaped incision in between, posteriomedially almost merged; apophysis anterioris with abrupt transition to anterior edge of sternite VIII, about length of sternite VIII; subostial sclerites narrow; colliculum very short, wide; ductus bursae with abrupt transition to clearly defined corpus bursae; corpus bursae spherical; signum on the right side at anterior part of ductus bursae; signum-hook short, strongly curved, basally with few additional spines. Differs from the related *S. disjectella* and *S. mixta* sp. n. particularly in the ventromedial depressions and the shape of the signum. Also similar to *S. artemisiella* but differs in the anteriorly broadly rounded and more strongly projected ventromedial depression.

DISTRIBUTION.– Only known from Russia (southern Ural Mountains) and Algeria.

BIONOMICS.– Host-plant and early stages unknown. The adults have been collected in mid-May and early September, indicating bivoltinism.

MATERIAL EXAMINED.– Holotype. ♂ 'RUSSIA S-Ural Orenburg oblast Pokrovka village 20 km S Schibendy valley 11.5.2005 K. Nupponen leg.' 'gen. Præparat No. 5457 ♂ *Scrobipalpa* H. Hendriksen' (NUPP).

Paratypes. 1 ♂, same data as holotype (ZMUC); 1 ♀, Russia, southern Ural, Orenburg oblast, Donskoje village, 6 km W mount Verbljushka, 9.ix.2002, leg. K. Nupponen, gen. slide 3993 ♀ Hendriksen (NUPP). Algeria: 2 ♂, Lambesa, Staudinger, gen. slides 4076 ♂ Hendriksen, GU 03/1187 ♂ Huemer (ZMHU, ZMUC).

ETYMOLOGY.– This new species is dedicated to Helmut Deutsch in recognition of his important contribution to European lepidopterology.

86. *Scrobipalpa disjectella* (Staudinger, 1859)

Gelechia disjectella Staudinger, 1859: 241.

DIAGNOSIS.– Adult. Wingspan. 12-15 mm. Labial palp cream-white, mottled with grey-brown on lower and outer surface of segment 2 and especially on segment 3. Antenna almost black. Face cream-coloured, becoming light grey-brown towards neck; thorax and tegula as forewing. Forewing light greyish brown from light-based and tipped black and brown scales, especially in dorsal half and along costa; fold brown with indistinct spots; a slender brown stripe in middle of wing from 2/5 to apex with longitudinal black spots surrounded with brown at 2/5 and 3/5 in middle of wing and as an apical stripe. Hindwing broad, light grey.

Variation. The two type specimens examined are in good condition, though faded. Compared with some recently collected, but more worn, specimens the former are much lighter, having the costal third and the outer margin covered with cream-white scales, whereas the latter specimens look more or less dark grey. One female has a prominent black streak from 2/5 to apex and distinct black dots.

Similar species. *S. disjectella* is characterized by its longitudinal pattern, especially the slender brown stripe with black streaks in the middle of the forewings from 2/5 to apex. *S. rebeli* q.v.; *S. deutschi* sp. n. q.v.; *S. phagnalella* q.v.; *S. mixta* sp. n. q.v.

MALE GENITALIA.– Uncus broadly subrectangular, apically truncate; gnathos-hook short, stout; valva stout, weakly curved, slightly shorter than uncus, apical part inflated; sacculus comparatively short and slender, about even width from base almost to apex, dorsoapically rounded with indistinct tooth turned inwards; vincular process almost same length and basally about same width as sacculus, distal part tapered, with outwardly turned tip, separated from sacculus by narrow gap; posterior margin of vinculum with narrow, moderately shallow V-shaped medial emargination; saccus broad at base, strongly tapered towards apex, extending slightly beyond apex of pedunculus; phallus short, stout, with distinct apical hook, coecum weakly inflated. Differs from *S. mixta* sp. n. in the shape of the sacculus and vincular process and the broader uncus.

FEMALE GENITALIA.– Segment VIII about same length and breadth, lateral edges straight, anterior edge membranous; subgenital plate with longitudinal depression, anterior part covered with small-sized foamy sculpture extending to base of apophysis anterioris, slightly oblique inner border long and straight; ventromedial depression broad, covered with microtrichia, depressions anteriorly separated by deep V-shaped emargination, posteriomedially almost merged; apophysis anterioris with gradual transition to anterior edge of sternite VIII, slightly shorter than segment VIII; subostial sclerites distinct, inwardly curved; colliculum very short; ductus bursae with abrupt transition to clearly defined corpus bursae; corpus bursae oval; signum on the right side at entrance of corpus bursae; signum-hook of moderate length, weakly curved near base, basal half with some additional spines. Very similar to *S. mixta* sp. n. differing in the anterior margin of the ventromedial depressions and slightly in the signum.

DISTRIBUTION.– In Europe restricted to southern Spain. Also recorded from North Africa (Algeria), Kyrgyzstan (Povolný, 1981: 391) and Russia (Ponomarenko, 2008: 96), but these records are dubious as additional and very similar taxa occur in these areas. Records from Macedonia (Povolný, 1967a: pl. 3, fig. 20; 2002a: pl. 31, fig. 280) and probably also Albania (Povolný, 1967a: 215) refer to *S. mixta* sp. n.

BIONOMICS.– The larva has been recorded in October and November feeding on *Artemisia barrelieri* Besser (Asteraceae) (Staudinger, 1859: 241). The adult is on the wing from March to late May.

REMARKS.– *Gelechia disjectella* was described from 3 specimens collected in Granada (Spain) (Staudinger, 1859: 241). Povolný (1967a: 215) without dissection selected a male as the lectotype, but labelled it as holotype. We have been able to study this specimen and a female paralectotype. The four genitalia drawings figured by Povolný (*op. cit.*, pl. 2-3, fig. 18-21) as belonging to *S. disjectella* are in fact a mixture of various species. The only female which belongs to *S. disjectella* is labelled 'Original' and was figured on pl. 3, fig. 19. According to Povolný (1967a: 215, 234) it comes from Lambèse (Algeria), but in fact it has no locality labels and Staudinger described the species exclusively from Granada. The second female comes from Ohrid (Macedonia) and was figured on pl. 3, fig. 20; it depicts *S. mixta* sp. n. However, later this figure and its locality were transposed with fig. 19 and consequently misidentified as the female of *S. disjectella* (Povolný, 2002a: pl. 73, fig. 754); furthermore the male genitalia in the latter work are referring to *S. mixta* sp. n. from Macedonia (Povolný, 2002a: pl. 31, fig. 280). Povolný (1967a: 215) moreover records *S. disjectella* from Tunisia, Speitla, based on a male in ZMUC. This specimen belongs to *S. forsteri* Povolný, 1971.

87. *Scrobipalpa mixta* sp. n.

DIAGNOSIS.– Adult. Male. Wingspan 15-17 mm. Labial palp light grey mottled with black. Antenna almost plain black. Head grey; thorax and tegula dark grey. Forewing brownish black from white-based black scales and mottled with light orange-brown, especially on fold, subcostal stripe and distal part of wing; a broad black stripe from base to 2/3 prolonged as a slender black stripe to apex; black spots indistinct; longitudinal black streaks along fold and costa; termen with black dots. Hindwing grey.

Female differs from male in: wingspan 14 mm. Labial palp cream-white mottled with grey and black on lower and outer surface of segment 2 and on segment 3. Antenna black, indistinctly lighter ringed. Face light grey from white-based dark grey scales. Forewing with elongate black spots and slender black streaks along fold and sub-costa edged with light brown. Hindwing light grey.

Variation. Based on the few available specimens *S. mixta* sp. n. seems to exhibit a moderate sexual dimorphism. Worn specimens look darker because they loose the light orange-brown scales.

Similar species. *S. mixta* sp. n. is characterized by the broad blackish stripe from base to 2/3, which is prolonged as a slender black stripe towards apex. Males resemble *S. stangei*, which has the central part of the thorax brownish, and *S. samadensis* which has the forewings with black, white-based scales along the veins. Females resemble *S. rebeli*, which has a light yellow stripe in the fold and *S. disjectella*, which has a slender brown stripe in middle of wing from 2/5 to apex.

MALE GENITALIA.– Uncus long, moderately slender, apically rounded; gnathos-hook short; valva stout, weakly curved, slightly shorter than uncus, apical part inflated; sacculus short and broadly subtriangular, apically rounded without distinct tip; vincular process moderately slender, subtriangular, about length and almost width of sacculus, slender, apically small tip, separated from sacculus by narrow subtriangular gap; posterior margin of vinculum with narrow, moderately shallow V-shaped medial emargination; saccus large, broad at base, strongly tapered towards apex, extending far beyond apex of pedunculus; phallus short, stout, with distinct apical hook, coecum inflated. Very similar to *S. disjectella* but differs in the comparatively larger sacculus and vincular process with slightly different shape and the narrower uncus.

FEMALE GENITALIA.– Segment VIII slightly broader than long, lateral edges straight, anterior edge moderately sclerotized; subgenital plate with longitudinal depression, anterior part covered with small-sized foamy sculpture extending to base of apophysis anterioris, medial part with some microtrichia, slightly oblique inner border long and straight; ventromedial depression broad, covered with microtrichia with the exception of smooth and convex anterior part, anteriomedially separated by deep V-shaped emargination, posteriomedially almost merged; apophysis anterioris with gradual transition to anterior edge of sternite VIII, slightly shorter than segment VIII; subostial sclerites distinct, inwardly curved; colliculum very short; ductus bursae with weak transition to oval corpus bursae; signum on the right side at entrance of corpus bursae; signum-hook of considerable length, weakly curved near middle, basal half with several additional spines. Very similar to *S. disjectella* but differs in the anteriorly convex and sclerotized ventromedial depressions and the heavily spined signum.

DISTRIBUTION.– Macedonia and probably also Albania (Povolný, 1967a: 215), the Ukraine.

BIONOMICS.– Host-plant and early stages unknown. The few adults known to date have been collected in April.

MATERIAL EXAMINED.– Holotype. ♂ 'Mazedonien Treskaschlucht I. Pinker IV.56' 'GU 04/1217 ♂ P. Huemer' (ZSM).

Paratypes. Macedonia. 1 ♂, 1 ♀, Ohrid, iv.1939, leg. Pinker, gen. slide GU 03/1186 ♀ Huemer, Klimesch 601 ♂ (ZSM). Ukraine. 1 ♂, Crimea, Karadagh, 24.iv.1989 leg. Budashkin, gen. slide GU 08/1270 ♂ Huemer (ZMUH).

ETYMOLOGY.– The specifies name refers to the confusion with other species, particularly *S. disjectella* and *S. rebeli* and is derived from the latin adjective *mixtus* (= mixed).

REMARKS.– See *S. disjectella*.

88. Scrobipalpa rebeli (Preissecker, 1914)

Gelechia (Lita) rebeli Preissecker, 1914: 16, fig.
Lita rebeli var. *fuscella* Klimesch, 1938b: 45.
Scrobipalpa japonica Povolný, 1977d: 222, figs. 6, 16, **syn. n.**

DIAGNOSIS.– Adult. Wingspan 12-14 mm. Labial palp cream-white mottled with grey-brown on lower and outer surface of segment 2 and 3. Antenna black, faintly lighter ringed. Head, thorax and tegula light brownish grey; face lighter. Forewing light greyish brown from light-based and tipped black and brown scales; a broad, longitudinal brownish edged stripe from base almost reaching apex; black spots surrounded with brown in fold and as double spots at 2/5 and 3/5 in middle of wing; subcostal stripe brown; a white subdorsal stripe; termen with black dots. Hindwing light grey, whitish towards base.

Variation. Specimens from Austria has darker grey-brown forewings compared with specimens from the Urals. Specimens of the form *fuscella* have darker labial palps and forewings, and the black longitudinal stripe and the spots thereby become indistinct.

Similar species. *S. rebeli* is characterized by its longitudinal forewing markings, especially the white subdorsal stripe, which is more or less missing in other species with this kind of forewing pattern. *S. disjectella* is similar and can have a smaller, light subdorsal stripe, but *S. rebeli* can be separated by its broader, black longitudinal stripe, edged with brown. *S. deutschi* sp. n. q.v.; *S. mixta* sp. n. q.v.

MALE GENITALIA.– Uncus moderately short and basally broad, distal part tapered with distinct lateral folds, apex truncate; gnathos-hook small; valva distinctly shorter than uncus, apically slightly curved and inflated, with distinct tip; sacculus stout, broad and short, basal width almost equal to length, distal part strongly tapered to inwardly directed apical tip, outer margin convex; vincular process slightly projected posteriad, sickle-shaped, stout, with outwardly directed and pointed tip, extending to about two-thirds length of sacculus, separated from sacculus by narrow gap; posterior margin of vinculum with moderately deep U-shaped emargination; saccus broad, distal part tapered to stout apex, projecting distinctly beyond apex of pedunculus; phallus moderately slender, with inflated coecum. The shape of the uncus, valva and the sacculus are unique within *Scrobipalpa*.

FEMALE GENITALIA.– Segment VIII slightly longer than broad; anterior margin of sternite

VIII membranous; subgenital plate covered with numerous microspines almost throughout, only posterior part smooth, with long and straight inner edge, turned outwards towards base of apophysis anterioris; ventromedial depressions covered with microspines, fused with sternite VIII to a large extent, reaching far beyond anterior edge of segment VIII, with sinusoid gap in between, posteriomedially merged; apophysis anterioris with gradual transition to anterior edge of sternite VIII, longer than segment VIII, basally with small zone of microspines; subostial sclerites distinct, inwardly curved; colliculum very short; ductus bursae with abrupt transition to clearly defined corpus bursae; corpus bursae spherical; signum on the right side of corpus bursae; signum-hook moderately stout and almost without curvation, basal half with several additional spines.

DISTRIBUTION.– Scattered records from Central Europe (Austria, northern Italy) to the Ukraine, southern Russia, Siberia and Japan. Recently recorded from China (Bidzilya & Li, 2010).

BIONOMICS.– The larva has been recorded from *Artemisia campestris* L. (Asteraceae) (Preissecker, 1914: 16). It feeds from a globular spun dwelling place on the seed panicle of the host-plant. Pupation takes place in a fragile cocoon on the ground surface and the pupa hibernates (Klimesch, 1938b: 45; Preissecker, *op. cit.*). The adults occur in a single generation from April to June. Preferred habitats are dry steppe slopes with xeric vegetation at low altidudes of up to about 500 m.

REMARKS.– *Lita rebeli* var. *fuscella* was described as a darker coloured local variation from Val Venosta (Südtirol, Italy) (Klimesch, 1938b: 45). However, examination of several syntypes gave no clear evidence of a subspecies state and we therefore treat this taxon in synonymy with nominotypical *S. rebeli*.

S. japonica was described from a single male collected on 24.v.1961 by Kumata on Hokkaido island (Povolný, 1977d: 222). Recently it was published from the Ukraine, based on two male specimens collected by Budashkin (Povolný, 1996b: 205). Povolný repeatedly stated that the moth and genitalia of *S. japonica* are similar to *S. rebeli*. The main differences are accordingly found in the more greyish forewings of *S. rebeli*, the more contrasting forewing markings, the shape of the uncus and particularly the short valva of *S. rebeli* and some minor 'characters' (Povolný, 2002a: 65). However, the various figures clearly show a considerable variation within the three known specimens of *S. japonica*, even from the same population (Povolný, 2002a: pl. 30, figs. 272-274) and no significant differences from *S. rebeli* (Povolný, *op. cit.*: pl. 30, fig. 271) can be observed. The two male specimens from Ukraine identified as *S. japonica* by this author and a further female from the southern Ural Mountains fully correspond with *S. rebeli* from the type locality in Austria and we therefore synonymize *S. japonica* with *S. rebeli*.

The year of description of *Gelechia* (*Lita*) *rebeli* was incorrectly stated as 1915 by Povolný (2002a: 64).

89. Scrobipalpa gallicella (Constant, 1885)

Lita gallicella Constant, 1885: 253, pl. 10, fig. 14.

DIAGNOSIS.– Adult. Wingspan 8-10 mm. Labial palp cream-coloured; segment 2 mottled with fuscous; segment 3 with black base and tip. Antenna black, ringed with light grey. Head, thorax and tegula light grey-brown. Forewing with white-tipped and white-based grey and black scales and longitudinal light brown stripes; black spots indistinct; black scales at tornus and on apex. Hindwing grey with light grey cilia.

Variation. There is some variation in the amount of light brown scales and in the distinctness of the black spots on the forewings. Specimens from the southern Ural Mountains have more black in the middle of the forewing, and a black streak towards apex.

Similar species. *S. gallicella* is characterized by its small size and the light grey forewings with longitudinal light brown stripes. *S. postulatella* sp. n. q.v.

MALE GENITALIA.– Uncus large, broadly subrectangular, apically flattened to slightly emarginate; gnathos-hook strong; valva extending slightly below tip of uncus, straight without inflated apex; sacculus short, outer margin strongly convex, apically narrowing to inwardly turned tip; vincular process shorter than sacculus, slender, sickle-shaped, distally narrowing, slightly curved outwards to a distinct tip, separated from sacculus by narrow gap, inner margin continuous with broad and deep U-shaped medial emargination of posterior margin of vinculum; saccus broad at base, moderately tapered distal part, extending far beyond apex of pedunculus; phallus stout, coecum weakly inflated, apical hook distinct. The shape of uncus and sacculus together with the stout saccus are quite characteristic for this species. The genitalia are very similar to *S. ustulatella* but differ in the more slender apex of the valva, the narrow gap between sacculus and vincular process and the apically stout saccus.

FEMALE GENITALIA.– Segment VIII slightly broader than long; subgenital plate with smooth posterior and lateral part, slightly oblique and straight inner edge, longitudinal depression covered with microtrichia posteriorly and with foamy sculpture anteriorly; ventromedial depression widened anteriorly, covered with microtrichia, anterior part with foamy sculpture, depressions anteriorly separated by U-shaped emargination, posteriomedially almost merged; apophysis anterioris about length of segment VIII, with abrupt transition to anterior edge of sternite VIII and ventromedial depression; subostial sclerites distinct, inwardly pointed; colliculum very short; ductus bursae broad, with abrupt transition to clearly defined corpus bursae; corpus bursae spherical; signum on the right side of ductus bursae near entrance to corpus bursae, surrounded by large field of microtrichia; signum-hook of moderate length, weakly curved near small base, basally with minute additional spines. Similar to *S. ustulatella* but differs in the anteriorly less separated ventromedial depressions and the shape of the signum.

DISTRIBUTION.– From Spain and France to Hungary. Recently also recorded from the Volga region in Russia (Piskunov & Anikin, 2001: 155) and the Ukraine (Bidzilya & Budashkin, 2009: 17). We were able to examine a small series of specimens from the southern Ural Mountains.

BIONOMICS.– The larva mines the leaves of *Artemisia caerulescens gallica* (Willd.) K. Persson and *A. alba* Turra (Asteraceae) (Povolný, 2002a: 43) which are spun together (Constant, 1885: 254). It has been recorded in three generations, from January to February, April to May and in June and July. The adults have been collected in June and early July at light.

REMARKS.– The signum of the holotype figured by Povolný (e.g. 2002a: pl. 67, fig. 664) is most probably deformed or broken since several additional females exmined by us have a typical signum-hook. In one male the genitalia were deformed especially in the vinculum (fig. 89a).

Specimens from eastern Europe have a broader sacculus and a slightly inflated apex of the valva whereas the only corresponding female from Hungary agrees with material from south-western Europe. We consider these populations conspecific for the time being.

90. Scrobipalpa ustulatella (Staudinger, 1871)

Gelechia ustulatella Staudinger, 1871: 307.

DIAGNOSIS.– Adult. Wingspan 9-10 mm. Labial palp white, mottled with brownish black. Antenna dark grey, indistinctly lighter ringed. Face whitish grey; neck, thorax and tegula light grey. Forewing light grey from white-based black scales; black patches at base and at 1/5 and 1/2 on costa, the latter two fused with black, orange-brown surrounded spots at 2/5 and 3/5 in middle of wing; a black streak along fold edged with orange-brown; a fine black streak from black spots at 3/5 to apex. Hindwing whitish grey.

Variation. The few specimens examined show some variation in the distinctness of the black markings and the amount of orange-brown scales in the fold and around the black spots in the middle of the wing. Similar species. *S. ustulatella* is characterized by its small size and light grey forewings with oblique black markings from costa to spots in the middle of the wings. *S. postulatella* sp. n. q.v.; *S. gallicella* q.v.

MALE GENITALIA.– Uncus large, broadly subrectangular, apically flattened; gnathos-hook strong; valva extending to about tip of uncus, stout, with weakly inflated apex; sacculus short, outer margin convex, apically narrowing to inwardly turned tip; vincular process shorter than sacculus, basally of about same width, distally narrowing, slightly curved outwards to a distinct tip, separated from sacculus by broad gap, inner margin continuous with broad and deep U-shaped medial emargination of posterior margin of vinculum; saccus broad at base, apically tapered, pointed, extending beyond apex of pedunculus; phallus stout, coecum weakly inflated, apical hook distinct. Very similar to *S. gallicella* but differs in characters such as the stout and inflated valva, the well separated sacculus and vincular process and the apically pointed saccus. Indistinguishable from *S. postulatella* sp. n.

FEMALE GENITALIA.– Segment VIII slightly broader than long; anterior margin of sternite VIII membranous; subgenital plate with smooth posteriolateral margin, slightly oblique and straight inner edge, longitudinal depression covered with foamy sculpture anteriorly, extending into base of apophysis anterioris; ventromedial depression covered with microtrichia in posterior half, anterior part digitate and covered with foamy sculpture, medially widely separated, reaching far beyond anterior margin of segment VIII, posteriomedially nearly merged; apophysis anterioris slightly shorter than segment VIII, with gradual transition to sternite VIII; subostial sclerites broad, inwardly directed; colliculum very short; signum a small spine-like sclerite. Differs from *S. gallicella* and from *S. postulatella* sp. n. in the digitate ventromedial depressions and the largely reduced signum.

DISTRIBUTION.– Only known from south-western Russia and the Ukraine (Bidzilya & Budashkin, 2009: 17). We were able to examine a few specimens from the southern Ural Mountains. Records from Spain (Povolný, 1987: 85) refer to *S. postulatella* sp. n.

BIONOMICS.– Host-plant and early stages unknown. Adults have been collected from mid-April to early June.

REMARKS.– *Gelechia ustulatella* was described from two specimens ('ein Pärchen') collected by H. Christoph in the surroundings of Sarepta (now Krasnoarmeysk) near Volgograd in southern Russia (Staudinger, 1871: 307). Povolný (1967a: 231, figs. 105-106) examined and figured a male and a female syntype and designated the male as the lectotype. The figures of the lectotype and the female paralectotype were later redrawn (Povolný, 1982: figs. 7-8). Finally Povolný (2002a: pl. 73, figs. 751-752) figured two female genitalia as different specimens which are in fact both based on the same paralectotype. The lectotype slide has been remounted and unrolled by us to show specific characters better. Furthermore we have been able to examine the female paralectotype preserved in the Povolný collection in MZMB.

The description of the female genitalia is based on a slide with largely destroyed ductus and corpus bursae. The remarkable shape of the signum may be an artefact, but in the absence of any further female from easternmost Europe this cannot be proved.

91. *Scrobipalpa postulatella* sp. n.

DIAGNOSIS.– Adult. Wingspan 10-12 mm. Labial palp cream-white; segment 2 mottled with dark grey on lower and outer surface; segment 3 mottled with black. Antenna black, ringed with light grey. Head cream-coloured mottled with light grey; thorax and tegula grey. Forewing grey from dark grey light based scales and mottled with black, brown and yellow; a black streak along fold edged by orange-brown and yellow; a black spot on costa at base; two black oblique streaks from 1/3 and 1/2 at costa towards black spots in middle of wing; a black streak in apex. Hindwing white with light grey margins and whitish-grey cilia.

Variation. Specimens can be lighter and darker grey, and the black spots in the middle of the wings and the black streak towards apex can be more or less prominent.

Similar species. *S. postulatella* sp. n. is characterized by its greyish forewings with black stripes in fold, at costa and in apex. It is very similar to *S. ustulatella* which is slightly smaller and with less contrasting markings on the forewings. *S. gallicella* has more brownish scales on the forewing and no distinct black oblique streaks from costa.

MALE GENITALIA.– Uncus large, broadly subrectangular, apically flattened; gnathos-hook strong; valva extending to about tip of uncus, stout, with weakly inflated apex; sacculus short, outer margin convex, apically narrowing to inwardly turned tip; vincular process shorter than sacculus, basally of about same width, distally narrowing, slightly curved outwards to a distinct tip, separated from sacculus by broad gap, inner margin continuous with broad and deep U-shaped medial emargination of posterior margin of vinculum; saccus broad at base, apically tapered, extending beyond apex of pedunculus; phallus stout, coecum weakly inflated, apical hook distinct. Differs from the similar *S. gallicella* in the stout and inflated valva, the well separated sacculus and vincular process and the apically pointed saccus. Indistinguishable from *S. ustulatella*.

FEMALE GENITALIA.– Segment VIII slightly broader than long; anterior margin of sternite

VIII membranous; subgenital plate with smooth posteriolateral margin, medial part covered with microtrichia, longitudinal depression covered with foamy sculpture extending to base of apopyhsis anterior, inner edge weakly separated from ventromedial depression; ventromedial depression covered with microtrichia in posterior half, anterior part broadly convex, medially widely separated, reaching far beyond anterior margin of segment VIII, posteriomedially nearly merged; apophysis anterioris slightly shorter than segment VIII, with gradual transition to sternite VIII; subostial sclerites broad, inwardly directed; colliculum very short; signum with small basal plate, signum-hook long and stout, distal half moderately curved, inner curvation with several strong spines reaching above middle of hook. Differs from *S. gallicella* and from *S. ustulatella* in the anteriorly broad ventromedial depressions and the shape of the signum.

DISTRIBUTION.– Restricted to southern Spain.

BIONOMICS.– Host-plant and early stages unknown. The adults have been collected from mid- to late April at altitudes from sea level to 1000 m.

MATERIAL EXAMINED.– Holotype. ♀ 'Spain Prov. Murcia 2 km S of Bolnuevo, 5 m, 28.-30.iv.2000 Peder Skou leg.' 'GU 01/1074 ♀ P. Huemer' (ZMUC).

Paratypes. Spain: 6 ♀, same data as holotype, but genitalia slide 3006 Hendriksen; 1 ♂, Almeria, Sierra Cabrera, 10 km S Mojacar, 100 m, 29.iii.1993, leg. Lingenhöle (ZMUC); 1 ♀, Almeria, Sierra Cabrera, 10 km S Gafarillos, 250 m, 21.iii.1994, leg. & coll. van der Wolf, gen. slide Wf 6196; 1 ♂, Murica, Sierra d'Espuna, C. F. de las Alguerias, 800 m, 4.iv.1997, leg. & coll. van der Wolf; 1 ♂, Murica, Ifae, 20.iv.1998, leg. & coll. C. Gielis, gen. slide 2701 Hendriksen; 3 ♂, 2 ♀, Murcia, Bolnuevo by Mazarrón, 10 m, 15.iv.2001, leg. Skou & Skule, gen. slide GU 09/1300 ♂ Huemer (ZMUC); 1 ♀, Bolnuevo, 5 m, 24.ix.2001, leg. Skou (ZMUC); 1 ♂, 4 ♀, Almeria, 10 km E Bedar, El Pinar, 325 m, 19.-27.iv.2001, leg. Skule & Skou (ZMUC); 6 ♂, 7 ♀, Almeria, Penon de Turillas, 1000 m, 23. & 24.iv.2001, leg. Hviid, Skou & Skule, gen. slides 3021 Hendriksen, GU 01/1073 ♂ Huemer (TLMF, ZMUC); 1 ♂, 3 km NW Torremerido, 27.iv.2009, leg. Tabel (ZMUC).

Material excluded from type series. Spain: 2 ♀, prov. Almeria, Penon de Turillas, 1000 m, 24.iv.2001, leg. Hviid, Skou & Skule, gen. slides 6280 ♀, 6343 ♀ Hendriksen (ZMUC).

ETYMOLOGY.– The name is derived from the past participle of *postulare* and refers to the precondition to describe a new species.

REMARKS.– *S. postulatella* sp. n. is the western vicariant of *S. ustulatella* differing mainly in the female genitalia. All data of *S. ustulatella* published from Spain so far in fact refer to this new species. Two female specimens differ in being larger (wingspan 13 mm) and darker. They have the antennae almost black, the forewings dark grey, and the oblique stripes from costa rather indistinct. In these specimens the black spots in the middle of the wing and the black streak towards apex are more prominent. They also show some differences in the female genitalia. Even though these specimens are collected together with some paratypes of *S. postulatella* sp. n. they are here excluded from the type series.

92. Scrobipalpa nitentella (Fuchs, 1902)

Lita nitentella Fuchs, 1902: 324.
Phthorimaea seminella Pierce & Metcalfe, 1935: 97.

DIAGNOSIS.– Adult. Wingspan 11-15 mm. Labial palp cream-white; segment 2 mottled with black on lower and outer surface; segment 3 with two, often confluent black rings. Antenna black, ringed with greyish. Head, thorax and tegula grey; face lighter. Forewing covered with grey, light-based scales and mottled with black and light brown, the latter especially as a subcostal stripe, in fold and surrounding the black spots at 2/5 and 3/5 in middle of the wing; additional small black spots at costa at 1/5, in middle of wing and in termen. Hindwing light grey.

Variation. *S. nitentella* exhibits both individual (also in specimens bred from the same locality) and geographical variation. Specimens can have almost plain light brown or light grey forewings with indistinct black spots, or they can have variegated forewings with few or many distinct black spots surrounded by orange scales. Some of the black spots can be confluent. Worn specimens tend to become light grey. Specimens from the coasts of north-western Europe have mostly light greyish brown forewings, often with reduced black spots. Specimens from the southern Ural Mountains are dark grey with many distinct black spots on the forewing, whereas specimens from Spain are predominantly light brown with little grey.

Similar species. Due to its variability *S. nitentella* may be confused with a number of other *Scrobipalpa* species. It is probably best recognized by the small black spots on the forewing (additional to the spots), when these are present. *S. nitentella* is very similar to *S. obsoletella*, together with which it often occurs. Even though *S. obsoletella* has less variegated forewings normally with only three black spots (spots), a safe identification will normally require examination of the genitalia.

MALE GENITALIA.– Coremata absent; uncus moderately slender, apically rounded; gnathos-hook short; valva extending to about or slightly beyond apex of uncus, weakly curved, with inflated apex; sacculus long and slender, about three times as long as wide, gradually tapered towards inwardly pointed apex, narrow gap between sacculus and vincular process; vinculum projected posteriorly, vincular process about two-thirds length of sacculus, digitate, curved inwards with blunt apex, continuous with shallow and broadly U-shaped medial emargination of posterior margin of vinculum; saccus broad at base, distal part strongly tapering to pointed apex, considerably exceeding apex of pedunculus; phallus long, moderately slender, coecum weakly inflated. Differs from related species in the long and slender sacculus and the shape of the vincular process.

FEMALE GENITALIA.– Segment VIII slightly longer than broad, anteriorly widened; subgenital plate covered with numerous microtrichia and small-sized foamy sculpture anteriorly, outer margin smooth, inner border long and straight, parallel-sided to outer border and turned outwards at base of apophysis anterioris, anterior edge membranous; ventromedial depressions covered with microtrichia, reaching far beyond anterior edge of segment VIII, with sinusoid gap in between, posteriomedially merged; apophysis anterioris with abrupt transition to anterior edge of sternite VIII, about length of segment VIII, basally with small zone of microtrichia; subostial sclerites inwardly curved and approaching in middle; colliculum rather broad and long; ductus bursae with abrupt transition to clearly defined corpus bursae; corpus bursae irregularly oval, dilated on the left side of ductus bursae; signum on the right side near entrance of corpus bursae; signum-hook large and stout, weakly curved, with numerous strong spines on concave side almost to apex.

DISTRIBUTION.– Widely distributed throughout Europe to the southern Ural Mountains, though mainly along the Mediterranean and Atlantic coasts; Ukraine (Bidzilya & Budashkin, 2009: 19). Outside Europe known from Tunisia, Turkey to Afghanistan (Povolný, 2002a: 58), Asian Russia (Transbaikal) (Bidzilya *et al.*, 1998: 47), Mongolia (Povolný, 1978b: 184) and China (Bidzilya & Li, 2010).

BIONOMICS.– The larva has been recorded from a large variety of Chenopodiaceae. Povolný (2002a: 58), following several authors, mentions *Atriplex hortensis* L. (=*A. nitens* Schkuhr), *A. prostrata* Bouch ex DC. (=*A. hastata* auctt.), *Halimione pedunculata* (L.) Aellen, *H. portulacoides* (L.) Aellen, *Salicornia europaea* agg., *Suaeda maritima* (L.) Dum. (= *Ch. maritimum* L.) and *Chenopodium album* agg. as host-plants. Further known host-plants are *Atriplex littoralis* L., *Beta vulgaris maritima* (L.) Arcang. (Michaelis, 1977) and *Suaeda altissima* (L.) Pallas (Falkovitsh & Bidzilya, 2006: 76). The young larva makes a short spiral leaf-mine, which soon becomes an irregular and branched greenish-white mine. Fully grown larvae live in spun galleries among the seeds. The larval period lasts from mid-August to late September. Pupation takes place in sand or debris. The pupal stage lasts from September to the following June (Jansen, 1999: 59-60). The adults have been collected from early June to early October when they are frequently attracted to light. In Denmark the species flies from late June to early August, rarely with a second generation in September. Hence, late records from September and October may indicate bivoltinism in other areas. *S. nitentella* is a characteristic species of halophytic habitats along coasts, preferably in salt marshes and storm flood strand lines within the tidal zone but also in inland tideless saline meadows (Jansen, *op. cit.*). Furthermore it occurs in deserts and steppe habitats with salty soils.

REMARKS.– Gaede (1937: 286) lists a '*Phthorimaea seminella* forma *spinosa* Pierce & Metcalfe, 1935: 97', and Povolný (2002: 58) a '*Lita seminella* (f.) *spinosa* Pierce & Metcalfe, 1935: 97'. No such name has been proposed in the paper in question by Pierce & Metcalfe.

93. *Scrobipalpa costella* (Humphreys & Westwood, 1845)

Anacampsis costella Humphreys & Westwood, 1845: 192, pl. 107, fig. 15.
Gelechia costimaculella Bruand d'Uzelle, 1859: 652; unnecessary replacement name for *Anacampsis costella* Humphreys & Westwood, 1845.

DIAGNOSIS.– Adult. Wingspan 13-16 mm. Labial palp cream-coloured, mottled with pink and black on lower and outer surface of segment 2; segment 3 with two broad, black rings. A small brush of pink scales at lower edge of the eye. Antenna black, indistinctly ringed with light grey. Face cream-coloured; neck brownish grey; thorax and tegula as forewing. Forewing in male grey-brown, in female light brown with a large, black, subtriangular block in middle of costa, reaching up to two thirds of wing breadth; black spots in fold and at 2/5 and 3/5 in middle of the wing, the former most distinct, surrounded by light brown; apical area with many black scales mixed with small spots of pinkish brown. Hindwing grey. Fore- and middle leg with the light rings at tibia pink.

Variation. Whereas the females seem rather constant there is some variation in the amount of black scales on the forewings of males. Some are almost covered with black, whereas these scales in other specimens are less prominent, leaving the black costal block more visible, but males are never as contrasting as females. The amount of pink scales on the labial palps, forewings and legs is variable, but less visible in worn specimens. Some specimens have a faint, light postmedial fascia.

Similar species. *S. costella* is, together with *S. hyoscyamella* q.v., characterized by the large, black, subtriangular costal block. Males especially can resemble some forms of *S. ocellatella* but, like all other European *Scrobipalpa*, it has no pink scales on the labial palps, forewings and legs.

MALE GENITALIA.– Coremata present; uncus comparatively slender, apically rounded; gnathos-hook strong; anterior margin of tegumen with comparatively shallow emargination; valva slightly shorter than uncus, weakly curved, with indistinctly inflated apex; sacculus long and slender, about four times as long as wide, gradually tapered towards pointed apex; vinculum strongly projected posteriorly; vincular process slightly shorter than sacculus, subtriangular, with narrow apex, continuous with shallow V-shaped medial emargination of posterior margin of vinculum; saccus broad at base, distal part strongly tapering to rounded apex, considerably exceeding apex of pedunculus; phallus comparatively long and slender, coecum weakly inflated. Differs from the most closely related *S. hyoscyamella* particularly in the more slender sacculus and the broader and longer saccus.

FEMALE GENITALIA.– Segment VIII about same length and breadth; subgenital plate laterally and posteriorly smooth, proximal depression densely covered with small-sized foamy sculpture, inwardly bordered by indistinct and straight edge, anteriorly a short transverse edge; ventromedial depression elongated with digitate anteriolateral projections, indistinctly separated from subgenital plate, posteriomedial area covered with microtrichia, anteriolateral part with small-sized foamy sculpture, not reaching smooth and rounded, medially emarginated anterior edge of segment VIII; apophysis anterioris with abrupt transition to anterior edge of sternite VIII, slightly longer than segment VIII; colliculum moderately long; ductus bursae long, with abrupt transition to clearly defined corpus bursae; corpus bursae almost spherical; signum on the right side near entrance of corpus bursae; signum-hook large and stout, weakly curved, with numerous strong spines on concave side almost to apex.

DISTRIBUTION.– Only known from north-western Europe, from Great Britain, Ireland and France to The Netherlands and Denmark, recently also found in Germany (Seliger, 2008: 45). Records from Austria (Elsner *et al.*, 1999: 44, pl. 25, fig. 223) are erroneous and refer to *S. hyoscyamella*. Records from Romania (Kovács & Kovács, 2000: 33) are most probably based on the incorrectly identified figure of a male specimen in Elsner *et al.* (*op. cit.*) and those from the Czech Republic (Povolný, 1996a: 114) and from Russia (Ponomarenko, 2008: 96) are dubious and very unlikely. Also a record from Spain (Requena, 2009: 44) is based on Elsner *et al.* (1999) and referable to *S. hyoscyamella* (Requena, *in litt.*). However, we know of one correctly identified specimen from Spain.

New record. Spain: 1 ♂, Gerona, Montseny by Coll de Rabell, 800 m, 13.vii.1988, leg. Fibiger, gen. slide 4010 Hendriksen (ZMUC).

BIONOMICS.– The larva has been recorded with certainty from *Solanum dulcamara* L. (Solanaceae) by several authors (Jansen, 1999: 53), whereas *Solanum tuberosum*

L. needs confirmation. It initially causes a large, slightly puckered, contorted blotch mine, usually near the leaf edge. Frass is irregularly deposited in the mine. Later the mine can be changed or the larva feeds between spun leaves, fruits or stems, frequently also in the stem (Schütze, 1931: 171; Stainton, 1867a: 78-81, pl. 9, fig. 3 a, b). The fully grown larva pupates between spun stems and leaves or on sandy ground surface. The larval stage was observed in a single generation from mid-September to June (Jansen, *op. cit.*), according to Schütze (1931, *op. cit*) in July and August. The adult, which is frequently observed at light, occurs from the middle of July to early October, exceptionally until mid-November. However in Britain there are records from February and March to November (Bland *et al.*, 2002) and even into December (Corley, *in litt.*). This suggests that it can hibernate in either larval, pupal or adult stage. The reported habitats are sandy dunes along coasts, but in Britain where the species is widespread and common it occurs inland in hedges and gardens, as well as on coastal shingle (Bland *et al.*, *op. cit*).

REMARKS.– *G. costimaculella* was unnecessarily proposed as a replacement name by Bruand d'Uzelle (1859: 652) who considered *Anacampsis costella* to be preoccupied by *Ypsolopha costella* Fabricius, 1775.

94. Scrobipalpa hyoscyamella (Stainton, 1869)

Lita hyoscyamella Stainton, 1869: 233.

DIAGNOSIS.– Adult. Wingspan 13-16 mm. Labial palp cream-coloured with pink suffusion; segment 2 mottled with black; segment 3 with two broad, black rings. A small brush of pink scales at lower edge of the eye. Antenna black, indistinctly ringed with light grey. Face cream-coloured; neck brownish grey; thorax and tegula as forewing. Forewing light brown; middle of costa with an oblong, irregular, black patch of one-third to half of the wing breadth; black spots in fold and at 2/5 and 3/5 in middle of the wing surrounded by light brown; apex with black scales. Hindwing grey. Fore- and middle leg with the light rings at tibia pink.

Variation. There is some variation in the amount of black on the costal part of the forewing. A series of specimens from southern Spain in the BMNH (bred from *Hyoscyamus* by Walsingham) differs in having the forewings almost light brown with only little black along costa, as an oblique streak from 1/4, in spots and along termen. Another male from Spain (prov. Zaragoza) does not match this unusual light brown form.

Similar species. *S. hyoscyamella* is similar to *S. costella*, but it can be separated by the more slender, black, costal block. Moreover it has apparently no sexual dimorphism and no pink scales on the forewings. Together with *S. costella* it is unique among European *Scrobipalpa* in having pinkish scales on the labial palps, lower edge of the eyes and on the light rings on the tibia of the fore- and midlegs (a character also found in *Cosmardia moritzella* and in a few species of *Caryocolum*).

MALE GENITALIA.– Coremata present; uncus comparatively broad, apically tapered, rounded; gnathos-hook strong; anterior margin of tegumen with comparatively deep emargination; valva slightly shorter than uncus, curved, with distinctly inflated apex; sacculus long and broad, about three times as long as wide, sickle-shaped, curved inwards and tapered towards pointed apex; vinculum strongly projected posteriorly; vincular process distinctly shorter than sacculus, broad, tapered towards apex; posterior margin of vinculum with shallow V-shaped medial emargination; saccus broad at base, distal part slender, slightly exceeding apex of pedunculus; phallus comparatively short and slender, coecum weakly inflated. The shape of the sacculus and the vinculum is characteristic and distinguishes *S. hyoscyamella* from related species such as *S. costella*.

FEMALE GENITALIA.– Segment VIII about same length and breadth; subgenital plate predominantly smooth, anterior margin a transverse edge, inner margin straight, anterioproximal area with small-sized foamy sculpture; ventromedial depression largely reduced, membranous; apophysis anterioris with abrupt transition to anterior edge of sternite VIII, slightly longer than segment VIII; colliculum short; ductus bursae long, gradually widened, with distinct transition to clearly defined corpus bursae; corpus bursae oval; signum on the right side near entrance of corpus bursae; basal plate of signum small, signum-hook short and stout, curved, with few basal spines on concave side.

DISTRIBUTION.– Only known from Spain, southern France, south-eastern Austria and probably Romania, recently also found in Portugal (Corley *et al.*, 2007: 327). The female figured by Povolný (2002a: pl. 10, fig. 19, pl. 75, fig. 786) and located from Jersey, Great Britain, is correctly identified but the locality is doubtful and needs confirmation (Bland *et al.*, 2002: 184). Even Povolný (2002a: 47) doubted an occurrence in England. Most if not all of the records from Russia and the Ukraine (Povolný, *op. cit.*; Ponomarenko, 2008: 96, 328) are probably based on misidentification of *S. hyoscyamivora* which is a junior subjective synonym of *S. indignella* q.v.

BIONOMICS.– The larva has been recorded with certainty from *Hyoscyamus* L. (Stainton, 1869: 233), *Solanum dulcamara* L. (Corley *et al.*, 2007: 327) and doubtfully from *Solanum lyratum* Thunb. (Solanaceae) (Povolný, 1990b: 196). According to Corley *et al.* (*op. cit.*) the larva was found in September between spun leaves and among fruits. However the larval habits are insufficiently described due to confusion with *S. costella* (Povolný, 2002a: 47). The adult occurs from October to March and hibernates in this stage, but Corley *et al.* (*op. cit.*) recorded adults in July so the species may be bivoltine. The adult is attracted to light.

REMARKS.– Povolný (2002a: 47) incorrectly synonymized *Gnorimoschema hyoscyamivora* Gerasimov, 1940, with *S. hyoscyamella*. This species is in fact a synonym of *S. indignella* q.v.

We examined a male specimen from the Ukraine (in ZMKU) with almost identical genitalia compared to *S. hyoscyamella*. The moth, however, is blackish grey with diffuse black spots on the forewing. It is uncertain if it represents a unicolorous dark form of *S. hyoscyamella* or an additional species. This can probably only be solved when the female becomes known.

95. Scrobipalpa portosanctana (Stainton, 1859)

Gelechia portosanctana Stainton, 1859: 212.
Gelechia eremaula Meyrick, 1891: 57, **syn. n.**
Gelechia lyciella Walsingham, 1900: 217, **syn. n.**
Lita desertella Rebel, 1901: 165, **syn. n.**
Phthorimaea bertramella Lucas, 1940: 228, **syn. n.**
Teleia leroyella Lucas 1950: 143, **syn. n.**

Gnorimoschema reisseri Povolný & Gregor, 1955: 85, fig. 4, **syn. n.**
Gnorimoschema philolycii Hering, 1957: (1): 644 (key), (3): 26-27, 138, pl. 45, fig. 388, 220, pl. 86, figs. 724-725, **syn. n.**
Gelechia gallincolella auctt., (*nec* Mann, 1872); misidentification.

DIAGNOSIS.– Adult. Wingspan 10-13 mm. Segment 2 of labial palp fuscous with cream-coloured upper and inner surface and a light ring at basal part; segment 3 fuscous with light patch at middle of inner surface. Antenna black with lighter rings. Head cream-coloured with light grey-brown neck; thorax and tegula as forewing. Forewing light grey-brown mottled with lighter and darker scales; a distinct, small, black patch on costa 1/5 from base, fused with a larger but less distinct black patch in middle of the wing 1/4 from base; black spots in fold and at 1/3 and 2/3 surrounded by brown scales; light costal and tornal spots indistinct, hardly forming a fascia; apical area with many black scales. Hindwing light grey, darker in apical part.

Variation. The forewing colour varies and can be light grey-brown, warm brown, grey or blackish brown. Some specimens have a white line running from near base of costa into the fold.

Similar species. The small distinct, black patch at costa near base and the, though variable, brownish forewings are characteristic for *S. portosanctana*. *S. vicaria* q.v.

MALE GENITALIA.– Uncus long, distally slightly tapered, apex with weak concave emargination; gnathos-hook strong, moderately short; valva slightly sigmoid in medial part, projecting slightly above tip of uncus, apically without distinct inflation; sacculus long and slender, about four times as long as wide, apically narrowing to inwardly turned tip, outer margin convex in apical part; vinculum strongly projected posteriad, with long and sickle-shaped vincular process, extending to about three-quarters length of sacculus, apically with outwardly directed tooth, weakly separated from sacculus by narrow gap; posterior margin of vinculum with shallow ovate medial emargination; saccus stout, moderately broad, apex with indistinct transverse sclerite, projecting slightly beyond apex of pedunculus; phallus comparatively long, with inflated coecum. The male genitalia are somewhat similar to those of *S. erichi* but differ in many characters such as the unique sacculus and vinculum.

FEMALE GENITALIA.– Segment VIII slightly broader than long, lateral margin slightly excavated; sternite VIII with transverse, sclerotized anterior margin; subgenital plate predominantly smooth, inner margin longitudinal and straight, depression with inwardly directed hump at anterior margin of sternite VIII, covered with small-sized foamy sculpture in this part; ventromedial depression anteriolaterally drop-shaped, medial part with some microtrichia, anteriomedial incision small; subostial sclerites broadly rounded and almost joined in middle; apophysis anterioris rod-like, abruptly broad-based, about length of segment VIII; colliculum short, closely attached to segment VIII; ductus bursae short, with gradual transition to large, pyriform corpus bursae; signum on the right side of corpus bursae; signum-hook large and stout, strongly curved at an almost right angle in basal half without additional spines, basal plate small. Very similar to *S. vicaria* but differs in the straight inner margin of the subgenital plate, the comparatively reduced foamy sculpture sculpture and the broadly rounded subostial sclerites.

DISTRIBUTION.– Local records from the Mediterranean part of Europe, ranging from Spain and southern France to Italy and Croatia; Madeira. More widely distributed in the Maghreb countries and the Near East.

New records. France (Corsica): 1 ♀, 4 km N Solenzara, Marine de Solaro, 3 m,

21.ix.2003, leg. Skou (ZMUC). Malta: 1 ♂, Mellieha Bay, Ghadira, 19.-31.viii.1992, leg. Petersen, gen. slide 2797 Hendriksen (ZMUC).

BIONOMICS.– The larva has been found feeding on *Lycium europaeum* L. (Walsingham, 1900: 217) and *L. barbarum* L. (Solanaceae) (Povolný, 1971a: 37). A record from galls of *Tamarix* L. (Tamaricaceae) (Mann, 1872: 38) refers to another species (see below). According to Povolný (2002a: 43) the species was also found in mines of *Rhamnus palaestina* Boiss. (Rhamnaceae) by Hering (1957) but in fact the latter author only bred the type-material of *Gnorimoschema philolycii*, currently considered a junior synonym of *S. portosanctana*, from *Lycium halimifolium* (Hering, 1957: 27). The larva inititally causes a large blotch-mine in the leaf, later it feeds between spun terminal shoots. It has been observed from May to October indicating polyvoltinism (Povolný, 2002a: 43). Pupation takes place outside the larval feeding place.

REMARKS.– *Gelechia portosanctana* was described from an unspecified number of specimens collected by T. V. Wollaston from the Madeiran island of Porto Santo. In BMNH are two syntypes, both without abdomen. A male labelled 'Lectotype' 'Madeira 1858, Wollaston, B.M. 1858-21' 'XXIV' '*Gelechia portosanctana* Stn., Ann. Mag. N. H. (3-5) III: 212, no. 16 (1859), Porto Santo, Type XXIV HD' 'Lectotype ♂ *Gelechia portosanctana*, select. K. Sattler 1961' is here designated as the lectotype. Even though the two syntypes are without abdomen their identity could easily be confirmed as only four species of *Scrobipalpa* occur in Porto Santo, viz. *S. ergasima*, *S. ocellatella*, *S. vasconiella*, *S. gallincolella*. The latter name has been in use for this species since Povolný (1964: 356), but *S. portosanctana* was still correctly used by Passos de Carvalho (1995: 566, 578).

A lectotype of *Gelechia eremaula* Meyrick was published by Clarke (1969: 150-151).

Gelechia lyciella was described from four specimens bred from *Lycium europaeum* in southern France (Walsingham, 1900: 217). Povolný (1966a: 133) claimed to figure the male genitalia of the holotype, but Walsingham did not select a holotype, only a male and a female type. According to ICZN (1999: Art. 74.5) that does not constitute a valid lectotype designation. In order to correct this and to preserve stability of nomenclature, we are here designating a lectotype, which was already labelled as such in the BMNH. It is labelled: 'Lectotype' '*Lycium europ*. 22.v. Perpigan Pyr. Or. France ex 14.vi.1899 Wals. 84427' 'Walsingham Collection 1910-427' '*Gelechia lyciella* Wlsh. Ent. Mo. Mag. XXXVI No. 1916(2) (1900) Type male descr.' 'BM ♂ Genitalia slide slide No. 7292'.

Gnorimoschema philolycii was described from a male holotype and a female paratype and the genitalia of both specimens are figured in detail in the original description (Hering, 1957: pl. 86, figs. 725-725). Later this species was synonymized with *S. gallincolella* sensu auctt. (Povolný, 1964: 356). Hering (1957: 27, 644) points to differences in the larvae which is grey with dark longitudinal lines and a blackish head and neck-shield in *G. philolycii* whereas the larva of *Gnorimoschema lyciella* (currently considered a junior synonym of *S. portosanctana*) is described as green with five reddish longitudinal lines and a reddish-brown head and neck-shield. Furthermore differences in the larval habits are described by this author. The larva of *G. philolycii* mines in 2-3 leaves which are gently spun together whereas the larva of *G. lyciella* mines several spun terminal leaves. However, the latter habits are probably described from material from North Africa since Hering mentions *G. lyciella* only from this area and hence may be attributed to a different species. The differences in the larva and its biology indicate two different species but that should be proved with new breedings.

Gelechia gallincolella was described from several specimens bred from galls of *Tamarix gallica* L. which were collected in the area of Spalato (nowadays Split, Croatia) in summer 1868. The type-material emerged in late September of the same year (Mann, 1872: 38). Povolný (1964: 356) examined two male specimen from the NHMW and figured and fixed a male from Ragusa as the lectotype (*op. cit.*, 342, pl. 5, fig. 26). We have been able to examine the genitalia slide of this male whereas the adult could not be traced. The genitalia agree with *S. portosanctana* but according to it´s collecting locality Ragusa (nowadays Dubrovnik, Croatia) it cannot be taken as part of the type series. According to ICZN (1999: Art. 74.2) a lectotype found not to have been a syntype loses its status. The second specimen examined by Povolný was collected at the type locality Spalato and according to the cited labels it should be one of the syntypes. However, this specimen was not figured by Povolný (*op. cit.*) and we have not been able to trace it in NHMW. The later published figure of the male genitalia of a specimen allegedly originating from Split (Povolný, 2002a: pl. 35, fig. 311) is in fact nothing but the lectotype of *Gelechia lyciella* from France already figured by Povolný himself in an earlier paper (1966a: 130, fig. 11). Since the original description of *Gelechia gallincolella* clearly contradicts the conspecifity with *S. portosanctana* and rather refers to a species of Teleiodini than to a Gnorimoschemini the hitherto accepted synonymy is doubtful and the true identity of the species cannot be stated for the time being.

Phthorimaea bertramella, described by Lucas (1940: 228) from Algeria has recently been considered as a species of *Scrobipalpa* (Sattler, 1985: 230). Unfortunately the genitalia of this species have been inadvertently mixed with a tortricid and the validity of *P. bertramella* remained doubtful. According to Povolný (2002a: 34, pl. 5, fig. 4) who figured the adult in colour, the species belongs to the *S. salinella*-complex. At the BMNH we were able to examine a colour transparency of the holotype belonging to the MNHN, and based on that we consider *bertramella* to be a synonym of *S. portosanctana*.

Teleia leroyella was described from an unspecified number of specimens from North Africa without further details. In the MNHN we were able to examine a male without abdomen labelled: 'Type' 'Tunis juin 1946' '*Lita leroyella* Dan. Lucas' 'Type, descrit. de le Buletin de la Ste' Entomologui de France en 1948'. The specimen matches *S. portosanctana* with which we here synonymize it.

96. Scrobipalpa vicaria (Meyrick, 1921)

Phthorimaea vicaria Meyrick, 1921: 74.
Scrobipalpa tineiformis Povolný, 1967a: 230, pl. 14, fig. 99, pl. 16, fig. 16.

DIAGNOSIS.– Adult. Wingspan 8-11 mm. Labial palp black, mottled with cream especially on upper and inner surface; segment 2 with light ring before middle and at tip; segment 3 with light ring at middle. Antenna black, more or less distinctly ringed lighter. Head cream-coloured with light grey-brown neck; thorax and tegula as forewing. Forewing brownish mottled with light and darker scales; a distinct, small, black patch on costa 1/5 from base, fused with a larger but less distinct black patch in middle of the wing 1/4 from base; black spots in fold and at 1/3 and 2/3 surrounded by brown scales; black patches at costa at half and 3/4; light costal and tornal spots indistinct, hardly forming a fascia; apical area with many black scales. Hindwing light grey, darker in apical part.

Variation. The forewing colour varies from light grey-brown to orange-brown, sometimes with many black scales. Specimens from the Atlantic coast of Morocco tend to be more orange-brown compared with more light grey-brown specimens from Spain.

Similar species. *S. portosanctana* is on average larger and has more uniformly brown forewings with less black patches, especially along the costa. The labial palps of *S. vicaria* are lighter than in *S. portosanctana*, with more distinct, lighter rings.

MALE GENITALIA.– Uncus long, distally strongly tapered, apex rounded; gnathos-hook moderately strong, short; valva almost straight, extending far beyond tip of uncus, apically without distinct inflation; sacculus long and slender, distal part strongly narrowing to inwardly pointed apex, outer margin convex in apical part; vinculum strongly projected posteriad, with moderately long and sickle-shaped vincular process, extending to about three-quarters length of sacculus, apically with outwardly directed tip, separated from sacculus by narrow gap; posterior margin of vinculum with shallow, ovate medial emargination; saccus stout, moderately broad, apex subrectangular, projecting distinctly beyond apex of pedunculus; phallus slender, with distinctly inflated coecum.

FEMALE GENITALIA.– Segment VIII slightly broader than long; sternite VIII with transverse, sclerotized anterior margin; subgenital plate predominantly smooth, inner margin weakly S-curved, with concave emargination anteriorly, depression with inwardly directed hump at anterior margin of sternite VIII, laterally covered with foamy sculpture; ventromedial depression posteriorly narrow, anteriorly drop-shaped, few microtrichia present, anteriolateral part with foamy sculpture, anteriomedial incision small; subostial sclerites very broad, pointed and almost joined in middle; apophysis anterioris moderately stout, gradually separated, almost length of segment VIII; colliculum short, closely attached to segment VIII; ductus bursae short, with gradual transition to large, pyriform corpus bursae; signum on the right side of corpus bursae; signum-hook moderately large and stout, strongly curved at an almost right angle in basal half with minute spines near base and small basal plate. Differs from the very similar *S. portosanctana* e.g. by the curved inner margin of the subgenital plate, the medially expanded foamy sculpture, the broadly rounded subostial sclerites and the smaller signum-hook.

DISTRIBUTION.– Unpublished records from southern Spain and Greece (Crete) and furthermore from the Canary Islands, Malta, Morocco, Libya and Cyprus. Widely distributed from southern Africa to Saudia Arabia, Iran and Pakistan (Povolný, 1971: 41, 2002a: 76). *S. vicaria* is probably a relatively new element in the European fauna.

New records. Spain: 1 ♂, Almeria, Carboneras, 10 m, 16.x.1992, leg. Fibiger, gen. slide 6277 Hendriksen (ZMUC). Numerous additional specimens from Spain in ZMUC. Cyprus: 2 ♂, 3 km W Dekeleia, 50 m, 9.v.2007, leg. Skule, gen. slide Hendriksen 6285, 6501 (ZMUC). Malta: 1 ♂, Rabat, 15.v.1999, leg. & coll. Sammut, gen. slide 5289 Hendriksen. Greece (Crete): 1 ♀, Mora Spakion, 30 m, 21.iv.1995, leg. Fibiger, gen. slide Hendriksen 3044 (ZMUC). Morocco: 7 ♂, 7 ♀, 8 km S Sidi Ifni, 50 m, 20.iii.2005, leg. Karsholt, gen. slide 5416 Hendriksen (ZMUC). Libya: 1

♂, Gharian, Wadi El Hira, 1.vi.1983, leg. Seneca, gen. slide 2907 Hendriksen (ZMUC). Spain (Canary Islands): 1 ♂, Gomera, Playa de Santiago, sea lev., 22.i.1981, leg. Stadel Nielsen, gen. slide 3084 Hendriksen (ZMUC).

BIONOMICS.– Host-plant and early stages unknown. Povolný (2002a: 76) suspects that the larva may be a miner of Solanaceae. The adult has been observed from January to July and from September to early November indicating at least bivoltinism. However, voltinism in Europe is unknown to a large extent. Preferred habitats are arid biotopes and *S. vicaria* primarily occurs in semi-deserts.

REMARKS.– The female genitalia figured by Povolný (2002a: figs. 784-785) are partially misidentified and fig. 785 is nothing but the original figure of female *S. laisinca* (Povolný, 1976: fig. 25).

97. *Scrobipalpa ocellatella* (Boyd, 1858)

Gelechia ocellatella Boyd, 1858: 143.
Gelechia ocellatella Stainton, 1859: 212; homonym and synonym of *Gelechia ocellatella* Boyd, 1858.
Gelechia submissella Stainton, 1859: 212.
Gelechia horticolella Rössler, 1866: 240.
Lita atriplicella var. *clarella* Caradja, 1920: 100.
Lita ocellatella ssp. *obscurior* Rebel, 1927: 187.
Gnorimoschema ocellatellum orientale Gregor & Povolný, 1954: 91, pl. 6, fig. 7, 8, pl. 7, fig. 4.
Gelechia portosanctana auctt., (*nec* Stainton, 1859); misidentification.

DIAGNOSIS.– Adult. Wingspan 11-14 mm. Labial palp cream-coloured; segment 2 with two indistinct and segment 3 with two distinct black rings. Antenna black, indistinctly ringed with grey. Head cream-coloured, neck mottled with grey and brown; thorax and tegula as forewing. Forewing variegated black and light or yellowish brown; black areas especially at base, as a block in costal two-thirds, in middle of wing and on the apex; black spots distinctly surrounded by light brown scales; costal and tornal spots light, nearly fused; a light, warm brown patch from before tornus to middle of wing. Hindwing light grey.

Variation. A rather variable species, especially in the amount of black scales on the forewing. Most specimens have the dorsal third lighter than the costal two-thirds. Rarely the forewings are almost completely black (f. *obscurior* Rebel).

Similar species. Due to its variability and wide distribution area *S. ocellatella* can be confused with a number of other *Scrobipalpa* species. *S. atriplicella* is of similar size, but has the dorsum concolorous with the forewing. From species with a lighter dorsum like, e.g. *S. suaedella*, *S. ocellatella* can normally be recognized by the light, warm brown patch from before tornus to middle of wing. *S. costella* q.v.

MALE GENITALIA.– Coremata absent; uncus broad with ventrolateral fold, apex truncate with strongly concave posterior margin; gnathos-hook distinct, short; tegumen deeply emarginate anteriorly; valva extending almost to apex of uncus, strongly bent at about one-third, distal part straight but directed inwards, weakly curved, with slightly inflated apex; sacculus broad and moderately long, same width or slightly widened from base to apical part, apex with broadly rounded outer margin and with distinct inwardly directed tooth at inner margin, gap between sacculus and vincular process broader than sacculus; vincular process shorter than sacculus, broadly projected posteriad, strongly narrowing towards outwardly directed tooth; posterior margin of vinculum with moderately narrow and deep emargination; saccus short and slender, strongly tapered towards apex, anteriorly extending to about apex of pedunculus; phallus short, coecum weakly inflated. The shape of the uncus and valva make this species unmistakable.

FEMALE GENITALIA.– Segment VIII of about same length and width; anterior margin of sternite VIII medially with distinct membranous opening; subgenital plate with oblique posterior edge, broadly convex proximally, patch like-depression of foamy sculpture posteriad of apophysis anterioris, extending to medial depression margin, band-like zone parallel to inner margin coverd with microtrichia; ventromedial depressions separated by V-shaped medial gap, densely covered with small-sized foamy sculpture in anterior third, foamy sculpture larger but increasingly replaced by microtrichia posteriorly; apophysis anterioris rod-like with abrupt transition to anterior edge of sternite VIII, at most length of segment VIII; colliculum short; ductus bursae rather short and broad, indistinct transition to small corpus bursae; signum on the right side of corpus bursae, basal plate small without additional tooth; signum-hook short, slender, strongly curved at an almost right angle in basal half.

DISTRIBUTION.– Widely distributed from Portugal to Greece and southern Russia. Very local records from inland localities of Central Europe. In the North-West along the coasts of the Atlantic Ocean and North Sea from Ireland to Denmark. In the southern Mediterranean recorded from the Canary Islands through North Africa to the Middle East.

New record. Serbia: 2 ♂, 7 ♀, Bačka Topola, 23.viii.1986, leg. Baraniak (ZMUC). Greece (Crete): 1 ♂, 7 km ESE Rethimno, Maroulas, 11-17.x.2003, leg. Skule (ZMUC). European Turkey: 2 ♂, 1 ♀, Kiyikoi, 26.ix.1989, leg. Baraniak (ZMUC).

BIONOMICS.– The larva has been recorded especially from *Beta vulgaris maritima* (L.) Arcang., and as a pest species on cultivated forms of *Beta vulgaris* L. (Chenopodiaceae). Other host-plant records include *Halimione portulacoides* (L.) Aellen, *Suaeda maritima* (L.) Dum. (= *Chenopodium maritimum* L.), *Suaeda vera* Forssk. ex J. Gmelin, *Salicornia europaea* agg. (Chenopodiaceae) and *Camphorosma monspeliaca* L. (Chenopodiaceae) (Gregor & Povolný, 1954: 91; Jansen, 1999: 68; Povolný, 1990b: 195). Up to eighty eggs are laid singly or in groups on the leaves, especially on the petioles close to the stem (Hammad *et al.*, 1968). The larva initially mines in the midrib, later the leaves are spun together with a silken web. Both the larva and the pupa have been reported as hibernating stage. However, in Europe the larva seems to overwinter. Pupation takes place on the ground surface within a loose silken cocoon spun among debris. The voltinism varies, depending on climatic conditions. *S. ocellatella* occurs in up to 5 generations in the Mediterranean and the Near East (Povolný, 2002a: 59), whereas it is bivoltine in the northern parts of Europe (Jansen, 1999: 68). Adults can be found almost throughout the year in subtropical and Mediterranean areas but only from June to September in central and northern Europe. In north-western Europe the species only occurs along the coast but preferably in almost non-saline environment at the edge of the dunes and it has never been recorded from inland arable fields with sugar beet (Jansen, *op. cit.*). However, in North Africa and the Mediterranean it is a serious pest of plantations of *Beta vulgaris* L.

REMARKS.– *Lita atriplicella* var. *clarella* was considered as an unavailable, infrasubspecific name and furthermore synonymized with *S. ocellatella* based on examination of the type-material (Park, 1996: 63). However, the name was unambiguously used subspecifically for populations from southern Spain and Algeria (Caradja, 1920: 100-101) and it is therefore available (IZCN, Art. 45.6.4).

Gelechia portosanctana Stainton, 1859 was erroneously treated as a synonym of *S. ocellatella* by Povolný (2002a: 59).

98. *Scrobipalpa pulchra* Povolný, 1967

Scrobipalpa (Euscrobipalpa) pulchra Povolný, 1967a: 223, pl. 9, figs. 64-68, pl.14, fig. 8.

DIAGNOSIS.– Adult. Wingspan 10-13 mm. Labial palpus white; segment 2 mottled with some brown, especially on outer surface; segment 3 with a slender, dark brown subbasal ring and a broader subapical one. Head cream white; thorax and tegula concolorous with forewing. Antenna greyish black, lighter ringed, especially near base. Forewing light yellowish brown with many white, blackish-tipped scales, especially in middle and subcostal part from base to two-thirds; one plical and two discal spots surrounded by brownish yellow scales; apical area yellowish brown, with black scales along the oblique termen; cilia light grey with black fringe line. Hindwing whitish grey, with grey cilia.

Variation. There is some variation in the black markings of the forewing. Specimens from Central Asia are darker, greyish.

Similar species. A rather characteristic species. The white ground colour of the labial palps and the light yellowish brown forewings, with its contrasting blackish-grey area in the middle of the wing are characteristic for *S. pulchra*, together with the oblique termen, which gives this species a pointed apex.

MALE GENITALIA.– Uncus long, distally weakly rounded, apex truncate with indistinct emargination; gnathos-hook strong, moderately short; valva extending almost to apex of uncus, weakly curved, apically slightly inflated; sacculus very long and slender, about four times as long as wide, widened below apex at about four-fifths, apically tapered to inwardly turned tip, outer margin convex in apical part; vincular process slightly projected posteriad, subtriangular shape without apical tooth, extending to about one-third length of sacculus, separated from sacculus by distinct gap; posterior margin of vinculum with moderately deep V-shaped emargination; saccus moderately broad basally, distal part strongly tapered to slender apex, projecting beyond apex of pedunculus; phallus stout, with inflated coecum. The shape of the sacculus and the comparatively short vincular process are distinctive characters.

FEMALE GENITALIA.– Segment VIII slightly broader than long; subgenital plate largely covered with granulated sculpture, extending into basal part of apophysis anterioris, posteriolateral margins smooth; ventromedial depression scarcely demarcated from subgenital plate, lateral and anterior part with granulated sculpture, anterior edge weakly convex, extending slightly beyond anterior edge of segment VIII, anteriomedial edge sclerotized, depressions separated by short incision, posterior half merged; subostial sclerites narrow, pointed; apophysis anterioris rod-like, with gradual transition to anterior edge of sternite VIII, slightly longer than segment VIII; colliculum short and small; ductus bursae long and broad, with abrupt transition to oval corpus bursae; signum on the right side of corpus bursae, signum-hook large and massive, weakly curved, basal third with several additional spines, basal plate large.

Distribution.– Southern Ural Mountains of Russia (Junnilainen *et al.*, 2010: 49) and south-eastern Latvia. Outside Europe known from the Middle East to Central Asia and Mongolia. Recently recorded from China (Bidzilya & Li, 2010).The single Latvian specimen was collected on a dry meadow, but it may have arrived with wind borne migration from south-east (Savenkov, pers. comm.).

New record. Israel: 1 ♂, 1 ♀, Jericho, Kalia, v.2004, leg. Kravchenko, gen. slide 5495 Hendriksen (TAU, ZMUC).

Bionomics.– The larva has been recorded from halophytic Chenopodiaceae: *Haloxylon persicum* Bunge, *Climacoptera crassa* (Bieb.) Botsch., *Gamanthus gamocarpus* (Moq.) Bunge (Falkovitsh & Bidzilya, 2003: 118), *Haloxylon ammodendron* (C. A. Meyer) Bunge and *Halothamnus subaphyllus* (C. A. Meyer) Botsch. (Falkovitsh & Bidzilya, 2006: 76). The pupa hibernates. Adults emerged from the end of April to late Ocotber, normally in not less than three generations (Falkovitsh & Bidzilya, 2006: 76). *S. pulchra* seems to be restricted to deserts with halophytic vegetation up to an altitude of about 1700 m.

99. Scrobipalpa gecko (Walsingham, 1911)

Lita gecko Walsingham, 1911: 12.

Diagnosis.– Adult. Wingspan 10 mm. Segment 2 of labial palp cream-white with scattered blackish brown scales; segment 3 white with two black rings. Antenna distinctly ringed black and white. Head, thorax and tegula white, mottled with some brownish and black scales. Forewing with white brownish or black-tipped scales, especially along costa and termen; at 1/3 and 2/3 black, elongate spots surrounded by orange-brown. Hindwing white with light grey cilia.

Variation. Only one specimen was examined.

Similar species. The white-grey forewings with elongate, black spots surrounded by orange-brown, and the white hindwings are characteristic for *S. gecko*.

Male genitalia.– Uncus long, apically tapered, rounded; gnathos-hook stout; valva long, slender, extending to about tip of uncus, slightly curved, distal part broader without distinct apical inflation; sacculus strong, long and broad, apically narrowing to inwardly turned tip, outer margin convex in apical part; vinculum projected posteriad, with small and finger-shaped vincular process, extending to about half length of sacculus, weakly separated from sacculus by narrow gap; posterior margin of vinculum with moderate V-shaped medial emargination; saccus stout, moderately broad, evenly tapered towards pointed apex, projecting beyond apex of pedunculus; phallus stout, with distinct apical hook, coecum moderately inflated.

Female genitalia.– Segment VIII longish, about same length and breadth, anterior margin with broad V-shaped emargination from middle of segment to base of apophysis anterioris; subgenital plate with weakly concave inner edge, anterior part with numerous short folds, small zone of foamy sculpture beside inner edge; longitudinal depression covered with minute microtrichia, anterior half with large and smooth sclerotization, anteriodistally bifurcate and pointed, medially rounded and merged; apophysis anterioris rod-like, slightly longer than segment VIII; colliculum short; ductus bursae long, narrow basal part evenly dilated to well separated oval corpus bursae; signum-hook slender, curved in middle, few additional spines at basal third, basal plate small.

DISTRIBUTION.– Described from Algeria and later recorded from Tunisia and the Middle East (Povolný, 2002a: 44). One unpublished specimen known from southern Greece.

New record. Greece: 1 ♀, Lakonia, 5 km S of Monemvasia, 18.iv.1981, leg. Skule, gen. slide GU 97/689 ♀ Huemer (ZMUC).

BIONOMICS.– The larva mines 'the leading articulations' of *Anabasis articulata* (Forskål) Moq. (Chenopodiaceae), a characteristic plant of desert sand dunes, in late December and January (Walsingham, 1911: 13). Pupation takes place outside the mine. The adult has been observed in April.

100. *Scrobipalpa hannemanni* Povolný, 1966

Scrobipalpa hannemanni Povolný, 1966b: 402, figs. 22, 5.
Scrobipalpa hannemanni furva Povolný, 1969a: 12, figs. 50-52, pl. 30, fig. 13. Subspecies.
Euscrobipalpa hannemanni gamanthi Falkovitsh & Bidzilya, 2006: 77. Subspecies.

DIAGNOSIS.- Adult. Wingspan 10-11 mm. Labial palp white, more or less mottled with black, especially on segment 3. Antenna black, ringed with light grey. Face cream-white, neck, thorax and tegula as forewings. Forewing grey from dark grey, light-based scales and mottled with some black and whitish yellow; base with many black scales; a distinct black spot on costa at 1/5; black spots in fold and at 2/5 and 3/5 in middle of wing only faintly surrounded with light brown; an indistinct light post-medial fascia; apical area with many black scales. Hindwing light grey, with grey margins.

Variation. There is some variation from lighter to darker grey on the forewings. The light scales can be almost white or light yellowish. The black spot at 3/5 in middle of the wing is sometimes divided into two, or it can be angulated.

Similar species. *S. hannemanni* is recognized by its relatively small size, its light grey forewings, and especially by its distinct black spot at 1/5 on costa. *S. erichi* has more variegated forewings and lighter hindwings.

MALE GENITALIA.- Uncus moderately broad, apex rounded; gnathos-hook comparatively small; valva long, exceeding uncus, medially curved, with moderately inflated apex; sacculus compact, long and evenly broad, about two times width of valva, outer margin convex, apex with distinct medial tooth; vincular process weakly developed, small, digitate, abruptly passing into moderately shallow and U-shaped medial emargination of posterior margin of vinculum; saccus short, broad at base, evenly tapered towards pointed apex, slightly exceeding apex of pedunculus; phallus comparatively short and slender, coecum strongly inflated. The compact sacculus and characters of the posterior margin of the vinculum make this species unmistakable.

FEMALE GENITALIA.- Segment VIII slightly longer than broad; subgenital plate with longitudinal inner margin, longitudinally sickle-shaped sclerotization, laterally membranous, posterior third of sclerotized zone smooth, anterior two-thirds depressed, covered with conspicuous foamy sculpture except for posterior area with some microtrichia; ventromedial depression narrow zones adjoined to inner margin of subgenital plate without projection, basal two-thirds merged and covered with few microtrichia, anterior third separated by V-shaped emargination, covered with foamy sculpture; apophysis anterioris gradually broad-based, rod-like distal part about length of segment VIII; colliculum short; ductus bursae moderately broad with abrupt transition to oval corpus bursae; signum on the right side at entrance of corpus bursae, with small basal plate; signum-hook large and stout, weakly curved, with numerous spines on concave side up to about middle of hook.

DISTRIBUTION.– In Europe with certainty only known from southern Russia and from the holotype which was probably collected in the area of Split (Croatia) (Povolný, 1966: 403). Outside Europe known from Siberia and Mongolia (Bidzilya, 2009: 5).

BIONOMICS.– Host-plant and early stages of the nominotypical subspecies are unknown. *S. hannemanni gamanthi* was bred from *Gamanthus gamocarpus* (Moq.) Bunge (Chenopodiaceae) (Falkovitsh & Bidzilya, 2006: 77). The holotype of *S. hannemanni* was collected in late May, other specimens have been recorded from June to August, at altitudes from 1120 to 1800 m (Povolný, 1969a: 13), in the southern Urals Mountains already at 200 m.

REMARKS.– *S. hannemanni furva* is regarded as darker and smaller subspecies restricted to Mongolia (Povolný, 2002a: 46). *S. hannemanni gamanthi* is a further subspecies recently described from Uzbekistan (Falkovitsh & Bidzilya, 2006: 77, 2009: 75). According to the original description it differs from the nominotypical subspecies by the larger size and the numerous fuzzy-greyish markings of the forwings. Furthermore the female genitalia are stated to have a broader and shorter signum. The name *hamanthi* used in the abstract of the original description is a misspelling.

101. Scrobipalpa erichi Povolný, 1964

Scrobipalpa erichi Povolný, 1964: 356, pl. 3, fig. 16, pl. 11, fig. 69, fig. 14.

DIAGNOSIS.– Adult. Wingspan 11 mm. Labial palp white; segments 2 and 3 each with two black rings. Antenna black, ringed with light grey. Head cream-white, neck grey; thorax and tegula as forewing. Forewing grey from black, white-based scales and mottled with light yellow, especially around black spots, as a postmedial fascia and in apical area; black spots at costa near base, at 1/4, half and 3/4, in fold and at 2/5 and 3/5 in middle of wing. Hindwing light grey, whitish towards base.

Variation. The few specimens examined by us show only slight variation.

Similar species. The relatively small size, the variegated appearance of the forewings and the light hindwings are characteristic for *S. erichi*. *S. vicaria* has normally light brown forewings and darker hindwings. *S. hannemanni* q.v.

MALE GENITALIA.– Uncus long, distally evenly tapered, apex with weak concave emargination; gnathos-hook very long; valva weakly curved, extending to about tip of uncus, apically without distinct inflation; sacculus comparatively long and broad, about two times as long as wide, distally narrowing to inwardly turned apical tooth, outer margin convex in distal part; vincular process subtriangular, short, extending to about middle of sacculus, sparsely separated from sacculus by small gap; posterior margin of vinculum with moderately shallow V-shaped medial emargination; saccus stout, moderately slender, tapered towards apex, apex with distinct transverse sclerite, distinctly projecting beyond apex of pedunculus; phallus comparatively long, with inflated coecum. The male genitalia are unmistakable within *Scrobipalpa*.

FEMALE GENITALIA.– Segment VIII about same length and breadth, anteriorly widened;

sternite VIII with indistinct, membranous anterior margin; subgenital plate with medially straight and anteriorly convex inner margin, covered with numerous microtrichia medially and increasingly with distinct foamy sculpture in anterior part, particularly along depression near inner edge; ventromedial depression rather narrow and distinctly shorter than subgenital plate, anteriorly evenly joined on anterior margin of subgenital plate, anteriomedially with distinct V-shaped emargination, depression largely covered with small-sized foamy sculpture in anterior part; apophysis anterioris rod-like, abruptly broad-based, shorter than segment VIII; colliculum short; ductus bursae anteriorly dilated; corpus bursae moderately separated from ductus bursae, rather small, oval; signum on the right side at entrance of corpus bursae; signum-hook moderately large, stout, distinctly curved in basal third, with short additional tooth near base, basal plate large.

DISTRIBUTION.– Eastern part of Central Europe, from Austria, Slovakia and the Czech Republic through Macedonia and European Turkey to the Russian Volga-Don region (Ponomarenko, 2008: 96). Outside Europe known from the Middle East to Mongolia and China, although earlier records by, e.g. Li & Yang (1996) refer to a related species (Bidzilya & Li, 2010).

BIONOMICS.– The larva feeds in three to four generations on leaves and flowers, and mines twigs and fruits of *Lycium barbarum* L. (Solanaceae). Also recorded from *Nicotiana tabacum* L. and *Solanum* L. (Elsner *et al.*, 1999: 43-44). Larvae on *Lycium* are twig-borers in April. The following two broods feed on the fruits in June and in August respectively, whereas the fourth generation infests fruits and flower buds from September to October. The larva hibernates in a web hidden among bark (Povolný, 2002a: 41). The adult occurs at light at altitudes up to 1500 m.

REMARKS.– The female genitalia figured by Povolný (2002a: pl. 75, figs. 778-780) are probably a mixture of two or even three species.

102. Scrobipalpa divisella (Rebel, 1936) **sp. rev.**

Lita divisella Rebel, 1936c: 74.

DIAGNOSIS.– Adult. Wingspan 14 mm. Segment 2 of labial palp cream-brown mottled with black on lower and outer surface; segment 3 cream-coloured with two black rings. Antenna black, indistinctly lighter ringed. Head cream-coloured with a light reddish brown tinge; thorax yellowish brown with darker margins; tegula dark brown. Forewing with dorsal third light brownish and costal two-thirds black, the separation between the two areas being waved; subcostal stripe light brown; black spots in fold and at 3/5 in middle of wing; apical area brown. Hindwing light grey.

Variation. The few specimens we have examined show only slight variation.

Similar species. *S. divisella* is characterized by its light head and by having the forewings separated into a light brownish dorsal third and a black costal two-thirds. It is similar to *S. voltinella*, but that species is on average smaller and lacks the black spots at 3/5 on the forewing. However, more material is needed to show if the two species can be separated on external characters. *S. dorsoflava* q.v.

MALE GENITALIA.– Uncus broad, subrectangular; tegumen distinctly widened anteriorly, with broad and moderately shallow emargination; gnathos-hook strong; valva slightly extending beyond uncus, weakly curved with slightly inflated apex; sacculus long and broad, tapering towards pointed apex, outer margin convex; vincular process largely reduced, small and digitate, separated from sacculus by small gap; posterior margin of vinculum with shallow V-shaped medial emargination; saccus broad at base, distally strongly tapered to moderately broad apex, scarcely exceeding apex of pedunculus; phallus short, with weakly inflated coecum. Differs from the similar *S. voltinella* by the distinctly larger sacculus and the smaller digitate process of the vinculum.

FEMALE GENITALIA.– Segment VIII subquadrate; subgenital plate with longitudinally folded inner and posterior edge, posterior and lateral part smooth, anteriomedial depression with large field of distinct foamy sculpture extending far into base of apophysis anterioris; ventromedial depression narrow, digitate, anteriorly weakly projected, membranous posterior part merged and covered with some microtrichia, anterior part covered with foamy sculpture, anteriomedial emargination V-shaped; apophysis anterioris gradually separated, distal two-thirds rod-like, slightly shorter than segment VIII; colliculum small; ductus bursae moderately broad with gradual transition to oval corpus bursae; signum on the right side at entrance of corpus bursae; signum-hook short, stout, curved at an almost right angle in middle, short additional tooth near base, basal plate small.

DISTRIBUTION.– With certainty only known from Spain, outside Europe from Turkey.

BIONOMICS.– Host-plant and early stages unknown. Adults have been collected from September to early November.

REMARKS.– *Lita divisella* was described from three males which were collected in former Syria (Marasch, nowadays Turkey) and in Akschehir (Turkey). Povolný (1967a: 215-216) found that the syntypes belonged to three different species. He selected and figured a lectotype (op. cit., pl. IV, fig. 22) and later he synonymized this species with *S. voltinella* (Povolný, 1981: 390). Despite our efforts we could not trace the lectotype in ZSM and hence our interpretation is solely based on the figure of the lectotype.

S. divisella and *S. voltinella* are indeed a taxonomically very difficult complex of species. Particularly the males are very similar and difficult to separate by genitalia. However, the males we have examined from Spain completely agree with the lectotype of *S. divisella* whereas unique male specimens from Morocco and from Sardinia respectively match neither species. In contrast to the males the female genitalia differ considerably, though we have only been able to examine females of alleged *S. divisella* from Spain and of *S. voltinella* from France and Croatia. Despite these shortcomings we attribute most specimens from Spain to *S. divisella* for the time being (but see below).

103. *Scrobipalpa voltinella* (Chrétien, 1898)

Lita voltinella Chrétien, 1898: 178.

DIAGNOSIS.– Adult. Wingspan 11-12 mm. Labial palp cream-coloured; segment 2 mottled with some blackish on lower and outer surface; segment 3 with two black rings. Antenna black, indistinctly lighter ringed. Face cream-coloured; neck, thorax and tegula light brown. Forewing with dorsal third light orange-brown and costal two-thirds black; subcostal stripe light brown; black spots in fold at 1/3, other spots indistinct; light brown costal and dorsal spots separated by black; apex black; termen with black scales. Hindwing light grey.

Variation. The few examined specimens are rather constant. Some have a black patch in the apical area.

Similar species. *S. voltinella* is characterized by its relatively small size and by having the forewings separated into a light orange-brown dorsal third and a black costal two-thirds. *S. divisella* q.v.; *S. dorsoflava* q.v.

MALE GENITALIA.– Uncus broad, subrectangular, slightly emarginate posteriomedially; tegumen distinctly widened anteriorly, with moderately shallow and broad emargination; gnathos-hook strong, short; valva extending to about tip of uncus, weakly curved with slightly inflated apex; sacculus moderately long and broad, tapering towards pointed apex, outer margin convex; vincular process short, broadly rounded, separated from sacculus by very small gap; posterior margin of vinculum with narrow and deep V-shaped medial emargination; saccus broader at base, distally strongly tapered, scarcely exceeding apex of pedunculus; phallus short, with inflated coecum. *S. divisella* q.v.

FEMALE GENITALIA.– Segment VIII slightly broader than long; subgenital plate largely smooth, inner edge indistinct, anteriomedial depression with large field of foamy sculpture extending to convex anterior margin; ventromedial depression largely reduced to narrow band-like zone adjoining subgenital plate, anteriorly separated by V-shaped emargination; apophysis anterioris rod-like, abruptly broad-based, slightly shorter than segment VIII; colliculum small; ductus bursae moderately narrow with gradual transition to oval corpus bursae; signum on the left side at entrance of corpus bursae; signum-hook moderately short, stout, curved at an almost right angle in basal half, short additional tooth near base, basal plate moderately large.

DISTRIBUTION.– Scattered records from Spain, southern France, Italy, Croatia (Povolný, 1967a: 231; 1996a: 115) and Turkey (Povolný, 2002a: 77). Recently recorded from the Ukraine (Bidzilya & Budashkin, 2004: 64). Records from Algeria (Povolný, 1967a: 231) and from Morocco (Povolný, 1971a: 43) probably refer to other species.

Confirmed record. Spain: 1 ♂, Teruel, Valdevecar Alb, 27.5.1984, leg. Cox, gen. slide 5037 ♂ Hendriksen (coll. Cox).

BIONOMICS.– The larva has been recorded mining the leaves of *Leuzia conifera* (L.) D. C. and *Tanacetum corymbosum* (L.) Schultz (Asteraceae) during March and April. It pupates in a small cocoon constructed of grains on the ground surface. Moths emerged in May (Chrétien, 1898: 178). Field observations date from late April to June and from September, indicating bivoltinism.

REMARKS.– *Lita voltinella* was described from an unspecified number of specimens bred from larvae which were collected in southern France (Chrétien, 1898: 178). Povolný (1967a: pl. XV, figs. 107-108) figured two males from NHMW and ZMHU which he considered as authentic. At a later date the same author designated and figured the female lectotype (Povolný, 1983a: 167-168, fig. 16). We have been able to study two female paralectotypes from MNHN which fully correspond with it. Conversely several other published data are based on misidentifications, e.g. the female genitalia figured by Povolný (1967a: pl. XV, figs. 109-110) refer to *S. suaedicola* whereas genitalia figures in Povolný (2002a: pl. 26, figs. 232-233, pl. 71, fig. 723) depict *S. divisella*. Hitherto published records of *S. voltinella* from Spain (e.g. Povolny, 2002a: 77) refer to *S. divisella*. See also *S. divisella*.

104. Scrobipalpa suaedicola (Mabille, 1906)

Lita suaedicola Mabille, 1906: 54, pl. 3, fig. 5.
Lita suaedicola Amsel, 1939: 72; homonym and synonym of *Lita suaedicola* Mabille, 1906.
Scrobipalpa (Euscrobipalpa) mabillei Povolný, 1969b: 75, figs. 6-8; unnecessary replacement name for *Lita suaedicola* Mabille, 1906.

DIAGNOSIS.– Adult. Wingspan 10-13 mm. Labial palp cream-coloured; segments 2 and 3 each with two black rings. Antenna black, ringed with light grey. Face cream-coloured; neck light grey; thorax and tegula as forewing. Forewing covered with white-based and white-tipped yellow, brown and black scales; black areas especially near base, as an ill-defined costal block and in apical part; black spots in fold and at 2/5 and 3/5 (the latter being double) surrounded by orange; dorsum and large tornal spot orange-brown; costal spot small, light yellow, separated from tornal spot by a small black streak. Hindwing grey.

Variation. There is some variation in the amount of yellow and orange-brown scales on the forewing, and specimens can be contrasting to almost unicolorous. In specimens with much orange-brown the black spots can be more or less covered orange.

Similar species. *S. suaedicola* is characterized by its variegated orange-brown and black forewings. *S. smithi* q.v.; *S. heretica* q.v.

MALE GENITALIA.– Uncus broad, apex slightly truncate with weak medial emargination; gnathos-hook short; valva slender, weakly curved, extending to about tip of uncus or slightly exceeding uncus, slightly curved with weakly inflated apex; sacculus comparatively short and broad, distodorsal part scarcely rounded with weak tip, inner margin nearly straight; vincular process wider and distinctly shorter than sacculus, broadly rounded at apex with indistinct outwardly directed tip; posterior margin of vinculum with narrow and moderately shallow emargination; saccus extremely short, about one half length of vinculum, tapered towards truncate apex, not reaching apex of pedunculus; phallus comparatively short and slender, with moderately inflated coecum. The shape of the saccus is quite characteristic for this species.

FEMALE GENITALIA.– Segment VIII subquadrate; subgenital plate largely smooth with small field of microtrichia anteriomedially, anterior quarter with thumb-like depression, extending dorsad beyond base of apophysis anterioris, covered with distinct foamy sculpture; ventromedial depressions longitudinal, narrow, smooth, anterior part sclerotized with convex inner margin, separated by V-shaped anteriomedial emargination, posteriomedially merged; subostial sclerite broad, band-like, medially adjoined; apophysis anterioris rod-like, stout, with abrupt transition to anterior edge of sternite VIII, slightly longer than segment VIII; colliculum short and small; ductus bursae gradually widened, with indistinct transition to oval corpus bursae; signum on the right side of corpus bursae; signum-hook relatively short, slender, strongly curved at an almost right angle in middle, minute tooth near base, basal plate small.

DISTRIBUTION.– Locally distributed along the Mediterranean coast of Spain and France. Also recorded from Madeira (Aguiar & Karsholt, 2006: 53).

BIONOMICS.– The larva feeds on *Suaeda vera* Forssk. ex J. Gmelin (Chenopodiaceae) (Amsel, 1939: 72). Adults have been collected in all months from April to October.

105. – 115. *Scrobipalpa salinella*-group

The following 11 species form a group of closely related taxa (the *S. salinella*-group) which can sometimes be difficult to separate. They can be grouped based on their male genitalia as follows: The first two species, *S. monochromella* and *S. samadensis* are characterized by their broad, curved vincular processes (they moreover differ from the other species by their characteristic wing pattern). *S. salinella, S. spergulariella* and *S. salicorniae* are characterized by their very short vincular processes, and by their relatively slender phallus (the phallus of *S. spergulariella* is the smallest in the group). The remaining species have longer and relatively slender vincular processes. They can be separated in a group with normally stout phallus (*S. halimioniella* sp.n. and *S. thymelaeae*), and a group having a large, stout phallus with a strong apical hook (*S. halymella, S. camphorosmella, S. instabilella* and *S. stabilis*). There are further differences in wing markings, female genitalia, and biology.

We have examined specimens belonging to this group which do not fit well to any of the specied treated below, and which probably belong to further, undescribed species. To clarify this situation studies of the biology and the DNA of such populations are recommended.

105. *Scrobipalpa monochromella* (Constant, 1895)

Gelechia monochromella Constant, 1895: 52.

DIAGNOSIS.– Adult. Wingspan 12-17 mm. Labial-palp cream-yellow; segment 2 mottled with blackish brown on lower and outer surface; segment 3 darker than segment 2. Antenna black, ringed with light grey. Face light yellowish grey; neck light grey; thorax and tegula as forewing. Forewing brown, mottled with orange-brown, especially along fold and as a subcostal stripe, and black, especially at base, along costa and in apical part; black spots indistinct. Hindwing grey.

Variation. Females are slightly larger and have lighter labial palps and forewings than males. Some males have uniformly black antennae. The veins of the forewings are sometimes covered with white-based black scales. Specimens from the spring generation are larger than specimens from the autumn generation.

Similar species. The relatively large size and the rather uniformly brownish forewings with indistinct black spots are characteristic for *S. monochromella*. Forms with the forewing veins thrown into relief by white-based black scales can resemble some forms of *S. samadensis*, but that species has the labial palps whitish and the forewings lighter.

MALE GENITALIA.– Uncus broad, apex slightly rounded; valva rather narrow, weakly curved, shorter than uncus, apically slightly inflated; gnathos-hook longish; sacculus comparatively short and broad, distodorsal part rounded with weak tip, inner margin straight; vincular process wider and shorter than sacculus, slightly curved outwards to an indistinct tip, continuous with comparatively shallow V-shaped medial emargination of posterior margin of vinculum; saccus broad and short, subtriangular, gradually tapered to pointed apex, extending to about apex of pedunculus; phallus broad, stout, with distinctly inflated coecum, large sclerotized arm apically. Very similar to *S. samadensis* but differs in the more slender sacculus without distinct tip and the stouter phallus.

FEMALE GENITALIA.– Segment VIII of about same length and breadth; anterior margin of sternite VIII with sclerotized rim and small membranous medial opening; subgenital plate moderately narrow, with broadly rounded posterior margin, lateromedial edge strongly folded, narrow longitudinal depression covered with foamy sculpture, extending from anterior part to about middle of subgenital plate, smooth area between foamy sculpture and anterior margin of sternite; ventromedial depressions relatively broad and almost subrectangular, medially largely fused, anteriorly extending almost to anterior edge of segment VIII, covered with microtrichia throughout; apophysis anterioris rod-like, with abrupt transition to sternite VIII, about 1.5 times length of segment VIII; colliculum short; ductus bursae broad, gradually widened towards corpus bursae; corpus bursae clearly demarcated from ductus bursae, oval; signum on the right side near entrance of corpus bursae, with small basal plate; signum-hook short, weakly curved, with indistinct additional spines near base, basal plate largely reduced. Differs from the very similar *S. samadensis* particularly in the short signum-hook.

DISTRIBUTION.– Only known from few localities in southern France and Italy. Recently recorded from Corsica (Billi *et al.*, 2010: 87), the Ukraine (Budashkin & Bidzilya, 2009: 17) and Siberia (Bidzilya, 2009: 6).

New record. Italy (Sicily): 5 ♂, 1 ♀, 5 km W Pachino, Pantano Longarini, sea lev., 12.ix.2002, leg. Karsholt *et al.*, gen. slide 3781 Hendriksen; 4 ♂, same data but 24.iv.2006, leg. Karsholt (ZMUC).

BIONOMICS.– The larva has been recorded from *Limonium vulgare* Mill. (Plumbaginaceae). It feeds in June between spun leaves (Constant, 1895: 52) or as a leaf-miner (Povolný, 2002a: 56). The adults emerged in July. In Sicily adults were collected in April and September. The adults are active from sunset onwards and later the species is easily attracted to artificial light sources. *S. monochromella* is restricted to halophytic habitats.

106. Scrobipalpa samadensis (Pfaffenzeller, 1870)

Gelechia samadensis Pfaffenzeller, 1870: 321.
Gelechia plantaginella Stainton, 1883: 253.
Lita brunhildae Schawerda, 1921: 134.
Lita zimmermani Zimmermann, 1923: (16); unavailable, nomen nudum.
Lita plantaginella var. *mariae* Zimmermann, 1926: 24.
Lita testacella Rebel, 1935a: 13.
Gelechia echo Meyrick, 1937: 122.

DIAGNOSIS.– Adult. Wingspan 11-18 mm. Labial palp whitish covered with blackish scales, especially on lower and outer surfaces. Antenna black, rather indistinctly lighter ringed. Face whitish grey, neck grey; thorax and tegula as forewing. Forewing light brown mottled with black, light-based scales, especially along veins; 1-2 small black spots in fold and 2-3 in middle of wing; termen with black dots. Hindwing grey.

Variation. *S. samadensis* exhibits both individual and geographical variation. The forewings can be almost plain light brown or they can be more or less covered with black scales. The black spots can be missing or they are sometimes distinct and confluent. In dark specimens the labial palps also become almost black, whereas in light specimens they may have a light ring in the middle of segment 3. Rarely the forewing has a black streak from base to apex. Specimens from salt marsh localities in NW Europe are more variable and on average darker than specimens from Central and South Europe and has been considered as ssp. *plantaginella* (see Povolný, 1967a: 225-226). However, the separation from the nominotypical subspecies is not very clear. The latter is also weakly defined and, e.g., specimens from Hungary have lighter brown and more variegated forewings compared with specimens from the mountains of Greece. On Iceland *S. samadensis* has the forewings more or less covered with light grey, with contrasting black spots.

Similar species. Despite its considerable variability *S. samadensis* can often be recognized by its relatively large size and the brownish forewings with white-based black scales along the veins. *S. stangei* q.v.; *S. monochromella* q.v.; *S. spumata* q. v; *S. mixta* sp. n., q.v.; *S. ergasima* q.v.

MALE GENITALIA.– Coremata present; uncus broad, apex slightly rounded; gnathos-hook distinct; valva rather narrow, weakly curved, shorter than uncus, apically inflated; sacculus comparatively short and broad, distodorsal part rounded with weak tip, inner margin straight; vincular process broadly hump-like, wider and shorter than sacculus, slightly curved outwards to an indistinct tip, continuous with comparatively shallow V-shaped medial emargination of posterior margin of vinculum; saccus broad and short, subtriangular, gradually tapered to pointed apex, extending to about apex of pedunculus; phallus broad, stout, with distinctly inflated coecum, distinct though short sclerotized arm apically. Differs from the similar *S. monochromella* in the broader and tipped sacculus and the more slender phallus with a rather small apical arm.

FEMALE GENITALIA.– Segment VIII slightly broader than long; anterior margin of sternite VIII with sclerotized rim and small membranous medial opening; subgenital plate with broadly rounded posterior margin, lateromedial edge strongly folded, narrow longitudinal depression covered with foamy sculpture, extending from anterior part to about middle of subgenital plate, smooth area between foamy sculpture and anterior margin of sternite; ventromedial depressions relatively broad and almost subrectangular, medially weakly separated, anteriorly reaching to anterior edge of segment VIII, covered with microtrichia throughout; apophysis anterioris rod-like, with abrupt transition to sternite VIII, almost two times length of segment VIII; colliculum short; ductus bursae broad, gradually widened towards corpus bursae; corpus bursae clearly demarcated from ductus bursae, pyriform; signum on the right side at entrance to corpus bursae, with small basal plate; signum-hook rather short, weakly curved, with indistinct additional tooth near base. *S. monochromella* q.v.

DISTRIBUTION.– Widely distributed in large parts of Europe, from Spain through Central Europe to Russia (southern Ural Mountains) and from southern Scandinavia to Greece. Also found in Turkey.

New records. Greece: 3 ♂, Ioannina, Katara Pass, 1600 m, 11.viii.1995, leg. Fibiger, gen. slide 2060 Hendriksen (ZMUC). Turkey: Agri, 10 km NE Diaydin, 800 m, 21.vii.1989, leg Fibiger & Esser, gen. slide 4372 Hendriksen (ZMUC).

BIONOMICS.– The larva has been recorded from *Plantago maritima* L., *P. coronopus*

L. and *P. lanceolata* L. (Plantaginaceae) (Jansen, 1991: 73). In mountainous regions such as the Alps further *Plantago* spp. should be considered as potential host-plants. Initially the larva feeds in a leaf of the host-plant, making a long and narrow mine in which it probably hibernates. Later on it lives in the upper part of the rootstock, where the frass is deposited. Pupation takes place inside a burrow in the rootstock (Jansen, 1991: 73). The voltinism depends on geographical latitude and altitude. In some parts of Europe such as Great Britain two generations are reported, with larva from October to May and from July to August and the adults occur from June to July and August to September (Bankes, 1894: 82). However, in northern Europe and in the Alps the species is reported as univoltine with adults from the second half of June to the first half of August (Kaitila, 1996: 94; Huemer, pers. obs.). *S. samadensis* is regularly attracted to light. It is found in halophytic coastal habitats and in dry grassland on the continent. Michaelis (1977: 220) furthermore reported sandhills and shingle beaches in Wales. Vertical distribution: from the lowlands up to about 2200 m in the Alps.

REMARKS.– *S. samadensis plantaginella* is occasionally considered as a distinct subspecies (e.g. Povolný, 2002a: 68). However, *S. samadensis* shows both individual variability and a tendency to geographical variation. Specimens from Great Britain (described as *Gelechia plantaginella*) and northern Europe frequently form an irregular, black, longitudinal streak through the forewing and they are often darker than those from the Alps (described as *Gelechia samadensis*). Specimens from Spain (described as *Lita testacella*) have on average more light brownish forewings. However, the individual variation is so extensive that it is impossible to separate material into subspecies.

107. *Scrobipalpa salinella* (Zeller, 1847)

Gelechia salinella Zeller, 1847: 853.
Teleia omachella auctt., (nec Oberthür, 1888: 43, pl. 6, fig. 28a, 28a′); misidentification.
Lita zernyella Rebel, 1918: 87.
Gnorimoschema corsicanum Gregor & Povolný, 1954: 94, pl. 6, fig. 3.
Gnorimoschema ignotum Gregor & Povolný, 1954: 94, pl. 6, fig. 1.
Scrobipalpa trebujenae Povolný, 1977c: 146, figs. 5, 8, **syn. n.**

DIAGNOSIS.– Adult. Wingspan 10-13 mm. Labial palp cream-coloured; segment 2 on lower and outer surface with black scales at base and on distal part; segment 3 with two black rings. Antenna black, ringed with light grey. Face cream-coloured, neck light grey; thorax and tegula as forewing. Forewing covered with black, white-based and white-tipped, and brown scales, the latter especially as a subcostal stripe, along fold and dorsum and as patches surrounding the more or less distinct black spots; a small black spot at 1/5 on costa; a faint postmedial fascia; veins in apical part greyish black with light brown in between. Hindwing light grey.

Variation. A variable species. There is some variation in the amount of black scales on the forewings, ranging from light brownish specimens with only slight mixture of black (especially near apex) and without distinct spots to specimens with a blackish area from base to apex and distinct black spots. Specimens with many dark scales and distinct black spots looks more grey than brown. Some specimens have white scales on the forewing, especially around the black spots and as a postmedial fascia.

Similar species. Due to its variability *S. salinella* can easily be confused with other species, especially the closely related *S. salicorniae* q.v., *S. spergulariella* q.v. and *S. halimioniella* sp. n. q.v. It can best be recognized by the rather large patches of light brown or orange scales surrounding the distinct (or missing) black spots on the forewings. *S. instabilella* q.v.; *S. halymella* q.v.; *S. thymelaeae* q.v.; *S. camphorosmella* q.v.

MALE GENITALIA.– Coremata present; uncus long, slender, distally tapered, apex rounded; gnathos-hook long; valva slender, evenly curved, extending slightly below tip of uncus, with inflated apex; sacculus moderately slender and short, slightly tapered towards apex, distal part with weakly convex outer margin, inner margin almost straight, apex constricted to inwardly directed tooth; vincular process weakly projected posteriad, shorter than sacculus, separated from sacculus by narrow space of about two-thirds width of sacculus, broadly hump-like to digitate with rounded to weakly pointed apex; posterior margin of vinculum with moderately broad and shallow V-shaped medial emargination, almost continuous with vincular process; saccus short, moderately broad at base, evenly tapered towards stout apex, extending to about apex of pedunculus; phallus stout, with weakly inflated coecum, distal part with sclerotized ventral part and strong, straight apical cornutus. The male genitalia are very similar to other species of the *S. salinella-instabilella* complex and can be reliably identified only by a combination of genitalia characters such as the shape of the saccculus, vincular process and the phallus.

FEMALE GENITALIA.– Segment VIII slightly broader than long, with almost parallel outer edges; anterior edge of sternite VIII well sclerotized lateromedially, medial part membranous; subgenital plate with convex inner edge, posteriomedially with some microtrichia, depression of anterior half largely covered with distinct foamy sculpture, anteriomedial edge folded and compact; ventromedial zone with longitudinal, weakly sclerotized and laterally folded depressions, not reaching anterior edge; apophysis anterioris about 1.5 times length of segment VIII, rod-like throughout with abrupt transition to segment VIII; colliculum short, strongly sclerotized ring; ductus bursae broad with small knob-like sclerotization near entrance; corpus bursae well distinguished, oval; signum on the right side at entrance of corpus bursae, basal plate small, signum-hook rather short, stout, curved, exceptionally with additional tooth. The female genitalia are similar to other species of the *S. salinella-instabilella* complex.

DISTRIBUTION.– Widely distributed in the Mediterranean parts of Europe, though local and mainly restricted to coasts and halophytic inland habitats. Outside Europe known from North Africa, the Middle East and Mongolia. However, several records may refer to other taxa such as *S. salicorniae*.

BIONOMICS.– The biology and host-plants are imperfectly known due to confusion with closely related species. According to Nel (1999: 457) the preferred host-plant is *Arthrocnemum fruticosum* (L.) Moq. (Chenopodiaceae) ('*Salicornia fruticosa*').

In Portugal it has been reared from *Arthrocnemum macrostachyum* (Moric.) C. Koch where it feeds in stems and shoot tips in May, producing adults in June (Corley, *in litt.*). Examined specimens were collected in all months from April to November, representing at least three generations. The moths are attracted to light and can be found flying freely at dusk. *S. salinella* is restricted to halophytic habitats both along the coast and in inland habitats.

REMARKS.– *Gelechia salinella* was described from an unspecified number of specimens collected in Sicily (Zeller, 1847: 853). The type series (in BMNH) is, as already described by Zeller (*op. cit.*), rather variable. Therefore, we herewith designate a lectotype in order to preserve stability of nomenclature. It is labelled: 'Lectotype' '*Salinella* Z. Syracus, 5. April' 'vidt. Hnm.' 'Type' 'Lectotype *Gelechia salinella* Z. Select. K. Sattler, 1961'. It is published here for the first time. The lectotype has light brown forewings mottled with grey and black scales (along costa, in the middle of the wing and along the veins in apical part of the wing) and distinct black spots. The unfortunately not unrolled genitalia slide is difficult to analyze but overall it agrees with the widely accepted species-concept of *S. salinella*.

Lita zernyella was described from two specimens collected in Durc (Albania) and an unspecified number of specimens from north-eastern Italy (Grado) (Rebel, 1918: 87). We have been able to examine a female syntype from the latter locality (coll. NHMW). It is a rather large specimen of about 13 mm wingspan and its genitalia fully correspond with *S. salinella*. The greyish forewing with reduced markings are characteristic but Rebel (*op. cit.*) describes the wings as ferrugineous.

Teleia omachella, described from Algeria (Oberthür, 1888: 43), was synonymized with *S. salinella* by Povolný (1966a: 134). The single syntype in the BMNH, which has been labelled as lectotype, differs in some details of the male genitalia (sacculus, saccus) from *S. salinella*, and is hardly conspecific with that species. We have not been able to assign it to any European species of this group.

Gnorimoschema corsicanum was described as separate species from Corsica (Gregor & Povolný, 1954: 94) and recently considered as a distinct subspecies by Leraut (1997: 123). We have been able to study a male paratype from MNHN which agrees with nominotypical *S. salinella*. Our study of the holotype of *Gnorimoschema ignotum* Gregor & Povolný, 1954 deposited in MNHN confirms its synonymy with *S. salinella*.

Scrobipalpa trebujenae was described from a single male collected in southern Spain (Prov. Sevilla). The holotype is rather worn and probably faded. Its remounted genitalia slide completely agrees with *S. salinella* and hence we formally synonymize *S. trebujenae*.

Male and female genitalia of *S. salinella* and other species of the *S. salinella-instabilella* are very similar. Characters of the adults and, if available, larval habits and appearance should be considered for a safe identification.

108. *Scrobipalpa spergulariella* (Chrétien, 1910) **sp. rev.**

Lita spergulariella Chrétien, 1910: 261.

DIAGNOSIS.– Adult. Wingspan 9-12 mm. Labial palp cream-coloured; segment 2 on lower and outer surface with two patches of black scales; segment 3 with two black rings. Antenna ringed black and cream. Head cream-coloured with some light grey in neck; thorax and tegula light brown. Forewing light orange-brown mottled with black, white-based scales, especially along costa, in apical part and between two patches of orange-brown in middle of wing; black spots indistinct. Hindwing light grey.

Variation. The few specimens examined show slight variation in the amount of black, white-based scales on the forewing, and in the distinctness of the black spots.

Similar species. *S. spergulariella* is very similar to light brownish specimens of *S.*

salinella, but it is smaller and has less blackish scales on the forewing. The small size also separates it from other related species.

MALE GENITALIA.– Very similar to *S. salinella* q.v. It differs mainly in the smaller phallus with a more delicate carina hook.

FEMALE GENITALIA.– Very similar to *S. salinella* q.v. Differs somewhat in the large-sized foamy sculpture with extended field of microtrichia, and furthermore in the smaller size, shorter apophysis posterioris and apophysis anterioris and the smaller corpus bursae.

DISTRIBUTION.– Only known from southern France and unpublished records from Italy (Sicily), Spain, Greece (incl. Crete) and Tunisia.

New records. Italy (Sicily): 2 ♂, 1 ♀, 5 km W Pachino, Pantano Longarini, sea lev., 12.ix.2002, leg. Karsholt *et al.*, gen. slides 3788, 3817, 6357 Hendriksen (ZMUC). Greece: 1 ♂, 1 ♀, Ipiros, Igumenitsa, 0 m, ult. vii.1994, leg. Selling, gen. slide 2749 Hendriksen. Greece (Crete): 1 ♀, Irakleon, Linoperamata, 17.vii.1984, leg. Baldizzone, gen. slide 3233 Hendriksen (ZMUC). Spain: 1 ♂, Catalonia, Rosas Salzgebiet, 20 m, 14.vi.1964, leg. Glaser, gen. slide Huemer 1015 (TLMF). Tunisia: 3 ♂, 1 ♀, Nefta area, 1.-4.v.1988, Karsholt leg., gen. slides 2941, 2959 Hendriksen (ZMUC).

BIONOMICS.– The larva has been recorded from *Spergularia media* (L.) C. Presl (Caryophyllaceae) (Nel, 1999a: 457). It is polyvoltine appearing throughout the year.

REMARKS.– *S. spergulariella* was recently considered a distinct species from *S. salinella* mainly because of its different host-plant together with a number of larval characters (Nel, 1999a: 457, figs. 19, 21).

109. Scrobipalpa salicorniae (Hering, 1889)

Lita salicorniae Hering, 1889: 302.
Lita caliacrae Caradja, 1932: 43, **syn. n.**

DIAGNOSIS.– Adult. Wingspan 13-15 mm. Labial palp cream-coloured; segment 2 with two patches of black on outer and lower surface; segment 3 with two black rings. Antenna black, ringed with light grey. Face cream-coloured, head light grey; thorax and tegula as forewing. Forewing with light brown, dark brown and black, white-based scales; light brown areas as a subcostal stripe, in fold, at dorsum and around black spots at 2/5 and 3/5; a black spot on costa at 1/5; apical area with many black scales. Hindwing light grey.

Variation. The labial palps, head, thorax, tegula and forewings are more or less covered with dark brown or fuscous scales or with white scales in lighter specimens. The black spots are often covered with blackish brown scales.

Similar species. Very similar to some forms of *S. salinella,* but besides the larger size differs in the white scales of light specimens or the dark mottling in darker specimens.

MALE GENITALIA.– Very similar to *S. salinella* q.v. The male genitalia of *S. salicorniae* are more stout, and the sacculus is longer, more slender and more pointed compared with *S. salinella*. Also the phallus is stouter. In *S. salinella* the sacculus is shorter and broader, almost square and hardly pointed and the phallus is more delicate.

FEMALE GENITALIA.– Very similar to *S. salinella* q.v. In the female genitalia of *S. salicorniae* the signum is more or less z-shaped; the foamy sculpture of sternite VIII is less developed, more restricted to medial part than in *S. salinella*. The signum of *S. salinella* is long, slender and bent. The foamy sculpture of *S. salinella* is broader and more rounded on sternite VIII.

DISTRIBUTION.– Scattered records from coastal areas of northern Europe and from halophytic inland localities in Central Europe. Recently published (as *S. salinella*) from Russia (southern Ural Mountains) (Junnilainen *et al.*, 2010: 50). A record from the Czech Republic (Povolný, 1996a: 114) is dubious.

BIONOMICS.– The larva has been recorded from *Salicornia europaea* agg. and less frequently from *Suaeda maritima* (L.) Dum., *Atriplex prostrata* Boucher ex. DC. (Chenopodiaceae), *Spergularia* spp. (including *S. salina* J. & C. Presl (Caryophyllaceae) and *Aster tripolium* L. (Asteraceae) (Bradford, 1979: 126; Courtois, 1988; Jansen, 1999: 71: Mutanen, *in litt.*). Young larvae are endophagous miners, later they make superficial galleries, spinning several shoots together with silken threads (Jansen, *op. cit.*), or when feeding on *Aster tripolium* they produce a series of galleries in the leaves (Bradford, *op. cit.*). The larva occurs from April to May, possibly after hibernation which is, however, unknown. Pupation takes place in a cocoon between spun parts of the host-plant and amongst plant debris and the pupal stage lasts from June to August. In The Netherlands the adult occurs in three peaks with the highest number of specimens from the end of July to the first half of August (Jansen, 1999: 73). Hering (1889: 302) suspected two generations. The moths are attracted to light and can be found flying freely at dusk. *S. salicorniae* occurs in halophytic habitats along the coast, but is also known from deserts, salt lakes and arid environments. According to Jansen (*op. cit.*) it occurs in Thero-Salicornion community both in tidal and non-tidal areas.

REMARKS.– *S. salicorniae* in our opinion differs sufficiently from *S. salinella* in size, colour and wing markings to be given status of a separate species. However, the genitalia are very similar to those of *S. salinella*. It will be important to study DNA from this and related taxa in future to prove the taxonomic status as separate species.

Lita caliacrae was described from two males collected on 14.vii.1931 in Romania: Balic. We studied a specimen labelled as holotype and published as such by Popescu-Gorj (1992: 140), an action which represents a valid lectotype designation. It is labelled: 'Balcie, 14/VII/[19]31/I' 'Holotyp *Lita caliacrae* Car. ♂' 'Type' '*Lita salicorniae* ssp. *caliacraella* Car Type' '*Scrobipalpa salinella* Zell.' '*Scrobipalpa salinella* Zell. Det. Povolný' [and on reverse side] '[genitalia slide Povolný] Car. 2192'. The specimen, which according to its frenulum is a male, is by a label 'Car. 2192' connected with a genitalia slide made by Povolný of female genitalia, probably of *caliacrae*. Referring to the 'holotype' of *caliacrae* and to the above mentioned genitalia slide as 'gen. Prep. No. 2192/Povolny' Park (1996: 63, 72, figs. 18-21) figures the genitalia of a male as '*Scrobipalpa caliacrae* Wocke'. Popescu-Gorj (1992: 140) also mentions a 'paratype' from 'Balic, Ak-Bunar, 15.VII.1931'. This specimen, which is a female without abdomen, is undoubtedly conspecific with the lectotype, but it is not a syntype, as it was not mentioned in the original description. It has no label referring to a genitalia slide. However, we are of the opinion that the genitalia slides of these two specimens have been exchanged, and that those figured by Park belong to the lectotype.

Lita caliacrae represents a light greyish form of *S. salicorniae* with contrasting patterns on the forewings.

110. *Scrobipalpa halimioniella* sp. n.

DIAGNOSIS.– Adult. Wingspan 11-14 mm. Labial palp cream-coloured; segment 2 mottled with light brown; segment 3 with two black rings. Antenna ringed with black and light grey. Head cream-coloured mottled with light grey towards neck; thorax and tegula as forewing. Forewing light grey from grey and black, white-based scales and mottled with light brown, especially along fold, as a sub-costal stripe, around black spots and along veins in apical part. Hindwing light grey.

Variation. Specimens can be lighter and darker, and there is variation in the amount of light brown scales on the forewing. The blackish spots at 3/5 in the middle of the wing can be double, sometimes appearing as a light brown ring without a black center. In some specimens the outer surface of segment 2 of the labial palps are more or less blackish.

Similar species. *S. halimioniella* sp. n. is characterized by its rather plain light grey-brown forewings almost without distinct black markings. It is similar to other species in the *S. salinella-instabilella* complex, especially the variable *S. salinella*, which is on average smaller and has more contrasting pattern. *S. halimioniella* sp. n. is also similar to weakly marked specimens of the more greyish *S. instabilella*. A safe identification should be based on examination of the genitalia.

MALE GENITALIA.– Coremata present; uncus long, slender, distally tapered, apex rounded; gnathos-hook long; valva slender, evenly curved, extending slightly to tip of uncus, with weakly inflated apex; sacculus moderately slender and short, slightly tapered towards apex, distal part with weakly convex outer margin, inner margin almost straight, apex constricted to inwardly directed tooth; vincular process weakly projected posteriad, shorter than sacculus, separated from sacculus by narrow space of about two-thirds width of sacculus, broadly hump-like to digitate with rounded to weakly pointed apex; posterior margin of vinculum with moderately broad and shallow V-shaped medial emargination, almost continuous with vincular process; saccus short, moderately broad at base, evenly tapered towards stout apex, extending to about apex of pedunculus; phallus stout, with weakly inflated coecum, distal part with sclerotized ventral part and strong, straight apical cornutus.

The male genitalia are similar to other species of the *S. salinella-instabilella* complex. *S. salinella* has a shorter and broader sacculus, shorter vincular processes, a shorter and less triangular saccus and a more delicate apical hook in phallus. *S. instabilella* has a deeper V-shaped posteriomedial emargination of the vinculum, a truncate (not triangular) saccus with rounded apex, and a much broader phallus with a large apical hook. *S. halymella* has a slightly more slender and pointed sacculus, longer and more slender vincular processes and deeper medial emargination; saccus is less tapered towards a rounded apex (almost triangular in *S. halimioniella* sp. n.); phallus is much broader (especially distal part) and has a larger apical hook.

FEMALE GENITALIA.– Segment VIII of about same length and breath, outer edges slightly oblique; anterior edge of sternite VIII well sclerotized lateromedially, small membranous opening in middle; subgenital plate with convex inner edge, anteriomedially folded and compact, bordering foamy sculpture which covers longitudinal depression in proximal and anterior part; ventromedial zone with longitudinal, weakly sclerotized and laterally folded depressions, not reaching anterior edge; apophysis anterioris slightly longer than segment VIII, rod-like; colliculum short and broad, strongly sclerotized ring; ductus bursae broad with small knob-like sclerotization near entrance; corpus bursae well distinguished, spherical; signum on the right side at entrance of corpus bursae; signum-hook short, stout, weakly curved, basal plate small, without additional spines.

The female genitalia are similar to other species of the *S. salinella-instabilella* complex. *S. salinella* has a slightly shorter and broader segment VIII, the subgenital plate is more convex and its foamy sculpture is more pronounced, the apophyses anteriores are relatively longer and the posterior part of the ductus bursae is more slender. *S. instabilella* has segment VIII anteriorly widened, the subgenital plate distinctly folded and with only a small field of foamy sculpture medially; the signum hook is large with numerous spines. *S. halymella* has a slightly shorter and broader segment VIII, the subgenital plate has less foamy sculpture, the medial part of its ventromedial depression is covered with microtrichia, the apophysis anteriores are relatively longer and the posterior part of the ductus bursae is slightly broader.

DISTRIBUTION.– Only known from southern France.

BIONOMICS.– The larva has been found on *Halimione portulacoides* (L.) Aellen (Chenopodiaceae). It lives between leaves which it spins together, mining them from the inside.

MATERIAL EXAMINED.– Holotype. ♂, 'FRANCE, Hernault [recte Herault] Frontignan, sea lev. la. 30.v.2004 *Halimione portulacoides* O. Karsholt' 'gen. Præparat No. 4739 ♂ H. Hendriksen' (ZMUC).

Paratypes. 11 ♂, 4 ♀, same data as holotype, but gen. slides 4740, 4790 Hendriksen, GU 09/1275 ♂, GU 09/1276 ♀ Huemer (TLMF, ZMUC); 1♀, same data but 26.-29.v.2004 (ZMUC); 1 ♂, Herault, Frontignan plage, 23.vi.1999, leg. & coll. Cox, gen. slide 3718 Hendriksen.

ETYMOLOGY.– The species name refers to the larval host-plant *Halimione*.

111. Scrobipalpa thymelaeae (Amsel, 1939) **sp. rev.**

Lita thymelaeae Amsel, 1939: 72, pl. 4, fig. 5.

DIAGNOSIS.– Adult. Wingspan 11-12 mm. Labial palp cream-coloured; each segment with two broad, brownish black rings. Antenna black, ringed with grey. Face cream-coloured with light greyish brown neck; thorax and tegula as forewing. Forewing light grey-brown from white-based black scales and partly covered with light yellow-brown, especially along dorsum, as a subcostal stripe and before light costal and tornal spots; black spots in fold and (double) at 2/5 and 3/5 in middle of wing; a distinct black streak in apical part of wing and a black spot in apex. Hindwing light grey. Underside of abdomen cream-white.

Variation. In the few examined specimens there is some variation in the density of light yellow-brown scales on the forewing. The black streak in the apical part of the forewing can be more or less distinct.

Similar species. *S. thymelaeae* is similar to some species of the *S. salinella* group. It can be separated from *S. instabilella* by its smaller size and more yellowish-brown scaling on the forewings. *S. salinella* and *S. salicorniae* lack the distinct black streak before apex. *S. camphorosmella* q.v.

MALE GENITALIA.– Coremata present; uncus long, slender, distally tapered, apex rounded; gnathos-hook long; valva slender, evenly curved, extending slightly below tip of uncus, with weakly inflated apex; sacculus moderately broad and short, almost even width from base to apex, with parallel inner and outer margins, apically convex with distinct inwardly directed tooth medially; vincular process distinctly projected posteriad, slightly shorter than sacculus, separated from sacculus by narrow space of about two-thirds width of sacculus, digitate and strongly tapered from base to tip, apex rounded with weakly crescent-shaped part turned outwards; posterior margin of vinculum with moderately broad and deep V-shaped medial emargination, almost continuous with vincular process; saccus moderately broad at base, evenly tapered towards stout apex, extending to about apex of pedunculus; phallus massive, with scarcely inflated coecum, distal part with strongly sclerotized ventral part and strong, slightly curved apical cornutus. The male genitalia are very similar to other species of the *S. salinella-instabilella* complex but differ e.g. by the characteristic shape of the sacculus.

FEMALE GENITALIA.– As in *S. salinella* q.v. Differs somewhat in the smaller-sized foamy sculpture of the subgenital plate.

DISTRIBUTION.– Hitherto only known from Sardinia. We also consider a small series from Bulgaria, which is similar to Sardinian specimens with only minor differences, e.g. in the more slender sacculus, as belonging to *S. thymelaeae*.

New record. Bulgaria: 1 ♂, 3 ♀, Tuzlata by Black Sea, 16.vi.2001, leg. Kaitila, gen. slides 6361, 6362 Hendriksen (KAIT, ZMUC).

BIONOMICS.– The larva has been found on *Thymelaea hirsuta* (L.) Endl. (Thymelaeaceae) in June (Amsel, 1939: 72). Contrary to its closests relatives, the biotope of *S. thymelaeae* is xeric meadoves. The adults emerged from July to August.

REMARKS.– *Lita thymelaeae* was described from a large series ('Typus und Cotypen: 130 Exemplare') of specimens collected in Sardinia. We studied some of these, including a male labelled 'Typus' in LNK. Povolný (1964: 352) synonymized *S. thymelaeae* with *S. salinella*, but we find the differences in genitalia and wing markings between these two taxa sufficiently different to consider them distinct species.

112. Scrobipalpa halymella (Millière, 1864) **sp. rev.**

Gelechia halymella Millière, 1864: 224, 242, pl. 42, figs. 4-8.

DIAGNOSIS.– Adult. Wingspan 11-12 mm. Labial palp cream-coloured; segment 2 mottled with black at base and on distal part of lower and outer surface; segment 3 with two black rings. Antenna ringed black and cream-grey. Face cream-coloured; neck, thorax and tegula greyish; distal part of tegula lighter. Forewing light greyish brown

from black, white-tipped and white-based scales and mottled with some yellow-orange, especially as a subcostal stripe, along fold and dorsum, surrounding black spots at 2/5 and 3/5 and in sub-apical area; black spots at base and at 1/5 at costa and near apex. Hindwing light grey.

Variation. There is slight variation in the greyish colour of the forewings, which can be mottled with yellowish, orange or light brown. The small black spots on the forewings can me more or less distinct and are occasionally fused into an ill-defined longitudinal black streak.

Similar species. The light greyish yellow forewings with small, distinct black spots are characteristic for *S. halymella*. *S. salicorniae* is larger, has more dark greyish forewings with less yellowish scales. *S. salinella* may be very similar, but has darker orange-brown and black forewings with more black in the apical area.

MALE GENITALIA.– Coremata present; uncus moderately slender, apically rounded; gnathos-hook long; valva almost extending to apex of uncus, weakly curved, with inflated apex; sacculus short and moderately slender, almost even width from base to truncate apex, small apical tooth directed inwards, gap between sacculus and vincular process almost two times width of sacculus; vincular process sickle-shaped, slightly shorter than sacculus, basally about same width, distally tapered towards outwardly directed tooth; vincular process continuous with broad and shallow V-shaped emargination of posterior margin of vinculum; saccus moderately broad, weakly tapered towards rounded apex, anteriorly projecting to about apex of pedunculus; phallus large, stout, with heavily sclerotized and broad distal part, large apical hook, coecum hardly inflated. Similar to other species of the *S. instabilella*-group, but differs in the broad gap between sacculus and vincular process and the distinctly pointed vincular process. However, these characters are also present in *S. stabilis*.

FEMALE GENITALIA.– Segment VIII slightly broader than long, posterior edge oblique; sternite VIII with transverse anterior edge with small membranous medial gap; subgenital plate with longitudinal and strongly sclerotized inner edge, medial and oblique depression covered with foamy sculpture, otherwise smooth; ventromedial depression narrow, subrectangular, largely covered with microtrichia, particularly in medial part; apophysis anterioris about 1.5 times length of segment VIII; colliculum broad and short; ductus bursae broad with small knob-like sclerotization near entrance; corpus bursae well distinguished, spherical; signum-hook small and stout, weakly curved, without additional spines. Very similar to *S. instabilella* q.v. and related species.

DISTRIBUTION.– Known from scattered and partially unpublished localities in southern France, Italy, Portugal, Canary Islands, Corsica and Malta; also Morocco, Tunisia and Israel.

New records. Portugal: 1 ♂, Algarve, Praia da Rocha, 8.x.1996, leg. Hendriksen, gen. slide Huemer GU 01/1037 (ZMUC). France (Corsica): 1 ♂, Calvi, 10 m, 18.-19.viii.1998, leg. Karsholt, gen. slide 2381 Hendriksen (ZMUC). Italy: 1 ♀, Toscana, Elba Isl., St. Andrea, 27.ix.1971, leg. Baldizzone, gen slide 3313 Hendriksen (BALD). Malta: 1 ♂, Mellieha Bay, Ghadira, 19.-31.viii.1992, leg. Petersen, gen. slide 2796 Hendriksen (ZMUC). Canary Islands: 5 ♂, 1 ♀, Grand Canaria, Maspalomas, 16.iii.1992, 19.v. & 3.x.1993, 21.iii., 26.vii. & 28.xii.1994, leg. Grotenfelt, gen. slides 5866, 5867, 5868, 5870, 5871, 5875 Hendriksen (ZMUH, ZMUC). Tunisia: 2 ♂, 3 ♀, Nefta area, 14.-16.iii.1986, leg. Karsholt, gen. slide 2942, 2943 Hendriksen (ZMUC). Israel: 6 ♂, 7 ♀, Jericho, Kalia, v.2004, leg. Kravchenko, gen. slides 5491, 5492, 5494 Hendriksen (TAU, ZMUC).

BIONOMICS.– The larva has been recorded from *Atriplex halimus* L. in its natural habitat but in the laboratory it also feeds freely on *Halimione portulacoides* (L.) Aellen (Chenopodiaceae) (Nel, 1999a: 460).

REMARKS.– *S. halymella* was until recently considered as a junior synonym of *S. salinella* (Leraut, 1997: 123) or *S. instabilella* (Povolný, 2002a: 49). However, larval characters clearly differ from related species (Nel, 1999a: 457, figs. 19-24), and also the adults show distinct characters.

113. *Scrobipalpa camphorosmella* Nel, 1999

Scrobipalpa camphorosmella Nel, 1999a: 462, figs. 5, 11, 17, 24.

DIAGNOSIS.– Adult. Wingspan 12 mm. Segment 2 of labial palp cream-coloured mottled with grey-brown on lower and outer surface; segment 3 white with two black rings. Antenna black, ringed with whitish grey. Face cream-coloured, neck light grey; thorax as forewings; tegula light yellow-brown. Forewing grey from white-based black scales and mottled with lighter and darker yellow-brown, especially along dorsum, as a subcostal stripe and around black spots in fold and (double) at 2/5 and 3/5 in middle of wing; black spots at base and 1/5 at costa; a short, black apical streak and a black spot in apex. Hindwing grey. Underside of abdomen white.

Variation. Only few specimens were examined, some of which have light costal and tornal spots separated by orange.

Similar species. *S. camphorosmella* is similar to some species of the *S. salinella*-complex. *S. instabilella* is larger and has normally less variegated forewings, and the underside of the abdomen is light yellowish grey. *S. salinella* has no black streak before the apex. *S. thymelaeae* has a more distinct apical streak and black dots in termen.

MALE GENITALIA.– Coremata present; uncus moderately slender, apically rounded or flattened; gnathos-hook long; valva almost extending to apex of uncus, weakly curved, with inflated apex; sacculus short and moderately slender, almost even width from base to apex, small medioapical tooth directed inwards, gap between sacculus and vincular process at most width of sacculus; vincular process moderately broad, shorter than sacculus, basally about same width, distally tapered towards indistinct outwardly directed tip; vincular process continuous with very broad V-shaped emargination of posterior margin of vinculum; saccus moderately slender, gradually tapered towards rounded apex, anteriorly projecting to about apex of pedunculus; phallus large, stout, with heavily sclerotized and broad distal part, large apical hook, coecum weakly inflated. Differs from other species of the *S. instabilella*-group, in the broadly emarginated posterior margin of the vinculum and in some details of the sacculus and the vincular process.

FEMALE GENITALIA.– Anterior margin of sternite VIII with transverse ridge and broad membranous medial gap; subgenital plate with extended sinusoid depression and large medial field of foamy sculpture, posterior part of depression covered with microspines; ventromedial depressions comparatively broad with convex inner and outer margins, sclerotized, smooth. Very similar to other species of the *S. instabilella*-group, differing slightly in the smooth and convex edged ventromedial depressions.

DISTRIBUTION.– Originally described from southern France. Recently also recorded from

Spain (Vives Moreno, 2003: 98); unpublished records from the eastern Mediterranean.

New records. Greece: 1 ♂, 1 ♀, Karpathos, Lefkos, 30 m, 21.v.1997, leg. Sutter. Greece (Crete): 1 ♀, Ag. Galini, 20 m, 25.v.1995 (LNK). Israel: 1 ♂, Jericho, Kalia, v.2004, leg. Kravchenko, gen. slide 5705 Hendriksen (TAU).

BIONOMICS.– The larva has been recorded from *Camphorosma monspeliaca* L. (Chenopodiaceae) (Nel, 1999a: 462). It differs from *S. instabilella* e.g. by the distinctly darker prothoracic shield (Nel, *op. cit.*, figs. 23-24).

114. Scrobipalpa stabilis Povolný, 1977 **stat. n.**

Scrobipalpa instabilella stabilis Povolný, 1977b: 192, figs. 16-19, 28.

DIAGNOSIS.– Adult. Wingspan 12-14 mm. Labial palp white; segment 2 mottled with black on lower and outer surface; segment 3 with two black rings. Antenna black, ringed with light grey. Face cream-white; neck grey; thorax greyish; tegula greyish at base, distally lighter. Forewing dark grey mottled with black and yellow-brown, the latter especially as a subcostal stripe, in fold and surrounding black spots at 2/5 and 3/5, and an elongate black streak at 4/5 in middle of wing; black spots at base and at 1/5 at costa dividing subcostal stripe; termen with black scales. Hindwing grey.

Variation. The forewings can be more or less variegated depending on the amount of black and yellow-brown. Apparently less variable than *S. instabilella*.

Similar species. Very similar to forms of *S. instabilella* with variegated forewings, and frequently not possible to be separated from that species without examination of the genitalia.

MALE GENITALIA.– As in *S. halymella* q.v. Differs from other species of the *S. instabilella*-group in the broad gap between sacculus and vincular process and the distinctly pointed vincular process.

FEMALE GENITALIA.– Anterior margin of sternite VIII a transverse ridge with broad membranous medial gap; subgenital plate with extended sinusoid depression and large medial field of foamy sculpture, posterior part of depression covered with microspines; ventromedial depressions comparatively broad, covered with microtrichia for the most part; ductus bursae broad with small knob-like sclerotization near entrance. Female genitalia hardly distinguishable from other species of the *S. instabilella*-group.

DISTRIBUTION.– Widely distributed in Spain, but not yet recorded elsewhere.

BIONOMICS.– Host-plant and early stages unknown. The adults have been collected in all months from April to October at light. *S. stabilis* prefers halophytic habitats along the coast and inland saline habitats.

REMARKS.– Povolný (1977b: 192) described *S. instabilella stabilis* from Spain (Granada: Baza), based on genitalic characters such as the knob-like sclerite of the ductus bursae, the large subgenital plate and signum and the stout phallus. The large subgenital plate and signum mentioned in the original description are merely due to the moderately large size of the adults. However, the geographical variation of *S. instabilella* does not cover other character states, particularly the unique sclerite of the ductus bursae, and *S. stabilis* is therefore considered as a species.

115. Scrobipalpa instabilella (Douglas, 1846)

Anacampsis instabilella Douglas, 1846: 1270, fig. 10.
Lita lagunella Chrétien, 1910: 272.
Lita strobilacella Caradja, 1920: 101; unavailable, nomen nudum.
Lita salsolella Amsel, 1935a: 301, fig. 139.
Phthorimaea halymiphaga Amsel, 1952: 127, fig. 32.

DIAGNOSIS.– Adult. Wingspan 13-16 mm. Labial palp cream-coloured; segment 2 mottled with dark grey on lower and outer surface; segment 3 with two black rings. Antenna dark grey, indistinctly lighter ringed. Face cream-coloured; neck, thorax and tegula greyish, the latter distally lighter. Forewing light grey mottled with dark grey, black and yellow, the latter especially as a subcostal stripe, in fold and surrounding black spots at 2/5 and 3/5; a black spot at 1/5 at costa. Hindwing grey.

Variation. A variable species. The labial palps are more or less covered with blackish brown scales, appearing from light with dark rings to almost black. Specimens can be almost plain light or dark grey, or they can have the forewings variegated from many black and yellow scales. The black spots can be distinct or almost absent. A distinct form has a more or less well marked, black streak from base to apex.

Similar species. Due to its variability *S. instabilella* may be confused with several other species. It is best recognized by its relatively large size, the greyish forewings with yellow markings and the more or less distinct black admixture. *S. salinella* is on average smaller and normally light brownish mottled with black, particularly in the central part of the forewings whereas the costal and dorsal parts of the forewings are brownish. *S. salicorniae* is almost of the same size as *S. instabilella*, but has the forewings dark brown (sometimes mottled with white). *S. halimioniella* sp. n. is very similar to light forms of *S. instabilella*, but it has the black scales on the forewings clearly white-based, whereas these in *S. instabilella* are with grey or yellowish base. *S. stabilis* q.v.; *S. suaedella* q.v.; *S. camphorosmella* q.v.; *S. thymelaeae* q.v.

MALE GENITALIA.– Coremata present; uncus moderately slender, apically rounded or flattened; gnathos-hook long; valva almost extending to apex of uncus, weakly curved, with inflated apex; sacculus short and moderately slender, almost even width from base to apex, small apical tooth directed inwards, gap between sacculus and vincular process about width of sacculus; vincular process broad, distinctly shorter than sacculus, basally about same width or broader, distally hump-like to weakly pointed; vincular process continuous with shallow V-shaped emargination of posterior margin of vinculum; saccus broad, evenly tapered towards rounded apex, anteriorly projecting slightly beyond apex of pedunculus; phallus large, stout, with heavily sclerotized and broad distal part, large apical hook, coecum weakly inflated. *S. instabilella* can be separated from the closely related *S. halymella* by the much narrower gap between sacculus and vincular process. It differs from *S. camphorosmella* in the shorter sacculus, the narrower medial emargination of the vinculum and the broader saccus, and from the very similar *S. stabilis* in the less pointed sacculus. *S. halimioniella* sp. n. q.v.

FEMALE GENITALIA.– Segment VIII longer than broad, anteriorly widened; anterior edge of sternite VIII sclerotized, transverse ridge with small membranous medial gap; subgenital plate with distinctly folded and slightly outwardly curved inner edge, longitudinal and slightly S-curved depression of subgenital plate with small field of foamy sculpture medially; medial depressions of sternite VIII smoothly sclerotized, extending almost to anterior margin of segment, anteriorly turned outwards with medial incision in between; apophysis anterioris rod-like with sinusoid outer margin anteriorly and with gradual transition to anterior edge of sternite VIII, slightly longer than segment VIII; colliculum very broad, short; ductus bursae broad, gradually widened with weak turning and abrupt transition to clearly defined corpus bursae; corpus bursae large, oval; signum on the right side near entrance of corpus bursae, with small basal plate; signum-hook large and stout, weakly curved, with numerous spines on concave side up to about middle of hook.

DISTRIBUTION.– Widely distributed along the coasts of western Europe and in the western Mediterranean; from the eastern Mediterranean we have only seen specimens from mainland Greece. Records from the Ukraine (Povolný, 2002a: 49; Bidzilya & Budashkin 2009: 17) are probably misidentifications (Bidzilya, *in litt.*). Records from Hungary (Szabóky *et al.*, 2006: 73) refer to *S. salicorniae*. Furthermore known from northern Africa to Palestine. Recently also recorded from Canada and the USA (Povolný, 1998; Powell & Povolný, 2001) (however, see below).

BIONOMICS.– The larva has been recorded from *Halimione portulacoides* (L.) Aellen (Chenopodiaceae) by several authors. In the laboratory it also feeds freely on *Atriplex halimus* L. (Nel, 1999a: 460). Bland *et al.* (2002: 172) stated records of *Atriplex littoralis* L. (Chenopodiaceae), *Plantago coronopus* L. (Plantaginaceae), *Aster tripolium* L. (Asteraceae), *Salicornia europaea* L. (Chenopodiaceae) and *Lycium barbarum* L. (Solanaceae) as probably accidental. However, at least the last host-plant is confirmed by breeding (Parsons, *in litt.*). Several other alleged host-plant records belonging to the families Chenopodiaceae, Plantaginaceae, Plumbaginaceae, Ericaceae and Thymelaeaceae (Povolný, 2002a: 49) are in strong need of verification. Most probably the majority of these records is based on misidentifications. The young larvae have been found mining the leaves of younger stalks. Later leaves and buds are spun together by a silky web and the leaves frequently turn greyish. The larva occurs univoltine in early September and after hibernation from March to early May. Pupation normally takes place in a cocoon between plant debris, occasionally between spun shoots of the host-plant (Jansen, 1999: 58). The adult is on the wing from early June to early September. During daytime it rests between the lowest branches of the host-plant and at night it can be attracted to light, frequently in numbers. Preferred habitats are coastal salt marshes which are only submerged at spring tide and storm flood (Jansen, *op. cit.*).

REMARKS.– Povolný (2002a: 49) treated several taxa as synonyms of *S. instabilella*. However, careful examination proved either distinct species or differing synonymy for some of them. *Gelechia halymella* Millière, 1864, is regarded as a valid species based on several adult and larval characters (q.v.). *Phthorimaea calaritanella* Amsel, 1951, is transferred into synonymy with *S. obsoletella* (q.v.). *S. stabilis*, initially described as a subspecies of *S. instabilella*, is regarded as separate species based on genitalic characters (q.v.).

We were able to examine a male published as *S. instabilella* from Canada in ZMUH (Povolný, 1998: 339). Both the moth and its genitalia show that it is a related species, which in our opinion is distinct from *S. instabilella*. A female published as *S. artemisiella* in the same paper represents probably the female of this species in the *salinella*-group.

S. instabilella was moreover recorded from a salt marsh locality in California

(Powell & Povolný, 2001: 22). We could examine a male from this series in coll. EME. It is indeed closely related to *S. instabilella*, but differs in several details of the genitalia (shorter sacculus, more curved vincular processes, much shorter saccus and less prominent apical hook of the phallus) and in colour and wing markings of the adult. In our opinion it represents a further undescribed species. The status and relation of the Nearctic members of the *S. salinella*-group with European taxa should be studied also with molecular data.

116. *Scrobipalpa peterseni* (Povolný, 1965)

Ilseopsis peterseni Povolný, 1965: 481, figs. 1-2.

DIAGNOSIS.– Adult. Wingspan: male 10-12 mm, female 8-9 mm. Head with frontal process. Labial palp white; segment 2 mottled with light brown; segments 2 and 3 with two black rings. Antenna black, indistinctly lighter ringed. Head white, more or less mottled with grey; thorax and tegula as forewing. Forewing slender, variegated white, light brown, grey and black; black spots indistinct; base blackish; a dark patch at 1/4 of costa fused with a black streak in middle of wing; light brown areas in costal part of wing. Hindwing white.

Variation. A highly variable species. Specimens can be white with black markings on forewings and labial palps, and they can have light brown suffusion on part or all of the forewings, or they can have thorax, tegula and larger parts of the forewings black. Females are slightly smaller than males.

Similar species. *S. peterseni* is the only species of *Scrobipalpa* (known to us) with a distinct frontal process of the head.

MALE GENITALIA.– Coremata absent; uncus subtriangular, apex slightly pointed; culcitula moderately large, gnathos hook strongly sclerotized, stout; tegumen broadly and deeply emarginated anteriorly; valva stout, basal part bent, distal part digitate, without distinct apical inflation, valva extending slightly beyond tip of uncus; sacculus widely separated from valva, broadly thumb-shaped and moderately short, distodorsal and distoventral part rounded, apex with strong inwardly directed tip; vincular process minute, adjacent to sacculus; posterior margin of vinculum with shallow V-shaped emargination; saccus moderately slender, short, tapered towards apex, extending moderately beyond apex of pedunculus; phallus slender, with distinctly inflated coecum, small sclerotized arm apically.

FEMALE GENITALIA.– Segment VIII subquadrate; subgenital plate with strongly convex inner edge, anterior part projected with convex outer and inner margins, posterior third covered with microtrichia, large zone of distinct foamy sculpture extending from about middle of sternite to basal part of apophysis anterior, lateral margin smooth; ventromedial depression largely reduced to narrow membrane separating subgenital plates, anteriomedially with deep V-shaped incision; rod-like apophysis anterioris extremely short, about one-fifth length of segment VIII, with distinct transition to projected subgenital plate; colliculum short and small; ductus bursae moderately broad, gradually widened to clearly defined corpus bursae; corpus bursae large, pyriform; signum-hook short, stout, curved at an almost right angle near middle, basal plate small.

DISTRIBUTION.– In Europe only known from an unpublished record from Sicily. Recorded from Algeria, Tunisia and Saudi Arabia (Sattler, 1988: 231).

New record. Italy (Sicily): 2 ♂, 5 km W Pachino, Pantano Longarini, sea lev., [collecting data unknown] leg. Bella, gen. slide GU 03/1183 ♂ Huemer (BELL, ZMUC).

BIONOMICS.– Host-plant and early stages unknown. Povolný (2002a: 81) suspected halophytic Chenopodiaceae and psammophytic Zygophyllaceae as host-plants. The adults have been collected in February and March and between August and November, probably occurring in several generations. It is attracted to light. Preferred habitats are sand dunes. The locality in Sicily is a salt marsh area.

REMARKS.– *S. peterseni* shows some rather unique characters within *Scrobipalpa* such as the broadly lanceolate and pointed, slightly brachypterous forewings. Furthermore the adult head possesses an irregular frontal process (Sattler, 1988: 228-229) and the genitalia show some specialised structures e.g. presence of a bent valva and very short apophysis anterior. However, these specialisations also occur to some extent in other species of *Scrobipalpa* and some transitions can be found. Considering the high diversity of genitalia structures and the variation in external appearance within *Scrobipalpa*, treatment of *S. peterseni* in a separate genus *Ilseopsis* is unsatisfactory and we follow Sattler (1988) in this respect.

117. Scrobipalpa ergasima (Meyrick, 1916) **comb. rev.**

Phthorimaea ergasima Meyrick, 1916: 568.
Teleia hyoscyamella Rebel, 1912: 89; secondary homonym of *Lita hyoscyamella* Stainton, 1869.
Lita mignatella Caradja, 1920: 100; unavailable, nomen nudum.
Phthorimaea intestina Meyrick, 1921: 74.
Gnorimoschema mirabile Povolný & Gregor, 1955: 85, fig. 5.
Gnorimoschema pervada Clarke, 1962: 101.

DIAGNOSIS.– Adult. Wingspan 10-14 mm. Labial palp cream-coloured; segment 2 mottled with dark brown on outer and upper surface; segment 3 with two black rings. Antenna ringed with black and dark grey. Head with cream-coloured face and grey neck; thorax and tegula as forewing. Forewing greyish brown from white-tipped black scales along the veins and light brown scales in between; indistinct black spots in fold and at 2/5 and 3/5 in middle of the wing surrounded with yellowish brown scales; a short, black, longitudinal streak before apical area; termen with many white-tipped black scales. Hindwing grey.

Variation. The forewings vary from light to dark grey and in the amount of yellowish brown scales. In worn specimens the small black spots become invisible.

Similar species. *S. ergasima* is characterized by its longitudinal streaks with several small black spots. It resembles small specimens of *S. samadensis*, but this species has broader forewings and more plain dark labial palps. *Keiferia lycopersicella* q.v.; *Phthorimaea operculella* q.v.

MALE GENITALIA.– Uncus comparatively slender, distally slightly tapered with rounded apex; gnathos-hook strong; anterior margin of tegumen with deep, V-shaped emargination; valva slightly shorter than uncus [appearing longer in unrolled slide], strongly bent at about one-third, apex without inflation; sacculus largely reduced, only a short hump; vinculum strongly projected posteriorly, with moderately shallow V-shaped emargination medially, vincular process broadly tooth-like, vinculum strongly projected posteriorly; saccus very broad, subrectangular, apically flattened, slightly exceeding apex of pedunculus; phallus long and stout, apical sclerotized arm distinct, coecum strongly inflated.

FEMALE GENITALIA.– Apophysis posterioris about 4-5 times length of segment VIII; segment VIII of subquadrate shape, posterior margin concave with medial incision; posterior half of sternite VIII densely covered with foamy sculpture, larger towards middle, small-scaled in distal part, medially with two drop shaped sclerotizations; anterior half of sternite VIII smooth, small ridge from posteriomedial part to base of apophysis anterioris; apophysis anterioris about length of segment VIII, basally broader with gradual transition to rod-like distal part; colliculum short, closely attached to sternite VIII; ductus bursae long and narrow, with abrupt transition to well separated, oval corpus bursae; signum on the right side of corpus bursae; signum-hook longish, evenly curved in middle, without additional tooth, basal plate of moderate size.

DISTRIBUTION.– Mediterranean area including eremic parts of Africa to South Africa and south-eastern Asia and Japan. Accidentally introduced into many countries.

New record. Malta: 1 ♂, Dwerja l/o Rabat, 13.vii.2004, leg. & coll Sammut.

BIONOMICS.– The larva is a leaf-miner on various Solanaceae and has been recorded from *Hyoscyamus albus* L., *Solanum nigrum* L. and *S. melongena* L. (Solanaceae) (Klimesch, 1984: 155). It is considered as a serious pest. The adult has been observed throughout the year, from January to December (Povolný, 2002a: 83).

REMARKS.– *Teleia hyoscyamella* was described from two females from Heluan in Egypt, both collected by Debski (Rebel, 1912: 89). A female from NHMW labelled 'LECTOTYPE' 'cuniculi *Hyoscyamus muticus* larva 4/4 1910 imago 18/4 1910' 'Heluan Debski ' '*Teleia hyoscyamella* Rbl Type' 'Mus.Vind. Gen.Präp. 16.642 ♀' is here designated as the lectotype in order to stabilize nomenclature. It fully agrees with *S. ergasima* and would principally have priority as the older species name. However, *Scrobipalpa hyoscyamella* (Rebel, 1912) becomes a junior secondary homonym of *S. hyoscyamella* (Stainton, 1869) and the species-group name is therefore invalid (ICZN, 1999: Art. 57.3.1). The other female syntype was bred from a pupa on *Tamarix* according to the original description, but we have not been able to trace this specimen.

According to Povolný (2002a: 83) *Lita epithymella* var. *glauca* Caradja, 1920 should be a further synonym of *S. ergasima*. Povolný refers to p. 26 in the paper by Caradja (1920), but the correct place is p. 100. However, there is no description, but a reference to "*L. epithymella* var. *glauca* Chrét.". This refers to *Lita thymifoliella* var. *glaucella* Chrétien, 1915 which is currently considered as belonging to the genus *Athrips* Billberg.

Turcopalpa Povolný, 1973

Turcopalpa Povolný, 1973b: 98.
Type species: *Turcopalpa glaseri* Povolný, 1973.

DIAGNOSIS.– See species description.

MALE GENITALIA.– Uncus small, slender, apically rounded; gnathos-hook short; tegumen comparatively small, anterior margin deeply emarginated; valva slightly shorter to longer than uncus, weakly curved, apex without inflation; sacculus short, pointed; vinculum broad, anterior margin with deep and broad U-shaped emargination, vincular process about length of sacculus, vinculum strongly projected posteriorly; saccus very broad, subrectangular, apically flattened, strongly exceeding apex of pedunculus; phallus long and stout, curved, distinct sclerotized arm near apex, coecum inflated.

FEMALE GENITALIA.– Apophyses posterioris moderately long; segment VIII subrectangular, longer than broad, laterally sclerotized with well separated subgenital plates, sternum VIII medially membranous with some microtrichia; apophysis anterioris rod-like, about length of segment VIII; intrasegmental membrane with subostial sclerite, extending from base of apophysis anterioris towards ostium; antrum exceptionally large, sub-oval; corpus bursae well defined; signum short and weakly curved, with several spines.

DISTRIBUTION.– Eremic parts of the western Palaearctic region, including northern Africa, Southwest Asia (Povolný, 2002a: 82) and the Ukraine, with two described species.

BIONOMICS.– Host-plant and early stages unknown. Povolný (2002a: 82) suspected halophytic and psammophytic Chenopodiaceae as host-plants. The adults have been collected from January to June and from August to September, probably occurring in several generations. Preferred habitats are semi-deserts.

REMARKS.– *Turkopalpa* (Bidzilya & Budashkin, 2009: 19; Falkovitsh & Bidzilya, 2009: 80) is a misspelling.

118. *Turcopalpa glaseri* Povolný, 1973

Turcopalpa glaseri Povolný, 1973b: 100, figs. 1, 16, 24.

DIAGNOSIS.– Adult. Wingspan 10-12 mm. Labial palp dirty-white; segment 2 mottled with light grey-brown and some blackish brown on lower and outer surface; segment 3 mottled with blackish brown. Antenna black; ringed with dirty white. Face cream-coloured; neck, thorax and tegula as forewing. Forewing light grey-brown from black, grey and light brown whitish-based scales; fold light yellow-brown with a black spot; two black spots at 2/5 and 3/5 in middle of wing, the latter angulated; indistinct black spots near base, at costa and in apex; indistinct light costal and tornal spots. Hindwing light grey, darker along margins; cilia light grey.

Variation. The above description is based on two specimens from the type locality in Turkey and one, rather worn specimen from the Ukraine. The latter has darker, grey-brown forewings with indistinct markings.

Similar species. *T. glaseri* is a rather characterless, greyish moth. It resembles several *Scrobipalpa* species, like *S. obsoletella*, and also *Scrobipalpula tussilaginis*. A safe identification is hardly possible without examination of the genitalia which show good diagnostic characters.

MALE GENITALIA.– Uncus small, apex broadly sub-oval; gnathos-hook short; tegumen slender, anterior margin with broad and deep emargination; valva stout, longer than uncus, apical third weakly curved, without inflation; sacculus short and slender, pointed; vinculum broad, anterior margin with deep and broad U-shaped emargination, vincular process slightly longer than sacculus, apically pointed; saccus massive, braodly subrectangular, exceeding apex of pedunculus; phallus long and stout, curved, distinct sclerotized arm near apex, coecum weakly inflated.

FEMALE GENITALIA.– Segment VIII subrectangular, longer than broad; subgenital plate well sclerotized, smooth, posterior and medial part of sternite VII membranous, posterior part with microtrichia; apophysis anterioris rod-like, about length of segment VIII; intrasegmental membrane with broad subostial sclerite, extending from base of apophysis anterioris towards ostium, with oblique anterior rim; antrum exceptionally large, sub-oval; corpus bursae well defined; signum with small basal plate; signum-hook stout, short and weakly curved, several spines on inner side of curvation from base almost to apex.

DISTRIBUTION.– Described from central Anatolia and recently recorded from a single female collected in the Ukraine (Bidzilya & Budashkin, 2009: 19).

BIONOMICS.– Host-plant and early stages unknown. The few adults known to date have been collected in August and September at light. In Anatolia the species was observed in eremic habitats at altitudes from 1000 to 2000 m (Povolný, 2002a: 82).

Scrobipalpula Povolný, 1964

Scrobipalpula Povolný, 1964: 339.
Type species: *Gelechia psilella* Herrich-Schäffer, 1854.

DIAGNOSIS. Adult. Wingspan 8-14 mm, with males mostly larger than females. Labial palp upturned, rather slender; segment 2 roughly scaled below forming two ridges with a groove in between, white with two indistinct blackish brown rings, lighter on upper and inner surface; segment 3 shorter than segment 2, smooth scaled, with a slender, black basal ring and a broad, black apical ring. Antenna dark greyish to black, more or less distinctly ringed. Forewing rather slender, mostly light greyish, in some species with a lighter costal part and a darker dorsal part separated by a longitudinal streak; three black spots; fold and sub-costal stripe often yellowish; apical area often darker than rest of the wing; cilia mostly grey with darker cilia line(s). Hindwing greyish.

MALE GENITALIA.– Sternite VIII broad, posteriorly rounded; tergite VIII tongue-shaped; uncus broadly rounded, normally broader than distal part of tegumen; gnathos with large culcitula and distinct, more or less broadly shovel-shaped, gnathos-hook; tegumen moderately widened from uncus to pedunculi, deeply emarginated anteriorly; pedunculi distinct, comparatively small; valva long, extending to about or beyond apex of uncus, broadly clavate, weakly curved ventrad (in slide inwards); sacculus extremely short, digitate, distinctly separated from valva; vinculum broad, posterior margin with prominent, paired process strongly projecting posteriad, medially with distinct incision; saccus moderately long and very broad, distally more or less rounded; phallus moderately to very long and slender, apex with sub-terminal sclerotization, coecum weakly inflated.

FEMALE GENITALIA.– Apophysis posterioris about 3 times length of segment VIII; segment VIII wide and longish, mediodorsally membranous; ventromedial zone of segment VIII with pair of longitudinal depressions, usually covered with foamy sculpture and extending to or beyond anterior edge of sternite into antrum; intermediate zone between depression membranous, weakly sculptured and covered with numerous microtrichia; apophysis anterioris rod-like, stout, slightly shorter to longer than segment VIII; antrum distinctly funnel-shaped, broad at base, distinctly tapered; ductus bursae with ring-shaped colliculum; corpus bursae usually well defined, covered with numerous microtrichia; signum a slender and long hook, with long and slender base.

DISTRIBUTION.– In the Palaearctic region represented by 5-6 species. Furthermore 16 Nearctic and 37 Neotropical species are described (Heckford & Sattler, 1999: 255; Lee *et al.*, 2009: 26; Povolný, 2002b: 43).

BIONOMICS.– The host-plants of European species are Asteraceae. Neotropical members are also known from Solanaceae. The larvae live between spun leaves or as leaf-miners.

REMARKS.– Members of this genus from the Palaearctic region belong to a complex of closely related taxa with disputed taxonomic state (Huemer & Karsholt, 1998; Povolný, 2000). The latter author formerly regarded the various taxa as part of a polytypical species with a wide distribution from South and North America to Europe and Asia. However, later Povolný (2000, 2002a) followed the argumentation of Huemer & Karsholt (*op. cit.*) to some degree and the European taxa were considered as 'semispecies'. Species of *Scrobipalpula* can be partially distinguished by external characters, differing ecology and by the male genitalia. Female genitalia are predominantly indistinguishable between species.

Males of *Scrobipalpula* differ from *Scrobipalpa* in the shape of the uncus, which can be examined by brushing off the tip of the abdomen.

119. Scrobipalpula psilella (Herrich-Schäffer, 1854)

[no genus] *psilella* Herrich-Schäffer, 1853: pl. 67, fig. 496.
Gelechia psilella Herrich-Schäffer, 1854: 171.
Gelechia nocturnella Staudinger, 1859: 241.
Lita pallidella Heinemann, 1870: 252.
Gelechia killiasii Frey, 1880: 362.
Gnorimoschema psilellum astericolellum Hering, 1957: 135; unavailable, nomen nudum.
Scrobipalpula psilella asiatica Povolný, 1968b: 17, fig. 72. Subspecies.

DIAGNOSIS.– Adult. Male. Wingspan 10-13 mm. Labial palp cream-white, segment 2 mottled with blackish brown on lower and outer surface; segment 3 with two black rings. Antenna black, indistinctly lighter ringed. Head grey with lighter face, thorax and tegula as forewing. Forewing with costal half light greyish brown and dorsal half blackish brown, separated by an ill-defined dark brownish streak from base through middle of wing almost to apex; black spots in fold indistinct; those at 2/5 and 3/5 in middle of wing more distinct, surrounded by some light orange-brown; fold, subcostal stripe and area between black spots at 2/5 and 3/5 light orange-brown. Hindwing grey. Female. Slightly smaller (9-12 mm), with slightly narrower wings and more contrast between dorsal and costal parts of the forewing.

Variation. A rather variable species. The forewings can be lighter or darker grey, and they can be more or less mottled with light orange-brown. The contrast between the light (costal) and dark (dorsal) parts of the forewings can be more or less pronounced, worn specimens becoming less contrasting.

Similar species. *S. psilella* is characterized by (normally) having the forewings separated into lighter costal and darker dorsal parts, separated by an ill-defined brownish streak from base through the middle of the wing almost to apex. A similar pattern is also found in the more contrasting *S. seniorum* q.v. and *S. ramosella* q.v.

MALE GENITALIA.– Uncus broadly rounded, sub-ovate; gnathos-hook spatulate, medially constricted to about half width of distal spatula; valva long, exceeding apex of uncus, distally slightly dilated with pointed apex; sacculus shortly digitate; vincular process basally broad, evenly narrowing towards outwardly pointed apex; phallus slightly curved at basal third, distal part long and slender, coecum weakly inflated. The male genitalia differ from other European *Scrobipalpula* in the tipped valva which also exceeds the uncus.

FEMALE GENITALIA.– Segment VIII ventromedially with comparatively extended foamy sculpture reaching to about anterior edge of segment; intermediate zone between depression membranous, weakly sculptured and covered with numerous microtrichia, extending into first half of antrum; apophysis anterioris rather slender, about length of segment VIII; antrum distinctly funnel-shaped, evenly tapered, comparatively long and slender, extending to about two-thirds length of apophysis anterioris; ductus bursae with short, ring-shaped colliculum; corpus bursae well defined, pear-shaped, covered with numerous microtrichia; signum a slender and long hook. The gradually and evenly tapered, moderately slender antrum is characteristic for *S. psilella*.

DISTRIBUTION.– Widely distributed throughout large parts of Europe. However, records from Great Britain (Povolný & Bradley, 1965: 16, figs. 1-2) are based on misidentification of *S. diffluella* (Heckford & Sattler, 1999); those from the Netherlands (Huismann & Koster, 1997: 56) refer to *S. tussilaginis*. The species is also known from North Africa and parts of Asia to Japan. Data from the New World most probably refer to additional species.

New record. France (Corsica): 1 ♂, 40 km S Bastia, Folelli, sea lev., 9-10.v.1999, leg. Karsholt (gen. slide 2445 Hendriksen) (ZMUC).

BIONOMICS.– The larva has been recorded from *Artemisia campestris* L. (Asteraceae) on several occasions. Additional host-plants of the family Asteraceae are *Artemisia maritima* L., *A. vulgaris* L. (Povolný & Bradley, 1965: 9), *Chrysanthemum* L., *Centaurea scabiosa* L., *Aster tripolium* L., (Elsner at al., 1999: 44), *A. amellus* L. (Hering, 1957), *Helichrysum arenarium* (L.) Moench (Herrich-Schäffer, 1854: 171) and *Erigeron acer* L. (Tabell, 2008: 137). Further records from *Cirsium* Miller, *Aster alpinus* L. (Elsner at al., *op. cit.*), *Achillea* L. (Povolný, 1964) and *Anthemis* L. (Hering, 1957) need confirmation. The larva is a leaf-miner or feeds between spun leaves. On the sea-shore it feeds on basal leaves from a long silken tube on the ground surface (Hering, 1891). The larval voltinism is insufficiently documented but seems to vary geographically from univoltine larvae occurring from April to May (Schütze, 1931: 190) to bivoltine populations with a second larval period in June and July. Accordingly the adults are on the wing from late May or June to mid-October, depending on climatic conditions of the locality. Vertical distribution: in the Alps up to an altitude of about 1200 m; reaching higher elevations of about 3000 m in Afghanistan and 3500 m in Nepal.

REMARKS.– *S. psilella* has been regarded as a polytypical and widely distributed species in the past (e.g. Povolný, 1987c: 3-7). In consequence it was repeatedly recorded from the New World, e.g. Patagonia and California (Povolný, 1987c: 3-7, figs. 1-4; Powell & Povolný, 2001: 16-17, figs. 48-51, 84-85). However, small genitalia differences in the available material indicate the probable occurrence of several additional sibling species. Powell & Povolný (*op. cit.*) support this interpretation with the description of sympatric and/or allopatric occurrence in California of phenotypically different populations with partially different host-plants and voltinism.

S. psilella asiatica was described from two males collected in Afghanistan (Povolný, 1968b: 17). The genitalia of this taxon fully agree with nominotypical *S. psilella* but the moth has ash-grey colour of the forewing with relatively large spots (Povolný, 2002a: pl. 12, fig. 13). In the absence of further material we follow the interpretation as a subspecies of *S. psilella* (Povolný, 2000: 87), though the separation of *S. psilella* into subspecies may be a matter of ongoing discussion.

S. (psilella) japonica was described as 'semispecies', based on a single male which was collected in Japan (Povolný, 2000: 88). The holotype is quite characteristic due to its uniformly whitish grey colouration of the forewing (Povolný, 2002a: pl. 12, fig. 14), but the genitalia seem to differ from *S. psilella* by the untipped apex of the valva. Further studies are required to clarify the taxonomic state of this taxon.

120. *Scrobipalpula ramosella* (Müller-Rutz, 1934)

Lita ramosella Müller-Rutz, 1934: 120.

DIAGNOSIS.– Adult. Wingspan, male: 13-14 mm, female: 12-13 mm. Labial palp white; segment 2 with two indistinct blackish brown rings; segment 3 with a slender, black basal ring and a broad, black apical ring. Antenna dark greyish brown, indistinctly ringed with lighter grey. Head light brownish grey with lighter face; thorax and tegula as forewing. Forewing with a whitish grey costal part and a blackish grey dorsal part, the two parts being separated by an orange-brown streak from base through the black spots at 2/5 and 3/5 in the middle of the wing towards apex; fold dark orange-brown; subcostal stripe light orange brown. Hindwing light grey.

Variation. The few examined specimens vary in the colour of the head (more or less mottled with grey), and in the amount of orange-brown, especially in the apical part of the forewings. Males are slightly larger than females.

Similar species. *S. ramosella* is characterized by having the costal part of the forewing whitish grey and the dorsal part blackish grey, and by its distinct light orange-brown subcostal stripe. It resembles *S. psilella*, but is on average larger, has more distinctly bi-coloured forewings with white stripes along costa and below the distinct light orange-brown subcostal stripe (the white scales are more scattered on the forewings of *S. psilella*). *S. seniorum* q.v.

MALE GENITALIA.– Gnathos-hook spatulate, medially strongly constricted to about one quarter width of distal spatula; valva moderately long, extending to about apex of uncus, distal part slightly and evenly dilated, apex rounded; saccus with rounded apex. Differs from other species of *Scrobipalpula*, except *S. seniorum* q.v., mainly in the strongly constricted gnathos-hook.

FEMALE GENITALIA.– Segment VIII ventromedially with distinct foamy sculpture, extending into anterior part of antrum; apophysis anterioris moderately stout, about length of segment VIII; antrum broadly funnel-shaped, anterior part abruptly tapered, comparatively short, extending to about half length of apophysis anterioris; signum-hook comparatively long and slender. Extremely similar to other species of European *Scrobipalpula*, with small differences in the shape of the antrum and particularly the long and slender signum-hook.

DISTRIBUTION.– Only known from the central part of the Alps (Switzerland), recently also recorded from the French Hautes-Alpes (Nel, 2003c: 398). Further records from Macedonia and Greece (Huemer & Karsholt, 1998: 51) are now attributed to *S. seniorum*. Recently also recorded from the Kuril Islands (Russia) (Piskunov, 2000: 117) and the Altai region (Ponomarenko, 2008: 97), but these records need verification.

BIONOMICS.– The larva has been recorded mining the leaves of *Erigeron* L. (Asteraceae) and spinning them together. Pupation took place between spun leaves or tissue in the laboratory (Huemer, pers. obs.). Schmid (*in litt.*) found the larva feeding gregariously between spun leaves of *Artemisia umbelliformis* Lam. (Asteraceae). The adult is on the wing from early July to early August. *S. ramosella* as far as known occurs on dry and rocky places on siliceous soil at high altitudes from about 2300 to 2700 m.

REMARKS.– *S. ramosella* is supported as a distinct species by external and genitalia characters as well as the sympatric occurrence with *S. diffluella* in the neighbourhood of Zermatt (Switzerland) (Sauter, 1961: 270).

121. *Scrobipalpula seniorum* Povolný, 2000

Scrobipalpula (psilella) seniorum Povolný, 2000: 87, pl. 1, fig. 1, pl. 5, fig. 19, pl. 8, figs. 11, 13.
Gnorimoschema ptarmicae Hering, 1957: 27; unavailable, nomen nudum.
Scrobipalpula psilella f. *compositella* Povolný, 1964: 343; unavailable, infrasubspecific.

DIAGNOSIS.– Adult. Wingspan 12-14 mm. Labial palp cream-white; segment 2 with two blackish brown rings; segment 3 with black ring at base (slender) and before tip (broad). Antenna black, ringed with grey. Head cream-white mottled with light grey in neck; thorax and tegula as forewing. Forewing whitish grey in costal and dorsal thirds (darkest in the latter); these two areas separated by a broad, irregular brownish black streak from base through distinct black (orange-brown surrounded) spots at 2/5 and 3/5 in the middle of the wing towards apex; fold dark orange-brown; subcostal streak light orange-brown.

Variation. A variable species. The whitish labial palps, head and forewings are more or less covered by grey and orange-brown scales. The irregular streak from base to apex can be black or orange-brown or a mixture of both colours. A female from Greece, Mt. Olympus has more slender hindwings, almost lacking the emargination at the outer margin.

Similar species. *S. seniorum* is characterized by its whitish grey forewings with a broad, irregular streak from base to apex. It resembles S. *ramosella*, which is darker, especially along dorsum, and has a less distinct separation between costal and dorsal parts of the forewings.

MALE GENITALIA.– Gnathos-hook spatulate, medially strongly constricted to about one-third width of distal spatula; valva long, extending to about apex of uncus, distal part dilated, apex rounded. Almost indistinguishable from *S. ramosella* q.v., and some alleged differences such as the broader valva may underly individual variation.

FEMALE GENITALIA.– As in *S. ramosella* q.v. Extremely similar to other species of European *Scrobipalpula*, with small differences in the shape of the antrum and the signum-hook.

DISTRIBUTION.– Only known from the Matka Treska Canyon (Macedonia) and Mt. Olympus (Greece).

BIONOMICS.– In Macedonia the larva has been recorded in May, mining the leaves of *Achillea ageratifolia* (Lindl.) Boiss. (=*Ptarmica ageratifolia serbica* (Lindl.) Nyman) (Asteraceae) and spinning them together. The adults did not emerge until September and October. At Mt. Olympus (Greece) the larva have been observed mining blotches in the basal leaves of *Centaurea pindicola* Griseb. (Asteraceae) during July and adults emerged in August (Klimesch, 1968: 122). According to Povolný (2000: 85) *Achillea holosericea* Sibth. & Sm. ought to be a further host-plant, but it is not authenticated by the type-material. The habitats are insufficiently described but from its host-plants *S. seniorum* seems to be restricted to steppe grassland on calcareous soil at altitudes up to about 2500 m.

REMARKS.– *S. seniorum* is extremely similar to *S. ramosella* as proved by the virtually indistinguishable genitalia. Populations from the Balkans were not given a separate taxonomic status by Huemer & Karsholt (1998: 53), but they have been subsumed as a further semispecies of the *S. psilella*-complex by Povolný (2000: 87), based on the quite characteristic external appearance and differences in ecology.

122. *Scrobipalpula diffluella* (Frey, 1870)

Gelechia diffluella Frey, 1870 [April]: 252.
Gelechia cacuminum Frey, 1870: 252.
Lita diffluella Heinemann, 1870 [Dec. 31]: 247.
Phthorimaea diffluella v. *bellidiastri* Klimesch, 1951b: 105.
Gnorimoschema uniflorellum Hering, 1957: 419, 542; unavailable, infrasubspecific.

DIAGNOSIS.– Adult. Wingspan 12-14 mm (male), 8-11 mm (female). Labial palp cream-white; segment 2 and 3 each with two more or less distinct blackish rings. Antenna blackish grey, indistinctly lighter ringed. Head light grey; thorax and tegula as forewing. Forewing grey mottled with white and orange-brown, the latter especially in fold, subcostal stripe and around and between spots in middle of wing; apical area with many black scales. Hindwing grey.

Variation. A variable species. Specimens can be almost plain whitish grey to blackish grey, and they can be variegated from longitudinal white, light orange-brown, grey and black lines. The black spots can be weak or distinct, and they can be more or less surrounded by orange-brown. Some specimens have a blackish brown streak from base to apex. This form is only known from the Alps and was named *S. bellidiastri* (Klimesch, 1951).

Similar species. In spite of its variability most specimens of *S. diffluella* can be recognized by their relatively small size, by having the neck and the thorax concolorous, and the forewings not separated into a dark dorsal and a light costal part. *S.*

psilella is normally larger and has the forewings separated into a lighter costal and a darker costal part. *S. tussilaginis* q.v.

MALE GENITALIA.– Genitalia armature comparatively short and stout; gnathos-hook spatulate, moderately constricted to about half width of distal spatula; valva moderately long, extending to about apex of uncus, distal part strongly dilated, apex rounded; saccus subrectangular. Differs from *S. psilella* in the shorter and apically rounded valva, from *S. tussilaginis* in the subrectangular saccus and from other *Scrobipalpula* mainly in the shape of the gnathos-hook.

FEMALE GENITALIA.– Segment VIII ventromedially with distinct foamy sculpture, extending far into antrum; apophysis anterioris moderately stout and short, about length of segment VIII; antrum broadly funnel-shaped, irregularly tapered, moderately short, extending to about half to two-thirds length of apophysis anterioris; corpus bursae short; signum-hook comparatively stout, distal part slightly dilated. Extremely similar to other species of European *Scrobipalpula*, particularly *S. tussilaginis*, with only small and somewhat variable differences in the shape of the antrum and the comparatively short signum-hook.

DISTRIBUTION.– Boreomontane species; locally distributed in the Alps (France to Austria and Slovenia); furthermore in Scandinavia and Latvia. Also reported from a few historical specimens in Great Britain (Heckford & Sattler, 1999). Recently recorded from Russia (southern Ural Mountains) (Junnilainen *et al.*, 2010: 51). The figure of female genitalia of a specimen from Hungary (Bükk mountains) (Povolný, 2000: 77, fig. 21) is correctly identified and indicates a wider distribution in the Carpathians. Nevertheless, a record from Romania needs confirmation (Kovács & Kovács, 2000: 33).

BIONOMICS.– Host-plant records include *Erigeron* L., *Homogyne alpina* (L.) Cass., *Aster alpinus* L. and *Aster bellidiastrum* (L.) Scop. (=*Bellidiastrum michelii* Cass.) (Klimesch, 1951) in the Alps, and *Erigeron acer* L. including *E. acer politus* Frr. (Asteraceae) in Scandinavia (Huemer & Karsholt, 1998: 59). Further host-plants are dubious and partially refer to *Scrobipalpa murinella* (Klimesch, 1943). The larva causes a blotch-like leaf-mine and it has been observed in June (Huemer, pers. obs.). Larvae living on *Erigeron acer* do not, contrary to *S. psilella*, feed from a silken, sand-covered tube (Tabell, 2008: 137). Heckford & Sattler (*op. cit.*) suspected that *S. diffluella* may be partially bivoltine with adults from mid-June to early July and again in late July in Scandinavia and from late May to mid-July and late July to mid-August in the Alps. However, at least in the Alps univoltinism is more likely and differences in flight period are mainly due to microclimatic conditions of the habitat and annual climatic changes. Vertical distribution: from low altitudes to about 1100 m in Norway; in the Alps from rarely below 2000 m to about 3000 m.

REMARKS.– *Gelechia cacuminum* was described on the same page as *Gelechia diffluella* (Frey, 1870: 252). Later the former taxon was synonymized with *Scrobipalpa murinella* by Povolný (1967a: 221) but in a correction leaflet to his paper he withdrew this nomenclatural act, based on the examination of type-material. In fact the lectotype of *G. cacuminum* is conspecific with *Scrobipalpula diffluella* but paralectotypes disagree and are misidentifications of *Scrobipalpa murinella* (Huemer & Karsholt, 1998: 59). Therefore records of *S. cacuminum* have to be reconsidered from case to case as they frequently refer to *S. murinella* (Klimesch, 1943; Burmann, 1951; Huemer & Tarmann, 1993).

123. Scrobipalpula tussilaginis (Stainton, 1867)

Gelechia tussilaginis Stainton, 1867a: 14.
Lita tussilaginella Heinemann, 1870: 251.
Xystophora retusella Rebel, 1891: 632.

DIAGNOSIS.– Adult. Wingspan 12-14 mm. Labial palp cream coloured; segment 2 mottled with some black on lower and outer surface: segment 3 with two black rings. Antenna black, ringed with light grey. Head cream coloured, darker towards neck; thorax and tegula light grey. Forewing light grey from white-based and white-tipped blackish grey scales and mottled with light yellow-orange, especially in fold, subcostal stripe and between black spots. Hindwing grey.

Variation. The forewings vary from light to darker grey. In dark specimens segment 2 of the labial palp is often heavily mottled with black. The black spots on the forewing vary in form and size, and they can occasionally be indistinct.

Similar species. *S. tussilaginis* is characterized by its rather plain light grey forewings with 3-4 distinct spots, and it is thereby separable from other European *Scrobipalpula* species. It may resemble light grey *Scrobipalpa* species like *S. obsoletella*, but that species has light yellow brownish scales scattered on the forewing.

MALE GENITALIA.– Genitalia armature comparatively short and stout; gnathos-hook spatulate, moderately constricted to about 0.4 to 0.7 times width of distal spatula; valva moderately long, extending to about apex of uncus, distal part moderately dilated, apex rounded; saccus subtriangular. Differs from *S. psilella* in the shorter and apically rounded valva, from *S. tussilaginis* in the subtriangular uncus and from the other European *Scrobipalpula* in the shape of the gnathos-hook.

FEMALE GENITALIA.– Segment VIII ventromedially with distinct foamy sculpture, extending into antrum; apophysis anterioris moderately stout and short, about length of segment VIII; antrum broadly funnel-shaped, distal part abruptly tapered, moderately short, extending to about two-thirds length of apophysis anterioris; corpus bursae long; signum-hook comparatively slender, distal part slightly dilated. Differs from *S. diffluella* slightly in the long corpus bursae and the abruptly tapered antrum and from other European *Scrobipalpula* in the shape of the antrum and the signum.

DISTRIBUTION.– Widely distributed in Central Europe, reaching Great Britain in the north-west and Greece and Turkey in the south-east. Recently also recorded from Denmark (Buhl *et al.*, 2009: 73-74), Sweden (Svensson, 2010: 21), Byelorussia (Piskunov, 2000: 117), European Russia (Ponomarenko, 2008: 97) and Siberia (Bidzilya, 2005: 15). Absent from the Iberian Peninsula and major parts of the Mediterranean.

BIONOMICS.– The larva has been recorded mining the leaves of *Tussilago farfara* L. and more rarely *Petasites* Mill. (Asteraceae) (Klimesch, 1958). Initially it produces a gallery which later becomes a large blotch-mine. Pupation takes place outside the mine on the underside of a leaf or among leaf-litter. Voltinism is insufficiently documented and seems to depend on latitude and altitude of the locality. Whereas *S. tussilaginis* is bivoltine in Great Britain (Pelham-Clinton, 1989) and in parts of northern and Central Europe, it is probably univoltine in montane areas. The adult has been observed from late April to mid-August, univoltine populations from late May to mid-June. Vertical distribution: from sea-level to about 1100 m in the Alps and in Greece.

Phthorimaea Meyrick, 1902

Phthorimaea Meyrick, 1902: 103.
Type species: *Gelechia (Bryotropha) operculella* Zeller, 1873.

DIAGNOSIS.– Hindwing of male with pencil of long hairs from base of costa (also present in *Keiferia lycopersicella*). See also species description.

MALE GENITALIA.– Sternite VIII long, basally broad, gradually narrowing towards rounded apex, with pair of long brushes of coremata; tergite VIII long, comparatively broad, tongue-shaped, with pair of long coremata; uncus broad and short; gnathos sub-oval without hook, culcitula large, cushion-like; tegumen gradually widened from uncus to pedunculi, emarginate; pedunculi distinct; valva broadly digitate, extremely long and projecting to or above uncus; sacculus largely reduced, rounded; vinculum short, with short and broad vincular processes, posterior margin with distinct medial incision; saccus large; phallus stout, long and slender, weakly curved, with small sclerotized apical arm, coecum weakly inflated.

FEMALE GENITALIA.– Apophysis posterioris about four times length of segment VIII; sternum VIII with large patches or transverse zone of foamy sculpture; ventromedial zone membranous, largely covered with microtrichia from posterior margin far into antrum; apophysis anterioris about length of or slightly longer than segment VIII; antrum broadly funnel-shaped, anteriorly narrowing and extending almost to apex of apophysis anterioris; colliculum long, sclerotized; ductus bursae short, indistinct; corpus bursae large, oval; signum on the left side of corpus bursae; signum-hook moderately stout, with small base, additional spines absent.

DISTRIBUTION.– Originally a Neotropical genus, but the type-species has been introduced to vast areas of the tropical and subtropical regions. According to Povolný (1989b: 93), the genus includes 6 species, but later the number was estimated at 3-4 species (Povolný, 2002a: 19).

BIONOMICS.– The larval host-plant of the only European species are various Solanaceae. Further details see species description.

REMARKS.– In former times numerous species of Gnorimoschemini were placed in the genus *Phthorimaea* (Meyrick, 1925: 91-96). Povolný (1967b: 104-107, 1994b: 10) restricted the genus to *P. operculella* and it nearest allies, a small group of Neotropical species with spezialized genitalia, and especially, having segment VIII in the male very long and with a pair of easily visible coremata.

124. *Phthorimaea operculella* (Zeller, 1873)

Gelechia (?Bryotropha) operculella Zeller, 1873: 262, pl. 3, fig. 17.
Gelechia terrella Walker, 1864b: 1024, n. obl.
Bryotropha solanella Boisduval, 1874: 713.
Gelechia tabacella Ragonot, 1879: cxlvi.
Gelechia sedata Butler, 1880: 560.
Phthorimaea argentinae Povolný, 1989: 95, figs. 43-44, 63, **syn. n.**
Gelechia piscipellis Howard, 1897: 76, (*nec* Zeller, 1873); misidentification.
Phthorimaea epicentra auctt., (*nec* Meyrick, 1909); misidentification.

DIAGNOSIS.– Adult. Wingspan 12-17 mm. Labial palps cream-coloured; segment 2 and 3 each with two black rings; segment 3 also with black tip. Antenna fuscous; indistinctly lighter ringed. Face cream-coloured, neck and tegula light grey-brown; thorax as forewing. Forewing variegated light brown and blackish grey, the latter especially along dorsum and costa; fold with light yellow scales between black spots; spots in middle of wing indistinct; apical area with many white-based black scales, especially along veins and in termen. Hindwing grey, with pencil of long hairs from base of costa in male. Abdominal segment VIII elongate in male, light yellow-brown; two lateral hair pencils from base of segment.

Variation. In some specimens segment 2 of the labial palps is light brown. Antennae are more distinctly ringed in females than in males. The forewings are variable with regard to the amount of light brown and blackish grey scales. When the former dominates specimens become less variegated. Occasionally there is a weak, light postmedial fascia.

Similar species. Males of *P. operculella* are unique among Palaearctic Gnorimoschemini because of the elongate abdominal segment VIII. Females may resemble *Scrobipalpa aptatella* and *Scrobipalpa ergasima*. The forewing pattern of *P. operculella* is characterized by the light brown fold having light yellow scales between the black spots, while the spots in the middle of the wing are indistinct.

MALE GENITALIA.– Uncus broadly subquadrate, apically nearly flattened with indistinct medial projection; gnathos with large, sub-oval sclerite, culcitula distinct; valva broadly digitate, distally narrowing, apical part inflated and curved, with pointed apex projecting above uncus; sacculus short, rounded hump; vincular process short with convex inner margin and outwardly directed apical tooth, posterior margin of vinculum with distinct medial incision; saccus large, subrectangular, about two-thirds width of vinculum; phallus long and moderately slender, weakly curved, with indistinct sclerotized apical arm, coecum indistinctly separated.

FEMALE GENITALIA.– Segment VIII about same length and breadth, distinct transverse zone of foamy sculpture in medial part of sternum VIII, covering about one-third of length of segment; subgenital plate posteriorly and anteriorly smooth, two pairs of rounded depressions in zone of foamy sculpture, outer depression with posteriorly pointed subtriangular sclerite; ventromedial zone membranous, posteriorly almost breadth of segment VII, anteriorly narrowing, covered with microtrichia except for foamy medial zone; apophysis anterioris rod-like, stout, longer than segment VIII; antrum broadly funnel-shaped with gradual transition to anterior margin of sternite VIII, apically extending to about three-quarters of apophysis anterioris; colliculum a rather long sclerotized band; ductus bursae very short with indistinct transition to large, oval corpus bursae; signum on the left side of corpus bursae; signum-hook smooth, strongly arched with small base.

DISTRIBUTION.– Neotropical species which has been introduced to many countries. In Europe mainly restricted to the southern parts. Records from Central Europe, e.g. Austria (Huemer & Mayr, 2000: 120) and Switzerland (Sauter & Whitebread, 2005: 65) are rare and exceptional. Records from northern Europe are probably based on introduced material, and the species cannot survive the winters there. Such records are not listed in the distribution table.

BIONOMICS.– The larva is a leaf-miner on various Solanaceae such as *Solanum melongena* L., *S. dulcamara* L., *S. tuberosum* L., *Nicotiana glauca* R.C. Graham and

N. tabacum L. (Klimesch, 1984: 154) and widely known as the Potato Tuber Moth; on some crops it is considered as a serious pest.

REMARKS.– *Phthorimaea epicentra* (Meyrick) is a species in the genus *Schizovalva* Janse, of which a paratype in BMNH belongs to *P. operculella* (Povolný, 1967b: 106).

Gelechia piscipellis Howard, 1897 has been listed as a synonym of *P. operculella* by, e.g., Povolný (2002a: 86), but it is a misidentification of *Gelechia piscipellis* Zeller, 1873, which is a synonym of *Isophrictis similiella* (Chambers, 1872) (Hodges, 1983: 19).

P. argentinae was described from a series of both sexes collected in several localities in Argentina and Chile (Povolný, 1989b: 99). It was stated to be the 'most striking and stoutest form of the genus', and to differ in details of the genitalia. We could examine the genitalia of the hitherto undissected male holotype in ZMUC (gen. slide 5160 Karsholt) and find them to be in complete agreement with those of *P. operculella*. The differences given in Povolný's description and the accompanying figure (*op. cit.*, p. 98, fig. 43) are due to differences in preparation techniques (not unrolling and very little pressing of the genitalia). We find it surprising that Povolný did not mention the widely distributed pest species *P. operculella* from Argentina and Chile, but instead described a new, similar looking species.

Also *P. robusta* as described and figured by (Povolný, 1989) from Argentina has similar genitalia to *P. operculella*, but it is on average smaller and differs in forewing colour and markings (Povolný, 1989b: 93-95), and it may represent a distinct species. However, the undissected male holotype belongs to another, not even congeneric species.

Tuta Kieffer & Jørgensen, 1910

Tuta Kieffer & Jørgensen, 1910: 363.
Type species: *Gnorimoschema (Tuta) atriplicella* Kieffer & Jørgensen, 1910.

DIAGNOSIS.– *Tuta absoluta* is a rather small (wingspan 10-12 mm) slender-winged moth with an apperance very similar to several species of *Scrobipalpa*. Hindwing of male without pencil of long hairs from base of costa (present in *Phthorimaea* and *Keiferia*). See also species description.

MALE GENITALIA.– See species description.

FEMALE GENITALIA.– See species description.

DISTRIBUTION.– Neotropical origin, introduced to Europe in 2007.

BIONOMICS.– The only species known from Europe is restricted to Solanaceae and its generic assignment is conflicting.

REMARKS.– The authorship of *Tuta* is disputed and we here follow Sattler (1973: 264) and subsequent authors such as Nye & Fletcher (1991: 317) who credited Kieffer & Jørgensen (1910) as authors of this and other taxa. This opinion was rejected by Povolný (1993: 86) who considered Strand as the author. In fact, Strand obviously was involved in the work of Kieffer & Jørgensen (1910) as an expert but he did not formally appear in the publication as author. Several of the taxa described in this work, including *Tuta*, were separately published by Strand (1911) a year later thus establishing new homonymies and objective synonymies.

Tuta was described as a subgenus of *Gnorimoschema* containing only *Gnorimoschema (Tuta) atriplicella* Kiefer & Jørgensen, 1910. Hodges & Becker (1990), revising the nomenclature of some Neotropical Gelechiidae, selected a neotype of the type species and considered it as a valid species in the genus *Phthorimaea*, thereby synonymizing *Tuta* with *Phthorimaea*. Povolný (1993) subsequently reinstated *Tuta* and argued that *G. (Tuta) atriplicella* is not a member of *Phthorimaea*, but belongs to the genus *Scrobipalpuloides* Povolný, 1987, which thereby became a synonym of *Tuta*. In fact the male genitalia of *atriplicella* Kiefer & Jørgensen are rather similar to those of *inapparens* Povolný, 1987, the type species of *Scrobipalpuloides*. Recently Lee *et al.* (2009), without further findings, followed Hodges & Becker (1990) and re-synonymized *Tuta* with *Phthorimaea* whereas they still treated *Scrobipalpuloides* as a distinct genus. However, the genitalia differences between these three taxa are small and in the absence of a phylogenetic analysis, based on morphological, ecological and molecular data, it is rather subjective to follow one or the other generic concepts.

The only European species, the recently introduced *Tuta absoluta*, was originally described in *Phthorimaea* by Meyrick, 1917 and later placed in different genera such as *Gnorimoschema* by Clarke (1965), in *Scrobipalpula* by Povolný (1967b) and in *Scrobipalpuloides* by Povolný (1987c). However, it is rather atypical in genitalia compared with the type species of these genera. Based on the synonymization of *Tuta* with *Phthorimaea* by Lee *et al.* (2009), *absoluta* would now again be combined with the latter genus which seems conflicting from genitalia morphology. Thus, its generic assignment remains questionable. Since *Tuta absoluta* is already a well known combination of an economically serious pest species, we hesitate to change the generic assignment and keep it in *Tuta* until further evidence reveals the correct genus.

125. *Tuta absoluta* (Meyrick, 1917)

Phthorimaea absoluta Meyrick, 1917: 44.

DIAGNOSIS.– Adult. Wingspan 10-12 mm. Labial palp cream-coloured; segment 2 with indistinct and segment 3 with two distinct black rings. Antenna black, ringed with light grey. Head grey, darkest towards neck; thorax and tegula as forewing. Forewing dark grey from white-based dark scales and mottled with orange-brown and white; black spots surrounded with some orange-brown or (in fold) yellow; costa blackish; faint costal and tornal spots separated by orange-brown; termen with many black scales. Underside of abdomen cream-coloured. Hindwing dark grey, lighter towards base.

Variation. There is some variation in the amount of orange-brown and white scales on the forewing. The black spots in the middle of the forewing can be double or drop-shaped. The costal and dorsal spots can be fused into a fascia.

Similar species. *Tuta absoluta* is characterized by its slender, variegated grey, black and orange-brown forewings. It is very similar to several common *Scrobipalpa* species, especially the normally larger and more broad-winged *S. atriplicella*. *Keiferia lycopersicella* q.v.

MALE GENITALIA.– Coremata absent; uncus broadly subquadrate, apically nearly flattened; gnathos with suboval sclerite, culcitula moderately distinct; tegumen comparatively short; valva broadly digitate, weakly curved, basal part smooth and with hump-like projection at about one-third, distal two-thirds less sclerotized, covered with long setae, apex extending almost to tip of uncus; sacculus reduced; vincular process short with convex inner margin and outwardly directed apical tooth, posterior margin of vinculum with V-shaped medial emargination; saccus longish subrectangular, about length of valva and vinculum, slightly smaller than uncus; phallus long and slender, straight, with indistinct sclerotized apical arm, short coecum weakly inflated.

FEMALE GENITALIA.– Segment VIII broader than long, ventromedial zone membranous; subgenital plate with strongly sclerotized, arched proximal margin, inner edge folded and covered with foamy sculpture, anterior part evenly passing over to outwardly directed apophysis anterioris; antrum extremely long, exceeding length of apophysis anterioris, broadly funnel-shaped, sclerotized; ductus bursae slightly longer than antrum, irregular sclerite in posterior third, membranous part of ductus bursae about one-third length of antrum, well separated from longish, oval corpus bursae; signum-hook small, straight, basal plate largely reduced.

DISTRIBUTION.– Neotropical species. 2007 detected in Spain, 2008 reported from Morocco, Algeria and The Netherlands (Sarto i Monteys and van der Straten, *in litt.*), 2009 from South France and Italy (Anonymous, 2009; Nel, 2009; Sannino & Espinosa, 2009), Malta (Dandria & Catania, 2009), England (Fera, 2009; Homan, 2010), Lithuania (Ostrauskas & Ivinskis, 2010), Denmark (Buhl *et al.*, in press), Greece (incl. Crete) (Roditakis *et al.*, 2010) and furthermore from Switzerland, Libya, Tunisia (Anonymous, 2010), The Canary Islands (Baéz, *in litt.*), Madeira (Aguiar, *in litt.*), Turkey, Syria (Agri Pest, 2010) and Egypt (Sattler, *in litt.*). In northern Europe *T. absoluta* has been found both in supermarkets, at tomato packing stations and in light traps operating in gardens and fields.

New record. Portugal: 2 ♂, 2 ♀, Algarve, 1 km N Salema, 30 m, 22.-23.ix.2009, leg. Skou (ZMUC). Cape Verde Islands, numerous specimens, Sal Island, December 2009, M. L. Lobo Lima leg. (ZMUC).

BIONOMICS.– *Tuta absoluta* is a serious pest on tomatos (*Lycopersicon esculentum* Mill.) but it also attacks several other cultivated Solanaceae, e.g. *Solanum muricatum* Aiton, *S. melongena* L., *S. nigrum* L., *S. puberulum* Nutt. ex Seem., *S. lyratum* Thunb., *S. elaeagnifolium* Cav., *Datura stramonium* L., *D. ferox* L., *Nicotiana glauca* Graham, *Lycium chilense* Bertero and *Lycopersicon hirsutum* Donal. The larva feeds on all parts of the host-plants and feeding habits vary from leaf-mining to boring into buds, stalks and fruits. The larvae do not enter diapause as long as food is available and there may be 10-12 generations per year by 25-30°C. *Tuta absoluta* has quarantine status in Europe (Anonymous, 2006).

REMARKS.– *Phthorimaea absoluta* was described from a single male from the Andes mountains in Peru (Meyrick, 1917: 45). Clarke (1965: 80) figured the holotype and its genitalia in lateral position. For the generic treatment see generic Remarks (above).

Keiferia Busck, 1939

Keiferia Busck, 1939: 568 [keys], 571.

Type species: *Phthorimaea lycopersicella* Busck, 1928.

DIAGNOSIS.– Small (wingspan 8-10 mm) greyish brown, rather characterless moths resembling many *Scrobipalpa* species. Hindwing of male with pencil of long hairs from base of costa (also present in *Phthorimaea*). See also species description.

MALE GENITALIA.– See species description.

FEMALE GENITALIA.– See species description.

DISTRIBUTION.– New world genus with the majority of species occurring in the Neotropical region. Recently, the type-species has been introduced to Europe. According to Povolný (2002a: 19), altogether 24 species are described.

BIONOMICS.– The larval host-plants of the only European species are various Solanaceae. Further details see species description.

126. *Keiferia lycopersicella* (Walsingham, 1897)

Eucatoptus lycopersicella Walsingham, 1897: 71.
Phthorimaea lycopersicella Busck, 1928: 171.

DIAGNOSIS.– Adult. Wingspan 8-10 mm. Labial palp white on upper and inner surface, cream-coloured on lower and outer surface; segment 2 with two black bands; segment 3 black with two cream-coloured rings. Antenna black, ringed with grey. Face cream-white; neck, thorax and tegula as forewing. Forewing variegated grey-brown from brown, grey and black lighter-based scales; small black spots in fold and at 2/5 and 3/5 in middle of wing; termen with many black scales. Hindwing grey.

Variation. The examined specimens show only slight variation.

Similar species. *K. lycopersicella* is characterized by its rather small size, and by the small black spots in the variegated brownish forewings. It resembles *Tuta absoluta*, but that species has the black spots in the fold followed by yellowish scales. It can also be confused with several species of *Scrobipalpa*, e.g. *S. ergasima*. Males of *K. lycopersicella* can definitively be recognized by the pencil of long hairs from the base on the hindwing costa.

MALE GENITALIA.– Uncus about length of tegumen, longitudinal, apex pointed; gnathos with long and broadly horseshoe-shaped medial sclerite; tegumen short, anteriorly strongly widened with broad and moderately deep emargination; valva long and slender, distinctly exceeding uncus, basal part straight, distinct curvature at two-thirds, distal third turned inwards, subapical process present, apical part turned posteriad, covered with numerous microtrichia, pointed; sacculus situated close to middle of vinculum, digitate, straight, about one-quarter length of valva; posterior margin of vinculum medially with longitudinal projection, apically incised, almost extending to tip of sacculus; Saccus long, about length of tegumen, stout; phallus long and slender, slightly curved, apical hooklet present, coecum inflated, about one-fifth length of phallus.

FEMALE GENITALIA.– Segment VIII with extremely short sclerotized part, laterally convex humps; ventromedial part of sternite VIII merged, trapezoid, posterior margin weakly emarginate; ostium bursae with ovate sclerotization, projected posteriad; apophysis anterioris long, slender, directed outwardly; antrum extremely long, almost length of apophysis anterioris, tubular, medially widened; ductus bursae slightly longer than antrum, irregular sclerite in posterior third, gradually widened to oval corpus bursae; signum-hook large, stout, weakly curved, basal plate distinct on proximal side of signum-hook.

DISTRIBUTION.– Widely distributed in South and Central America and also in the southern part of the U.S.A. In 2008 introduced to Europe (Albenga, Liguria, Italy) (Espinosa & Sannino, 2009; Sannino & Espinosa, 2009). Acording to Sannino (*in litt.*), the infestation has disappeared in 2009 probably not only due to frost during winter, but also because of strong competition with *Tuta absoluta*.

BIONOMICS.– The larva feeds on a large variety of Solanaceae and is a well known pest of *Lycopersicon esculentum* L., known as the Tomato Pinworm (Elmore & Howland, 1943). Furthermore *Solanum melongena* L. var *'esculentum'* Nees, *S. tuberosum* L., *S. carolinense* L., *S. xanthii* A. Gray, *S. bahamese* L., and *S. umbelliferum* Esch. are recorded hosts (http://edis.ifas.ufl.edu/IN231). Eggs are laid singly or in small groups on a leaf. Initially the larva causes a blotch-like leaf-mine, later it feeds between spun leaves or enters the stem or fruits. Pupation takes place outside of the larval feeding place in a web on the ground surface. Seven to eight or even more generations per year have been reported outside Europe.

REMARKS.– The homology of the sacculus compared with other genera of Gnorimoschemini is uncertain.

Ephysteris Meyrick, 1908

Ephysteris Meyrick, 1908: 724.
  Type species: *Ephysteris chersaea* Meyrick, 1908.
Microcraspedus Janse, 1958: 137.
  Type species: *Megacraspedus brachypogon* Meyrick, 1937.
Opacopsis Povolný, 1964: 344.
  Type species: *Gelechia inustella* Herrich-Schäffer, 1854.
Echinoglossa Clarke, 1965: 85.
  Type species: *Echinoglossa trinota* Clarke, 1965.

DIAGNOSIS.– Wingspan 8-15 mm. Segment 3 of labial palp shorter than segment 2. Antenna more or less distinctly lighter and darker ringed. Forewing more or less slender, white to blackish brown; the outer plical spots is situated more dorsally than in most other gnorimoschemine genera. Termen and basal part of cilia in forewing with black-tipped white scales. Hindwing greyish.

MALE GENITALIA.– Sternite VIII broader than long, posterior margin with medial emargination; tergite VIII smaller than sternite, broad, posterior margin rounded or exceptionally with bifurcate process; uncus longer than broad, distally narrowing, apically curved; gnathos with strong base, hardly curved hook long and stout, culcitula largely reduced; tegumen slender, anteriorly deeply emarginate; pedunculi of moderate size; valva curved, broadly digitate, distally widened, extending at most to base of gnathos; sacculus weakly curved to straight, digitate, slender, rarely clubbed, at most two-thirds length of valva; vinculum broad and short, posterior margin with short to moderately long lateromedial projections, separated by shallow to deep medial emargination; saccus of variable shape, slender to broad, moderately to extremely long; phallus long and slender with small and weakly swollen coecum, rarely moderately short and stout with larger coecum, apical sclerite distinct to largely reduced.

FEMALE GENITALIA.– Segment VIII with almost evenly sclerotized subgenital plates, smooth, ventromedial zone membranous; ostium bursae frequently projected anteriorly and distinctly edged; antrum tubular, short to very long, up to more than two times length of apophysis anterioris; apophysis anterioris stout, rod-like, slightly longer than segment VIII; ductus bursae moderately slender, longer than apophysis anterioris; corpus bursae clearly separated, oval; signum arrow-shaped with three pointed lobes.

DISTRIBUTION.– Almost worldwide distribution with a concentration of species in the eremic parts of the Eastern hemisphere, particularly in the Afrotropical and Oriental region. Only two species are known from the Nearctic.

BIONOMICS.– Host-plants of *Ephysteris* are poorly known. The few records on host-plants of European *Ephysteris* belong to the Poaceae, whereas records on Zygophyllaceae refer to *Ochrodia* species (Karsholt & Sattler, 1998: 36).

REMARKS.– *Ephysteris* and *Microcraspedus* are here treated as synonyms, following the earlier discussion in Karsholt & Sattler (1998). In contrast to Povolný (2002a) we do not regard the comparatively small differences in the shape of the sacculus, the anterior margin of the vinculum or the female antrum as of subgeneric or even generic level. Otherwise *Ephysteris* s. str. would only include *E. promptella*, leaving a vast majority of species in a polyphyletic genus (or subgenus) *Microcraspedus* (see e.g. figures in Povolný (2002a) which depict various examples of genitalia variation within this group).

Several taxa have been described in this taxonomically difficult genus, frequently without giving an informative differential diagnosis, or based on insufficient material such as unique specimens. Furthermore, drawings of conventional slides with various pressure on the cover slides are extremely difficult and sometimes almost hopeless to interpret (see again figs. in Povolný (2002a)). Of particular help for a safe identification are, besides the often minute genitalia differences, frequently the moths themselves which can show quite distinct characters.

127. ***Ephysteris promptella*** (Staudinger, 1859)

Gelechia (Lita) promptella Staudinger, 1859: 241.
Tinea despectella Walker, 1863: 477.
Gelechia petiginella Mann, 1867: 843.
Doryphora parvula Staudinger, 1879: 323.
Aristotelia cacomicra Walsingham, 1908: 931.
Ephysteris chersaea Meyrick, 1908: 725.
Epithectis oschophora Meyrick, 1910: 219.
Anapatetris crystallista Meyrick, 1911: 229.
Phthorimaea dispensata Meyrick, 1921: 73.
Phthorimaea fanatica Meyrick, 1921: 73.
Metzneria xanthorhabda Gozmány, 1951: 20, fig.
Ephysteris (Ephysteris) promptella australiae Povolný, 1977d: 423, figs. 36-37, 55.

DIAGNOSIS.– Adult. Wingspan 8-12 mm. Labial palp cream-white; two black rings in segment 2 (blurred) and in segment 3 (distinct). Antenna black, rather indistinctly lighter ringed. Neck, thorax and tegula as forewing; face lighter. Forewing slender, light to dark grey-brown from white-based blackish scales, mottled with light orange-brown; fold and subcostal stripe orange-brown; small black spots at 2/5 and 3/5 in middle of wing. Hindwing grey.

Variation. A variable species. The forewings can be almost plain light brown, almost blackish grey or variegated yellowish brown to blackish grey. The black spots can be prominent or more or less missing. The black-tipped scales in apex can also be found along the veins of parts of or all of the forewings. *E. promptella* is also variable in size.

Similar species. Due to its variability *E. promptella* can be confused with other *Ephysteris* species. *E. diminutella* has broader forewings. *E. olympica* q.v.

MALE GENITALIA.– Uncus slender, apex rounded; gnathos-hook prominent, about length of uncus, stout; valva curved, broadly digitate, distally widened, extending almost to base of gnathos; sacculus broadly spatulate, extending slightly beyond middle of valva; posterior margin of vinculum with medial hump and narrow, short incision; saccus shorter than valva, stout, distally tapered with broad tip; phallus distinctly shorter than tegumen, distally with slight S-curvature, moderately stout, weakly inflated coecum about two-fifths length of phallus, apical sclerite reduced.

FEMALE GENITALIA.– Segment VIII short, about 1.5 times broader than long; subgenital plates evenly sclerotized, smooth, anteriorly with field of coremata; anterior margin of sternite VIII with sclerotized edge; antrum weakly separated, very short, tubular; apophysis anterioris stout, distal part rod-like, slightly longer than segment VIII; ductus bursae slender, almost two times length of apophysis anterioris; corpus bursae large, oval; signum broadly arrow- to horseshoe-shaped with two short lateral and one long medial lobe anteriorly.

DISTRIBUTION.– Southern parts of Europe to southern Russia. Also recorded from South and East Africa, Canary Islands and Madeira to the Middle East and Pakistan, Indoaustralian region. Recently also found in China (Li & Bidzilya, 2008: 47).

New records. France (Corsica): 5 ♂, 1 ♀, 4 km N Solenzara, Marine de Solaro, 3 m, 21.ix.2003, leg. Skou, gen. slide 4773 Hendriksen (ZMUC). Malta: 1 ♀, Bingemma, 25.vi.1999, leg. & coll. Sammut. European Turkey: 1 ♂, Kiyikoi, 26.ix.1989, leg. Baraniak (ZMUC).

BIONOMICS.– The larva has been recorded from *Andropogon* L., *Oryza* L., *Panicum* L., *Saccharum* L., *Sorghum* Moench, *Stipa* L. and *Triticum* L. (Poaceae), feeding in the stems of the host-plant (Karsholt & Sattler, 1998: 36). In subtropical and tropical areas the adults have been observed from February to late November/early December, occurring in several generations (Povolný, 2002a: 87). Vertical distribution from sea leavel up to 2500 m (Li & Bidzilya, 2008: 47).

REMARKS.– The extensive synonymy of *E. promptella* was discussed in detail by Sattler (1992: 110). According to (Klimesch, 1984: 157) specimens from the Canary Islands are darker than those of the nominotypical populations, but specimens from there examined by us fall into the range of variation described above.

128. Ephysteris tenuisaccus Nupponen, 2010

Ephysteris tenuisaccus Nupponen *in* Junnilainen & Nupponen, 2010: 24, figs. 50-54.

DIAGNOSIS.– Adult. Wingspan 11-12 mm. Labial palp white; tip of segment 3 slightly mottled with black. Antenna ringed with black and white. Head white, mottled with some light brown in neck; thorax and tegula as forewing. Forewing rather broad, cream-coloured, mottled with some brown scales; a large blackish brown spot in fold, a smaller at 2/5 in middle of wing, and a large, drop-shaped at 3/5; black spots near base and at 1/3 and 2/3 at costa; termen and basal half of cilia with brown-tipped white scales. Hindwing light grey.

Variation. There is some variation in the amount of brown scales on the forewing, and these can be blackish brown. In specimens with many brown scales thorax and tegulae are darker, and the apical part of the forewing becomes almost blackish brown.

Similar species. *E. tenuisaccus* is characterized by its rather broad, cream-coloured forewings with brownish markings, especially the large spots in fold and at 3/5, which are not surrounded by orange scales. *Microlechia karsholti* q.v.

MALE GENITALIA.– Uncus moderately broad, subtriangular, apex rounded; gnathos-hook prominent, about length of uncus, apically pointed; tegumen anteriorly deeply emarginated; valva curved, broadly digitate, distally widened with rounded apex, extending almost to middle of uncus, largely covered with microsetae; sacculus narrow, digitate, weakly curved, extending to about middle of valva; posterior margin of vinculum with short and broad lateromedial projections, medially with narrow U-shaped emargination; saccus about length of valva, slender, distally tapered with acute tip; phallus about length of tegumen, straight, moderately stout, weakly inflated coecum of about one-third length of phallus, apical sclerite distinct. Differs from related species such as *E. deserticolella* particularly in the slender saccus.

FEMALE GENITALIA.– Segment VIII with almost evenly sclerotized subgenital plates, smooth; antrum tubular, posteriorly indented, short, about one-third length of apophysis anterioris; apophysis anterioris rod-like, about 1.5 times length of segment VIII; ductus bursae slender, almost two times length of apopyhsis anterior; corpus bursae large, oval; signum horseshoe-shaped with three pointed lobes anteriorly. Somewhat similar to *E. deserticolella* but differs in the shape of the signum and the short and slender antrum.

DISTRIBUTION.– Only known from the type locality in Russia (southern Ural Mountains).

BIONOMICS.– Host-plant and early stages unknown. The adults were collected in June and mainly swept from *Hedysarum* sp. and *Nanophyton erinaceum* (Pallas) in afternoon sunshine. Specimens were rarely attracted to light (Junnilainen & Nupponen, 2010). The habitat is a steppe on limestone.

129. Ephysteris deserticolella (Staudinger, 1871)

Gelechia deserticolella Staudinger, 1871: 305.
Lita albocapitella Rebel, 1928: (33).
Opacopsis buvati Povolný, 1992b: 332, figs. 2, 4, **syn. n.**

DIAGNOSIS.– Adult. Wingspan 9-12 mm. Labial palp white, segments 2 and 3 each with two rather indistinct dark brown rings. Antenna black, indistinctly lighter ringed. Head, thorax and tegula cream-white, the latter mottled with some brown. Forewing light brownish grey from white-tipped brown and grey scales; some yellow-orange in fold and around black spots; a black spot near base; termen and basal half of cilia with many black-tipped white scales. Hindwing whitish grey.

Variation. There is some variation in the amount of brown, grey and sometimes black scales of the forewing, which can vary from almost white through light brown to light grey-brown.

Similar species. *E. deserticolella* is characterized by its cream-white head and thorax, and by its light forewings. *E. insulella* q.v., *E. iberica* q.v., *Microlechia karsholti* q.v.

MALE GENITALIA.– Uncus small, apex rounded; gnathos-hook prominent, about length of uncus; valva curved, broadly digitate, distally widened, extending almost to base of gnathos; sacculus narrow, digitate, weakly curved, extending to about middle of valva; posterior margin of vinculum with short lateromedial projections, medially with deep U-shaped emargination; saccus slightly longer than valva, stout, basally broad, distally tapered with truncate tip; phallus about length of tegumen, straight, moderately stout, weakly inflated coecum of about one-third length of phallus, apical sclerite distinct. Differs from the similar *E. insulella* and *E. diminutella* in the slightly broader and shorter sacculus the broadly emarginate vinculum with less developed projections and the comparatively stout saccus and phallus. *E. tenuisaccus* q.v.

FEMALE GENITALIA.– Segment VIII with almost evenly sclerotized subgenital plates, smooth; antrum tubular, broad and short, about one quarter length of apophysis anterioris; apophysis anterioris stout, distal part rod-like, slightly longer than segment VIII; ductus bursae moderately broad; corpus bursae large, oval; signum horseshoe- to arrow-shaped with three pointed lobes anteriorly. *E. tenuisaccus* q.v.

DISTRIBUTION.– Described from south-western Russia. Also recorded from the Ukraine and known from an unpublished specimen from Greece; Cyprus, Turkey, Iran, Afghanistan. According to Povolný (1968b: 6, 2002a: pl. 80, fig. 846) also on Malta, but this record is a misidentification of *E. iberica*. An isolated population occurs in southern France.

New record. Greece: 1 ♂, Rodopi, Fanari, sea lev., 19.viii.1985, leg. Fibiger, gen. slide 1407 Hendriksen (ZMUC).

BIONOMICS.– Host-plant and early stages unknown. The adults are on the wing from April to September, probably in at least two generations. They were collected in halophytic and eremic habitats from lowland localities up to altitudes of about 1600 m (Nel, 2003c: 395; Povolný, 1992b: 332, 2002a: 90).

REMARKS.– *E. buvati* was described from one male specimen only (Povolný, 1992b: 332). We have been able to examine the holotype and a further male collected by Nel (2003c: 395), whereas the female figured in this paper is a misidentification of *E. promptella*. The specimens of *E. buvati* are slightly darker than most *E. deserticolella*, but besides that we cannot find any differences. The moth is stated as small by Povolný (2002a: 89), whereas *E. deserticolella* should be middle-sized with 4.4-5.3 mm forewing-length. However, already in the original description the forewing length of *E. buvati* is correctly given as 5 mm (Povolný, 1992b: 332), thus just falling within the variation of *E. deserticolella*. Additional material from southern France may prove a separate subspecies but considering the geographical isolation it is not surprising to find such differences in forewing colour. *E. deserticolella* generally shows a high degree of variation with very light coloured specimens in Turkey.

130. *Ephysteris insulella* (Heinemann, 1870)

Lita insulella Heinemann, 1870: 251.
Gelechia insularis Staudinger, 1871: 304.
Doryphora? praticolella Christoph, 1872: 25, pl. 2, fig. 22. Subspecies, **comb. n., syn. n., stat. nov.**
Opacopsis gallica Povolný, 1992b: 330, figs. 1, 3, **syn. n.**

DIAGNOSIS.– Adult. Wingspan 12-15 mm. Labial palp white; segment 2 with two indistinct blackish brown rings; segment 3 with a slender basal and a broader apical black ring. Antenna black, ringed with white and grey. Head, thorax and tegula cream-white mottled with blackish brown; face lighter. Forewing slender, light grey from white-based blackish grey scales, mottled with some yellow and brown; fold and subcostal stripe yellow; three distinct small, black spots; termen and basal half of cilia with many black-tipped white scales. Hindwing grey.

Variation. There is some variation in the amount of yellow and brown scales on the forewings. The black spots can be more or less covered with yellow. One specimen has a slender black streak at 4/5 on the forewing. The above description refers to the nominal subspecies. Ssp. *praticolella* has lighter, cream-white (instead of light grey) labial palps, head, thorax, tegulae and forewings, which are more or less covered with light yellow, but with only little light grey, and then especially along apical margin and (sometimes) along veins in apical part of the wing.

Similar species. *E. insulella* is characterized by its relatively long and slender light grey or cream-coloured forewings with small, but distinct black spots. It can resemble *E. deserticolella*, but that species is smaller and has less distinctly ringed antennae and larger, black spots.

MALE GENITALIA.– Uncus small, apex rounded; gnathos-hook about length of uncus, slender; valva curved, broadly digitate, extending almost to base of gnathos; sacculus narrow, digitate, weakly curved, extending beyond middle of valva; posterior margin of vinculum with distinctly projecting lateromedial humps, medially with deep V-shaped emargination; saccus longer than valva, moderately stout, distally tapered to acute tip; phallus slightly longer than tegumen, straight, slender, weakly inflated coecum of about one-fifth length of phallus, apical sclerite distinct. Very similar to *E. diminutella* but normally with more distinct projection of the vinculum. *E. deserticolella* q.v.

FEMALE GENITALIA.– Segment VIII with almost evenly sclerotized subgenital plates, smooth; ostium bursae projected anteriorly with semi-circular edge; antrum tubular, long, from about 1.5 to almost two times length of apophysis anterioris; apophysis anterioris stout, rod-like, hardly exceeding length of segment VIII; ductus bursae moderately slender; corpus bursae well distinguished from ductus bursae, large, oval; signum large, subtriangular, medially with longitudinal ridge, anteriolateral lobes comparatively broad, anteriomedial lobe long and pointed. Practically indistinguishable from *E. diminutella* in characters of the female genitalia. *E. inustella* q.v.

DISTRIBUTION.– The nominotypical subspecies is known from isolated historical localities by the river Rhine in Germany (during the 19th century) and Switzerland, and from recent records from south-eastern France; a record from Austria (Povolný, 2002a: 91) is unconfirmed. Ssp. *praticolella* is known from Italy, the Balkans and south-western Russia (Christoph, 1872: 25; Ponomarenko, 2008: 97), Tuva region (Siberia (Povolný, 2001a: 206). as well as China and Mongolia (Li & Bidzilya, 2008: 55).

BIONOMICS.– Host-plant and early stages unknown. The species may be associated with *Stipa pennata* L. (Poaceae) as it was collected among this grass both in Russia (Christoph, 1873: 27) and France (Povolný, 1992b: 332). Reutti (1898: 247) collected the species in May and June along river banks in places without vegetation on the younger islands of the Rhine river. The adults have been observed from late April to early September, indicating bivoltinism, at least in southern Europe. The species is occasionally attracted to light. *E. insulella* prefers sandy habitats such as coastal dunes and sand banks of large rivers and steppe slopes.

REMARKS.– Genitalia characters such as the length of the antrum and the shape of the signum underly a considerable individual variation and largely agree with those of related taxa, particularly *E. diminutella*. Therefore *E. insulella* is more reliably identified externally than from genitalia structures.

E. insulella was described as '*Lita insulella* HS.' by Heinemann (1870: 251) based on an unspecified number of specimens from the 'Rheininseln' in June (see below). There is no description of an *insulella* in the work by Herrich-Schäffer (1847-55) and Heinemann is the author of *insulella*.

E.insularis was described as '*Gelechia insularis* Reutti *in litt.*' by Staudinger (1871: 304) based on one male and four females collected by Reutti in June 1857, also on the 'Rheininseln'. Reutti did not describe the species himself, but he dispersed specimens under the name of *insulella* and /or *insularis* (Reutti, 1898: 247), and Heinemann and Staudinger were probably unaware of their almost simultaneous description of this species. The type locality is an island of the river Rhine near Wittenweier, Ortenaukreis in Baden-Württenberg, Germany (Hausenblass, 2009: 104; Reutti, 1898: 247).

Doryphora? praticolella was a Zeller name which was made nomenclatorially available by Christoph (1872: 25-27), based on specimens from southern Russia ('Sarepta'). – Ponomarenko (2008: 97) listed this taxon as *Phthorimaea practicolella* [sic!] which is a misspelling.

The type series of *Gelechia furfurella* (Staudinger, 1871: 308) is a mixture of *Caulastrocecis furfurella* and *Ephysteris insulella*. We follow Povolný (1964: 348) in the attribution of the name to the species in the genus *Caulastrocecis*.

Opacopsis gallica was described from a unique specimen collected in May at Les Vigneaux (Hautes-Alpes, France) on a dry steppe slope with *Stipa pennata* L. (Poaceae) (Povolný, 1992b: 332). The description does not give a differential diagnosis but Povolný particularly mentions that the valva is obliquely pointed outwards. We have been able to study the genitalia slide of the holotype from MZMB and found that the genitalia are simply strongly pressed and hence distorted. Based on additional material collected in the Hautes-Alpes the genitalia characters completely fall within the range of variation of *E. insulella* and *E. diminutella*. The adults largely agree with specimens from the type locality of *E. insulella* from BMNH and we therefore consider *E. gallica* as a junior subjective synonym of the latter species.

131. *Ephysteris diminutella* (Zeller, 1847)

Gelechia diminutella Zeller, 1847: 855.
Phthorimaea lunaki Hartig, 1941: 157, pl. 6, fig. 6.

Ephysteris (Opacopsis) treskensis Povolný, 1964: 348, pl. 8, fig. 53, pl. 10, fig. 60, fig. 17., **syn. n.**
Ephysteris (Opacopsis) hispanica Povolný, 1981: 371, figs. 6-7, 29-30, 54, **syn. n.**
Ephysteris (Opacopsis) foulonsensis Povolný, 1981: 375, figs. 12-13, 34-35, 52, **syn. n.**

DIAGNOSIS.– Adult. Wingspan 8-12 mm. Labial palp with segment 3 almost as long as segment 2, cream-white; segments 2 and 3 each with two black rings. Antenna black, ringed with grey. Head cream-coloured, mottled with grey towards neck; thorax and tegula as forewing. Forewing dark grey from white-based black scales and mottled with much orange-brown, especially in dorsal third, including fold, subcostal stripe, around black spots and in apical part of wing; termen and basal part of cilia with many black-tipped light grey scales. Hindwing light grey.

Variation. A rather variable species both in size and forewing colour. The dark areas vary from light grey to almost black, and the areas with orange-brown scales can be lighter or darker and more or less dominant. Some specimens have a faint black streak in the apical part and / or at the base of the forewing.

Similar species. *E. diminutella* is characterized by its grey and orange-brown forewings with small but distinct black spots. Due to its variability it can be confused with several other *Ephysteris* species, and it is also similar to some *Scrobipalpa* (e.g. *S. suaedivorella*), but these have the black spots at 3/5 on the forewings closer to costa. *E. promptella* q.v., *E. inustella* q.v., *Vladimirea glebicolorella* q.v.

MALE GENITALIA.– Uncus small, apex rounded; gnathos-hook prominent, about length of uncus, weakly pointed; valva curved, broadly digitate, distally widened, extending to about base of gnathos; sacculus slender, digitate, weakly curved, extending from middle to about two-thirds of valva; posterior margin of vinculum with moderately broad and short lateromedial projections, inwardly with small hump, medially with deep V-shaped emargination; saccus longer than valva, stout, basally broader, distally tapered; phallus distinctly longer than tegumen, straight, moderately slender, weakly inflated coecum of about one-quarter length of phallus, apical sclerite distinct. Very similar and almost indistinguishable from the closely related *E. insulella* with weak differences in the less developed projections of the vinculum. *E. deserticolella* q.v.

FEMALE GENITALIA.– Segment VIII with almost evenly sclerotized subgenital plates, smooth, longitudinal inner edge of subgenital plate distinct; ostium bursae projected anteriorly with semi-circular edge; antrum tubular, moderately slender to broad, long, from about 1.5 to more than two times length of apophysis anterioris; apophysis anterioris stout, rod-like, about length of segment VIII; ductus bursae moderately slender; corpus bursae well distinguished from ductus bursae, large, oval; signum large, subtriangular, medially with longitudinal ridge, anteriolateral lobes comparatively broad, anteriomedial lobe long and pointed. The signum has a tendency to be more slender than that of *E. insulella* but the infraspecific variation frequently does not allow a safe identification of the female genitalia. *E. inustella* q.v.

DISTRIBUTION.– Widely though locally distributed in southern Europe, ranging from Spain to the Balkans. Recently found in Romania (Kovács & Kovács, 2000: 33). Male genitalia of an unidentified species from the southern Ural Mountains (Russia) (Junnilainen *et al.*, 2010: 51) largely agree with those from Sicily and should be considered conspecific. Probably also in North Africa (Morocco) (see below).

A male specimen from north-eastern Italy figured by Povolný (2002a: pl. 42, fig. 374) is a misidentification.

BIONOMICS.– Host-plant and early stages unknown. The adults have been observed from May to late September. They are attracted to light.

REMARKS.– A variable species which is, however, most reliably identified and separated from related species externally. We have been able to study rich material ranging from Spain to the Urals and found that several characters regarded as distinct at species level by Povolný underly a considerable individual and geographical variation. Particularly the length of the male sacculus, saccus and the phallus varies individually and between populations. These structures have a tendency to become increasingly stout in populations at the western and southern border of the distribution (Spain, Sicily) whereas they are on average longer in the north and east. However, e.g. material from Krk island (Croatia) is intermediate in these characters. Similarly female genitalia vary in some characters, particularly the length of the antrum which is shorter in western and southern parts of the distribution (Spain, Sicily) than in other areas. Therefore we came to the conclusion that most of the described taxa should not be kept separate for the time being. The possibility that there are hidden species within *E. diminutella* should be tested with molecular data in addition to morphology.

E. diminutella was described from a few specimens collected by Zeller in late July near Messina in Sicily. We were able to examine both type material and recently collected specimens. The population in Sicily is rather constant in the colour and wing markings.

Ephysteris treskensis was described from a male holotype and a female paratype collected in Macedonia by Kasy (Povolný, 1964: 348). The species was characterized by a long and slender phallus and particularly long signum. We have been able to study the genitalia slides of both type specimens from NHMW. They completely agree with several populations e.g. from the western Alps and fall into the infraspecific variation of the widely distributed *E. diminutella*. Furthermore external characters allow no separation of *E. treskensis* and hence we formally synonymize it with *E. diminutella*.

Ephysteris hispanica was described from three males and three females from Spain. The male holotype was collected by Klimesch in Albarracin (Povolný, 1981: 375). Two specimens with the same labels from the Klimesch collection (ZSM) are labelled 'holotype'. However, only the male with the slide number 'Kl. 5641 ♂' fits the holotype as the other specimen is a female. The female genitalia of populations from Spain are in general more stout than those of other populations, with a comparatively short saccus, phallus and antrum. However, considering the general variation within *E. diminutella* we synonymize *E. hispanica* with the latter species.

Ephysteris foulonsensis was described from six males and seven females collected in the middle of August in southern France (Povolný, 1981: 375). One further paratype from late September is labelled "Afrat". Povolný & Luquet (1983: 70) were unable to fix this name to a place in France, and we find it more likely that it refers to Afrat in northern Morocco. Povolný points to 'qualitative differences' compared to *E. treskensis*, such as the considerably more slender and subtle genitalia and the sclerotized part of the ductus bursae (antrum) which is stated to be about one-third longer in *E. foulonsensis*. We have been able to examine genitalia slides of three male and three female type-specimens from MZMB which exhibit a small degree of individual variation. Compared to the overall variability within different populations from southern Europe we consider this variation purely infrasubspecific. This interpretation is furthermore supported by a paratype without abdomen which largely agrees with *E. treskensis*. We therefore formally synonymize *E. foulonsensis* with *E. diminutella*.

132. *Ephysteris inustella* (Zeller, 1847)

Gelechia inustella Zeller, 1847: 853.
Epithectis delminiella Rebel, 1904: 353.
Lita gredosensis Rebel, 1935a: 14. Subspecies, **stat. rev.**

DIAGNOSIS.– Adult. Wingspan 10-12 mm. Labial palp pure white; segment 2 with two dark brown rings; segment 3 with two black rings. Antenna black, ringed with whitish grey. Face pure white; neck, thorax and tegula white mottled with black-tipped scales. Forewing grey from white-based black scales and mottled with orange especially in fold, apical area and around two pairs of distinct black spots situated above each other at 2/5 and 3/5; white scales in middle of wing and as faint costal and tornal spots; termen and basal half of cilia with many black-tipped scales. Hindwing grey.

Variation. The forewings vary from lighter to darker grey. The black spots can be more or less distinct, and the upper one at 3/5 is sometimes covered with orange scales. Also the neck can be pure white. Worn specimens become darker. *E. inustella* ssp. *gredosensis* is slightly smaller than the nominate form (wingspan 10-11 mm). It has darker labial palps, antennae, head, thorax and tegula. The forewings are grey from white-based black scales and mottled with orange-brown in fold, subcostal stripe, around distinct black spots and in apical part. One specimen has a distinct black patch near base.

Similar species. *E. inustella* is separated from other *Ephysteris* by the pure white colour of the head and labial palps, and by the distinct black spots which are broadly surrounded by orange. *E. inustella* ssp. *gredosensis* is hardly separable on external characters from some forms of the variable *E. diminutella*.

MALE GENITALIA.– Uncus small, apex rounded; gnathos-hook distinct, about length of uncus; valva curved, broadly digitate, distally obovate, extending almost to base of gnathos; sacculus slender, digitate, curved, extending to about middle of valva; posterior margin of vinculum lateromedially distinctly projected with pointed hump, medially with extremely deep V-shaped emargination; saccus longer than valva, stout, basally broad, distally tapered with rounded tip; phallus slender, longer than tegumen, slightly curved, weakly inflated coecum of about one-fifth length of phallus, apical sclerite distinct. The posterior margin of the vinculum and the shape of the valva are characteristic for this species.

FEMALE GENITALIA.– Segment VIII with almost evenly sclerotized subgenital plates, smooth, longitudinal inner edge of subgenital plate distinct; ostium bursae distinctly projected with oval anterior edge; antrum tubular, moderately slender, about 2.5 times length of apophysis anterioris; apophysis anterioris stout, basally widened, distal part rod-like, about length of segment VIII; corpus bursae well distinguished from ductus bursae, large, oval; signum moderately large, subtriangular with slender posterior lobe, medially with longitudinal ridge, anteriolateral lobes broad, anteriomedial lobe long and pointed. Differs from *E. insulella* and *E. diminutella* in the shape of the ostium bursae combined with the long and comparatively slender antrum and the smaller signum.

DISTRIBUTION.– Scattered distribution in Central and eastern Europe, reaching Estonia (Saaremaa Isl.) in the north and the Russian Volga-Don region in the east (Ponomarenko, 2008: 97). A record from Switzerland formerly attributed to *E. diminutella* with some doubts belongs to *E. inustella* (Sauter & Whitebread, 2005: 65); the occurrence of *E. inustella* in Switzerland has recently been confirmed (Schmid, 2007: 127-128). Recently also recorded from the Ukraine (Povolný, 2001a: 206). Most records from southern Europe (e.g., Povolný, 1964: 346) are probably referable to *E. diminutella. E. inustella* ssp. *gredosensis* occurs in mountains of Portugal and Spain at altitudes between 1600 and 2500 m.

BIONOMICS.– Host-plant and early stages unknown. The adults have been collected from late May to early June, at high altitudes in central Spain in late July. Further data remain unreliable due to the mixture with other species of *Ephysteris*.

REMARKS.– *Gelechia inustella* was first mentioned as '*Inustella* Zell. *in litt.*' as a synonym (and a misidentification) of *Gelechia* (now *Scrobipalpa*) *artemisiella* Zeller (1839: 201), and it was made nomenclatorially available by Zeller (1847: 853). Zeller did not give the exact locality for *inustella*, but it is assumed to be Silesia (Poland), from where it was also figured by Herrich-Schäffer (1854: 171).

Epithectis delminiella was described from a male specimen collected on Prenj mountain (Bosnia and Herzegovina) (Rebel, 1904: 354). We have not been able to trace the whereabouts of the holotype.

Lita gredosensis was described from two males and one female specimen collected in central Spain (Sierra de Gredos) by Reisser (Rebel, 1935a: 14). Povolný later examined and figured the three syntypes and synonymized *E. gredosensis* with *E. inustella* stating that is might be a subspecies of the latter (Povolný, 1964: 346, 1968b: 7). However, more recently he treated both as pure synonyms (Povolný, 2002a: 91). We have been able to study both male syntypes from NHMW - the figure of the 'holotype' (Povolný, 1968: pl. VI, fig. 27) does not constitute a valid lectotype designation (ICZN, 1999: Art. 74.5) - which largely agree with *E. inustella* in genitalia. The whereabouts of the only known female syntype is unknown but according to Povolný (1968b: pl. VII, fig. 34) it has a slightly more narrow antrum than *E. inustella*. Furthermore, differences in external appearance of specimens from Spain and Portugal seem to support a geographical differentiation. We therefore prefer to keep *E. inustella gredosensis* **stat. rev.** as a separate subspecies for the time being.

The figure of the adult *E. inustella* in Povolný (2002a: pl. 12, fig. 17) is possibly a misidentification.

133. Ephysteris olympica Povolný, 1968

Ephysteris (*Opacopsis*) *olympica* Povolný, 1968b: 8, fig. 36.
Ephysteris (*Opacopsis*) *olympica* ssp. *monticola* Povolný, 1981: 379, figs. 31-32, 61, **syn. n.**

DIAGNOSIS.– Adult. Wingspan 9-10 mm. Labial palp blackish brown mottled with white. Antenna black. Head, thorax and tegula as forewing; face lighter. Forewing with blackish brown, lighter-based scales and mottled with some light brown, especially along subcostal stripe and in fold; black spots indistinct, surrounded by a few light brown scales; dorsal part of termen with grey scales, apical part with whitish scales. Hindwing grey.

Variation. The few specimens examined show some variation on the forewing colour (from greyish brown to almost black) and in the distinctness of the black markings.

Similar species. *E. olympica* is characterized by its small size, and its dark, almost blackish colour. It may resemble dark specimens of *E. promptella*, but that species has lighter labial palps and normally also more brownish forewings.

MALE GENITALIA.– Uncus small, apex rounded; gnathos-hook distinct, about length of uncus; valva curved, broadly digitate, distally widened, extending almost to base of gnathos; sacculus moderately slender, digitate, curved, about two-thirds length of valva; posterior margin of vinculum lateromedially moderately projected with pointed hump, medially with deep V-shaped emargination; saccus about length of valva, stout, basally broad, distally tapered; phallus stout, short, about length of tegumen, with sclerotized longitudinal ridge, apex broadly rounded, apical sclerite distinct, weakly inflated coecum of about one-quarter to one-third length of phallus. Differs from other species of *Ephysteris* particularly in the phallus.

FEMALE GENITALIA.– Segment VIII with almost evenly sclerotized subgenital plates, smooth with weak longitudinal fold and moderately distinct inner edge; ostium bursae broadly projected anteriorly with ovate anterior edge; antrum funnel-shaped, anterior part weakly tapered, slightly exceeding apex of apophysis anterioris; apophysis anterioris stout, rod-like, slightly longer than segment VIII; signum small, arrow-shaped, anteriolateral lobes short and broad, anteriomedial lobe long and pointed. Differs from other species of the genus in the characteristic shape of the antrum.

DISTRIBUTION.– Disjunct distribution in the Olympus and Chelmos mountains (Greece), the western Alps and the Pyrenees (Spain, France).

BIONOMICS.– Host-plant and early stages unknown. The adults have been collected in July in alpine habitats up to altitudes of about 2700 m in Greece.

REMARKS.– The confusing mixture of the *E. olympica* type-material which included a specimen of *Carpatolechia decorella* (Haworth, 1812) was clarified by Sattler (1987a) who also selected a lectotype of *E. olympica*.

We have not been able to examine the female and its genitalia and our description is solely based on the figure published by Povolný (2002a: pl. 80, fig. 852).

Ephysteris (*Opacopsis*) *olympica* ssp. *monticola* was described from three specimens from La Grave (Hautes Alpes) and Vernet (Pyrenees) in France (Povolný, 1981: 370, 378-379). The type material is not specified in the description, and the localities of the specimens can only be seen from the legends to the genitalia figures. All three are stated to be paratypes, but a specimen from Vernet was later figured by Povolny (2002a, pl. 13, fig. 11) as the holotype of ssp. *monticola*. No differential diagnosis is given in the original description, but Povolny (2002a: 93) later stated spp. *monticola* to be smaller and with greyish (instead of black) forewings and more indistinct markings. We were able to examine a few specimens of *E. olympica* from Greece (Mt. Olympos and Mt. Chelmos), and a series from France (Hautes Alpes) and they agree in size and colour. We cannot find differences which could justify keeping the Greek and French populations as different subspecies, and they are therefore synonymized.

134. *Ephysteris iberica* Povolný, 1977

Ephysteris (*Opacopsis*) *iberica* Povolný, 1977b: 195, figs. 3, 33.

DIAGNOSIS.– Adult. Wingspan 9-11 mm. Labial palp white, mottled with blackish brown on lower and outer surface. Antenna greyish brown, indistinctly lighter ringed. Head cream-white, mottled with grey-brown towards neck; thorax and tegula as forewing. Forewing rather slender, light greyish from white-based black scales and mottled with light yellow and brown, especially in fold, subcostal stripe and along veins in apical half; black spots indistinct, surrounded by a few lighter scales; termen and basal half of cilia with many black-tipped white scales. Hindwing light grey.

Variation. The number of white-based black scales of the forewing is variable, and the forewings may look from almost whitish to light grey-brown.

Similar species. *E. iberica* is characterized by its rather slender, light grey forewings, mottled with yellow and brown, and with indistinct spots. *E. deserticolella* is similar, but has head and thorax white, less yellow and brown scales on the forewings and a black spot near base.

MALE GENITALIA.– Uncus small, apex rounded; gnathos-hook prominent, almost length of uncus; valva strongly curved in apical third, broadly digitate, extending almost to base of gnathos; sacculus narrow, digitate, straight, extending to apical third of valva; posterior margin of vinculum with short lateromedial projections, medially with deep U-shaped emargination; saccus about two times length of valva, slender, basally broad, distally strongly tapered with pointed tip; phallus much longer than tegumen, straight, slender, inflated coecum of about one-fifth length of phallus, apical sclerite reduced.

FEMALE GENITALIA.– Segment VIII with almost evenly sclerotized subgenital plates, smooth and without folds, medial part membranous and covered with microtrichia; ostium bursae projected anteriorly with short and broadly rounded antrum, extending to about one-quarter length of apophysis anterioris; basal third of apophysis anterioris broad, distal part rod-like, about length of segment VIII; ductus bursae basally narrow, gradually widening; corpus bursae well distinguished, oval; signum small, arrow-shaped, anteriolateral lobes short and broad, anteriomedial lobe longer, pointed.

DISTRIBUTION.– Widely distributed in the Mediterranean region from France and Spain (common) through Croatia to Crete and Israel but not yet published from most of these countries.

New records. France: 1 ♂, Aude, 5 km N St. Paul-de-Fenouillet, 450 m, 11.viii.2001, Skou leg. (ZMUC). Italy (Sicily): 1 ♂, 18 km E Gela, Acate valley, 50 m, 1.v.1987, leg. Fibiger, gen. slide 2540 Hendriksen (ZMUC). Greece: 1 ♂, 1 ♀, Lakonia, 5 km S Monemvasia, 7.viii.1980, leg. Christensen, gen. slide 2559 Hendriksen (ZMUC). Greece (Crete): 3 ♀, Lassithi distr., Makrigialos, 11-19.vi.1988, leg. Johansson (ZMUC). Croatia: 2 ♂, 1 ♀, Drvenik, 7.ix.2008, leg. & coll. I. Richter. Israel: 2 ♂, Jerusalem, 6.vi.2004, leg. Kravchenko, gen. slide 5498 Hendriksen (TAU, ZMUC).

BIONOMICS.– Host-plant and early stages unknown. The adults have been collected in all months from April to October. They are attracted to light.

REMARKS.– Data about the holotype were omitted in the original description. According to Povolný (2002a: 91) only known from the male holotype. However, we have been able to study series of both sexes from various localities.

Ochrodia Povolný, 1966

Ochrodia Povolný, 1966a: 142.
Type species: *Gelechia subdiminutella* Stainton, 1867.

DIAGNOSIS.– See species description.

MALE GENITALIA.– Sternite VIII much larger than tergite, posterior margin deeply excavated; tergite VIII short, rounded; uncus moderately broad, mitre-shaped with convex apex; gnathos-hook distinct, shorter than uncus; tegumen slender, anteriomedially with longitudinal ridge, anterior margin with V-shaped emargination; pedunculi of moderate size, subtriangular; valva short and evenly curved, extending to base of gnathos-hook, broadly thumb-shaped with mucronate apex; sacculus slender digitate, about one-third width of valva, slightly shorter than valva; vinculum moderately long, posterior margin with broad and shallow subrectangular emargination, short lateromedial projections with two humps; saccus about two to three times length of valva, slender, rod-like; phallus moderately to very long, straight to weakly curved, distal part slender, inflated coecum of about one-third length of phallus, apical sclerite reduced.

FEMALE GENITALIA.– Segment VIII with almost evenly sclerotized subgenital plates, smooth, posteriorly with several strong and stiff setae, posterior and inner margins convex; ostium bursae projected anteriorly, indistinct and very short antrum ring-shaped; apophysis anterioris stout, rod-like, basally converged and originating from subostial area, rod-like part obliquely turned outwards and ditally curved anteriad, about two to three times length of subgenital plate; ductus bursae narrow, gradually widening; corpus bursae well distinguished, oval; signum large, basal plate subtriangular with broad lateral lobes, medially with longitudinal and slightly denticulate ridge, signum-hook stout, weakly curved.

DISTRIBUTION.– Monotypic genus with a vast distribution in the eremic parts of the Palaearctic, Afrotropical, Australasian and Indo-Malayan region. Furthermore recorded from Jamaica where it was probably introduced.

BIONOMICS.– The larva has been recorded from *Balanites* Del., *Tribulus* L. and *Zygophyllum* L. (Zygophyllaceae) (Karsholt & Sattler, 1998: 36), according to Povolný (2002a: 94) also on *Zizyphus* Miller (Rhamnaceae). Further details see species description.

135. *Ochrodia subdiminutella* (Stainton, 1867)

Gelechia subdiminutella Stainton, 1867b: 45.
Gelechia jamaicensis Walsingham, 1897: 76.
Gnorimoschema bucolica Meyrick, 1904: 317.
Lita zygophyllella Rebel, 1912: 89.
Phthorimaea ericnista Meyrick, 1914: 245.
Aristotelia ferritincta Turner, 1919: 115. (Subspecies).
Phthorimaea ochrodeta Meyrick, 1923: 23.

Phthorimaea extorris Meyrick, 1923: 50.
Phthorimaea crocoleuca Meyrick, 1923: 51.
Lita unitella Turati, 1930: 78, pl. 2.
Aristotelia tribulivora Dumont, 1931: 150, fig. 1.
Aristotelia pulverea Janse, 1950: 115, pl. 50, figs. 7-8, pl. 54, figs. 10-11, pl. 55, fig. 9, pl. 62, figs. 5-6.
Scrobipalpa turgida Janse, 1951: for: 218, pl. 87, fig. 5, pl. 88, fig. 4, pl. 90, fig. 6, pl. 97, figs. 13-14, pl. 99, fig. 12
Ephysteris pentamacula Janse, 1958: 107, pls 16-18, pl. 39h, pl. 52d, pl. 63i-j.
Gnorimoschema infallax Gozmány, 1960: 417.
Gnorimoschema tractatum Gozmány, 1960: 418.

DIAGNOSIS.– Adult. Wingspan 7-11 mm. Labial palp white; segments 2 and 3 each with a black ring. Antenna black, ringed with grey. Head cream-coloured; thorax and tegula as forewing. Forewing variegated from white-based black scales and orange-brown markings, the latter especially near base, by broad patches surrounding or covering two pairs of black spots at 2/5 and 3/5 and as a streak separating light costal and tornal spots. Hindwing with costal third grey and dorsal two thirds whitish grey.

Variation. A highly variable species. The above description is based on specimens from southern Spain (probably representing one species). Their forewings vary from cream-coloured mottled with yellow-brown or orange-brown to blackish grey with black spots surrounded by brown. Also the size of the moths is variable. Specimens from Malta are larger (wingspan 11-13 mm) and lighter than those from Spain. Specimens from Cyprus are on average slightly larger (10-12 mm) than Spanish ones and have a darker head and grey olive-brown forewings with the black spots surrounded with less orange-brown and have also darker hindwings. They may represent a separate species (see remarks below).

Similar species. Due to its variation *O. subdiminutella* can be confused with several *Scrobipalpa* and *Ephysteris* species. Most specimens can be recognized by the many white-tipped scales and the large orange markings on the forewings, and especially by the hindwings which are darker in the costal third than in the dorsal two-thirds.

MALE GENITALIA.– See genus description.

FEMALE GENITALIA.– See genus description.

DISTRIBUTION.– In Europe known from Spain (several provinces) and unpublished records from Malta and Crete. Also recorded from Italy (Basilicata, Abruzzi) and Croatia (Dalmatia) (Povolný, 1987: 86; 2002a: 94). Subtropical and tropical Africa and Asia, from the Mediterranean, the eremic parts of Africa, the Middle East to Australia.

New records. Greece (Crete): 1 ♂, Drepanides by Kissamos, 100 m, 27.iii-2.iv.1999, leg. Fibiger & Jeppesen *et al.*, gen. slide 3188 Hendriksen (ZMUC). Malta: 1 ♂, 1 ♀, Mellieha, 2.-4.vii.2002, leg. Hendriksen (ZMUC).

BIONOMICS.– The larva (of *O. subdiminutella* s.l.) has been recorded from *Zygophyllum album* L. and various other species of *Zygophyllum* L. and from *Balanites aegyptiaca* (L.) Del. (Zygophyllaceae) (Povolný, 2002a: 94). In Australia specimens were bred from the seed-capsules of *Tribulus terrestris* L. (Zygophyllaceae) (Povolný, 1977d). The larva initially mines a short gallery and later it causes a large reticulate blotch-mine with or without frass inside the mine. The initial leaf-mine may be changed at a later stage and frequently several leaves are spun together and mined. Finally pupation takes place outside the leaf-mine (Povolný, 2002a: 94). Rebel (1912: 89) found the larva during late March and early April on *Zygophyllum guyotii* Kneucker & Muschl. and the adults emerged in April.

REMARKS.– As already stated in the original description of *Ochrodia*, *O. subdiminutella* is one of the taxonomically most confused species of the whole tribe Gnorimoschemini (Povolný, 1966: 145). Being unable to resolve its taxonomy, Povolný (2002a: 93-94) synonymized all *Ochrodia* taxa with *O. subdiminutella*. The genus *Ochrodia* is most diverse in the drier parts of the Old World subtropical and tropical zones, reaching only the warmest parts of Europe. A revision of *Ochrodia* therefore falls outside the scope of the present volume, and we tentatively follow Povolný's opinion. Preliminary studies of Afrotropical *Ochrodia* indicate that *O. subdiminutella* is a species complex, with several species distributed, e.g., in the Arabian Peninsula. Also European *O. subdiminutella* seems to cover more than one species, none of which are clearly conspecific with the type of *Gelechia subdiminutella* from Israel. Specimens from Cyprus are larger and darker that specimens from Spain, and the former are probably referable to *O. extorris* (Meyrick).

Vladimirea Povolný, 1967

Vladimirea Povolný, 1967c: 148.
Type species: *Vladimirea wiltshirei* Povolný, 1967.
Distinxia Povolný, 1967c: 157.
Type species: *Vladimirea (Distinxia) amseli* Povolný, 1967.

DIAGNOSIS.– Rather small to medium sized moths (wingspan 9-12 mm). Segment 3 of the labial palps almost as long as segment 2. The outer black spots closer to dorsum than the middle one. See species description.

Outside Europe *Vladimirea* species are rather diverse and include large and strongly marked species.

MALE GENITALIA.– Sternite VIII broader than long, posterior margin flattened and with or without medial emargination; tergite VIII tongue-shaped; uncus variable, broad to moderately narrow, apex convex, with or without medial emargination; gnathos-hook small to long; tegumen moderately slender, anterior margin with broad emargination; pedunculi of moderate size, subtriangular; valva moderately short to long, at most slightly exceeding tip of uncus, evenly curved, broadly digitate with hardly inflated, rounded apex; sacculus rod-shaped to slender digitate or bulb-shaped, about half to two-thirds length of valva; vinculum moderately long, posterior margin frequently with broad and shallow emargination, short lateromedial projections present or reduced; saccus stout, moderately slender to broad, of moderate length hardly ever exceeding length of valva; phallus stout and short, rarely long and slender, straight to weakly curved, inflated coecum of about one-third to one-quarter length of phallus, apex frequently with strong sclerotized hook and additional sclerotizations.

FEMALE GENITALIA.– Segment VIII short, broader than long, subgenital plates frequently smooth, occasionally folded, rarely with distinct medioproximal field of small-sized foamy sculpture; ostium bursae in anterior part of sternite VIII, frequently edged by sclerotized rim; apophysis anterioris stout, rod-like, exceeding length of segment VIII;

antrum and colliculum absent; ductus bursae moderately narrow; corpus bursae well distinguished, oval; signum variable and of moderate size, basal plate usually subtriangular, signum-hook small to long.

DISTRIBUTION.– Palaearctic distribution, ranging from North Africa to Mongolia. 16 species have been described so far.

BIONOMICS.– The host-plants are various species of Zygophyllaceae. The larval habits vary considerably from leaf-miners to gall-makers or they spin leaves and/or seeds together.

REMARKS.– The subgenus *Distinxia* with the only described species *Vladimirea (Distinxia) amseli* Povolný, 1967 was considered as intermediate between *Vladimirea* and *Ephysteris* (Povolný, 1967c: 157). In fact the genitalia characters of both sexes are very close to species of *Ephysteris* such as *E. tenuisaccus*. However, in the absence of a generic revision a possible synonymization is postponed.

136. Vladimirea glebicolorella (Erschoff, 1874)

Bryotropha glebicolorella Erschoff, 1874: 101, pl. 6, fig. 114.
Vladimirea submaculata Povolný, 1967c: 154, figs. 16-17.

DIAGNOSIS.– Adult. Wingspan 9-11 mm. Labial palp cream-white mottled with blackish brown, especially on lower and outer surfaces. Antenna ringed with black and dirty-white. Face cream-white; neck, thorax and tegula as forewing. Forewing light brownish grey from black and brown, white-based scales; a rather large black spot in fold and two such spots in middle of wing at 2/5 and 3/5, the latter closest to dorsum; cream-white costal and tornal spots weakly fused forming an indistinct bent fascia; apical area and termen with many black scales. Hindwing light grey with whitish grey cilia.

Variation. The above description is based on specimens from Crimea which are slightly variable. One specimen has the black spots on the forewings followed by many cream-white and light brown scales. A specimen from Afghanistan is larger (wingspan 12 mm), with cream-white head and light brown forewings with small and rather indistinct black spots. Its genitalia are identical with those of specimens from Crimea.

Similar species. *V. glebicolorella* is characterized by having the forewings covered by black and brown, white-tipped scales, and by the three rather large black spots with the outer one closer to dorsum than the middle one. It may resemble specimens of the variable *Ephysteris diminutella*, but that species has segment 3 of the labial palps shorter than segment 2.

MALE GENITALIA.– Uncus with convex apex; gnathos-hook very small; valva distinctly shorter than uncus, evenly curved, digitate with weakly rounded apex; sacculus broadly digitate, almost size of valva, short, about half length of valva; vinculum moderately long, posterior margin with broadly U-shaped emargination, lateromedial hump present; saccus stout, basally broad, distal part evenly tapered to pointed apex, about length of valva; phallus stout and short, straight, inflated coecum of about one-third length of phallus, apex with distinct sclerotized hook.

FEMALE GENITALIA.– Segment VIII subrectangular, about two times broader than long, smooth; sternite VIII with transverse, sclerotized anterior margin; ostium bursae at anterior margin of sternite VIII, edged by short lateral sclerite; apophysis anterioris rod-like, about two times length of segment VIII; ductus bursae moderately narrow; corpus bursae well distinguished, oval; signum small, basal plate subtriangular, signum-hook distinct.

DISTRIBUTION.– Only known from the Ukraine (Crimea) (Bidzilya & Budashkin, 2009: 19). Outside Europe recorded from Armenia, Uzbekistan, Turkmenistan, Iran and Afghanistan (Piskunov, 1988: 360; Povolný, 2002a: 95).

BIONOMICS.– The larva mines on *Zygophyllum fabago* L. (Zygophyllaceae), later it also feeds freely on the leaves or causes stem-galls (Piskunov, 1988: 360). It has been observed in June and July, and adults occur from April to June (Povolný, 2002a: 95).

REMARKS.– The identification of *V. glebicolorella* from the Ukraine is based on Bidzilya & Budashkin (2009: 19) and we follow these authors, despite some discrepancies in the male genitalia compared to figures of *V. glebicolorella* in literature. Particularly the short valva and sacculus differ from the two figures of male genitalia in Povolný (2002a: pl. 48, figs. 428-429) which, however, indicate a considerable individual variation in several characters such as the shape of the uncus and the sacculus. Fig. 428 depicts the holotype of *V. submaculata* which was collected in Sharud in Armenia, whereas the other specimen allegedly also from Armenia is mismatched and depicts a paratype from Afghanistan (see Povolný, 1967c: fig. 17). Considering the genitalia differences it cannot even be excluded that these two specimens are not conspecific.

Microlechia Turati, 1924 **gen. rev.**

Microlechia Turati, 1924: 162.
Type species: *Microlechia chretieni* Turati, 1924.
Hedma Dumont, 1932: 714, **syn. n.**
Type species: *Hedma abzacella* Dumont, 1932.
Megalocypha Janse, 1960: 196 **syn. n.**
Type species: *Megalocypha polioptera* Janse, 1960.

DIAGNOSIS.– Small moths (wingspan 6-9 mm). Labial palps slender; segment 2 only slightly broader than segment 3; both segments of about same length. Antenna more or less distinctly lighter and darker ringed. Forewing in *M. chretieni* and *M. rhamnifoliae* with small patches of raised scales. Hindwing whitish grey to grey.

MALE GENITALIA.– Sternite VIII broad, posteriorly rounded, with pair of coremata; tergite VIII broadly tongue-shaped; uncus elongated oval, without apical emargination; gnathos-hook moderately curved, slender, extraordinarily long, about length of uncus; tegumen anteriorly widened, anterior margin with broadly rounded emargination; pedunculi of moderate size, distally rounded; valva distinctly shorter than uncus, weakly curved, digitate with inflated and usually rounded apex; sacculus needle-shaped, about half to almost length of valva, distal part curved inwards; posterior margin of vinculum with moderately broad and shallow emargination, short to moderate lateromedial humps present; saccus slender to broad,of moderate length hardly exceeding length of valva; phallus long and slender, straight to weakly curved, inflated coecum of about one-fifth to one-quarter length of phallus, apex frequently with strong, sclerotized hook and additional spines.

FEMALE GENITALIA.– Apophysis posterioris moderately long; segment VIII with smoothly sclerotized subgenital plates, occasionally folded; anterior margin frequently with short lateromedial humps; apophysis anterioris stout, rod-like, about length of segment VIII; antrum about length of segment VIII, moderately short funnel-shaped to long and tubular; ductus bursae with or without lateral sclerites; corpus bursae large, oval; signum-hook large, slender, curved, without additional spines, basal plate small.

DISTRIBUTION.– Palaearctic distribution, ranging from the Canary islands and North Africa to Central Asia and India. 8 species have been described so far.

BIONOMICS.– The host-plants are various species of Solanaceae and Rhamnaceae. The larvae are leaf-miners.

REMARKS.– *Microlechia* was hitherto considered as a junior synonym of *Recurvaria* Haworth, 1828 (Huemer & Karsholt, 1999: 35). However, a careful re-examination of the original description provided evidence that *Microlechia* is congeneric with *Hedma* and *Megalocypha* and we formally synonymise these genus group names.

137. Microlechia rhamnifoliae (Amsel & Hering, 1931) **comb. n.**

Lita rhamnifoliae Amsel & Hering, 1931: 142, pl. 1, fig. 1.
Lita rhamnifoliae Amsel, 1935b: 301, pl. 12, figs. 144-145.

DIAGNOSIS.– Adult. Wingspan 7-8 mm. Labial palp white; segment 2 with a broad black ring; segment 3 with two slender black rings at base and near tip. Antenna dark grey, indistinctly lighter ringed. Head cream-white mottled with some grey-brown towards neck; thorax and tegula as forewing. Forewing light grey from black, light-based and light-tipped scales; a large black patch in middle of wing; small patches of orange-brown raised scales in fold, middle of wing and near tornus; black spots near base and along costa; apical part with many black scales. Hindwing light grey with light grey cilia.

Variation. The few examined specimens show some variation in the amount of black scales on the forewing. In one specimen the antennae are uniformly dark grey.

Similar species. *M. rhamnifoliae* is characterized by its small size and by the greyish forewings with a large black patch which is missing in *M. chretieni*.

MALE GENITALIA.– Uncus large, suboval, broadest at base; gnathos-hook long and stout, longer than uncus, curved, with pointed apex; valva digitate, extending beyond base of uncus, weakly curved, distal fifth inflated with rounded apex; sacculus basally about half width of valva, extending to distal fifth of valva, basal part straight, subapically with distinct curvature, apex pointed; vincular process hump-like; posterior margin of vinculum with indistinct and shallow medial emargination; saccus shorter than valva, broad at base, distally narrowing to pointed apex; phallus about length of valva, slender, weakly curved, apex with subterminal hooklet, coecum inflated and about one-quarter length of phallus.

FEMALE GENITALIA.– Apophysis posterioris short, about three times length of segment VIII; segment VIII with almost evenly sclerotized subgenital plate, smooth; anterior margin with rounded lateromedial humps; antrum tubular, anteriorly obliquely edged, shorter than apophysis anterioris; ductus bursae anteriorly widened, almost three times length of apophysis anterior, small field of microspines in basal third; corpus bursae large, oval; signum-hook moderately large, slender, curved at an almost right angle in middle, without additional spines, basal plate small.

DISTRIBUTION.– In Europe only known from Greece; furthermore recorded from Cyprus, Canary Islands and North Africa (Morocco) to Palestine and Saudi Arabia. Also in South Africa.

New records. Greece: 1 ♀, I. Karpathos, Lefkos, 30 m, 23.v.1997, leg. Sutter, gen. slide GU 09/1308 ♀ Huemer (ZMUC). Cyprus: 2 ♀, Alethrikó, 11.-13.iv.1991, leg. Johansson, gen. slide 2584 Hendriksen (ZMUC). Morocco: 2 ♂, 5 ♀, 8 km S Sidi Ifni, 50 m, 20.iii.2005, leg. Karsholt, gen. slides 5056, 5110, 5116, 5117 Hendriksen (ZMUC). Spain (Canary Islands): 1 ♂, Teneriffe, Callao Salvaje, 1.-5.i.1985, leg. Schnack, gen. slide 3985 Hendriksen (ZMUC).

BIONOMICS.– The larva causes characteristic mines in the leaves of *Rhamnus palaestina* Boiss. (Rhamnaceae) during February and March. It starts as a short gallery or blotch and later the leaf blades of two or three leaves are spun together and usually the whole leaves are mined. Frass is irregularly deposited in the mine. Pupation takes place between spun leaves in a slight cocoon and adults emerged after about 4 weeks (Amsel & Hering, 1931: 141-142). The species also occurs outside the range of *Rhamnus palaestina* Boiss. and additional host-plants are to be expected (Povolný 2002a: 100).

REMARKS.– *Lita rhamnifoliae* was described twice, first by Amsel & Hering (1931: 142) and then by Amsel (1935b: 301), without mentioning the description from 1931. The original description is solely based on the larva.

138. Microlechia chretieni Turati, 1924

Microlechia chretieni Turati, 1924: 163, pl. 6, fig. 5.
Phthorimaea microcasis Meyrick, 1929: 532, **syn. n.**
Gelechia micradelpha Walsingham, 1900: 217; homonym of *Gelechia micradelpha* Lower, 1897.
Teleia hyoscyamella Amsel & Hering, 1931: 132, pl. 1, fig. 5; homonym of *Teleia hyoscyamella* Rebel, 1912.
Hedma abzacella Dumont, 1932: 714, **syn. n.**
Teleia hyoscyami Amsel, 1935b: 299, pl. 12, fig. 147, **syn. n.**
Megalocypha polioptera Janse, 1960: 197, pl. 81, fig. d, pl. 93, fig. a, pl. 98, fig. h, pl. 102, figs., **syn. n.**
Ephysteris aellographa Janse, 1960: 220, pl. 91, fig. d, pl. 106, figs., pl. 126, fig. h,i, pl. 129, fig. g, **syn. n.**

DIAGNOSIS.– Adult. Wingspan 6-9 mm. Labial palp white; segments 2 and 3 each with one or two black rings. Antenna indistinctly ringed black and grey. Head cream-white, mottled with grey-tipped scales; thorax and tegula as forewing. Forewing grey from white-based blackish grey scales and mottled with orange-brown; small groups of black and orange-brown raised scales in fold at 1/3, and at 1/2 and 3/4 in middle of wing. Hindwing grey.

Variation. There is some variation in the amount of orange-brown scales and in the number and size of black spots and patches on the forewing.

Similar species. *M. chretieni* is characterized by its small size and the greyish forewings with tufts of raised black and orange-brown scales. *M. rhamnifoliae* q.v.

MALE GENITALIA.– Uncus large, oval, medially about same width as length; gnathos-hook long and stout, longer than uncus, curved, with pointed apex; valva slender digitate, extending to middle of uncus, distal fifth distinctly inflated with rounded apex; base of sacculus almost width of valva, extending to distal fifth of valva, basal part straight, distinct curvature at three-quarters, distal quarter turned inwards, apex pointed; vincular process posteriorly projected, pointed, inner margin convex; posterior margin of vinculum with deep, almost U-shaped emargination with sclerotized margins; saccus moderately long, about length of valva, slender, broader at base; phallus about twice length of valva, slender, straight, about half a dozen of distinct spines in posterior half, apex with bent subterminal hooklet; coecum inflated, spherical, about one-quarter length of phallus. Differs from the similar *M. karsholti* e.g. by the shape of the uncus, the emargination of the vinculum and the distinctly longer saccus and phallus.

FEMALE GENITALIA.– Apophysis posterioris short, about three times length of segment VIII; segment VIII with smoothly sclerotized subgenital plate; anterior margin with weak lateromedial humps; antrum cylindrical, medially weakly constricted; slightly longer than apophysis anterioris; ductus bursae short, with lateral sclerites; corpus bursae large, oval; signum-hook large, evenly curved, basal plate small.

DISTRIBUTION.– Widely distributed in the Mediterranean, ranging from the Canary Islands and Madeira (Porto Santo) to an unpublished record from Greece (Crete); North Africa, South Africa and the Middle East (Povolný, 1978a).

New record. Greece (Crete): 1 ♂, 1 ♀, Lassithi distr., Makrigialos, 11.-19.vi.1998, 1 ♀, 20.v.1993, leg. Johansson, gen. slides 2607, 2608 Hendriksen (ZMUC).

BIONOMICS.– The larva mines the leaves, flower-buds and stems of Solanaceae and has been recorded from *Lycium europaeum* L. (Walsingham, 1900: 218; Hartig & Amsel, 1952: 85), *Lycium intricatum* Boiss. (Klimesch, 1984: 156) and other *Lycium* spp. and *Hyoscyamus* L. (Povolný, 1978a: 44). It causes small mines in the leaves in late May. Preferred habitats are coastal dunes and rocky slopes.

REMARKS.– *Microlechia chretieni* was described from 3 specimens collected at Bengasi (Libya) on the 15th of February (Turati, 1924: 147). The whereabouts of the type-material are dubious but probably at least two specimens were destroyed together with a majority of Turati´s collection. A further syntype which should be in the collection of Chrétien, now MNHN, could not be traced by Leraut (*in litt.*) and is probably lost. However, following the detailed original description including a colour photograph of the adult we have no doubt that *Microlechia chretieni* is nothing but a senior synonym of *Phthorimaea microcasis* and we therefore formally synonymize both taxa.

139. Microlechia karsholti (Nupponen, 2010) **comb. n.**

Hedma karsholti Nupponen *in* Junnilainen & Nupponen, 2010: 26, figs. 55-59.

DIAGNOSIS.– Adult. Wingspan 8-9 mm. Labial palp white; apical half of segment 2 faintly mottled with brown. Antenna black, ringed with greyish white. Head, thorax and tegula white. Forewing cream-white; an elongate black spot in fold and two circular black spots at 2/5 and 3/5 in middle of wing, the latter larger and situated closer to dorsum than the middle one; black scales at costa near base and at 3/5, and in apical part of the wing. Hindwing whitish grey.

Variation. The above description is based on the holotype.

Similar species. *M. karsholti* is characterized by its white palps, head, thorax and tegulae, and by its cream-white forewings with distinct black spots. It may resemble whitish *Ephysteris* species like *E. deserticolella* and *E. tenuisaccus*, but *Ephysteris* species have segment 3 of the labial palps shorter than segment 2.

MALE GENITALIA.– Uncus large, oval, elongated; gnathos-hook long and stout, longer than uncus, curved, with pointed apex; valva slender digitate, extending to middle of uncus, distal fifth weakly inflated with slightly pointed apex; sacculus more slender than valva, extending to distal fifth of valva, basal part straight, distinct curvature at three-quarters, distal quarter turned inwards, apex pointed; vincular process posteriorly projected, pointed, inner margin convex; posterior margin of vinculum with deep, oval emargination with sclerotized margins; saccus moderately short, about two-thirds length of valva, stout, broader at base, apically pointed; phallus about length of valva, slender, straight, several minute spines in posterior half, apex with small subterminal hooklet; coecum inflated, spherical, about one-third length of phallus. *M. chretieni* q.v.

FEMALE GENITALIA.– Apophysis posterioris long, about four times length of segment VIII; segment VIII with smoothly sclerotized, weakly folded subgenital plate, anterior margin heavily sclerotized, without lateromedial humps; apophysis anterioris with elongated base, distal part rod-like; antrum short, about one-third length of apophysis anterioris, tubular, posterior and anterior margin concave; ductus bursae anteriorly widened, sub-basally wrinkled; corpus bursae distinct, sub-oval; signum-hook large, slender, evenly curved, basal plate small.

DISTRIBUTION.– Only known from the type locality in Russia (southern Ural Mountains).

BIONOMICS.– Host-plant and early stages unknown. The few adults known to date have been collected in June by sweeping low vegetation just before dusk and at light during night. The habitat is a steppe on limestone.

Cosmardia Povolný, 1965

Cosmardia Povolný, 1965: 484.

Type species: [*Tinea*] *morizella* [sic !] Geyer, 1836.

DIAGNOSIS.– See species description.

MALE GENITALIA.– Sternite VIII broad, posteriorly rounded; tergite VIII broadly tongue-shaped; uncus short, broad, narrower than tegumen; gnathos without strong hook, culcitula comparatively large; tegumen widened from uncus to pedunculi, shallowly emarginate; pedunculi distinct; valva digitate, reaching to about base of gnathos; sacculus digitate, slightly shorter than valva; vinculum broad, weakly incised posteriomedially and lateromedially, resulting in two pairs of short processes; saccus long, slender; anellus without separate sclerites; phallus slender, straight, with sclerotized apical arm; coecum weakly separate.

FEMALE GENITALIA.– Apophysis posterioris long; segment VIII short, broader than long, without specialized modifications, anteriorly with sclerotized margin; apophysis anterioris comparatively short, rod-like; antrum funnel-shaped, about length of segment VIII, extending slightly beyond apex of apophysis anterioris; corpus bursae well defined; signum a strong and large hook with rounded basal plate.

DISTRIBUTION.– Monotypic genus, restricted to the Western Palaearctic region, extending into western Siberia.

BIONOMICS.– The host-plants of the larva are *Silene* spp. and *Lychnis* L. (Caryophyllaceae). Further details see species description.

140. Cosmardia moritzella (Treitschke, 1835)

Oecophora moritzella Treitschke, 1835: 214.
[*Tinea*] *morizella* [sic !] Geyer, 1836: pl. 71, figs. 476-477.
Lita roseella Zetterstedt, 1839: 1005, **syn. n.**

DIAGNOSIS.– Adult. Wingspan 14-17 mm. Segment 2 of labial palp light brown, more or less mottled with dark brown on lower and outer surface; segment 3 black. Head with light brown face and dark brown neck; thorax and tegula black; a group of erect, red scales between base of proboscis and compound eye. Antenna black, very indistinctly lighter ringed. Forewing black in dorsal and apical part, light whitish brown to yellowish brown in costal part; at the border of the two colours in middle of wing two black spots; costal and tornal spots whitish, indistinct; cilia dark grey, lighter beyond black cilia line. Hindwing grey, with grey cilia. Lighter areas in medial parts of fore- and middle legs red.

Variation. The lighter, costal part of the forewings varies from uniform light brown to cream-white mottled with brownish. Specimens may be more or less reddish: besides the red scales below the eye and on fore- and middle legs some specimens also have the light parts of the hindlegs red, and the labial palps and costa of the forewing can have a reddish tinge.

Similar species. The bicoloured forewings are characteristic for *C. moritzella*.

MALE GENITALIA.– See genus description.

FEMALE GENITALIA.– See genus description.

DISTRIBUTION.– Locally recorded from various Central European countries, ranging from France to Romania and according to Povolný (2002a: 104) to the Ukraine and the southern Ural Mountains (Russia) (Junnilainen *et al.*, 2010: 52); northwards through north-eastern Germany and northern Poland to Latvia and southern Finland; an isolated population in central Norway; recently also recorded from Portugal (Corley, 2009: 170) but absent in the Mediterranean area and from much of Scandinavia. Extra-European records are known from western and southern Siberia.

BIONOMICS.– The larva and biology were described in some details already in the original description (Treitschke, 1835: 215-216). The larva has been found on *Silene dioica* (L.) Clairv. (Schütze, 1931: 89), *Silene latifolia* Poirr . (= *S. pratensis* (Rafn.) Godr., = *Melandrium album* Garcke) (Thomann, 1956: 433), *S. bellidifolia* Juss. ex Jacq. (Povolný, 2002a: 104) and *Lychnis flos-jovis* (L.) Desv. (Huemer & Erlebach, 2003). A record from *Silene vulgaris* (Moench) Garcke (Povolný, *op. cit.*) requires confirmation. The larvae live more or less gregariously among the flowers and unripe seed-capsules, which are spun together. One can often find small and nearly full fed larvae together (Treitschke, *op. cit.*), probably because the females appear during a long period after hibernation. In the laboratory pupation took place in a slight coccon amonst debris, spun leaves or between tissue. Adults can be found during most of the year, but there is probably only one generation per year.

REMARKS.– *Oecophora moritzella* was described from an unspecified number of specimens found as larvae by Moritz in Brandenburg, Germany (Treitschke, 1835: 214-215). We examined two syntypes from the Treitschke collection in the HNHM, and in order to stabilize nomenclature one of these is here published as lectotype. It is a female, labelled: 'TREITS. 3703'. 'Lectotype *Oecophora moritzella* Treitschke, 1835 O. Karsholt, design. 2004'.

Lita roseella was described from an unspecified number of specimens collected by Boheman in Dovre, Norway. We examined a male in the NHRS, and in order to stabilize nomenclature this specimen is here designated as lectotype. It is male labelled: 'Dv. [Dovre] Bhn [Boheman] Zr. [Zetterstedt]' 'Lectotype *Lita roseella* Zetterstedt, 1839 O. Karsholt, design. 2003'.

Wocke (1862: 235-236) examined a syntype of *roseella*, but probably another one, because he stated it to be a female, having the costal part of the forewing rose-red (light greyish red in the lectotype). Wocke concluded that the specimen examined by him differed so much from usual *moritzella* that he considered them a distinct species. *Lita roseella* was not discussed again in the literature, and is only listed, as a distinct species, in a few catalogues.

Specimens from the isolated population in middle Norway, where the larva feeds on flowers of *Silene latifolia*, are large and generally with red suffusion. They have similar genitalia to specimens from Latvia, northern Germany, Central Europe and the Altai Mts, and we consider them to be conspecific, and even hardly separable at subspecies level.

The name *moritzella* has been erroneously attributed to Hübner by numerous authors (Sattler, 1973: 188). Even though Povolný (2001: 104) cites the paper by Sattler (1973), he still used the wrong spelling *morizella*.

Lutilabria Povolný, 1965

Lutilabria Povolný, 1965: 482.
Type species: *Gelechia lutilabrella* Mann, 1857.

DIAGNOSIS.– Medium sized to rather large (wingspan 14-21 mm), more or less broad-winged moths. Labial palps rather long and slender, segment 3 distinctly shorter than segment 2. Underside of antenna in male with very short cilia. Forewing almost unicolorous yellowish to greyish brown with only indistinct black spots; cilia light grey beyond indistinct, darker cilia line. Hindwing almost without emargination on outer margin before apex.

MALE GENITALIA.– Sternite VIII broad, posteriorly rounded; tergite VIII broadly tongue-shaped; uncus large, broad; gnathos with paired basal sclerites, apically with small sclerite but without hook, culcitula large; tegumen widened from uncus to pedunculi, deeply emarginate; pedunculi distinct; valva digitate, comparatively broad and long, almost reaching tip of uncus; sacculus broadly digitate, about three-quarters length of valva; vinculum broad, weakly incised posteriomedially, with one pair of short lateromedial processes; saccus long, stout, distally slightly tapered; anellus without specialised sclerites; phallus slender, straight to curved, with sclerotized apical arm; coecum weakly to distinctly inflated.

FEMALE GENITALIA.– Apophysis posterioris very long, about 6-7 times length of segment VIII; segment VIII longer than broad, posterior margin with several strong setae, medial region with broad longitudinal zone of microspines; tergite VIII much longer than sternite VIII with oblique and well sclerotized anterior margin, sclerotized modifications of segment VIII absent, several longitudinal folds present posterioventrally; apophysis anterioris comparatively long, rod-like; antrum funnel-shaped with lateral sclerotization; corpus bursae distinctly separated from ductus bursae; signum a large hook with moderate curvation, basal plate distinct to reduced.

DISTRIBUTION.– Extremely scattered records from Slovenia and Croatia to the Ukraine, southern Russia and Mongolia.

BIONOMICS.– Host-plants and early stages unknown. Adults have been collected in steppe habitats, mainly in mountainous regions.

141. *Lutilabria lutilabrella* (Mann, 1857)

Gelechia lutilabrella Mann, 1857: 179.
Lutilabria olympica Huemer, 1993: 341, figs. 1-2, 6, 8, 10, 14-16. Subspecies, **stat. n.**

DIAGNOSIS.– Adult. Wingspan 15-17 mm. Labial palp light yellow mottled with brown on lower and outer surface. Antenna black, ringed with whitish grey. Head and tegula light brown; thorax dark brown. Forewing rather broad, covered with yellow-brown, light-based scales. Hindwing grey with lighter cilia.

Variation. The forewing colour varies from light yellow-brown to light brown. – Ssp. *olympica* is larger (wingspan 17-21 mm) and more broad-winged. The labial palps, head, thorax, tegulae and forewings in males are darker, greyish brown. Females of ssp. *olympica* have pure yellow labial palps and yellowish brown forewings.

Similar species.The nominotypical subspecies of *L. lutilabrella* resembles females of *Agonochaetia intermedia*, but it can be recognized by its distinctly ringed antennae and by its shorter and lighter labial palps.

MALE GENITALIA.– Uncus posteriorly evenly curved, without medial emargination; gnathos with moderately large, horseshoe-shaped medial sclerite, culcitula large; tegumen anteriorly with deep U-shaped emargination; valva weakly curved, distal part sword-like, apically pointed, reaching tip of uncus; sacculus about three-quarters length of valva, strongly curved in basal half, distal part clubbed, apically a weak point; vinculum with largely reduced posteriomedial humps, medially with shallow convexity, oblique and setose lateromedial margin of vinculum inwards with short hump; saccus moderately long, stout; phallus slender, distal part slightly curved, with sclerotized apical arm; coecum distinctly swollen, almost one-third length of phallus.

FEMALE GENITALIA.– Segment VIII longer than broad, posteriorly tapered, with setae along posterior margin; sternite VIII with ventral folds present, transverse sclerotized edge from middle towards anteriolateral margin, anterior edge moderately sclerotized, oblique; apophysis anterioris basally widened and evenly joined to oblique anterior edge, distal four-fifths rod-like; antrum funnel-shaped, posterior margin concave, anterior margin indented, lateral part broadly sclerotized; ductus bursae slender, hardly

longer than apophysis anterioris; corpus bursae distinctly separated, oval; signum horn-shaped, basal plate ovate to largely reduced, signum-hook long, weakly curved. The sclerotized antrum, short ductus bursae and the differently shaped basal plate of the signum differ from *L. volgensis*.

DISTRIBUTION.– Until recently only known from the type-material which was collected in Croatia and Greece respectively. It has now been found in the Ukraine (Bidzilya, 1997: 124). Also recorded from Bulgaria (Salin, 2004: 80-81) (as *L. olympica*) and an unpublished record from Slovenia.

New record. Slovenia: 1 ♀, Karst, Kozina – Materija, 30.v.2001. leg. Habeler, gen. slide GEL 1060 ♀ Huemer (TLMF).

BIONOMICS.– Host-plant and early stages unknown. In Slovenia the adult was observed in late May, in the Ukraine it was collected from late June to mid-July among grass and bushes of *Amygdalus nana* L. (Rosaceae) around 7 a.m. Search at other times gave no results, and no moths were attracted to light (Bidzilya, *in litt.*).

REMARKS.– Povolný (2002a: 104) synonymized *L. olympica* with *L. lutilabrella* and argued that differences figured in the original description were merely based on different preparation techniques and age of specimens. Compared with other species of the genus the differences in genitalia are weak and may be based on some individual variation, but the limited material and the lack of molecular data does not allow a final decision. However, besides of small observed differences e.g. in the basal plate of the signum, *L. olympica* differs from *L. lutilabrella* in the larger size. We therefore prefer to keep it as a subspecies of the latter for the time being.

142. *Lutilabria volgensis* Anikin & Piskunov, 1996

Lutilabria volgensis Anikin & Piskunov, 1996: 172, figs. 1-2.

DIAGNOSIS.– Adult. Male. Wingspan 14-16 mm. Labial palp yellow; outer and lower surface (especially of segment 2) densely mottled with blackish brown. Antenna, head, thorax and tegula grey-brown. Forewing grey-brown; fold with some orange scales and an indistinct black spot; termen with black scales. Hindwing dark grey with light grey cilia. Female. Slightly smaller than male (wingspan 13-14 mm) and with less blackish brown scales in the yellow labial palps.

Variation. The above description is based on males from the southern Ural Mountains which, apart from the slight sexual dimorphism, are rather constant. The holotype of *L. volgensis*, which is without labial palps and the right forewing, is slightly larger (wingspan 17 mm) and has lighter, yellowish brown head and forewing.

Similar species. *L. volgensis* is characterized by its rather small size, relatively slender dark forewings and yellowish labial palps. The holotype of *L. volgensis* is very similar to the specimen of *L. lutilabrella* from the Ukraine figured on the colour plate. *L. prolata* q.v.

MALE GENITALIA.– Uncus posteriorly with broadly rounded lateral flaps, separated by medial emargination; gnathos with small medial sclerite, culcitula large; tegumen anteriorly with moderately deep U-shaped emargination; valva weakly curved, distal part sword-like, apically pointed, reaching tip of uncus; sacculus about three-quarters length of valva, strongly curved in basal half, distal part clubbed, apically a weak point; vinculum with pair of short posteriomedial humps, medially separated by short incision, oblique and setose lateromedial margin of vinculum slightly projected posteriad; saccus long, slender; phallus slender, distal part slightly curved, with sclerotized apical arm; coecum distinctly swollen, nearly one-third length of phallus. *L. prolata* q.v.

FEMALE GENITALIA.– Segment VIII longer than broad, posteriorly tapered, with distinct setae along posterior margin, ventral folds present, anterior edge strongly sclerotized, oblique; is longer than segment VIII, basally widened and evenly joined to oblique anterior edge, distal four-fifths rod-like; antrum funnel-shaped, membranous, laterally with weakly sclerotized margins; ductus bursae slender, about two times length of apophysis anterioris; corpus bursae distinctly separated, oval; signum horn-shaped, basal plate distinct, evenly rounded, signum-hook long. *L. lutilabrella* q. v.

DISTRIBUTION.– Only known from Russia (Saratovskaya Oblast (type locality) and the southern Ural Mountains).

BIONOMICS.– Host-plant and early stages unknown. The holotype was collected in late May on limestone hills. Additional material from the southern Ural Mountains was collected from late May to mid-June in steppe habitats (Junnilainen *et al.*, 2010: 52).

REMARKS.– *L. volgensis* was described from a single male from Saratov province. The original figure of the genitalia depicts an inwardly projected sclerotized hump of the posterior margin of the vinculum and a straight phallus (Anikin & Piskunov, 1996: 172, figs. 1-2). During reexamination of the genitalia of the holotype which were preserved in crystallised sugar, we found the phallus to be lost and the structures of the vinculum destroyed. All other characters of diagnostic value we could observe agree with material from the southern Ural Mountains which we consider conspecific for the time being.

143. *Lutilabria prolata* Junnilainen & Nupponen, 2010

Lutilabria prolata Junnilainen & Nupponen, 2010: 28, figs. 63-64.

DIAGNOSIS.– Adult. Wingspan 17-19 mm. Labial palp whitish; segment 2 densely covered with black; segment 3 mottled with some black scales. Antenna, head, thorax and tegula fuscous. Forewing covered with black, grey-based scales; fold with some brownish scales and an indistinct black spot. Hindwing dark grey with black marking and light grey cilia.

Variation. The few more or less worn specimens examined show only slight variation.

Similar species. *L. prolata* is characterized by its dark fuscous forewings and by the whitish ground colour of the labial palps. *L. volgensis* is smaller and has yellowish labial palps.

MALE GENITALIA.– Uncus posteriorly with broadly rounded lateral flaps, separated by medial emargination; gnathos with small medial sclerite, culcitula large; tegumen anteriorly with very broad and moderately deep emargination; valva strongly curved, distal part broad, folded, apically pointed, reaching tip of uncus; sacculus about three-quarters length of valva, strongly curved in basal half, distal part clubbed, apically projected with distinct point; vinculum with pair of rounded posteriomedial humps, medially separated by short incision, medial area of vinculum with transverse sclerite, oblique and setose lateromedial margin slightly projected and folded; saccus moderately long, stout; phallus slender, distal part curved, with sclerotized apical arm; coecum distinctly swollen, nearly one-third length of phallus. Differs from the similar *L. volgensis* e.g. by the broader and strongly curved valva, the structures of the vinculum and the shorter saccus.

Female genitalia.– Unknown.

Distribution.– Described from few localities, ranging from Russia (southern Ural Mountains) to the Ukraine and Romania.

Bionomics.– Host-plant and early stages unknown. The few adults known to date have been collected from late May to early June. *L. prolata* seems to be restricted to steppe habitats.

Klimeschiopsis Povolný, 1967

Klimeschiopsis Povolný, 1967d: 191.
Type species: *Gelechia discontinuella* Rebel, 1899.

Diagnosis.– Adult. Wingspan 9-14 mm. Labial palps slender; underside of segment 2 without a groove. Apex of hindwing with slight emargination. Antenna black, indistinctly lighter ringed. Forewing black mottled with white; two light, transverse fasciae and light costal and tornal spots.

Male genitalia.– Sternite VIII broad, posteriorly rounded; tergite VIII broadly tongue-shaped; uncus broad at base, distal half narrow; gnathos without hook, culcitula large; tegumen widened from uncus to pedunculi, deeply emarginate; pedunculi small; valva digitate, comparatively broad and long, extending beyond base of gnathos; sacculus broadly digitate, slightly curved apically, about four-fifths length of valva; vinculum broad and long, with deep, usually V-shaped posteromedial emargination, without separate lateromedial processes; saccus long, stout, distally slightly tapered; anellus without specialised sclerites; phallus stout, straight, with sclerotized apical arm; coecum weakly inflated, usually with basal sclerite.

Female genitalia.– Apophysis posterioris long, about 4 to 5 times length of segment VIII; segment VIII posteriorly with short setae, without modifications ventrally but with numerous weak folds; apophysis anterioris long, rod-like; antrum short, funnel-shaped; posterior part of ductus bursae without or with weak lateral sclerite; corpus bursae well defined; signum a strong, moderately short hook with largely reduced basal plate.

Distribution.– Holarctic distribution, with four species occurring in the Palaearctic region. Three of them occur in Europe, one species in Afghanistan. The taxonomy of the Nearctic species is not yet resolved.

Bionomics.– Host-plants and early stages are unknown. However, probably larvae feed on various Caryophyllaceae.

Remarks.– Genitalia structures are extremely similar within *Klimeschiopsis* and safe identification is a lot easier when based on examination of external characters.

144. *Klimeschiopsis kiningerella* (Duponchel, 1843)

Lita kiningerella Duponchel, 1843: 456, pl. 85, fig. 5.
Phthorimaea atralbella Palm, 1947: 36, fig. 1 a, e.

DIAGNOSIS.– Adult. Wingspan 11-13 mm. Labial palp cream-white, mottled with black. Antenna black, indistinctly ringed with grey. Face cream-coloured, mottled with black in neck; thorax as forewing; tegula black with white tip. Forewing black mottled with white; an oblique white stripe from near base of costa to before dorsum at 1/3 and one from 1/3 at costa to 2/3 in middle of wing; distinct costal and tornal spots white, separated by a black streak; indistinct black spots in fold and at 2/5 and 3/5 in middle of wing; apical area black. Hindwing whitish grey.

Variation. There is some variation in the extent of white markings of the forewing.

Similar species. *K. kiningerella* is characterized by its black forewings with two blurred white oblique streaks from costa of the forewings. It is similar to *K. discontinuella*, which has a cream-white head and broad white bands on the forewing, and to *K. terroris*, which has more clear black and white markings and less oblique white markings on the forewings.

MALE GENITALIA.– Uncus broadly subrectangular with rounded apex; gnathos with rounded basal sclerite, culcitula large; tegumen with deep U-shaped anterior emargination; valva digitate, long, extending to about apex of uncus, weakly curved, apex weakly inflated; sacculus broadly digitate, about four-fifths length of valva, apex curved, pointed; vinculum broad and long, with deep and narrow posteromedial emargination, lateromedially with rounded projections; saccus long and slender; phallus stout, straight, with sclerotized apical arm and weakly inflated coecum, basal sclerite reduced. Very similar to *K. discontinuella* and *K. terroris*, differing mainly by the rounded lateromedial projections of the vinculum.

FEMALE GENITALIA.– Segment VIII posteriorly with short setae, sternite VIII medially membranous, intensively covered with microtrichia, subgenital plate smooth, laterally weakly folded; apophysis anterioris about length of segment VIII, basal part weakly widened, distal four-fifths rod-like; antrum short, broadly funnel-shaped, about one-fifth length of apophysis anterioris; ductus bursae with moderately abrupt transition to longish-oval corpus bursae; signum-hook long and strong, about half length of apophysis anterioris, weakly curved, basally with distinct lateral projections, basal plate small. Differs from the other species of *Klimeschiopsis* particularly in the long signum-hook.

DISTRIBUTION.– Mainly restricted to lower mountains of Central Europe, from France to Romania and the Ukraine (Carpathian mts.), but also in Scandinavia. Recently recorded from Russia (southern Ural Mountains) (Junnilainen *et al.*, 2010: 52). Absent from the European and African Mediterranean. A specimen from Spain (Albarracin) figured by Povolný 2002a (p. 15, fig. 12) but not mentioned under distribution in the text most probably belongs to *K. terroris*.

New record. Turkey: 1 ♀, Ankara, 10 km NW Kizilkahaman, 1150-1250 m, 6-7.viii.1989, leg. Fibiger & Esser (ZMUC).

BIONOMICS.– Host-plant and early stages unknown. According to Povolný (2002a: 103) the larva feeds on *Silene dioica* (L.) Clairv. but no evidence to the source is given. Kaitila (1996: 103) suspects *Lychnis viscaria* L. and *L. alpina* L. (= *S. suecica*

(Lodd.) Greut. et Burd.) (Caryophyllaceae) as possible host-plants. Adults occur in June and July, according to Povolný (*op. cit.*) in May and August. Preferred habitats are rocks covered in mosses and lichens. Vertical distribution: only known from altitudes up to about 1000 m.

145. *Klimeschiopsis discontinuella* (Rebel, 1899)

Gelechia discontinuella Rebel, 1899: 178.

DIAGNOSIS.– Adult. Wingspan 12-14 mm. Labial palp cream-white; segment 2 mottled with some black on lower and outer surface; segment 3 with two black rings. Antenna black, indistinctly ringed with grey. Head cream-coloured; thorax and tegula black, the latter with white tip. Forewing blackish grey from white-based black scales; a broad greyish white oblique fascia near base connected along dorsum with a similar curved fascia in middle of wing; white costal and tornal spots separated by black; indistinct black spots in fold and at 2/3 and 3/5 in middle of wing. Hindwing grey.

Variation. The white markings of the forewing can be of different size and more or less mottled with white-based black scales.

Similar species. *K. discontinuella* is characterized by its cream-white head and by the broad greyish white fascia on the forewings. *K. terroris* is smaller, with darker head and more distinct black and white markings on the forewings. *K. kiningerella* q.v.

MALE GENITALIA.– Posterior margin of vinculum with deep V-shaped emargination, broad and slightly pointed lateromedial projections; saccus stout. Otherwise as in *K. kiningerella* q.v. and *K. terroris* q.v. The shape of the posterior margin of the vinculum and the stout saccus are the most reliable diagnostic characters compared with other species of *Klimeschiopsis*.

FEMALE GENITALIA.– Antrum broadly funnel-shaped, with concave outer margin; signum-hook moderately short, strong, about one-quarter length of apophysis anterioris, weakly curved, basally with long lateral projections. Differs from the otherwise similar *K. kiningerella* in the shorter signum, and from *K. terroris* in the slightly longer and laterally concave antrum as well as in the longer signum hook.

DISTRIBUTION.– Only known from a restricted area of the southern Alps, ranging from northern Italy to parts of Switzerland. Records from Austria (Povolný, 2002a: 103) are erroneous and refer to the former Austrian Südtirol, which is now part of Italy. Recently also recorded from the Altai Mountains in Central Asia (Bidzilya, 2005: 15).

BIONOMICS.– Host-plant and early stages unknown. The adult occurs from May to late September, pending on the altitude of its habitat. It has been observed in steppe habitats with rock and scree up to altitudes of about 2100 m and is attracted to light.

REMARKS.– Povolný (2002a: 103) treats *K. terroris* as a junior synonym of *K. discontinuella*. However, the two taxa are here regarded as distinct due to morphological differences and biogeographic reasons (see below).

146. *Klimeschiopsis terroris* (Hartig, 1938) **sp. rev.**

Phthorimaea (Lita) terroris Hartig, 1938: 89, pl. 7, fig. 5.

DIAGNOSIS.– Adult. Wingspan 9-12 mm. Labial palp white; segment 2 mottled with black; segment 3 with two ill-defined black rings. Antenna black, ringed with light grey. Head cream-white mottled with black towards neck; thorax and tegula black, the latter with white tips. Forewing black with three irregular cream-white fasciae near base, in middle of wing and from costal to tornal spots; black spots ill-defined. Hindwing dark grey.

Variation. The white fasciae of the forewing are variable in form and distinctness. In one specimen the middle one is reduced.

Similar species. *K. terroris* is characterized by its relatively small size and its black forewings with three irregular cream-white fasciae. *K. kiningerella* q.v., *K. discontinuella* q.v.

MALE GENITALIA.– Posterior margin of vinculum with deep V-shaped emargination, moderately slender and slightly pointed lateromedial projections; saccus slender. Otherwise as in *K. kiningerella* q.v. and *K. discontinuella* q.v. The slender shape of the posterior margin of the vinculum and the slender saccus are characteristic features of *K. terroris*.

FEMALE GENITALIA.– Ventromedial area covered with microtrichia comparatively small; antrum broad and short, funnel-shaped; corpus bursae small; signum-hook short, about one-sixth length of apophysis anterioris, curved, basally with lateral projections. Otherwise as in *K. kiningerella* q.v. and *K. discontinuella* q.v. The small corpus bursae and signum are the most important diagnostic characters compared with other species of the genus.

DISTRIBUTION.– Only known from Teruel (Albarracin), Castellón and Andalusia (Spain).

BIONOMICS.– Host-plant and early stages unknown. The few existing specimens were collected from middle to late August. In Castellón (Mt. Penyagolosa) the adult has been observed at light near a rock-surface dominated by Mediterranean steppe vegetation and thermophilous shrubs such as *Amelanchier* Medik. (Rosaceae) at altitudes of about 1500 m (Huemer, pers. obs.).

REMARKS.– Povolný (2002a: 103) treats *K. terroris* and *K. discontinuella* in synonymy. However, the remote distributional areas of the two taxa remain unexplained and differences of the adults and genitalia are distinct, indicating two separate species.

Caryocolum Gregor & Povolný, 1954

Caryocolum Gregor & Povolný, 1954: 87.
Type species: *Gelechia leucomelanella* Zeller, 1839.

Diagnosis.– Diagnosis. Small to medium sized (8-18 mm) species. Segment 2 and segment 3 of labial palp of about the same length; in *C. marmorea* and *C. pullatella* segment 2 is longer than segment 3. Forewing with black, brown, grey, orange or white scales. Cilia of forewing in most species with grey basal part and light grey distal part beyond a black cilia line. Cilia of hindwing lighter or darker grey. As the colour of the cilia are of only little importance in the identification they are not dealt with in the diagnosis of the single species. In a few species (*C. proxima, C. fibigerium, C. junctella*) the tarsi of the forelegs and especially the midlegs with light reddish scales. The female of *C. laceratella* is slightly brachypterous.

MALE GENITALIA.– Sternite VIII broad, posteriorly rounded, frequently with pair of coremata; tergite VIII broadly tongue-shaped; uncus short, broad, narrower than tegumen; gnathos without hook, culcitula large; tegumen widened from uncus to pedunculi, shallowly to deeply emarginate anteriorly; pedunculi distinct; transtilla sclerotized, with or without spines; valva variably shaped, never extending beyond uncus; sacculus separate or rarely fused with valva, usually shorter than valva; vinculum broad, incised or weakly emarginate posteriomedially, frequently also emarginate lateromedially, resulting in up to two pairs of processes; saccus long, slender to extremely broadened; anellus with or without pair of peg-like to needle-shaped sclerites; phallus slender, frequently straight, rarely curved, with sclerotized apical arm and with or without minute cornuti at apex; coecum weakly to distinctly swollen.

FEMALE GENITALIA.– Apophysis posterioris short to long; segment VIII posteriorly with long setae, usually with sclerotized modifications ventrally, subgenital plate frequently with folds, occasionally with drop-shaped, flap-like or digitate processes; apophysis anterioris rod-like, variable in length; antrum tubular to funnel-shaped, extending to about apex of apophysis anterioris, rarely reduced; posterior part of ductus bursae with or without lateral sclerite; corpus bursae normally well defined, frequently with microtrichia; signum strongly hook-like with basal plate, exceptionally reduced to small sclerotized plate.

DISTRIBUTION.– Holarctic genus with approximately 70 species mainly from temperate Europe to the Middle East and with few Nearctic species (Huemer, 1988).

BIONOMICS.– The host-plants exclusively belong to the Caryophyllaceae and additional records are doubtful or may refer to xenophagy. The larvae usually feed between spun shoots or in stems, occasionally also in flowers. Numerous species occur in mountainous habitats.

REMARKS.– *Caryocolum* can be subdivided into two divisions: division A is characterized by the presence of tiny cornuti and a pair of lateral sclerites of the ductus bursae; division B by the large pedunculi, a medially strongly narrowing tegumen and prolongations of the ventral margin of the tegumen towards transtilla. Within these divisions species-groups are strongly reflected by host-plant selection, indicating a co-evolution. The genus *Caryocolum* was revised by Klimesch (1953-1954) and Huemer (1988) and only few species have been described since.

147. *Caryocolum fischerella* (Treitschke, 1833)

Lita fischerella Treitschke, 1833: 84.

DIAGNOSIS.– Adult. Wingspan 12-14 mm. Segment 2 of labial palp blackish brown, upper part of inner surface and tip almost white; segment 3 black. Antenna ringed brown and black. Head light fuscous with cream frons; thorax and tegula light brown, posterior tip of thorax black. Forewing light brownish, mottled with white and blackish; black spots at base on costa and dorsum; two large black patches in middle of the wing at 2/5 and 3/5; costal and tornal spots white, nearly reaching each other; apical area with many black scales. Hindwing grey.

Variation. A fairly constant species. Worn specimens look more greyish and have lighter labial palps.

Similar species. Due to the large black spot in the light brown forewings *C. fischerella* should not be misidentified.

MALE GENITALIA.– Uncus broadly rounded; tegumen shallowly emarginate anteriorly; transtilla broad, band-like, without spines; valva long, digitate, slightly tapered and curved towards apex; sacculus shorter, broad, apically pointed; posterior margin of vinculum with slight incision medially, lateromedial projections indistinct; saccus stumpy, frequently with transverse sclerite apically; phallus short, straight, with few minute cornuti apically.

FEMALE GENITALIA.– Segment VIII without processes; apophysis anterioris about 1.5 times length of segment VIII; antrum short, about one-fifth length of apophysis anterioris, funnel-shaped, with lateral folds; posterior part of ductus bursae with pair of lateral sclerites, extending to about two-fifths of apophysis anterioris; signum an extremely prominent hook, slightly bent at apex, base of signum largely reduced.

DISTRIBUTION.– Mainly recorded from Central and eastern Europe, reaching southern Scandinavia and European Russia. Absent from the Iberian Peninsula and the British Isles.

New record. Macedonia: 1 ♀, Prosečnik, 7.viii.1986, leg. Baraniak (ZMUC).

BIONOMICS.– The larvae have been recorded feeding gregariously on *Saponaria officinalis* L. (Treitschke, 1833: 85) and they are also mentioned from *S. ocymoides* L. (Klimesch, 1953: 315). Other records of host-plants, e.g. *Silene bellidifolia* Juss ex Jacq. (Klimesch, *op. cit.*), probably refer to different species. The larva occurs in May and early June and spins the terminal shoots and leaves together. Pupation takes place in a web between the leaves (Treitschke, *op. cit.*). The adults emerge from mid-June to late July, in the wild they have been observed until early August. *C. fischerella* mainly occurs on riversides at low altitudes, in northern Europe also in wastelands and gardens.

148. Caryocolum tischeriella (Zeller, 1839)

Gelechia tischeriella Zeller, 1839: 11.

DIAGNOSIS.– Adult. Wingspan 11-12 mm. Segment 2 of labial palp black, with upper part of inner surface white; segment 3 black. Antenna black, indistinctly lighter ringed; lower surface of antennal scape white. Head with cream-white frons and black neck; thorax and tegula blackish, more or less mottled with brown. Forewing black; an oblique white spot from costa near base; a white patch in middle of wing; costal and tornal spot white, nearly reaching each other. Hindwing grey.

Variation. The white spots on the forewings are variable in size, and some specimens have an additional, small one in the fold. Sometimes thorax and tegula are almost brown.

Similar species. The cream-white head and black neck, the antennal scape with white lower surface and particularly the white patch in the middle of the forewing are characteristic for *C. tischeriella*.

MALE GENITALIA.– Uncus broad, subrectangular; tegumen shallowly emarginate anteriorly; transtilla weakly sclerotized, without spines; valva long, broad at base, abruptly tapered to long digitate distal part, apex curved; sacculus broadened medially, apically pointed; posterior margin of vinculum with slight incision medially, lateral pair of projections small, distinct; saccus long, slender; phallus long, curved, with numerous minute cornuti apically.

FEMALE GENITALIA.– Apophysis posterioris short; segment VIII without processes, pair of narrow lateromedial folds ventrally; apophysis anterioris short, about length of segment VIII; antrum short, about one-third length of apophysis anterioris, funnel-shaped, with lateral folds; posterior part of ductus bursae with pair of lateral sclerites, anteriorly extending to apex of apophysis anterioris, posteriorly fused with antrum; signum with distinct base and comparatively large, curved hook.

DISTRIBUTION.– Widely distributed in Europe, except the British Isles, the far north and parts of the Mediterranean. Furthermore not yet recorded e.g. from Denmark, The Netherlands, Belgium and Poland. A record from Hungary (Klimesch, 1953: 315) is considered doubtful (Fazekas, 2002: 310) and needs confirmation. Occurs in North Africa (Morocco) and recently also found in Central Caucasus and in Siberia (Ponomarenko, 2008: 98).

New records. Italy (Sicily): 3 ♂, Ragaha, Bosco de Patemò, Monte Arso, 1000 m, 9.ix.2002, leg. Karsholt *et al.* (ZMUC). Greece (Crete): 1 ♂, 4 km S. Topolia, 300 m, 25-29.vi.2000, leg. Fibiger *et al.*, gen. slide 3274 Hendriksen (ZMUC). 'Russia (Caucasus): 1 ♂, Kabardino-Balkarskij zap., 35 km SE mt. Elbrus, subalp. meadows, 2300 m, 20.vii.1990, leg. Jalava, gen. slide GU 94/479 ♂ Huemer (ZMUH).

BIONOMICS.– The larva has been recorded from *Silene nutans* L. (Schmid, 1887: 132). It overwinters at a very early stage (Wörz, 1954: 130) and can be found from early May to late June between two leaves which are spun together. The larva eats the inner surface of the leaves whereas the outer epidermis remains intact (Schmid, *op. cit.*; Huemer, pers. obs.). Pupation takes place outside the larval dwelling among debris on the ground. The adult occurs from mid-June to mid-September and is attracted to light. Preferred habitats are sunny and thermophilous localities such as sunny forest-edges, meadows and pastures. Vertical distribution: from lowlands to about 1200 m in the Alps, 1600 m in the Sierra Nevada '(Spain) and up to 2300 m in Caucasus.

149. *Caryocolum alsinella* (Zeller, 1868)

Gelechia alsinella Zeller, 1868: 145.
Lita albifrontella Heinemann, 1870: 266.
Lita tristella Heinemann, 1870: 267.
Lita semidecandriella Tutt, 1887a: 29.
Gelechia semidecandrella Threlfall & Stainton, 1887: 233.

DIAGNOSIS.– Adult. Wingspan 9-11 mm. Segment 2 of labial palp cream-coloured on inner surface, mottled with brown and black on outer surface; segment 3 black. Antenna black, indistinctly lighter ringed. Head fuscous with lighter frons; thorax and tegula concolorous with forewing. Forewing with a mixture of white, grey, light brown and black scales; black spots at base; a distinct black patch from 2/5 at costa almost reaching dorsum; another at 3/5 from middle towards tornus; costal and tornal spots white, sometimes fused; apical area blackish. Hindwing light grey to grey.

Variation. The colour of the forewings varies from brownish to greyish or blackish. Whereas the number of brownish scales on the forewing are variable even within specimens bred from the same locality, specimens from sand dune areas tend to be lighter grey or lighter brown.

Similar species. *C. viscariella* is larger and has the angulated fascia from costal to tornal spot light yellowish brown; in *C. alsinella* it is white. *C. blandulella* and *C. arenbergeri* q.v.

MALE GENITALIA.– Uncus broad, subrectangular; tegumen shallowly emarginate anteriorly; transtilla weakly sclerotized, without spines; valva long, slender, digitate, apex nearly straight; sacculus short, thumb-shaped, distoventrally weakly concave, broadened medially, apically pointed; posterior margin of vinculum with deep medial and pair of shallow lateromedial incisions, resulting in medial pair of long, digitate processes, reaching to about posterior margin of vinculum, lateromedial processes of vinculum weakly developed; saccus long, slender; phallus long and slender, nearly straight, with numerous minute cornuti apically, coecum with tiny sinistral sclerite. Extremely similar to *C. viscariella* and *C. albifaciella* with minor differences from the latter in the more distinct lateromedial processes of the posterior margin of vinculum and the tiny sinistral sclerite of the coecum which is absent in *C. albifaciella* and well developed in *C. viscariella*. However, these characters are quite difficult to observe.

FEMALE GENITALIA.– Segment VIII without processes, two pairs of short folds surrounding ostium bursae; apophysis anterioris about 1.5 times length of segment VIII; antrum absent; posterior part of ductus bursae with pair of narrow to comparatively broad lateral sclerites of variable length, anteriorly extending from beyond middle almost to apex of apophysis anterioris; signum with crescent-shaped base and stout, strongly bent hook. The genitalia are practically indistinguishable from those of *C. viscariella* and *C. albifaciella*.

DISTRIBUTION.– Locally distributed in most of the European countries, only absent from parts of Scandinavia; newly recorded recorded from some large Mediterranean islands. Also recorded from North Africa, Turkey and Central Asia to the Lower-Amur region (Ponomarenko, 2008: 98).

New records. Italy (Sardinia): 1 ♂, Belvi, Mte. Istiddi, 700 m, 12.ix.1974, leg. Hartig, gen. slide 3468 Karsholt (ZMUC). Greece (Crete): 1 ♂, Omalos, 1100-1200 m, 28.vii.-2.viii.2001, leg. Fibiger *et al.*, gen. slide 3276 Hendriksen (ZMUC).

BIONOMICS.– The larva has been found feeding on *Cerastium arvense* L. (Huemer, 1988: 461), *C. semidecandrum* L. (Stainton, 1867: 154), *C. fontanum* Baumg. (Parsons, 1995: 74), *Arenaria montana* L. (Corley *et. al.*, 2006) and *Minuartia verna* (L.) Hiern (Zeller, 1868: 146). In spring, the larva is a leaf-miner and later feeds on spun shoots, flowers and seed-capsules. Pupation takes place up to the end of June, in a small firm cocoon, covered with grains of sand (Karsholt, 1981: 261; Stainton, 1867: 154). The adult occurs from late June to early October and can be collected around the host-plant in the early evening or at light. Preferred habitats are sandy coasts and similar dry and sandy inland localities. Vertical distribution: from lowland localities to about 2000 m in southern Europe and 2700 m in Morocco.

REMARKS.– *C. alsinella* is very similar to *C. viscariella* in genital characters indicating a possible conspecificity. However, host-plants and habitat of both taxa differ and the adults of *C. alsinella* are distinctly smaller than *C. viscariella*. Furthermore, Stainton (1867) illustrates the larva of *C. alsinella* (as *C. maculiferella*) as yellow and according to Bland *et al.* (2002: 193-194) it is light green, whereas that of *C. viscariella* is dull green with faint rosy tinge anteriorly and posteriorly. The taxonomic state of the former subspecies *C. a. semidecandrella* from north-western Europe with more orange-brown scales has been discussed earlier (Karsholt, 1981: 261). There is no clear border between populations of lighter and darker specimens, and we consider it more likely that the light form, which often occurs in sand dune areas, is due to ecological conditions.

150. *Caryocolum viscariella* (Stainton, 1855)

Gelechia viscariella Stainton, 1855: 43.
Lita crepusculella Teich, 1889: 111, **syn. n.**

DIAGNOSIS.– Adult. Wingspan 11-15 mm. Segment 2 of labial palp yellowish cream-coloured on inner and upper surface, heavily mottled with blackish brown on lower and outer surface; segment 3 blackish grey. Antenna black, indistinctly lighter ringed. Head with light yellowish face and blackish grey neck; anterior part of thorax and tegula almost black, posterior part reddish brown. Forewing blackish grey, more or less mottled with brownish; three black patches: at base, from costa at 2/5 towards middle of the wing, and from middle of wing towards tornus; costal and tornal spots light yellowish brown, often forming an angulated fascia; apical area with many black scales. Hindwing light grey.

Variation. The number of black scales on the forewings are variable, resulting in some specimens looking almost black. In specimens with less black the brown colour becomes more dominant, often forming a distinct patch at 3/5. The angulated, subapical fascia may be more or less distinct.

Similar species. *C. alsinella* and *C. albifaciella* q.v.

MALE GENITALIA.– Coecum with distinct sinistral sclerite. Genitalia very similar to those of *C. alsinella* and *C. albifaciella* but larger than those of the former (corresponding to the size differences of adults). From *C. albifaciella* it differs in the presence of a sinistral sclerite of the coecum.

FEMALE GENITALIA.– Medial part of anterior margin of segment VIII frequently slightly projected anteriad, otherwise as in *C. alsinella* q.v. and *C. albifaciella* q.v.

DISTRIBUTION.– Central and north-western Europe, also found in Italy and recently in European Russia (Piskunov & Uskov, 2006: 3). A record from Spain (Vives Moreno, 1991) is most probably based on the incorrect synonymization of *C. viscariella* and *C. albifaciella* and the latter species was listed for this country earlier by Agenjo (1968). Asian Russia (Transbaikal) (Bidzilya *et al.*, 1998: 47).

BIONOMICS.– The larva has been recorded feeding on *Silene dioica* (L.) Clariv., *S. viscaria* (L.) Jessen (Huemer, 1988: 462) and *S. latifolia alba* (Mill.) Greut. et Burd. (Bland *et al.*, 2002: 194). A record from *Cucubalus baccifer* L. (Lhomme, 1946: 628) needs verification. The alleged host-plant *Stellaria uliginosa* Murray (Bland *et al.*, 2002: 194) is based on a misidentification of *C. albifaciella*. The larva feeds in April and May inside a spun shoot and between spun terminal leaves. The black frass is deposited inside the larval dwelling. Later it regularly bores into the stem which becomes slightly swollen. Pupation takes place on the ground surface, in captivity amongst spun leaves and tissue paper. The adult occurs from late June to early August. It can be found by sweeping stands of the host-plants during the day or at night when it is attracted to light (Bland *et al.*, 2002: 194). Preferred habitats are various types of dry and extensively managed grassland, hedgerows, forest edges and glades at comparatively low altitudes up to about 1000 m.

REMARKS.– *Lita crepusculella* was described from two specimens collected by Teich in Latvia at Kurtenhof [Salaspils 20 km E Riga] and Kemmern [Kemeri, about 50 km W Riga] on 27th June and 2nd July, respectively [between 1885 and 1889]. The type-material seems to be lost, and the synonymy of *C. viscariella* is based on the detailed original description. No further specimens of *C. viscariella* are known from Latvia.

151. *Caryocolum albifaciella* (Heinemann, 1870)

Gelechia albifaciella Heinemann, 1870: 205.
Lita behenella Constant, 1890a: 125, pl. 1, fig. 3.

DIAGNOSIS.- Adult. Wingspan 14-16 mm. Labial palp fuscous, lighter on inner surface. Antenna black, indistinctly lighter ringed. Head with cream-white frons and dark neck; thorax and tegula concolorous with forewing. Forewing dark brown; indistinct black patches at base, 2/5 and 3/5; costal and tornal spots white; apical area with many black scales. Hindwing grey.

Variation. The examined material shows only slight variation. The forewing may be lighter or darker brown, and the costal and tornal spots more or less distinct.

Similar species. *C. viscariella* is on average smaller and has normally lighter forewings, which are more marbled with brown. *C. inflativorella* q.v.

MALE GENITALIA.- Very similar to *C. alsinella* and *C. viscariella*. Differs from both species in the complete lack of a sinistral sclerite of the coecum.

FEMALE GENITALIA.- As in *C. alsinella* q.v. and *C. viscariella* q.v.

DISTRIBUTION.- Only known from the Alps (France, Switzerland, Austria, Germany) and the Massive Central (France); firstly recorded from the Pyrenees (Andorra). Probably also in Spain (Agenjo, 1968).

New record. Andorra: 1 ♂, Arnisal, 1500 m, 1.viii.1997, leg. Baungaard, gen. slide 4938 Karsholt (ZMUC).

BIONOMICS.- The larva has been recorded exclusively from *Silene vulgaris* (Moench) Garcke (Huemer, 1993b). A doubtful host-plant record from *Stellaria uliginosa* Murray (Huemer, 1988: 462), erroneously attributed to *C. viscariella* (Bland *et al.*, 2002: 194), refers to *C. albifaciella*. The adult occurs from July to late August and is attracted to light. Preferred habitats are alpine meadows and pastures both on calcareous and siliceous soil.

REMARKS.- *C. albifaciella* is very similar to *C. viscariella* and the two species were synonymized by Huemer (1988: 461). However, later *C. albifaciella* was recalled from synonymy due to its different biology, appearance of the adults and also small, though diagnostic characters in the genitalia (Huemer, 1993b).

152. Caryocolum vicinella (Douglas, 1851)

Gelechia vicinella Douglas, 1851: 102.
Lita inflatella Chrétien, 1901: 17.
Gelechia leucomelanella ab. *albescens* Bankes, 1909: 265; unavailable, infrasubspecific; misidentification.
Gelechia leucomelanella ab. *suffusa* Bankes, 1909: 265; unavailable, infrasubspecific; misidentification.

DIAGNOSIS.– Adult. Wingspan 11-15 mm. Labial palp black; segment 2 white on inner and upper surface. Antenna black, indistinctly lighter ringed. Head, thorax and tegula black; face cream-coloured. Forewing black, mottled with white along dorsum; a triangular white patch at 1/5, often fused along dorsum with a less distinct white patch at 3/5, giving the forewing a *Plutella*-like impression; costal and tornal spots distinct, white, often fused. Hindwing grey.

Variation. In some specimens the black forewings are mottled with white and light brown. Worn specimens look more greyish. Sometimes most of the head is cream-coloured.

Similar species. The almost *Plutella*-looking markings of the forewings are characteristic for *C. vicinella*. *C. bosalella* and *C. schleichi* q.v.

MALE GENITALIA.– Uncus subrectangular; transtilla weakly sclerotized, without spines; valva long, slender, digitate, apex weakly curved; sacculus long, thumb-shaped, distoventrally with concave emargination; posterior margin of vinculum with deep medial and pair of shallow lateromedial incisions, resulting in medial pair of digitate processes, reaching below posterior margin of vinculum, lateromedial part of vinculum hump-like; saccus long, slender; anellus with pair of short needle-shaped sclerites; phallus long and slender, weakly curved, with numerous minute cornuti apically. Genitalia very similar to those of *C. bosalella* and only differing in the more shallow lateral humps of the vinculum.

FEMALE GENITALIA.– Apophysis posterioris long, about 6 times length of segment VIII; segment VIII without processes, with pair of narrow ventromedial folds converging towards posterior margin; apophysis anterioris about 1.2 to 1.4 times length of segment VIII; antrum short, about one-fifth length of apophysis anterioris, broadly funnel-shaped, with lateral folds; posterior part of ductus bursae with pair of lateral sclerites, extending to about three-fifths of apophysis anterioris; signum with crescent-shaped base and stout, strongly bent hook. Differs from the closely related *C. bosalella* mainly in the longer antrum and from *C. amaurella* e.g. in the distinct folds and the lack of a ventromedial sclerotization of segment VIII, and the more stout signum.

DISTRIBUTION.– Widely distributed in Europe, not yet recorded from most of the large Mediterranean islands.

New record. Italy (Sicily): 14 ♂, above Lingublosa, Bocca Racabo, 1450 m, 8.-9.ix.2002, leg. Karsholt et al., gen. slides 3807, 3808 Hendriksen (ZMUC).

BIONOMICS.– Reliable host-plant records are *Spergularia rubra* (L.) J. & C. Presl (Benander, 1965: 17), *Stellaria media* (L.) Vill. (Schütze, 1931: 91), *Cerastium arvense* (Pröse, 1979: 66), *Lychnis alpina* L. (= *Silene suecica* (Lodd.) Greut. et Burd.) (Parsons, 1995: 73), *S. vulgaris* (Moench) Garcke (incl. *S. maritima* With.) (Stainton,

1867a: 60), *S. nutans* L. and *Petrorhagia saxifraga* (L.) Link (Klimesch, 1953: 312). Other host-plant records are erroneous or doubtful (Huemer, 1988: 463). The larva feeds from April to June within spun young shoots, in larger plants it bores into the stem. Pupation takes place in a whitish cocoon on the ground surface, frequently on a stone. The adult occurs from late June to mid-September. It flies by night and is attracted to light. Preferred habitats are coastal shingle, pebbly shores, alpine scree and rock formations. Vertical distribution: from coastal localities up to 2500 m in southern Spain.

153. *Caryocolum bosalella* (Rebel, 1936)

Lita bosalella Rebel, 1936a: 94.

DIAGNOSIS.– Adult. Wingspan 10-12 mm. Segment 2 of labial palp whitish, on lower and outer surface mottled with blackish brown; segment 3 black. Antenna black, more or less distinctly ringed with whitish. Head, thorax and tegula black, mottled with some brownish; face cream. Forewing dark grey with black and white markings, base black; a black patch from costa at 2/5 nearly reaching dorsum; a black fascia at 3/5; a white patch at 1/5 and one in middle of wing; costal and tornal spots distinct, white, separate. Hindwing grey.

Similar species. Very similar to *C. vicinella*, but on average smaller; the white patch at 1/5 of dorsum is less triangular, and the white spot in the middle of the wing is more distinct in *C. bosalella*.

Variation. The white patch in the middle of the wing may be mottled with fuscous. In specimens from Corsica the light markings of the forewing tend to be yellowish white. One specimen has a cream-coloured head.

MALE GENITALIA.– Very similar to *C. vicinella* but differing in the more projected posteriolateral margin of the vinculum.

FEMALE GENITALIA.– Antrum largely reduced; posterior part of ductus bursae with pair of lateral sclerites, extending to about middle of apophysis anterioris. Otherwise indistinguishable from *C. vicinella*.

DISTRIBUTION.– Endemic to Sardinia and Corsica.

BIONOMICS.– Host-plant and early stages unknown. The few adults known to date have been collected from early August to early September at light up to altitudes of about 1780 m (Huemer, 1988: 464).

154. *Caryocolum amaurella* (Hering, 1924)

Lita amaurella Hering, 1924: 82, figs. 11-12.
Lita viscariae Schütze, 1926: 171.

DIAGNOSIS.– Adult. Wingspan 10-12 mm. Segment 2 bone-white on inner and upper surface, blackish grey on outer and lower surface; segment 3 almost black with light tip. Antenna black, indistinctly lighter ringed. Head with light yellow frons and black neck; thorax and tegula black mottled with brown. Forewing blackish grey mottled with some light brown; base black; two indistinct black spots in fold; one oblique spot above it and one in cell; some white scales before and after these spots; costal and tornal spot small, white, rarely fused. Hindwing light grey.

Variation. The colour of the forewings varies from greyish to blackish. Worn specimens look lighter than fresh ones. Sometimes there are no white scales in the middle of the wing.

Similar species. The more or less uniformly blackish forewings with small white costal and tornal spots are characteristic for *C. amaurella*. *C. leucothoracellum* q.v.

MALE GENITALIA.– Uncus subovate; tegumen stout; transtilla membranous; valva short, digitate, apex rounded; sacculus short, with angular ventral and convex dorsal margin, apically pointed; posterior margin of vinculum with deep medial emargination and slight medial incision, two pairs of short processes developed; saccus long, comparatively broad at base, distal part gradually tapered; phallus long and slender, weakly curved, with some minute cornuti apically. The characteristic shape of the sacculus makes this species unmistakable.

FEMALE GENITALIA.– Segment VIII without processes, numerous narrow folds, ventromedially with large sub-ovate sclerotization; apophysis anterioris about 1.5 times length of segment VIII; antrum short, about one-fifth length of apophysis anterioris, broadly funnel-shaped; posterior part of ductus bursae with pair of lateral sclerites, extending to about middle of apophysis anterioris and two tiny sclerites anteriorly; signum with crescent-shaped base and slender, strongly bent hook. *C. amaurella* shows some resemblance to *C. vicinella* q.v. but differs in several characters.

DISTRIBUTION.– Scattered records from northern and Central Europe, mountains of southern Europe and Turkey (Huemer, 1989c: 37; Nel, 2003a: 84). Recently recorded from the Ukraine (Bidzilya & Budashkin, 2009: 19) and Russia (southern Ural Mountains) (Junnilainen *et al.*, 2010: 53).

BIONOMICS.– The larva has been recorded feeding on *Lychnis viscaria* L. (Schütze, 1931: 88), *Silene otites* (L.) Wibel (Burmann, 1990: 175) and *S. nutans* L. (Huemer, 1989c: 37). It feeds in April and May between spun terminal shoots. The later stages bore into the stem and the young shoots often become swollen and stunted (Schütze, 1931: 88; Huemer, pers. obs.). Pupation takes place on the ground in a cocoon among debris. The adult occurs from late June to early September and it is attracted to light. *C. amaurella* is restricted to warm and sunny open habitats such as dry meadows, pastures and rock formations. Vertical distribution: from lowland localities to about 1600 m in the Alps.

155. *Caryocolum oculatella* (Thomann, 1930)

Lita oculatella Thomann, 1930: 24; replacement name for *Lita ochraceella* Thomann, 1929.

Lita ochraceella Thomann, 1929: 191; homonym of *Lita ochraceella* Chrétien, 1915 (Lepidoptera, Gelechiidae).

DIAGNOSIS.– Adult. Wingspan 11-13 mm. Labial palp cream-coloured; segment 3 with black tip. Antenna black. Head, thorax and tegula light yellow-brown. Forewing black mottled with some light grey-brown, especially along middle of costa and dorsum; an oblique white fascia from 1/6 on costa to 1/5 on dorsum; a white patch in middle of wing; a distinct white subapical fascia. Hindwing grey.

Variation. The few specimens examined show only minor variation.

Similar species. The cream to light yellow-brown labial palps, head, thorax and tegula, and the distinct white fasciae on the forewings are characteristic for *C. oculatella*. *C. confluens* q.v.

MALE GENITALIA.– Uncus broadly rounded; culcitula large; tegumen very short, stout; transtilla membranous; valva short, digitate, dorsal margin curved, apex weakly pointed; sacculus short, semicircular, broader than valva; posterior margin of vinculum with deep subrectangular emargination and slight medial incision, pair of small lateral processes developed; saccus very long, weakly broadened at base, distal part evenly tapered; anellus with pair of short sclerites; phallus comparatively short, straight, with few minute cornuti apically. Similar to *C. leucofasciatum* but differs in the broader sacculus and the larger culcitula.

FEMALE GENITALIA.– Apophysis posterioris about 2.5 times length of apophysis anterioris; segment VIII long and broad, without processes, ventromedial part with large ovate sclerotized plate; apophysis anterioris long, about 1.5 times length of segment VIII; antrum short, ring-shaped; posterior part of ductus bursae with pair of lateral sclerites, extending to about middle of apophysis anterioris; corpus bursae well defined, small; signum a stout, bent, medium-sized hook with sickle-shaped base. *C. oculatella* can be distinguished from *C. leucofasciatum* by the longer segment VIII.

DISTRIBUTION.– Only known from two nearby localities in the Alps (Switzerland, Austria) (Huemer, 1988: 471). Outside Europe single specimens have been collected in Transbaikalia and Mongolia (Bidzilya *et al.*, 1998: 47; Huemer, 1989b: 202).

BIONOMICS.– The larva has been recorded exclusively from *Gypsophila repens* L. (Thomann, 1929: 191). During May it spins several terminal leaves of a shoot together into a pod-like dwelling. Within the pod the larva eats the parenchyma filling the upper part of the dwelling with frass. The fully grown larva finally leaves the dwelling and pupates on the ground surface or between dead stalks (Huemer, pers. obs.). The adults emerged from late June to early July. Only a few specimens have been observed in the wild, flushed out of stands of the host-plant with a bee-smoker or attracted to light. In the Alps the habitats are steep rocks with stands of the host-plant.

156. *Caryocolum leucofasciatum* Huemer, 1989

Caryocolum leucofasciatum Huemer, 1989b: 197, figs. 1, 4, 7, 10.

DIAGNOSIS.– Adult. Wingspan 11 mm. Labial palps almost white. Antenna ringed black and white. Head, thorax and tegula white. Forewing white with black patches at base, at 2/5 from costa to middle of wing, as a fascia at 3/4 and in apical area. Hindwing light grey.

Variation. The few specimens examined are rather uniform.

Similar species. *C. confluens* q.v.

MALE GENITALIA.– Uncus broadly rounded; culcitula medium-sized; transtilla membranous; valva short, digitate, dorsal margin curved, apex weakly pointed; sacculus short, semicircular, about breadth of valva; posterior margin of vinculum with deep subrectangular emargination and slight medial incision, pair of small lateral processes and indisctint medial humps developed; saccus very long, weakly broadened at base, distal part evenly tapered; anellus with pair of short sclerites; phallus comparatively short, straight, with some minute cornuti apically. Distinguished from *C. oculatella* mainly by the smaller culcitula and the narrower sacculus.

FEMALE GENITALIA.– Very similar to *C. oculatella* but apophysis anterioris shorter (corresponding to the size differences of adults) and segment VIII distinctly shorter. Apophysis anterioris therefore relatively longer, of about 2 to 2.5 times length of segment VIII.

DISTRIBUTION.– Restricted to southern Spain (Granada: Baza, Sierra Nevada).

BIONOMICS.– Host-plant and early stages unknown. Possibly the larva feeds on *Gypsophila* L. which is the host-plant of the closely related *C. oculatella*. The adults have been observed from late May to early June and again in early October at low altitudes, indicating two generations. However, in the higher regions of the Sierra Nevada at about 1900 m, the moth has been collected in late August (Huemer, 1989b: 197).

157. *Caryocolum petryi* (Hofmann, 1899)

Lita petryi Hofmann, 1899: 139.
Lita repentella Chrétien, 1908b: 258.
Lita rougemonti Rebel, 1907: 236.
Lita petryi benanderi Hering, 1933: 86, figs. 1-2.

DIAGNOSIS.– Adult. Wingspan 10-13 mm. Segment 2 of labial palp bone white on inner and upper surface, light brown on lower and outer surface, darkest towards tip; segment 3 black, more or less mottled with bone white. Antenna black, indistinctly lighter ringed. Head light brown, darkest towards neck; thorax and tegula concolorous with forewing. Forewing light grey to blackish grey, sometimes with some yellow; black patches at 2/3, 3/4 and in apical area; middle of dorsum light grey; an indistinct light grey patch from 2/5 at costa and one such patch in middle of wing; costal and tornal spots weak, separated by yellow scales. Hindwing grey.

Variation. *C. petryi* exhibits both geographical and individual variation. Specimens from the Swedish island of Öland are small and have the labial palps, head, thorax and tegulae covered with charcoal grey scales, and the forewings are hardly contrasting and without yellow scales. This population has been described as ssp. *benanderi*. Similar dark, but larger specimens also occur in the Baltic countries and furthermore in higher regions of the Alps.

Similar species. Specimens of the nominotypical form from Central Europe are rather characterless and may be confused with, e.g., dark specimens of *C. alsinella*, but that species can be recognized by the dark spot on the forewing at 3/5, pointing towards tornus. *C. baischi* sp. n. q.v.

MALE GENITALIA.– Uncus subrectangular; tegumen distinctly emarginate anteriorly; transtilla weakly sclerotized, spinous; valva long, slender, evenly tapered and weakly curved apically; sacculus long and broad, up to three times breadth of valva, apical part broadly rounded; posterior margin of vinculum with pair of broadly rounded medial processes and pair of small lateral processes; saccus long, slender; phallus long, slender, weakly S-curved. *C. petryi* differs from the similar *C. repentis* mainly in the broadly rounded apex of the sacculus and the less curved phallus and from *C. baischi* sp. n. e.g. in the longer valva, the more slender sacculus and the shorter saccus.

FEMALE GENITALIA.– Apophysis posterioris about 4 times length of apophysis anterioris; segment VIII without processes, two pairs of distinct longitudinal folds ventrolaterally; apophysis anterioris almost length of segment VIII; antrum very broad, funnel-shaped, extending to about middle of apophysis anterioris; posterior part of ductus bursae with pair of long lateral sclerites, extending to apex of apophysis anterioris; signum a slender, medium-sized hook with sickle-shaped base. *C. petryi* differs from *C. repentis* and from *C. baischi* sp. n. mainly in the much shorter antrum. *C. siculum* q.v.

DISTRIBUTION.– Scattered records from the Pyrenees, the Alps, Germany, Hungary and some localities in northern Europe (Finland, Sweden, Latvia and Lithuania). Also known from Mongolia (Huemer, 1989c: 38; Povolný, 1973a: 18), southern Ural Mountains (Junnilainen *et al.*, 2010: 53) and from the Transbaikal region (Bidzilya, 2005: 16).

BIONOMICS.– The host-plants are *Gypsophila repens* L. (Müller-Rutz, 1934: 123), *G. fastigiata* L. (Hofmann, 1899: 143) and *G. paniculata* L. (Klimesch, 1954: 275). The first instar larva is unknown, though probably a leaf-miner. Later the larva spins several terminal leaves of a shoot together to form a pod-like dwelling. The larva has been observed from May to mid-July, depending on the elevation of the biotope. It pupates to a blackish pupa within the dwelling. In the laboratory adults emerged from early June to mid-August. Preferred habitats are alpine scree and rock formations. *C. petryi* seems to be restricted to limestone areas. Vertical distribution: in the Alps from about 550 to 2000 m, outside also in the lowlands with stands of the host-plant.

REMARKS.– Specimens from Sweden are very dark and have been regarded as the subspecies *benanderi*. However, similar populations also occur in the higher regions of the Alps and are probably due to climatic conditions.

The larval habits of *C. petryi* match those of *C. repentis* and *Tila capsophilella*. However, the larvae of the former pupate inside the dwelling-place whereas those of the latter two species pupate outside on the ground surface or between dead stalks. The final instar of the larva has a characteristic reddish ventrum on the first thoracic segment. Furthermore the pupa is distinctly darker brown than in the above mentioned species (Huemer, 1998: 53).

158. *Caryocolum baischi* sp. n.

DIAGNOSIS.– Adult. Wingspan 12-13 mm. Labial palp cream-coloured; segment 2 mottled with black, especially on lower and outer surface; segment 3 almost black. Head cream-coloured, darker towards neck; thorax and tegula black, posteriorly light reddish brown. Forewing covered with black, light-based scales; light grey-brown oblique fasciae from costa at 1/5, 1/2 and 4/5, almost reaching dorsum; an elongate orange patch in fold; black spots indistinct, followed by orange-brown scales; costal and tornal spot indistinct, fused into a weak, broad, light brown fascia; cilia around tornus dark grey. Hindwing light grey, darker towards margins; cilia grey.

Variation. The type series varies in the colour of the oblique transverse fascia from costa which can vary from light grey-brown to yellow-brown.

Similar species. *C. baischi* sp. n. is characterized by the dark grey-brown forewings with more or less indistinct markings, blackish labial palps and cream-white head. It is very similar to some forms of *C. petryi*, but it can be recognized by the more prominent elongate orange patch in the fold.

MALE GENITALIA.– Uncus subrectangular, apically rounded; tegumen distinctly emarginate anteriorly; transtilla weakly sclerotized, spinous; valva rather short, straight, evenly tapered towards apex; sacculus broad and slightly longer than valva, distal part gradually widened towards apex, shovel-shaped, up to four times breadth of valva; posterior margin of vinculum with pair of broadly rounded medial processes and pair of small lateral processes; saccus very long, slender; phallus long, slender, weakly S-curved. *C. baischi* sp. n. differs from the similar *C. petryi* and *C. repentis* particularly in the peculiar shape of the sacculus.

FEMALE GENITALIA.– Apophysis posterioris about 5 times length of apophysis anterioris; segment VIII without processes, two pairs of distinct longitudinal folds ventrolaterally; apophysis anterioris slightly longer than segment VIII; antrum very broad, funnel-shaped, extending to last third of apophysis anterioris; posterior part of ductus bursae with pair of long lateral sclerites, extending distinctly beyond apex of apophysis anterioris; signum a slender, medium-sized hook with sickle-shaped base. The female genitalia are similar to other species of the group but differ in the length of the antrum.

DISTRIBUTION.– Restricted to southern Greece.

BIONOMICS.– Host-plant and early stages unknown. Probably the larva feeds on *Gypsophila* L. which is the host-plant of all closely related taxa such as *C. petryi*. The adults have been observed rather late in the season, from early September to mid-October at light.

MATERIAL EXAMINED.– Holotype. ♂ 'Griechenland Peloponnes Leonidon Kloster Elona 7.9.2002 Licht 500 m G. Baisch' 'P. Huemer GEL 1121 ♂' (TLMF).

Paratypes. 1 ♂, same data as holotype; 2 ♂, ditto, but 13.x.1994; 1 ♂, ditto, but 11.x.1995 (coll. Baisch; TLMF); 1 ♂, 1 ♀, 5 km N Galaxidi, 600 m, 2.x.2005, leg. Schmitz, gen. slides 5919 ♀, 5920 ♂ Hendriksen (coll. Schmitz; ZMUC).

ETYMOLOGY.– The name is dedicated to Günter Baisch who collected the first specimens.

159. Caryocolum repentis Huemer & Luquet, 1992

Caryocolum repentis Huemer & Luquet, 1992: 35, figs. 1, 6, 10-12, 14.
Lita repentella auctt., (*nec* Chrétien, 1908); misidentification.

DIAGNOSIS.– Adult. Wingspan 9-12 mm. Segment 2 of labial palp cream-white on inner and upper surface, on outer and lower surface light brown at base, grey-brown towards tip; segment 3 almost black. Antenna black, indistinctly lighter ringed. Head light brown, darker towards neck; thorax and tegula anteriorly blackish grey, posteriorly light brown. Forewing light brown along basal 2/3 of dorsum, black along costa, interrupted by a slender oblique whitish fascia from 2/3 of costa; a distinct orange-brown patch in middle of wing, interrupting white costal and dorsal spots; apical area black, mottled with white scales. Hindwing light grey.

Variation. The orange-brown patch in the middle of the forewings is sometimes reduced.

Similar species. *C. repentis* is characterized by the light brown colour at the dorsum of the forewings. *C. siculum* q.v.

MALE GENITALIA.– Uncus subrectangular; tegumen distinctly emarginate anteriorly; transtilla weakly sclerotized; valva long, slender, evenly tapered and weakly curved towards apex; sacculus long and broad, up to three times breadth of valva, apical part broadly rounded; posterior margin of vinculum with pair of broadly rounded medial processes and pair of short digitate processes; saccus long, slender; phallus long, slender, distinctly S-curved. *C. repentis* differs from *C. petryi* in the pointed apex of the sacculus and the distinctly curved phallus.

FEMALE GENITALIA.– Apophysis posterioris about 4 times length of apophysis anterioris; segment VIII without processes, two pairs of distinct longitudinal folds ventrolaterally; apophysis anterioris slightly shorter than segment VIII; antrum about breadth of segment, broadly funnel-shaped, extending to about apex of apophysis anterioris; posterior part of ductus bursae with pair of long lateral sclerites, situated beyond apex of apophysis anterioris; signum a stout, bent hook with sickle-shaped base. Differs from the related *C. petryi* in the much larger antrum which is projected to the apex of apophysis anterioris.

DISTRIBUTION.– Scattered records from the Alps (Austria, Italy, Switzerland, France) and from various parts of Spain; recently found in Russia (South Ural) (Junnilainen *et al.*, 2010: 53).

BIONOMICS.– The larva has been found from early May to late June, feeding exclusively on *Gypsophila repens* L. (Huemer, 1988: 473). It initially mines a leaf, causing a short gallery towards the tip. The frass is protruded out of a hole at the base of the leaf. Later several terminal leaves of a shoot are spun together into a pod-like dwelling. Within the pod the larva eats the parenchyma filling the upper part of the dwelling with frass. The fully grown larva finally leaves the dwelling and pupates on the ground surface or between dead stalks. The colour of the pupa is brown. In the laboratory adults emerged from mid-June to late July, in the field they have been observed from mid-July to early September (Huemer, 1998: 52). Preferred habitats are alpine scree and rock formations. *C. repentis* seems to be restricted to limestone areas. Vertical distribution: from about 700 to 2400 m.

REMARKS.– *C. repentis* was formerly misidentified with *C. repentella*. However, examination of the type-material of *Lita repentella* proved its conspecificity with *C. petryi* (Huemer & Luquet, 1992) and other records of the former before 1992 refer to *C. repentis*.

160. *Caryocolum siculum* Bella, 2008

Caryocolum siculum Bella, 2008: 69, figs. 1-13.

DIAGNOSIS.– Adult. Wingspan 13-16 mm. Segment 2 of labial palp greyish black, mottled with cream, at inner and upper surface almost cream-coloured; segment 3 almost black. Antenna black, indistinctly lighter ringed. Face cream-white, neck blackish grey; thorax and tegula black at margins, brown in the middle. Forewing bicoloured, with sinuously separated, black costal and apical part and light brown (darkest at base) dorsal part; a whitish brown bent fascia from costal to tornal spot.

Variation. The brownish areas vary from whitish brown or light yellow-brown to light brown.

Similar species. *C. repentis* is smaller, and has two whitish brown fasciae reaching from dorsum almost to costa, and the costal and tornal spots separated by orange scales.

MALE GENITALIA.– Uncus broad; tegumen with shallow emargination anteriorly; transtilla with several spines; valva moderately slender, evenly tapered, curved, with rounded apex; sacculus two-thirds length of valva, broad with convex dorsal and ventral margin, apicomedially pointed; posterior margin of vinculum medially with weak incision, with pair of lateral processes, medial process largely reduced; saccus long and slender, about same length or slightly longer than distance from anterior margin of vinculum to tip of valva; phallus long and slender, slightly S-curved. Differs from *C. saginella* and species such as *C. petryi* in the length of the valva which distinctly overtops the sacculus which is also of different shape compared to related species.

FEMALE GENITALIA.– Apophysis posterioris about 3 times length of apophysis anterioris; segment VIII without processes, two pairs of distinct longitudinal folds and several indistinct folds ventrolaterally; apophysis anterioris slightly longer than segment VIII; antrum broadly funnel-shaped, short, extending to about one-fifth length of apophysis anterioris; posterior part of ductus bursae with pair of moderately long lateral sclerites, extending to about two-thirds length of apophysis anterioris; signum a slender, medium-sized hook with sickle-shaped base. Differs from the similar *C. petryi* particularly in the shorter antrum and the short lateral sclerites of the ductus bursae.

DISTRIBUTION.– Only known from Sicily (Italy).

BIONOMICS.– The larva has been recorded exclusively from *Gypsophila arrostii* Guss. (Caryophyllaceae) (Bella, 2008: 73). It feeds in the stems of the host-plant from April to late June, causing an internodal gall of 4-6 mm width and 15-18 mm length. In the last stage the larva prepares an emerging hole for the adult and pupation takes place inside the gall (Bella, 2008: 71; Karsholt, pers. obs.). The adults emerged from late May to early July. In the field the moth has not yet been observed, despite searching with light traps. The species occurs in xeric steppe grassland at altitudes from about 300 to 500 m.

161. *Caryocolum inflativorella* (Klimesch, 1938)

Lita inflativorella Klimesch, 1938a: 85, figs. 1-3.
Lita xuthella Rebel, 1941: 4, fig. 6.
Gnorimoschema census Gozmány, 1954: 278.

DIAGNOSIS.– Adult. Wingspan 11-14 mm. Labial palps blackish brown, mottled slightly lighter. Antenna black, indistinctly lighter ringed. Head cream-white, darker towards neck; thorax and tegula concolorous with forewing. Forewing rather broad, grey-brown, mottled with white and black; base, a large costal patch, and distal 2/3 (separated by a slender white fascia) black; a whitish brown patch at 2/3 and 1/2. Hindwing grey, lighter towards base.

Variation. In some of the few specimens examined the labial palps are lighter, with cream-coloured areas, but the tip of segment 3 is still black. Specimens from the

Balkans, initially described under the name *Lita xuthella*, are in general lighter yellowish brown.

Similar species. *C. cauligenella* has white head and thorax. *C. albifaciella* has more brownish (less blackish) and less variegated forewings, and the upper part of the neck is darker than in *C. inflativorella*. *C. amaurella* has more slender and deeper black forewings.

MALE GENITALIA.– Uncus broad; tegumen distinctly emarginate anteriorly; transtilla with numerous spines; valva broad basally, evenly tapered, straight; sacculus broad, ventrodistal margin slightly serrated, apex pointed; posterior margin of vinculum with pair of broadly rounded medial processes and indistinct lateral processes; saccus strong, comparatively short, evenly tapered towards apex; phallus short and stout, apex weakly curved. The shape of the sacculus and vinculum distinguish this species from related species.

FEMALE GENITALIA.– Segment VIII without processes, ostium surrounded by short longitudinal folds ventrolaterally; apophysis anterioris long, about 1.5 times length of segment VIII; antrum short, broad, cup-shaped, weakly tapered anteriorly with straight anterior margin, extending to about one-fourth of apophysis anterioris; posterior part of ductus bursae with pair of very short lateral sclerites, extending to about middle of apophysis anterioris; signum a large, strongly curved hook with sickle-shaped base. Differs from otherwise similar species by the characteristic shape of the antrum.

DISTRIBUTION.– Known from Slovakia, Hungary, Macedonia and Turkey; recently recorded from Romania (Kovács& Kovács, 2000: 33).

BIONOMICS.– The larva has been recorded exclusively from *Silene vulgaris* (Moench) Garcke (Klimesch, 1938a: 86). It feeds in the stems of the host-plant from the middle of June to early July, causing a large internodal gall. Up to 4 galls have been observed in a single plant which then becomes stunted. The frass is deposited in the upper and lower part of the gall. Before pupation the larva prepares a hole for the emerging adult at slightly about the middle of the gall. Furthermore it separates the frass from the pupal chamber with a firm spinning before pupation takes place within the gall (Klimesch, 1938a: 86-87). The adults were bred from late July to early August. In the field the moth has been observed from early July to early October. It is attracted to light. Preferred habitats are shady places under old oak trees.

REMARKS.– Specimens from Macedonia are usually more variegated with distinct yellowish brown forewing markings. Such specimens were described as *C. xuthella*. However, the genitalia are indistinguishable from typical *C. inflativorella* (see also Huemer, 1988: 476) and similar looking specimens could also be found among populations from Hungary.

162. *Caryocolum saginella* (Zeller, 1868)

Gelechia saginella Zeller, 1868: 146.
Lita coussonella Chrétien, 1908a: 245.

DIAGNOSIS.– Adult. Wingspan 10-11 mm. Labial palp white, segment 2 with some brown on outside near base; segment 3 with a few dark brown scales in the middle, forming a ring. Antenna black, ringed with white. Head, most of thorax and tegula white; posterior part of thorax blackish brown. Forewing white with black markings: at base, from 2/5 of costa towards middle of wing, a spot in middle of wing, a fascia at 3/4, and apical part of wing. Hindwing light grey.

Variation. In some populations (especially from the Balkans) the white forewings are more covered with black scales, leaving only small areas of white between the black areas.

Similar species. *C. cauligenella*, which has similar genitalia, differs in being on average larger, and by having head, thorax, tegulae and light marking of the forewing cream coloured (white in *saginella*).

MALE GENITALIA.– Uncus broad; tegumen shallowly emarginate anteriorly; transtilla with numerous spines; valva broad basally, evenly tapered, apically curved; sacculus broad, pointed, dorsal margin rounded, ventral margin slightly serrated; posterior margin of vinculum with pair of comparatively broad medial processes and weakly developed lateral processes; saccus long and slender, about 1.2 to 1.5 times length of distance from anterior margin of vinculum to tip of valva; phallus long and slender, apex weakly curved. Maybe apart from the slightly longer saccus no reliable differences from *C. cauligenella*. *C. siculum* q.v.

FEMALE GENITALIA.– Segment VIII without processes, ventromedial part with sub-ovate sclerotization, ostium surrounded by longitudinal folds laterally; apophysis anterioris comparatively short, about length of segment VIII; antrum small, anterior margin distinctly emarginate medially; posterior part of ductus bursae with pair of characteristic, posteriorly dilated, narrow lateral sclerites, slightly separated from antrum, extending to about two-thirds length of apophysis anterioris; signum short, slightly curved hook with sickle-shaped base. The genitalia are practically indistinguishable from those of *C. cauligenella*. A specimen from Croatia has a smaller signum and slightly shorter sclerites of ductus bursae.

DISTRIBUTION.– Restricted to the southern Alps (France, Italy and Austria) and the Balkan Peninsula (Macedonia, Greece). Recently recorded from Croatia.

New record. Croatia: 1 ♀, South Velebit, 15.viii.2003, leg. Srnka, gen. slide in tube (TOKA).

BIONOMICS.– The larva has been found feeding on *Silene saxifraga* L. (Chrétien, 1908a). It feeds in May and June in the shoots and causes a terminal gall. The plant becomes stunted, the terminal leaves are broadened and form a bud-like gall. Pupation takes place outside the larval dwelling in a slight cocoon on the ground surface (Chrétien, 1908a: 246; Klimesch, 1951: 192; Huemer, pers. obs.). The adult occurs from early June to late July and it is frequently found resting on rocks. Preferred habitats are rock formations with stands of the host-plant. *C. saginella* is restricted to limestone areas at altitudes up to about 1700 m, reaching 1900 m in Greece (Huemer, 1989c: 38).

REMARKS.– *C. saginella* is extremely similar to *C. cauligenella* in genital characters indicating a possible conspecificity. However, the adults are smaller and more whitish and the host-plants and larval behaviour of both taxa differ considerably. We therefore treat them as separate species for the time being.

163. Caryocolum cauligenella (Schmid, 1863)

Gelechia cauligenella Schmid, 1863: 73.

DIAGNOSIS.– Adult. Wingspan 11-14 mm. Segment 2 of labial palp cream white, with black on outer surface near base; segment 3 cream-white, with a few black scales. Antenna black, very indistinctly lighter ringed. Head cream-white to light brown; thorax and tegula light brown, mottled with a few black scales. Forewing dark grey to black, mottled with cream and light brown; a black patch at base and one from 2/5 of costa towards middle of wing; a black fascia at 3/4; between these patches cream or light brown; costal and tornal spots cream-white, sometimes fused into a subapical fascia. Hindwing grey.

Variation. In some specimens the light markings of the forewing are more or less overlaid by black scales. Bred specimens have a tendency to become greasy, which may change the colour so that, e.g., the thorax becomes brownish.

Similar species. *C. saginella* q.v.

MALE GENITALIA.– Saccus length equal to distance from anterior margin of vinculum to apex of valva. Otherwise as in *C. saginella*.

FEMALE GENITALIA.– Signum with longer hook, corresponding to the larger size of the adults. Otherwise as in *C. saginella*.

DISTRIBUTION.– Locally distributed in Europe, except most of Scandinavia and the British Isles; the Middle East.

Confirmed record. Italy (Sicily): 4 ♂, above Lingublosa, Bocca Racabo, 1450 m, 8.-9.ix.2002, leg. Karsholt *et al.*, gen. slide 3809 Hendriksen (ZMUC).

BIONOMICS.– The larva has been found feeding on *Silene nutans* (Schmid, 1863: 63) and *S. otites* (L.) Wibel (Snellen, 1882: 657). Zerkowitz (1946: 134) adds *S. gallica* L., *S. portensis* L., *S. inaperta* L. and *S. longicilia* Hoth., but no moths were reared from these host-plants (Mendes, 1904) so they must be considered questionable. A record from *Silene vulgaris* (Moench) Garcke (Wörz, 1954: 130) needs confirmation. The larva feeds from April to July, depending on the climatic situation of its habitat. It bores in the stem and causes a large gall, usually in the terminal internode of the central shoot, rarely in a lateral stem. Affected plants become stunted. Pupation takes place on the ground surface in a weak cocoon. The adult occurs from June to late August and it is rarely attracted to light. Preferred habitats are sunny edges of oak and pine forests but also dry meadows. In the Mediterranean collected up to altitudes of about 1600 m.

164. Caryocolum trauniella (Zeller, 1868)

Gelechia trauniella Zeller, 1868: 147.

DIAGNOSIS.– Adult. Wingspan 14 mm. Segment 2 of labial palp clear white; segment 3 white mottled with black, especially towards tip. Antenna black with some white on basal segments. Head and thorax clear white; neck and tegulae white with some dark scales. Forewing black with three clear white fasciae, the two basal ones not reaching dorsum. Hindwing dark grey.

Variation. Only one specimen was examined.

Similar species. In the larger *C. peregrinella* the basal white fascia of the forewings reaches dorsum, and it is fused with the white fascia in the middle of the wing. *C. trauniella* is also a smaller species with particularly dark and not shiny hindwings.

MALE GENITALIA.– Uncus broad; tegumen shallowly emarginate anteriorly; transtilla membranous; valva lanceolate, evenly tapered towards pointed and weakly curved apex; sacculus comparatively slender, distally tapered; posterior margin of vinculum with pair of broadly rounded medial processes and short, hump-like lateral processes; saccus strong, comparatively short, evenly tapered towards apex; phallus stout, straight, entrance of ductus ejaculatorius parallel to longitudinal axis of phallus. Differs from the otherwise very similar *C. peregrinella* and *C. delphinatella* in the entrance angle of the ductus ejaculatorius and also the shape of the saccus.

FEMALE GENITALIA.– Segment VIII without processes, ventromedial part with subovate sclerotization, ostium surrounded by distinct longitudinal folds laterally; apophysis anterioris comparatively long, about 1.5 times length of segment VIII; antrum broadly conical, short, anteriorly without emargination; posterior part of ductus bursae with pair of short and narrow lateral sclerites, extending to about one-third length of apophysis anterioris; signum a very large, strongly bent hook with small base. The female genitalia are particularly characterized by the large signum.

DISTRIBUTION.– Endemic to the south-eastern Alps (Italy, Austria, Slovenia). Records from the Middle Volga region (Ponomarenko, 2008: 98) and from Hungary (Pastorális, 2008: 117) are most probably based on misidentifications.

BIONOMICS.– Host-plant and early stages unknown. The probable host-plants are *Silene* spp. The adults have been collected in very low numbers from late June to the middle of August both during the day and at light. Preferred habitats are alpine scree and rock formations, exclusively in limestone areas. *C. trauniella* is only known from a few mountain localities from about 900 to 1800 m.

165. Caryocolum peregrinella (Herrich-Schäffer, 1854)

Gelechia peregrinella Herrich-Schäffer, 1854: 106.
Gelechia melantypella Mann, 1877: 498.

DIAGNOSIS.– Adult. Wingspan 15-18 mm. Segment 2 of labial palp white, black at base on outer surface; segment 3 white, mottled with some black. Antenna black, indistinctly lighter ringed. Head, thorax and tegula cream-white; proximal part of thorax black. Forewing white and black; white areas as transverse fasciae at 1/5, 3/5 and as costal and tornal patches. Hindwing light grey.

Variation. There is only slight variation. Some specimens have a black spot in the middle of the white fascia at 3/5. A slight geographical variation in the extent of the white forewing markings is here regarded as infrasubspecific. Particularly specimens from the south-eastern and south-western Alps are more whitish than those from the Balkans.

Similar species. *C. peregrinella* is characterized by its large size and its distinct black and white markings on the forewing. *C. delphinatella* q.v.; *C. trauniella* q.v.

MALE GENITALIA.– Uncus broad; tegumen shallowly emarginate anteriorly; transtilla membranous; valva lanceolate, evenly tapered towards pointed and weakly curved apex; sacculus comparatively broad, distally tapered; posterior margin of vinculum with pair of broadly rounded medial processes and short, hump-like lateral processes; saccus strong, evenly tapered towards apex; phallus stout, straight, entrance of ductus ejaculatorius at acute angle to longitudinal axis of phallus. Distinguished from *C. trauniella* mainly by broader sacculus and the entrance of the ductus ejaculatorius, from *C. delphinatella* by the shape of the sacculus.

FEMALE GENITALIA.– Segment VIII without processes, ventromedial part with subtriangular sclerotization, ostium surrounded by distinct longitudinal folds laterally; apophysis anterioris comparatively long, about 1.3 times length of segment VIII; antrum broadly conical, short, anteriorly emarginate; posterior part of ductus bursae with pair of short and rather broad lateral sclerites, extending to about one-third length of apophysis anterioris; signum small to medium-sized hook, comparatively small base.

DISTRIBUTION.– Only known from the southern Alps (France, Italy, Slovenia and Austria) and the Balkan Peninsula: Bosnia and Herzegovina (Rebel, 1904: 350), Macedonia and Greece; recently also found in Spain (Heckford, *in litt.*).

BIONOMICS.– The larva has been recorded from *Silene* sp. (Huemer, pers. obs.), *S. ciliata* Pourr., *Petrocoptis pyrenaica* (J. Bergeret) A. Braun ex Walpers, probably subspecies *glaucifolia* (Lag.) P. Monts. & Fernández Casas (Heckford, *in litt.*) and an unidentified Caryophyllaceae (Klimesch, 1968: 124). It lives in a silken tube covered with green frass at the base of the host-plant and feeds on the young leaves and shoots. The larval stage has been observed from the middle of May to mid-June. Pupation takes place in a slight cocoon on the ground surface. The adult occurs from mid-July to mid-September and is attracted to light. *C. peregrinella* is restricted to rock formations and alpine scree on limestone. Vertical distribution: mountain areas from about 1000 to 2500 m.

REMARKS.– *Gelechia melantypella* was described from the male holotype collected near Schluderbach in the Dolomites (South Tyrol, Italy) (Mann, 1877:498). Later this species was synonymized with *C. peregrinella* (Huemer, 1988: 480). However, recent sequencing of the DNA-barcode suggests cryptic diversity within *C. peregrinella*.

166. Caryocolum delphinatella (Constant, 1890)

Lita delphinatella Constant, 1890b: 6, pl. 1, fig. 2.
Gnormioschema fiorii Klimesch, 1953: 277, figs. 23-24.

DIAGNOSIS.– Adult. Wingspan 15-16 mm. Labial palp white; segment 3 mottled with some black scales. Antenna black, indistinctly lighter ringed. Head, thorax and tegula white; posterior tip of thorax black. Forewing white and black mottled with some light brown; a white fascia near base of costa; a distinct white patch in middle of wing; these two connected by a light brown streak along dorsum; costal and tornal patches white, distinct and separated. Hindwing grey.

Variation. The examined material shows only slight variation.

Similar species. The relatively large wingspan together with the clear white labial palps, head, thorax and tegulae are characteristic for *C. delphinatella*. *C. peregrinella* is larger and has more distinct black and white markings on the forewings.

MALE GENITALIA.– Uncus broad; tegumen shallowly emarginate anteriorly; transtilla membranous; valva lanceolate, evenly tapered towards pointed and curved apex; sacculus broad, rounded towards pointed apex; posterior margin of vinculum with pair of broadly rounded medial processes and short, hump-like lateral processes; saccus strong, evenly tapered towards apex; phallus stout, straight, entrance of ductus ejaculatorius at acute angle to longitudinal axis of phallus. Differs from the similar *C. trauniella* and *C. peregrinella* particularly in the shape of the saccus, from the former species also in the entrance of the ductus ejaculatorius.

FEMALE GENITALIA.– Apophysis posterioris about 4 times length of apophysis anterioris; segment VIII without processes, ostium bursae surrounded by folds; apophysis anterioris slightly longer than segment VIII; antrum comparatively long, funnel-shaped, posteriorly fused with pair of short lateral sclerites, these extending to about tip of apophysis anterioris; corpus bursae well defined, pyriform; signum long, hook-shaped. Similar to *C. trauniella* and *C. peregrinella* but differs in the distinctly longer antrum and the anteriorly extended sclerites of the ductus bursae.

DISTRIBUTION.– Widely scattered localities in the south-western Alps of France, Switzerland and Italy (Alpes Maritimes, Alpes de Haute Provence, Hautes Alpes, Isère, Walliser Alpen, Alpi Cozie), the French Pyrenees (Haute-Garonne) and the Abruzzi mountains in central Italy.

BIONOMICS.– Host-plant and early stages unknown, though *Silene* spp. and/or *Minuartia* spp. are suspected as larval host-plants (Huemer, 1991: 31). The adults have been collected in July and August. They are attracted to light and can be disturbed from rocks during the day. Preferred habitats are alpine scree and rock formations. *C. delphinatella* is probably restricted to limestone mountains. Vertical distribution: mountains from about 1600 to 2500 m.

REMARKS.– The identity of *C. delphinatella* was discussed in detail by Huemer (1991).

167. Caryocolum provinciella (Stainton, 1869)

Gelechia provinciella Stainton, 1869: 221.

DIAGNOSIS.– Adult. Wingspan 12 mm. Segment 2 of labial palp black, lighter on upper and inner surface, with cream-white tip; segment 3 black mottled with white, especially towards tip. Antenna black. Head cream-brown, darker towards neck; thorax and tegula light brown slightly mottled with black. Forwing separated into a blackish brown costal part and light orange-brown dorsal part by a sinuous, interrupted black line from base to tornal spot; no transverse fascia at 4/5; apical area with many black scales. Hindwing light grey.

Variation. In the few specimens examined the colour of the head varies from cream-brown to dark grey.

Similar species. In *C. marmorea* segment 2 of the labial palp is without a light tip and furthermore longer than segment 3. It is also missing the black streak, which separates the forewings of *C. provinciella* into a costal dark part and a dorsal light part.

MALE GENITALIA.– Uncus broadly rounded; tegumen shallowly emarginate anteriorly; transtilla membranous; valva long, slender lanceolate, evenly tapered towards pointed and straight apex; sacculus semiovate, dorsal margin convex, ventral margin straight;

posterior margin of vinculum with deep U-shaped emargination medially and pair of hump-like lateral processes; saccus stout, gradually tapered towards apex; anellus with pair of long and needle-shaped sclerites; phallus short, stout, apically curved.

FEMALE GENITALIA.– Segment VIII without processes, ventromedial part with sub-ovate sclerotization, surrounded by distinct lateral folds, two additional folds branching off towards antrum; apophysis anterioris shorter than segment VIII; antrum ring-shaped, posterior margin sinuate; posterior part of ductus bursae with pair of short and rather broad lateral sclerites, fused with antrum posteriorly, slightly exceeding base of apophysis anterioris anteriorly; signum a strongly bent hook with basal spines, large sickle-shaped base.

DISTRIBUTION.– Only known from comparatively few localities in the Mediterranean: Portugal, Spain, France, Morocco and hitherto unpublished records from Greece and Tunisia.

New records. Greece: 1 ♂, Peloponnes, Gythion, dunes near Elos, 26.x.1998, leg. Grünewald, gen. slide GEL 980 Huemer (TLMF). 2 ♂, Tunisia, 21 km E Tabarka, 7.-13.v.1988, leg. Karsholt, gen. slide Hendriksen 3000 (ZMUC).

BIONOMICS.– The larva has been recorded from *Silene nicaensis* All. (Stainton, 1869: 221) and *Herniaria* L. (Huemer, 1988: 482). It lives in sand tubes at the base of the host-plant and feeds on the young shoots. Pupation takes place in a sand-cocoon and the adults emerged from early April to beginning of May (Stainton, *op. cit.*). According to Lhomme (1946: 627), *C. provinciella* occurs in two generations with larvae from September to March and in June, and adults from April to May and in August. The adult is attracted to light. The species is exclusively restricted to sandy coasts.

168. Caryocolum mucronatella (Chrétien, 1900)

Lita mucronatella Chrétien, 1900a: 138.
Lita poschiavensis Rebel, 1936b: 2, pl. 1, fig. 5.

DIAGNOSIS.– Adult. Wingspan 10-12 mm. Segment 2 of labial palp bone-white on inner and upper surface, on outer and lower surface mottled with grey-brown; segment 3 black mottled with whitish. Antenna black, indistinctly lighter ringed. Head grey, lighter towards face, darker towards neck; thorax and tegula black mottled with red brown. Forewing white, more or less overlaid with black; black patches at base, from costa at 1/3, before the distinct white costal and tornal patch, and in apical area. Hindwing grey.

Variation. A variable species, especially in the south-eastern part of its distribution area. It is especially the number of black scales on the forewing which varies, and some specimens are almost black, whereas others have distinct whitish markings. Thorax and tegulae are sometimes almost red-brown, but this colour is rarely extended to the base of the forewings.

Similar species. Due to its variation *C. mucronatella* may be confused with several other species, especially *C. alsinella* and *C. leucomelanella*. The former has normally orange brown scales in the middle of the forewings; in the latter the red-brown colour at thorax and tegulae is extended to the base of the forewings, and the costal and tornal patches are less distinct. *C. mazeli* q.v.; *C. fibigerium* q.v.

MALE GENITALIA.– Uncus broadly rounded; tegumen shallowly emarginate anteriorly; transtilla membranous; valva short, evenly tapered distally; sacculus broad and short, pointed at apex, dorsal margin convex, distoventral margin with concave excavation; posterior margin of vinculum with pair of broadly rounded medial processes and short, hump-like lateral processes; saccus long, slender; phallus slender, straight. Differs from somewhat similar species such as *C. peregrinella* by the characteristic shape of the sacculus.

FEMALE GENITALIA.– Segment VIII without processes, ostium bursae pentagonally surrounded by strong folds, pair of narrow ventromedial folds converging towards posterior margin and a further pair of parallel folds extending into antrum; apophysis anterioris about 1.5 times length of segment VIII; antrum short, comparatively narrow, funnel-shaped, about one-fifth to one-third length of apophysis anterioris; posterior part of ductus bursae with pair of long lateral sclerites, extending to about apex of apophysis anterioris; signum with sickle-shaped base and slender, strongly bent hook. *C. mucronatella* shows a remarkable variation in the length of the antrum. The long lateral sclerites of the antrum combined with the posteriorly extended folds of segment VIII which reach the posterior margin make the species unmistakable.

DISTRIBUTION.– Locally distributed in mountainous regions mainly of southern Europe, from Portugal to Greece; towards the north extending into southern Germany. Turkey and recently collected in Egypt.

New record. Egypt: 1 ♀, Sinai, near mt. Moyses, 13.iv.1999, leg. Skule, gen. slide Hendriksen 2873 (ZMUC).

BIONOMICS.– The larval host-plants are *Minuartia mutabilis* Schinz & Thell. ex Becherer (Lhomme, 1946: 639), *M. setacea* (Thuill.) Hayek (Schmid, 1887: 132), *M. verna attica* (Boiss. & Spruner) Hayek (Huemer, 1988: 483) and *M. laricifolia* (L.) Schinz & Thell. (Burmann, 1949: 71); furthermore bred from *Cerastium* sp. in Greece (Karsholt, pers. obs.). The larva has been observed from late May to mid-July. It constructs a silken web which is extended from the root-stock to the terminal parts of the host-plant and feeds on the leaves (Schmid, 1887: 132; Burmann, 1949: 71). On *Cerastium* it has been found feeding between spun shoots (Karsholt, pers. obs.). Pupation takes place between dry vegetation or between tissue in the laboratory. The adult occurs from late June to mid-September (in the Near East as early as April) and is frequently attracted to light. Preferred habitats are xeromontane localities including alpine scree and rock formations. Vertical distribution: from the valleys to about 1200 m in the Alps; reaching elevations of about 2400 m in southern Europe.

169. *Caryocolum leucomelanella* (Zeller, 1839)

Gelechia leucomelanella Zeller, 1839: 138.
Gelechia gypsophilae Stainton, 1869: 210, fig.

DIAGNOSIS.– Adult. Wingspan 9-12 mm. Segment 2 whitish on inner and upper surface, heavily mottled with greyish black on lower and outer surface; segment 3 almost black. Antenna black, finely lighter ringed. Head with face cream and neck dark grey; thorax and tegula greyish black on anterior and posterior part, brown in middle. Forewing greyish black; a brownish or reddish brown streak from base, connected with a white oblique fascia from costa at 1/5; a small white patch in middle of wing surrounded by black spots; costal and tornal spots white. Hindwing grey.

Variation. A rather variable species. The greyish black colour of the forewing may be more or less mottled with light grey. Sometimes the brownish streak from the base is reduced or absent. The white costal and tornal spots may be more or less distinct.

Similar species. *C. leucomelanella* is most easily recognized by the brownish, medial part of thorax, which extends along costa to 1/5 of the forewings. When it is absent identification is problematic without examination of the genitalia. *C. mucronatella* q.v.; *C. schleichi* q.v.; *C. mazeli* q.v.

MALE GENITALIA.– Uncus broad; culcitula large; tegumen shallowly emarginate anteriorly; transtilla sclerotized, without spines; valva broad at base, dorsally rounded, distal part abruptly tapered, turned ventrad; sacculus slender and short, knife-shaped; posterior margin of vinculum with pair of broadly rounded medial processes and short, hump-like lateral processes; saccus stout, comparatively short; phallus long, slender, straight. Differs from the similar *C. mazeli* and *C. leucothoracellum* mainly by the absence of a dorsal hump and the shorter distal part of the valva and the size of the sacculus.

FEMALE GENITALIA.– Segment VIII without processes, ostium bursae pentagonally surrounded by strong folds, pair of short folds converging towards posterior margin of segment VIII, a further pair of parallel folds extending into antrum; apophysis anterioris about length of segment VIII; antrum short, conical, anteriorly indented, about one-third length of apophysis anterioris; posterior part of ductus bursae with pair of long lateral sclerites, extending to about apex of apophysis anterioris; signum a large, strongly bent hook, sickle-shaped base. *C. leucomelanella* is practically indistinguishable from *C. mazeli* in female genitalia. It differs from the closely related *C. leucothoracellum* in the much shorter apophysis anterioris and the relatively longer sclerites of the ductus bursae.

DISTRIBUTION.– Central and southern parts of Europe, absent from the north-west and north. Turkey and China, Asian Russia (Transbaikal) (Bidzilya *et al.*, 1998: 47); probably more widely distributed in Asia.

New record. Italy (Sicily): 13 ♂, Ragaha, Bosco de Patemò, Monte Arso, 1000 m, 9.ix.2002, leg. Karsholt *et al.*, gen. slide 3932 Hendriksen (ZMUC).

BIONOMICS.– The larval stage has been recorded from *Petrorhagia saxifraga* (L.) Link, *Dianthus carthusianorum* L. (Klimesch, 1953: 230), *D. sylvestris* Wulfen (Huemer, 1988: 485) and probably also *D. gratianopolitanus* Vill. and *D. seguieri* Vill. (Pröse, 1979: 66). Records from *Lychnis* L. and *Minuartia* Loefl. ex L. (Elsner *et al.*, 1999: 46) need verification. The larva has been found from April to June, in the Mediterranean as early as March (Stainton, 1869: 210). It feeds in young shoots or bores in the stem. The latter behaviour is obligatory when *Petrorhagia* (Ser. ex Dc.) Lk. is attacked. This small plant reacts to the larval infestation with the development of a distinct gall. The shoots of the foodplant become stunted and the galls are finally left by the larva. *Dianthus* feeding larvae preferably attack the shoots but they can also bore in the stems. Pupation takes place on the ground surface. The adults are on the wing from early June to September and occasionally early October. They are frequently attracted to light. *C. leucomelanella* is restricted to xeric habitats such as dry grassland and sunny rock formations. Vertical distribution from lowland to about 2200 m.

REMARKS.– Specimens bred from *Petrorhagia saxifraga* were formerly treated as a separate species, viz. *C. gypsophilae*, by many authors. However, according to more recent investigations. *C. gypsophilae* is regarded as a host-race of *C. leucomelanella* (Huemer, 1988: 485).

C. quinlingensis Li & Zheng, 1995 from China was compared to *C. schleichi* in the original description. However, the genitalia figures of the holotype (Li & Zheng, 1995: 294, figs. 1-2) show a male specimen closely related to *C. leucomelanella* with the valvae turned inside and therefore somewhat difficult to interpret. The species is larger with a wingspan of 14.5 mm and therefore we hesitate to syonymize it with *C. leucomelanella*.

Additional closely related species have been described from the Middle East, viz. *C. immixtum* Huemer, 1988, and *C. similellum* Huemer, 1989.

170. *Caryocolum mazeli* Huemer & Nel, 2005

Caryocolum mazeli Huemer & Nel, 2005: 125, figs. 1a-b, 3, 5a-b.

DIAGNOSIS.– Adult. Wingspan 12-13 mm. Segment 2 of labial palp cream-coloured mottled with black; segment 3 almost black. Antenna ringed with black and light grey. Head cream-white, darker towards neck; thorax and tegula concolorous with forewing. Forewing blackish grey, with indistinct black patches and spots; base, an oblique patch at 1/4 from base, 1-2 spots in middle of wing and apical area black; an oblique streak at 1/5 from base, a small patch in middle of wing and a fascia from costal to tornal spots white. Hindwing light grey.

Variation. Only a few, worn specimens were examined.

Similar species. *C. leucomelanella* has normally brown scales on thorax and tegula extending into a brownish patch at costa near base. *C. mucronatella* has more distinct black spots on the forewings. *C. mazeli* is hardly distinguishable from some forms of *C. schleichi*, though the costal and tornal patches are normally separate in that species.

MALE GENITALIA.– Uncus broad; tegumen shallowly emarginate anteriorly; transtilla sclerotized, without spines; valva broad at base, dorsally with rounded hump, distal part abruptly tapered, comparatively long, turned ventrad; sacculus slender and long, knife-shaped; posterior margin of vinculum with pair of broadly rounded medial processes and short, hump-like lateral processes; saccus stout, comparatively short; phallus long, slender, straight. Very similar to *C. leucomelanella* but the dorsal hump of the valva is more distinct and the sacculus is longer and more slender.

FEMALE GENITALIA.– As in *C. leucomelanella* q.v.

DISTRIBUTION.– Only recorded from the Pyrenees and the south-western Alps (France), occurring sympatrically with *C. leucomelanella* in these regions.

New record. Andorra: 1 ♂, Camillo, 4.-23.vii.1979, leg. Schreurs, gen. slide 6123 Hendriksen (ZMUC).

BIONOMICS.– Host-plant and early stages unknown. Most probably the larva feeds on *Dianthus* L. and/or *Petrorhagia* (Ser.) Link which are the only known host-plants within the *leucomelanella*-group. The few adults known to date have been collected from early to late August at light. Vertical distribution: from about 450 to 1800 m.

REMARKS.– The diagnostic characters of the genitalia weakly support this species; its status should be proved by molecular data.

171. Caryocolum leucothoracellum (Klimesch, 1953)

Gnorimoschema leucothoracellum Klimesch, 1953: 233, figs. 7-9.

DIAGNOSIS.– Adult. Wingspan 10-13 mm. Labial palp with segment 2 white at upper and inner surface; lower and outer surface black with white tip; segment 3 black on basal half, white on apical half. Antenna black, ringed with white. Head white; thorax and tegula cream-white mottled with light brown. Forewing black, lighter along middle of costa and dorsum, and with white markings: an oblique fascia from 1/5 of costa and a patch in middle of wing at 3/5; a light brown patch at costa near base; costal and tornal spots white, distinct, often fused into a fascia. Hindwing grey.

Variation. The few specimens examined show only slight variation.

Similar species. The whitish head and thorax separate *C. leucothoracellum* from most other *Caryocolum* species. *C. peregrinella* and *C. delphinatella* are larger, and like *C. cauligenella* and *C. saginella* they have the lower and outer surface of segment 2 of labial palp light (not black).

MALE GENITALIA.– Uncus broad; tegumen shallowly emarginate anteriorly; transtilla sclerotized, without spines; valva broad at base, dorsally with rounded hump, distal part abruptly tapered, comparatively long, turned ventrad; sacculus slender and comparatively long, knife-shaped; posterior margin of vinculum with pair of broadly rounded medial processes and short and comparatively broad lateral processes; saccus stout, comparatively short; phallus long, straight. Very similar to *C. leucomelanella* but the dorsal hump of the valva is more distinct and the sacculus is longer and more slender.

FEMALE GENITALIA.– Segment VIII without processes, ostium bursae pentagonally surrounded by strong folds, pair of short folds converging towards posterior margin of segment VIII, a further pair of parallel folds extending into antrum; apophysis anterioris about 1.5 times length of segment VIII; antrum short, conical, anteriorly indented, about one-sixth length of apophysis anterioris; posterior part of ductus bursae with pair of relatively short lateral sclerites, slightly exceeding middle of apophysis anterioris; signum a small hook, indistinct base. Differs from related species in the longer apophysis anterioris, the relatively short sclerites of the ductus bursae, and a short signum hook.

DISTRIBUTION.– Mediterranean and submediterranean area, from France and Italy to Austria and Hungary. Recently also found in Switzerland (Bryner *et al.*, 2005: 128). A single record from Morocco (Huemer, 1988: 487) indicates a wider distribution in the Mediterranean.

BIONOMICS.– The larva has been found in mid-April feeding in the young shoots and boring in the stem of *Dianthus* sp. (Burmann, 1990: 178; Huemer, pers. obs.), according to Nel (2003b: 395) on *D. sylvestris longicaulis* (Tenore) Greuter et Burdet. Pupation took place between spun tissues in the laboratory. The adults have been observed from mid-June to late October and they are attracted to light. *C. leucothoracellum* is restricted to xeric grassland and rock formations. Vertical distribution: from comparatively low elevations of 100 to about 1500 m in the Alps.

172. Caryocolum schleichi (Christoph, 1872)

Lita schleichi Christoph, 1872: 22, pl. 1, fig. 19.
Caryocolum syriacum Povolný, 1977a: 171, figs. 1-2, 6.
Lita dianthella Chrétien, 1925: 246. Subspecies.
Caryocolum hackeri Derra, 1985: 373, figs. 1, 4-6.
Lita improvisella Rebel, 1936b: 3, pl. 1, fig. 7. Subspecies.
Lita arenariella Benander, 1937: 31, figs. A, B. Subspecies.

DIAGNOSIS.– Adult. Wingspan 9-13 mm. Segment 2 of labial palp bone-white to cream-white on inner and upper surface, mottled with greyish brown on lower and outer surface; segment 3 black, mottled with some white. Antenna black, indistinctly lighter ringed. Head light yellowish brown, grey-brown towards neck; thorax and tegula greyish black on anterior and posterior part, brown in middle. Forewing greyish black with white markings: a fascia at 1/5 and a patch in middle of wing; costal and tornal patches white, normally separated. Hindwing grey. Ventral side of abdomen cream-coloured.

Variation. *C. schleichi* exhibits both individual and geographical variation. The light areas of the forewing may be more or less distinct, varying in colour from clear white to light brown. Ssp. *dianthella* from Spain is rather small (9-11 mm) and has clear white markings on the forewings. Ssp. *arenariella* from Sweden and the Baltic countries has the forewings blackish, and apart from the distinct white costal and tornal spots the 'light' markings are hardly visible. Ssp. *improvisella* from the Alps has lighter, more greyish forewings, and the white markings are fused along dorsum. The colour of the head, thorax and tegulae varies from white to dark brown between the different subspecies. Specimens from Turkey resemble the nominotypical subspecies, but have often darker head and thorax.

Similar species. *C. schleichi* is, due to its geographical and individual variation, easily confused with several other species of *Caryocolum*. It is normally without the light brown base at the forewing costa found in most specimens of *C. leucomelanella*, and the head is not black as in *C. vicinella*. *C. mucronatella* has more distinct black spots on the forewing. *C. mazeli* q.v. It will often be necessary to examine the genitalia for a safe identification.

MALE GENITALIA.– Uncus broad, subrectangular; tegumen shallowly emarginate anteriorly; transtilla sclerotized, without spines; valva almost straight, broadly shovel-shaped, apex with two processes, with or without weak distinct hump dorsally, dorsal process longer than ventral one; sacculus short, knife-shaped; posterior margin of vinculum deeply incised with pair of long medial processes and short, hump-like lateral processes; saccus long, slender; phallus long, slender, almost straight.

FEMALE GENITALIA.– Segment VIII without processes, anterior margin of tergite VIII with deep subrectangular emargination; sternite VIII with pair of narrow and posteriorly converging folds surrounding ostium bursae; apophysis anterioris about length to slightly longer than segment VIII; antrum short, tubular, about one-fifth to one-sixth length of apophysis anterioris; posterior part of ductus bursae with pair of long lateral sclerites, extending to about apex of apophysis anterioris; signum a slender, strongly curved hook, sickle-shaped base.

DISTRIBUTION.– *C. schleichi* is distributed through much of the Palaearctic region: ssp. *dianthella* occurs in the western Mediterranean (Portugal, Spain, France (Pyrenees),

Morocco); ssp. *improvisella* in Central Europe (mainly the Alps) and Macedonia. ssp. *arenariella* in southern Scandinavia, the Baltic states, Mongolia and Asian Russia (Transbaikal) (Bidzilya *et al.*, 1998: 47); and ssp. *schleichi* in southern Russia, Turkey, Syria and Jordan (Huemer, 1989c: 39), and after a recent record also in Romania (Kovács & Kovács, 2000: 35). Further records from Syria and Afghanistan may represent a distinct subspecies (Huemer, 1988: 489).

New records. Croatia: 1 ♂, Dalmatia, Baska Voda, 18.-28.ix.1981, leg. Wolschrijn, gen. slide 6184 Hendriksen (ZMUC). Greece: 3 ♂, 2 ♀, Flórina, 2 km W Kariés, Little Prespa Lake, 970 m, 28.viii.2008, leg. Skou, gen. slide 5132 Karsholt (ZMUC).

BIONOMICS.– The larva has been recorded from *Dianthus scaber toletanus* (Boiss & Reuter) Tutin, *D. deltoides* L. (*schleichi dianthella*) (Chrétien, 1925: 246), *D. sylvestris* Wulfen (ssp. *improvisella*) (Klimesch, 1953: 236), *D. lusitanicus* Brot. (Corley, *in litt.*) and *D. arenarius* L. (ssp. *arenariella*). The host-plant of the nominotypical subspecies is unknown. In the Austrian Alps and the French Pyrenees it feeds in the shoots of the host-plant during May and June (Huemer, 1988: 488), in northern Europe the larva bores into the stem which becomes stout and gall-like. Pupation takes place outside the larval feeding place, on the ground (Benander, 1926: 173). The adult occurs from mid-May to early October, depending on local climatic conditions and the altitude and latitude of its habitat. It inhabits various steppe and xeric biotopes on sandy or rocky surface and can be attracted to light. Vertical distribution: in northern and south-eastern Europe at low altitudes, otherwise mainly in mountainous areas up to about 2600 m in North Africa and 2000 m in the Alps.

REMARKS.– *C. schleichi* shows a considerable geographical variation in male genitalia structures and to a certain degree in forewing markings and colour of head and thorax. These geographically separated populations were initially described as separate species but are nowadays treated as subspecies. Four subspecies are separated: *C. s. schleichi*, *C. s. dianthella*, *C. s. improvisella* and *C. s. arenariella*. *C. s. schleichi* can be recognized by the almost pure white colour of the head and thorax, occasionally mottled with some light brown. The other three subspecies have a grey-brown head and thorax, with only the frons whitish. Specimens from Greece and Turkey are not clearly referable to any of the above mentioned subspecies. *C. s. dianthella* differs from the remaining subspecies in the small dorsal hump and the short distodorsal process of the valva; *C. s. improvisella* is characterized by the broad valva with a comparatively long distodorsal hump, and *C. s. arenariella* can be identified by the broad valva and particularly the very long distodorsal process of the valva. The female genitalia of these subspecies are indistinguishable and the observed differences, particularly in the length of the slerites of the ductus bursae, seem to be based on individual variation.

173. *Caryocolum marmorea* (Haworth, 1828)

Recurvaria marmorea Haworth, 1828: 553.
Gelechia manniella Zeller, 1839: 198.
Gelechia marmorella Doubleday, 1859: 30; unjustified emendation of *Recurvaria marmorea* Haworth, 1828.
Caryocolum marmoreum auctt.

Gelechia pulchra Wollaston, 1858: 121. Subspecies.
Caryocolum pulchrum auctt.

DIAGNOSIS.– Adult. Wingspan 10-12 mm. Segment 2 of labial palp yellowish white, more or less covered with greyish brown; segment 3 shorter than segment 2, greyish black. Antenna black, weakly lighter ringed. Head, thorax and tegula concolorous with forewing, frons whitish grey. Forewing light brown, mottled with white, yellow, grey and black scales; black patches and spots in middle of wing at 2/5, 3/5 and 4/5 at tornus and in apical area; white and yellowish patches along dorsum, between black patches and as a postmedial fascia. Hindwing light grey.

Variation. *C. marmorea* is both individually and geographically a variable species. Specimens from coastal areas along the North Sea are white to light brown along dorsum and light brown along costa, often with a well-defined, white block in the middle of the forewings. In such specimens segment 2 of the labial palps are often pure yellowish white. Specimens from mountain areas often have the forewings more or less covered with greyish black, and lighter markings tend to be orange. The populations in Madeira and in the Canary Islands differ in habitus from mainland European populations.

Similar species. *C. marmorea* can (together with *C. pullatella* q.v.) be recognized by having the dark segment 3 of the labial palps shorter than the light segment 2, and by having the M-shape at the dorsum of the forewings covered with many light scales, giving the impression that the wing is divided into a darker costal and a lighter dorsal part. *C. provinciella* q.v.

MALE GENITALIA.– Tergite VIII deeply emarginate anteriorly; sternite VIII comparatively long. Uncus broad, subrectangular; tegumen with deep V-shaped emargination anteriorly; transtilla band-like; valva short, slender, digitate; sacculus longer than valva, subrhomboid, distoventrally rounded, distodorsally pointed; vinculum very short, posterior margin with pair of comparatively broad and rounded medial processes and short lateral processes; saccus very long, slender; phallus long, stout, almost straight to slightly curved, distinctly inflated basally, with numerous minute cornuti apically. Differs from the very similar *C. pullatella* in the comparatively long sternite VIII, and the distinctly emarginate tergite VIII and anterior margin of the tegumen.

FEMALE GENITALIA.– Segment VIII without processes, lateroposteriorly with pair of prominent vaulted semiovate sclerotizations, anterior margin of segment VIII strongly sclerotized; ostium bursae with triangular sclerite, converging posteriorly; apophysis anterioris slightly longer than segment VIII; antrum very short, ring-shaped; posterior part of ductus bursae with pair of narrow, short lateral sclerites, extending from about one-third to middle of apophysis anterioris; signum a stout, strongly curved hook, with comparatively small base. The female genitalia are indistinguishable from *C. pullatella*.

DISTRIBUTION.– From the Canary Islands, Madeira and north-western Africa through most of Europe except the north-east; Turkey; a single record from Canada (British Columbia) (Huemer, 1988: 494).

BIONOMICS.– The larva has been recorded from *Cerastium fontanum vulgare* (Hartm.) Greut et Burd. (Stainton, 1867a: 142), *C. semidecandrum* L. (Huemer, 1988: 493) and *Silene nocteolens* Webb & Berth. (Klimesch, 1984: 158). It initially causes irregular, semitransparent upper surface leaf-mines on *C. semidecandrum*, usually extending the length of the leaf near the midrib or running along the edge of a leaf at the tip (Heckford, 2000). In February the early part of the gallery had been filled with greenish brown frass. The larva only mines a single leaf from which it enters into the stem. Later it has been reported feeding on leaves touching the ground from a silken tube in the sand on the ground surface (Stainton, 1867: 142). On larger plants such as *Silene* spp. the larva has been found between spun leaves (Klimesch, 1984: 159) or boring into the stem (Huemer, pers. obs.). It is unknown whether or not the larva is an initial leaf-miner when it feeds on these host-plants. The early instar larva of *C. marmorea* has been observed as early as late January (Heckford, 2000), mid- or fully-grown larval stages are usually found from March to May. Pupation takes place within the larval tube or in the laboratory also between spun leaves. The adults have been recorded from May to October in a single generation. They are frequently attracted to light. Preferred habitats are sandbanks but also dry and thermophilous inland localities. It has been recorded from altitudes between sea level and 2100 m in Spain and Turkey.

REMARKS.– *C. marmorea* shows a considerable individual and geographical variation in colour and wing markings. Particularly populations from the Desertas Islets in the Madeira Islands (Portugal) can be readily separated by a white head, thorax and tegulae. Initially described as separate species, these populations are nowadays treated as subspecies *C. m. pulchra* (Wollaston, 1858) (Huemer, 1988: 494).

This species is closely related to *C. pullatella* and *C. protectum* (Braun, 1965) from the U.S.A. Only minute differences in the male genitalia and pregenital segments could be observed in these taxa and it is disputable whether they are distinct species.

174. *Caryocolum pullatella* (Tengström, 1848)

Gelechia pullatella Tengström, 1848: 126.
Gelechia pulla Tengström, 1848: 162.
Gelechia subtractella Walker, 1864a: 592.
Lita alsinella var. *livoniella* Teich, 1889: 88.
Gelechia agricolaris Meyrick, 1933: 448.

DIAGNOSIS.– Adult. Wingspan 11-13 mm. Labial palp blackish brown, mottled with lighter brown, especially on segment 2. Antenna black. Head, thorax and tegula blackish brown; frons lighter. Forewing blackish grey with black patches at base, 2/5 and 4/5, and mottled with white and yellowish brown scales near base, in middle on costal part of the wing and in and beyond white postmedial fascia; a white patch in middle of wing. Hindwing dark grey.

Variation. The white patch in the middle of the forewing is sometimes mottled with yellowish brown and dark grey scales, thereby becoming less conspicuous. The black patches at 2/5 and 4/5 are sometimes reduced to black spots.

Similar species. Dark specimens of *C. marmorea*, especially from mountain areas, may be confused with *C. pullatella*, but in such specimens the white or light greyish brown patch in the middle of the forewing is less conspicuous, and they have the costal part of the forewing more mottled with orange-brown scales. *C. cassella* has the white patch in the middle of the forewing extended to an oblique, white fascia.

MALE GENITALIA.– Tergite VIII shallowly emarginate anteriorly; sternite VIII comparatively short. Tegumen with shallow emargination anteriorly. Otherwise indistinguishable from *C. marmorea*.

FEMALE GENITALIA.– As in *C. marmorea* q.v.

DISTRIBUTION.– Holarctic. In Central Europe only known from two localities in the Alps (western part of Austria); widely distributed from Scandinavia, Latvia, Poland, European Russia (mainly northern parts) to Siberia, Korea and Japan; in the Nearctic region numerous records from the U.S.A. and Canada (Huemer, 1988: 492).

BIONOMICS.– The only reliable host-plant record refers to *Dianthus superbus* L. (Kaitila, 1996: 96). However, additional host-plants are to be expected. The larva feeds in June inside a flowering shoot which becomes withered. The infested shoot can be changed and a totally dry flowering shoot indicates a larva inside the closest neighbouring shoot (Kaitila, 1996: 103). The adults are found in July and August in Europe. They are on the wing from mid-May to mid-October in North America.

REMARKS.– *C. pullatella* is indistinguishable from the North American *C. protectum* in genitalia characters but the two species differ strongly externally.

175. Caryocolum stramentella (Rebel, 1935)

Bryotropha stramentella Rebel, 1935b: 39.
Caryocolum emarginatum Huemer, 1988: 497, figs. 69, 139, **syn. n.**

DIAGNOSIS.– Adult. Wingspan 13-14 mm. Labial palp cream-brown, mottled with blackish brown, especially on segment 3. Antenna ringed with dark and light grey. Head cream-coloured; thorax and tegula concolorous with forewing. Forewing rather broad, creamy brown mottled with black especially at base, along dorsum, before the light postmedian fascia and in apical area; three black spots at 2/5 form a triangle. Hindwing whitish grey.

Variation. The few known specimens vary in the mixture of light and dark scales on thorax and tegulae, and in the amount of dark scales on the forewing.

Similar species. Though it is a rather characterless species, *C. stramentella* is not exactly similar to other European *Caryocolum* species. The relatively broad creamy brown (not white) forewings separates it from its relatives.

MALE GENITALIA.– Uncus subrectangular, with pair of transverse folds dorsally; tegumen emarginate anteriorly; transtilla with numerous spines; valva of moderate length, slender, evenly tapered and weakly curved apically; sacculus shorter and broader than valva, apically tapered; posterior margin of vinculum with extremely deep, V-shaped lateromedial emarginations, medially a pair of long and largely fused, needle-shaped processes; saccus long and stout, distal part tapered; phallus long, stout, slightly curved.

FEMALE GENITALIA.– Unknown.

DISTRIBUTION.– Italy (Piedmont) and Turkey.

BIONOMICS.– Host-plant and early stages unknown. The few specimens known to date have been collected from mid-July to early October in xeromontane habitats.

REMARKS.– The type series of *Bryotropha stramentella* consists of two specimens representing two species: *C. stramentella* and *C. inflativorella*. Due to a lectotype designation by Huemer (1988: 497) the former specimen was fixed as the name bearing type.

C. emarginatum was described from a single male collected close to the border between Italy and France and differing from *C. stramentella* in the broader valva and sacculus (Huemer, 1988: 497). However, further specimens from Turkey demonstrated some variation in these characters and hence *C. emarginatum* is here synonymized with *C. stramentella*.

176. *Caryocolum hispanicum* Huemer, 1988

Caryocolum hispanicum Huemer, 1988: 498, figs. 70, 140, 203.

DIAGNOSIS.– Adult. Wingspan 11-12 mm. Labial palp greyish brown; segment 3 darker than segment 2. Antenna black, indistinctly lighter ringed. Head light grey; thorax and tegula concolorous with forewing. Forewing dark grey with black patches at 2/5, 3/5 and in apical area; orange scales in fold and between black patches; a weak whitish postmedial fascia.

Variation. Specimens from Turkey have lighter grey forewings, often mottled with white scales.

Similar species. *C. fraternella* has warm yellow labial palps. Very similar to dark forms of *C. marmorea*, but in that species segment 3 of the labial palps is shorter than segment 2.

MALE GENITALIA.– Uncus comparatively narrow; tegumen shallowly emarginate anteriorly; transtilla sclerotized, with numerous spines; valva broadly shovel-shaped, with slightly pointed distodorsal hump and broadly rounded distoventral hump; sacculus short, digitate, well separated from valva; posterior margin of vinculum narrowly emarginate medially, with pair of short lateromedial processes; saccus comparatively short, broad at base, gradually tapered towards middle, distal part slender; phallus slender, almost straight.

FEMALE GENITALIA.– Segment VIII without processes or stronger folds; apophysis anterioris long, about 1.5 times length of segment VIII; antrum tubular, long, extending to about two-thirds of apophysis anterioris; posterior part of ductus bursae with pair of short lateral sclerites; corpus bursae well defined; signum a large, slender, hook. Differs from the very similar *C. gallagenellum* and *C. fraternella* in the intermediate length of the antrum and the more slender signum.

DISTRIBUTION.– In Europe only known from Central Spain and Greece. Turkey (Huemer, 1989c: 39)

BIONOMICS.– Host-plant and early stages unknown. The adult has been collected in July at light, in Turkey until mid-September. Vertical distribution: only known from higher mountains between 1200 and 1800 m.

REMARKS.– Specimens from Greece slightly differ in the shape of the valva but they are here considered conspecific with *C. hispanicum*.

177. *Caryocolum confluens* Huemer, 1988

Caryocolum confluens Huemer, 1988: 498, figs. 71, 141, 204.

DIAGNOSIS.– Adult. Wingspan 12-13 mm. Labial palp cream-coloured, mottled with blackish brown, especially on segment 3. Antenna black, indistinctly lighter ringed. Head cream-coloured, on neck mottled with grey; thorax and tegula concolorous with forewing. Forewing black, with white patches at 1/5, in middle of wing and as costal and tornal spots. Hindwing grey.

Variation. The few specimens available show only slight variation. A specimen from Montenegro (Durmitor) is slightly larger (wingspan 14 mm), has the labial palps almost blackish brown, the face light grey, and the neck, thorax and tegula black. The forewings are black with less distinct white patches.

Similar species. The pure black and white forewings, without yellow or brown mixture are characteristic for *C. confluens*. *C. leucofasciatum* and *C. oculatella* have segment 3 of the labial palps almost white.

MALE GENITALIA.– Uncus comparatively broad; tegumen distinctly emarginate anteriorly; transtilla weakly sclerotized, with few spines; valva broadly shovel-shaped, with broadly rounded distodorsal and distoventral humps; sacculus largely fused with valva, only apical part somewhat separate; posterior margin of vinculum with large ovoid emargination, pair of lateromedial processes distinct; saccus comparatively short and slender; phallus slender, almost straight, basally strongly inflated.

FEMALE GENITALIA.– Segment VIII with some weak lateromedial folds; apophysis anterioris comparatively long, about 1.5 times length of segment VIII; antrum short, ring-shaped; posterior part of ductus bursae with pair of short lateral sclerites; corpus bursae well defined; signum a large, strongly curved hook. Differs from the related species in the distinctly shorter antrum.

DISTRIBUTION.– Endemic to Greece (Mt. Taygetos, Mt. Chelmos). The identity of the above mentioned specimen from Montenegro (Durmitor) remains dubious (Huemer & Jakšic, 1996: 86) and it may belong to a further undescribed species.

BIONOMICS.– Host-plant and early stages unknown. The few adults known to date have been collected from early June to late July mostly at light. Vertical distribution: only known from higher mountains between 1500 and 2300 m.

178. *Caryocolum gallagenellum* Huemer, 1989

Caryocolum gallagenellum Huemer, 1989a: 22, figs. 1-4, 6, 8, 10.

DIAGNOSIS.– Adult. Wingspan 11 mm. Segment 2 of labial palp yellowish grey, mottled with blackish grey; segment 3 blackish brown with light yellow tip. Antenna black with fine light grey rings. Head with cream-yellow frons and dark grey neck; thorax and tegula concolorous with forewing. Forewing black mottled with white and light brown scales, especially as transverse fasciae (not reaching costa) at 2/5 and 3/5, and along dorsum; costal and tornal spots white, separated by a light brown patch. Hindwing grey, becoming lighter towards base.

Variation. The specimens available for study show only minor variation.

Similar species. In *C. fraternella* the light brown scales are placed in longitudinal stripes, whereas in *C. gallagenellum* they form transverse fasciae. *C. fraternella* can moreover be recognized by the warm yellow lower surface of segment 2 of the labial palps.

MALE GENITALIA.– Uncus comparatively broad; tegumen shallowly emarginate anteriorly; transtilla weakly sclerotized, with several small spines; valva broadly shovel-shaped, with rounded distodorsal and distoventral humps; sacculus almost completely fused with valva, only minute apical part with some setae discernible; vinculum comparatively short, posterior margin projected posteriad, with broadly V-shaped emargination, pair of lateromedial processes indistinct; saccus comparatively short and slender; phallus slender, almost straight. Differs from the closely related *C. confluens* particularly in the strongly reduced sacculus and the subtriangular projection of the posterior margin of the vinculum. *C. gallagenellum* can be distinguished from the related *C. fraternella* by several characters such as the broader uncus, rounded apices of the valva and the different vinculum.

FEMALE GENITALIA.– Apophysis posterioris about 1.5 mm long; segment VIII without processes or distinct folds; apophysis anterioris long, about 1.5 times length of segment VIII; antrum tubular, medium-sized, extending to about middle of apophysis anterioris; posterior part of ductus bursae without lateral sclerites; corpus bursae well defined; signum a large and stout, strongly bent hook. Differs from the very similar *C. hispanicum* and *C. fraternella* e.g. in the shorter antrum and stout, bent signum.

DISTRIBUTION.– Widely separated localities in France, Central Italy and southern Germany.

BIONOMICS.– The larva has been found on *Cerastium arvense* L. and probably also on other *Cerastium* spp. (Huemer, 1989a: 21). It spins several small leaves of a terminal shoot together into a spherical dwelling. The leaves become swollen and the 7-8 mm large and gall-like larval dwelling turns downwards. The inner wall of the gall is covered with silken web. Frass is mainly deposited inside the gall, partially also outside through a small hole at the tip of the dwelling. Pupation takes place outside the larval dwelling, probably on the ground surface. In South France the larva has been observed in April and the adults emerged after 4 to 5 weeks of pupal stage in late May. However, field observations indicate a flight period between early July and beginning of September. *C. gallagenellum* prefers xeric biotopes such as limestone scree with sparse vegetation. In their habitat adults are also attracted to light.

179. Caryocolum fraternella (Douglas, 1851)

Gelechia fraternella Douglas, 1851: 101.
Lita intermediella Hodgkinson, 1897: 36.

DIAGNOSIS.– Adult. Wingspan 10-13 mm. Labial palp warm yellow, lighter at inner surface, mottled with blackish brown on outer surface and on segment 3. Antenna blackish brown, weakly lighter ringed. Head with cream-yellow frons and light grey neck; thorax and tegula concolorous with forewing. Forewing black, mottled with white, light brown and grey scales; light brown stripes along fold and in middle of wing from 1/3 to 4/5; costal and tornal spots white. Hindwing grey.

Variation. The amount of light brown scales is variable. Some specimens have the central part of the wing almost light brown. Worn specimens become lighter. Specimens in collections have a tendency to become greasy.

Similar species. The warm yellow lower surface of segment 2 of the labial palp is characteristic for *C. fraternella*. *C. gallagenellum* q.v.; *C. hispanicum* q.v.

MALE GENITALIA.– Uncus comparatively narrow; tegumen shallowly emarginate anteriorly; transtilla sclerotized, with numerous spines; valva broadly shovel-shaped, with pointed distodorsal and distoventral humps; sacculus largely fused with valva, only minute apical part with some setae discernible; vinculum comparatively long, posterior margin with distinct U-shaped emargination medially, lateromedial area with two pairs of characteristic short projections; saccus comparatively short and slender; phallus slender, almost straight.

FEMALE GENITALIA.– Apophysis posterioris about 1.0 mm long; segment VIII without processes or distinct folds; apophysis anterioris short, reaching at most length of segment VIII; antrum tubular, long, extending to about apex of apophysis anterioris; posterior part of ductus bursae without lateral sclerites; corpus bursae well defined; signum a large and stout, weakly curved hook. Differs from the similar *C. gallagenellum* and other related species e.g. in the shorter apophyses anteriores and posteriores and in the longer antrum.

DISTRIBUTION.– Widely distributed in north-western Europe, but also known from the northern part of the Iberian Peninsula and the Ukraine, becoming scarce towards the south and largely absent in the Mediterranean.

BIONOMICS.– The larva has been recorded from *Stellaria holostea* L., *S. uliginosa* Murray (Schütze, 1931: 92), *S. graminea* L. (Bradford, 1979: 128), *Cerastium fontanum* Baumg. (Klimesch, 1954: 281) and *C. arvense* L. (Huemer, 1993: 149). *Cerastium semidecandrum* L. (Huemer, 1988: 499) needs confirmation as a hostplant. The larva feeds in a spun young shoot from April to the end of May (Stainton, 1867: 122). Pupation takes place in a slight cocoon on the ground surface. The adult emerges from June to July but in the wild it has been observed mainly in August. It flies by night, but is rarely observed at light (Bland *et al.*, 2002: 196). Preferred habitats are various grassland, heaths, rough meadows and forest edges at low elevations.

180. Caryocolum klosi (Rebel, 1917)

Gelechia klosi Rebel, 1917: (30).

DIAGNOSIS.– Wingspan 14-16 mm. Labial palp cream-yellow mottled with blackish brown, especially on segment 3. Antenna black, ringed with whitish grey. Head with cream-white frons and light grey neck; anterior part of thorax and tegula blackish grey, posterior part whitish brown. Forewing black, mottled with white and light brown; white costal spots reaching middle of wing at 2/5, 1/2 and 4/5, the latter fused with tornal spot to form a bent fascia; diffuse light brown streaks along fold and in middle of wing; 1-2 black spots in middle of wing. Hindwing grey.

Variation. The specimens available for study show only minor variation.

Similar species. *C. interalbicella* has a cream-white frons and neck, forewing less clear black and white, the medial white costal block smaller and the black spots more distinct.

MALE GENITALIA.– Uncus long, narrow; tegumen very broad and deeply emarginate anteriorly, strongly constricted medially; transtilla weakly sclerotized, with some spines; valva long, comparatively broad, distal part slightly tapered, curved ventrad, apex truncate, with some stiff setae; sacculus short and broadly digitate, distal part tapered; posterior margin of vinculum slightly incised medially, lateromedial projections indistinct; saccus comparatively short, basal half broader than rod-like distal part; phallus slender, almost straight. Differs from the very similar *C. interalbicella* in the broader valva.

FEMALE GENITALIA.– Apophysis posterioris about 2.5 mm; segment VIII with pair of longitudinal, sclerotized ridges dorsally; ventromedial part of segment VIII membranous; apophysis anterioris about 1.5 times length of segment VIII; antrum tubular, short, about one-third length of apophysis anterioris; posterior part of ductus bursae membranous; signum a large hook, with distinct base and short spines basally. The female genitalia differ from *C. interalbicella* particularly in the larger antrum and signum.

DISTRIBUTION.– Only recorded from few and scattered localities in Central Europe: France, southern Germany, Austria, Czech Republic and a doubtful record from Romania. Recently also found in the Italian Alps (Karsholt, 2005: 129).

BIONOMICS.– The larva has been recorded from *Stellaria nemorum* L. (Süssner, 1966). It feeds in April between spun shoots in a similar manner to *C. cassella* which however, occurs two weeks later. The adults have been bred in the second half of May in the laboratory. In the wild they occur from late June to late July. Preferred habitats are shady mountain forests and forest clearings.

181. *Caryocolum interalbicella* (Herrich-Schäffer, 1854)

Gelechia interalbicella Herrich-Schäffer, 1854: 187.
Tinea quadrella Fabricius, 1794: 298; homonym of *Tinea quadrella* [Denis & Schiffermüller], 1775 (Lepidoptera, Pyralidae).

DIAGNOSIS.– Adult. Wingspan 14-17 mm. Labial palp cream-white, mottled with black on outer surface of segment 2 and on segment 3. Antenna black, ringed with grey. Head cream-white; thorax and tegula concolorous with forewing. Forewing black with white markings and scattered orange-grey scales; white costal spots reaching middle of wing at 2/5, 1/2 and 4/5, the latter often fused with tornal spot to form a bent fascia; black spots in fold and on each side of the white patch in middle of wing; scattered orange scales, especially in fold. Hindwing grey.

Variation. The two inner white spots may be less distinct due to many grey scales. In some specimens there are many orange to light brown scales on the forewing, and they may form a patch before costal and tornal spots, sometimes separating them. Other, more clear black and white specimens may have white scales along termen.

Similar species. *C. klosi* q.v.

MALE GENITALIA.– Very similar to *C. klosi* but the valva is more slender in the basal half and the sacculus is slender with only a slightly broadened medial part.

FEMALE GENITALIA.– Apophysis posterioris about 2.8 mm; antrum very short, about one-fifth length of apophysis anteriois; signum is a medium-sized hook, with comparatively small base, without additional spines. Otherwise as in *C. klosi*.

DISTRIBUTION.– Endemic to the Alps, widely distributed in the central and southern part of the alpine arc.

BIONOMICS.– The larva has been recorded from *Cerastium arvense* L. (Burmann, 1954b: 203). Almost fully grown larvae have been found at high altitudes already from mid-April to mid-May, indicating hibernation in the larval stage. The larva constructs a silken tube which is hidden between the roots of the host-plant. It usually is well hidden at the base of the larval tube which is only left when the larva feeds on the young shoots and buds. Occasionally it bores into the shoots. Moths emerged from late May to mid-June (Burmann, 1954b: 204). The adult has been collected from late June to late August in the field. Preferred habitats are sunny rock formations. Vertical distribution: from about 1000 to 2500 m, rarely at lower elevation.

182. Caryocolum laceratella (Zeller, 1868)

Gelechia laceratella Zeller, 1868: 143.
Lita thurneri Pinker, 1953: 180, figs. 1-2.

DIAGNOSIS.– Adult. Male. Wingspan 18 mm. Labial palp white, mottled with brown on outer surface of segment 2 and on tip of segment 3. Antenna black. Head white; thorax white with some light brown scales; tegula black at base, whitish yellow at tip. Forewing white, mottled with yellowish brown and with black more or less oblique markings as follows: at base of costa, along fold, at 1/4 of costa, as a patch in middle of wing, and as another patch before costal spot; apical area mottled white, black and yellowish brown. Hindwing light grey. Female. Wingspan 14 mm. Similar to male but smaller and slightly brachypterous.

Variation. The specimens available for study show only minor variation.

Similar species. The white head and labial paps, and the white forewings with oblique black markings are characteristic. *C. laceratella* is moreover unique in *Caryocolum* in the distinct difference in wingspan between males and females.

MALE GENITALIA.– Uncus comparatively short, broad; tegumen very broad and deeply emarginate anteriorly, strongly constricted medially; transtilla pear-shaped with numerous spines; valva long, comparatively broad, distal part slightly broadened, apex truncate, with some stiff setae; sacculus digitate, slightly pointed; posterior margin of vinculum straight with weak incision medially; saccus comparatively slender; phallus stout, almost straight. The broad and short valva and the straight posterior margin of the vinculum without projections are characteristic for *C. laceratella*.

FEMALE GENITALIA.– Apophysis posterioris about 1.6 to 1.7 mm; segment VIII with pair of longitudinal, sclerotized ridges dorsally; ventromedial part of segment VIII membranous; apophysis anterioris slightly shorter than segment VIII; antrum tubular, short, about one quarter length of apophysis anterioris, anterior margin deeply incised; posterior part of ductus bursae membranous; signum a medium-sized hook, with distinct base. Readily distinguished from related species such as *C. klosi* and *C. interalbicella* by the distinctly shorter apophysis anterioris and the different antrum.

DISTRIBUTION.– Endemic to the south-eastern Alps (Italy, Slovenia).

BIONOMICS.– The larva has been recorded from *Moehringia ciliata* (Scop.) Dalla Torre, *Minuartia sedoides* (L.) Hiern and *Cerastium* sp. (*cf. C. latifolium* L.) (Huemer & Sattler, 1989: 263). The main host-plant seems to be *Moehringia ciliata*, a small Caryophyllaceae growing between debris and preferring alpine scree on limestone soil. The larva lives in a silken tube covered with dark green frass and hidden among basal stalks of the host-plant. It occurs very late in the year, depending on the elevation of the locality and duration of cover with snow. In the Alpi Giulie (Italy) larvae have been recorded as late as early August. Pupation takes place between withered parts of the plant near the ground surface (Huemer & Sattler, *op. cit.*). Pinker (1953: 180) found pupae between late August and mid-September. The adults have been collected in late July at lower altitudes and from late August to September at higher elevation. *C. laceratella* is restricted to alpine scree and similar habitats with scarce vegetation such as dry river-beds, exclusively in limestone areas. Vertical distribution: from about 1000 to 2100 m.

183. Caryocolum blandella (Douglas, 1852)

Gelechia blandella Douglas 1852a: 246.
Recurvaria maculea Haworth, 1828: 552; unjustified emendation of *Tinea maculella* Fabricius, 1794; misidentification.
Phthorimaea horoscopa Meyrick, 1926: 279. Subspecies, **syn. n., stat. n.**

DIAGNOSIS.– Adult. Wingspan 12-15 mm. Labial palp cream-white, outer surface of segment 2 and segment 3 mottled with blackish brown. Antenna ringed with black and cream-white. Head and tegula white; thorax white mottled with brown. Forewing brownish white mottled with grey and black scales; a distinct oblique black patch from 1/4 on costa; 2-3 black spots in middle of wing at 2/5 and 3/5; a faint outwards bent fascia at 4/5; apical area a mixture of white, brown and black scales. Hindwing grey.

Variation. The forewings may look light brown or whitish grey depending on the amount of light brownish scales. The black patch on costa varies in size and may be confluent with the black spot at 2/5 in the middle of the wing. The two anterior black spots may be confluent. Specimens of spp. *horoscopa* are smaller (wingspan 10-11 mm) and distinctly darker, having the labial palps, neck, thorax, tegulae and light areas of the forewing covered by grey scales.

Similar species. *C. blandelloides* is slightly smaller and shows less contrast in the markings of the forewings. The black patch on costa is up to twice as wide in *C. blandella* as in *C. blandelloides*. The antenna of *C. blandella* is more distinctly ringed with light and dark compared to *C. blandelloides*. Spp. *horoscopa* is similar to *C. blandulella*, but that species has a more regular black costal block, which is normally not confluent with the first black spot in the middle of the forewings. The forewings of ssp. *horoscopa* have moreover more orange-brown scales, which interrupt the white fascia at 3/4. It is also similar to specimens of *C. blandelloides* from Crete, but the black costal block of the latter species is more slender and less distinct; spp. *horoscopa* moreover has shorter and broader forewings than *C. blandelloides*.

MALE GENITALIA.– Uncus long, narrow; tegumen very broad and deeply emarginate anteriorly, strongly constricted towards uncus; transtilla with numerous spines; valva curved ventrad, much longer than sacculus, slender, narrowing medially, apical part a rounded bulge with some stiff setae; sacculus slender, knife-shaped; posterior margin of vinculum slightly projected with small medial incision; saccus subtriangular, broad at base, gradually tapered towards apex; phallus slender, basally inflated, almost straight. Very similar to *C. blandelloides*, differing mainly in the rounded apex of the valva and to a certain degree in the shape of the saccus. Differs from *C. proxima* in the much longer valva with a larger apical bulge.

FEMALE GENITALIA.– Segment VIII with pair of broad, digitate dorsal processes, pointed ventrad; ventromedial part of segment VIII with large ovate sclerotization, almost extended to posterior margin of segment; apophysis anterioris about twice as long as segment VIII; antrum funnel-shaped, short, about one quarter length of apophysis anterioris, anterior and posterior margins emarginate; posterior part of ductus bursae with indistinct sclerotized plate and pair of tiny lateral sclerites; signum slender, slightly curved hook, with comparatively small and sickle-shaped base. Differs from the very similar *C. blandelloides* mainly in the shorter ventral plate of segment VIII and the larger antrum.

DISTRIBUTION.– North-western Europe through Central Europe to Russia, extending to the Altai Mountains in the east (Ponomarenko, 2008: 98). Absent from the Iberian Peninsula and the Mediterranean apart from a recent record from north Portugal (Corley *et al.*, 2006). Ssp. *horoscopa* occurs from Greece and Turkey to Afghanistan and India (Kashmir).

BIONOMICS.– The larva has been observed feeding on *Stellaria holostea* L. (Douglas, 1852b: 77). A record from *Stellaria media* (L.) Vill. (Wörz, 1954: 129) needs confirmation. In the early spring the larva is a leaf-miner. Later it feeds between a spun shoot or within the flower-buds and finally within the seed-capsules (Douglas, *op. cit.*; Stainton, 1867: 90; Schütze, 1931: 92). In June, when the larva is fully grown, it pupates in a cocoon on the ground surface among debris. The adult occurs from early July to the beginning of September. It is rarely attracted to light but can be easily found resting on trunks (Karsholt, 1981: 257). Preferred habitats are deciduous forests, mainly with oaks. The few adults of ssp. *horoscopa* known to date have been collected from July to late August at altitudes of about 500 m in Greece, reaching elevations of up to 2200 m in Afghanistan.

REMARKS.– Specimens from Greece and Turkey are smaller and mottled with dark grey on head and thorax. They have with some doubts been identified as *C. horoscopa* (Meyrick, 1926) which is otherwise known from a few specimens collected in Kashmir and Afghanistan (Huemer, 1989d: 133-134). However, differences in genitalia characters are insufficient to separate these populations from nominotypical *C. blandella*. We therefore treat *C. horoscopa* as a subspecies of *C. blandella* for the time being.

184. Caryocolum blandelloides Karsholt, 1981

Caryoclum blandelloides Karsholt, 1981: 252, figs. 1-2, 13-14, 23.

DIAGNOSIS.– Wingspan 10-13 mm. Segment 2 of labial palp cream-white mottled with blackish brown on outer and lower surface; segment 3 blackish brown with some cream-white. Antenna ringed black and light brown. Head with cream-white face and whitish brown neck; thorax and tegula concolorous with forewing. Forewing whitish brown mottled with grey and black; a rather slender black oblique streak from 1/4 of costa, a small black spot at 2/5 and a larger comma-shaped spot at 3/5; an outward bent white fascia at 4/5; apical area with many black scales. Hindwing grey.

Variation. Males are slightly larger than females. The black streak from 1/4 on costa may be divided into two, and the black spot in the middle of the wing at 3/5 may be straight or angled. Specimens from Corsica and Crete have less brown scales on the forewings, giving them a more contrasting look. In these populations the antennae are also ringed black and white, and some specimens have much black on the forewings. The specimens from Crete moreover have the black costal streak broader.

Similar species. *C. blandelloides* may be confused with several other *Caryocolum* species, e.g. *C. blandella* q.v.; *C. blandulella* q.v.; *C. moehringiae* q.v.; *C. petrophila* q.v.; *C. kroesmanniella* q.v.; *C. huebneri* q.v.

MALE GENITALIA.– Very similar to *C. blandella* but the apical bulge of the valva is of triangular shape and the saccus is more irregularly tapered in *C. blandelloides*.

FEMALE GENITALIA.– Very similar to *C. blandella* but differs in the smaller digitate process of segment VIII, the posteriorly extended ventral plate and the shorter and smaller antrum.

DISTRIBUTION.– Initially described from Scandinavia this species is now known from scattered localities in the Mediterranean, Great Britain and Central Europe to north-western Russia and the Ukraine (Bidzilya & Budashkin, 2009: 19).

New records: Greece: 2 ♂, 16 ♀, Dráma, Falakró above Vókalás, 1650 m, 31.viii.2008, leg. Skou (ZMUC). Greece (Crete): 4 ♂, 1 ♀, Kallergi Mts., 1450-1550 m, 28-30.vii.2001, leg Fibiger, Jeppesen, Madsen & Nilsson, gen slides 3275, 3277, 4735 Hendriksen (ZMUC).

BIONOMICS.– The larva has been recorded from *Cerastium arvense strictum* (Haenke) Gaudin (Huemer, 1987: 207) and *C. fontanum* Baumg. (Hoare *et al.* 1999: 153); it also feeds on *C. semidecandrum* L. (Buhl *et al.*, 1998: 111). It has been observed from mid- to late June feeding inside a seed capsule of the host-plant. Frass is either ejected or deposited in the capsule itself. Capsules are changed but the larva never spins adjacent capsules together (Hoare *et al.*, *op. cit.*; Huemer, *op. cit.*). Pupation took place in a slight cocoon on the ground surface and moths emerged from mid-July to early August. In the wild adults have been collected from July to late August. Preferred habitats are sandy coasts but also sandy soil on alpine scree.

185. Caryocolum jaspidella (Chrétien, 1908)

Lita jaspidella Chrétien, 1908c: 231.

DIAGNOSIS.– Adult. Wingspan 8-10 mm. Segment 2 of labial palp bone-white on inner and upper surface, on lower and outer surface light grey mottled with black, especially at distal end; segment 3 almost white. Antenna black, faintly lighter ringed. Head with face cream-coloured and neck covered with light grey dark-tipped scales; thorax and tegula concolorous with forewing. Forewing covered with white, light brown and black scales; base white mottled with light brown; an inwards bent black fascia from costa at 1/5 to dorsum at 1/6; from this a black streak along costa; middle of wing with 3 black spots; between these and an almost straight white fascia from costal to tornal spots a brown patch; apical area with many black scales. Hindwing grey.

Variation. Specimens vary in the density of black scales on the forewings.

Similar species. *C. jaspidella* is unique within *Caryocolum* in having a black fascia from costa at 1/5 to dorsum at 1/6, giving the impression that it is directed towards the base of the forewings.

Male genitalia.– Uncus broadly rounded; tegumen distinctly widened anteriorly with deep medial emargination; transtilla with numerous spines; valva extremely short and almost completely reduced, with setae; sacculus short but more than twice as long as valva, subtriangular, evenly tapered to pointed apex; posterior margin of vinculum weakly projected with shallow medial incision; saccus strong, comparatively short, tapered towards apex; phallus basally inflated, short and stout. Differs from the otherwise similar *C. arenbergeri* mainly by the almost completely reduced valva and the weakly projected vinculum.

Female genitalia.– Segment VIII with pair of broad, digitate ventrolateral processes, pointing medially; ventromedial part of segment VIII with large ovate sclerotization; apophysis anterioris about 2.5 times length of segment VIII; antrum ring-shaped, short, anterior and posterior margins emarginate; posterior part of ductus bursae with indistinct sclerotization; signum stout, slightly curved hook, with comparatively small base. Differs from *C. proxima* in the longer apophysis anterioris and the larger signum.

Distribution.– Only known from the western Mediterranean area, Portugal (Corley *et al.*, 2006: 418), Spain (Karsholt, 1992: 307) and North Africa (Algeria, Morocco).

Bionomics.– The larva has been recorded from *Cerastium gibraltaricum* Boiss. (Karsholt, 1992: 304). It is initially a leaf-miner and later it feeds between spun leaves of a shoot. Pupation takes place in the soil in a cocoon covered with grains of sand (Karsholt, 1992: 304). The larval stage has been found in mid-April at an elevation of about 2000 m. Moths emerged in May in the laboratory. In its natural habitat the adult has been collected in April and from August to early September, indicating two generations. In the Middle Atlas Mountains *C. jaspidella* has been observed in the upper part of a cedar forest, preferring warm spots among spiny bushes which were not eaten by goats and sheep. Vertical distribution: known from about 200 m in Portugal to 2100 m in North Africa.

186. Caryocolum proxima (Haworth, 1828)

Recurvaria proxima Haworth, 1828: 552.
Gelechia maculiferella Douglas, 1851: 102.
Gelechia maculivicinella Bruand d'Uzelle, 1859: 652.
Gelechia horticolla Peyerimhoff, 1871: 411.
Caryocolum proximum auctt.

Diagnosis.– Adult. Wingspan 10-13 mm. Segment 2 of labial palp cream-coloured mottled with blackish brown on outer and lower surface; segment 3 blackish brown with some lighter scales. Antenna ringed with black and grey-brown. Head, thorax and tegula concolorous with forewing; face cream-yellow. Forewing greyish brown mottled with white, orange and black; a distinct black block at 1/4 of costa, almost reaching dorsum and edged with yellow-white towards base; a black spot in middle of wing at 2/5 and a larger such at 3/5, the latter often fused with a black patch near tornus; light costal and tornal spots separated by orange scales; apical area with many black scales. Hindwing grey. Light rings of tibia of forelegs, and especially midlegs with red scales.

Variation. The black patch from costa at 1/4 may vary in form and size. Some specimens have many orange scales, especially in the middle of the forewings.

Similar species. *C. proxima* is most easily recognized by the light reddish scales on the tibia of the forelegs and especially of the mid-legs. This character is also found in many specimens of *C. junctella* (which has many white-tipped scales along the dorsum of the forewing, and the black costal block is less prominent) and in specimens of *C. fibigerium* from south-western Europe (but that species has black labial palps, thorax and tegulae).

MALE GENITALIA.– Uncus long, comparatively broad; tegumen very broad and deeply emarginate anteriorly, strongly constricted towards uncus; transtilla with numerous spines; valva curved ventrad, short, slightly longer than sacculus, slender, strongly narrowing medially, apical part a small rounded bulge with some stiff setae; sacculus slender, knife-shaped; posterior margin of vinculum slightly projected with small medial incision; saccus subtriangular, broad at base, gradually tapered towards apex; phallus slender, basally inflated, almost straight. Differs from the otherwise similar *C. blandella* and *C. blandelloides* in the distinctly shorter valva.

FEMALE GENITALIA.– Segment VIII with pair of broad and long, digitate ventrolateral processes, pointing medially; ventromedial part of segment VIII with large ovate sclerotization, not extending to posterior margin of segment VIII; apophysis anterioris about 1.5 times length of segment VIII; antrum ring-shaped, short, anterior and posterior margins emarginate; posterior part of ductus bursae with indistinct sclerotization; signum a stout, short hook, with comparatively small base. Very similar to *C. jaspidella* but differs in the shorter apophysis anterioris and the signum.

DISTRIBUTION.– Widely distributed in Europe except the northernmost parts. Recently also found in Turkey and Israel. Recorded from the U.S.A., where it was possibly introduced.

New records. Spain: 1 ♂, prov. Madrid, Sierra de Guadarama, Lozoya, 1000 m, 20-21.vii.1995, leg. Skou, gen. slide 2053 Hendriksen (ZMUC). Italy (Sicily): 3 ♂, above Lingublosa, Bocca Racabo, 1450 m, 8.-9.ix.2002, leg. Karsholt *et al.*, gen. slidea 3806, 5481 Hendriksen (ZMUC). Greece (Crete): 2 ♂, 4 km S Topolia, 300 m, 25.-29.vi.2000, leg. Fibiger *et al.*, gen. slide 5406 Hendriksen (ZMUC). Israel: 1 ♂, slopes by Yflach, 31.v.2004, leg. Kravchenko, gen. slide 5703 Hendriksen (TAU). Turkey: 1 ♀, Elazig, Hazar, Gölü N, 1250 m, 25.vii.1989, leg. Fibiger & Esser, gen. slide GU 94/492 ♀ Huemer (ZMUC).

BIONOMICS.– The larva has been recorded from *Stellaria media* (L.) Vill. (Karsholt, 1981: 258) and *Cerastium fontanum* (Bradford, 1979: 128). It feeds in May on the flowers and seeds of the host-plant (Bradford, *op. cit.*; Huemer, pers. obs.) and pupates in a slight cocoon among debris on the ground surface. The adult occurs from late June to mid-September, but mainly in August. It has been found during the day resting in cracks on the trunks of old oaks but is also quite frequently attracted to light (Huemer, pers. obs.). Preferred habitats are margins of *Quercus*-forests and open grassland, but also gardens with lawns (Karsholt, *op. cit.*).

REMARKS.– *Gelechia maculiferella* Douglas is an unnecessary objective replacement name for *Recurvaria proxima* Haworth, 1828, which is not an emendation of *Tinea proximella* Hübner, 1796.

The supposed synonymy of *Gelechia maculivicinella* is based on Leraut (1997: 124).

There is a high degree of confusion concerning both biological and distributional data between *C. proxima* and *C. alsinella*. Most of the classical literature on *C. proxima* (under the name *maculiferella*), mainly based on Stainton's (1867: 155) treatment, refers to *C. alsinella*. These misidentifications were later corrected by Karsholt (1981: 258).

187. Caryocolum blandulella (Tutt, 1887)

Gelechia blandulella Tutt, 1887b: 105.

DIAGNOSIS.– Adult. Wingspan 9-11 mm. Labial palp cream-coloured mottled with black, especially on segment 3. Antenna ringed black and cream-grey. Head, thorax and tegulae with light grey, dark-tipped scales; frons cream-white. Forewing light grey mottled with white, light brown and black; base with many black scales; a prominent black block from costa at 1/4 nearly reaching dorsum; 2-3 black spots in middle of wing, the outer one fused with a black patch near tornus; an almost straight white fascia between costal and tornal spots, not interrupted by orange-brown; apical area almost black. Hindwing grey.

Variation. The coverage of black scales on the forewing is variable. Rarely there are many orange-brown scales on the forewings, and these may then separate the white costal and tornal spots. Specimens from sand dune localities on the North Sea coast have lighter forewings with many light orange-brown (and less grey and black) scales, and the neck is almost cream-white.

Similar species. The dark form may resemble *C. alsinella*, which has dark forewings with a less distinct costal block, and costal and tornal spot often separated by orange-brown. The light form resembles *C. blandelloides*, which is slightly larger and has a less distinct black costal block than in *C. blandulella*. *C. blandella* spp. *horoscopa* q.v.

MALE GENITALIA.– Uncus sub-ovate; tegumen strongly widened anteriorly, with deep medial emargination; transtilla with numerous spines; valva very short and slender, weakly tapered towards apex, with comparatively indistinct setae apically; sacculus knife-shaped, proximal part broader than valva, about length of valva; posterior margin of vinculum with deep V-shaped medial emargination and broadly projected lateromedial humps; saccus subtriangular, broad at base, gradually tapered towards apex; phallus weakly curved, stout. The short valva without apical inflation is particularly characteristic for this species.

FEMALE GENITALIA.– Segment VIII short, with pair of broad and short digitate ventrolateral processes, pointing medially; ventromedial part of segment VIII membranous; apophysis anterioris about three times length of segment VIII; antrum ring-shaped, reaching about two-thirds length of segment VIII, anterior and posterior margins scarcely emarginate; posterior part of ductus bursae with indistinct sclerotization; signum short and stout hook, with distinct base. Differs from related species such as *C. jaspidella* and *C. proxima* e.g. in the membranous ventrum of segment VIII and the shape of the antrum. *C. arenbergeri* q.v.

DISTRIBUTION.– From Spain and Great Britain through the southern parts of northern Europe, Central Europe and the Balkan Peninsula to Turkey (Huemer, 1989c: 39). Recently also found in Switzerland (Bryner *et al.*, 2005: 128) and Finland (Kullberg *et al.*, 2002: 148).

New records. Croatia: 1 ♂, 1 ♀, Krk, Misučaynica, 6-8.viii.1975, leg. Jäckh, gen. slides 2785, 2786 Karsholt (ZMUC). France (Corsica): 1 ♀, Col de Vergio, 1400-1500 m, 14-15.viii.1998, leg. Karsholt, gen. slide 2204 Hendriksen (ZMUC).

BIONOMICS.– The larva has been found feeding on *Cerastium pumilum* Curt. (Benander, 1965: 17) and *C. semidecandrum* L. (Huemer, 1993: 151). It lives from early May to June in the seed-capsules which are spun together (Benander, *op. cit.*). The adult occurs from the middle of July to late August. *C. blandulella* is a characteristic species of sandy coasts and similar inland habitats. Vertical distribution: from coastal areas to 1600 m in Greece and 2000 m in Turkey.

REMARKS.– The type of *C. blandulella* is of the light form with lighter forewings and head. Karsholt (1981: 260) suggested that the dark form may represent a subspecies, but it is now considered more likely that the variation is correlated with the habitat, viz. the light form occur in localities with white sand.

188. Caryocolum arenbergeri Huemer, 1989

Caryocolum arenbergeri Huemer, 1989b: 202, figs. 3, 6, 9.

DIAGNOSIS.– Adult. Wingspan 9 mm. Segment 2 of labial palp cream-white mottled with black on outer and lower surface; segment 3 almost black. Antenna black, faintly lighter ringed. Head with white frons and grey neck; thorax and tegula concolorous with forewing. Forewing light grey, mottled with black, white and especially brown; an oblique black block from costa at 1/4; a small black spot in middle of wing at 2/5 and a larger oblique black spot at 3/5; an almost straight white fascia at 3/4; apical area a mixture of brown, black and white. Hindwing grey.

Variation. The intensity of brown scales on the forewing is variable. They are most prominent in fresh specimens.

Similar species. *C. arenbergeri* can be recognized by its small size and its more or less brownish forewings with a few distinct black markings. It may be mistaken for *C. alsinella*, but that species has more black (and less brown) scales on the forewing.

MALE GENITALIA.– Uncus broadly rounded; tegumen distinctly widened anteriorly, with deep medial emargination; transtilla with numerous spines; valva very short, pointed towards apex, with apical brush of setae; sacculus short, slightly longer than valva, with convex dorsal and concave ventral margin, apex pointed; posterior margin of vinculum with deep V-shaped medial emargination and broadly projected lateromedial humps; saccus strong, comparatively short, abruptly tapered towards apex; phallus short and stout. Differs from the related *C. jaspidella* in the long valva and the medial emargination of the vinculum. Genitalia furthermore similar to those of *C. blandulella* but differing e.g. in the shorter and strongly tapered valva, the longer sacculus and the more slender saccus.

FEMALE GENITALIA.– Segment VIII short, with pair of broad and short digitate ventrolateral processes, pointing medially, medial part of subgenital plate with small angular ridge; ventromedial part of segment VIII membranous; apophysis anterioris about three times length of segment VIII; antrum ring-shaped, reaching about two-thirds length of segment VIII, anterior and posterior margins scarcely emarginate; posterior part of ductus bursae with indistinct sclerotization; signum short and stout hook, with broad base. Particularly the angular ridge is characteristic and differentiates *C. arenbergeri* from the otherwise very similar *C. blandulella*.

DISTRIBUTION.– Endemic to the Iberian Peninsula where it occurs in the Sierra de Gredos and the Sierra de Guadarama (Spain).

BIONOMICS.– Host-plant and early stages unknown, but probably the larva feeds on species of Alsinoideae (*Cerastium* L. and/or *Stellaria* L.). The few adults known to date have been collected in mid July at light. Vertical distribution: known from mountainous areas from about 1000 to 1700 m.

189. *Caryocolum tricolorella* (Haworth, 1812)

Tinea tricolorella Haworth, 1812: 338
Recurvaria contigua Haworth, 1828: 552.
Gelechia acernella Herrich-Schäffer, 1854: 185, pl. 77, fig. 580.

DIAGNOSIS.– Adult. Wingspan 13-15 mm. Segment 2 of labial palp bone-white on upper and inner surface, black mottled with white on lower and outer surface; segment 3 almost black. Antenna black, faintly lighter ringed. Head with cream-white frons and blackish grey neck; thorax and tegula almost black. Forewing covered with brown, black and white scales; basal and dorsal part light brown; a large black block at costa, inwardly bordered with white; a large brown patch in middle of wing between half and 3/4 separating white costal and tornal spots; apical area almost black. Hindwing grey.

Variation. There is some variation in the intensity of black scales on the forewings. In worn specimens thorax and tegulae become lighter.

Similar species. The large black block on costa of the forewings contrasting with the light brown base and dorsum is characteristic for *C. tricolorella*.

MALE GENITALIA.– Uncus long, narrow; tegumen very broad and deeply emarginate anteriorly, strongly constricted medially; transtilla weakly sclerotized, with numerous spines; valva long, slender, distal part slightly broadened, curved ventrad, apex truncate, with some stiff setae; sacculus long, knife-shaped; vinculum comparatively short, posterior margin of vinculum weakly emarginate medially, lateromedial projections indistinct; saccus slender, evenly tapered towards pointed apex; anellus with pair of long and needle-shaped sclerites; phallus stout, almost straight. Differs from the very similar *C. fibigerium* in the posterior margin of the vinculum which is less emarginated and without lateral projections. The genitalia are furthermore similar to those of *C. klosi* and *C. interalbicella* with a number of differing characters such as the more slender valva, the posterior margin of the vinculum and the shorter saccus.

FEMALE GENITALIA.– Segment VIII with pair of small flaps dorsally; ventromedial part of segment VIII membranous with numerous microtrichia; apophysis anterioris short, about length of segment VIII; antrum broadly funnel-shaped, about one-third breadth of segment VIII, short, reaching about half length of segment VIII; posterior part of ductus bursae with indistinct sclerotization; signum a long and slender hook, with large base. Differs from the similar *C. fibigerium* e.g. in the shorter and comparatively narrow antrum.

DISTRIBUTION.– Northern, Central and eastern Europe. From the Mediterranean only a recent record from northern Spain. A single record from Sicily (Mariani, 1943: 169) most probably refers to *C. fibigerium*.

New record. Spain: 1 ♀, Lleida, Castellbò, 850 m, 11.vii.2009, leg. F. Groenen, gen. slide FG1782 (coll. Groenen).

BIONOMICS.– The larva has been recorded from *Stellaria holostea* L. (main host-plant), *S. alsine* Grimm (Sorhagen, 1886: 194), *S. media* (L.) Vill. (Hering, 1937: 512) (doubtful record) and *Cerastium arvense* (Huemer, 1988: 509). It initially produces a gallery-like leaf-mine which has been observed as early as September. In Britain the mining larva occurs from December to January. Later it feeds on buds or between spun terminal shoots. Pupation takes place in a cocoon among detritus on the ground surface in May. The adult occurs from June to mid-September and it is attracted to light and sugar (Stainton, 1867: 102). Preferred habitats are thermophilous deciduous forests, particularly oak forests.

190. Caryocolum fibigerium Huemer, 1988

Caryoclum fibigerium Huemer, 1988: 510, figs. 86, 153, 214.

DIAGNOSIS.– Adult. Wingspan 10-13 mm. Labial palp black, slightly mottled with lighter scales. Antenna black, faintly lighter ringed. Head with cream-coloured frons and black neck; thorax and tegula black. Forewing covered with black, grey, brown and white scales; a black costal block at 1/4, separated from base by a white and light brown area; distal 2/5 of wing black with contrasting white costal and tornal spots. Hindwing grey.

Variation. In some specimens black scales dominate on the forewings making them look blackish. Sometimes the dorsum is more or less brown. In specimens from SW Europe the light rings of tibia of the forelegs and mid-legs have red scales.

Similar species. Specimens from south-western Europe can be recognized by the red scales of the forelegs and mid-legs (the blackish *C. fibigerium* should not be confused with the two other species having this character, viz. *C. proxima* and *C. junctella*). *C. mucronatella* has lighter labial palps, thorax and tegulae compared to *C. fibigerium*.

MALE GENITALIA.– Very similar to *C. tricolorella* but the valva and sacculus are somewhat more slender and the posterior margin of the vinculum is more distinctly emarginated and exhibits a pair of weak lateral projections.

FEMALE GENITALIA.– Similar to *C. tricolorella* but the antrum is much broader and longer, reaching about half width of segment VIII and extending to apex of apophysis anterioris.

DISTRIBUTION.– Mediterranean, from the Iberian Peninsula, southern France and Italy to south-eastern Europe; Morocco, Turkey (Huemer, 1989c: 39).

BIONOMICS.– The larva has been found feeding on *Arenaria montana* L. (Corley, 2002a). It initially probably produces a short linear leaf-mine and has been reported to feed from November to mid-December between two spun leaves, usually at the growing tip of a shoot. Afterwards the larva constructs a cocoon where it remains inactive until pupation in spring (Corley, *op. cit.*). The adult has been observed in April at lower altitudes and from mid-June to early October at higher elevations. It is attracted to light. Preferred habitats are xeric grassland and rocky places. Vertical distribution: from lowland localities to about 2300 m in Spain (2400–2800 m in Morocco).

191. Caryocolum junctella (Douglas, 1851)

Gelechia junctella Douglas, 1851: 103.
Phthorimaea aganocarpa Meyrick, 1935: 585.

DIAGNOSIS.– Adult. Wingspan 10-11 mm. Segment 2 of labial palp cream-coloured mottled with grey brown; segment 3 almost black. Antenna black, faintly lighter ringed. Face pale, neck metallic shining dark fuscous; thorax and tegula greyish brown, the latter with light tip; a small tuft of slender light red scales at the ventral margin of the compound eye. Forewing with a mixture of grey, yellow, reddish, brown and black scales. An oblique black block from 1/4 on costa; 2-3 black spots in middle of wing; white costal and tornal spots separated by yellow scales; areas with yellowish scales in fold and between outer black spot; apical area with many black scales. Hindwing grey.

Variation. *C. junctella* is a variable species. The forewings are often covered with many dark-tipped scales. Fresh specimens, especially, tend to be reddish from the forewings more or less covered with reddish brown scales, and also by having the labial palps and the light rings at the tibia of the fore- and midlegs reddish. Specimens from Denmark are without red legs and the small tuft of slender light red scales at the ventral margin of the compound eye. Hibernated specimens are often worn, looking more or less uniformly grey.

Similar species. Due to its variability *C. junctella* is often confused with several other small *Caryocolum* species. Red tibia of the fore- and mid-leg, which are only found in some specimens of *C. junctella*, are otherwise only known in this genus in *C. proxima* q.v. and *C. fibigerium* q.v. *C. junctella* is moreover the only *Caryocolum* known to hibernate in the adult stage, and specimens found in spring should therefore belong to this species. The patches of yellowish scales in the fold and before the white costa are also characteristic for *C. junctella*.

MALE GENITALIA.– Uncus long, narrow; tegumen very broad and deeply emarginate anteriorly, strongly constricted medially; transtilla with numerous spines; valva long, slender, distal part slightly broadened, curved ventrad, apex rounded with some stiff setae; sacculus shorter than valva, apex hook-shaped; vinculum short, posterior margin straight with a weak incision; saccus stout, comparatively slender, evenly tapered towards pointed apex; anellus with pair of semicircular sclerites; phallus comparatively slender, basally inflated, almost straight.

FEMALE GENITALIA.– Segment VIII short, with pair of small flaps dorsally; ventromedial part of segment VIII with numerous microtrichia; apophysis anterioris long, about two times length of segment VIII; antrum funnel-shaped, basally broad, long, extending almost to apex of apophysis anterioris; signum a prominent, long and slender hook, with five short spines basally.

DISTRIBUTION.– Palaearctic. From France and Great Britain throughout northern, Central and eastern Europe to Asian Russia (Transbaikalia) (Bidzilya *et al.*, 1998: 47), China, Korea and Japan. Records from the Mediterranean, e.g. from Spain (Agenjo, 1968), are rare and need verification.

BIONOMICS.– The biology of *C. junctella* is as yet insufficiently described. The larva has been recorded from *Stellaria graminea* L. (Parsons, 1995: 76), *S. media* (L.) Vill. (Huemer, 1988: 511) and *Cerastium arvense* (Benander, 1928: 79). *Cerastium pauciflorum* Steven ex Ser. (Liu & Pai, 1979: 276) and *Polygonum aviculare* L. (Polygonaceae) (Park, 1993: 18) are stated as further host-plants from eastern Asia. Particularly the latter record is extremely surprising and may be due to xenophagy. The larva is initially a leaf-miner (Parsons, 1995: 76). The feeding habits of later stages are insufficiently described, according to Park (1993: 18) the leaves are attacked. The larva occurs in June. *C. junctella* is the only species of the genus known to hibernate as an adult. The moth occurs from July to May, particularly along forest edges and lanes. It has been found on the trunks of oaks, pines and fir and can be beaten out of the vegetation. The adult is attracted to light.

192. *Caryocolum cassella* (Walker, 1864)

Gelechia cassella Walker, 1864a: 594.
Gelechia melanotephrella Erschoff, 1877: 345.
Lita albifasciella Toll, 1936: 407, pl. 49, fig. 20.
Phthorimaea subvicinella Hackman, 1946: 63, figs. 9-10.
Caryocolum falellum Piskunov, 1975: 869, fig. 17.

DIAGNOSIS.– Adult. Wingspan 11-13 mm. Segment 2 of labial palp cream-coloured mottled with blackish brown; segment 3 almost black with white tip. Antenna ringed with black and cream-brown. Head with cream-coloured face and greyish neck; thorax light and dark grey; tegula with cream-coloured tip. Forewing with white, grey, black and light brownish scales; base, an oblique costal block at 1/4 and one at 3/4, dorsum and apical area greyish black; patches of white mottled with light brown between base and first and second costal blocks; costal and tornal spots cream-white, interrupted by black. Hindwing grey.

Variation. *C. cassella* is only slightly variable, but the light areas on the forewings can be more or less covered with light brown scales. The basal black block can be divided in two patches.

Similar species. The two outwards oblique black costal blocks are characteristic for *C. cassella*. In other species they are either less oblique or only the inner one is oblique. *C. pullatella* q.v.

MALE GENITALIA.– Uncus long, narrow; tegumen very broad and deeply emarginate anteriorly, strongly constricted medially; transtilla strongly sclerotized, large subovate part turned posteriorly, without spines; valva short and broad, knife-shaped, evenly tapered towards apex, with indistinct apical brush of setae; sacculus about length of valva, slender, knife-shaped; vinculum prominent, posterior margin slightly incised medially, lateromedial part weakly projected; saccus very broad at base, distally gradually tapered, apex rod-like; phallus stout, weakly curved.

FEMALE GENITALIA.– Segment VIII with pair of distinct oblique folds ventrolaterally, additionally a pair of drop-shaped processes developed; ventromedial part of segment VIII ovate, well sclerotized, with numerous microtrichia; apophysis anterioris long, nearly two times length of segment VIII; antrum tubular, short, extending almost to middle of apophysis anterioris, posteriorly deeply emarginate; posterior part of ductus bursae with pair of short sclerites; signum a distinct hook, with broad base.

Distribution.– Holarctic. In Europe restricted to the northern parts and mountains of Central Europe; a single record from Romania (Kovács & Kovács, 2000: 35); not yet recorded from the Iberian Peninsula, the British Isles and the Mediterranean. Widely distributed in Asia, from Siberia, Transbaikalia and Korea to Japan; numerous records from the U.S.A. and Canada (Huemer, 1988: 514).

Bionomics.– The larva has been found feeding on *Stellaria nemorum* L. (Petry, 1912: 117), *S. graminea* L. (Kaitila, 1996: 96), *S. alsine* Grimm (Schütze, 1931: 92) and *Moehringia trinervia* (L.) Clairv. (Elsner *et al.*, 1999: 48). A record from *Silene vulgaris* (Moench) Garcke (Elsner *et al., op. cit.*) is dubious and probably refers to *C. vicinella*. The larva lives in May and June, first in the flower-buds, later between spun shoots (Petry, *op. cit.;* Schütze, *op. cit.*). The adult occurs from late June to late August. It can be beaten from trunks and is also attracted to light (Karsholt, 1981: 268; Huemer, pers. obs.). Preferred habitats are damp shady forests, edges of deciduous woods and clearings. Vertical distribution: in Europe restricted to lowland and mountain localities up to about 1000 m, in North America reaching 2500 m.

193. Caryocolum moehringiae (Klimesch, 1954)

Gnorimoschema moehringiae Klimesch, 1954: 338, figs. 36-38.

Diagnosis.– Adult. Wingspan 12-13 mm. Segment 2 of labial palp black mottled with white, lighter on inner surface; segment 3 darker than segment 2. Antenna ringed black and whitish grey. Head cream-white, neck mottled with grey; thorax blackish grey mottled with some light grey; tegula lighter than thorax. Forewing white, overlaid with brown and black scales; an oblique fascia, often interrupted by brown scales, at 1/4 on costa; black spots before white costal and tornal spots, the latter confluent with a black comma-shaped spot in middle of wing; apical area with many black scales. Hindwing grey.

Variation. The black spots on the forewings are variable in size. Worn specimens become whitish.

Similar species. Very similar to *C. blandelloides* and not always separable from that species on external features. Most specimens of *C. moehringiae* are slightly larger and have the oblique fascia at 1/4 broader. *C. petrophila*, *C. kroesmanniella*, *C. huebneri* q.v.

Male genitalia.– Uncus long, narrow, smooth; tegumen very broad and deeply emarginate anteriorly, strongly narrowing medially; transtilla with numerous spines; valva short, digitate; sacculus shorter than valva, digitate; vinculum long, posterior margin slightly incised medially, lateromedial part broadly rounded and projected, lateral part with unique pliers-shaped process, almost reaching apex of valva; saccus subtriangular, evenly tapered towards slender apex; phallus stout, basally inflated, weakly curved.

Female genitalia.– Segment VIII comparatively long, with pair of small flaps ventrolaterally; ventromedial part of segment VIII membranous, with numerous microtrichia; apophysis anterioris long, about two times length of segment VIII; antrum funnel-shaped, extending to about middle of apophysis anterioris; signum a spiny disk, without hook.

Distribution.– Extremely scattered records from the Alps (Switzerland, Germany, Austria, Italy, Slovenia, France (Nel, 2006: 51)) and Greece (Huemer, 1989c: 40).

BIONOMICS.– The larva has been recorded feeding on *Moehringia muscosa* L. and *M. bavarica* (L.) Gren (Klimesch, 1954: 341). In the early stages in April it mines short galleries where the frass is deposited within. Later in May the larva feeds between spun terminal shoots or on spun leaves (Klimesch, *op. cit.*). Moths were bred from early June to mid-July. In the wild the adult has been observed only in a few specimens which were collected in July and August. *C. moehringiae* is preferably found in shady coniferous woodland.

194. Caryocolum petrophila (Preissecker, 1914)

Gelechia (Lita) petrophila Preissecker, 1914: 18, fig.
Phthorimaea kemnerella Palm, 1947: 39, figs. 1c, 1f.
Caryocolum petrophilum auctt.

DIAGNOSIS.– Adult. Wingspan 11-13 mm. Labial palp similar to that of *C. moehringiae*. Antenna black, ringed with light grey. Head cream-white with grey-brown neck; thorax and tegula black, sometimes mottled with light grey. Forewing white, overlaid by brown and black scales; a broad black oblique streak from 1/4 on costa to fold; two black spots in middle of wing; black patches (confluent) before white costal and tornal spots; apical area almost black. Hindwing grey.

Variation. The forewings can, to some extent, be more or less covered with black or brown scales.

Similar species. *C. moehringiae* is very similar, but the oblique black streak from 1/4 on costa is more slender in that species. This is even more the case in the otherwise similar *C. blandelloides*. *C. kroesmanniella* and *C. huebneri* q.v.

MALE GENITALIA.– Uncus long, narrow; tegumen very broad and distinctly emarginate anteriorly, strongly constricted medially; transtilla with numerous spines; valva short, slender, digitate; sacculus broader than valva basally, subtriangular, knife-shaped; posterior margin of vinculum with broadly V-shaped emargination and distinct subtriangular projection lateromedially; saccus prominent, subtriangular, very broad at base, distally gradually tapered, apex rounded; phallus stout, weakly curved.

FEMALE GENITALIA.– Segment VIII short, with pair of small flaps ventrolaterally; ventromedial part of segment VIII membranous; apophysis anterioris very long, about three times length of segment VIII; antrum funnel-shaped, longer than segment VIII, extending to about middle of apophysis anterioris, with few microtrichia; signum a long, slender hook. Differs from the otherwise similar *C. huebneri* e.g. in the shape and length of the antrum, and the shorter signum of *C. huebneri*.

DISTRIBUTION.– Only known from scattered localities in Scandinavia (Norway, Sweden, Finland), the Alps (France, Switzerland, Austria, Italy) and the Apennines, the Balkan Peninsula and Turkey. Recently also found in Central Asia and in the southern Ural Mountains (Russia) (Junnilainen *et al.*, 2010: 54). A record from Sicily (Huemer, 1988: 516) is erroneous.

New record. Russia: 1 ♂, SW Altai, Katun valley, 10 km W Katanda, 1200 m, 26.-27.viii.1983, leg. Mikkola, Hippa & Jalava, gen. slide GU 94/472 ♂ Huemer (ZMUH).

BIONOMICS.– The larva has been recorded from *Cerastium arvense* L. and *Stellaria graminea* L. (Kaitila, 1996: 97), probably also on additional species of *Cerastium*

L. (Huemer, pers. obs.). According to Skala (1950: 115) the early larval stage is a leaf-miner. Later, in May and June, it feeds between spun flowers and leaves (Kaitila, *op. cit.*). The adult occurs from June to mid-September and it is attracted to light. Preferred habitats are thermophilous, rocky places and dry meadows.

195. *Caryocolum huebneri* (Haworth, 1828)

Recurvaria huebneri Haworth, 1828: 551.
Gelechia hubnerella Doubleday, 1859: 30; unjustified emendation of *Recurvaria huebneri* Haworth, 1828.
Gelechia knaggsiella Stainton, 1866: 167.

DIAGNOSIS.– Adult. Wingspan 11-12 mm. Segment 2 of labial palp white mottled with black; segment 3 almost black. Antenna greyish white, more or less distinctly ringed with black. Head white mottled with grey towards neck; thorax and tegula with light grey dark-tipped scales. Forewing white mottled with many light brown, grey and black scales; black patches at 1/5 of costa and at 1/4 in middle of wing; in middle of wing at 1/3 a small black spot and a comma-shaped spot at 2/3, the latter often confluent with a black tornal patch; a bent white fascia at 4/5; apical area with many black scales. Hindwing grey.

Variation. There is some variation in the amount of brown and black scales on the forewings. The black patches near the base are sometimes fused into an oblique streak. Worn specimens become more whitish

Similar species. *C. kroesmanniella* is larger and on average lighter. Also *C. blandelloides* has on average lighter forewings, and the black patches at the basal part of the wings are fused into a slender streak.

MALE GENITALIA.– Uncus long, comparatively broad; tegumen very broad and distinctly emarginate anteriorly, strongly constricted medially, long processes towards transtilla; transtilla sclerotized, with few minute spines; valva short, thumb-shaped; sacculus short, knife-shaped; posterior margin of vinculum with broadly V-shaped emargination, lateromedial part with broadly rounded projection; saccus prominent, extremely broad, weakly tapered towards rounded apex; phallus stout, curved apex.

FEMALE GENITALIA.– Segment VIII comparatively short, with pair of small flaps ventrolaterally; ventromedial part of segment VIII membranous; apophysis anterioris long, about two times length of segment VIII; antrum broadly tubular, almost reaching length of segment VIII, extending to about one-third of apophysis anterioris; signum a short hook. Differs from the similar *C. petrophila* in the longer segment VIII with shorter apophysis anterior, the tubular antrum and the shorter signum-hook.

DISTRIBUTION.– Scattered records from the greater part part of western and Central Europe, extending into western Russia in the east (Ponomarenko, 2008: 98); not recorded from the Mediterranean. The distribution of *C. huebneri* is insufficiently known due to its confusion with *C. kroesmanniella*.

New record. Serbia: 2 ♀, Beograd, Rakovica, 24.-26.viii.1964, leg. Zouhar, gen. slide 3842 Karsholt (ZMUC).

BIONOMICS.– The larva is exclusively recorded from *Stellaria holostea* L. (Bradford, 1979: 129). According to Bradford (*op. cit.*) it occurs in May, feeding between spun shoots. Schütze (1931: 92) mentions the larva feeding in the seed-capsules of the host-plant in June. Moths emerged from mid-July to late July and have been collected until late August in the wild. In its preferred habitats such as deciduous forests it is attracted to light.

REMARKS.– *C. huebneri* was for decades confused with *C. kroesmanniella* (Ridout, 1977) and literature sources should be carefully considered.

196. Caryocolum kroesmanniella (Herrich-Schäffer, 1854)

Gelechia kroesmanniella Herrich-Schäffer, 1854, pl. 77, fig. 581.
Recurvaria huebneri auctt., (*nec* Haworth, 1828); misidentification.

DIAGNOSIS.– Adult. Wingspan 13-14 mm. Segment 2 of labial palp cream-coloured mottled with brown, especially on outer surface; segment 3 blackish brown, slightly mottled with lighter scales. Antenna ringed with light and blackish brown. Head white mottled with light brown towards neck; thorax and tegula cream-coloured mottled with grey and brown. Forewing white mottled with light brown and black; black patches at 1/5 on costa and at 1/4 in middle of wing; a black spot at 1/3 in middle of wing; a comma-shaped spot at 2/3 in middle of wing often confluent with a black tornal patch; a weak white bent fascia at 4/5; apical area with many black scales. Hindwing grey.

Variation. A rather constant species. The comma-shaped black spots on the forewings can be of different shape. Worn specimens become whitish grey.

Similar species. *C. blandelloides* is on an average smaller and normally has the two black patches near the base of the forewing fused into an oblique streak. The antenna of *C. kroesmanniella* is more distinctly ringed. *C. huebneri* q.v.

MALE GENITALIA.– Uncus long, comparatively broad; tegumen very broad and distinctly emarginate anteriorly, strongly constricted medially, long processes towards transtilla; transtilla sclerotized, with largely reduced spines; valva short, slender, digitate, thumb-shaped; sacculus short, about length of valva, digitate; posterior margin of vinculum with extremely broad subrectangular emargination, lateromedial part with large triangular process; saccus prominent, extremely broad, almost subrectangular; phallus stout, weakly curved apically.

FEMALE GENITALIA.– Segment VIII comparatively long, with pair of drop-shaped processes dorsally; ventromedial part of segment VIII membranous; apophysis anterioris comparatively short, about length of segment VIII; antrum long, funnel-shaped, extending to about apex of apophysis anterioris; signum a small plate, without hook.

DISTRIBUTION.– Scattered records from parts of Scandinavia, the British Isles, Central and eastern Europe to European Russia (Ponomarenko, 2008: 98); only recently found in Switzerland (Schmid, 2006). Not recorded from the Mediterranean.

BIONOMICS.– The larva has been recorded from *Stellaria holostea* L., *S. alsine* Grimm (Klimesch, 1954: 358), *S. media* (L.) Vill. and exceptionally *S. nemorum* L. (Huemer, 1988: 518); furthermore *Stellaria graminea* L. may be a host-plant (Pröse, 1979: 67). The larva initially is a leaf-miner during autumn. After hibernation it feeds between spun leaves until late May (Huemer, 1993: 152). The adult occurs from early July to the beginning of September. It may be found on trunks of oak and is also attracted to light. *C. kroesmanniella* is preferably found in open woodland at low altitudes.

Tila Povolný, 1965

Tila Povolný, 1965: 485.
Type species: *Lita capsophilella* Chrétien, 1900.

DIAGNOSIS.– Segment 2 of the labial palps with lateral bristles. Males with lateral brushes of long scales on abdominal tergites III-VI. See also species description.

MALE GENITALIA.– Sternite VIII broad and long, posteriorly rounded, with pair of long coremata; tergite VIII moderately slender, short; uncus subrectangular to subtriangular, apically flattened to pointed; gnathos with irregularly shaped sclerite, culcitula small; tegumen gradually widened from uncus to pedunculi, anteriorly with shallow emargination; pedunculi moderately small, rounded; valva extending to about base of gnathos, basal two-thirds slender, rod-like, weakly curved, distal part broadly subtriangular; sacculus slightly shorter and basally broader than valva, digitate, apically pointed; vinculum short, posterior margin with subtriangular medial projection and small medial incision, separating pair of indistinct lateromedial humps; saccus moderately long, basal half broad, distal part evenly or abruptly tapered; anellus with pair of long, basally fused needle-shaped sclerites, slightly shorter than phallus; phallus short, slender, almost straight, coecum hardly swollen.

FEMALE GENITALIA.– Apophysis posterioris moderately short, about three times length of segment VIII; segment VIII subrectangular.

DISTRIBUTION.– Restricted to isolated localities in central Europe and Turkey.

BIONOMICS.– The larval host-plant of the only European species is *Gypsophila repens* L. (Caryophyllaceae), whereas the bionomy of *Tila sequanda* Povolný, 1974 is unknown. Further details see species description.

REMARKS.– According to Povolný (2002a: 102) the signum of the female genitalia should be absent in *Tila*. However, the author himself (Povolný, *op. cit.*, pl. 85, fig. 914) depicts a signum in one of the two figured specimens of *T. capsophilella*. We have not been able to examine the female of *T. sequanda* which may well also have a signum.

197. *Tila capsophilella* (Chrétien, 1900)

Lita capsophilella Chrétien, 1900b: 223.

DIAGNOSIS.– Adult. Wingspan 11-13 mm. Segment 2 of labial palp cream-coloured mottled with black on lower and outer surface; segment 3 white mottled with black. Antenna black, ringed with light grey. Head light grey from white-based grey scales and mottled with black, especially at base and in apical area; black spots in fold and at 2/5 and 3/5 in middle of wing, the latter drop-shaped towards tornus; whitish grey areas before and after black spots at 3/5. Hindwing whitish grey, darker towards margin.

Variation. There is some variation in the amount of black scales on the forewings, which may look from whitish grey to blackish grey.

Similar species. *T. capsophilella* is characterized by the lateral bristles on segment

2 of the labial palps. Outside of *Tila* these are only found in the related genus *Pogochaetia*, and in most genera of Tineidae. *T. capsophilella* can be separated from *Pogochaetia solitaria* by its greyish head and forewings; those of *P. solitaria* being brown and black.

MALE GENITALIA.– Uncus subrectangular, apically flattened or weakly convex; gnathos with broad sclerite; valva shorter than uncus, basal two-thirds slender, rod-like, distal part bird's head shaped, with strong and long setae; sacculus slender digitate, weakly curved, slightly shorter and basally broader than valva, apically pointed; posterior margin of vinculum with subtriangular medial projection and small medial incision, pair of indistinct lateromedial humps; saccus with broadly U-shaped base, distal part abruptly tapered, rod-like; anellus with pair of long, needle-shaped sclerites; phallus short, slender, almost straight, coecum hardly discernible.

FEMALE GENITALIA.– Segment VIII subrectangular, slightly broader than long, anteriodorsal edge distinctly sclerotized band, posteriad of concave anterioventral edge; subgenital plate smooth, medial part covered with microtrichia, anteriomedial edge with rounded sclerotized flap; apophysis anterioris almost length of segment VIII, basally widened, tapered to rod-like distal part; antrum short, funnel-shaped, laterally distinctly sclerotized, medially membranous with incised anterior margin incised; ductus bursae narrow, distinctly separated from transversely suboval corpus bursae; signum a longish plate.

DISTRIBUTION.– Only known from scattered localities in the Alps (France, Switzerland, Austria) and Germany.

BIONOMICS.– The larva has been found from May to early July, feeding exclusively on *Gypsophila repens* L. (Caryophyllaceae) (Chrétien, 1900b: 224; Huemer, pers. obs.). It spins the terminal leaves of a shoot together into a pod-like dwelling. Within the pod the larva eats the parenchyma, filling the upper part of the dwelling with frass. The fully grown larva leaves the dwelling and pupates on the ground surface or between dead stalks (Huemer, 1998: 51). The colour of the pupa is brown. In the laboratory adults emerged from mid-June to early August. Preferred habitats are alpine scree and rock formations, exclusively in limestone areas. Vertical distribution: from about 700 to 2000 m.

Pogochaetia Staudinger, 1879

Pogochaetia Staudinger, 1879: 310.
Type species: *Pogochaetia solitaria* Staudinger, 1879.
Pogonochaetia Rye, 1881: 9; unjustified emendation.
Type species: *Pogochaetia solitaria* Staudinger, 1879.
Chaetopogon Rye, 1881: 9; unjustified emendation.
Type species: *Pogochaetia solitaria* Staudinger, 1879.

DIAGNOSIS.– Segment 2 of the labial palps with lateral bristles. Males with lateral brushes of long scales on abdominal tergites III-VI. See also species description.

MALE GENITALIA.– Sternite VIII broad and long, posteriorly rounded, with pair of long brushes of coremata; tergite VIII broad, short; uncus sub-oval, apically rounded with medial projection or incision; gnathos with a claw-like sclerite, culcitula moderately small; tegumen gradually widened from uncus to pedunculi, weakly emarginate; pedunculi indistinct; valva long, broadly digitate to bar-shaped, apically inflated or weakly broadened, reaching tip of uncus, basally curved, distal part straight; sacculus broader and about half length of valva, apically pointed; vinculum broad, extremely short, posterior margin with medial incision, lateromedial projections present; saccus long, slender; anellus with pair of long, needle-shaped sclerites, about length of phallus; phallus short, slender, almost straight, coecum weakly separated.

FEMALE GENITALIA.– Apophysis posterioris long; segment VIII short, distomedially membranous with pair of short sclerotized edging ridges, ventrally and ventrolaterally strongly sclerotized; apophysis anterioris longer than segment VIII, basally broad, gradually narrowing towards rod-like distal half; antrum about length of segment VIII, extending to about two-fifths length of apophysis anterioris, funnel-shaped, laterally distinctly sclerotized, medially membranous; ductus bursae comparatively long; corpus bursae well separate from ductus bursae, suboval, with distinct suboval signum plate.

DISTRIBUTION.– Restricted to the mountains of the western Palaearctic region, from the Pyrenees and the Alps to Turkey, northern Iran and Central Asia.

BIONOMICS.– The larval host-plant of the European *P. solitaria ocymoidella* is *Saponaria ocymoides* L. (Caryophyllaceae), whereas the bionomy of related taxa is unknown. Further details see species description.

REMARKS.– Until recently only known from a single species. Bidzilya (2005: 16) described a second species, viz. *P. dmitrii*, from the Altai mountains. The female of this species is unknown.

198. Pogochaetia solitaria Staudinger, 1879

Pogochaetia solitaria Staudinger, 1879: 310.
Gelechia (?) *ocymoidella* Walsingham, 1900: 218. Subspecies.
Pogochaetia solitaria ssp. *cabreretsi* Povolný, 1981: 385, figs. 36-38, 45, 47, 58.

DIAGNOSIS.– Adult. Wingspan 12-14 mm. Labial palp cream-coloured, heavily mottled with black, especially at lower and outer surface of segment 2. Antenna black, indistinctly ringed with grey. Head light brown, darker towards neck; thorax and tegula black, the latter with light brown tip. Forewing black with blackish grey dorsum and mottled with light grey-brown, especially on dorsal part of basal area, in fold, between black spots and as confluent costal and tornal spots. Hindwing dark grey, lighter towards base.

Variation. The amount of light grey-brown areas on the forewings is variable, and in some specimens they are almost uniformly black. Specimens from Spain are on average lighter than specimens from the Alps. See also remarks.

Similar species. *P. solitaria* is characterized by the lateral bristles on segment 2 of the labial palps. Males with specialized scales on the dorsal side of the abdomen.

MALE GENITALIA.– Uncus suboval, apically rounded, with medial incision; valva long, broadly digitate, strongly curved at about one-fifth, distal part slightly widened, reaching tip of uncus; sacculus about 1.5 times width and half length of valva, apex distinctly pointed; posterior margin of vinculum weakly projected posteriorly, with short lateromedial humps and small medial incision; saccus long, slender; anellus with pair of long, basally fused and needle-shaped sclerites; phallus short, slender, almost straight, coecum weakly swollen.

FEMALE GENITALIA.– See genus description.

DISTRIBUTION.– Restricted to highly isolated localities in central Spain, the Pyrenees and the Alps (France, Switzerland, Italy, Austria) (ssp. *ocymoidella*). The nominotypical subspecies is known from hitherto unpublished records from Greece, and from Turkey to northern Iran (Elburs Mountains).

New records. Greece: 1 ♂, Aegina I., Marathonas, 10 m, 3.vi.1995, leg Sutter, gen slide. 4794 Sutter (LNK). Spain: 1 ♀, Teruel, 6 km S Albarracin, 1400 m, 22.viii.2001, leg. Skou & Skule; 1 ♂, 5 km W Albarracin, 1200 m, 2.viii.2003, leg. Skou (ZMUC).

BIONOMICS.– The larva has been found from late May to early June, feeding in the flowers of *Saponaria ocymoides* L. (Caryophyllaceae) (Walsingham, 1900: 218; Burmann, 1949: 70; Huemer, pers. obs.). It bores from the base of the flower into the ovary and eats the seeds. Sometimes it spins two to three flowers together, but normally the larva is well hidden and difficult to find. Burmann (*op. cit.*) found up to 8 larvae in one cushion of the host-plant. Pupation takes place in a webbing on the ground surface or between dead stalks. In the laboratory adults emerged from the second half of June to July. Preferred habitats are xeric scree and rock formations up to altitudes of about 1300 m.

REMARKS.– *P. solitaria* shows some variation in colour and wing markings and Povolný (2002a: figs. 5-6) separates two subspecies: specimens of the nominotypical populations from Turkey are less variegated with a darker dorsum than those from Europe. The latter populations are nowadays treated as subspecies *P. solitaria ocymoidella*. *P. solitaria cabreretsi* Povolný, 1981, described as a further separate subspecies from France (Lot: Cabrerets) shows no significant differences from other populations from Central Europe and is therefore considered as a synonym (Sattler, 1992: 111).

Agonochaetia Povolný, 1965

Agonochaetia Povolný, 1965: 487.
Type species: *Agonochaetia incredibilis* Povolný, 1965.
Sautereopsis Povolný, 1965: 488.
Type species: *Gelechia terrestrella* Zeller, 1872.

DIAGNOSIS.– Rather large (wingspan 18-20 mm), broad-winged moths. Labial palps rather long and slender without dark rings; segment 3 slightly shorter than segment 2. Underside of antenna in male with short cilia. Forewing almost unicolorous with indistinct black spots and terminal dots; cilia light grey beyond indistinct black cilia line. Hindwing almost without emargination on outer margin before apex.

MALE GENITALIA.– Sternite VIII broad, posteriorly rounded; tergite VIII broadly tongue-shaped; uncus subrectangular to suboval, short, basally broad, apically frequently constricted; gnathos without hook, culcitula moderately large; tegumen widened from uncus to pedunculi, deeply emarginate; pedunculi distinct; valva long, digitate, frequently extending far beyond uncus, apical part curved, medial part with additional widening or hump; sacculus digitate, distally curved, with bulbous apex, about three-quarters length of valva; vinculum broad, posterior margin usually without distinct incision or emargination, pair of digitate lateral humps developed; saccus long, basally comparatively broad, gradually tapered to slender apical part; anellus with pair of long and very distinct needle-shaped sclerites, exceeding length of phallus; phallus slender, frequently straight, rarely curved, with sclerotized apical arm; coecum weakly to distinctly swollen.

FEMALE GENITALIA.– Apophysis posterioris long; segment VIII longer than broad, without distinct sclerotized modifications, only short longitudinal folds ventrally; apophysis anterioris about length of segment VIII, rod-like, broader at base; antrum large, funnel-shaped; posterior part of ductus bursae without sclerotization; corpus bursae well defined; signum with extraordinarily large subtriangular basal plate with a strong hook.

DISTRIBUTION.– Holarctic, with a single species in North America. Recently two additional species have been described from Asia (Bidzilya, 2000; 2001), increasing the number of species to six.

BIONOMICS.– Host-plant and early stages unknown, possibly feeding on Caryophyllaceae. All species inhabit xeric steppe-slopes, in Europe up to altitudes of about 2000 m.

REMARKS.– The serious lack of material is a major hindrance for the validation of various taxa in the genus *Agonochaetia*.

199. Agonochaetia terrestrella (Zeller, 1872)

Gelechia terrestrella Zeller, 1872: 111.
Gelechia muestairella Müller-Rutz, 1922: 241.

DIAGNOSIS.– Adult. Wingspan 20 mm. Labial palp light cream-yellow mottled with brown, especially on lower and outer surface. Antenna grey-brown, indistinctly lighter ringed. Head, thorax and tegula light brownish. Forewing broad, covered with dark brown, lighter-based scales; an indistinct black spot in fold and two such spots at 2/3 and 3/5 in middle of wing, the latter angulated; termen with indistinct black spots. Hindwing dark grey with light grey cilia.

Variation. The above description is based on two males from Switzerland, one of which lacks dark spots in termen. Females are smaller (wingspan 18 mm), slightly less broad-winged and have lighter grey-brown forewings with even more indistinct black spots. The indistinct light fascia at 3/4 shown in the colour figure by Povolný (2002a: pl. 15, fig. 17) is not present in the few specimens examined by us.

Similar species. *A. quartana* q.v.; *A. intermedia* q.v.

MALE GENITALIA.– Uncus short, subtriangular, apex with minute medial indention; culcitula distinct; valva extending far beyond apex of uncus, distal half curved inwards, medial part with large subtriangular projection; sacculus digitate, strongly curved near middle, with bulbous distal part and pointed apex, about three-quarters length of valva; vinculum broad and short, posterior margin with pair of weakly developed sinusoid medial humps, indistinct medial incision, lateral humps slender and digitate; saccus long, basally comparatively broad, distal part tapered, rod-like; anellus with pair of long and very distinct needle-shaped sclerites, exceeding length of phallus;

phallus slender, straight, with sclerotized apical arm; coecum distinctly swollen, about one-fifth length of phallus. Differs from other species of the genus mainly in the large projection of the valva.

FEMALE GENITALIA.– Segment VIII longer than broad, without distinct sclerotized modifications, only short longitudinal folds ventrally; apophysis anterioris about length of segment VIII, rod-like, broader at base; antrum slightly funnel-shaped, entrance between middle and anterior half of sternite VIII, moderately widened and concave, distal part of antrum extending to about or beyond middle of apophysis anterioris; signum large with broadly subtriangular basal plate, strong and straight distal hook of about length of basal plate. Differs from *A. intermedia* in the longer and less constricted antrum and from *A. quartana* in the broader basal plate of the signum with longer hook.

DISTRIBUTION.– Only known from isolated localities in south-eastern Switzerland (Graubünden) and Albania (Elsner *et al.*, 1999: 49), and an unpublished record from Romania.

New record. Romania: 1 ♀, Dobrogea, Baneasa, Rez. Canaraua Fetii, 17.v.1994, leg. Kovác, gen. slide GU 04/1216 ♀ Huemer (coll. Kovác).

BIONOMICS.– Host-plant and early stages unknown. The adults have been collected from late June to July, according to Povolný (2002a: 106) already in May. Preferred habitats are mountainous steppe slopes at altitudes up to above 2000 m.

200. Agonochaetia intermedia Sattler, 1968

Agonochaetia intermedia Sattler, 1968: 120.

DIAGNOSIS.– Adult. Wingspan 18-19 mm. Male. Labial palp light cream-yellow; lower and outer surface of segment 2 densely mottled with dark grey-brown. Antenna blackish grey-brown. Head light brown; thorax and tegula as forewing. Forewing rather broad, grey-brown mottled with yellow-brown (especially in fold) and black (especially along costa and termen); indistinct black spots in fold and at 2/3 and 3/5 in middle of wing. Hindwing dark grey with light cilia. Female. Similar to male, but with labial palp, antenna, head, thorax, tegula and forewing light yellow-brown mottled with only some black scales and without black spots.

Variation. Apart from being sexual dimorphic the few examined specimens show only slight variation.

Similar species. *A. intermedia* is characterized by its sexual dimorphism, with the female being light yellow-brown. The male resembles *A. terrestrella* and *A. quartana*, but is more slender-winged and has segment 2 of the labial palps darker.

MALE GENITALIA.– Uncus short, subtriangular; culcitula distinct; valva extending beyond apex of uncus, curved inwards, medial part widened, apical area slender; sacculus digitate, strongly curved near middle, with weakly bulbous distal part and pointed apex, about three-quarters length of valva; vinculum broad and short, posterior margin with pair of weakly developed sinusoid medial humps, indistinct medial incision, lateral humps slender and digitate; saccus moderately long, basally broadened, distal part tapered, rod-like; anellus with pair of long and distinct needle-shaped sclerites, about length of phallus; phallus slender, straight, with sclerotized apical arm; coecum distinctly swollen, about one-quarter length of phallus. Differs from the otherwise similar *A. terrestrella* mainly in the hump-like projection of the valva.

FEMALE GENITALIA.– Segment VIII longer than broad, without distinct sclerotized modifications, short longitudinal folds ventrally; apophysis anterioris about length of segment VIII, rod-like, broader at base; antrum broadly funnel-shaped, entrance in anterior half of sternite VIII, distal part extending slightly beyond base of apophysis anterioris; signum large with broadly subtriangular basal plate, strong and almost straight distal hook of about length of basal plate. Differs from other species of the genus particularly in the broad and short antrum.

DISTRIBUTION.– Isolated records from southern Switzerland, western Austria (Huemer, 1989e: 387) and Russia (Volga Don region and southern Ural Mountains) (Junnilainen *et al.*, 2010: 54).

BIONOMICS.– Host-plant and early stages unknown. The adults are on the wing from late May to July. *A. intermedia* is restricted to xeromontane habitats up to altitudes of about 2000 m. It flies freely at dusk and is attracted to light.

201. Agonochaetia quartana Povolný, 1990

Agonochaetia quartana Povolný, 1990a: 150, figs. 3, 11.

DIAGNOSIS.– Adult. Wingspan 20 mm. Labial palp cream-yellow mottled with brown, especially on lower and outer surface. Antenna grey-brown, ringed with light yellowish. [Head denuded], thorax and tegula as forewing. Forewing broad, covered with grey-brown, lighter-based scales; an indistinct black spot at 3/5 in middle of wing; termen with some black scales. Hindwing dark grey with light grey cilia.

Variation. Only the somewhat worn holotype was examined.

Similar species. *A. quartana* is characterized by its broad, almost unicolorous wings. It is similar to *A. terrestrella*, but lacks dark spots on the forewings and has the antennae more distinctly ringed. *A. intermedia* q.v.

MALE GENITALIA.– Unknown.

FEMALE GENITALIA.– Segment VIII longer than broad, without distinct sclerotized modifications, short longitudinal folds posterioventrally; apophysis anterioris about length of segment VIII, rod-like, broader at base; antrum slightly funnel-shaped, entrance in middle of sternite VIII, moderately widened and concave, distal part of antrum extending to about middle of apophysis anterioris; signum large with longitudinal subtriangular basal plate, straight distal hook distinctly shorter than basal plate. Differs from *A. intermedia* in the longer and less constricted antrum and from *A. terrestrella* in the longer than broad basal plate of the signum with shorter distal hook.

DISTRIBUTION.– Only known from Bulgaria.

BIONOMICS.– Host-plant and early stages unknown. The holotype was collected in June.

REMARKS.– *A. quartana* was described from a single female collected in Bulgaria (Slivno) (Povolný, 1990a: 150) and no additional material has become available. It is distinct only based on the signum of the holotype, whereas the shape of the antrum, considered to be characteristic for the species by Povolný (*op. cit.*), falls within the individual variation of *A. terrestrella*. Therefore these two taxa may well prove to be conspecific in future.

Sattleria Povolný, 1965

Sattleria Povolný, 1965: 490.
Type species: *Gelechia dzieduszyckii* Nowicki, 1864.

DIAGNOSIS.– Adult. Male. Wingspan, male 15-24 mm. Labial palp recurved, cream-coloured; basal part of segment 2 and all of segment 3 mottled with blackish brown; segment 3 acute, about as long as segment 2. Antenna black, in some species more or less distinctly lighter ringed. Forewing rather slender with rounded apex, light grey brown mottled with yellowish, white and black; one or two elongate black spots in fold and one (dot-shaped) at 1/2 and one (often angulated) at 3/5; termen often with black dots. Hindwing light grey. Female. Brachypterous. Wingspan 10-19 mm. Labial palp and antenna as in male. Forewing ellipsoid, on average darker than in male. Hindwing short (up to two-third length of forewing) and slender, light grey.

MALE GENITALIA.– Sternite VIII broad, posteriorly rounded, occasionally with weak median emargination; tergite VIII tongue-shaped, narrow; uncus short, narrower than tegumen; gnathos hook strong, culcitula comparatively small; tegumen widened from uncus to pedunculi, broadly emarginate; pedunculi long, slender; valva long, slender; sacculus separate, shorter than valva; vinculum broad, deeply emarginate posteriomedially with pair of long, strongly sclerotized, frequently forked processes; saccus comparatively long with almost parallel margins; anellus with pair of rounded, centrally perforated sclerites; phallus long, slender, straight with sclerotized apical arm and sometimes ventromedial projection; coecum weakly separate.

FEMALE GENITALIA.– Apophysis posterioris very long; segment VIII dorsally membranous, ventrally and ventrolaterally strongly sclerotized with pair of small pockets; sternite VIII medially with irregular folds or ridges; apophysis anterioris rod-like, about one-third length of apophysis posterioris, basally strongly sclerotized and extended into ostium bursae and to sternal pockets; antrum long and tubular, strongly sclerotized; ductus bursae usually shorter than antrum; corpus bursae spherical or oval, without or rarely with barely discernible signum.

DISTRIBUTION.– Members of *Sattleria* are restricted to the European alpine system. The largest number of species occurs in the Alps and Pyrenees, with single species extending to the Carpathians, mountains of the Balkan Peninsula and the Appenines, respectively.

BIONOMICS.– Information on the biology is scarce. The ovum and oviposition sites are unknown. Host-plant records include Caryophyllaceae (*Silene* spp., *Moehringia* sp.) and Saxifragaceae (*Saxifraga* spp.). The larva has been found in July and August, feeding from a loosely spun silken tube that is often extended from the upper parts of the host-plant to the roots or nearby stones (Pitkin & Sattler, 1991: 212). Frass is deposited outside the larval tube. The presence of young and mature larvae together indicates a biennial development. The usual overwintering stage is most probably the larva. However, early snowfall, such as occasionally observed shortly after the flight season had started, indicates a possible hibernation of the pupa (Huemer, pers. obs.). Pupation takes place on the ground surface in a dense cocoon, preferably under stones or amongst the host-plant. The adults are on the wing between late June and mid-September but the timing reflects local climatic factors. Mating behaviour is largely unknown. Burmann (1954: 349) found a mating pair in the early afternoon. However, Pitkin & Sattler (*op. cit.*) suspect that increased male activity about midnight may be related to search for the females. Males of several species can be collected at light both with a mercury vapour light in front of a white sheet or with a Petromax paraffin lantern. According to personal observations (P.H.) the males are also active in the early morning indicating mating behaviour. Preferred habitats are alpine grassland and open slopes with coarse scree and sparse vegetation. Vertical distribution: from about 1500 to 3500 m.

REMARKS.– The status of species within this genus was disputed by Povolný (1986; 2001b). This author repeatedly accentuated a high degree of individual variation both in genitalia and in the adult habitus and concluded that *Sattleria* contains only a single polymorphic species. However, as already demonstrated by Pitkin & Sattler (1991: 207) this interpretation is mainly due to a misinterpretation of interspecific characters. In particular Povolný´s last mentioned paper is not based on the author´s own field experience nor on additional material. Numerous inaccuracies such as mixture of figures and localities, ignorance of characters figured and discussed in earlier publications such as the sacculus and the phallus, inaccurate redrawings of original figures and the emotional style of this paper cannot be regarded as a substantial contribution to solve problems within *Sattleria*.

Species from the eastern part of the Alps or the Carpathians and the Balkans, namely *S. styriaca* and *S. dzieduszyckii*, have never been observed sympatrically with other taxa of *Sattleria*. However, our own field experiences have shown that several other taxa occur sympatrically and even syntopically without any intermediate forms. Such a sympatric occurrence has been observed in *S. pyrenaica* – *S. arcuata* and *S. pyrenaica* – *S. angustispina* in the Pyrenees, and in *S. melaleucella* – *S. triglavica*, *S. melaleucella* – *S. breviramus, S. melaleucella* – *S. izoardi* and *S. melaleucella* – *S. marguareisi* in the western Alps. These taxa can readily be distinguished by genitalic characters but also by wing markings and size and hence they are regarded as separate species. Observations on the adult behaviour and larval host-plants furthermore support this view. Last but not least, examination of mtDNA, e.g. of sympatric *S. melaleucella* and *S. izoardi*, clearly supports several distinct species (Huemer, unpubl.). Molecular data should also help to resolve the taxonomy of the more widely distributed *S. triglavica* and *S. melaleucella* in a foreseeable future.

202. *Sattleria arcuata* Pitkin & Sattler, 1991

Sattleria arcuata Pitkin & Sattler, 1991: 215, figs. 1, 4-5, 17, 29, 48, 51.

DIAGNOSIS.– Adult. Male. Wingspan 20-22 mm. Labial palp cream-coloured mottled with blackish brown, especially at base of segment 2 and at tip of segment 3. Antenna blackish brown. Face cream-coloured; neck light brown; thorax and tegula blackish brown, the latter with lighter tip. Forewing with base, costa, a dorsal stripe and termen black and central part lighter, cream-coloured; elongate black spots in fold and middle of wing; the one at 3/5 from base angulated; blackish areas from 1/4 on costa to second spots in fold, and before light postmedial fascia. Hindwing light grey. Female. Wingspan 15 mm. Labial palps, antennae, head, thorax and tegula similar to male. Forewing fuscous, black at base, at 1/4 on costa and in fold; a black spot in middle of wing at 3/5; some yellowish scales, especially in fold and in basal part of costa and dorsum. Hindwing grey.

Variation. A rather constant species. Specimens can have lighter or darker forewings, but that is mostly due to their condition with worn specimens becoming lighter.

Similar species. *S. arcuata* is a medium-sized *Sattleria* with slender, rather dark forewings. It occurs in the same area as *S. pyrenaica*, which is smaller and has a brownish lustre on the forewings.

MALE GENITALIA.– Sacculus gradually broadened at base; apex of primary process of vinculum almost level with apex of sacculus, with or without fine setae on outer surface; secondary process of vinculum a short spine with broad base, arising in basal quarter of primary process; saccus comparatively broad and long; phallus without median projection, broad medial arch developed, apical arm comparatively large, curved.

FEMALE GENITALIA.– Sternite VIII medially without distinct folds or ridges, pair of comparatively small pockets in anterior third; apophysis anterioris with broadly sclerotized base, projecting distinctly above sternal pockets, inner edge evenly curved; antrum strongly sclerotized, long and tubular, extending far beyond tip of apophysis anterioris, widely separate from sclerotized parts of sternite VIII.

DISTRIBUTION.– Endemic to the Hautes-Pyrenees (France, Spain).

BIONOMICS.– Host-plant and early stages unknown. The adult occurs from mid-July to early August. The preferred habitats are more or less heavily grazed alpine pastures (Pitkin & Sattler, 1991: 216). Vertical distribution from about 2100 to altitudes of 2700 m.

203. *Sattleria pyrenaica* (Petry, 1904)

Gelechia pyrenaica Petry, 1904b: 3.

DIAGNOSIS.– Adult. Male. Wingspan 16 mm. Labial palp cream-white mottled with blackish brown on lower and outer surface of segment 2 and at tip of segment 3. Antenna black. Face cream-coloured; thorax and tegula light grey-brown. Forewing light grey mottled with dark grey (especially along costa, dorsum and in apical area) and light brown (especially in central part of the wing); an elongate black spot in fold, a dot-shaped one at 1/2 and an angulated one at 3/5 in middle of wing; termen lined black. Hindwing light grey. Female. Wingspan 11 mm. Labial palp almost cream-coloured. Antenna black. Head cream-white; thorax and tegula light brown. Forewing light grey-brown mottled with orange-brown, especially in fold and subcostal stripe; an elongate black spots in fold and two in middle of wing. Hindwing light grey.

Variation. The examined material shows only slight variation.

Similar species. *S. pyrenaica* is characterized by its small size and by its light grey-brown forewings. *S. arcuata* q.v.; *S. izoardi* q.v.

MALE GENITALIA.– Sternite VIII deeply emarginate posteriorly; sacculus not broadened into lobe basally; apex of primary process of vinculum almost level with apex of sacculus, with or without up to three fine setae on outer surface; secondary process of vinculum comparatively long blade with weakly serrated or uneven edge, arising at right angles to primary process between about one-third and middle; saccus comparatively narrow and short; phallus basally strongly splayed, without median projection, narrow medial arch sometimes developed, apical arm comparatively small.

FEMALE GENITALIA.– Sternite VIII medially with comparatively weak folds, pair of distinct pockets in posterior third with long and straight anterior edge; apophysis anterioris with broadly rounded sclerite at base, inner edge of apophysis anterioris weakly curved; antrum shorter than apophysis anterioris, meeting base of sclerotized parts of sternite VIII, irregularly sclerotized at posterior end.

DISTRIBUTION.– Endemic to the Hautes-Pyrenees and the Pyrenees Orientales (France, Andorra, Spain).

BIONOMICS.– Host-plant and early stages unknown. The adults have been observed on more or less heavily grazed alpine pastures flying at dawn and dusk above low-growing *Rhododendron* scrub (Pitkin & Sattler, 1991: 216). They are also attracted to light at about midnight. Vertical distribution from 1600 to about 2700 m.

204. Sattleria angustispina Pitkin & Sattler, 1991

Sattleria angustispina Pitkin & Sattler, 1991: 217, figs. 4, 7, 19, 33, 50, 54-55.

DIAGNOSIS.– Adult. Male. Wingspan 18 mm. Labial palp cream-coloured mottled with blackish brown on basal part of segment 2 and on segment 3. Antenna black, indistinctly lighter ringed. Head cream-coloured, darker towards neck; thorax and tegula as forewing. Forewing variegated white, yellow, brown and black; black areas separated by white near base, from costa at 1/4 towards second black spot in fold, as a large block on costa and dorsum before white postmedial fascia and in apical area; yellow scales in fold, in short subcostal stripe and around black spots; termen 'checkered' with black and white dots. Hindwing light grey. Female. Wingspan 14 mm. Segment 2 of labial palp cream-coloured; segment 3 almost black. Antenna and head as in male; thorax blackish brown mottled with some light scales; tegula lighter. Basal half of forewing similar to male; apical half with reduced markings: a black block before angulated, white postmedial fascia reduced; black apical area reduced; termen without scales. Hindwing whitish grey.

Variation. Only one male and one female were examined.

Similar species. *S. angustispina* is characterized by its cream-coloured head and by its variegated forewings with clear white markings.

MALE GENITALIA.– Sacculus strongly broadened basally, distal part long and slender; apex of primary process of vinculum about level with apex of sacculus, without setae; secondary process of vinculum long and slender, spine-like, arising at acute angle in basal fifth of primary process; saccus tapered; phallus with indistinct median projection, apical arm small and almost straight.

FEMALE GENITALIA.– Sternite VIII medially with indistinct folds, pair of comparatively small pockets on anterior margin; apophysis anterioris without specialised sclerite at base, inner edge almost at right angle to sternite VIII; antrum about same length as apophysis anterioris, meeting base of sclerotized parts of sternite VIII, irregularly sclerotized at posterior end.

DISTRIBUTION.– Endemic to Mt. Canigou (Pyrénées Orientales, France).

BIONOMICS.– Host-plant and early stages unknown. The few adults known to date have been collected from mid-July to early August at an altitude of about 2600 m.

205. *Sattleria izoardi* Huemer & Sattler, 1992

Sattleria izoardi Huemer & Sattler, 1992: 71, figs. 7-8, 17, 24.

DIAGNOSIS.– Adult. Male. Wingspan 14-17 mm. Labial palp cream-coloured mottled with some blackish brown on basal part of segment 2; segment 3 with dark tip. Antenna black. Head cream-coloured, darkest at neck; thorax and tegula light grey-brown. Forewing light greyish mottled with white (especially in costal half of middle part), light brown (especially in fold and subcostal stripe), light grey (especially in dorsal part) and black (especially at base, along costa and before white postmedial fascia); black spots in fold elongate, a dot-shaped one at 1/2 and angulated one at 3/5 in middle of wing; termen with black dots. Hindwing light grey. Female. Wingspan 10 mm. Labial palp, antenna, thorax and tegula as in male. Head cream-white. Forewing light grey-brown with black costa and termen; black spots in fold fused, followed by yellow scales; a black patch between second spot in fold and first spot in middle of wing; second spot in middle of wing dot-shaped; white postmedial fascia near apex angulated. Hindwing light grey.

Variation. The examined specimens are rather constant. Worn specimens look lighter than fresh ones.

Similar species. *S. izoardi* is characterized by its small size and its light greyish forewings with an indistinct broad, black submedial fascia followed by a similar white postmedial fascia. It resembles *S. pyrenaica*, but that species has more light brown forewings without white areas. The tip of segment 3 of the labial palp is darker in *S. pyrenaica*. *S. marguareisi* q.v.

MALE GENITALIA.– Sacculus not broadened into lobe basally; apex of primary process of vinculum almost level with apex of sacculus, with or without up to three fine setae on outer surface; secondary process of vinculum very long, spine-like, with weakly serrated edge, arising at acute angle in basal third of primary process; saccus comparatively short; phallus short, with weak basal sclerites, apical arm small and almost straight.

FEMALE GENITALIA.– Sternite VIII medially with indistinct folds, pair of comparatively small pockets in posterior third with long and straight anterior edge; Apophysis anterioris without specialised sclerite at base, inner edge evenly curved; antrum shorter than apophysis anterioris, posteriodorsally with large sub-oval sclerites.

DISTRIBUTION.– Endemic to the south-western Alps (France, Italy).

BIONOMICS.– Host-plant and early stages unknown. The adult occurs from late July to late August, preferably on alpine pastures. Vertical distribution from 2400 up to an altitude of about 2700 m.

206. *Sattleria breviramus* Pitkin & Sattler, 1991

Sattleria breviramus Pitkin & Sattler, 1991: 218, figs. 4, 8, 47.

DIAGNOSIS.– Adult. Male. Wingspan 21 mm. Labial palp cream-white mottled with blackish brown on basal half of segment 2 and on apical half of segment 3. Antenna black, indistinctly lighter ringed. Head cream-coloured; thorax and tegula blackish brown mottled with light brown. Forewing grey; costal part of base, fold, spots in middle of wing and terminal line black; yellow scales in fold, subcostal stripe and around black spots; a white streak in middle of wing from 1/5 to 3/5; apical third fuscous, separated by light, almost fused, costal and tornal spots. Hindwing light grey. Female. Wingspan 17 mm. Labial palp and head cream-white; thorax dark brown; tegula light brown. Forewing light grey mottled with yellow; base of costa black; a distinct black, yellow edged streak in fold; a white area around large black spots in middle of wing; apical area with white, yellow and black scales. Hindwing light grey.

Variation. Only one male and one female were examined.

Similar species. *S. breviramus* is a medium sized *Sattleria*, which is characterized by its distinct black streak in the fold and by having the costal part of the base black. Differs from the partially sympatric *S. melaleucella* particularly by the longitudinal streak in fold.

MALE GENITALIA.– Sacculus strongly broadened basally, distal part comparatively long and slender, curved; primary process of vinculum largely reduced, hair-like; secondary process of vinculum slender, spine-like; saccus short, almost parallel-sided; phallus short, without projections, apical arm small and almost straight.

FEMALE GENITALIA.– Sternite VIII weakly folded without pair of pockets, anterior margin a sclerotized transverse ridge, medial part between subgenital plates sclerotized and smooth; apophysis anterioris simple rod-like, about length of segment; antrum short tubular, about half length of apophysis anterioris, distinctly separate from sclerotized parts of sternite VIII; signum largely reduced to minute hook-like sclerite.

DISTRIBUTION.– Endemic to the Alpes Maritimes and the Alpes-de-Haute-Provence (France).

BIONOMICS.– Host-plant and early stages unknown. The adults have been collected in July both at light and flying freely in the early evening. Preferred habitats are alpine scree with sparse vegetation on siliceous soil. Vertical distribution: from 2400 to about 2750 m.

REMARKS.– The single female known to date was first described by Nel & Varenne (2003: 87-88, fig. 2). Its identification was based on evidence such as differences in genitalia and external appearance compared to the known alpine species.

207. Sattleria melaleucella (Constant, 1865)

Gelechia melaleucella Constant, 1865: 197, pl. 7, figs. 14a, b.
Gelechia mariae Frey, 1867: 302; unavailable, nomen nudum.
Gelechia dzieduszyckii ssp. *fusca* Burmann, 1954a: 345, pl. 18, figs. 1-6, text-fig. 1.

DIAGNOSIS.– Adult. Male. Wingspan 19-24 mm. Labial palp cream-coloured; basal part of segment 2 and all of segment 3 mottled with blackish brown. Antenna black, ringed with grey brown. Head and tegula light brown; thorax dark brown. Forewing light grey mottled with yellowish brown, white and black; costal half of base and costa black; two elongate black spots in fold and one (dot-shaped) at 1/2 and one (elongate) at 3/5; costal and tornal spots black, fused with spot at 3/5; subcostal stripe and streak through fold to tornus yellow-brown; many white scales in middle of wing; veins in apical area white; termen with black dots. Hindwing light grey. Female. Wingspan 10-19 mm. Labial palp cream-yellow; segment 3 mottled with some black. Antenna, face and thorax as in male. Neck brown; tegula grey-brown with light tip. Forewing fuscous mottled with white; fold with one square and one elongate spot, separated by yellow; black spot at 1/2 elongate, small, the one at 3/5 large. Hindwing rather long (up to one-third of forewing) and slender, whitish grey.

Variation. *S. melaleucella* is a geographically variable species. The description of males above is based on material from Switzerland, whereas specimens from the north-western Alps of Austria are generally larger and have more or less fuscous forewings with rather indistinct markings. These populations have been described as ssp. *fusca* (Burmann, 1954: 345). However, similar specimens also occur in other areas, e.g. in the south-western Alps. The description above of females is based on such dark individuals. Specimens from the southern Alps and especially from Switzerland have lighter and more variegated forewings. A specimen from Italy (Monte Baldo) has a broad whitish submedial fascia.

Similar species. Due to its variability *S. melaleucella* can be confused with other *Sattleria* species. It is mostly characterized by its large size and numerous white scales on the forewing (not in f. *fusca*). The rather long, slender hindwing of the female is also characteristic. Differs from other large species such as *S. triglavica*, *S. styriaca* and *S. dzieduszyckii* by the oblique dark spot on the fold which is a longitudinal streak in the other species.

MALE GENITALIA.– Sternite VIII slightly to deeply emarginate posteriorly; sacculus evenly tapered; apex of primary process of vinculum nearly level with apex of sacculus, without setae; secondary process of vinculum comparatively long, slender to subtriangular, spine-like, arising at right angles to primary process between about one-third and one quarter; saccus tapered; phallus with large to small median projection, basally more or less distinctly splayed, apical arm small and almost straight.

FEMALE GENITALIA.– Sternite VIII medially with pair of distinct and widely separate longitudinal ridges and frequently transverse folds in between, pair of comparatively large pockets in middle of segment VIII; apophysis anterioris with broad subtriangular sclerotized base, projecting slightly above sternal pockets, inner edge evenly curved with sternite VIII; antrum exceeding tip of apophysis anterioris, distinctly separate from sclerotized parts of sternite VIII, strongly sclerotized dorsally, ventral part membranous, posterior part of sclerite convex to inverted V-shaped.

DISTRIBUTION.– Scattered records from the Alps (Alpes Maritimes to western Austria) and the Appenines (Gran Sasso). Records in the northern and central Alps are restricted to nunataks.

BIONOMICS.– The larva has been found feeding on leaves of *Cerastium latifolium* L. (Caryophyllaceae) (Huemer, 1986: 152), *C. uniflorum* Clairv., probably *Moehringia* sp. (Caryophyllaceae) and *Saxifraga biflora macropetala* (Kern.) Rouay & Camus (Saxifragaceae) (Pitkin & Sattler, 1991: 220). It lives in a silken tube among the basal leaves and stalks of the host-plant, extending its housing to the shoots and to nearby stones. Disturbed larvae quickly retreat to the base of the tube. Pupation takes place in a cocoon amongst debris or under stones. During July and August larvae of different stage, pupae and adults may be observed simultaneously, indicating that the development cycle is at least biennial (Burmann, 1954: 350; 1977: 146; Huemer, pers. obs.). The adults have been observed from late June to mid-September. The males are attracted to light, but they have also been observed flying freely during the day, particularly in the morning hours. *S. melaleucella* occurs on alpine scree with scattered vegetation at altitudes from about 2000 to 3500 m, exceptionally at 1600 m.

REMARKS.– Pitkin & Sattler (1991: 219-220) distinguished 3 geographically separate forms which mainly differ in small characters of the secondary process of the vinculum. First molecular data give some hints of cryptic species but in the absence of females of the most separate form from the southern Alps (Monte Baldo, Adamello), a taxonomic re-assessment has to be postponed.

208. Sattleria triglavica Povolný, 1987

Sattleria dzieduszyckii ssp. *triglavica* Povolný, 1987a: 91, figs. 3-4, 19, pl., figs. 4-5.
Gelechia dzieduszykii [sic] f. *basistrigella* Müller-Rutz, 1934: 121, pl. 1, fig. 7; unavailable, infrasubspecific.
Sattleria triglavica ssp. *basistrigella* Huemer, 1997: 286, figs. 1-2, 5. Subspecies.

DIAGNOSIS.– Adult. Male. Wingspan 20-21 mm. Labial palp cream-coloured, slightly mottled with brown on basal half of segment 2 and on segment 3. Antenna black, indistinctly lighter ringed. Head light brown, darkest on neck; thorax and tegula dark brown, the latter with white tips. Forewing light brown mottled with white (especially in costal parts and at veins in apical area), grey (especially in middle of wing) and black (especially on costa, as dots around termen and as a costal spot at 3/4); a black spot through fold; a dot-shaped black spot at 1/2 and an angulated one at 3/5. Hindwing light grey. Female. Wingspan14 mm. Labial palp and antenna as in male. Head, thorax and tegula white mottled with light brown. Forewing with a white streak from near base to 2/3 on costa and white along base to 1/2; a distinct black streak along fold; black spots in middle of wing distinct, separated by light orange; subcostal stripe light orange-brown; apical area variegated black, orange-brown and white. Hindwing light grey.

Variation. The above description is based on two males of ssp. *basistrigella*. One of these is lighter than the other, but this specimen may be faded. In the other specimen the two black spots in the middle of the forewing are fused in one wing. A slightly worn male belonging to the nominal subspecies ssp. *triglavica* from Montenegro differs in having many white (instead of light brown) scales on the forewings, especially in the apical area of the forewings. It has moreover a black patch in the middle of the wing at 3/4.

Similar species. *S. triglavica* is characterized by its rather large size, by the light ground colour of the forewings, and by the distinct black streak through the fold. *S. styriaca* q.v.

MALE GENITALIA.– Sacculus distinctly broadened medially; apex of primary process of vinculum about level with apex of sacculus; secondary process of vinculum subrectangular to broadly rounded, with very broad base, extending from base to about middle of primary process; saccus comparatively narrow to broad, usually weakly tapered; phallus without or rarely with median projection, apical arm short.

FEMALE GENITALIA.– Sternite VIII with pair of distinct longitudinal ridges close together, pair of comparatively small pockets in anterior third (central and southern Alps) or posterior half (south-western Alps) of segment VIII; apophysis anterioris with broad sclerotized base, projecting slightly above sternal pockets, inner edge evenly curved with sternite VIII; antrum usually slightly exceeding tip of apophysis anterioris, separate from sclerotized parts of sternite VIII, posterior end irregularly shaped.

DISTRIBUTION.– Restricted to the highest mountains of the central and southern Alps (France, Switzerland, Italy and Slovenia), from Savoie to Julian Alps; also in the Balkans: Korab (Albania) (Povolný, 1987a: 92), Durmitor (Montenegro) (Huemer & Jakšic, 1996: 84).

BIONOMICS.– The larva has been observed during July, living in a silken tube on *Silene acaulis* (L.) Jacqu. (Caryophyllaceae) (Pitkin & Sattler, 1991: 221). In the laboratory adults emerged from late September to early October. However, in the field they have been collected from July to August. The male is attracted to light. Vertical distribution from about 2400 to 3100 m in the Alps and at 1900 m on Durmitor.

REMARKS.– *S. triglavica* shows a comparatively high degree of geographical variation. Hence, two subspecies are distinguished by some authors: *S. triglavica triglavica* from the south-eastern Alps (Slovenia), Montenegro and Albania and *S. triglavica basistrigella* from parts of the southern and central Alps (Povolný, 1987a: 91; Pitkin & Sattler, 1991: 221). Nominotypical populations from the Julian Alps and from Durmitor differ from other populations by the less swollen sacculus. Specimens from the western and south-eastern range can be separated from material originating from the central Alps by their smaller size, and by the larger and more rounded secondary process of the vinculum. Furthermore the south-western form is characterized by a median projection of the phallus and the comparatively broad pockets of sternite VIII which are situated in the middle of sternite VIII. In specimens from the central and southern Alps the pockets lie more anteriorly. Females of the nominotypical subspecies are unknown (see also Pitkin & Sattler, 1991).

209. Sattleria styriaca Pitkin & Sattler, 1991

Sattleria styriaca Pitkin & Sattler, 1991: 223, figs. 4, 15, 24, 44, 61.

DIAGNOSIS.– Adult. Male. Wingspan 23 mm. Labial palp cream-white mottled with light grey-brown on lower and outer surface of segment 2 and on segment 3. Antenna black, indistinctly ringed with brown. Head and tip of tegula whitish brown; thorax and base of tegula light grey-brown. Forewing white and light brown with grey and black markings; a black streak in fold through two black spots separated by orange-brown; a small, elongate spot at 1/2 and an angulated one at 3/5 in middle of wing; costa grey; a broad blackish grey fascia from 3/4 on costa followed by a broad white fascia in apical part of wing; termen with black dots. Hindwing whitish grey. Female. Wingspan 15 mm. Segment 2 of labial palp bone-white mottled with brown on lower and outer surface; segment 3 grey-brown. Head, thorax and tegula light grey-brown; face lighter. Antenna dark brown. Forewing dirty white mottled with light and dark brown; subcostal stripe light brown; a distinct black streak along fold, edged with brown; black spots at 2/5 and 3/5 in middle of wing, the latter being largest; subapical part of wing dark brown; apical part white; termen with blackish brown scales. Hindwing light grey.

Variation. Only one, probably somewhat faded male and one female were examined.

Similar species. *S. styriaca* is characterized by its large wingspan, and by its rather variegated forewings, especially the broad blackish grey and white fascia before apex. It is larger and more contrasting than *S. triglavica*, but has a less prominent basal streak through the fold.

MALE GENITALIA.– Sacculus strongly broadened basally, distal part comparatively long and slender, curved; primary process of vinculum short, about level with middle of sacculus; secondary process of vinculum with broad base and slender distal part, arising at base of primary process; saccus distinctly tapered; phallus weakly splayed at base, with indistinct median projection, apical arm small.

FEMALE GENITALIA.– Sternite VIII with pair of comparatively large pockets in anterior third of segment VIII; apophysis anterioris with broad sclerotized base, projecting slightly above sternal pockets, inner edge evenly curved with sternite VIII; antrum extending to about tip of apophysis anterioris, separate from sclerotized parts of sternite VIII, posterior end irregularly shaped.

DISTRIBUTION.– Endemic to the north-eastern and south-eastern Alps (Austria).

BIONOMICS.– Host-plant and early stages unknown. The few adults known to date have been collected from late June to early August. Vertical distribution from 1850 to about 2300 m.

210. Sattleria marguareisi Huemer & Sattler, 1992

Sattleria marguareisi Huemer & Sattler, 1992: 63, figs. 5, 10.

DIAGNOSIS.– Adult. Male. Wingspan 15-16 mm. Labial palp bone-white mottled with fuscous on segment 2 and at tip of segment 3. Antenna black, ringed with brown. Head cream-white, darker towards neck; thorax blackish brown; tegula lighter. Forewing greyish with dark grey base and costa; fold yellow between two elongate spots; a black patch from fold to black dot-shaped spots at 1/2 in middle of wing; a slightly angulated black spot at 3/5; a broad dark grey shadow before indistinct light tornal and costal spots; a distinct black line along termen. Hindwing light grey. Female. Unknown.

Variation. The examined specimens show only slight variation.

Similar species. *S. marguareisi* is characterized by its small size and its greyish forewings. *S. pyrenaica* has more light brown forewings, and *S. izoardi* has light areas on the forewings.

MALE GENITALIA.– Sacculus narrow, evenly tapered distally; primary process of vinculum short, about level with middle of sacculus; secondary process of vinculum broadly suboval with weakly serrated outer margin, extending from about one-third to two-thirds of primary process; saccus comparatively short, distinctly tapered; phallus short, without basal or median projections, apical arm long and curved.

FEMALE GENITALIA.– Unknown.

DISTRIBUTION.– Only known from Marguareis (Alpes Maritimes, France).

BIONOMICS.– Host-plant and early stages unknown. *S. marguareisi* seems to be restricted to limestone areas. The adults have been observed in the second half of July.

They can be flushed out of stands of the possible host-plant *Saxifraga moschata* Wulf. (Saxifragaceae) with a bee-smoker and are attracted to light. Vertical distribution from about 2100 to 2600m.

REMARKS.– *S. marguareisi* occurs sympatrically with *S. melaleucella* but it inhabits different microhabitats. The latter species seems to be restricted to alpine scree with stands of various Caryophyllaceae.

211. Sattleria dzieduszyckii (Nowicki, 1864)

Gelechia (Anacampsis) dzieduszyckii Nowicki, 1864: 20, fig. 4.
Gelechia dzieduszyckii tatrica Gregor & Povolný, 1955: 120, 127, phot. 2, fig. 4.

DIAGNOSIS.– Adult. Male. Wingspan 19-20 mm. Labial palp almost pure white. Antenna black, indistinctly lighter ringed. Head cream-white; thorax light brown; tegula with light tips. Forewing light grey-brown, more or less overlaid with white scales; an elongate black streak in the otherwise orange fold; rather indistinct black spots at 1/2 (dot-shaped) and 3/5 (angulated); a black spot at 3/4 on costa; termen with black dots. Hindwing light grey. Female. Wingspan 13 mm. Labial palps white. Antenna brown. Head, tegula and thorax white, the latter mottled with light brown. Forewing whitish grey mottled with light brown and black especially in apical part; fold with a blackish strak (most distinct in distal part) edged with yellow brown; black spots at 2/5 and 3/5 in middle of wing, the latter largest; veins in apical area and termen with black scales. Hindwing whitish grey.

Variation. The few examined specimens show some (probably geographical) variation. A specimen from the Polish Tatra mountains has almost plain whitish brown forewings and white veins in the apical part. Two specimens from the Carpathians of Romania have more distinct markings and the apical area almost whitish. One of these has also segment 2 of the labial palps cream-coloured.

Similar species. *S. dzieduszyckii* is characterized by its light labial palps, head and forewings, the latter with many white scales with more or less indistinct black markings. Compared with the other *Sattleria* species, *S. dzieduszyckii* is the lightest and least strongly marked. However, worn specimens of other *Sattleria* species often become lighter.

MALE GENITALIA.– Sternite VIII emarginate posteriorly; sacculus narrow, evenly tapered distally; apex of primary process of vinculum about level with apex of sacculus, basally broadened, basoventral margin slightly serrated; secondary process of vinculum broadly subovate with uneven or slightly serrated margin, extending from near base to about middle of primary process; saccus rarely slightly tapered; phallus weakly splayed at base, with indistinct median projection, apical arm short.

FEMALE GENITALIA.– Sternite VIII with pair of well separate, distinct longitudinal ridges, pair of broad pockets in posterior third of segment VIII; apophysis anterioris without broadly rounded sclerotized base, inner edge evenly curved with sternite VIII; antrum usually extending to about tip of apophysis anterioris, posterior end irregularly convex, without distinct separation from sclerotized parts of sternite VIII.

DISTRIBUTION.– Restricted to the higher peaks of the Carpathians, from Tatra mountains (Poland, Slovakia) in the north to the southern Carpathians (Romania). Records from

Rila Planina (Bulgaria) (Rebel, 1903: 329) are unconfirmed, whereas those from the Appenines, Alps and Pyrenees apply to other species of *Sattleria*.

BIONOMICS.– Host-plant and early stages insufficiently known. According to Povolný (1987a: 88) the larva lives on *Saxifraga* sp. (Saxifragaceae) but adults have never been reared. Pitkin & Sattler (1991: 224) suspect various Caryophyllaceae as possible host-plants. The adults have been collected from late June to mid-August. They have been observed on heavily grazed alpine pastures and on sparsely vegetated ground. The male is attracted to light. Vertical distribution from about 1800 to 2600 m.

REMARKS.– *Gelechia* (*Anacampsis*) *dzieduszyckii* was described from the Tatra mountains (Nowicki, 1864: 20). Records from the Alps, which are in fact all based on misidentifications, misled many authors to transpose the type locality into the Alps. In consequence of this error, Gregor & Povolný (1955: 120) described a separate subspecies, viz. *Gelechia dzieduszyckii tatrica*, again from the Tatra mountains.

Incorrect and doubtful records for the European fauna

The following taxa are not treated in this book due to various reasons:

A. *Scrobipalpa brandti* Povolný, 1972

REMARKS.– *Scrobipalpa brandti*, initially described from Iran, was recorded from a single male collected in southern Russia (Anikin & Piskunov, 1998: 69-70, figs. 1, 3-4). This specimen is the only one published from Europe. However, the figures of the adult show a dark species with different markings compared to the distinctly lighter coloured *S. brandti* (Povolný, 2002a: pl. 7, fig. 19). Furthermore, *S. brandti* is characterized by short valvae which never reach the tip of the uncus, and a comparatively broad saccus (Povolný, 1972: 189, 2002a: pl. 25, figs. 224-225). All these characters are different in the species figured by Anikin & Piskunov (1998). Though this species cannot be safely identified from the figures alone, its conspecificity with *S. brandti* can be denied.

B. *Scrobipalpa eremica* Povolný, 1967

REMARKS.– *Scrobipalpa eremica*, initially described from Iran and Syria (Povolný, 1967a: 217) was listed in *Fauna Europaea* (Karsholt, 2004-2009) without further data. The record was based on a specimen from the Ukraine which later turned out to be misidentified.

C. *Scrobipalpa filia* Povolný, 1969

REMARKS.– *Scrobipalpa filia* was described from Mongolia and later published from Iran (Povolný, 1969a: 7; 1972: 195). Recently, the species was also recorded from south-western Russia, based on a single male, collected from 25.-30.v.2000 (Anikin & Piskunov, 2003: 70). Unfortunately, no figures of the adult and its genitalia were given, leaving doubts about the identity of this finding. Through the courtesy of V. Anikin we received the genitalia of this specimen in glycerine and it seems to be correctly identified. However, since we have not been able to examine the moth itself or the female sex the occurrence of this species in Europe needs further confirmation.

D. *Scrobipalpa gregori* Povolný, 1967

REMARKS.– *Scrobipalpa gregori* Povolný, 1967, was described from Afghanistan (Povolný, 1967a: 218), later recorded from southern Russia (Anikin & Piskunov, 1998: 69-70, figs. 1, 3-4) and with some doubts from Italy (Povolný, 1981: 372, figs. 14-15, 2002a: pl. 7, fig. 24, pl. 26, fig. 236). The figures of the single male specimen from Russia clearly contradict important characters of *S. gregori* such as the slightly concave uncus and the wing markings and it must belong to another species which cannot be identified safely from the figures alone. The single male specimen from Italy, preserved in ZMUC, is a misidentification of *S. artemisiella*.

E. *Scrobipalpa maniaca* Povolný, 1969

REMARKS.– *Scrobipalpa maniaca* was described from Mongolia (Povolný, 1969a: 13) and it is widely distributed through Central Asia to the Transbaikal area (Povolný, 1972: 195; 2002a: 54). Recently the species was recorded from south-western Russia, based on a single male collected from 25.-30.v.2000 (Anikin & Piskunov, 2003: 70). Similar to the aforementioned species, no figures of the adult and its genitalia were published. The genitalia capsule in glycerine sent to us by V. Anikin belongs to a species which is probably not conspecific with *S. maniaca*. Again, further material is needed to clarify the status of this material.

F. *Scrobipalpa meyricki* Povolný, 1971

REMARKS.– *Scrobipalpa meyricki* Povolný, 1971, was described from 3 males and 3 females from Algeria. We studied the male holotype and a female paratype in the NHMW, which are conspecific with a series of specimens from Tunisia in ZMUC. The genitalia slide of the holotype is embedded in a semilateral position, and this probably misled Povolný (1981: 388-389) to synonymize *S. glaserorum* with the otherwise different looking *S. meyricki*. The former species is here treated as a synonym of *S. bradleyi* q.v.

G. *Scrobipalpa optima* Povolný, 1969

REMARKS.– A record of *Scrobipalpa optima*, initially described from Iran (Povolný, 1969d: 4) and later recorded from Russia is most probably due to confusion with *S. grossa* which is considered as a junior synonym (Ponomarenko, 2008: 96). However, the latter taxon in fact is conspecific with *S. indignella*, a species which occurs in Russia.

H. *Scrobipalpa remota* Povolný, 1972

REMARKS.– *Scrobipalpa remota* was described from seven specimens collected in the Middle East and in Algeria and a male from North Iran was selected as the holotype (Povolný, 1972: 191). Later this species was also published from Spain (Povolný, 1977b: 196). A record from France, probably based on a female figured in Povolný & Luquet (1983: 67-68, fig. 11) derives from inadequate labelling of an old specimen. Hence it is doubted by the authors themselves and also in the most recent French catalogue (Leraut, 1997: 52). All other records from western Europe and those from north-western Africa are based on male specimens. Since the males are similar to those of *S. superstes*, a species widespread in the western Mediterranean, identification of *S. remota* should be confirmed with females which are well differentiated. All known females from western Europe we have been able to examine in fact belong to *S. superstes*. Therefore the alleged occurrence of *S. remota* in this part of Europe is extremely doubtful and unlikely. A further record of *S. remota* from the Ukraine is based on two male specimens from ZMKU (Povolný, 2001a: 205). We have been able to study both specimens and their genitalia slides which show a superficial similarity to *S. remota*. However, the adults are very different from this species and fully agree with *S. clintoni* q.v., a species also known from the Ukraine.

I. Scrobipalpa rjabovi Piskunov, 1990

REMARKS.– *Scrobipalpa rjabovi* was described from Mongolia and subspecies *S. rjabovi kizilcumica* from Uzbekistan (Piskunov, 1990: 301-303). Later, this species was recorded from Saratov district in south-western Russia by Anikin *et al.* (1999: 316). We have not been able to study material from this area and no figures prove the correct identification. Therefore, the occurrence of this species in Europe needs confirmation.

J. Scrobipalpa vladimiri Povolný, 1966

REMARKS.– *Scrobipalpa vladimiri* Povolný, 1966, initially described from the Middle East and later found in Germany (Thüringen) as a single specimen (Povolný, 1971b: 73) is most probably mislabelled and this species is not regarded as a member of the European fauna (Karsholt, 1995: 148-149; Povolný, 2002a: 77).

Taxa incertae sedis

K. *Lita melanella* Heinemann, 1870

REMARKS.– *Lita melanella* was described from an unspecified number of specimens from Regensburg in southern Germany. It was stated to have black, violet shining forewings with indistinct markings, and black labial palps with segment 2 longer than segment 3 (Heinemann, 1870: 258). We have been unable to trace type material or otherwise authentically identified specimens of *L. melanella*. Based on the original description we cannot refer it to any species described in this volume, and we are not even convinced that it belongs in the Gnorimoschemini.

L. *Lita nigripalpella* Heinemann, 1870

REMARKS.– *Lita nigripalpella* was described from an unspecified number of specimens from Regensburg in southern Germany. It was stated to be similar to *Lita melanella*, but slightly lighter and by having more slender and pointed forewings and somewhat lighter labial palps (Heinemann, 1870: 258). We have been unable to trace type material or otherwise authentically identified specimens of *L. nigripalpella*. Based on the original description we cannot refer it to any species described in this volume, and we are not even convinced that it belongs in the Gnorimoschemini. It may well be a synonym of *L. melanella* as suggested by Gaede (1937: 272).

M. *Lita pygmaeella* Heinemann, 1870

REMARKS.– *Lita pygmaeella* was described from an unspecified number of specimens from Ober-Engadin in Switzerland. It was named from its small size and compared to and separated from *Lita melanella, L. nigripalpella, Scrobipalpa acuminatella* and *S. murinella* (Heinemann, 1870: 259 – not 359 as printed in the book). We have been unable to trace type material or otherwise authentically identified specimens of *L. pygmaeella*. Based on the original description we cannot certainly refer it to any species described in this volume, although the rusty yellow surrounding of the black spots make it possible that it belongs in the Gnorimoschemini and maybe it is conspecific with either *Scrobipalpa murinella* or *Scrobipalpula diffluella*.

N. *Lita trochilella* Heinemann, 1870

REMARKS.– *Lita trochilella* was described from an unspecified number of specimens from Braunschweig in northern Germany. It was stated to have a superficial resemblance to [*Ephysteris*] *diminutella*, differing in the form of the hindwings and the longer segment 3 of the labial palps (Heinemann, 1870: 253). We have been unable to trace type material or otherwise authentically identified specimens of *L. trochilella*. Based on the original description we cannot refer it to any species described in this volume, and we are not even convinced that it belongs in the Gnorimoschemini.

Distribution catalogue

Similar to the first volume on Gelechiidae published in *Microlepidoptera of Europe* (Huemer & Karsholt, 1999) we include in Europe the countries and larger islands listed below. Contrary to, e.g. *Fauna Europaea* (Karsholt 2004-2009) we do not include the Canary Islands, Madeira, the Azores Islands and Cyprus. To the East we include (contrary to *Fauna Europaea*) also the eastern parts of the Ural Mountains.

The distribution catalogue mainly follows Karsholt & Razowski (1996). However, former Yugoslavia is now separated into 7 countries: Bosnia-Herzegovina, Croatia, Kosovo, Macedonia, Montenegro, Serbia and Slovenia, and we follow this new political scenery. Distributional data from the former Soviet Union are primarily based on Bidzilya (*in litt.* and in several papers) who kindly prepared check-lists for European Russia, Byelorussia, Ukraine and Moldavia, and Ponomarenko (2008). Important papers by Anikin & Piskunov (2001) and Junnilainen *et al.* (2010) deal with the Volga region and the southern Ural Mountains, respectively.

Additional information has been obtained from following sources: Portugal (Corley *et al.*, 2000; Corley, 2002b), Andorra (Pitkin & Sattler, 1991), Finland (Kullberg *et al.*, 2002), Germany (Gaedike & Heinicke, 1999), Austria (Huemer & Tarmann, 1993), Switzerland (Bryner *et al.*, 2005; Sauter & Whitebread, 2005), Poland (Buszko *et al.*, 2000), Czech Republic, Slovakia and Hungary (Elsner, 1998; Elsner *et al.*, 1999; Fazekas, 2002; Pastorális, 2008), France (Leraut, 1997), former Yugoslavia (Gomboc & Carnelutti, *in litt.*; Huemer & Jakšic, 1996; Huemer & Sattler, 1995; Klimesch, 1968; Rebel, 1891; Thurner, 1941), Ukraine (Povolný, 2001a), Romania (Kovács & Kovács, 2000), Estonia (Jürivete & Õunap, 2008), Latvia (Savenkov & Šulcs, in press) and Lithuania (Ivinskis, 2004). Also data from *Fauna Europaea* (Karsholt, 2004-2009) have been used.

Furthermore several hitherto unpublished records, mainly from the collections of ZMUC, are included.

No data about Gnorimoschemini are available from Kosovo, Moldavia, and the Principality of Liechtenstein so far.

? - doubtful records

Acronyms of the countries

Acronyms of countries follow
http://publications.europa.eu/code/de/de-5000600.htm,
except for the larger Mediterranean Islands. Slovenia is listed as SL instead of SI to avoid confusion with Si (Sicily).

PO Portugal
ES Spain
AD Andorra
FR France
Co Corsica
Sa Sardinia
Si Sicily
IT Italy
MA Malta
IE Ireland
UK Great Britain
BE Belgium
NL The Netherlands
LU Luxemburg
DE Germany
CH Switzerland
AT Austria
PL Poland
CZ Czech Republic
SK Slovakia
HU Hungary
RO Romania
SL Slovenia
HR Croatia
RS Serbia
ME Montenegro
BA Bosnia-Herzegovina
MK Macedonia
AL Albania
BG Bulgaria
EL Greece
Cr Crete
TR Turkey (European part)
IC Iceland
DK Denmark
NO Norway
SE Sweden
SF Finland
EE Estonia
LV Latvia
LT Lithuania
RU Russia (European part)
BY Byelorussia
UA Ukraine

1.	soffneri		ES			Co		Si	IT												HU	
2.	herbichii	PO	ES		FR								NL		DE			PL		SK	HU	RO
3.	bodillum														DE							
4.	nupponeni																					
5.	robustella																					
6.	steueri				FR				IT						DE		AT		CZ	SK		
7.	epithymella		ES		FR		Sa		IT							CH	AT					
8.	nordlandicolella																					
9.	nilsi				FR				IT								AT					
10.	valesiella		ES		FR				IT						?	CH	AT					
11.	streliciella				?						UK	?	NL		DE			PL				
12.	hoefneri								IT								AT					
13.	petasitis				?				IT						DE	CH	AT		?			RO
14.	aptatella				?				?													
15.	kasyi																					
16.	notata																					
17.	acuminatella	PO	ES	AD	FR	Co		Si	IT	IE	UK	BE	NL	LU	DE	CH	AT	PL	CZ	SK	HU	RO
18.	skulei																					
19.	hungariae																AT				HU	
20.	adaptata																AT		CZ	SK		
21.	brahmiella				FR							BE			DE					SK		RO
22.	vasconiella	PO	ES		FR		Sa	Si	IT													
23.	dorsolutea																					
24.	amseli		ES		FR																	
25.	hyssopi				FR																	
26.	montanella				FR			Si														
27.	corleyi	PO																				
28.	chrysanthemella								IT						DE	CH	AT		CZ	SK	HU	RO
29.	proclivella		ES		FR				IT			BE	NL		DE	CH	AT	PL	CZ	SK	HU	RO
30.	frugifera																					
31.	oleksiyella																					
32.	smithi		ES																			
33.	occulta																					
34.	grisea																					
35.	usingeri																					
36.	clintoni				?						UK				DE							
37.	reiprichi																			SK	HU	
38.	obsoletella	PO	ES		FR	Co	Sa	Si	IT	IE	UK	BE	NL	LU	DE	CH	AT	PL	CZ	SK	HU	RO
39.	feralella								IT							CH	AT					
40.	halonella				FR				IT								AT	?		SK	HU	
41.	perinii				FR				IT													
42.	phagnalella	PO	ES		FR	Co	Sa															
43.	tokari																					
44.	karadaghi																					
45.	heimi																					
46.	acuta																					
47.	soffneri																					
48.	jarii																					
49.	murinella								IT	IE	UK				DE	CH	AT			SK		RO
50.	wiltshirei	PO	ES																			
51.	caucasica																					
52.	pauperella		ES		FR				IT		UK			LU	DE	CH	AT	PL	CZ	SK	HU	RO
53.	spumata																					
54.	arenbergeri								IT								AT		CZ	SK	HU	
55.	nana																					
56.	heretica		ES																			
57.	bigoti		ES		FR			Si														
58.	dorsoflava																					

						BG		Cr													soffneri	1.
				MK							DK	NO	SE	SF	EE	LV	LT	RU		UA	herbichii	2.
											DK										bodillum	3.
																		RU		UA	nupponeni	4.
																		RU			robustella	5.
																					steueri	6.
												NO	SE	SF	EE	LV		RU	BY		epithymella	7.
												NO	SE	SF	EE			RU			nordlandicolella	8.
																					nilsi	9.
										IS		NO	SE	SF		LV		RU			valesiella	10.
											DK	NO	?	SF	EE	LV		RU			streliciella	11.
SL																					hoefneri	12.
SL												NO	SE	SF							petasitis	13.
						?															aptatella	14.
				MK		BG	EL														kasyi	15.
																		RU		UA	notata	16.
SL	HR					BG	EL		TR		DK	NO	SE	SF	EE	LV	LT	RU	BY	UA	acuminatella	17.
SL																					skulei	18.
	HR																			UA	hungariae	19.
																		RU		UA	adaptata	20.
							EL											RU		UA	brahmiella	21.
				MK			EL											RU		?	vasconiella	22.
																		RU			dorsolutea	23.
																					amseli	24.
																					hyssopi	25.
																					montanella	26.
																					corleyi	27.
	?	?	BA															RU		?	chrysanthemella	28.
						BG	EL				DK		SE	SF	EE	LV	LT	RU	BY	UA	proclivella	29.
																		RU			frugifera	30.
																		RU			oleksiyella	31.
																				UA	smithi	32.
																		RU			occulta	33.
																		RU			grisea	34.
																		RU			usingeri	35.
											DK	NO	SE		EE	LV		RU	BY	UA	clintoni	36.
							?					NO									reiprichi	37.
	?	?			AL	BG	EL	Cr	TR		DK	NO	SE	SF	EE	LV	LT	RU	BY	UA	obsoletella	38.
																					feralella	39.
																		RU		?	halonella	40.
SL				MK			EL														perinii	41.
								Cr										?			phagnalella	42.
?	HR																				tokari	43.
																		RU		UA	karadaghi	44.
																		RU			heimi	45.
																		RU		UA	acuta	46.
						BG												RU			soffneri	47.
						BG															jarii	48.
SL												NO	SE	SF		LV	LT	RU	BY	UA	murinella	49.
																					wiltshirei	50.
																		RU			caucasica	51.
	?	?											SE	SF	EE	LV		RU	BY	UA	pauperella	52.
																		RU			spumata	53.
	HR																	RU		UA	arenbergeri	54.
																		RU			nana	55.
																		RU			heretica	56.
							EL														bigoti	57.
																		RU			dorsoflava	58.

59.	magnificella																						
60.	abstrusa																						
61.	superstes	PO	ES		FR		Sa	Si															
62.	plesiopicta																						
63.	bradleyi		ES		FR			Si															
64.	selectella																						
65.	alterna																						
66.	lutea																						
67.	niveifacies	PO	ES																				
68.	indignella																						
69.	punctata																						
70.	lagodes																						
71.	deluccae																						
72.	atriplicella		ES		FR	Co	Sa	Si	IT			UK	BE	NL	LU	DE	CH	AT	PL	CZ	SK	HU	RO
73.	suaedella	PO	ES		FR	Co	Sa		IT			UK				DE							
74.	solitaria																						
75.	dagmaris								?														
76.	suasella	PO	ES		FR																		
77.	hendrikseni																						
78.	traganella	PO	ES		FR					MA													
79.	bazae		ES		FR																		
80.	artemisiella		ES	AD	FR	Co	Sa	Si	IT		IE	UK	BE	NL	LU	DE	CH	AT	PL	CZ	SK	HU	RO
81.	stangei											UK				DE		AT	PL			HU	?
82.	suaedivorella		ES																				
83.	bryophiloides																						
84.	algeriensis		ES																				
85.	deutschi																						
86.	disjectella		ES																				
87.	mixta																						
88.	rebeli								IT									AT					
89.	gallicella		ES		FR				IT													HU	
90.	ustulatella																						
91.	postulatella		ES																				
92.	nitentella		ES		FR						IE	UK	BE	NL		DE		AT	PL	CZ	SK	HU	RO
93.	costella		ES		FR						IE	UK	BE	NL		DE							
94.	hyoscyamella	PO	ES		FR													AT					RO
95.	portosanctana		ES		FR	Co	Sa	Si	IT	MA													
96.	vicaria		ES							MA													
97.	ocellatella	PO	ES		FR	Co	Sa	Si	IT	MA	IE	UK	BE	NL		DE	CH	AT		CZ	SK	HU	RO
98.	pulchra																						
99.	gecko																						
100.	hannemanni																						
101.	erichi																	AT		CZ	SK	HU	RO
102.	divisella		ES																				
103.	voltinella		ES		FR																		
104.	suaedicola		ES		FR		Sa																
105.	monochromella				FR	Co		Si	IT														
106.	samadensis		ES		FR				IT		IE	UK	BE	NL		DE	CH	AT	PL	CZ	SK	HU	RO
107.	salinella	PO	ES		FR	Co	Sa	Si	IT														
108.	spergulariella		ES		FR			Si															
109.	salicorniae				FR							UK	BE	NL		DE		AT	PL				RO
110.	halimoniella				FR																		
111.	thymelaeae						Sa																
112.	halymella	PO			FR	Co			IT	MA													
113.	camphorosmella		ES		FR																		
114.	stabilis		ES																				
115.	instabilella	PO	ES		FR		Sa	Si	IT		IE	UK	BE	NL		DE		?					
116.	peterseni							Si															

																		RU		UA	magnificella	59.
																		RU			abstrusa	60.
																					superstes	61.
																		RU			plesiopicta	62.
																					bradleyi	63.
							EL											RU		UA	selectella	64.
																		RU			alterna	65.
																		RU			lutea	66.
																					niveifacies	67.
																		RU		UA	indignella	68.
																		RU			punctata	69.
																					lagodes	70.
																					deluccae	71.
SL				MK		BG			TR	IS	DK	NO	SE	SF	EE	LV	LT	RU	BY	UA	atriplicella	72.
																					suaedella	73.
																		RU		UA	solitaria	74.
																					dagmaris	75.
																					suasella	76.
							EL														hendrikseni	77.
																					traganella	78.
																					bazae	79.
SL	HR	?	?	MK	AL	BG	EL				DK		SE	SF	EE	LV	LT	RU	BY	UA	artemisiella	80.
											DK	NO	SE	SF	EE	LV	LT	RU			stangei	81.
																		RU			suaedivorella	82.
							EL							SF		LV		RU		UA	bryophiloides	83.
																					algeriensis	84.
																		RU			deutschi	85.
																		?			disjectella	86.
				MK	?															UA	mixta	87.
																		RU		UA	rebeli	88.
																		RU		UA	gallicella	89.
																		RU		UA	ustulatella	90.
																					postulatella	91.
											DK	NO	SE	SF	EE	LV		RU		UA	nitentella	92.
											DK										costella	93.
																					hyoscyamella	94.
	HR																				portosanctana	95.
																					vicaria	96.
SL	HR	RS	BA	MK	AL	BG	EL	Cr	TR		DK							RU		UA	ocellatella	97.
																LV		RU			pulchra	98.
							EL														gecko	99.
	?																	RU			hannemanni	100.
									TR									RU		UA	erichi	101.
																					divisella	102.
	HR																			UA	voltinella	103.
																					suaedicola	104.
																				UA	monochromella	105.
	HR						EL			IS	DK	NO	SE	SF	EE	LV		RU		UA	samadensis	106.
	HR				AL	BG	EL											?		?	salinella	107.
							EL														spergulariella	108.
											DK		SE	SF		LV		RU		UA	salicorniae	109.
																					halimoniella	110.
						BG															thymelaeae	111.
																					halymella	112.
							EL	Cr													camphorosmella	113.
																					stabilis	114.
							EL				DK									?	instabilella	115.
																					peterseni	116.

117.	ergasima	PO	ES		FR	Co	Sa	Si		MA												
118.	glaseri																					
119.	psilella	PO	ES		FR	Co		?	IT				?		DE	CH	AT	PL	CZ	SK	HU	RO
120.	ramosella				FR											CH						
121.	seniorum																					
122.	diffluella				FR				IT			UK			DE	CH	AT				HU	?
123.	tussilaginis				FR				IT			UK		NL	DE	CH	AT	PL	CZ	SK	HU	RO
124.	operculella	PO	ES		FR	Co		Si	IT	MA		UK	BE	NL	DE	CH	AT				HU	RO
125.	absoluta	PO	ES		FR			Si	IT	MA		UK		NL	DE	CH						
126.	lycopersicella								IT													
127.	promptella	PO	ES		FR	Co		Si	IT	MA										SK	HU	RO
128.	tenuisaccus																					
129.	deserticolella				FR																	
130.	insulella				FR				IT						DE	CH						RO
131.	diminutella		ES		FR			Si	IT													RO
132.	inustella	PO	ES		FR				?						DE	CH	AT	PL	CZ	SK	HU	
133.	olympica		ES		FR																	
134.	iberica		ES		FR			Si		MA												
135.	subdiminutella		ES						IT	MA												
136.	glebicolorella																					
137.	rhamnifoliae																					
138.	chretieni	PO	ES		FR		Sa															
139.	karsholti																					
140.	moritzella	PO			FR				IT						DE	CH	AT	PL	CZ	SK	HU	RO
141.	lutilabrella																					
142.	volgensis																					
143.	prolata																					RO
144.	kiningerella				FR				IT						DE	CH	AT	PL	CZ		HU	RO
145.	discontinuella								IT							CH	AT					
146.	terroris		ES																			
147.	fischerella				FR										DE	CH	AT	PL	CZ	SK	HU	RO
148.	tischeriella	PO	ES		FR			Si	IT						DE	CH	AT		CZ	SK		RO
149.	alsinella	PO	ES		FR	Co	Sa		IT		IE	UK	BE	NL	DE	CH	AT	PL	CZ	SK	HU	
150.	viscariella				FR				IT		IE	UK			DE	CH	AT				HU	
151.	albifaciella		?	AD	FR										DE	CH	AT					
152.	vicinella		ES		FR			Si	IT		IE	UK			DE	CH	AT		CZ	SK	HU	RO
153.	bosalella					Co	Sa															
154.	amaurella				FR				IT						DE	CH	AT		CZ	SK	HU	
155.	oculatella															CH	AT					
156.	leucofasciatum		ES																			
157.	petryi		?		FR										DE	CH	AT				HU	
158.	baischi																					
159.	repentis		ES		FR				IT						DE	CH	AT					
160.	siculum							Si														
161.	inflativorella																			SK	HU	RO
162.	saginella				FR				IT							CH	AT					
163.	cauligenella		ES		FR			Si	IT				BE	NL	DE	CH	AT	PL	CZ	SK	HU	
164.	trauniella								IT								AT					
165.	peregrinella		ES		FR				IT								AT					
166.	delphinatella				FR				IT							CH						
167.	provinciella	PO	ES		FR																	
168.	mucronatella	PO	ES		FR				IT						DE	CH	AT					
169.	leucomelanella		ES	AD	FR			Si	IT						DE	CH	AT	PL	CZ	SK	HU	RO
170.	mazeli		?	AD	FR																	
171.	leucothoracellum				FR				IT							CH	AT				HU	
172.	schleichi	PO	ES		FR				IT						DE	CH	AT		CZ			RO
173.	marmorea	PO	ES		FR	Co	Sa		IT		IE	UK	BE	NL	DE	CH	AT	PL	CZ	SK	HU	
174.	pullatella																AT	PL				

							EL	Cr												ergasima	117.
																			UA	glaseri	118.
							EL		TR	DK	NO	SE	SF	EE	LV	LT	RU	BY	UA	psilella	119.
																	?			ramosella	120.
				MK			EL													seniorum	121.
											NO	SE	SF	EE	LV		RU			diffluella	122.
SL				MK						DK		SE					RU	BY	UA	tussilaginis	123.
				MK			EL	Cr									RU	BY	UA	operculella	124.
								Cr		DK						LT				absoluta	125.
																				lycopersicella	126.
				MK		BG	EL	Cr	TR								RU		UA	promptella	127.
																	RU			tenuisaccus	128.
							EL										RU		UA	deserticolella	129.
						BG											RU			insulella	130.
				MK			EL										RU		?	diminutella	131.
			BA											EE	LV	LT	RU		UA	inustella	132.
							EL													olympica	133.
	HR						EL	Cr												iberica	134.
	HR							Cr												subdiminutella	135.
																			UA	glebicolorella	136.
							EL													rhamnifoliae	137.
								Cr												chretieni	138.
																	RU			karsholti	139.
	?	?									NO		SF	EE	LV	LT	RU	BY	UA	moritzella	140.
SL	HR					BG	EL												UA	lutilabrella	141.
																	RU			volgensis	142.
																	RU		UA	prolata	143.
				MK		BG	EL				NO	SE	SF	EE	LV		RU		UA	kiningerella	144.
																				discontinuella	145.
																				terroris	146.
			BA	MK		BG				DK		SE	SF	EE	LV	LT	RU	BY	UA	fischerella	147.
SL		?	?	MK		BG	EL	Cr			NO	SE	SF	EE	LV	LT	RU			tischeriella	148.
SL				MK			EL	Cr		DK		SE	SF	EE	LV				UA	alsinella	149.
										DK	NO	SE	SF	EE	LV	LT	RU			viscariella	150.
																				albifaciella	151.
SL							EL			DK	NO	SE	SF	EE	LV	LT	RU	BY	?	vicinella	152.
																				bosalella	153.
	?	?					EL			DK	NO	SE	SF	EE	LV		RU		UA	amaurella	154.
																				oculatella	155.
																				leucofasciatum	156.
SL		?	?									SE	SF		LV	LT	RU			petryi	157.
							EL													baischi	158.
SL																				repentis	159.
																				siculum	160.
				MK																inflativorella	161.
SL	HR			MK			EL													saginella	162.
SL				MK			EL			DK		SE	SF	EE	LV					cauligenella	163.
SL																				trauniella	164.
SL			BA	MK			EL												UA	peregrinella	165.
																				delphinatella	166.
							EL													provinciella	167.
SL							EL													mucronatella	168.
SL	HR	ME	BA	MK	AL	BG	EL										RU		UA	leucomelanella	169.
																				mazeli	170.
																				leucothoracellum	171.
SL	HR			MK			EL					SE	SF	EE	LV	LT	RU			schleichi	172.
SL				MK	AL		EL	Cr		DK	NO	SE							?	marmorea	173.
										DK	NO	SE	SF	EE	LV		RU		UA	pullatella	174.

175. stramentella								IT													
176. hispanicum		ES																			
177. confluens																					
178. gallagenellum				FR				IT						DE							
179. fraternella		ES		FR					IE	UK		NL	LU	DE	CH		PL				
180. klosi				FR				IT						DE		AT		CZ			?
181. interalbicella				FR				IT						DE	CH	AT					
182. laceratella								IT													
183. blandella	PO			FR					IE	UK	BE	NL	LU	DE	CH	AT	PL	CZ	SK	HU	
184. blandelloides		ES			Co	Sa		IT		UK				DE	CH	AT	PL	CZ	SK		
185. jaspidella	PO	ES																			
186. proxima	PO	ES		FR	Co	Sa		IT		UK		NL		DE	CH	AT	PL	CZ	SK	HU	RO
187. blandulella		ES		FR	Co		Si	IT		UK		NL		DE	CH			CZ	SK	HU	
188. arenbergeri		ES																			
189. tricolorella		ES		FR					IE	UK	BE	NL	LU	DE	CH	AT	PL	CZ	SK	HU	RO
190. fibigerium	PO	ES		FR			?	IT													
191. junctella		?		FR				IT		UK	BE			DE	CH	AT	PL	CZ	SK		RO
192. cassella				FR										DE	CH	AT	PL	CZ	SK	HU	RO
193. moehringiae				FR				IT						DE	CH	AT					
194. petrophila				FR				IT							CH	AT					
195. huebneri				FR				IT		UK	BE			DE	CH	AT	PL	CZ	SK	HU	RO
196. kroesmanniella				FR						UK	BE	NL	LU	DE	CH	AT	PL	CZ	SK	HU	RO
197. capsophilella				FR										DE	CH	AT					
198. solitaria		ES		FR				IT							CH	AT					
199. terrestrella															CH						RO
200. intermedia															CH	AT					
201. quartana																					
202. arcuata		ES		FR																	
203. pyrenaica		ES	AD	FR																	
204. angustispina				FR																	
205. izoardi				FR				IT													
206. breviramus				FR																	
207. melaleucella				FR				IT						DE	CH	AT					
208. triglavica				FR				IT							CH						
209. styriaca																AT					
210. marguareisi				FR																	
211. dzieduszyckii																	PL		SK		RO

																			stramentella	175.
							EL												hispanicum	176.
		?					EL												confluens	177.
																			gallagenellum	178.
									DK	NO	SE	SF	EE	LV				UA	fraternella	179.
																			klosi	180.
																			interalbicella	181.
SL																			laceratella	182.
SL							EL		DK	NO	SE	SF	EE	LV	LT	RU	BY	UA	blandella	183.
							EL	Cr	DK	NO	SE	SF	EE			RU		UA	blandelloides	184.
																			jaspidella	185.
SL		ME		MK	AL		EL	Cr	DK		SE					RU		UA	proxima	186.
	HR						EL		DK		SE	SF						UA	blandulella	187.
																			arenbergeri	188.
SL									DK		SE	SF	EE	LV	LT	RU	BY	UA	tricolorella	189.
						BG	EL												fibigerium	190.
SL			BA			BG	EL		DK	NO	SE	SF	EE	LV	LT	RU		UA	junctella	191.
									DK	NO	SE	SF	EE	LV	LT	RU	BY	UA	cassella	192.
SL							EL												moehringiae	193.
				MK						NO	SE	SF	EE						petrophila	194.
		RS							DK		SE			LV	?	RU	?	UA	huebneri	195.
									DK		SE	SF	EE	LV	LT	RU	BY	UA	kroesmanniella	196.
																			capsophilella	197.
							EL												solitaria	198.
					AL														terrestrella	199.
																RU			intermedia	200.
						BG													quartana	201.
																			arcuata	202.
																			pyrenaica	203.
																			angustispina	204.
																			izoardi	205.
																			breviramus	206.
																			melaleucella	207.
SL		ME			AL														triglavica	208.
																			styriaca	209.
																			marguareisi	210.
																		UA	dzieduszyckii	211.

Colour plates

Plate 1 All specimens 3 x natural size

1. *Gnorimoschema soffneri* (Riedl) –
   a) ♂, Greece, Crete, Lassithi distr., Makrigialos, 20.v.1993, leg. J. Johansson (ZMUC).
   b) ♂, Spain, Granada, Camino de Capileira, 1250 m, 13.vii.1985, leg. E. Traugott-Olsen (ZMUC).
2. *G. herbichii* (Nowicki) –
   a) ♀, Denmark, EJ, Anholt, 17.-18.viii.1979, leg. O. Karsholt (ZMUC).
   b) ♀, Denmark, EJ, Anholt, 14.vi.1996, leg. & coll. P. Falck.
   c) ♀, Denmark, SZ, Jungshoved, 6.vi.1959, leg. N. L. Wolff (ZMUC).
   d) ♂, Finland, PP, Oulu, 13.v.1978, leg. J. Kyrki, gen. slide Karsholt 3230 (ZMUC).
   e) ♂, Netherlands, Gld., Heerde, 10.vii.1982, leg. J. B. Wolschrijn, gen. slide van der Wolf 1798 (ZMUC).
3. *G. bodillum* Karsholt & Nielsen –
   a) ♂, Denmark, NEJ, Råbjerg Mile, 20 m, e.l. 3.-16.vi.1996, leg. A. Scholz (ZMUC).
   b) ♀, paratype, Denmark, NEJ, Råbjerg Mile, e.p. 13.vi.1973, leg. E. S. Nielsen (ZMUC).
4. *G. nupponeni* sp. n. –
   a) ♀, paratype, Russia, South Ural, Orenburg district, 6 km W Donskoje village, mount Verbljushka, 27.viii.2000, leg. K. Nupponen, gen. slide Hendriksen 6426 (NUPP).
   b) ♂, paratype, Russia South Ural, Orenburg district, 6 km W Donskoje village, mount Verbljushka, 27.viii.2000, leg. K. Nupponen (NUPP).
   c) ♀, paratype, Russia, South Ural, Orenburg district, 40 km W Orsk, near Guberlja village, 9.ix.2009, leg. K. Nupponen (NUPP).
5. *G. robustella* (Staudinger) –
   ♂, Russia, Sarepta, 3.vii.1875, leg. H. Christoph (BMNH).
6. *G. steueri* Povolný –
   a) ♀, paratype, Germany, Thuringia, Blankenburg, Pörzmühle, 30.viii.1974, leg. H. Steuer (ZMUC).
   b) ♂, Germany, Thuringia, Blankenburg, 28.viii.1984, leg. M. K. Gleitz (ZMUC).
7. *G. epithymella* (Staudinger) –
   a) ♀, Spain, Huesca, Benasque, 1500 m, 31.vii.-2.viii.1981, leg. A. Cox, gen. Slide Hendriksen 5002 (ZMUC).
   b) ♀, Austria, Nordtirol, Finstermünz, 1200 m, 30.vii.1982, leg. Burmann, genitalia in tube (TLMF).
   ssp. *boerneri* Amsel –
   c) ♂, Italy, Sardinia, Gennargentu, Arcu Tascussi, 1000 m, 25.vii.1981, leg. G. Baldizzone (ZMUC).
8. *G. nordlandicolella* (Strand) –
   a) ♂, Finland, PPN, Tornio, 25.vi.1996, leg. M. Mutanen (ZMUC).
   b) ♀, Finland, PPN, Tornio, 25.vi.1996, leg. M. Mutanen (ZMUC).
   c) ♂, Finland, PK, Liperi, Ylämylly, 24.vi.2006, leg. J.-P. Kaitila, gen. slide Karsholt 5141 (ZMUC).
   d) ♂, Finland, PK, Liperi, Ylämylly, 24.vi.2006, leg. J.-P. Kaitila (ZMUC).
   e) ♂, Turkey, Kayseri, 5 km NW Erciyos Dag, 2000 m, 4.vii.1987, leg. M. Fibiger (ZMUC).
9. *G. nilsi* Huemer –
   ♂, paratype, Austria, Osttirol, Venedigergruppe, Virgental, Sajathütte W, 2500 m, 4.viii.1993, leg. N. Ryrholm (ZMUC).

1a
1b
2a
2b
2c
2d
2e
3a
3b
4a
4b
4c
5
6a
6b
7a
7b
7c
8a
8b
8c
8d
8e
9

Plate 2 All specimens 3 x natural size

10. *Gnorimoschema valesiella* (Staudinger) –
    a) ♂, Russia, Pskov region, Sebezh, Bondarevo, ex l. 15.-22.v.2002, leg. N. Savenkov (ZMUC).
    b) ♂, Iceland (east), Hallormssaður, 12.vii.1950, leg. N. L. Wolff (ZMUC).
11. *G. streliciella* (Herrich-Schäffer) –
    a) ♂, Finland, V, Kiikala, 18.vi.1994, leg. J. Junnilainen (ZMUC).
    b) ♀, Finland, V, Kiikala, 17.vi.2000, leg. T. Mutanen (ZMUC).
12. *G. hoefneri* (Rebel) –
    a) ♂, Austria, Osttirol, Lavant, Weg zu Lavanter Alm, 1200 m, 2.viii.1991, leg. H. Deutsch, P. Huemer & O. Karsholt (ZMUC).
    b) ♂, Austria, Lienzer Dolomiten, Hochstadel, 2000 m, vii.1950, leg. R. Pinker, abdomen missing (TLMF).
13. *Scrobipalpopsis petasitis* (Pfaffenzeller) –
    a) ♀, Austria, Nordtirol, Halltal, 1400 m, 31.v.1958, leg. K. Burmann (TLMF).
    b) ♂, Austria, Nordtirol, Halltal, 1400 m, 31.v.1958, leg. K. Burmann, gen. prep. Tokár 2788 (in tube) (TLMF).
    c) ♂, Romania, Carpathian Mts. Mt. Harghita, Tinovul Lucs1100 m, 11.vii.1988, leg. S. & Z. Kovács (ZMUC).
    d) ♂, Finland, Ks, Kuusamo, 9.vii.1991, leg. K. Saloranta, gen. slide Karsholt 4881 (ZMUC).
14. *Scrobipalpa aptatella* (Walker) –
    ♂, United Arabian Emirates, 10 km NNE Huwaylat, 450 m, 17.iv.2006, leg. C. Gielis, gen. slide Hendriksen 6019 (RMNH).
15. *S. kasyi* Povolný –
    ♂, Greece, Larissa, Olympos, 700-2100 m, 21.-26.v.1990, leg. O. Karsholt (ZMUC).
16. *S. notata* (Povolný) –
    a) ♀, Russia, Bashkira, Sakmara river, Jantyshevo village, 21.vi.1996, leg. K. Nupponen, J.-P. Kaitila, J. Junnilainen & M. Ahola, gen. slide Huemer GU 03/1196 (NUPP).
    b) ♂, Russia, South Ural, Orenburg district, 6 km W Donskoje village, mount Verbljushka, 14.vii.1998, leg. K. Nupponen (NUPP).
17. *S. acuminatella* (Sircom) –
    a) ♂, Switzerland, Vaud, Col de la Croix, 1770 m, 5.vii.1989, leg. B. Å. Bengtsson, gen. slide Karsholt 4617 (ZMUC).
    b) ♂, Finland, EH, Hartola, la. 10.viii.2004, leg. J. Tabel, gen. slide Hendriksen 5376 (ZMUC).
    c) ♂, Denmark, SZ, Skibinge, 6.vi.1978, leg. O. Karsholt (ZMUC).
    d) ♀, Italy, Piemonte (CN), Parco Natur. Reg. Alpi Marittime, Palanfré, Lago di Alberghi, 1600-1800 m, 12.vi.1999, leg. G. Baldizzone (ZMUC).
    e) ♀, Germany, Württemberg, Marbach, Neckar, e.l. 23.vii.1970, leg. L. Süssner (TLMF).
18. *S. skulei* sp. n. –
    ♀, paratype, Slovenia (SW), Stavnik Mts., 11 km above Kozina, 950 m, 30.vi.2003, leg. C. Hviid & B. Skule, gen. slide Hendriksen 4764 (ZMUC).
19. *S. hungariae* (Staudinger) –
    ♂, Hungary, Kaposvár, 14.vii.1952, leg. S. Pazsiczky (ZMUC).
20. *S. adaptata* (Povolný) –
    a) ♂, holotype, Ukraine, Khomutovskaya Steppe nature preserve, 8.v.1996, leg. A. Bidzilya, gen. slide Huemer GU 02/1178 (ZMKU).
    b) ♂, Slovakia, Devinska Kobyla, 2.v.1993, leg. J. Liška, gen. slide GU 02/1166 (ELSN).
    c) ♂, Russia, South Ural, Orenburg district, 20 km S Pokrovka village, Schibendy valley, 11.v.2005, leg. K. Nupponen, gen. slide Hendriksen 5468 (NUPP).

10a
10b
11a
11b
12a
12b
13a
13c
13b
13d
14
15
16a
16b
17a
17b
17c
17d
17e
18
19
20a
20b
20c

Plate 3 All specimens 3 x natural size

21. *Scrobipalpa brahmiella* (Heyden) –
    a) ♂, Romania, Transsylvania, Viisoara, Cămpia Turzii, leg. S. & Z. Kovács, gen. slide Karsholt 5147 (ZMUC).
    b) ♂, Russia, Volgograd district, 20 km SW Frolovo village, 16.v.2005, leg. K. Nupponen, gen. slide Hendriksen 6415 (NUPP).
22. *S. vasconiella* (Rössler) –
    a) ♂, Spain, Malaga, 12 km W Estepona, Casares, 350 m, 26.-27.vii.2003, leg. P. Skou, gen. slide Hendriksen 4674 (ZMUC).
    b) ♀, Portugal, Madeira Isl., Porto Santo, 14.iv.1996, leg. O. Karsholt (ZMUC).
    c) ♂, Spain, Teruel, Albarracin, Val de Vecar, 1150 m, 1.viii.2003, leg. P. Skou, gen. slide Karsholt 5146 (ZMUC).
23. *S. dorsolutea* sp. n. –
    ♀, paratype, Russia, South Ural, Orenburg district, 6 km W Donskoje village, mount Verbljushka, 13.v.1999, leg. K. Nupponen, gen. slide Hendriksen 6428 (NUPP).
24. *S. amseli* Povolný –
    a) ♂, France, Hautes-Alpes, Reotier, 1000 m, 8.vi.1995, leg. A. Cox, gen. slide Hendriksen 3255 (ZMUC).
    b) ♂, holotype, France, Hautes-Alpes, Durance Valley, St. Crépin, 900 m, 24.-30.vi.1959, gen. slide Povolný Am. 1736 (LNK).
25. *S. hyssopi* Nel –
    ♂, holotype, France, Hautes-Alpes, Villars de Saint Crépin, e.l. 17.vii.2002, leg. J. Nel, gen. slide Nel 14755 (TLMF).
26. *S. montanella* (Chrétien) –
    ♂, Italy, Sicily, Ragalna, Monte Arso, Bosco di Paternò,1000 m, 9.ix.2002, leg. M. Fibiger, G. Jeppesen & O. Karsholt (ZMUC).
27. *S. corleyi* sp. n. –
    a) ♂, holotype, Portugal, Beira Baxa, Serra da Estréla, Torre, 1600-1950 m, 10.-11.vii.1986, leg. O. Karsholt, gen. slide Huemer GU 01/1093 (ZMUC).
    b) ♀, paratype, Portugal, Albergaria, Serra da Estréla, la. ix.2004, leg. M. F. V. Corley, abdomen missing (TLMF).
    c) ♂, paratype, Portugal, Albergaria, Serra da Estréla, la. ix.2004, leg. M. F. V. Corley, gen. slide Huemer GEL 1108 (TLMF).
28. *S. chrysanthemella* (Hofmann) –
    a) ♂, Germany, Bayern, Schwäbische Alb, Urach, Nägelesfälsen, 700 m, la. 21.x.1991, leg. A. Scholz (ZMUC).
    b) ♀, Italy, Piemonte (AL), Parco N. R. Capanne di Marcarolo, Loc. Magnoni, 600 m, 27.v.2003, leg. G. Baldizzone, gen. slide Hendriksen 4349 (ZMUC).
29. *S. proclivella* (Fuchs) –
    a) ♂, Germany, Bayern, Schwäbische Alb, Urach, Giengen, Eselsburger Tal, 550 m, la. 10.x.1993, leg. A. Scholz (ZMUC).
    b) ♀, Denmark, B, Boderne, la. 8.vii.1978, leg. O. Karsholt (ZMUC).
    c) ♀, Germany, Bayern, Schwäbische Alb, Urach, Giengen, Eselsburger Tal, 550 m, la. 10.x.1993, leg. A. Scholz (ZMUC).
30. *S. frugifera* Povolný –
    ♀, Russia, South Siberia, Krasnoyarsk kraj, Jenisej, Minussinsk, Tagarskij Isl., 18.vi.1927, leg. S. Tzygankov, gen. slide. Bidzilya 6 (ZMKU).
31. *S. oleksiyella* sp. n. –
    a) ♀, paratype, Russia, South Ural, Orenburg district, 20 km S Pokrovka village, Schibendy valley, 7.vi.1998, leg. K. & T. Nupponen (NUPP).
    b) ♀, paratype, Russia South Ural, Orenburg district, 20 km S Pokrovka village, Schibendy valley, 2.viii.2005, leg. K. Nupponen (ZMUC).
    c) ♂, paratype, Russia, South Ural, Orenburg district, 20 km S Pokrovka village, Schibendy valley, 29.vi.2003, leg. K. & T. Nupponen (NUPP).
32. *S. smithi* Povolný & Bradley –
    a) ♂, Spain, Zaragoza, Tosos, 600 m, 16.vi.1999, leg. P. Skou, gen. slide Hendriksen 3026 (ZMUC).
    b) ♀, Spain, Almeria, Sierra Alhamilla, above Turrillas, Colativi, 1000-1350 m, 3.ix.2001, leg. C. Hviid & B. Skule, gen. slide Hendriksen 3529 (ZMUC).
33. *S. occulta* (Povolný) –
    a) ♂, Russia, South Ural, Cheliabinsk district, Arkaim reserve, near Amurskii, 18.vi.1996, leg. K. Nupponen, J.-P. Kaitila, J. Junnilainen & M. Ahola, gen. slide 2933 Hendriksen (NUPP).
    b) ♀, Russia, Altai, Kosh-Agachskij district, Dzhazator village, 1500 m, 28.vii.2001, leg. A. Bidzilya, gen. slide Bidzilya 144/07 (ZMKU).

21a
21b
22a
22b
22c
23
24a
24b
25
26
27a
27b
27c
28a
28b
29a
29b
29c
30
31a
31b
31c
32a
32b
33a
33b

Plate 4 All specimens 3 x natural size

34. *Scrobipalpa grisea* Povolný –
   a) ♂, Russia, South Ural, Cheliabinsk district, Ajat river, near Nikolaevska village, Sokolovskaja gora, 5.vii.1997, leg. K. Nupponen & J. Junnilainen, gen. slide 3986 (NUPP).
   b) ♂, Russia, South Ural, Orenburg district, 6 km W Donskoje village, mount Verbljushka, 28.vi.2003, leg. K. Nupponen (NUPP).
   c) ♀, Russia, South Ural, Cheliabinsk district, Arkaim reserve near Amurskii village, 14.vi.1996, leg. K. Nupponen, J.-P. Kaitila, J. Junnilainen & M. Ahola (NUPP).
   d) ♂, Russia South Ural, Orenburg district, 20 km S Pokrovka village, Schibendy valley, 3.vi.1998, leg. K. & T. Nupponen (NUPP).
35. *S. usingeri* Povolný –
   ♂, holotype, Mongolia, Sout Gobi aimak, 100 km W Ovot Chuural, 1250 m, 22.vi.1967, leg. Z. Kaszab, gen. slide Povolný Gz.3593 (HNHM).
36. *S. clintoni* Povolný –
   a) ♀, Denmark, NWZ, Gudmindrup Lyng, pup. 17.viii.1970, leg. O. Karsholt, gen. slide Karsholt 688 (ZMUC).
   b) ♂, Denmark, EJ, Ebeltoft, Boeslum, pup. 11.iv.1971, leg. G. Pallesen (ZMUC).
37. *S. reiprichi* Povolný –
   a) ♀, Norway, ON, Vinstra, 11.vi.1985, leg. O. Karsholt & V. Michelsen (ZMUC).
   b) ♂, Norway, ON, Vinstra, 11.vi.1985, leg. O. Karsholt & V. Michelsen, gen. slide Huemer GU 01/1021 (ZMUC).
   c) ♂, Greece, Larissa, Olympos S, 30 km NW Karia, 1700 m, 10.vii.1990, leg. M. Fibiger, gen. slide Hendriksen 1011 (ZMUC).
38. *S. obsoletella* (Fischer von Röslerstamm) –
   a) ♂, France, Corse, Cargèse, 0-30 m, 15.-16.v.1999, leg. O. Karsholt, gen. slide Hendriksen 2562 (ZMUC).
   b) ♀, Denmark, LFM, Mellemskoven, 30.vi.1992, leg. O. Karsholt, gen. slide Hendriksen 916 (ZMUC).
   c) ♀, Denmark, NEZ, Kongelunden, pup. 11.iii.1969, leg. O. Karsholt, gen. slide Karsholt 624 (ZMUC).
   d) ♂, Deamark, LFM, Mandemarke, 11.-16.viii.1996, leg. O. Karsholt (ZMUC).
39. *S. feralella* (Zeller) –
   a) ♂, Austria, Oberösterreich, Nat. Park Kalkalpen, Gr. Bach, Wasserboden area, 460 m, 10.viii.2002, leg. J. Wimmer (TLMF).
   b) ♀, Austria, Oberösterreich, Nat. Park Kalkalpen, Gr. Bach, Wasserboden area, 460 m, 5.viii.2002, leg. J. Wimmer (ZMUC).
40. *S. halonella* (Herrich-Schäffer) –
   ♂, Italy, Friuli Venezia Giulia, Pordenone, Magredi di Cordenons, 100 m, 29.iv.2001, leg. L. Morin, gen. slide Huemer GU 02/1137 (TLMF).
41. *S. perinii* (Klimesch) –
   ♀, Italy, Trentino, Pietramurata, 250 m, e.l. 6.viii.1963, leg. K. Burmann (ZMUC).
42. *S. phagnalella* (Constant) –
   a) ♀, Morocco, 25 km S Essaouria, Sidi Kaouki, 100 m, 28.iii.2005, leg. O. Karsholt, gen. slide Hendriksen 5068 (ZMUC).
   b) ♂, Greece, Crete, 10 km W ag. Varvara, 500 m, 7.xi. 1991, leg. M. Fibiger (ZMUC).
43. *S. tokari* sp. n. –
   ♂, paratype, Croatia, Zaostrog, 6.ix.2002, leg. Z. Tokár, gen. prep. Tokár 7061 (TOKA).
44. *S. karadaghi* (Povolný) –
   ♂, holotype, Ukraine, Crimea, Karadagh, 2.viii.1990, leg. Yu. Budashkin, gen. slide Huemer GU02/1179 (ZMKU).
45. *S. heimi* sp. n. –
   a) ♂, paratype, Russia, South Ural, Cheliabinsk district, near Moskovo village, 26.v.1998, leg. K. & T. Nupponen (NUPP).
   b) ♂, holotype, Russia, South Ural, Cheliabinsk district, near Moskovo village, 26.v.1998, leg. K. & T. Nupponen, gen. slide Hendriksen 3177 (NUPP).
46. *S. acuta* (Povolný) –
   ♂, holotype, Ukraine, Kamennye Mogily nature preserve, 24.viii.1995, leg. A. Bidzilya, gen. slide Huemer GU 01/1175 (ZMKU).

34a
34b
34c
34d
35
36a
36b
37a
37b
37c
38a
38b
38c
38d
39a
39b
40
41
42a
42b
43
44
45a
45b
46

Plate 5 All specimens 3 x natural size

47. *Scrobipalpa soffneri* Povolný –
    a) ♂, Russia, South Ural, Orenburg district, near Burannoe village, 12.v.2005, leg. Nupponen, gen. slide Huemer 08/1266 (NUPP).
    b) ♂, Russia, Volgograd district, 20 km SW Frolovo village, 17.v.2005, leg. K. Nupponen, gen. slide Nupponen 1/11.x.2005 (NUPP).
48. *S. jariorum* sp. n. –
    a) ♂, paratype, Bulgaria (NE), Black Sea coast, Tuzlata, 26.v.2002, leg. J.-P. Kaitila (KAIT).
    b) ♂, paratype, Bulgaria (NE), Black Sea coast, Tuzlata, 26.v.2002, leg. J.-P. Kaitila (ZMUC).
49. *S. murinella* (Duponchel) –
    a) ♂, Austria, Nordtirol, Franz-Sennhütte, 2700 m, 17.vi.1949, leg. K. Burmann (ZMUC).
    b) ♀, Sweden, Öl, Möcklemossen, 30.v.1981, leg. O. Karsholt (ZMUC).
    c) ♂, Austria, Kärnten, Grossglockner, Gamsgrube, 2600 m, e.l. 30.viii.1943, leg. J. Klimesch (ZSM).
    d) ♀, Austria, Nordtirol, Ötztal, Rofenberg, 2600 m, la. 10.viii.1942, leg. J. Klimesch (ZMUC).
50. *S. wiltshirei* Povolný –
    a) ♂, Spain, Alicante, La Marina, 12.ix.1998, leg. J. B. Wolschrijn, gen. slide Hendriksen 2778 (ZMUC).
    b) ♂, Spain, Murcia, 4 km SW Aguilas, 12.-14.ix.1995, leg. H. van der Wolf (ZMUC).
51. *S. caucasica* (Povolný) –
    a) ♂, holotype, Russia, Caucasus, Bezengi, 3.vi.1997, leg. I. Kostjuk, gen. slide Povolný Bdz.18 (ZMKU).
    b) ♂, Russia, Caucasus, Bezengi, 3.vi.1997, leg. I. Kostjuk, gen. slide Povolný Bdz.26 (ZMKU), (holotype of *S. bezengensis* (Povolný)).
52. *S. pauperella* (Heinemann) –
    a) ♀, Austria, Burgenland, Hainburger Berge, Hundsheimer Kogel, 16.v.1964, leg. W. Glaser, gen. slide Hendriksen 1319 (ZMUC).
    b) ♀, Finland, EH, Sysmå, la. 25.vii.2004, leg. J. Tabel, gen. slide Hendriksen 5381 (ZMUC).
    c) ♂, Finland, EH, Sysmå, la. 25.vii.2004, leg. J. Tabel, gen. slide Hendriksen 5377 (ZMUC).
    d) ♂, Estonia (E), Alatskivi, 14.-25.v.2000, leg. J. Viidalep, gen. slide Hendriksen 5417 (ZMUC).
    e) ♂, Slovenia, below Nanos Mts., near Vipava, 150 m, 1.vi.2003, leg. C. Hviid & B. Skule, gen. slide Hendriksen 4811 (ZMUC).
53. *S. spumata* (Povolný) –
    a) ♂, Russia, South Ural, Orenburg district, 6 km W Donskoje village, mount Verbljushka, 11.vi.1998, leg. K. & T. Nupponen (NUPP).
    b) ♂, Russia, South Ural, Orenburg district, 6 km W Donskoje village, mount Verbljushka, 17.vi.1999, leg. K. & T. Nupponen (NUPP).
54. *S. arenbergeri* Povolný –
    ♂, Croatia, Istria, Rovinj area, Koukuletovica, 20 m, 20.vii.2000, leg. T. Mayr, gen. prep. without number (in tube) (TLMF).
55. *S. nana* Povolný –
    a) ♂, Russia, South Ural, Orenburg district, 20 km S Pokrovka village, Schibendy valley, 17.vii.1998, leg. K. Nupponen, gen. slide Karsholt 5155 (ZMUC).
    b) ♂, Russia South Ural, Orenburg district, 20 km S Pokrovka village, Schibendy valley, 26.v.2004, leg. K. Nupponen, gen. slide Huemer GU09/1309 (NUPP).
56. *S. heretica* Povolný –
    a) ♂, Spain, Teruel, Albarracin, 11.-24.vii.1987, leg. J. B. Wolschrijn, gen. slide Hendriksen 4066 (ZMUC).
    b) ♀, Kazakhstan, Dehambulskaya district, 70 km NNE Frunze, 950 m, 18.vii.1990, leg. L. Kaila (ZMUH).
57. *S. bigoti* Povolný –
    a) ♀, Spain, Murcia, 4 km SW Aguilas, 5.-6.ix.1997, leg. H. van der Wolf (ZMUC).
    b) ♀, France, Herault, Frontignan, 17.-20.viii.2001, leg. A. Cox, gen. slide Hendriksen 5028 (ZMUC).
    c) ♂, Italy, Sicily, 5 km W Pachino, Pantano Longarini, sea lev., 12.ix.2002, leg. O. Karsholt *et al.* (ZMUC).
58. *S. dorsoflava* (Povolný) –
    a) ♂, Russia, South Ural, Cheliabinsk district, near Moskovo village, 2.vi.2004, leg. K. Nupponen, gen. slide Hendriksen 5472 (NUPP).
    b) ♂, Russia, South Ural, Cheliabinsk district, near Moskovo village, 26.v.1998, leg. K. & T. Nupponen, gen. slide Hendriksen 3161 (ZMUC).
59. *S. magnificella* Povolný –
    a) ♂, Russia, South Ural, Orenburg district, 40 km W Orsk, near Guberlja village, 25.vi.2003, leg. K. Nupponen, gen. slide Hendriksen 5019 (NUPP).
    b) ♂, Russia South Ural, Orenburg district, 20 km S Pokrovka village, Schibendy valley, 29.vi.2003, leg. K. Nupponen, gen. slide Hendriksen 5020 (NUPP).
    c) ♂, Russia, South Ural, Orenburg district, 6 km W Donskoje village, mount Verbljushka, 13.v.1999, leg. K. Nupponen, gen. slide Hendriksen 5018 (NUPP).

47a
47b
48a
48b
49a
49b
49c
49d
50a
50b
51a
51b
52a
52b
52c
52d
52e
53a
53b
54
55a
55b
56a
56b
57a
57b
57c
58a
58b
59a
59b
59c

Plate 6 All specimens 3 x natural size

60. *Scrobipalpa abstrusa* sp. n. –
   a) ♂, paratype, Russia South Ural, Orenburg district, 20 km S Pokrovka village, Schibendy valley, 11.v.2005, leg. K. Nupponen (NUPP).
   b) ♂, paratype, Russia South Ural, Orenburg district, 20 km S Pokrovka village, Schibendy valley, 11.v.2005, leg. K. Nupponen (NUPP).
61. *S. superstes* Povolný –
   a) ♂, Spain, Alicante, La Marina, 7.ix.1998, leg. J. B. Wolschrijn (ZMUC).
   b) ♀, Spain, Alicante, La Marina, 10.ix.1998, leg. J. B. Wolschrijn, gen. slide Hendriksen 2775 (ZMUC).
   c) ♀, Spain, Almeria, Cabo de Gata, 19.iv.2000, leg. H. van der Wolf (ZMUC).
62. *S. plesiopicta* Povolný –
   ♂, Russia, South Ural, Cheliabinsk district, Ajat river, near Nikolaevska village, Sokolovskaja gora, 25.vii.1998, leg. K. Nupponen (NUPP).
63. *S. bradleyi* Povolný –
   a) ♂, Spain, Huelva, Las Marismas, W of Belavista, 16.viii.1988, leg. E. Traugott-Olsen.
   b) ♀, Italy, Sicily, 5 km W Pachino, Pantano Longarini, sea lev., 12.ix.2002, leg. O. Karsholt *et al.* (ZMUC).
64. *S. selectella* (Caradja) –
   a) ♂, Greece, Thassos, 10.vii.2003, leg. N. J. Aachmann-Andersen, gen. slide Hendriksen 4819 (ZMUC).
   b) ♀, Kazakhstan (SE), valley of river Charyn, 15 km W Chundzhi, Yasenevaya Rostscha, 7.vii.1994, leg. I. Kostjuk, gen. slide Hendriksen 2061 (ZMUC).
   c) ♀, Russia, South Ural, Orenburg district, 6 km W Donskoje village, mount Verbljushka, 18.vi.1999, leg. K. & T. Nupponen, gen. slide Hendriksen 3157 (NUPP).
65. *S. alterna* (Falkovitsh & Bidzilya) –
   a) ♀, Russia, South Ural, Orenburg district, 6 km W Donskoje village, mount Verbljushka, 13.v.1999, leg. K. Nupponen, gen. slide Hendriksen 3175 (NUPP).
   b) ♂, Russia, South Ural, Orenburg district, 6 km W Donskoje village, mount Verbljushka, 13.v.1999, leg. K. Nupponen, abdomen missing (NUPP).
66. *S. lutea* Povolný –
   a) ♂, Russia, South Ural, Orenburg district, 20 km S Pokrovka village, Schibendy valley, 23.vi.1999, leg. K. & T. Nupponen (NUPP).
   b) ♂, Russia, South Ural, Orenburg district, 20 km S Pokrovka village, Schibendy valley, 2.vi.2003, leg. K. Nupponen (NUPP).
   c) ♂, Russia, South Ural, Orenburg district, 20 km S Pokrovka village, Schibendy valley, 24.vi.1999, leg. K. & T. Nupponen (NUPP).
67. *S. niveifacies* Povolný –
   a) ♀, Spain, Huesca, 8 km S Candasnos, Barranco de Valcuerna, 175 m, 7.ix.2001, leg. C. Hviid & B. Skule (ZMUC).
   b) ♂, Spain, Almeria, 5 km SW Tabernas, Rambla de Tabernas, 18.-25.iv.2000, leg. C. Hviid & B. Skule, gen. slide Hendriksen 3247 (ZMUC).
   c) ♀, Portugal, Algarve, Praia da Rocha, 9.x.1996, leg. H. Hendriksen (ZMUC).
68. *S. indignella* (Staudinger) –
   a) ♂, Russia South Ural, Orenburg district, 20 km S Pokrovka village, Schibendy valley, 30.vi.2003, leg. K. Nupponen (NUPP).
   b) ♀, Turkey, Nevsehir, Göreme by Ürgüp, Love Valley, 1300 m, 2.vii.1987, leg. M. Fibiger (ZMUC).
   c) ♂, Russia South Ural, Orenburg district, 20 km S Pokrovka village, Schibendy valley, 22.vi.1999, leg. K. & T. Nupponen, gen. slide Hendriksen 3182 (NUPP).
69. *S. punctata* (Povolný) –
   a) ♂, Russia, South Ural, Orenburg district, 20 km S Pokrovka village, Schibendy valley, 19.vii.2003, leg. K. Nupponen, gen. slide Hendriksen 012 (NUPP).
   b) ♂, Russia, South Ural, Orenburg district, 6 km W Donskoje village, mount Verbljushka, 19.vi.1999, leg. K. & T. Nupponen (NUPP).
70. *S. lagodes* (Meyrick) –
   ♂, Kazakhstan, Kokshetau hills, near river Terisakkan, Kok-pek, la. 13.vii.1958, leg. Falkovich, gen. slide Huemer GU 09/1317 (ZMKU).

60a
60b
61a
61b
61c
62
63a
63b
64a
64b
64c
65a
65b
66a
66b
66c
67a
67b
68a
68b
68c
69a
69b
70

Plate 7 All specimens 3 x natural size

71. *Scrobipalpa deluccae* Povolný –
 ♂, Libanon, Bescherre, 1900 m, 13.-14.v.1961, leg. F. Kasy & E. Vartian, gen. slide Mus. Vind. 16525 (NHMW).
72. *S. atriplicella* (Fischer von Röslerstamm) –
 a) ♂, Denmark, LFM, Møns Klint S, la. 16.ix.2007, leg. O. Karsholt (ZMUC).
 b) ♀, Denmark, NEZ, Tåstrup, la. Viii.2006, leg. J. Martin & L. Sigsgaard (ZMUC).
 c) ♀, Germany, Sachsen-Anhalt, Aschersleben, e.l. 15.v.1971, leg. H. Patzak, gen. slide Hendriksen 6290 (ZMUC).
 d) ♀, Denmark, LFM, Møns Klint S, la. 16.ix.2007, leg. O. Karsholt (ZMUC).
73. *S. suaedella* (Richardson) –
 a) ♀, Gr. Britain, Essex, Bradwell-on-Sea, 9.vii.1988, leg. O. Karsholt, gen. slide Karsholt 5156 (ZMUC).
 b) ♀, Gr. Britain, Essex, Bradwell-on-Sea, 9.vii.1988, leg. O. Karsholt, gen. slide Hendriksen 2002 (ZMUC).
 c) ♀, Spain, Almeria, Capo de Gata 16.-17.iv.2001, leg. P. Skou & B. Skule (ZMUC).
 d) ♂, Spain, Murcia, Bolnuevo by Mazarrón, 10 m, 15.iv.2001, leg. P. Skou, gen. slide Karsholt 5157 (ZMUC).
74. *S. solitaria* Povolný –
 a) ♂, Russia, South Ural, Orenburg district, 6 km W Donskoje village, mount Verbljushka, 18.vi.1999, leg. K. & T. Nupponen (NUPP).
 b) ♂,Ukraine, Lugansk District, Melovoj district, Streltsovskaya Steppe nature preserve, 5.vii.2002, leg. A. Bidzilya, gen. slide Bidzilya 254/08 (ZMKU).
75. *S. dagmaris* Povolný –
 a) ♂, holotype, Italy [?], Brescia, Lago di Garda, Gargano, 4.viii.1972, leg. E. Heiss, gen. slide Povolný Ht.5793 (MZMB).
 b) ♀, Uzbekistan, Kyzylkum, 40 km E Dzhingildy, Ayakguzhumdy, 1.vi.1965, leg. Pastukhov, gen. slide Bidzilya 44/05 (ZMKU).
76. *S. suasella* (Constant) –
 a) ♂, France, Vaucluse, Pertuis, 7.vii.1992, leg. A. Lăstůvka (ZMUC).
 b) ♀, Spain, Huesca, Puerto del Sarrablo, 1200 m, 9.viii.2003, leg. J. Viehmann (ZMUC).
 c) ♂, Spain, Zaragoza, Caspe, 22.vi.1992, , leg. A. Lăstůvka (ZMUC).
77. *S. hendrikseni* sp. n. –
 ♀, holotype, Greece, Ioánina, 4 km W Agia Paraskevi, 750 m, 26.viii.2008, leg. P. Skou, gen. slide Huemer GU 09/1286 (ZMUC).
78. *S. traganella* (Chrétien) –
 a) ♀, Malta (N), Mellieha Bay, 25 m, 5.-11.iv.2004, leg. B. Skule, gen. slide Hendriksen 4973 (ZMUC).
 b) ♀, Spain, Almeria, 6 km S Tabernas, Mini Hollywood, 400 m, 12.iv.1997, leg. H. van der Wolf (ZMUC).
 c) ♂, Morocco, 8 km S Sidi Ifni, 20.iii.2005, leg. O. Karsholt, gen. slide Hendriksen 5065 (ZMUC).
79. *S. bazae* Povolný –
 a) ♂, Spain, Almeria, 6 km S Roquetas de Mar, 10 m, 17.x.1992, leg. M. Fibiger (ZMUC).
 b) ♂, Spain, Almeria, Mini Hollywood, 230 m, 14.-15.x.1992, leg. M. Fibiger (ZMUC).
80. *S. artemisiella* (Treitschke) –
 a) ♂, Denmark, WJ, Skallingen, 28.vi.1982, leg. O. Karsholt (ZMUC).
 b) ♂, Denmark, SJ, Stensbæk Plantage, 25.vii.1978, leg. O. Karsholt (ZMUC).
 c) ♂, Italy, Verona, Monte Baldo, Bocca di Navene, 1400 m, 18.vi.1976, leg. G. Baldizzone, gen. slide Karsholt 4813 (ZMUC).
 d) ♂, Italy, Piemonte (CN), Parco Natur. Reg. Alpi Marittime, S. Giacomo di Entraque, Gias d. Siula, Rifugio Soria, 1700-1800 m, 22.vii.1997, leg. G. Baldizzone, gen. slide Hendriksen 2031 (ZMUC).
 e) ♂, Italy, Piemonte (CN), Parco Natur. Reg. Alpi Marittime, S. Giacomo di Entraque, Gias d. Siula, Rifugio Soria, 1600 m, 25.vii.1996, leg. G. Baldizzone, gen. slide Hendriksen 1865 (ZMUC).
 ssp. *oreocyrniella* (Petry) –
 f) ♀, Italy, Sardinia (C), Gennargentu, Arcu Tascussi, 1000 m, 28.vii.1981, leg. G. Baldizzone (ZMUC).

71
72a
72b
72c
72d
73a
73b
73c
73d
74a
74b
75a
75b
76a
76b
76c
77
78a
78b
78c
79a
79b
80a
80b
80c
80d
80e
80f

Plate 8 All specimens 3 x natural size

81 *Scrobipalpa stangei* (Hering) –
- a) ♀, Denmark, NEZ, Amager, 31.vii.1924, leg. F. Gudmann, abdomen missing (ZMUC).
- b) ♂, Denmark, NEJ, Læsø, Bovet, 13.vii.1982, leg. O. Karsholt (ZMUC).
- c) ♀, Poland, KP, Inowrocław, Mątwy, 14.vii.1995, leg. T. Rynarzewski, gen. slide Hendriksen 1832 (ZMUC).
- d) ♂, Denmark, SZ, Skibinge, 23.vi.-5.vii.1985, leg. O. Karsholt, gen. slide Hendriksen 914 (ZMUC).

82. *S. suaedivorella* (Chrétien) –
- ♀, Spain, Zaragoza, Rio del Huerva, 3 km E Cerveruela, 800 m, 7.vii.2002, leg. B. Skule, gen. slide Hendriksen 4011 (ZMUC).

83. *S. bryophiloides* Povolný –
- a) ♂, Russia, South Ural, Orenburg district, 40 km W Orsk,near Guberlja village, 26.vi.2003, leg. K. Nupponen, gen. slide Hendriksen 5014 (NUPP).
- b) ♀, Turkey, Ankara, Tuz Gölü N, 900 m, 5.viii.1989, leg. M. Fibiger & N. Esser, gen. slide Hendriksen 4389 (ZMUC).
- c) ♀, Turkey, Konya, 19 km SE Eregli, Bolkar Dagi, Halkapinar, 1300 m, 28.vii.1985, leg. B. Van Oorschot & W. De Prins, gen. slide Hendriksen 3090 (ZMUC).

84. *S. algeriensis* Povolný & Bradley –
- a) ♂, Morocco, Tinerhir area, 1400-2000 m, 12.iv.1989, leg. O. Karsholt, gen. slide Hendriksen 2906 (ZMUC).
- b) Spain, Granada, Baza, 650 m, 28.-29.ix.1992, leg. F. Schepler, abdomen missing (ZMUC).

85. *S. deutschi* sp. n. –
- a) ♂, paratype, Russia, South Ural, Orenburg district, 20 km S Pokrovka village, Schibendy valley, 11.v.2005, leg. K. Nupponen (ZMUC).
- b) ♂, paratype, Algeria, Lambesa, gen. slide Huemer GU 03/1187 (ZMUC).

86. *S. disjectella* (Staudinger) –
- a) ♂, lectotype, Spain, Granada, leg. O. Staudinger (?), abdomen missing (ZMHU).
- b) ♀, Spain, Huesca, 3 km E Penalba, 15.iv.2003, leg. J. Viehmann, gen. slide Hendriksen 6346 (ZMUC).
- c) ♂, Spain, Murcia, Sierra de Espunas, C. F. De las Alguerias, 800 m, 20.iii.1994, leg. & coll. H. van der Wolf, gen. slide van der Wolf 6121.

87. *S. mixta* sp. n. –
- a) ♂, paratype, Macedonia, Ohrid, iv.1939, leg. R. Pinker, gen. slide Klimesch 601 (ZSM).
- b) ♀, paratype, Macedonia, Ohrid, iv.1939, leg. R. Pinker, gen. slide Huemer GU03/1186 (ZSM).
- c) ♂, Ukraine, Crimea, Karadagh, 24.iv.1989, leg. Yu. Budaskhin, gen. slide Huemer GU 08/1270 (ZMKU).

88. *S. rebeli* (Preissecker) –
- a) ♂, Austria, Niederösterreich, Krems, Kuhberg, z. 2.vi.1927, leg. F. Preissecker, gen. slide Hendriksen 3101 (ZMUC).
- b) ♀, Russia, South Ural, Cheliabinsk district, near Moskovo village, 26.v.1998, leg. K. & T. Nupponen, gen. slide Huemer GU 03/1189 (NUPP).
- c) ♂, Italy, Südtirol, Meran, Naturns, e.l. 10.iv.1935, leg. J. Klimesch (ZSM).

89. *S. gallicella* (Constant) –
- a) ♂, Italy, Liguria, Strada Triora-Verdeggia, 1000 m, 9.vii.1985, leg. U. Parenti, gen. slide Karsholt 4805 (ZMUC).
- b) ♀, France, Alpes Maritimes, S of Tende, Saorge, 482 m, 3.-4.vii.2008, leg. O. Karsholt (ZMUC).

90. *S. ustulatella* (Staudinger) –
- a) ♂, paralectotype, Russia, Volgograd district, Sarepta, leg. H. Christoph, gen. slide Povolný St. 739 (ZMHU).
- b) ♂, Russia, South Ural, Orenburg district, 20 km S Pokrovka village, Schibendy valley, 25.v.2004, leg. K. Nupponen, gen. slide Hendriksen 6385 (ZMUC).

81a
81b
81c
81d
82
83a
83b
83c
84a
84b
85a
85b
86a
86b
86c
87a
87b
87c
88a
88b
88c
89a
89b
90a
90b

Plate 9 All specimens 3 x natural size

91. *Scrobipalpa postulatella* sp. n. –
    a) ♀, paratype, Spain, Almeria, Penon de Turillas, 1000 m, 24.iv.2001, leg. C. Hviid, P. Skou & B. Skule (ZMUC).
    b) ♂, paratype, Spain, Almeria, Sierra Cabrera, 10 km S Mojacar, 100 m, 29.iii.1993, leg. Lingehöle (ZMUC).
    c) ♀, Spain, Almeria, Penon de Turrillas, 1000 m, 24.iv.2001, leg. C. Hviid, P. Skou & B. Skule, gen. slide Hendriksen 6280 (ZMUC).
92. *S. nitentella* (Fuchs) –
    a) ♂, Denmark, SZ, Stavreby, la. 4.x.1978, leg. O. Karsholt (ZMUC).
    b) ♂, Poland, KP, Ciechocinek, e.l. 1.ii.1984, leg. J. Buszko, gen. slide Karsholt 4602 (ZMUC).
    c) ♀, Denmark, LFM, Blands, 16.vi.1964, leg. K. Pedersen (ZMUC).
93. *S. costella* (Humphreys & Westwood) –
    a) ♀, Denmark, WJ, Nr. Lyngvig, la. 4.v.1995, leg. O. Karsholt (ZMUC).
    b) ♂, Denmark, WJ, Nr. Lyngvig, la. 4.v.1995, leg. O. Karsholt (ZMUC).
94. *S. hyoscyamella* (Stainton) –
    a) ♂, Austria, Steiermark, Rein, Mühlbachgraben, 20.ix.1988, leg. H. Habeler (TLMF).
    b) ♀, Spain, Andalusia, Malaga, 10.iii.1001, leg. T. Walsingham (BMNH).
95. *S. portosanctana* (Stainton) –
    a) ♂, Spain, Almeria, 7 km S Tabernas, Mini Hollywood, 650 m, 17.-18.iv.2000, leg. H. van der Wolf (ZMUC).
    b) ♀, Malta (N), Mellieha Bay, 25 m, 5.-11.iv.2004, leg. B. Skule (ZMUC).
    c) ♀, Portugal, Madeira Isl., Porto Santo, 15.iv.1996, leg. O. Karsholt, gen. slide Huemer GU 01/1057 (ZMUC).
96. *S. vicaria* (Meyrick) –
    a) ♂, Spain, Almeria, El Pozo de Esparto 20 m, 22.-26.iv.2001, leg. C. Hviid, P. Skou & B. Skule (ZMUC).
    b) ♀, Morocco, 8 km S Sidi Ifni, 20.iii.2005, leg. O. Karsholt (ZMUC).
    c) ♂, Libya, Gharian, Wadi El Hira, 1.iv.1983, leg. U. Seneca, gen. slide Hendriksen 2907) (ZMUC).
97. *S. ocellatella* (Boyd) –
    a) ♂, Spain, Granada, Baza, 7.x.1975, leg. M. & W. Glaser (ZMUC).
    b) ♂, Spain, Zaragoza, Lake Caspe, 3.x.2004, leg. A. Cox (ZMUC).
    c) ♂, Spain, Sevilla, Mazagón, 1.-8.x.1975, , leg. M. & W. Glaser (ZMUC).
    d) ♂, Macedonia, Treskaschlucht, iv.1956, leg. R. Pinker, gen. slide Huemer GU 00/899 (TLMF).
98. *S. pulchra* Povolný –
    a) ♀, Uzbekistan, Kyzylkum, 40 km E Dzhingildy, Ayakguzhumdy, 27.v.1965, leg. Pastukhov, (ZMUC)
    b) ♂, Turkey, Konya, Taurus, road Bozkir-Hadim, 5 km SE Öcpinar, 1100 m, 13.vii.1986, leg. M. Fibiger (ZMUC).
99. *S. gecko* (Walsingham) –
    ♀, Greece, Lakonia, 5 km S Monemvasia, 18.iv.1981, leg. B. Skule, gen. slide Huemer GU 97/689 (ZMUC).
100. *S. hannemanni* Povolný –
    a) ♂, Russia South Ural, Orenburg district, 20 km S Pokrovka village, Schibendy valley, 5.vi.1998, leg. K. & T. Nupponen, gen. slide Hendriksen 3153 (ZMUC).
    b) ♀, Russia South Ural, Orenburg district, 20 km S Pokrovka village, Schibendy valley, 29.vi.2003, leg. K. Nupponen (NUPP).

91a
91b
91c
92a
92b
92c
93a
93b
94a
94b
95a
95b
95c
96a
96b
96c
97a
97b
97c
97d
98a
98b
99
100a
100b

Plate 10 All specimens 3 x natural size

101. *Scrobipalpa erichi* Povolný –
   a) ♂, Slovakia, Komárno, 10.v.1985, leg. G. Pastorális (ZMUC).
   b) ♂, Slovakia, Komárno, 26.iv.1983, leg. G. Pastorális, gen. prep. Pastoralis 1394 (in tube) (ZMUC).
102. *S. divisella* (Rebel) –
   ♂, Spain, Almeria, Sierra del Almagro, Huercal Overa, 300 m, 29.x.1989, leg. T. Grünewald, gen. slide Huemer GU 01/986 (TLMF).
103. *S. voltinella* (Chrétien) –
   a) ♀, lectotype, France, Ardéche, La Voulte-sur-Rhône,v.1897, leg. P. Chrétien, gen. slide without number (MNHN).
   b) ♂, Spain, Teruel, 1 km E Tramacastilla, 1250 m, 24.viii.2001, leg. P. Skou & B. Skule, gen. slide Karsholt 5158 (ZMUC).
   c) ♂, Huesca, Biescas, Toria, 30.viii.1989, leg. J. B. Wolschrijn, gen. slide Karsholt 5159 (ZMUC).
104. *S. suaedicola* (Mabille) –
   a) ♂, Spain, Alicante, La Marina, 10.ix.1998, leg. J. B. Wolschrijn, gen. slide Hendriksen 2606 (ZMUC).
   b) ♀, Spain, Alicante, La Marina, 15.-22.ix.1999, leg. J. B. Wolschrijn (ZMUC).
   c) ♂, Spain, Alicante, La Marina, Elche, Platje el Pinet, 22.-25.ix.2000, leg. J. B. Wolschrijn (ZMUC).
   d) ♂, Spain, Zaragoza, Rio del Huerva, 3 km E Cerveruela, 800 m, 7.vii.2002, leg. B. Skule, gen. slide Hendriksen 4004 (ZMUC).
   e) ♂, Spain, Almeria, 2 km SW Tabernas, Rambla de Tabernas, 24.-25.v.2006, leg. P. Skou, gen. slide Hendriksen 6292 (ZMUC).
105. *S. monochromella* (Constant) –
   a) ♂, Italy, Toscana, Foce F. Ombrone, e.l. 10.vii.1977, leg. G. Baldizzone, gen. slide Karsholt 4814 (ZMUC).
   b) ♀, France, Herault, Frontignan, sea lev., 26.-29.v.2004, leg. O. Karsholt (ZMUC).
106. *S. samadensis* (Pfaffenzeller) –
   a) ♀, Denmark, B, Randkløve, 4.vii.1980, leg. O. Karsholt (ZMUC).
   b) ♀, Greece, 18 km SSE Varia, 1550 m, 18.vii.1997, leg. M. Fibiger (ZMUC).
   c) ♂, Denmark, NEJ, Ålbæk, 26.-30.vii.1981, leg. O. Karsholt (ZMUC).
   d) ♀, Slovakia, Tvrdošovce, 23.vi.2004, leg. Z. Tokár (ZMUC).
   e) ♂, Denmark, WJ, Blåvand, 2.viii.1973, leg. O. Karsholt, gen. slide Karsholt 1495 (ZMUC)
   f) ♀, Denmark, NEJ, Læsø, Vesterø, 15.vii.1982, leg. O. Karsholt (ZMUC).
   g) ♀, Iceland (W), Reykjavik, 18.vii.1950, leg. N. L. Wolff, gen. slide Wolff 2603 (ZMUC).
107. *S. salinella* (Zeller) –
   a) ♂, Croatia, Krk Isl., Soline, ex la. 4.vii.1978, leg. G. Baldizzone (ZMUC).
   b) ♀, Spain, Alicante, La Marina, 16.ix.1998, leg. J. B. Wolschrijn, gen. slide Hendriksen 2777 (ZMUC).
   c) ♂, Spain, Almeria, Capo de Gata, 5 m, 30.vii.2003, leg. P. Skou, gen. slide Hendriksen 4760 (ZMUC).
   d) ♂, Spain, Sevilla, Guadalquivir, Trebujena, 4.x.1975, leg. M. & W. Glaser, gen. slide Povolný Gl. 5016 (MZMB) (holotype of *Scrobipalpa trebujenae* Povolný).
108. *S. spergulariella* (Chrétien) –
   a) ♂, France, Var, Giens Isl., e.l. 25.ix.1996, leg. J. Nel, gen. slide Huemer GEL 984 (TLMF).
   b) ♂, Spain, Catalonia, salt marshes near Rosas, 20 m, 14.vi.1964, leg. M. & W. Glaser, gen. slide Huemer GEL 1015 (TLMF).

101a
101b
102
103a
103b
103c
104a
104b
104c
104d
104e
105a
105b
106a
106b
106c
106d
106e
106f
106g
107a
107b
107c
107d
108a
108b

Plate 11 All specimens 3 x natural size

109. *Scrobipalpa salicorniae* (Hering) –
   a) ♂, Denmark, WJ, Skallingen, 25.vi.1982, leg. O. Karsholt (ZMUC).
   b) ♂, Poland, KP, Inowrocław, Mątwy, 19.v.1993, leg. J. Buszko (ZMUC).
   c) ♂, Denmark, NEZ, Amager, la. Vi.1920, leg. F. Gudmann, gen. slide Karsholt 1363 (ZMUC).

110. *S. halimioniella* sp. n. –
   a) ♂, paratype, France, Herault, Frontignan, sea lev., la. 30.v.2004, leg. O. Karsholt (ZMUC).
   b) ♂, paratype, France, Herault, Frontignan, sea lev., la. 30.v.2004, leg. O. Karsholt (ZMUC).

111. *S. thymelaeae* (Amsel) –
   a) ♂, paratype, Italy, Sardinia, S. Caterina, ex l. vii.-viii.1936, leg. H. G. Amsel, gen. slide Karsholt 5043 (LNK).
   b) ♀, paratype, Italy, Sardinia, S. Caterina, ex l. vii.-viii.1936, leg. H. G. Amsel (LNK).

112. *S. halymella* (Millière) –
   a) ♂, France, Corse, Calvi, 10 m, 18.-19.viii.1998, leg. O. Karsholt, gen. slide Hendriksen 2381 (ZMUC).
   b) ♂, Spain, Huesca, 8 km S Candasnos, Barranco de Valcuerna, 175 m, 13.-14.ix.2002, leg. P. Skou (ZMUC).

113. *S. camphorosmella* Nel –
   a) ♂, paratype, France, Var, Giens Isl., e.l. 26.iv.1995, leg. J. Nel, gen. slide Huemer GEL 982 (TLMF).
   b) ♂, France, Alpes Maritimes, Beaulieu, la. 8.iii.1893, leg. T. Walsingham, gen. slide. BMNH 20108 (BMNH).

114. *S. stabilis* Povolný –
   a) ♂, Spain, Zaragoza, Rio del Huerva,Tosos, 550 m, 15.ix.2002, leg. P. Skou (ZMUC).
   b) ♂, paratype, Spain, Granada, Baza, 2.vi.1975, leg. M. & W. Glaser, gen. slide Povolný Gl. 4978 (LNK).

115. *S. instabilella* (Douglas) –
   a) ♂, Denmark, WJ, Oksby, 23.vii.1969, leg. O. Karsholt, gen. slide Karsholt 671 (ZMUC).
   b) ♀, Denmark, WJ, Skallingen, 8.-13.viii.1983, leg. O. Karsholt (ZMUC).
   c) ♂, Great Britain, England, Essex, Bradwell-on-Sea, 9.vii.1988, leg. O. Karsholt (ZMUC).
   d) ♀, Netherlands, Zld, Ellewoutsdijk, 1.viii.1994, leg. A. Cox, gen. slide Karsholt 5148 (ZMUC).
   e) ♂, France, Herault, Frontignan, sea lev., 26.-29.v.2004, leg. O. Karsholt, gen. slide Karsholt 5150 (ZMUC).

116. *S. peterseni* (Povolný) –
   a) ♂, Tunisia, Nefta, 14.-16.iii.1986, leg. O. Karsholt (ZMUC).
   b) ♂, Italy, Sicily, 5 km W Pachino, Pantano Longarini, sea lev., leg. S. Bella, gen. slide Huemer GU 03/1183 (BELL).

117. *S. ergasima* (Meyrick) –
   ♂, Portugal Madeira, Funchal Lido, 15.-17.ix.1973, leg. N. L. Wolff (ZMUC).

118. *Turcopalpa glaseri* Povolný –
   ♀, Turkey, Ankara, Tuz Gölü N, 3.ix.1969, leg. M. & W. Glaser, gen. slide Povolný Gl. 4733 (MZMB).

119. *Scrobipalpula psilella* (Herrich-Schäffer) –
   a) ♂, Italy, Piemonte, Val Susa, Villar Dora, 500 m, 4.vii, 198, leg. G. Bassi, gen. slide Hendriksen 1406 (ZMUC).
   b) ♀, Finland, EH, Hartola, la. 15.-16.ix.2004, leg. J. Tabel, gen. slide Hendriksen 5379 (ZMUC).
   c) ♂, Denmark, NEZ, København, 19.vii.1873, leg. ? (ZMUC).

120. *S. ramosella* (Müller-Rutz) –
   a) ♂, Switzerland, Valais, Zermatt, 2000 m, 5.viii.1934, leg. P. Weber, gen. prep. Tokár 4118 (in tube) (TLMF).
   b) ♂, Switzerland, Valais, Zermatt, Kühberg, 2700 m, 8.vii.1934, leg. P. Weber, gen. slide Klimesch 589 (ZSM).

109a
109b
109c
110a
110b
111a
111b
112a
112b
113a
113b
114a
114b
115a
115b
115c
115d
115e
116a
116b
117
118
119a
119b
119c
120a
120b

Plate 12 All specimens 3 x natural size

121. *Scrobipalpa seniorum* Povolný –
  a) ♂, Macedonia, Matka, Treska Schlucht, ex l. 17.ix.1963, leg. J. Klimesch (ZMUC).
  b) ♂, Macedonia, Matka, Treska Schlucht, ex l. 21.ix.-10.x.1955, leg. J. Klimesch (ZSM).
  c) ♀, Greece, Larissa, Olympus Mt., above Kataphygion (A1), 2500 m, e.l. 10.viii.1962, leg. F. Kasy (ZSM).
122. *S. diffluella* (Frey) –
  a) ♂, Sweden, Upl., Älvkarleby, e.l. vii.1948, leg. W. Brandt, gen. slide Hendriksen 1660 (ZMUC).
  b) ♂, Austria, Nordtirol, Ötztal, Rofenberg, 2600 m, e.l. iii.1943, leg. J. Klimesch (TLMF).
  c) ♂, Norway, ST(i), Kongsvold, 900-1100 m, 12.-20.vi.1985, leg. O. Karsholt (ZMUC).
  d) ♀, Austria, Nordtirol, Ötztal, Rofenberg, 2600 m, e.l. iii.1943, leg. J. Klimesch, gen. slide Huemer GEL 619 (TLMF).
  e) ♀, Finland, EH, Heinola, la. 15.ix.2004, leg. J. Tabel, gen. slide Hendriksen 5380 (ZMUC).
123. *S. tussilaginis* (Stainton) –
  a) ♀,Great Britain, England, Dorset, Lyme Regis, sea lev., la. 17.vii.1988, leg. O. Karsholt (ZMUC).
  b) ♀, Great Britain, England, Dorset, Lyme Regis, sea lev., la. 17.vii.1988, leg. O. Karsholt (ZMUC).
  c) ♂, Greece, Katara pass, 1650 m, 28.vii.1990, leg. M. Fibiger (ZMUC).
124. *Phthorimaea operculella* (Zeller) –
  a) ♂, Portugal, Madeira, Funchal, 50 m, 22.ii.1979, leg. O. Karsholt (ZMUC).
  b) ♂, Spain, Andalusia, Marbella, 20.-21.vii.1983, leg. G. Baldizzone & P. Triberti (ZMUC).
  c) ♀, Italy, Liguria, Savona, dint. Noli, v.1954, leg. U. Parenti (ZMUC).
125. *Tuta absoluta* (Meyrick) –
  a) ♂, Argentina, Neuquen, San Martin de los Andes, 640 m, 13.x.1981, leg. O. Karsholt & E. S. Nielsen, gen. slide Povolný Dk. 0251 (ZMUC).
  b) ♂, Chile, Aconcagua, Los Andes Curimon, 700 m, 28.iii.1979, leg. Mission Cientica Danesa, gen. slide Povolný Dk. 0331 (ZMUC).
  c) ♂, Spain, Granada, Benamaurel, 700 m, 15.ix.2009, leg. P. Skou (ZMUC).
126. *Keiferia lycopersicella* (Walsingham) –
  ♀, Dominican Republic, Bani, el.l 13.-17.ii.1974, leg. J. Klapperich, gen. slide Povolný Am. 5245 (ZMUC).
127. *Ephysteris promptella* (Staudinger) –
  a) ♂, Italy, Aeolian Islands, Lipari, 4.v.1979, leg. U. Parenti (ZMUC).
  b) ♂, Greece, Crete, Kato Stalos, 19.-27.iv.1995, leg. J. P. Baungaard, gen. slide Hendriksen 2438 (ZMUC).
  c) ♂, Spain, Andalusia, Camino de Istan, 400 m, 25.vi.1975, leg. E. Traugott-Olsen, gen. slide Karsholt 3531 (ZMUC).
  d) ♂, Turkey, Agri, 10 km NE Diyadin, 1800 m, 21.vii.1989, leg. M. Fibiger & N. Esser, gen. slide Hendriksen 4558 (ZMUC).
  e) ♀, Turkey, Gaziantep, 16 km NE Kadirli, 700 m, 10.vii.1987, leg. M. Fibiger (ZMUC).
128. *E. tenuisaccus* Nupponen –
  a) ♂, Russia, South Ural, Orenburg district, 20 km S Pokrovka village, Schibendy valley, 7.vi.1998, leg. K. & T. Nupponen (ZMUC).
  b) ♀, Russia, South Ural, Orenburg district, 20 km S Pokrovka village, Schibendy valley, 7.vi.1998, leg. K. & T. Nupponen (ZMUC).
129. *E. deserticolella* (Staudinger) –
  a) ♂, Turkey, Ankara, Tuz Gölü, northern bank, 30.viii.-12.ix.1968, leg. N. & U. Glaser (ZMUC).
  b) ♂, Turkey, Kars, 10 km NW Aralik, 825 m, 22.vii.1989, leg. M. Fibiger & N. Esser (ZMUC).
  c) ♂, France, Var, Embouchurc de l'Argens, 20.ix.1994, gen. slide Nel 7601 (TLMF).

121a
121b
121c
122a
122b
122c
122d
122e
123a
123b
123c
124a
124b
124c
125a
125b
125c
126
127a
127b
127c
127d
127e
128a
128b
129a
129b
129c

Plate 13 All specimens 3 x natural size

130. *Ephysteris insulella* (Heinemann) –
    a) ♀, France, Hautes-Alpes, Reotier, 1000 m, 8.-15.viii.1990, leg. A. Cox, gen. slide Huemer GEL 1124 (TLMF).
    b) ♂, France, Hautes-Alpes, Reotier, 1000 m, 8.-15.viii.1990, leg. A. Cox (ZMUC).
    ssp. *praticolella* (Christoph) –
    c) ♂, Italy, Fano, Marche, Torette, 5.ix.1956, leg. A. Fiori (ZSM).
    d) ♂, Russia, Sarepta, leg. ?, gen. slide Povolný Co. 2084 (ZMUC).
131. *E. diminutella* (Zeller) –
    a) ♀, Italy, Sicily, Ragalna, Monte Arso, Bosco di Paternò,1000 m, 9.ix.2002, leg. M. Fibiger, G. Jeppesen & O. Karsholt (ZMUC).
    b) ♂, Italy, Sicily, Ragalna, Monte Arso,Bosco di Paternò, 1000 m, 9.ix.2002, leg. M. Fibiger, G. Jeppesen & O. Karsholt (ZMUC).
    c) ♂, Italy, Sicily, Ragalna, Monte Arso, Bosco di Paternò, 1000 m, 9.ix.2002, leg. M. Fibiger, G. Jeppesen & O. Karsholt (ZMUC).
    d) ♀, Italy, Sicily, Ragalna, Monte Arso, Bosco di Paternò, 1000 m, 9.ix.2002, leg. M. Fibiger, G. Jeppesen & O. Karsholt (ZMUC).
    e) ♀, Morocco(?), Afrat, 27.ix.1900, leg. ?, gen. slide Povolný Pr. 5298 (missing) (MNHN) (paratype of *Ephysteris foulonensis* Povolný).
    f) ♂, Spain, Teruel, Albarracin, 12.-22.v.1953, leg. J. Klimesch, gen. slide Povolný Kl. 5640 (ZSM) (paratype ["false holotype"] of *Ephysteris hispanica* Povolný).
    g) ♂, Spain, Teruel, Albarracin, 12.-22.v.1953, leg. J. Klimesch, gen. slide Povolný Kl. 5641 (ZSM) (holotype of *Ephysteris hispanica* Povolný).
    h) ♂, Italy, Piemonte, S. Anna di Valderi, 1000 m, 7.viii.2007, leg. M. Fibiger & O. Karsholt (ZMUC).
    i) ♀, Macedonia, Drenovo pr. Kavadar, 7.-10.vi.1956, leg. J. Klimesch, gen. slide Hendriksen 3855 (ZMS).
    j) ♂, Macedonia, Drenovo pr. Kavadar, 10.-20.vi.1956, leg. J. Klimesch, gen. slide Povolný Kl. 5624 (ZMS).
    k) ♂, Greece, Larissa, Olympos S, 30 km NW Karia, 1850 m, 11.vii.1990, leg. M. Fibiger (ZMUC)
    l) ♂, Italy, Piemonte (CN), Parco Natur. Reg. Alpi Marittime, Entracque, Mt. Ray, 1800 m, 20.vii.1999, leg. G. Baldizzone (BALD).
    m) ♂, Italy, Piemonte (CN), Parco Natur. Reg. Alpi Marittime, Entracque, Mt. Ray, 1800 m, 22.vii.1999, leg. G. Baldizzone, gen. slide Hendriksen 2848 (ZMUC).
    n) ♂, Greece, Florina, Agios Ilias, 900 m, 15.ix.1991, leg. F. Schepler, gen. slide Hendriksen 1168 (ZMUC).
132. *E. inustella* (Zeller) –
    a) ♂, Estonia, Pidula, 4.vii.1994, leg. J. Junnilainen (ZMUC).
    b) ♀, Latvia, Lps., Pape, 24.vi.1994, N. Savenkov (ZMUC). ssp. *gredosensis* (Rebel) –
    c) ♂, Portugal, BB, Serra da Estréla, Torre, 1600-1950 m, 10.-11.vii.1986, leg. O. Karsholt, gen. slide Hendriksen 1421 (ZMUC).
    d) ♂, syntype, Spain, Avila, Sierra de Gredos, Garg. Pozas, 1900 m, 15.vii.1934, leg. H. Reisser, gen. slide Mus. Vind. 3901 (NHMW).
133. *E. olympica* Povolný –
    a) ♂, Greece, Peloponnese, Chelmos, 2300 m, 24.vi.1958, leg. J. Klimesch, gen. slide Hendriksen 3856 (ZSM).
    b) ♂, France, Hautes-Alpes, Col d'Izoard, 2250 m, 28.vii.2001, leg. J. Nel, gen. slide Nel 12913 (TLMF).
134. *E. iberica* Povolný, 1977 –
    a) ♂, Spain, Murcia, Bolnuevo by Mazarrón, 10 m, 1.vi.1998, leg. P. Skou, gen. slide Hendriksen 2537 (ZMUC).
    b) ♂, Spain, Almeria, 10 km E Bedar, El Pinar, 325 m, 19.-27.iv.2001, leg. P. Skou & B. Skule (ZMUC).
    c) ♂, Greece, Crete (SW), Paleohora, 20 m, 1.iv.1999, leg. G. Jeppesen *et al.* (ZMUC).
135. *Ochrodia subdiminutella* (Stainton) –
    a) ♂, Spain, Almeria, La Marina del Pinet, 7.v.2000, leg. H. van der Wolf (ZMUC).
    b) ♀, Spain, Almeria, San Jose, Nihar, 6.-20.x.2000, leg. J. B. Wolschrijn, gen. slide Karsholt 5142 (ZMUC).
    c) ♂, Spain, Murcia, Bolnuevu by Mazarrón, 10 m, 1.vi.1998, leg. P. Skou & B. Skule (ZMUC).
    d) ♂, Spain, Murcia, Bolnuevu by Mazarrón, 10 m, 1.vi.1998, leg. P. Skou & B. Skule (ZMUC).
    e) ♀, Spain, Almeria, 10 km E Bedar, El Pinar, 325 m, 19.-27.iv.2001, leg. B. Skou & P. Skule, gen. slide Karsholt 5143 (ZMUC).
    f) ♂, Cyprus, Platres, N Limassol, 1150 m, 23.-26.vi.1997, leg. M. Fibiger *et al.* (ZMUC)
136. *Vladimirea glebicolorella* (Erschoff) –
    a) ♂, Ukraine, Crimea, Meganom, e.l. 27.-28.v.2004, leg. Yu. Budaskhin, gen. slide Bedzilya 527/07 (ZMKU).
    b) ♀, Ukraine, Crimea, Meganom, e.l. 27.-28.v.2004, leg. Yu. Budaskhin, gen. slide Bedzilya 521/07 (ZMKU).

130a
130b
130c
130d
131a
131b
131c
131d
131e
131f
131g
131h
131i
131j
131k
131l
131m
131n
132a
132b
132c
132d
133a
133b
134a
134b
134c
135a
135b
135c
135d
135e
135f
136a
136b

Plate 14 All specimens 3 x natural size

137. *Microlechia rhamnifoliae* (Amsel & Hering) –
   a) ♀, Greece, Karpathos, Lefkos, 30 m, 23.v.1997, leg. R. Sutter, gen. slide Huemer GU 09/1308 (ZMUC).
   b) ♀, Cyprus, Alethrikó, aa.iv.1991, leg. R. Johansson, gen. slide Hendriksen 2584 (ZMUC).
138. *M. chretieni* Turati –
   a) ♂, Spain, Almeria, San Jose, Nihar, 6.-20.x.2000, leg. J. B. Wolschrijn (ZMUC).
   b) ♀, Greece, Crete, Lassithi distr., Makrygialos, 11.-19.vi.1988, leg. R. Johansson, gen. slide Hendriksen 2607 (ZMUC)
139. *M. karsholti* (Nupponen) –
   ♂, hootype, Russia, Russia, South Ural, Orenburg district, 20 km S Pokrovka village, Schibendy valley, 16.v.2004, leg. K. Nupponen (NUPP).
140. *Cosmardia moritzella* (Treitschke) –
   a) ♀, Norway, On, Vinstra, la. 20.vi.1981, leg. O. Karsholt (ZMUC).
   b) ♂, Italy, Südtirol, Sesvenna group, Val Monastero, Eselweg, 1700 m, la. 8.vii.2004, leg. O. Karsholt (ZMUC).
141. *Lutilabria lutilabrella* (Mann) –
   a) ♂, Ukraine (SE), Donetsk reg., Kamennye Mogily Nat. Res., 29.v.1999, leg. A. Bidzilya, gen. slide Huemer GU 02/1154 (ZMUC).
   b) ♂, Croatia, Fiume (Rijeka), vi.1840, leg. J. Mann, gen. slide Huemer GU 90/192 (ZIAP).
   c) ♂, Croatia, Fiume (Rijeka), 1840, leg. J. Mann, gen. slide Huemer GU 90/191 (ZIAP).
   ssp. *olympica* Huemer –
   d) ♂, Greece, Larissa, Olympos Mts., Kataphygion, 2000 m, 27.-30.vi.1962, leg. J. Klimesch (ZMM).
   e) ♀, Greece, Larissa, Olympos Mts., 11 km NE Kalivia, below Skolio, 1850 m, 3.vii.2004, leg. B. Skule (ZMUC).
142. *L. volgensis* Anikin & Piskunov –
   a) ♂, holotype, Russia, Saratov, Khvalynsk distr., 5 km N Khvalynsk, 25.v.1990, leg. B. V. Anikin, gen. slide Huemer GU 03/1207 (ZIAP).
   b) ♂, Russia, South Ural, Cheliabinsk district, 20 km S Mednogorsk, near Kidriasovo village, 16.vi.1999, leg. K. & T. Nupponen, gen. slide Huemer GU 09/ 1303 (NUPP).
   c) ♀, Russia, South Ural, Orenburg district, 6 km W Donskoje village, mount Verbljushka, 10.vi.1998, leg. K. & T. Nupponen, gen. slide Huemer GU 09/1304 (NUPP).
143. *L. prolata* Junnilainen & Nupponen –
   a) ♂, paratype, Russia, South Ural, Orenburg district, near Zesnokovka village, 9.vi.2001, leg. K. Nupponen (NUPP).
   b) ♂, Romania, Dobrogea, Macin G. - Greci, 100-300 m, 3.vi.1997, leg. C. Wieser, gen. slide Huemer GU 99/866 (LMKÄ).
   c) ♂, Ukraine, Khomutovskaya Steppe nature preserve, 6.-7.vi.2004, leg. E. Rutyan, gen. slide Huemer GU 05/1233 (ZMKU).
144. *Klimeschiopsis kiningerella* (Duponchel) –
   a) ♀, Norway, On, Vinstra, 4.-5.vii.1984, leg. O. Karsholt, abdomen in capsule (ZMUC).
   b) ♂, Finland, Ab, Lojo, 30.vi.1972, leg. H. Krogerus, gen. slide Karsholt 1687 (ZMUC).
145. *K. discontinuella* (Rebel) –
   a) ♀, Italy, Südtirol, Laatsch, 1000 m, 19.viii.1972, leg. K. Burmann (TLMF).
   b) ♂, Italy, Südtirol, Laatsch, 1000 m, 19.viii.1972, leg. K. Burmann, gen. prep. Tokár 3784 (in tube) (TLMF).
146. *K. terroris* (Hartig) –
   a) ♀, Spain, Teruel, 1 km E Tramacastilla, 1250 m, 7.viii.2007, leg. P. Skou & B. Skule, gen. slide Karsholt 5144 (ZMUC).
   b) ♀, Spain, Andalucia, Camino de Ronda, Urb. Madronal, Loma de Colmenas, 500 m, 3.ix.1988, leg. E. Traugott-Olsen, gen. slide Karsholt 5145 (ZMUC).

137a
137b
138a
138b
139
140a
140b
141a
141b
141c
141d
141e
142a
142b
142c
143a
143b
143c
144a
144b
145a
145b
146a
146b

Plate 15 All specimens 3 x natural size

147. *Caryocolum fischerella* (Treitschke) –
   a) ♀, Denmark, SZ, Fakse, la. 29.v.1978, leg. O. Karsholt (ZMUC).
   b) ♀, Sweden, Öl, Böda, la. 1-5.vi.1980, leg. R. Johansson (ZMUC).
148. *C. tischeriella* (Zeller) –
   a) ♂, Sweden, Ög, Kolmården, 21.vii.1971, leg. I. Svensson (ZMUC).
   b) ♂, Italy, Torino, Valle di Susa, Rocci Amelone, e. l. 10.vi.1980, leg. G. Baldizzone (ZMUC).
149. *C. alsinella* (Zeller) –
   a) ♀, Spain, Granada, Sierra Nevada, Ruta del Valeta, 2000 m, 23.vii.2003, leg. P. Skou (ZMUC).
   b) ♂, Denmark, EJ, Øer, la. 3.vi.1996, leg. O. Karsholt (ZMUC).
   c) ♀, Denmark, LFM, Busene, la. 30.iv.2000, leg. O. Karsholt (ZMUC).
   d) ♀, Denmark, EJ, Øer, la. 3.vi.1996, leg. O. Karsholt (ZMUC).
   e) ♀, Sweden, Öl, Böda, Längalvaret, 1.viii.1985, leg. O. Karsholt (ZMUC).
150. *C. viscariella* (Stainton) –
   a) ♀, Denmark, EJ, Lovnkær Skov, la. 12.v.1991, leg. H. Hendriksen (ZMUC).
   b) ♂, Denmark, EJ, Lovnkær Skov, la. 16.iv.1990, leg. P. Falck (ZMUC).
151. *C. albifaciella* (Heinemann) –
   ♀, Austria, Nordirol, Vennatal, 1550 m, e. l. 13.vii.1988, leg. P. Huemer, gen. slide Huemer GEL 428 (TLMF).
152. *C. vicinella* (Douglas) –
   a) ♂, Denmark, B, Randkløve, 4.vii.1980, leg. O. Karsholt (ZMUC).
   b) ♀, Italy, Abruzzo, Mte la Rocca, 6 km SW Pescasséroli, 1600-1950 m, 13-14.viii.1972, leg. R. Johansson (ZMUC).
153. *C. bosalella* (Rebel) –
   a) ♂, Italy, Sardinia, Mte Limbara, Pt. Balistreri, 1150 m, 3.vii.1998, leg. F. Iversen, gen. slide Hendriksen 3967 (ZMUC).
   b) ♂, France, Corsica, Bocca di Vezzu, Desert des Agriates, 300 m, 28.vi.2005, leg. B. Skule, gen. slide Hendriksen 5366 (ZMUC).
154. *C. amaurella* (Hering) –
   a) ♂, Denmark, B, Hammeren, 18.vii.1977, leg. O. Karsholt leg (ZMUC).
   b) ♀, Denmark, B, Hammeren, 18.vii.1977, leg. O. Karsholt leg (ZMUC).
   c) ♂, France, Alpes Maritimes, Col de la Cayolle, 2200-2300 m, 29-30.vii.2005, leg. P. Skou (ZMUC).
155. *C. oculatella* (Thomann) –
   ♂, Austria, Nordtirol, Finstermünz, 1200 m, e. l. 12.vi.1989, leg. P. Huemer (TLMF).
156. *C. leucofasciatum* Huemer –
   ♂, paratype, Spain, Granada, Baza, 22-26.v.1979, leg. M. & W. Glaser, gen. slide Huemer 87/073 (TLMF).

147a
147b
148a
148b
149a
149b
149c
149d
149e
150a
150b
151
152a
152b
153a
153b
154a
154b
154c
155
156

Plate 16 All specimens 3 x natural size

157. *Caryocolum petryi* (Hofmann) –
a) ♂, Germany, S. Kyffhäuser, Kattenburg, e. l. 20.vii.1938, leg. E. Jäckh (ZMUC).
b) ♀, Sweden, Öl, Möcklemossen, la. 30.v.1981, leg. O. Karsholt (ZMUC).
c), ♀ Sweden, Öl, Möcklemossen, la. 30.v.1981, leg. O. Karsholt (ZMUC).
d) ♀, Slovenia, Juliske Alpi, Soca, 550 m, e. l. 29.vi.1990, leg. P. Huemer (TLMF).
e) ♂, Austria, Tirol, Venedigergr., Dorfertal W, 1990 m, e. l. vii.1993, leg. P. Huemer (TLMF).
f) ♂, Russia, South Ural, Cheliabinsk distr., Mias, Ilmel State Res., 26.vi.1997, K. Nupponen & J. Junnilainen, gen. prep. Junnilainen without number (in tube) (NUPP).

158. *C. baischi* sp. n. –
♂, paratype, Greece, Peloponnese, Leonidon, Bona Monastry, 13.x.1994, 500 m, leg. G. Baisch (TLMF).

159. *C. repentis* Huemer & Luquet –
a) ♀, paratype, Italy, Trentino, Piz Ciavazes S, 2150 m, la. 3.vii.1988, leg. P. Huemer (TLMF).
b) ♀, Austria, Steiermark, Altaussee, 750 m, la. 4.v.1947, leg. J. Klimesch (ZMUC).

160. *C. siculum* Bella –
a) ♀, paratype, Italy, Sicily, 6 km S Vizzini, E Lago Dirillo, la. 27.iv.2006, leg. O. Karsholt (ZMUC).
b) ♀, paratype, Italy, Sicily, 6 km S Vizzini, E Lago Dirillo, la. 27.iv.2006, leg. O. Karsholt (ZMUC).

161. *C. inflativorella* (Klimesch) –
a) ♀, paratype, Hungary, Cserkút, Pécs, la. 19.vi.1936, leg. J. Klimesch (TLMF).
b) ♂, Macedonia, Ochrid, Petrina Planina, 1600 m, 23.viii.1955, leg. F. Kasy, abdomen missing (TLMF).
c) ♂, Macedonia, Ochrid, Petrina Planina, 1600 m, 4.x.1953, leg. J. Thurner (TLMF).

162. *C. saginella* (Zeller) –
a) ♂, Italy, Brescia, Lavenone, Corno Barzo S, 1650 m, la. medio vi.1987, leg. P. Huemer (TLMF).
b) ♂, Italy, Trentino, Loc. Vezzano, 600 m, la. vi.1983, leg. K. Burmann (ZMUC).

163. *C. cauligenella* (Schmid) –
a) ♂, Macedonia, Ohrid, Petrina Plan, ex la. 20.vii.1959, leg. J. Klimesch (ZMUC).
b) ♀, Denmark, LFM, Høvblege, la. 13.vi.1971, leg. O. Karsholt (ZMUC).
c) ♀, Denmark, LFM, Høvblege, la. 20.vi.1978, leg. O. Karsholt (ZMUC).
d) ♀, Austria, Nordtirol, Umhausen, 1.viii.1944, leg. K. Burmann, abdomen missing (TLMF).

164. *C. trauniella* (Zeller) –
♀, Slovenia, Juliiske Alpe, Alta Val Trenta, Isonzo (Soca), 950 m, 11.viii.2000, leg. L. Morin (TLMF).

165. *C. peregrinella* (Herrich-Schäffer) –
a) ♂, Italy, Trentino, Sella-Gruppe, Piz Ciavazes S, 2150 m, 7.viii.1991, leg. P. Huemer (TLMF).
b) ♀, Greece, Larissa, Olymp SW, Kataphygion B, 1900 m, ex la. 25.vi.1988, leg. P. Huemer (TLMF).

157a
157b
157c
157d
157e
157f
158
159a
159b
160a
160b
161a
161b
161c
162a
162b
163a
163b
163c
163d
164
165a
165b

Plate 17 All specimens 3 x natural size

166. *Caryocolum delphinatella* (Constant) –
  a) ♀, Italy, Imperia, Punta Marguraeis S, 2250-2400 m, 19.vii.1991, leg. P. Huemer & G. Tarmann (TLMF).
  b) ♂, France, Alpes Maritimes, Tete Chaudon, NE Col de la Boira, 2200 m, 20.vii.1991, leg. P. Huemer & G. Tarmann (TLMF).
167. *C. provinciella* (Stainton) –
  a) ♂, Portugal, Algarve, Carrapateira, 6.x.1993, leg. M. Corley (ZMUC).
  b) ♂, Spain, Huelva, Torre la Higuera, 23.ix.-5.x.1994, leg. J. B. Wolschrijn (ZMUC).
168. *C. mucronatella* (Chrétien) –
  a) ♂, Greece, Parnassos, N Arakhova, 1900 m, 24.vii.1984, leg. M. & E. Arenberger (ZMUC).
  b) ♂, Austria, Nordtirol, Umhausen, ex la. 10.vii.1947, leg.K. Burmann (TLMF).
  c) ♂, Greece, Arkadia, Parnon Oros, at Prof. Ilias, 1650 m, 24.vii.1998, leg. B. Skule & D. Nilsson (ZMUC).
169. *C. leucomelanella* (Zeller) –
  a) ♂, Ukraine, Crimea, Karadagh, 30.v.1994, leg. I. Kostjuk (ZMUC).
  b) ♀, Italy, Udine, 3 km NW Arta Terme, T. But, la. primo v.1987, leg. P. Huemer (ZMUC).
  c) ♂, Poland, Gacki, 16.vii.1997, leg. E. Baraniak (ZMUC).
170. *C. mazeli* Huemer & Nel –
  a) ♂, France, Pyrénées-Orientales, Eyne, 1950 m, 23.vii.2001, leg. R. Mazel (TLMF).
  b) ♀, France, Pyrénées-Orientales, Eyne, 1700 m, 22.vii.2001, leg. R. Mazel (TLMF).
171. *C. leucothoracellum* (Klimesch) –
  ♀, Italy, Verona, Monte, 300 m, la. 18.iv.1987, leg. P. Huemer (TLMF).
172. *C. schleichi* (Christoph) –
  a) ♂, Turkey, Sivas, 10 km W Gürün, 1650 m, 27.vii.1989, leg. M. Fibiger & N. Esser (ZMUC).
  ssp. *dianthella* (Chrétien) –
  b) ♂, Spain, Almeria, Sierra de los Filabres, Calar Alto, 2050 m, 29.viii.2001, leg. B. Skule (ZMUC).
  c) ♂, Spain, Teruel, Puerto de Orihuela, 1650 m, 23.viii.2001, leg. B. Skule & P. Skou (ZMUC).
  ssp. *improvisella* (Rebel) –
  d) ♂, Italy, Südtirol, Laas, la. 13.v.1987, leg. P. Huemer (ZMUC).
  e) ♂, Austria, Nordtirol, Vennatal, 1500 m, ex la. 6.viii.1954, leg. K. Burmann (ZMUC).
  ssp. *arenariella* (Benander) –
  f) ♂, Sweden, Sk, Kivik, ex la. 1.vii.1974, leg. R. Johansson (ZMUC).
173. *C. marmorea* (Haworth) –
  a) ♂, Denmark, WJ, Skallingen, 30.vi.1982, leg. O. Karsholt (ZMUC).
  b) ♂, Spain, Almeria, Sierra de los Filabres, Calar Alto, 2150 m, 29.vii.2003, leg. P. Skou (ZMUC).
  c) ♀, Greece, Rhodos, 5 km S Rhodos City, 250 m, 8.vii.2000, leg. M. Fibiger (ZMUC).
  d) ♀, Portugal, Madeira, Porto Moniz, sea level, la. 21.iv.1996, leg. O. Karsholt (ZMUC).
  ssp. *pulchra* (Wollaston) –
  e) ♂, Portugal, Madeira Isl., Deserta Grande Isl., 17.-19.vi.1962, leg. E. W. Classey, gen. slide BMNH 24262 (BMNH).
174. *C. pullatella* (Tengström) –
  a) ♂, Sweden, Sm, 15 km NE Växjö, N. Åreda, 20.vi.1975, leg. R. Johansson (ZMUC).
  b) ♂, Austria, Nordtirol, Hochgurgl, 2100 m, 30.viii.1992, leg. Dobler (TLMF).
175. *C. stramentella* (Rebel) –
  a) ♂, Turkey, Nevsehir, 10 km W Ürgüp, 1300 m, 17-20.vi.1996, leg. F. Schepler (ZMUC).
  b) ♂, Turkey, Mersin, Taurus, 10 km SE Arslanköy, 1300 m, 11.vii.1987, leg. M. Fibiger (ZMUC).
176. *C. hispanicum* Huemer –
  a) ♂, Spain, Avila, Sierra de Gredos, 15 km S-SW Hoyos del Espinos, 1720 m, 27-28.vii.1988, leg. M. Fibiger (ZMUC).
  b) ♀, Turkey, Nevsehir, 7 km SE Avanos, 1200 m, 2.viii.1989, leg. M. Fibiger & N. Esser (ZMUC).

166a
166b
167a
167b
168a
168b
168c
169a
169b
169c
170a
170b
171
172a
172b
172c
172d
172e
172f
173a
173b
173c
173d
173e
174a
174b
175a
175b
176a
176b

Plate 18 All specimens 3 x natural size

177. *Caryocolum confluens* Huemer –
  a) ♂, Greece, Chelmos, 2300 m, 1.vii.1990, leg. F. Schepler (ZMUC).
  b) ♂, Montenegro, Durmitor, Crna Gora, Sedlo pass, 1907 m, 25.vii.1984, leg. P. Jakcić, gen. slide Huemer GU92/329 (TLMF).
178. *C. gallagenellum* Huemer –
  a) ♀, Italy, Umbria, above Assisi, 1000-1100 m, la. 7.v.1999, leg. O. Karsholt (ZMUC).
  b) ♂, paratype, France, Alpes Maritimes, Col de vence, 1050 m, la. 11.iv.1988, leg. P. Huemer & G. Tarmann (ZMUC).
179. *C. fraternella* (Douglas) –
  a) ♂, Denmark, SZ, Skibinge, la. 19.v.1986, leg. O. Karsholt (ZMUC).
  b) ♂, UK, Sussex, Burgess Hill, 31.vii.-2.viii.1988, leg. O. Karsholt (ZMUC).
  c) ♀, Denmark, LFM, Møns Klint S, 3.-8.viii.2008, leg. O. Karsholt (ZMUC).
180. *C. klosi* (Rebel) –
  ♀, Germany, Württemberg, Schwarzwald, Wildbad, 530 m, ex la. 26.v.1976, leg. L. Süssner (ZMUC).
181. *C. interalbicella* (Herrich-Schäffer) –
  a) ♂, France, Alpes Maritimes, Col de la Cayolle, Esteng, 1850 m, 9.vii.1988, leg. M. Fibiger (ZMUC).
  b) ♀, Italy, Piemonte (CN), Parco Natur. Reg. Alpi Maritime, 1900-2200 m, 26.vii.2000, leg. G. Baldizzone (ZMUC).
  c) ♀, Switzerland, Valais, Laquintal, 29.vii.1924, leg. P. Weber (ZMUC).
182. *C. laceratella* (Zeller) –
  a) ♂, Italy, Udine, Alpi Giulie, Montasio, Malga Pecol, 1800 m, la. viii.1987, leg. P. Huemer (ZMUC).
  b) ♀, Italy, Udine, Alpi Giulie, Montasio, Malga Pecol, 1800 m, la. viii.1987, leg. P. Huemer (ZMUC).
183. *C. blandella* (Douglas) –
  a) ♂, Denmark, NEJ, Læsø, Skoven, 22.vii.-3.viii.1984, leg. O. Karsholt (ZMUC).
  b) ♂, Denmark, SZ, Vemmetofte, 31.vii.1974, leg. O. Karsholt (ZMUC).
  ssp. *horoscopa* (Meyrick) –
  c) ♂, Turkey, Ankara, 20 km NW Kizilcahaman, 1200 m, 24.vii.1986, leg. M. Fibiger (ZMUC).
  d) ♀, Greece, Komotini, Kürü river by Lasmos, 19.vii.1987, leg. M. Fibiger (ZMUC).
184. *C. blandelloides* Karsholt –
  a) ♂, paratype, Sweden, Öl, Seberneby, 23.vii.1975, leg. O. Karsholt (ZMUC).
  b) ♀, Italy, Lucania, Mt. Pollino, sotto Rif. Fasanelli, 1200 m, 13.vii.1991, leg. G. Baldizzone (ZMUC).
  c) ♀, France, Corsica, Col de Vergio, 1400-1500 m, leg. O. Karsholt (ZMUC).
185. *C. jaspidella* (Chrétien) –
  a) ♂, Morocco, Itzèr area, la. 16.iv.1989, 2100 m, leg. O. Karsholt (ZMUC).
  b) ♀, Morocco, Azrou / Ifrane area, 35 km SW Azrou, la. 17-19.iv.1989, 1400-2000 m, leg. O. Karsholt (ZMUC).
186. *C. proximum* (Haworth) –
  a) ♂, Denmark, NEZ, Ermelunden, 4.viii.1978, leg. O. Karsholt (ZMUC).
  b) ♀, Hungary, Hajdu-Bihar, 5 km N Oüspökladany, 75 m, 27.vi.2002, leg. B. Skule & C. Hviid (ZMUC).
187. *C. blandulella* (Tutt) –
  a) ♀, Sweden, Öl, Seberneby, 19.vii.1975, leg. O. Karsholt (ZMUC).
  b) ♀, Denmark, WJ, Skallingen, 23.vii.1978, leg. O. Karsholt (ZMUC).
  c) ♂, Spain, Teruel, 1 km E Tramacastillo, 1250 m, 24.viii.2001, leg. B. Skule & P. Skou (ZMUC).
188. *C. arenbergeri* Huemer –
  ♀, Spain, Madrid, Sierra de Guadarama, Lozoya, 1000 m, 20-21.vii.1995, leg. P. Skou, gen. slide Huemer GU 10/1323 (ZMUC).

177a
177b
178a
178b
179a
179b
179c
180
181a
181b
181c
182a
182b
183a
183b
183c
183d
184a
184b
184c
185a
185b
186a
186b
187a
187b
187c
188

Plate 19 All specimens 3 x natural size

189. *Caryocolum tricolorella* (Haworth) –
   a) ♂, Denmark, SZ, Vemmetofte, la. 9.v.1987, leg. O. Karsholt (ZMUC).
   b) ♀, Denmark, SZ, Vemmetofte, la. 9.v.1987, leg. O. Karsholt (ZMUC).
190. *C. fibigerium* Huemer –
   a) ♂, Spain, Granada, Rio de Baza, 800 m, 27.ix.1987, leg. P. Skou (ZMUC).
   b) ♂, Greece, Larissa, Olympos S, 30 km NW Karia, 1850 m, 11.vii.1990, leg. M. Fibiger (ZMUC).
   c) ♂, Greece, Katara pass, 1650 m, 28.vii.1990, leg. M. Fibiger (ZMUC).
191. *C. junctella* (Douglas) –
   a) ♂, Slovakia, Slovenský Raj, 7.viii.1990, leg. Z. Tokár (ZMUC).
   b) ♂, Poland, „Dirsdorf", 12.viii.1892, leg. ? (ZMUC).
192. *C. cassella* (Walker) –
   a) ♂, Denmark, LFM, Bøtø, la. 23.v.1999, leg. H. Hendriksen (ZMUC).
   b) ♂, Sweden, Sm, Högsby, 8.vii.1971, leg. R. Johansson (ZMUC).
193. *C. moehringiae* (Klimesch) –
   a) ♀, Greece, Drama, Mt. Phalakron, above Volas, 1700 m, 20.vii.1987, leg M. Fibiger (ZMUC).
   b) ♀, paratype, Austria, Oberösterreich, Partenstein, Mühltal, ex la. 15.vi.1952, leg. J. Klimesch (ZMUC).
194. *C. petrophila* (Preissecker) –
   a) ♀, Finland, St, Rauma, ex la. 1985, leg. J. Itämies (ZMUC).
   b) ♂, France, Alpes Maritimes, Col de la Cayolle, Esteng, 1850 m, 3-4.viii.1988, leg. M. Fibiger (ZMUC).
195. *C. huebneri* (Haworth) –
   a) ♂, Hungary, Pecs, 5 km N Pecs center, 600 m, 3.vii.1999, leg. B. Skule (ZMUC).
   b) ♀, Poland, Wroclaw, leg. ? (ZMUC).
196. *C. kroesmanniella* (Herrich-Schäffer) –
   a) ♀, Denmark, SJ, Draved Skov, 26.vii.1978, leg. O. Karsholt (ZMUC).
   b) ♀, Denmark, SZ, Vemmetofte, la. 9.v.1987, leg. O. Karsholt (ZMUC).
197. *Tila capsophilella* (Chrétien) –
   a) ♂, Austria, Tirol, Brenner, Vennatal, 1800 m, e.l. 13.-21.viii.1947, leg. J. Klimesch (ZMUC).
   b) ♀, Austria, Nordtirol, Brenner, Vennatal, 1500 m, e.l. 29.viii.1947, leg. K. Burmann (TLMF).
198. *Pogochaetia solitaria* Staudinger –
   a) ♂, Italy, Trentino, Pietramurata, e.l. 26.vii.1987, leg. Huemer (TLMF).
   b) ♀, Italy, Trentino, Pietramurata, e.l. 26.vii.1987, leg. Huemer (TLMF).
   c) ♀, Spain, Teruel, 5 km SE Albarracin, 1400 m, 18.vii.1995, leg. P. Skou (ZMUC).
199. *Agonochaetia terrestrella* (Zeller) –
   a) ♀, Switzerland, Graubünden, St. Maria im Münstertal, 29.vi.1923, leg. H. Thomann (BNMC).
   b) ♀, Switzerland, Graubünden, St. Maria im Münstertal, 29.vi.1923, leg. H. Thomann, gen. slide Huemer GU 06/1254 (BNMC).
   c) ♀, Romania, Dobrogea, Băneasa, Rez. Canarăua Fetii, leg. & coll. S. & Z. Kovacs, gen. slide Huemer GU 04/1216 (TLMF).

189a
189b
190a
190b
190c
191a
191b
192a
192b
193a
193b
194a
194b
195a
195b
196a
196b
197a
197b
198a
198b
198c
199a
199c
199b

Plate 20 All specimens 3 x natural size

200. *Agonochaetia intermedia* Sattler –
   a) ♂, Austria, Nordtirol, Fliess, 1000 m, 3.vi.1983, leg. K. Burmann (TLMF).
   b) ♀, Austria, Nordtirol, Fliess, 1000 m, 10.vi.1987, leg. P. Huemer, gen. slide Huemer GU 87/107 (TLMF).
201. *A. quartana* Povolný –
   ♀, holotype, Bulgaria, Slivno, vi.1896, leg. H. Rebel, gen. slide Povolný Mn. 3222 (SZM).
202. *Sattleria arcuata* Pitkin & Sattler –
   a) ♂, Spain, Lerida, 23 km E. Vielha, Porto de Bonaigua, 2050 m, 31.vii.1988, leg. M. Fibiger (ZMUC).
   b) ♀, France, Pyrenees Central, Pic du Midi de Bigorre, 2650 m, 4.viii.1981, leg. G. Robinson, K. Sattler & K. Tuck, gen. slide Huemer GEL 330 (TLMF).
203. *S. pyrenaica* (Petry) –
   a) ♂, Spain, Huesca, 3 km W Laco Urdiceto, 11 km NE Bielsa, 2150 m, 20.vii.1992, leg. M. Fibiger (ZMUC).
   b) ♀, France, Pyrenees Central, Pic du Midi de Bigorre, 2650 m, 2.viii.1981, leg. G. Robinson, K. Sattler & K. Tuck, gen. slide Huemer GEL 331 (TLMF).
204. *S.angustispina* Pitkin & Sattler –
   a) ♂, paratype, France, Pyrénées Orientales, Mt. Canigou, 15.vii.1894, leg. ?, gen. slide Gibeaux 84 (MNHN).
   b) ♀, paratype, France, Pyrénées Orientales, Mt. Canigou, 15.vii.1894, leg. ?, gen. slide Pitkin 53 (MNHN).
205. *S. izoardi* Huemer & Sattler –
   a) ♂, France, Hautes-Alpes, Izoard, Clôt la Cime, 2400-2650 m, 2.viii.1984, leg. C. Gibeaux (ZMUC).
   b) ♀, Italy, Piemonte, Cuneo, Colle dell'Agnello, 2640 m, 20.vii.2001, leg. P. Huemer (TLMF).
   c) ♂, Italy, Piemonte, Cuneo, Colle della Fauniera, 2500 m, 28.vii.2009, leg. P. Skou & B. Skule (ZMUC).
206. *S. breviramus* Pitkin & Sattler –
   a) ♂, France, Alpes-de-Haute-Provence, SW Castel de Restfont, Roche Chevalier, 2480 m, 25.vii.1990, leg. P. Huemer & G. Tarmann (TLMF).
   b) ♀, France, Alpes Maritimes, Valdeblore, Teras des Bresses, 2750 m, 23.viii.2001, leg. T. Varenne, gen. slide Nel 13696 (TLMF).

200a
200b
201
202a
203a
202b
203b
204a
204b
205a
205b
205c
206a
206b

Plate 21 All specimens 3 x natural size

207. *Sattleria melaleucella* (Constant) –
   a) ♂, Switzerland, Appenzell, Säntis, 2300 m, 22.vii.1989, leg. P. Huemer (TLMF).
   b) ♂, Austria, Nordtirol, Muttekopf, 2700 m, 5.viii.1953, leg. K. Burmann, abdomen in capsule (TLMF) (paratype of *Gelechia dzieduszyckii* ssp. *fusca* Burmann).
   c) ♂, Italy, Verona, Monte Baldo, Longino, 2200 m, 29.vi.1985, leg. G. Tarmann (TLMF).
   d) ♂, Italy, Piemonte, Colle della Lombardia, 2300 m, 27.vii.2009, leg. P. Skou & B. Skule (ZMUC).
   e) ♀, Austria, Nordtirol, Muttekopf, 2700 m, 5.viii.1953, leg. K. Burmann (TLMF). (paratype of *Gelechia dzieduszyckii* ssp. *fusca* Burmann).
208. *S. triglavica* Povolný –
   a) ♂, Montenegro, Durmitor, Velika Kalica, 1900 m, 26.vii.1985, leg. P. Jakšić, gen. slide Huemer GEL 323 (TLMF).
   ssp. *basistrigella* Huemer –
   b) ♂, Switzerland, Valais, Zermatt,Trift, 2800 m, 1.viii.1981, leg. S. E. Whitebread (ZMUC).
   c) ♀, paratype, Switzerland, Valais, Zermatt,Triftkummen, 30.vii.1934, leg. P. Weber, gen. slide L. Pitkin 88 (TLMF).
   d) ♂, paratype, Switzerland, Valais, Zermatt,Triftkummen, 30.vii.1934, leg. P. Weber (TLMF).
209. *S. styriaca* Pitkin & Sattler –
   a) ♂, Austria, Steiermark, Zirbitzkogel, 1700, ex l. V.1954, leg. R. Pinker, gen. slide Huemer GEL 160 (TLMF).
   b) ♀, holotype, Austria, Styria, Seetaler Alpen, Zirbitzkogel, 2200 m, e.l. 1.viii.1911, leg. ?, gen. slide Mus. Vind. 11240 (NHMW).
210. *S. marguareisi* Huemer & Sattler –
   ♂, paratype, France, Alpes Maritimes, Marguareis, western slope, Navela, 2100-2200 m, 18.-19. vii.1991, leg. P. Huemer & G. Tarmann (ZMUC).
211. *S. dzieduszyckii* (Nowicki) –
   a) ♂, Poland, Tatra Mts., Trzydniowlański W, 1650 m, 11.viii.1987, leg. J. Buszko (ZMUC).
   b) ♀, Poland, Tatra Mts., Liljowe, 1950 m, 5.vii.1951, leg. S. Błeszyński, gen. slide Pitkin 98 (TLMF).
   c) ♂, Romania, Central Carpathian Mts., Mt. Făgarăs, Lac Capra, 2100 m, 8.viii.1992, leg. S. & Z. Kovács (ZMUC).

207a
207b
207c
207d
207e
208a
208b
208c
208d
209a
209b
210
211a
211b
211c

Male genitalia illustrations

1

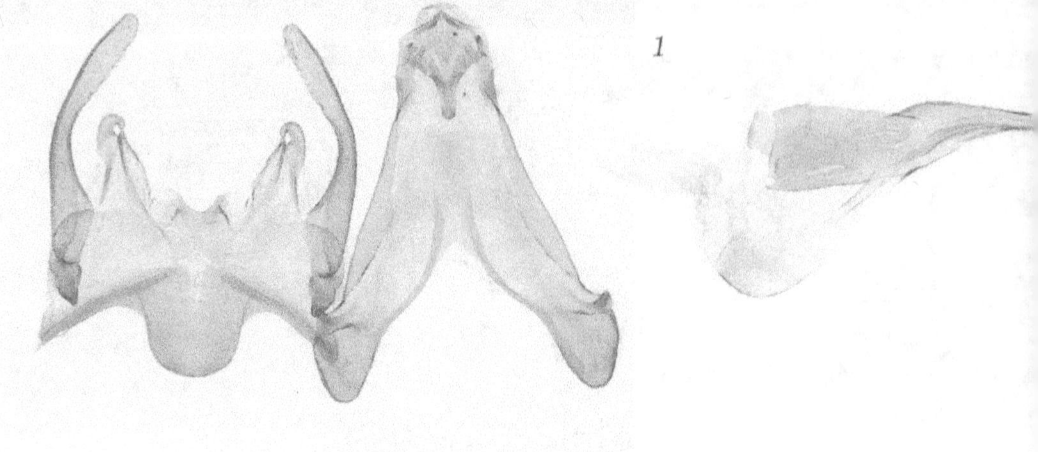

2

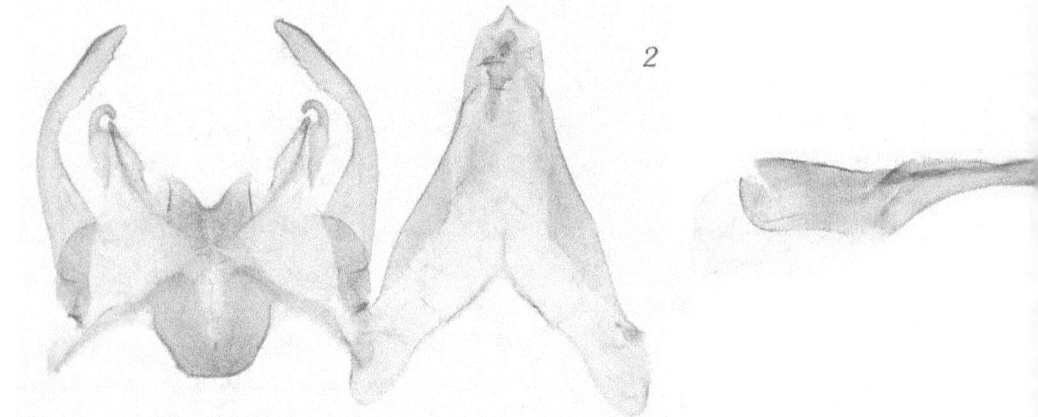

3

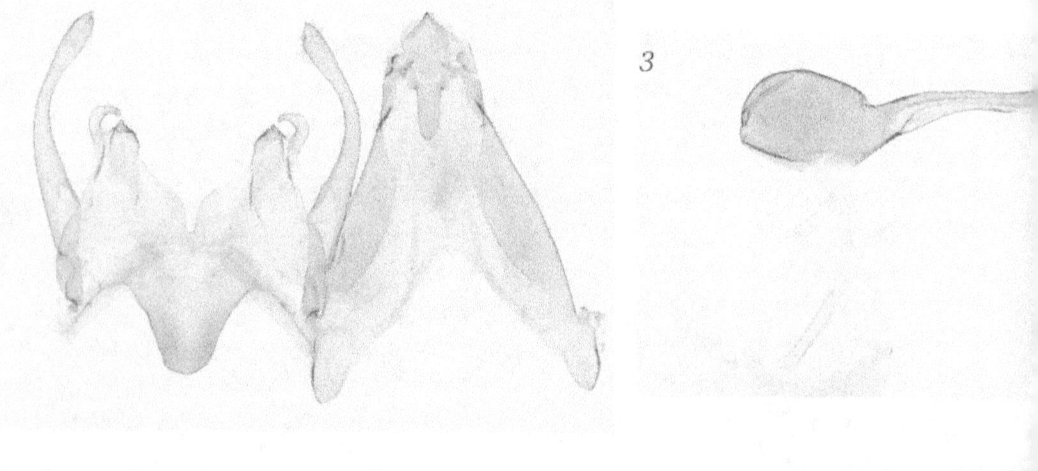

MALE GENITALIA

1. *Gnorimoschema soffneri* (Riedl) – Greece, GEL 635 P.H. (TLMF).
2. *Gnorimoschema herbichii* (Nowicki) – Russia, GU 94/497 P.H. (TLMF).
3. *Gnorimoschema bodillum* Karsholt & Nielsen – Denmark, GEL 53 P.H. (TLMF).

4

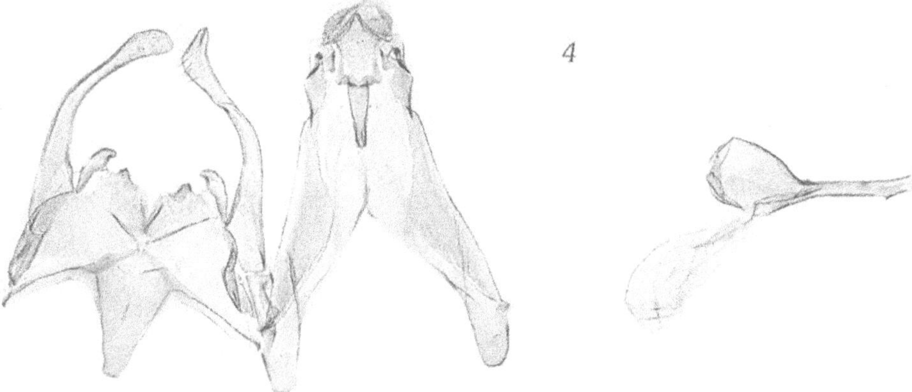

5

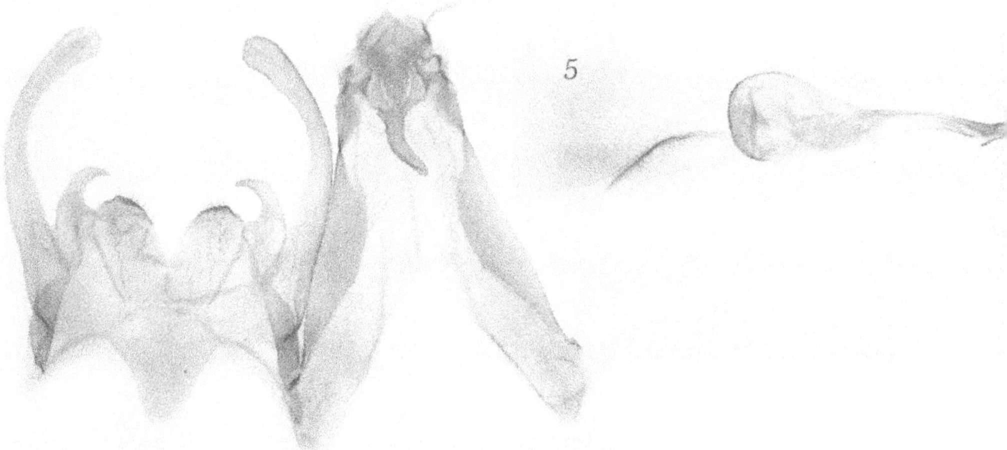

6

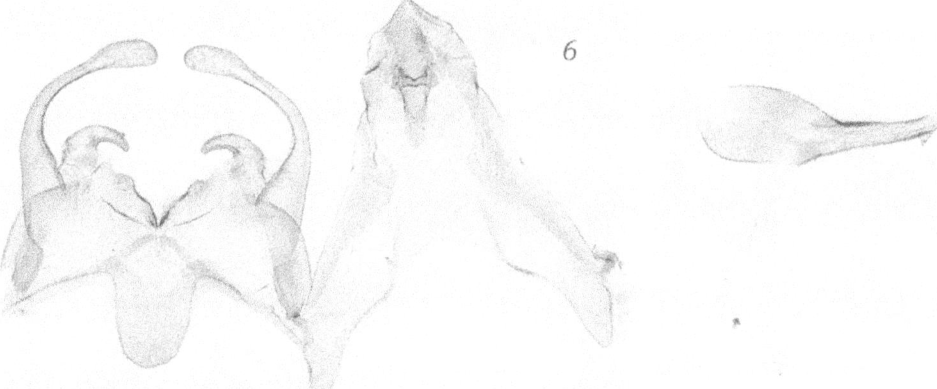

Male Genitalia

4. *Gnorimoschema nupponeni* sp. n. – Paratype, Russia, GU 09/1280 P.H. (NUPP).
5. *Gnorimoschema robustella* (Staudinger) – Paralectotype GU 08/1261 (ZMHU).
6. *Gnorimoschema steueri* Povolný – Germany, GU 89/016 P.H. (TLMF).

7

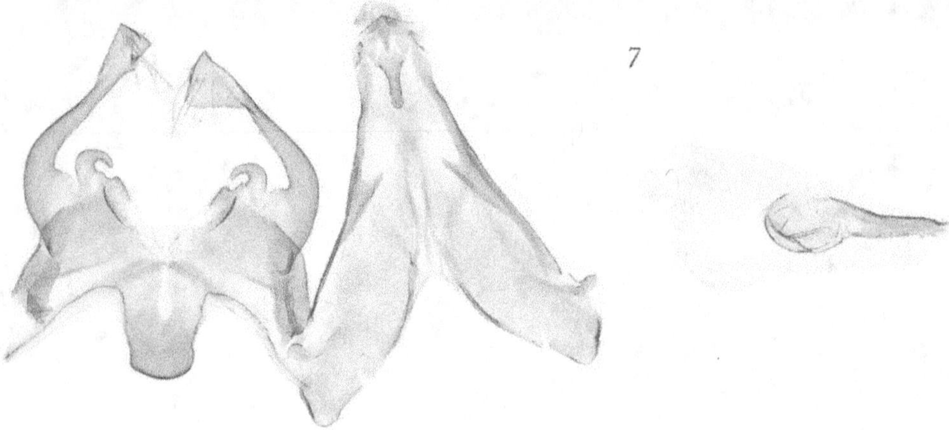

8

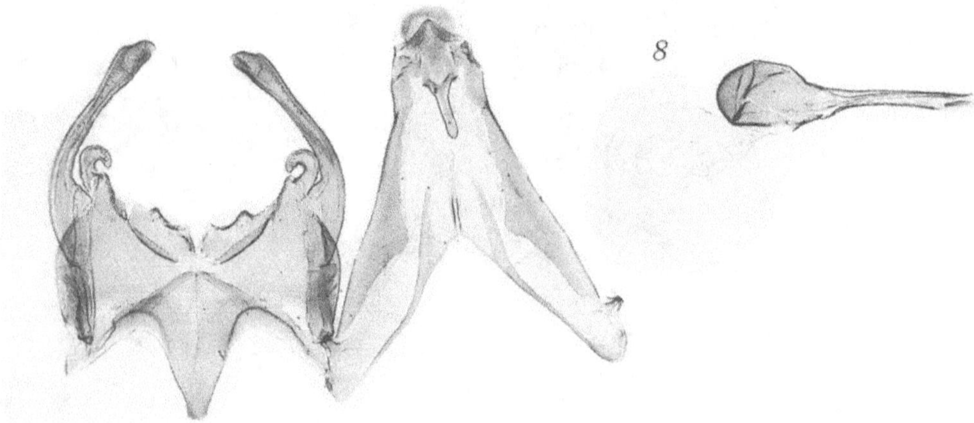

9

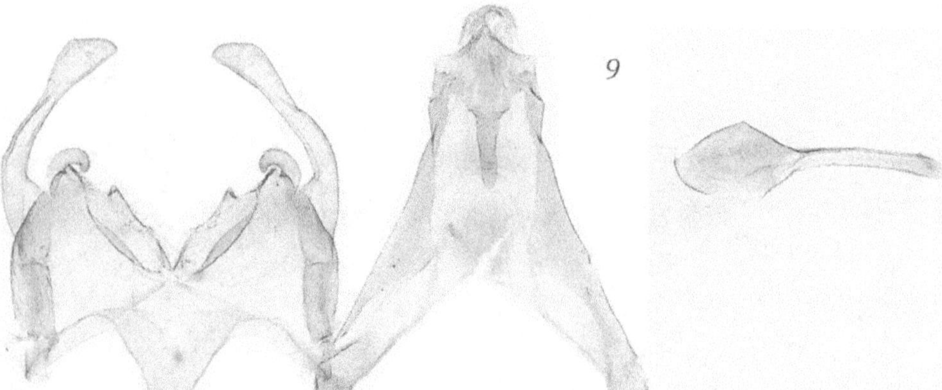

MALE GENITALIA

7. *Gnorimoschema epithymella* (Staudinger) – Finland, GU 89/037 P.H. (TLMF).
8. *Gnorimoschema nordlandicolella* (Strand) – Finland, GU 94/473 P.H. (ZMUC).
9. *Gnorimoschema nilsi* Huemer – Paratype, Austria, GEL 518 P.H. (TLMF).

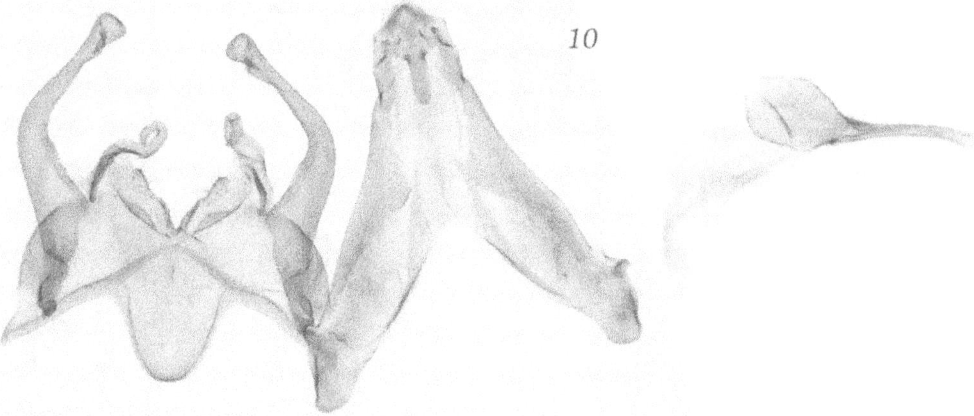

10

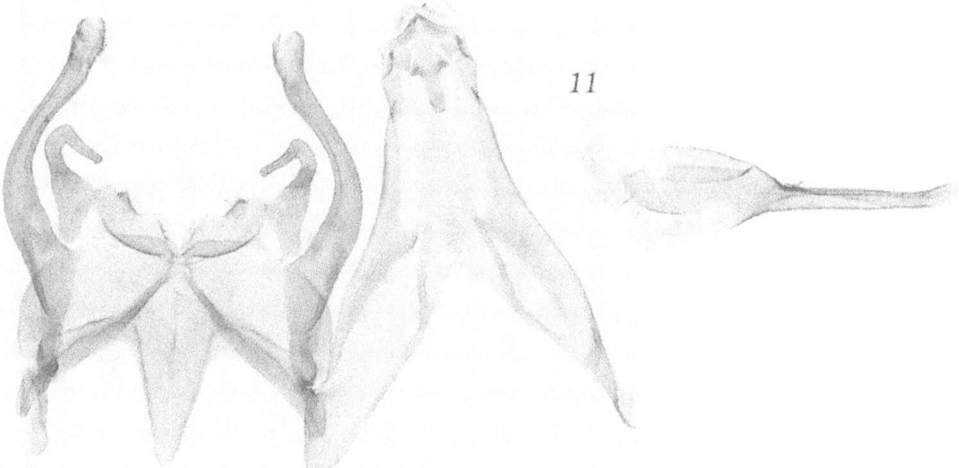

11

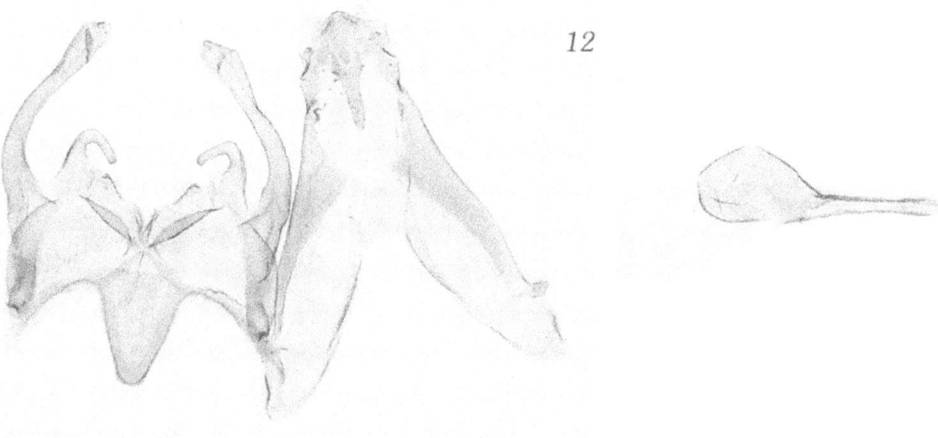

12

MALE GENITALIA

10. *Gnorimoschema valesiella* (Staudinger) – Andorra, GU 01/1046 P.H. (ZMUC).
11. *Gnorimoschema streliciella* (Herrich-Schäffer) – Finland, GEL 553 P.H. (TLMF).
12. *Gnorimoschema hoefneri* (Rebel) – Austria, GEL 50 P.H. (TLMF).

13

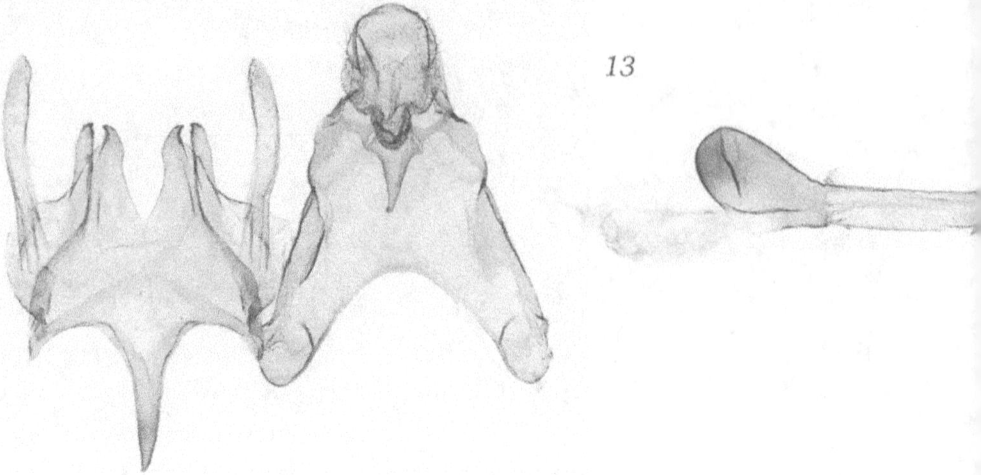

14

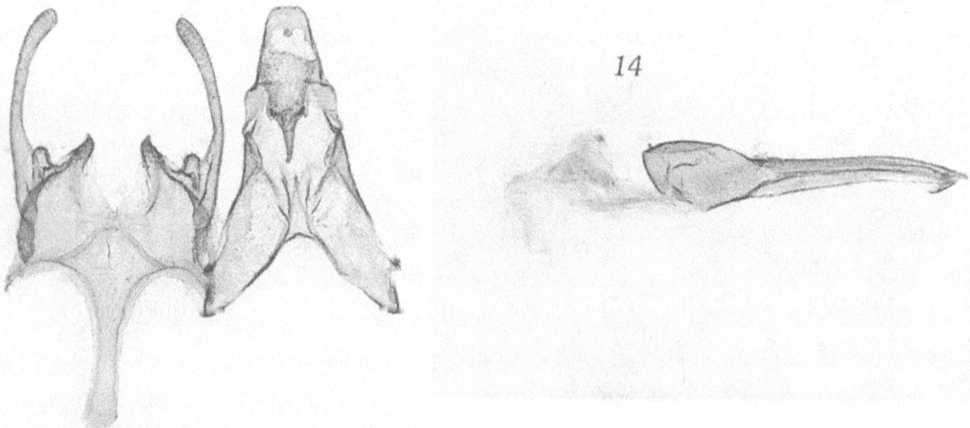

15

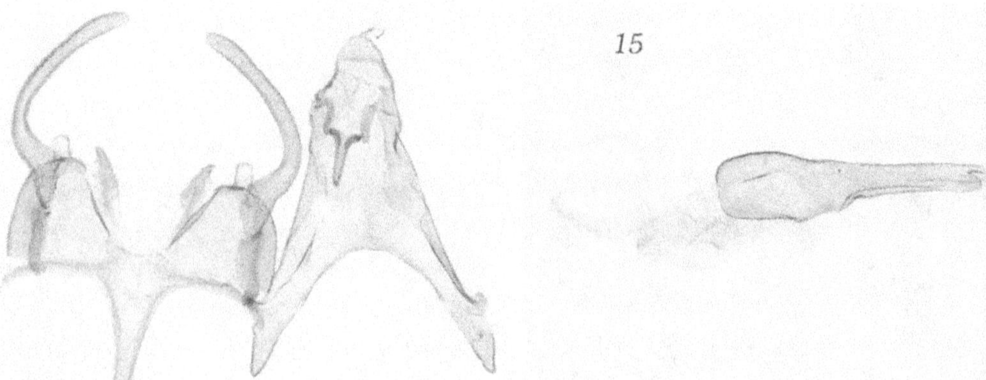

Male Genitalia

13. *Scrobipalpopsis petasitis* (Pfaffenzeller) – Austria, GEL 1077 P.H. (TLMF).
14. *Scrobipalpa aptatella* (Walker) – Egypt, gen. slide 26582 (BMNH).
15. *Scrobipalpa kasyi* Povolný – Macedonia, gen. slide 16517 (NHMW).

16

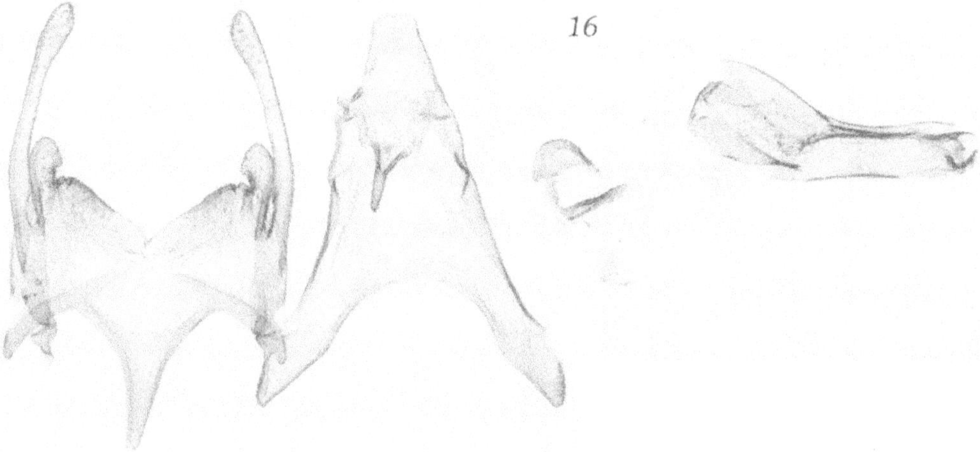

17

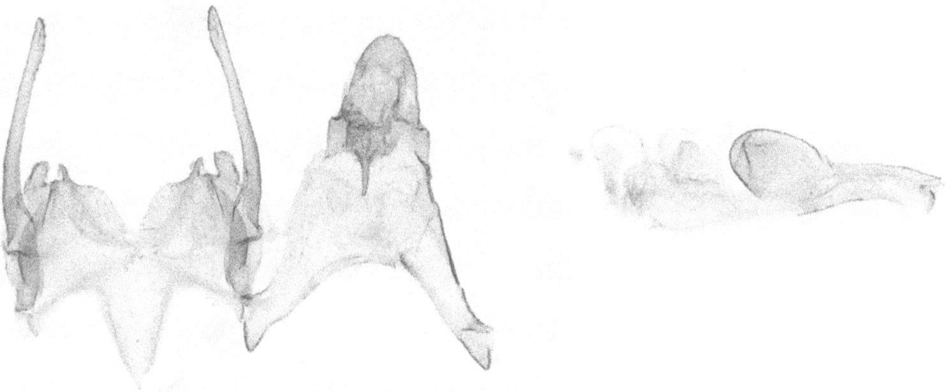

18

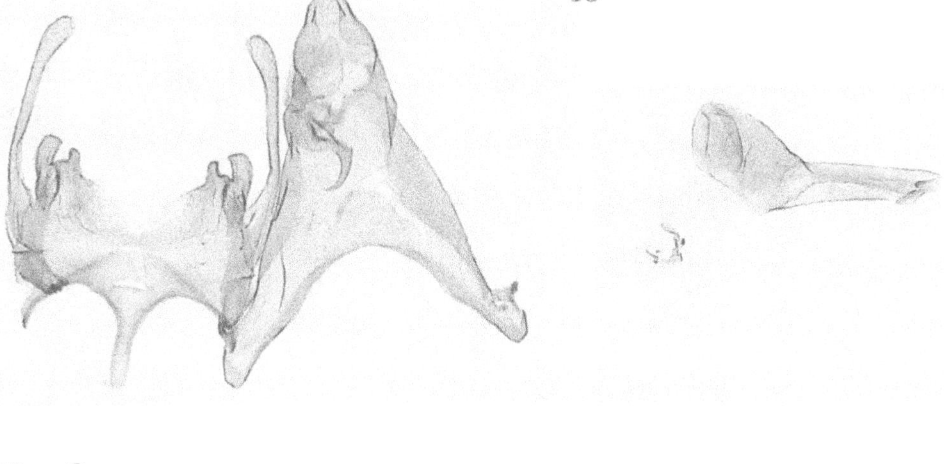

MALE GENITALIA

16. *Scrobipalpa notata* (Povolný) – Russia, GU 03/1193 P.H. (NUPP).
17. *Scrobipalpa acuminatella* (Sircom) – Austria, GEL 389 P.H. (TLMF).
18. *Scrobipalpa skulei* sp. n. – Paratype, Slovenia, gen. slide 4763 H.H. (ZMUC).

19

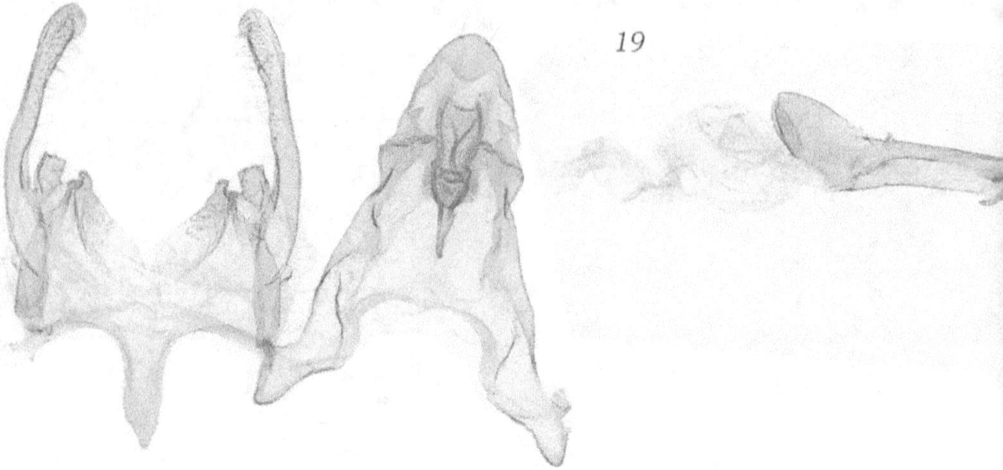

20

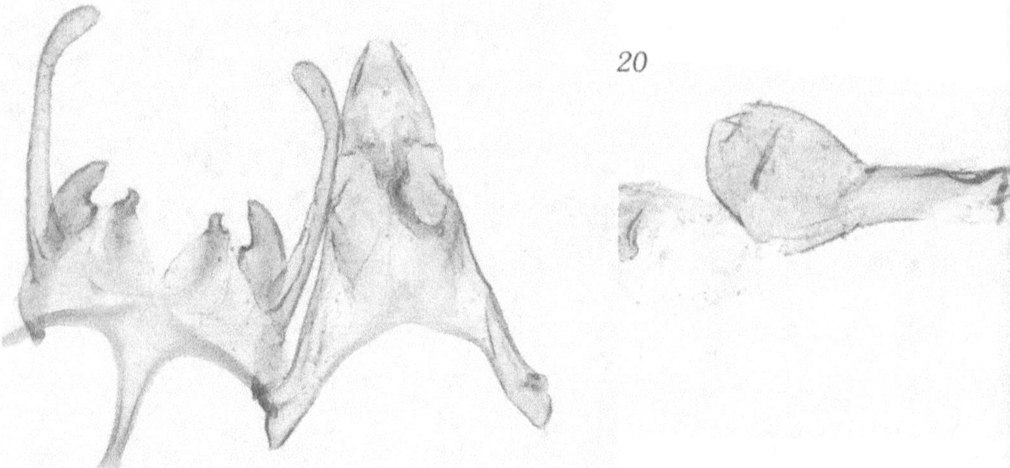

21

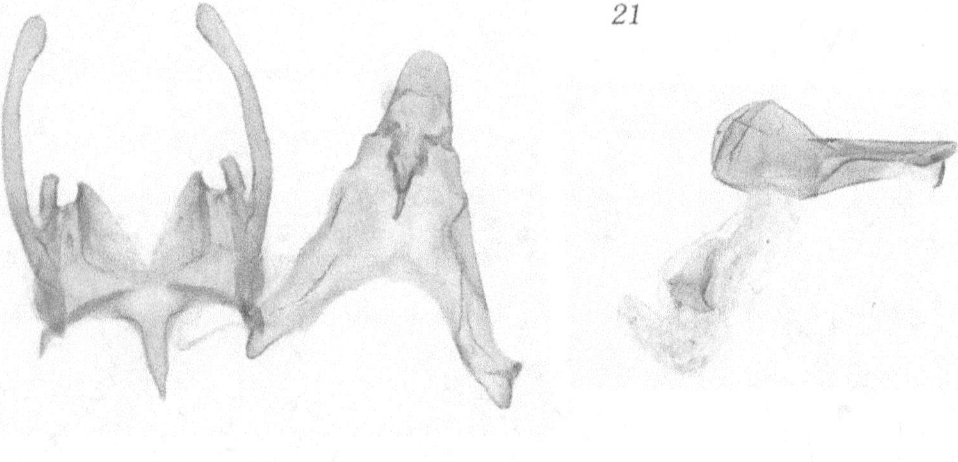

MALE GENITALIA

19. *Scrobipalpa hungariae* (Staudinger) – Hungary, gen. slide 16531 (NHMW).
20. *Scrobipalpa adaptata* (Povolný) – Holotype, Russia, GU 03/1178 P.H. (ZMKU).
21. *Scrobipalpa brahmiella* (Heyden) – Germany, GU 00/901 P.H. (ZSM).

22

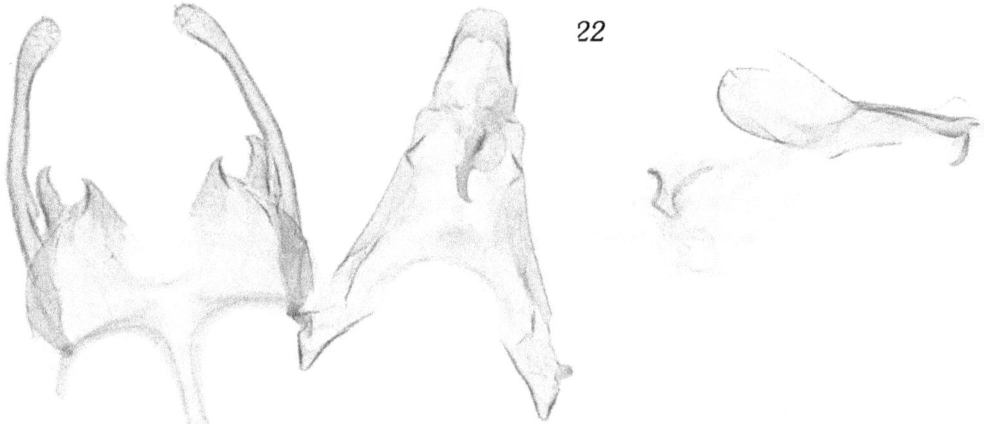

23

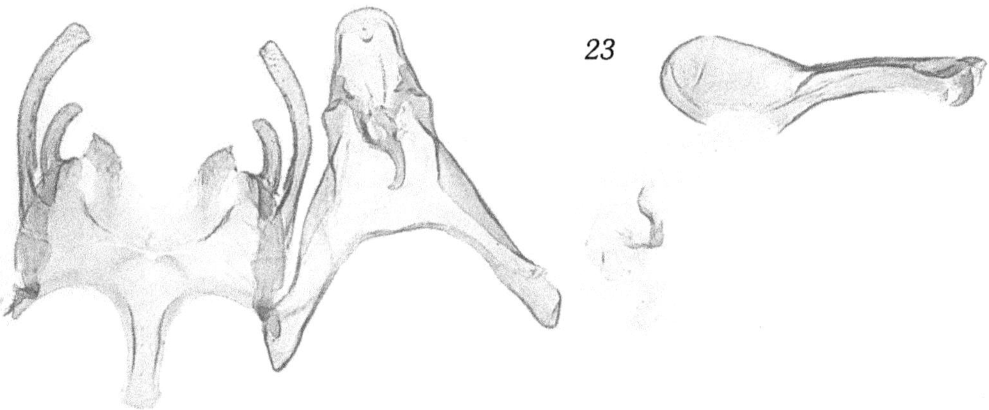

24

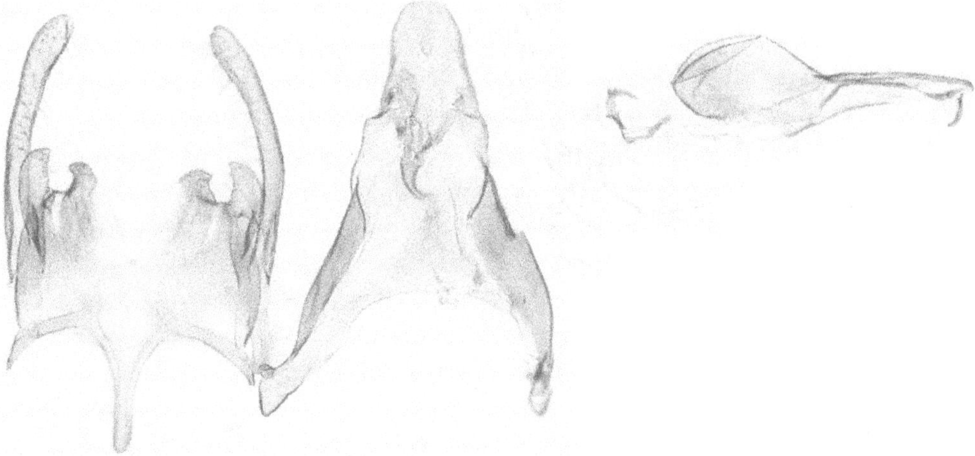

Male Genitalia

22. *Scrobipalpa vasconiella* (Rössler) – France, GU 01/1013 P.H. (ZMUC).
23. *Scrobipalpa dorsolutea* sp. n. – Holotype, Russia, gen. slide 09/1288 P.H. (NUPP).
24. *Scrobipalpa amseli* Povolný – Spain, gen. slide 2566 H.H. (ZMUC).

25

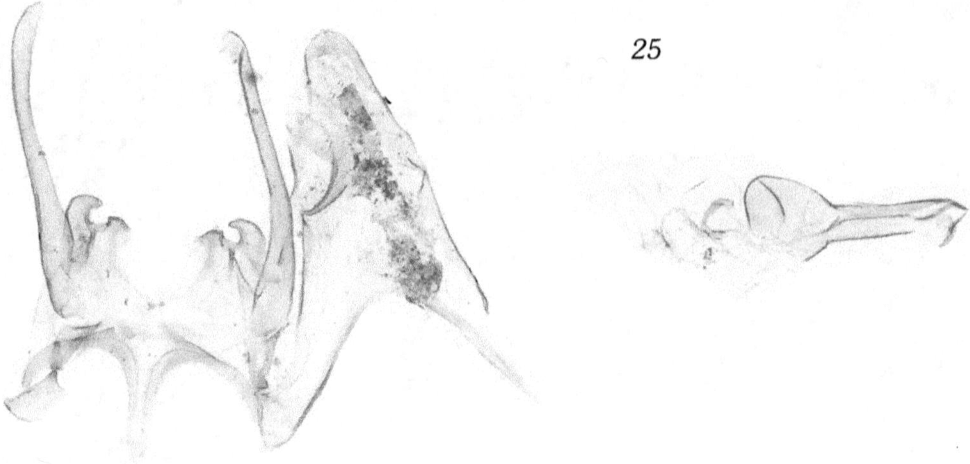

26

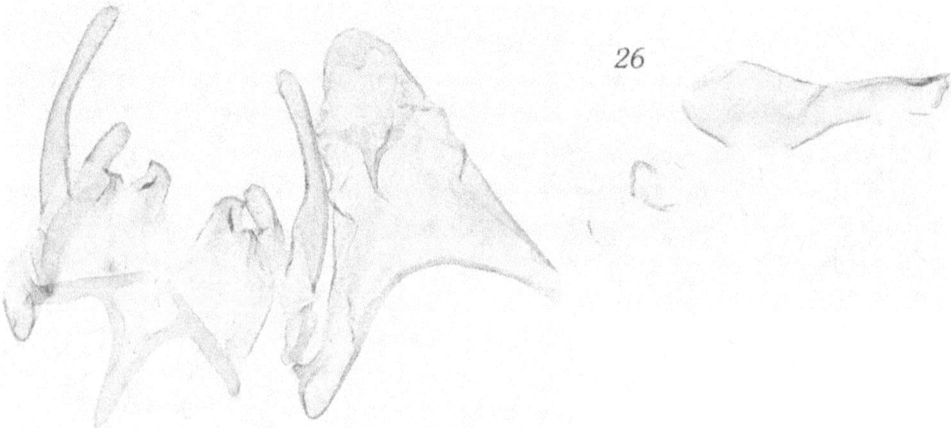

27

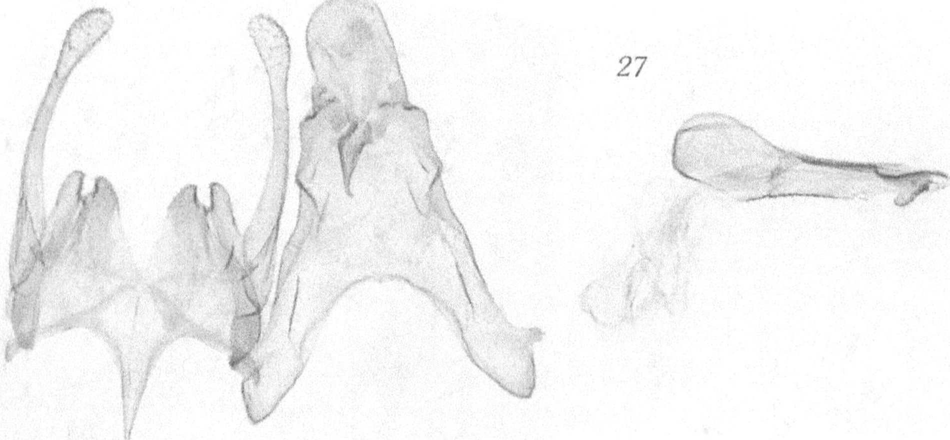

MALE GENITALIA

25. *Scrobipalpa hyssopi* Nel – Holotype, France, gen. slide 14755 J.N. (TLMF).
26. *Scrobipalpa montanella* (Chrétien) – Syntype, France, GU 06/1247 (MNHN).
27. *Scrobipalpa corleyi* sp. n. – Holotype, Portugal, GU 01/1093 P.H. (ZMUC).

28

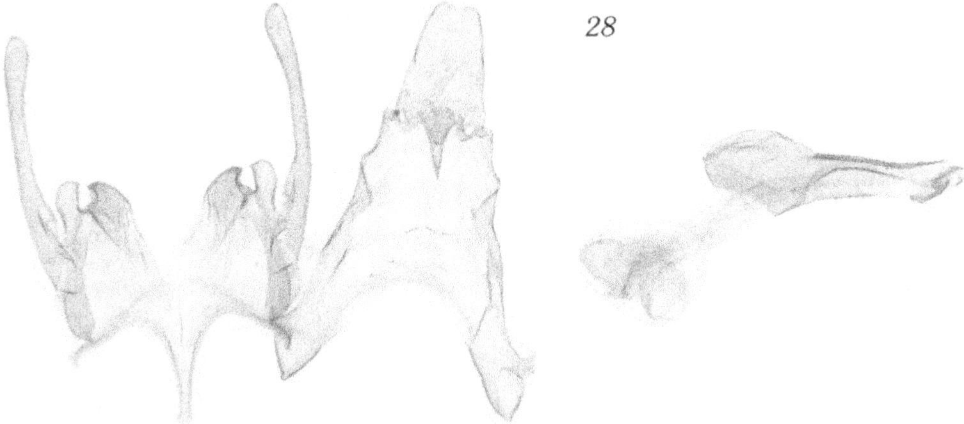

29

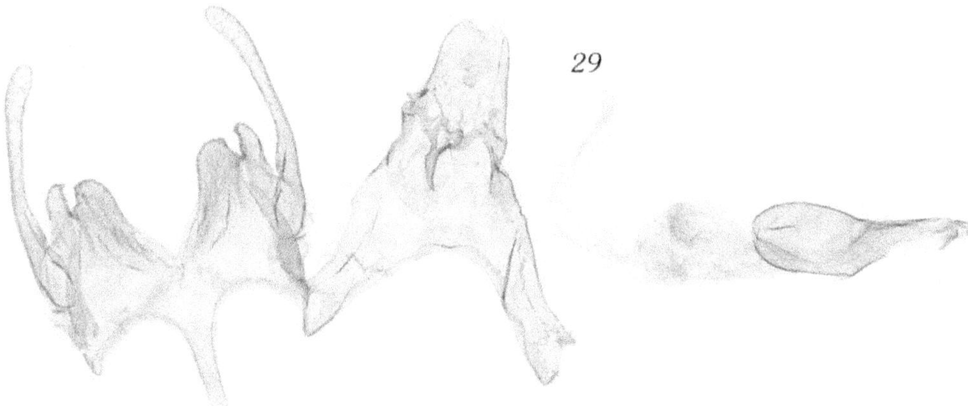

30

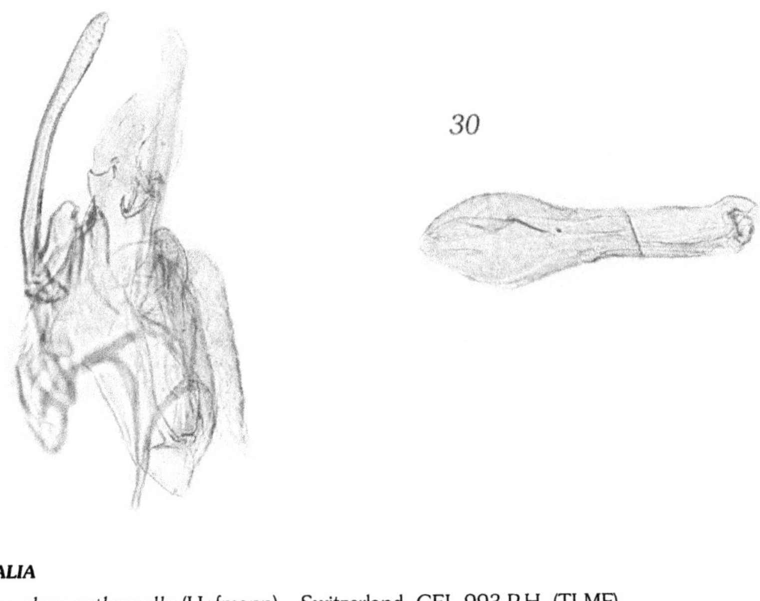

Male Genitalia

28. *Scrobipalpa chrysanthemella* (Hofmann) – Switzerland, GEL 993 P.H. (TLMF).
29. *Scrobipalpa proclivella* (Fuchs) – Austria, GEL 995 P.H. (TLMF).
30. *Scrobipalpa frugifera* Povolný – Paratype, Mongolia, gen. slide 3340 Gz. Povolný (HNHM).

31

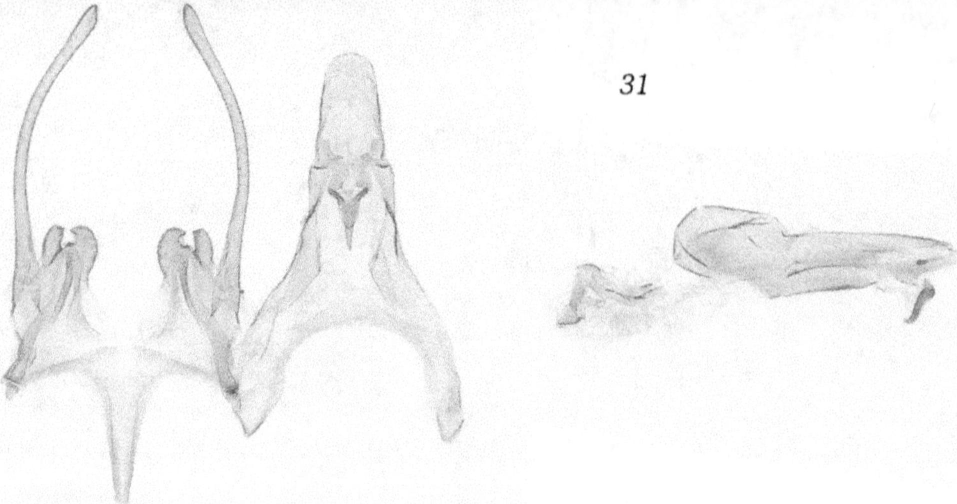

32

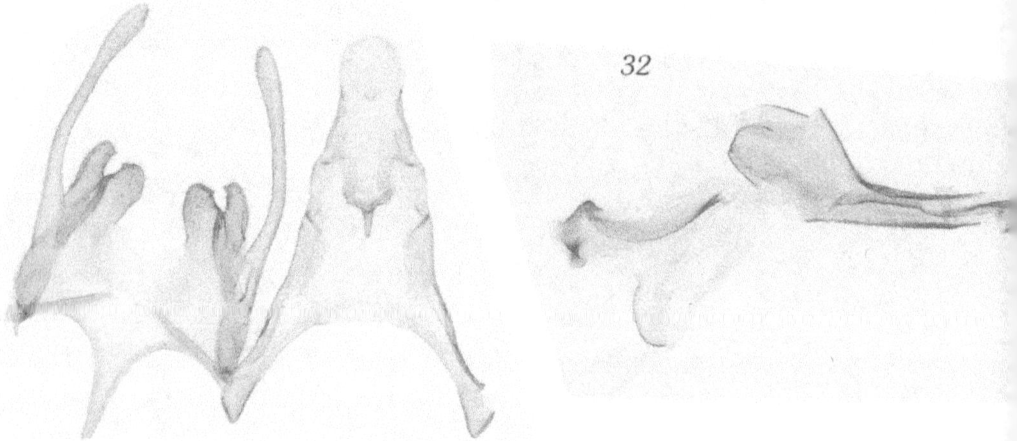

33

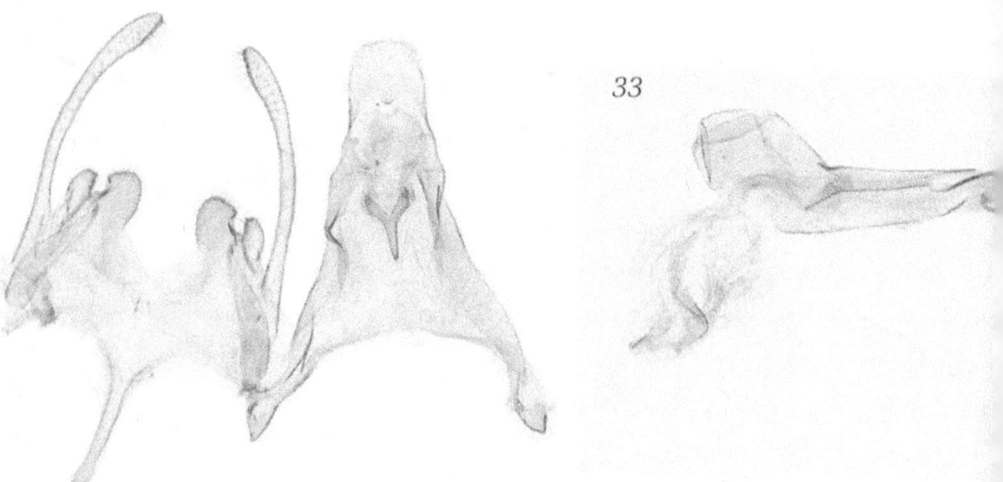

Male Genitalia

31. *Scrobipalpa oleksiyella* sp. n. – Paratype, Russia, GU 02/1121 P.H. (JUNN).
32. *Scrobipalpa smithi* Povolný & Bradley – Lebanon, gen. slide 16521 (NHMW).
33. *Scrobipalpa occulta* (Povolný) – Russia, GU 01/1104 P.H. (JUNN).

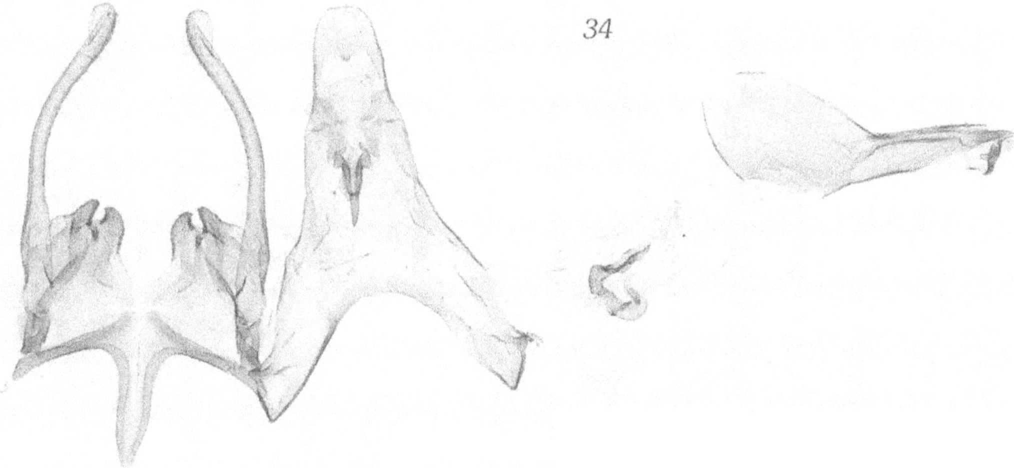

34

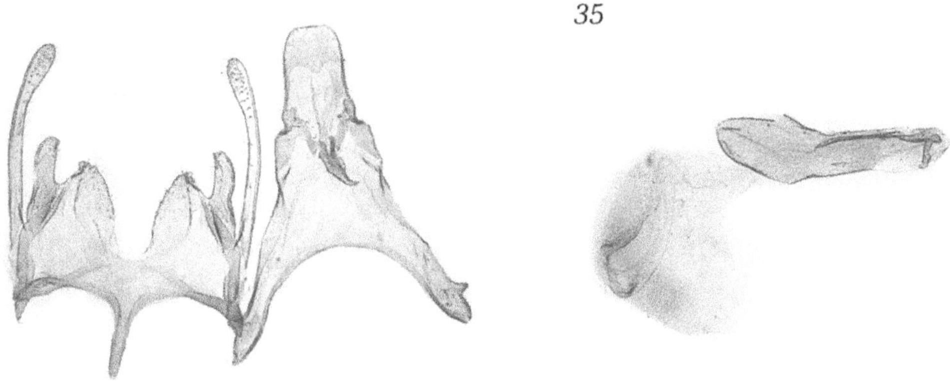

35

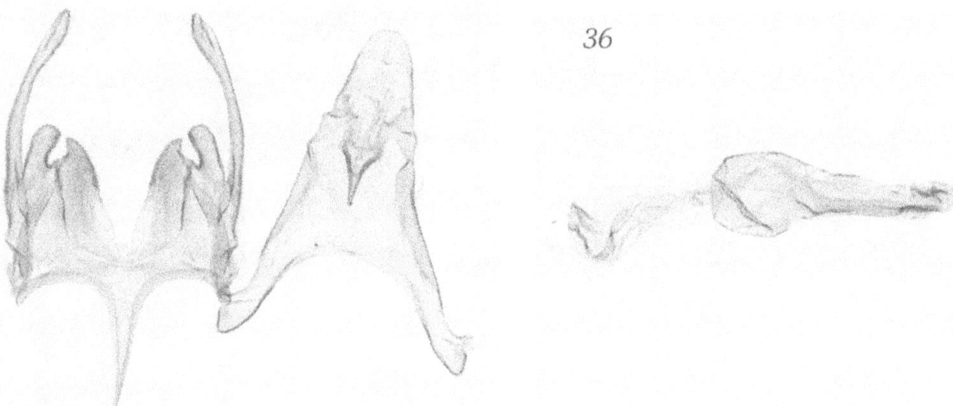

36

Male Genitalia

34. *Scrobipalpa grisea* Povolný – Russia, GU 03/1195 P.H. (NUPP).
35. *Scrobipalpa usingeri* Povolný – Mongolia, GU 06/1294 P.H. (HNHM).
36. *Scrobipalpa clintoni* Povolný – Denmark, GEL 991 P.H. (TLMF).

37

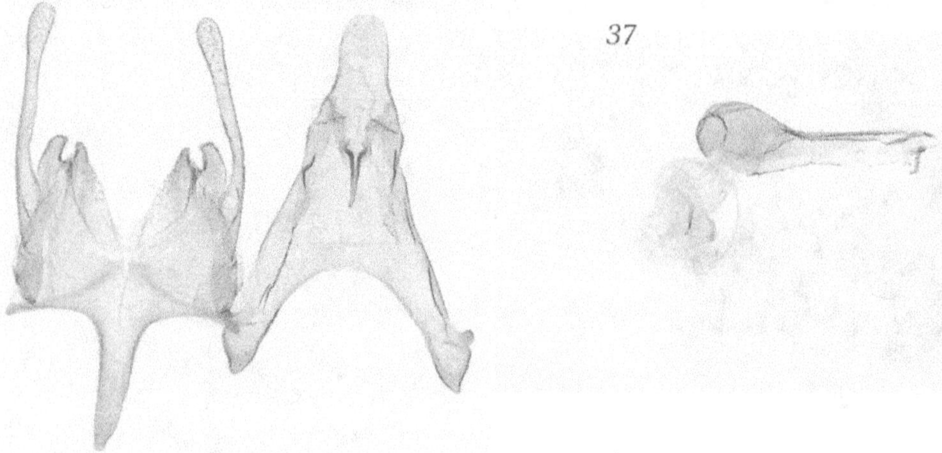

38

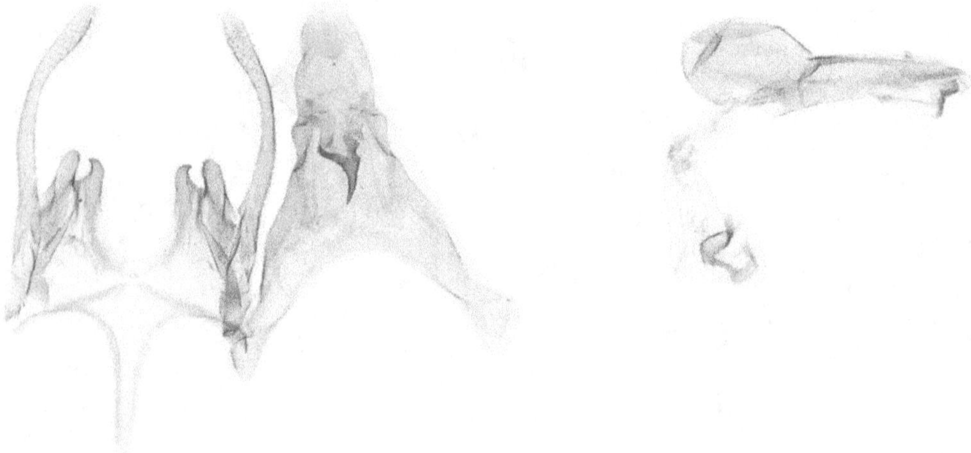

39

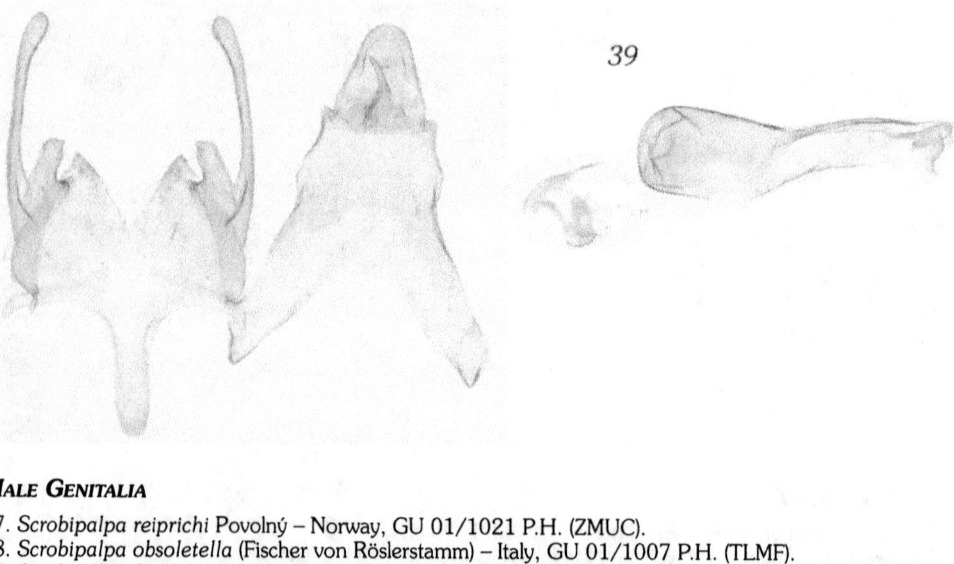

Male Genitalia

37. *Scrobipalpa reiprichi* Povolný – Norway, GU 01/1021 P.H. (ZMUC).
38. *Scrobipalpa obsoletella* (Fischer von Röslerstamm) – Italy, GU 01/1007 P.H. (TLMF).
39. *Scrobipalpa feralella* (Zeller) – Italy, GEL 408 P.H. (TLMF).

40

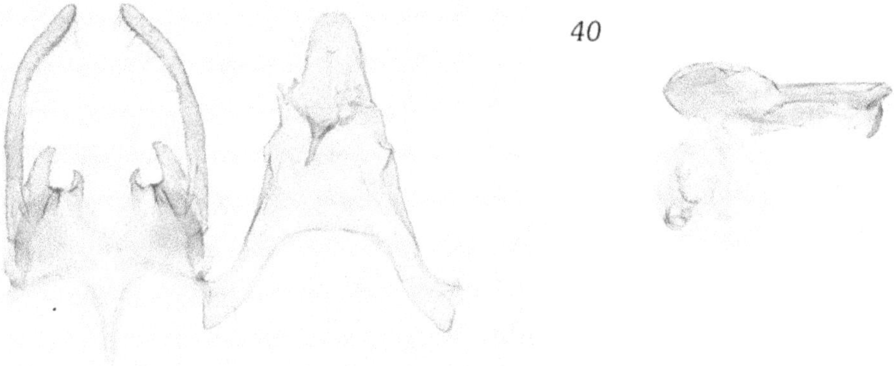

41

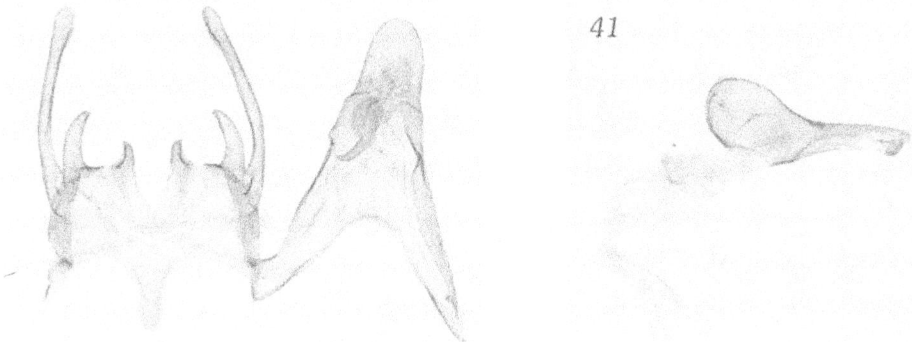

42

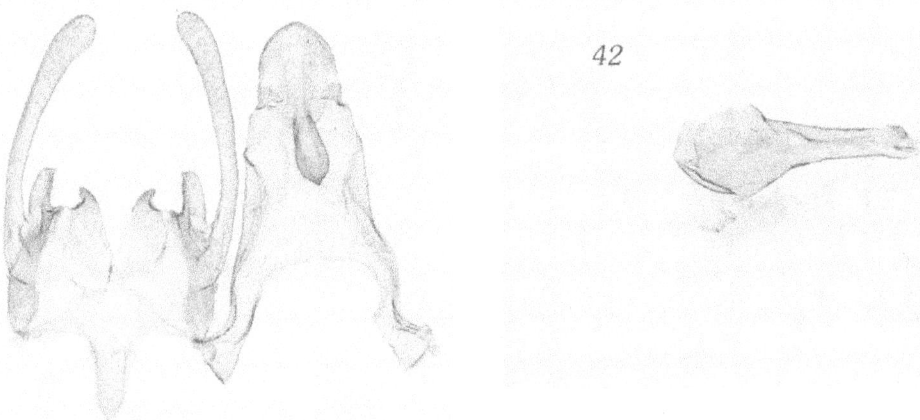

MALE GENITALIA

40. *Scrobipalpa halonella* (Herrich-Schäffer) – Italy, GU 02/1137 P.H. (MFSN).
41. *Scrobipalpa perinii* (Klimesch) – Italy, GEL 946 P.H. (TLMF).
42. *Scrobipalpa phagnalella* (Constant)*ww*Spain, GU 00/896 P.H. (ZSM).

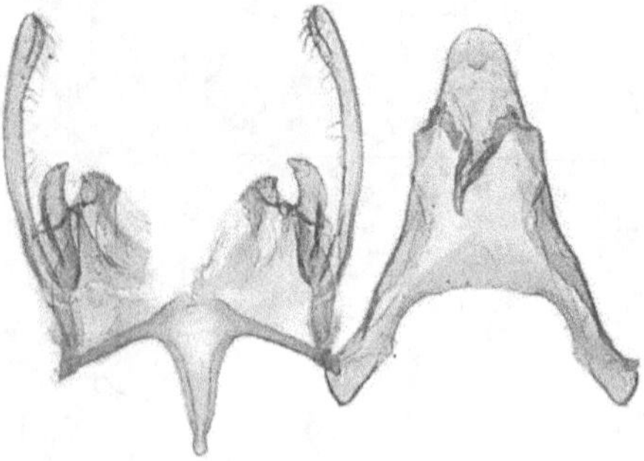

43

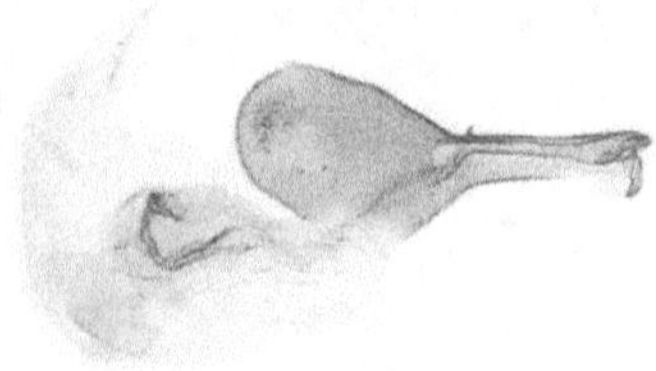

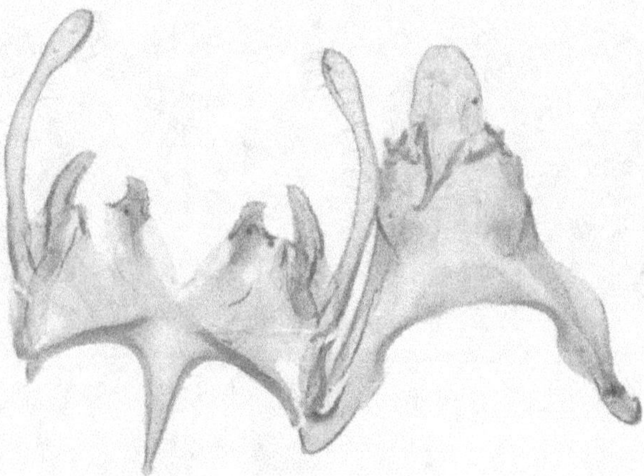

44

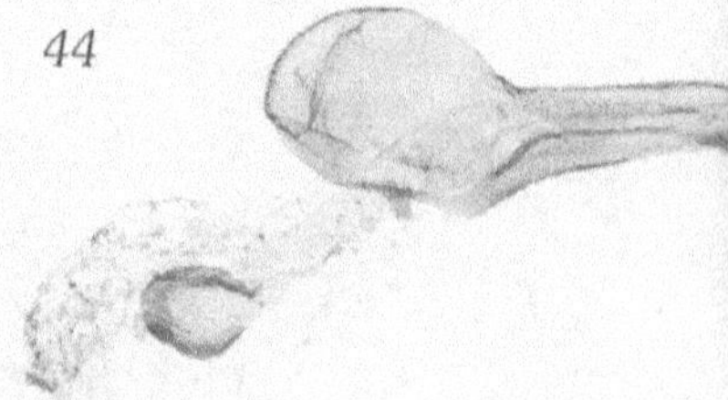

45

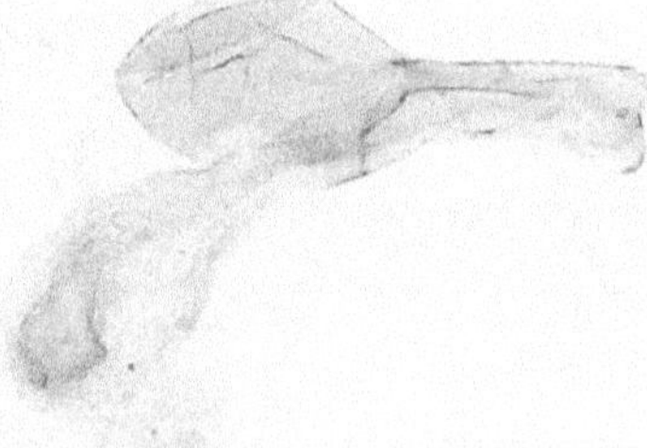

MALE GENITALIA

43. *Scrobipalpa tokari* sp. n. – Paratype, Croatia, GU 09/1289 P.H. (TOKA).
44. *Scrobipalpa karadaghi* (Povolný) – Holotype, Russia, GU 03/1179 P.H. (ZMKU).
45. *Scrobipalpa heimi* sp. n. – Paratype, Russia, GU 01/1110 P.H. (JUNN).

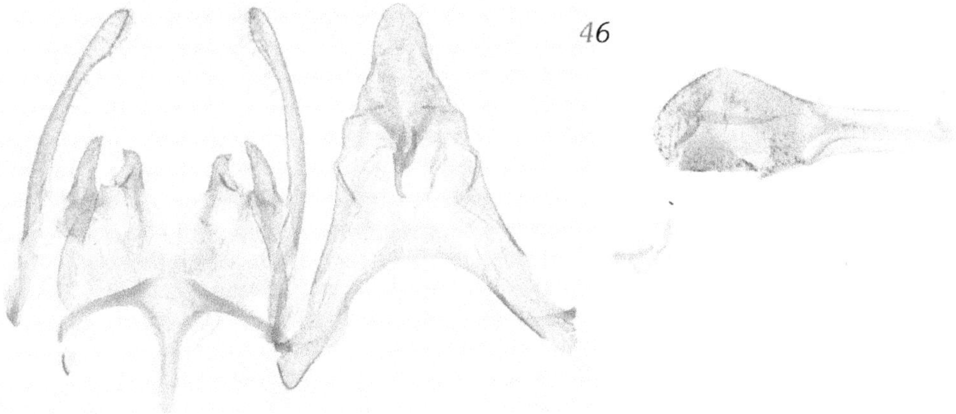
46

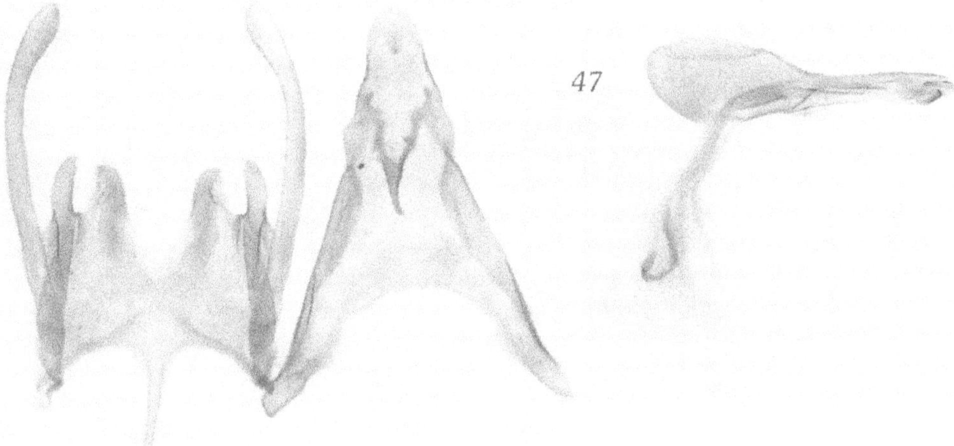
47

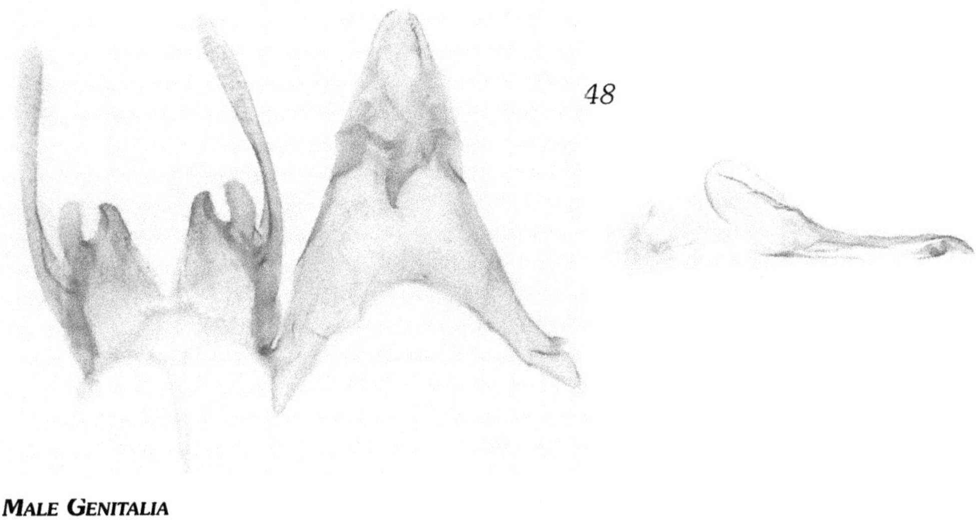
48

MALE GENITALIA

46. *Scrobipalpa acuta* (Povolný) – Holotype, Russia, GU 03/1175 P.H.; Ukraine, write gen. slide 2876 H.H. (phallus) (ZMKU).
47. *Scrobipalpa soffneri* Povolný – Russia, GU 08/1266 P.H. (NUPP).
48. *Scrobipalpa jariorum* sp. n. – Paratype, Bulgaria, GU 08/1272 P.H. (KAIT).

49

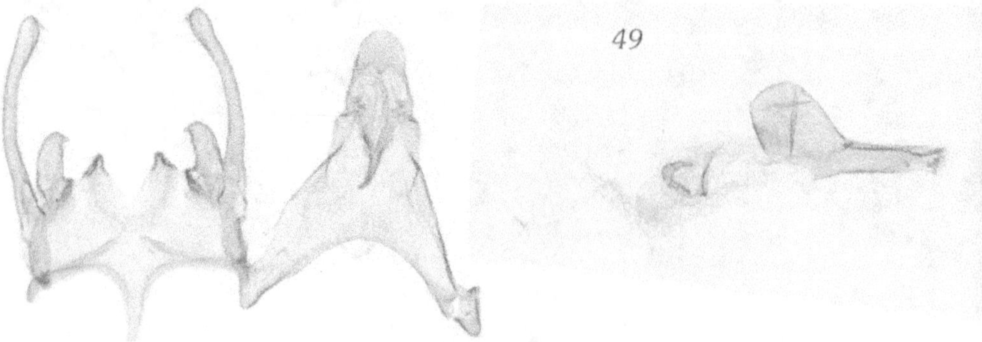

50

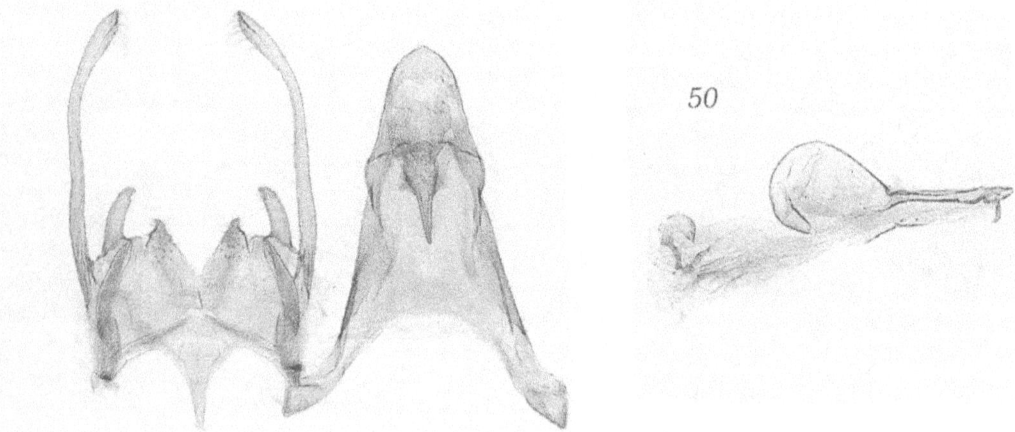

51

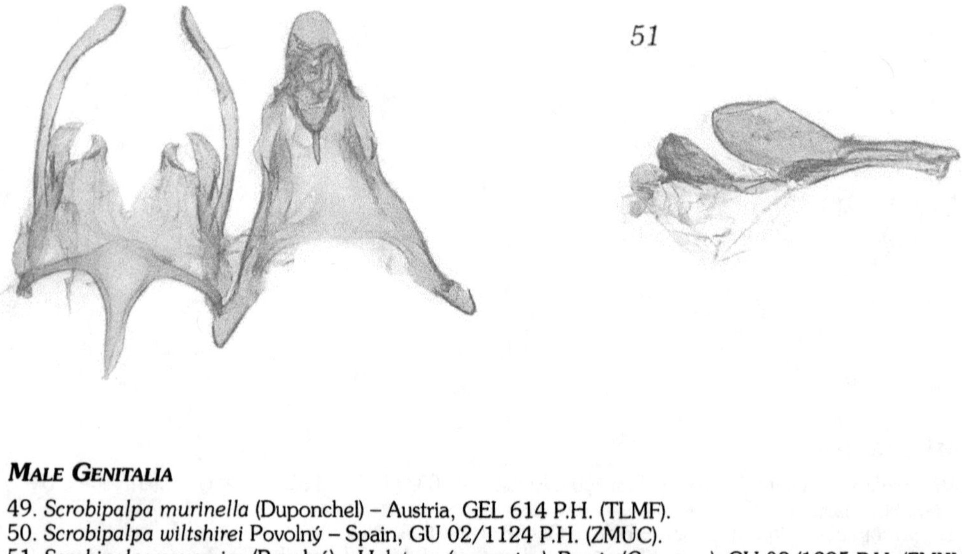

Male Genitalia

49. *Scrobipalpa murinella* (Duponchel) – Austria, GEL 614 P.H. (TLMF).
50. *Scrobipalpa wiltshirei* Povolný – Spain, GU 02/1124 P.H. (ZMUC).
51. *Scrobipalpa caucasica* (Povolný) – Holotype *(caucasica)*, Russia (Caucasus), GU 09/1295 P.H. (ZMKU).

52

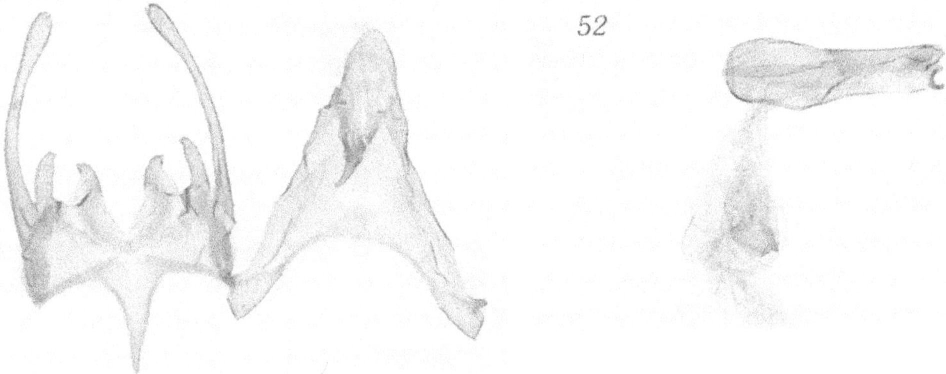

53

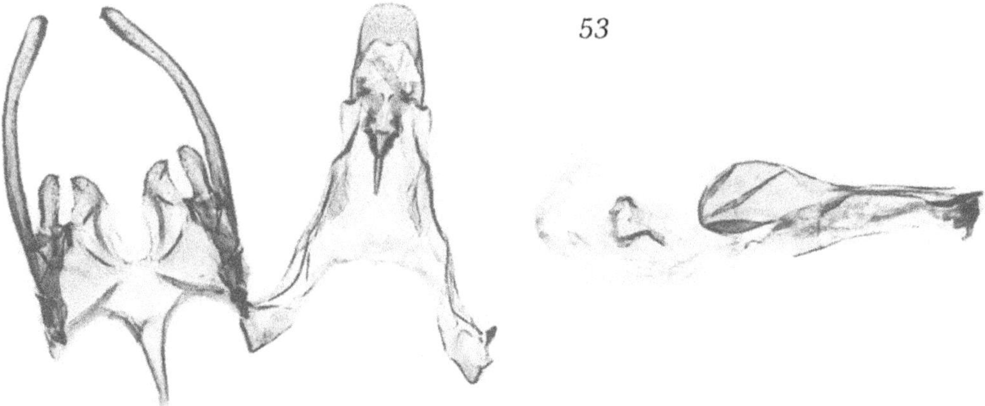

54

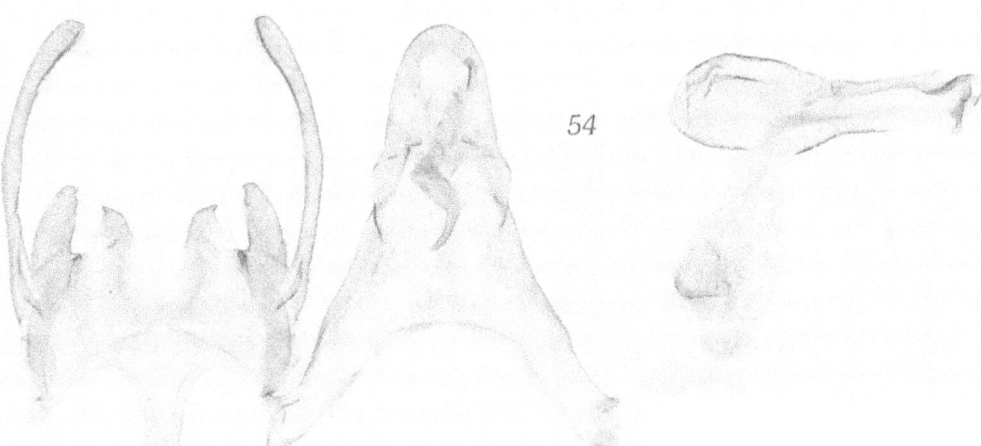

Male Genitalia

52. *Scrobipalpa pauperella* (Heinemann) – Czechia, GEL 1064 P.H. (TLMF).
53. *Scrobipalpa spumata* (Povolný) – Russia, GU 01/1084 P.H. (JUNN).
54. *Scrobipalpa arenbergeri* Povolný – Croatia, GEL 973 P.H. (TLMF).

55

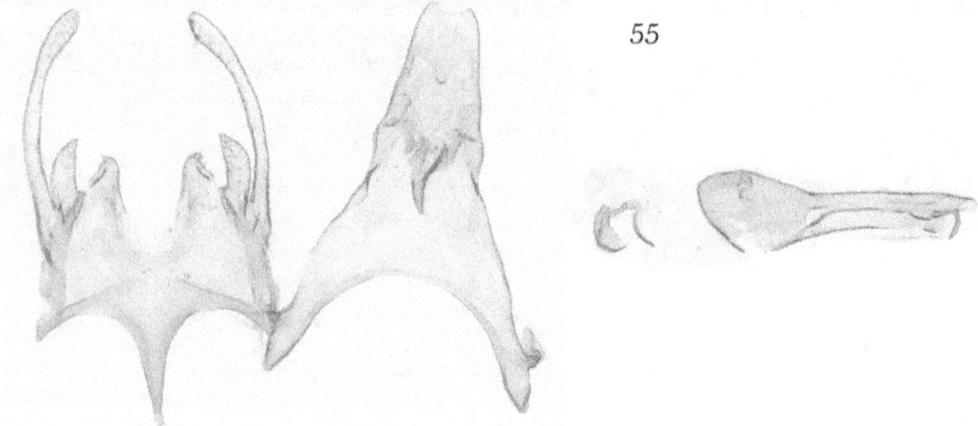

56

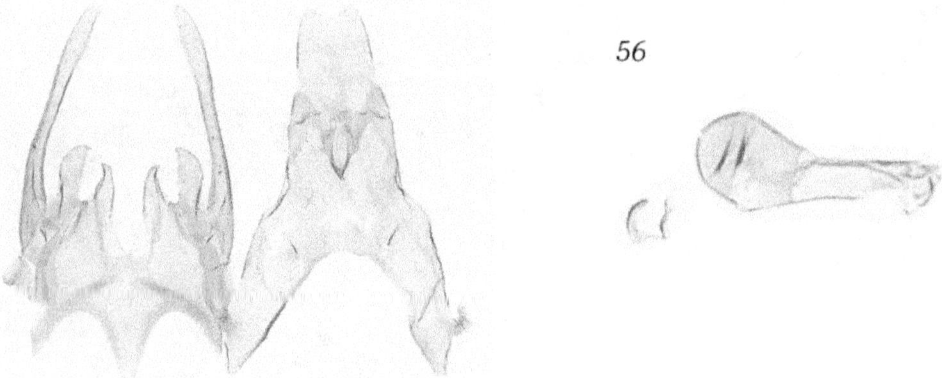

57

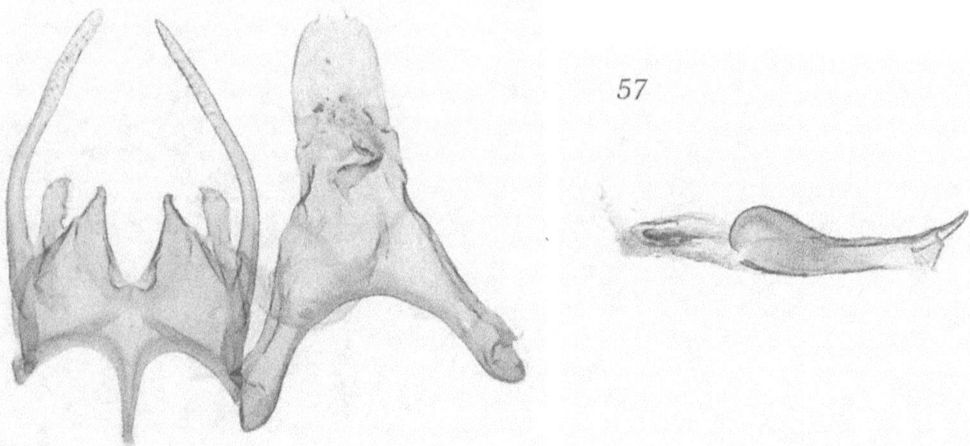

Male Genitalia

55. *Scrobipalpa nana* Povolný – Russia, GU 03/1200 P.H. (ZMUC).
56. *Scrobipalpa heretica* Povolný – 4066 H.H. (ZMUC).
57. *Scrobipalpa bigoti* Povolný – Cyprus, GEL 1102 P.H. (TLMF).

58

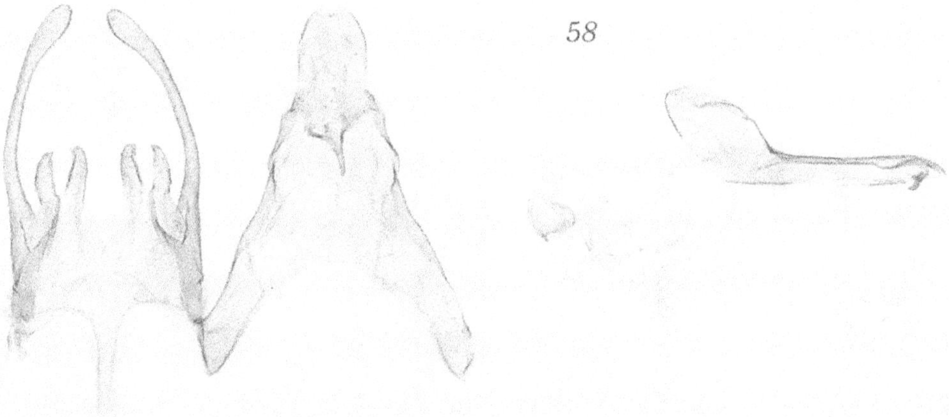

59

60

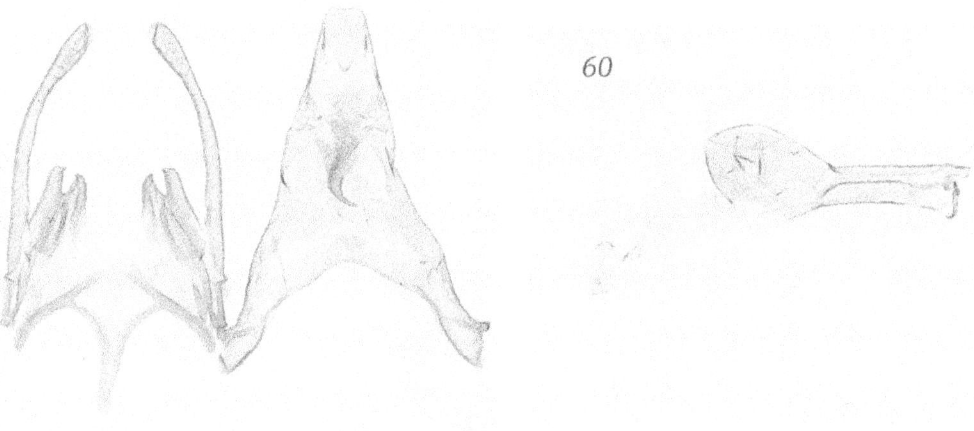

MALE GENITALIA

58. *Scrobipalpa dorsoflava* (Povolný) – Russia, GU 02/1126 P.H. (JUNN).
59. *Scrobipalpa magnificella* Povolný – Russia, GU 02/1116 P.H. (ZMUC).
60. *Scrobipalpa abstrusa* sp. n. – Paratype, Russia, gen. slide 2886 H.H. (ZMUC).

61

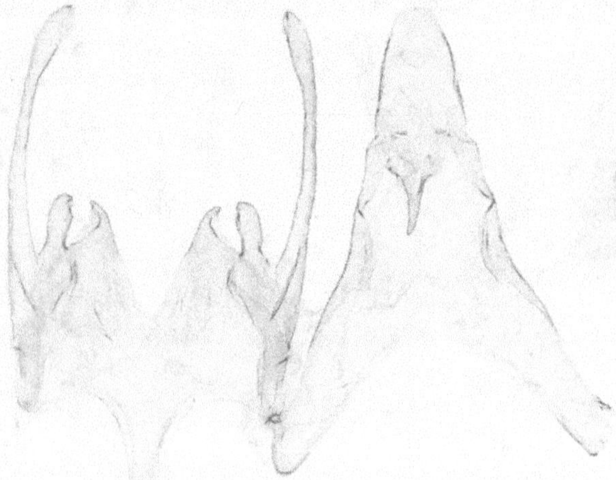

62

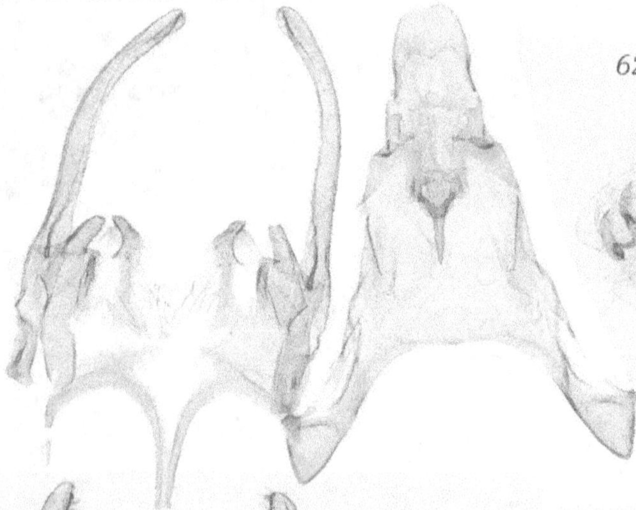

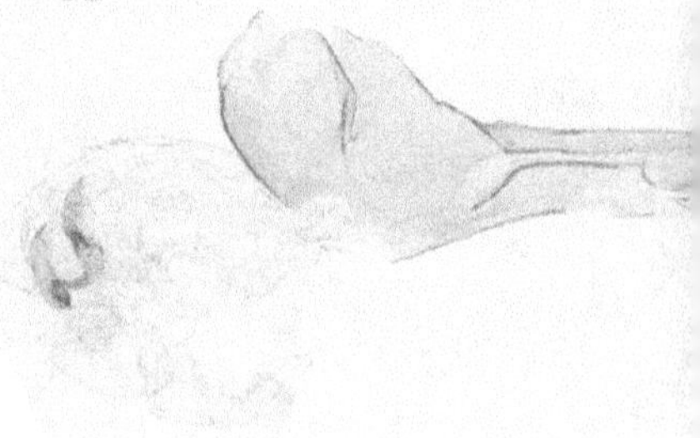

63

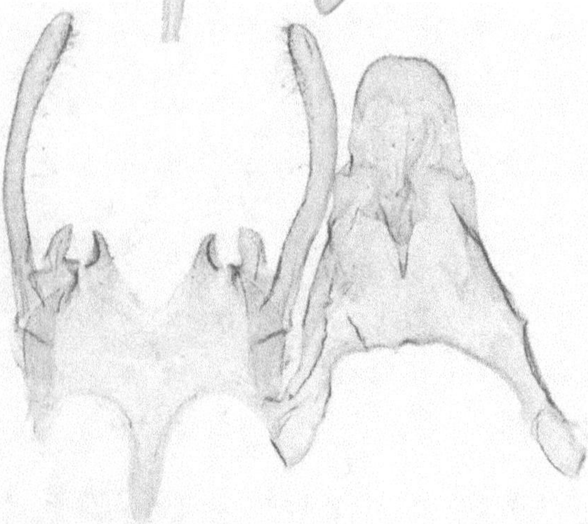

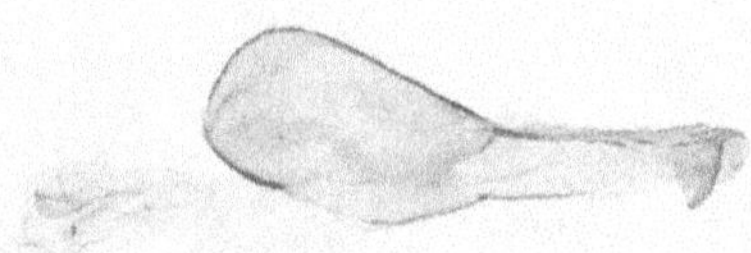

MALE GENITALIA

61. *Scrobipalpa superstes* Povolný – Portugal, GEL 1055 P.H. (TLMF).
62. *Scrobipalpa plesiopicta* Povolný – Russia, GU 03/1191 P.H. (NUPP).
63. *Scrobipalpa bradleyi* Povolný – France, GEL 1104 P.H. (TLMF).

64

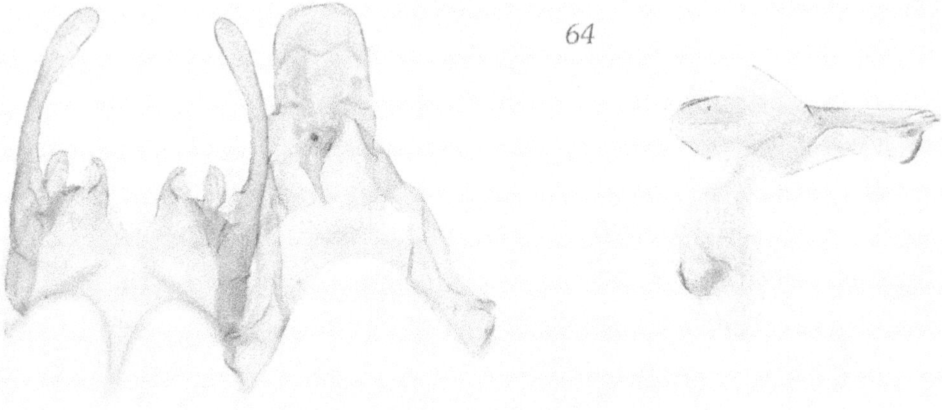

65

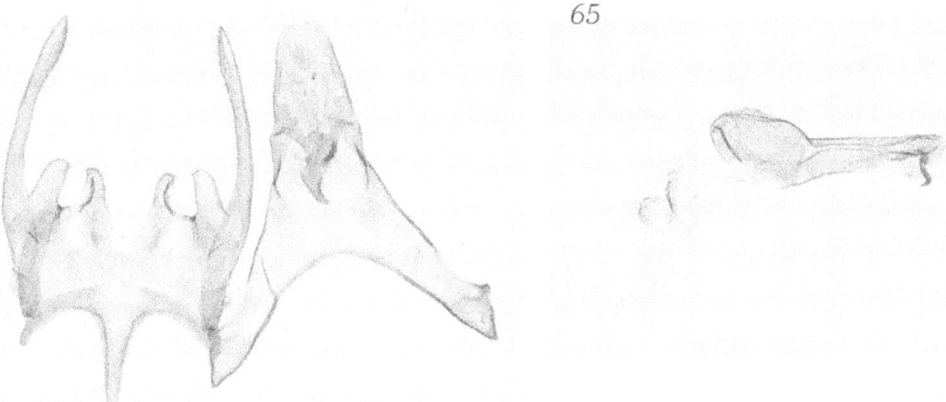

66

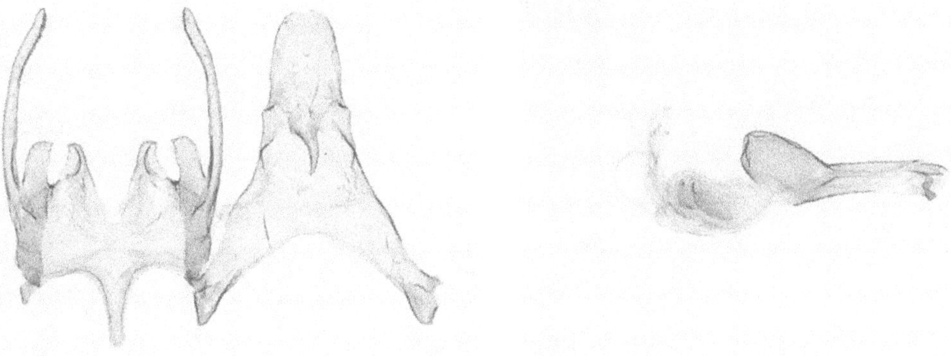

Male Genitalia

64. *Scrobipalpa selectella* (Caradja) – Kazakhstan, GU 01/1058 P.H. (ZMUC).
65. *Scrobipalpa alterna* (Falkovitsh & Bidzilya) – Russia, GU 03/1190 P.H. (NUPP).
66. *Scrobipalpa lutea* Povolný – Russia, GU 06/1258 P.H. (NUPP).

67

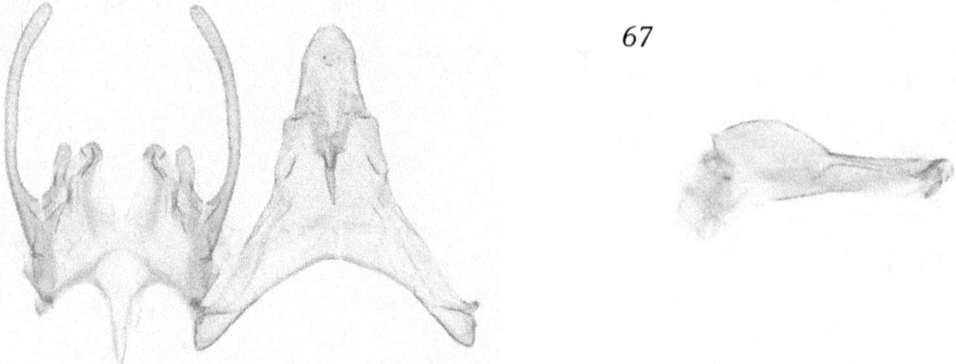

68

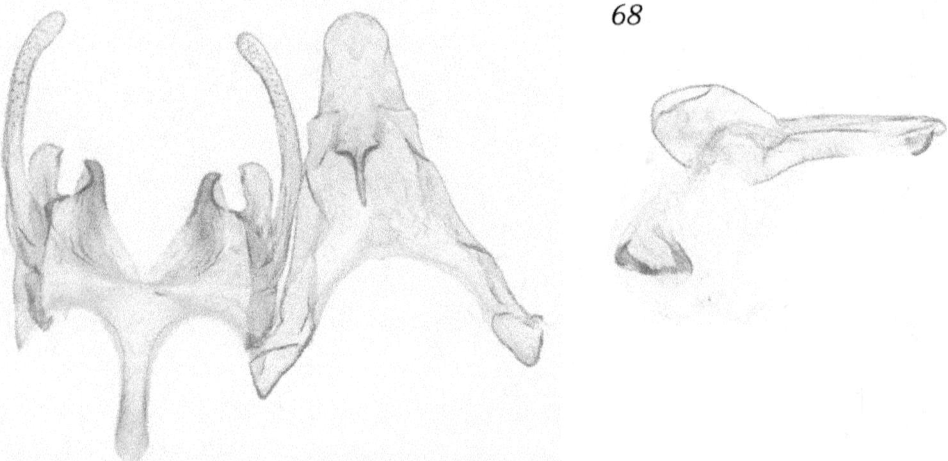

69

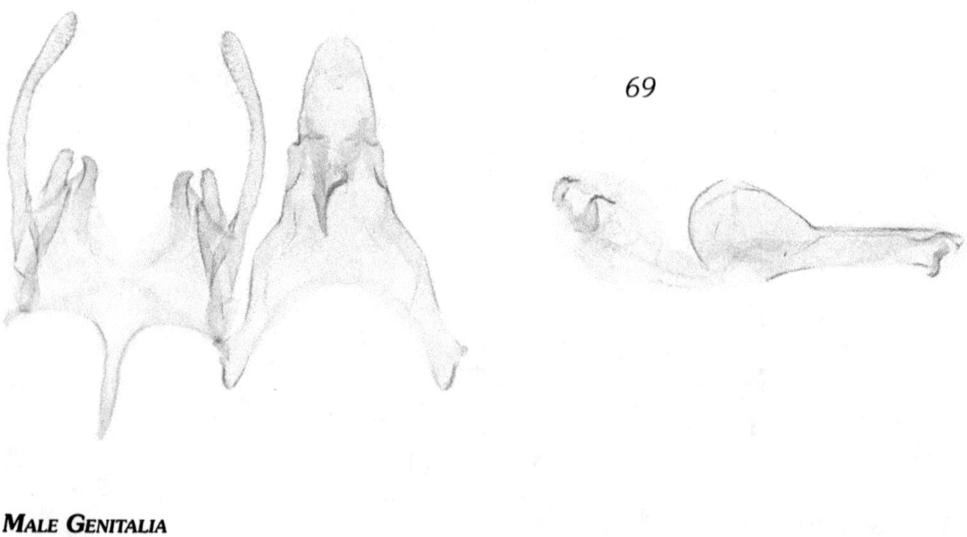

MALE GENITALIA

67. *Scrobipalpa niveifacies* Povolný – Spain, GU 01/1023 P.H. (BALD).
68. *Scrobipalpa indignella* (Staudinger) – Syria, gen. slide 16519 (NHMW).
69. *Scrobipalpa punctata* (Povolný) – Russia, GU 01/1092 P.H. (JUNN).

70

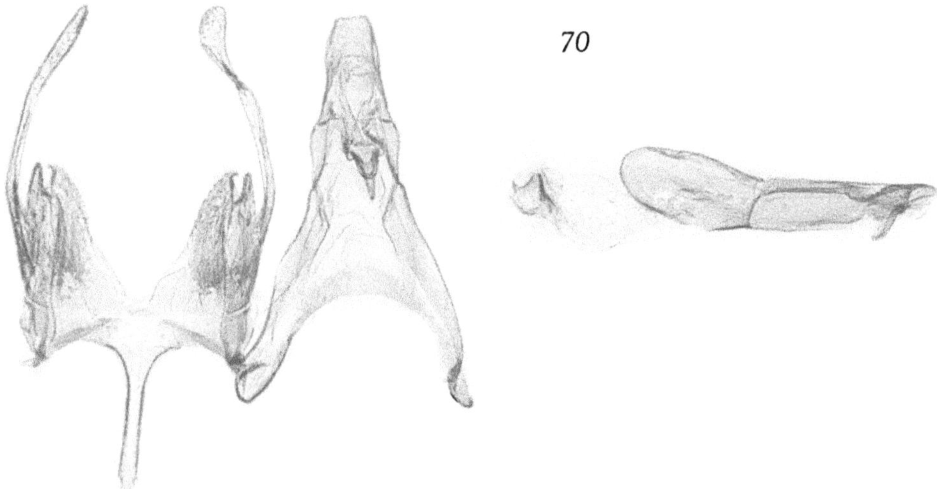

71

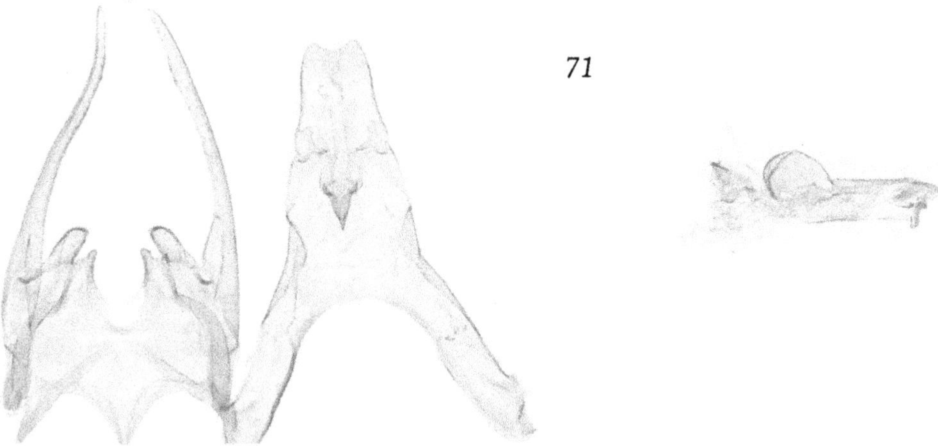

72

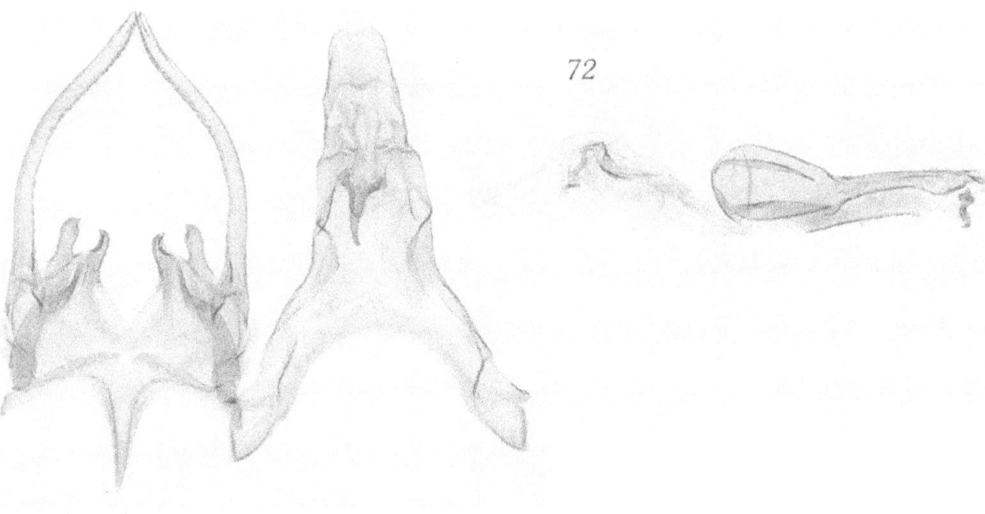

MALE GENITALIA

70. *Scrobipalpa lagodes* (Meyrick) – Kazakhstan, GU 09/11317 P.H. (ZMKU).
71. *Scrobipalpa deluccae* Povolný – Lebanon, gen. slide 16525 (NHMW).
72. *Scrobipalpa atriplicella* (Fischer von Röslerstamm) – Germany, GEL 997 P.H. (TLMF).

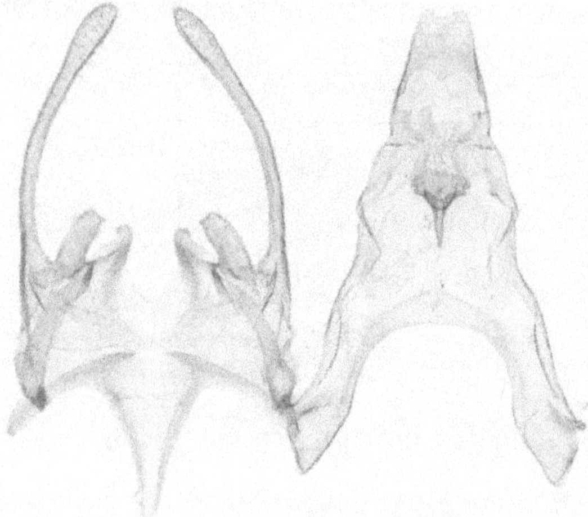

73

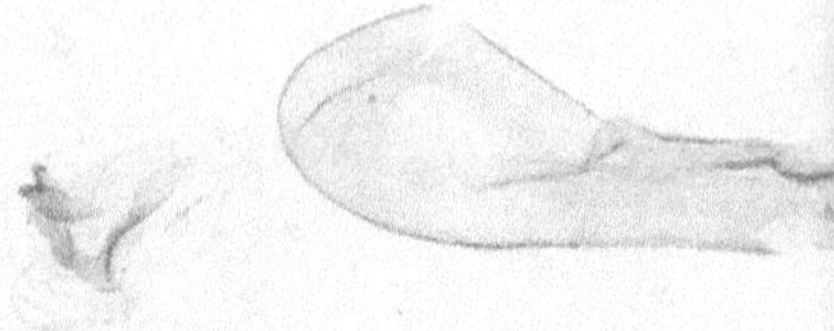

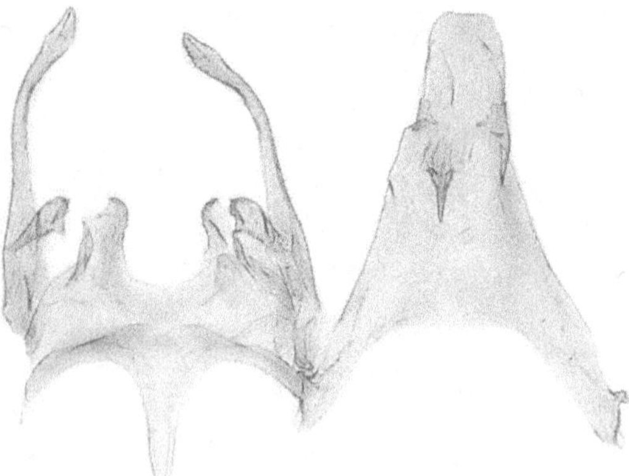

74

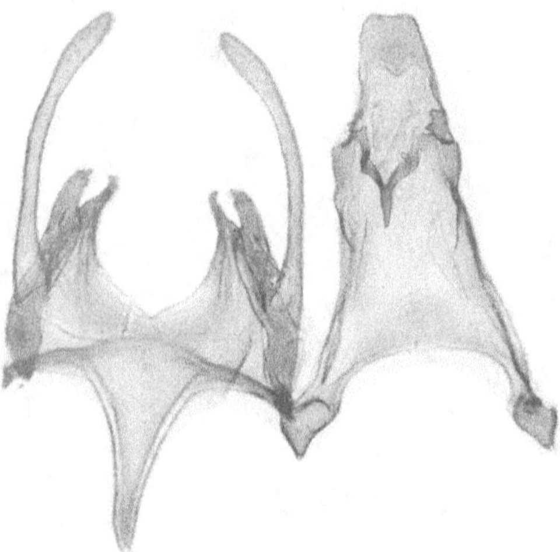

75

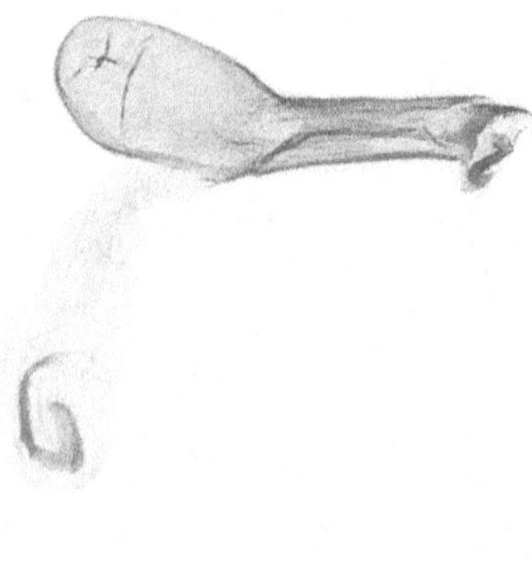

Male Genitalia

73. *Scrobipalpa suaedella* (Richardson) – Great Britain, GEL 976 P.H. (TLMF).
74. *Scrobipalpa solitaria* Povolný – Russia, gen. slide 3159 H.H. (NUPP).
75. *Scrobipalpa dagmaris* Povolný – Azerbaijan, GU 09/1318 P.H. (ZMKU).

76

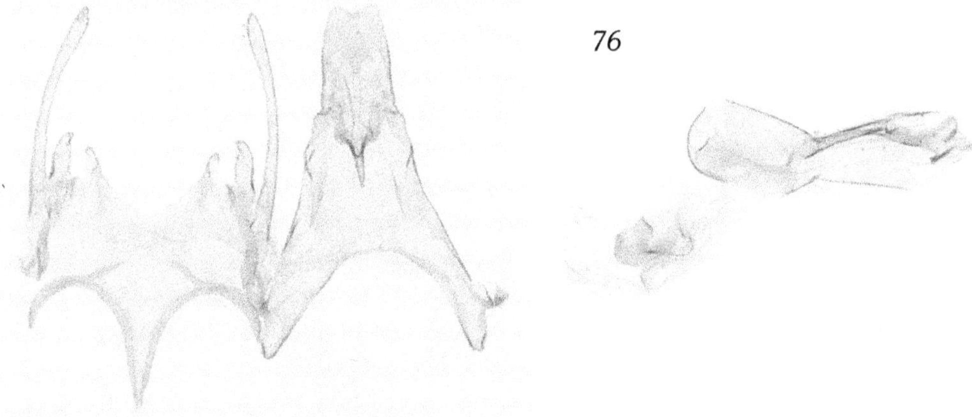

77

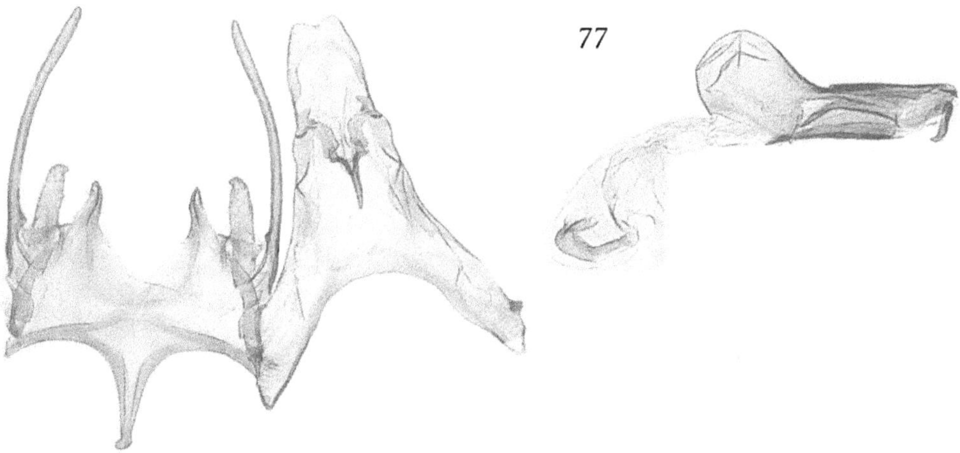

78

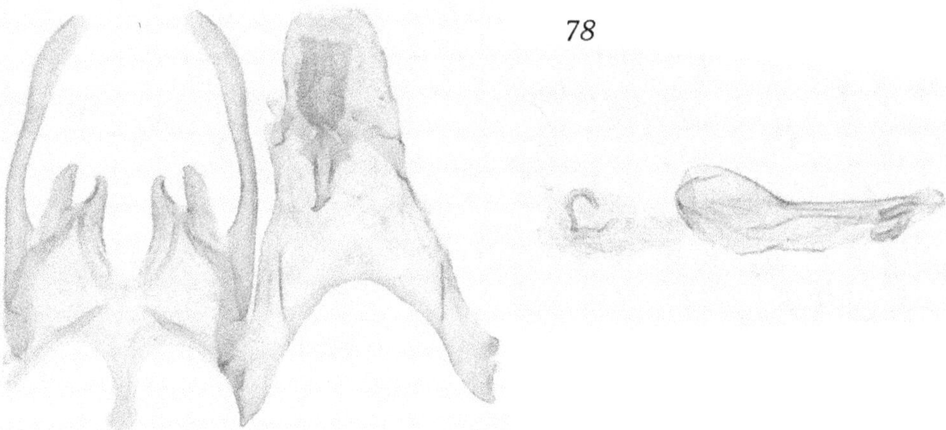

Male Genitalia

76. *Scrobipalpa suasella* (Constant) – France, GU 01/1053 P.H. (ZMUC).
77. *Scrobipalpa hendrikseni* sp. n. – Paratype, Greece, GU 09/1285 P.H. (ZMUC).
78. *Scrobipalpa traganella* (Chrétien) – Spain, GU 01/1035 P.H. (GRÜN)

79

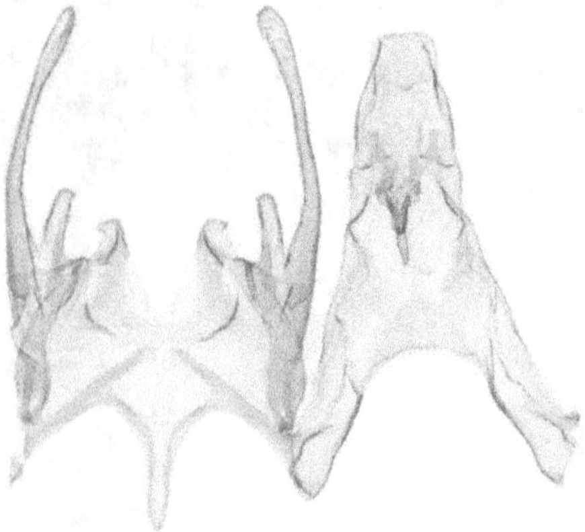

80

81

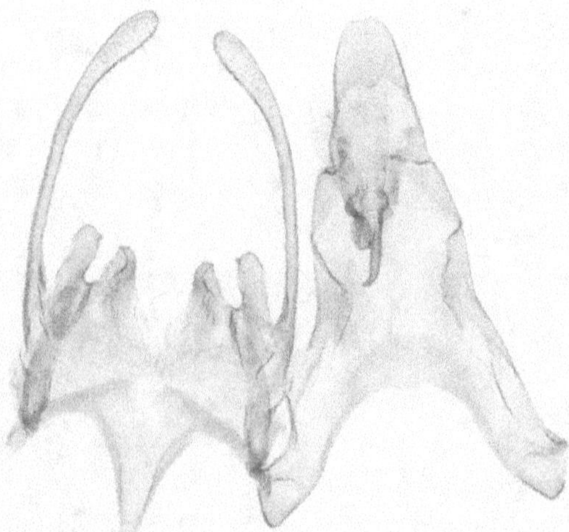

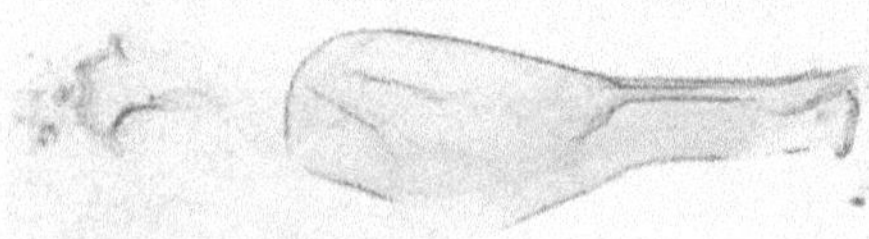

Male Genitalia

79. *Scrobipalpa bazae* Povolný – Spain, GEL 1004 P.H. (TLMF).
80. *Scrobipalpa artemisiella* (Treitschke) – Italy, GEL 57 P.H. (TLMF).
81. *Scrobipalpa stangei* (Hering) – Denmark, GEL 979 P.H. (TLMF).

82

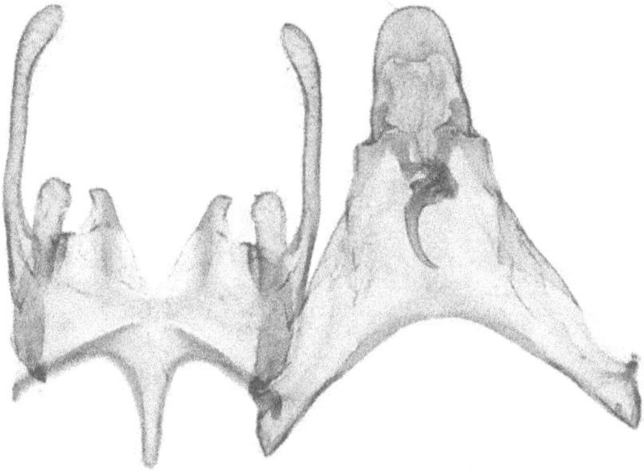
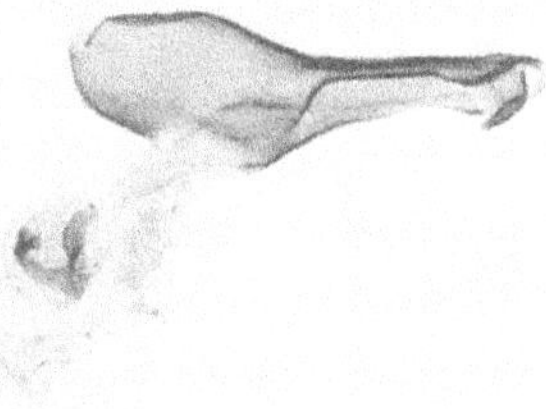

83

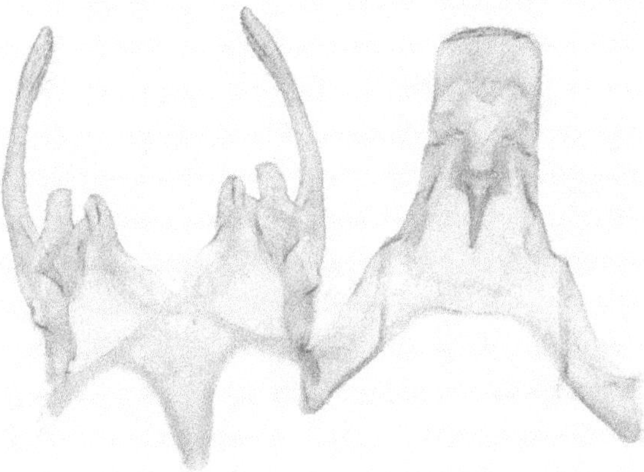
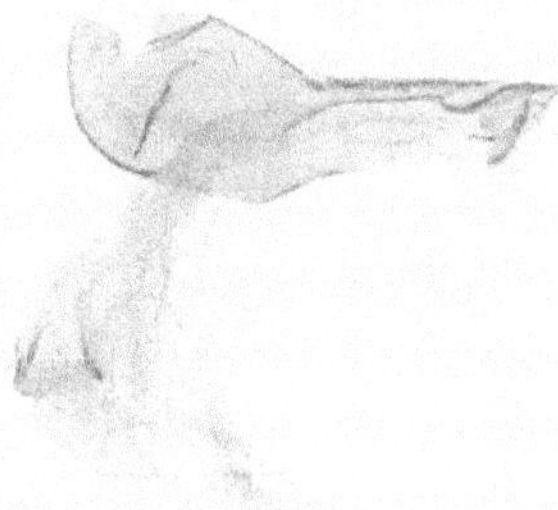

84

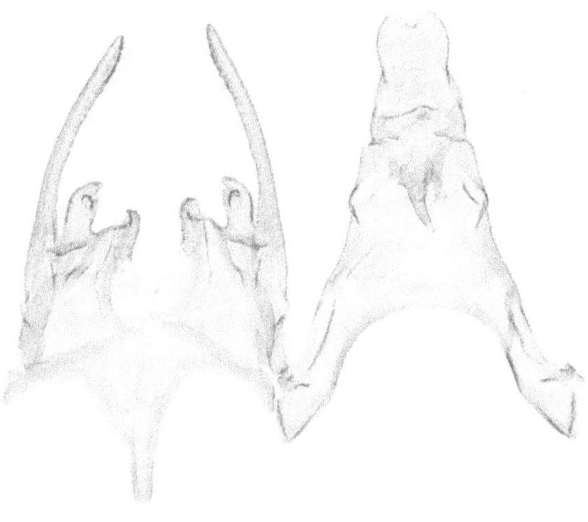
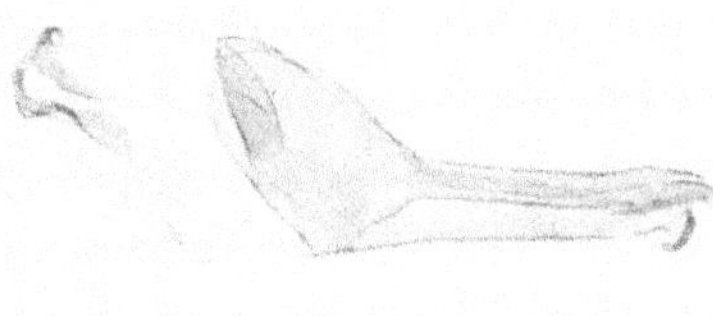

MALE GENITALIA

82. *Scrobipalpa suaedivorella* (Chrétien) – Russia, GU 06/1244 P.H. (ZMKU).
83. *Scrobipalpa bryophiloides* Povolný – Turkey, gen. slide 16530 (NHMW).
84. *Scrobipalpa algeriensis* Povolný & Bradley – Spain, gen. slide 999 H.H. (ZMUC).

85

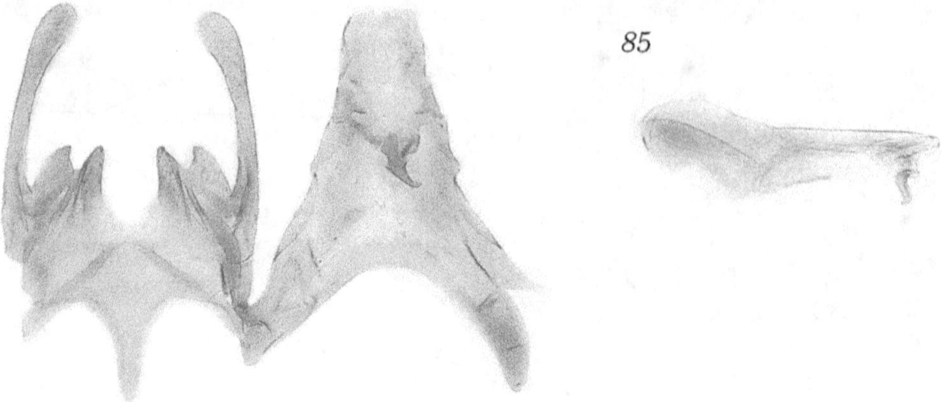

86

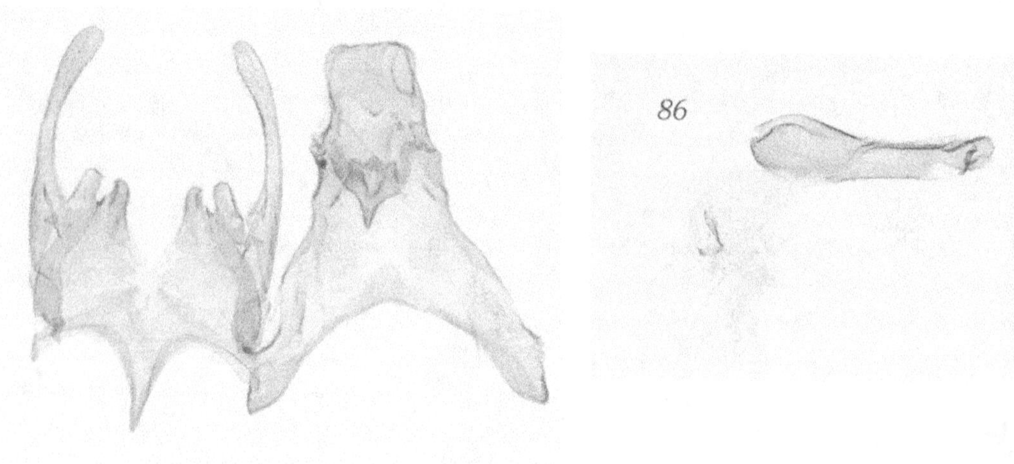

87

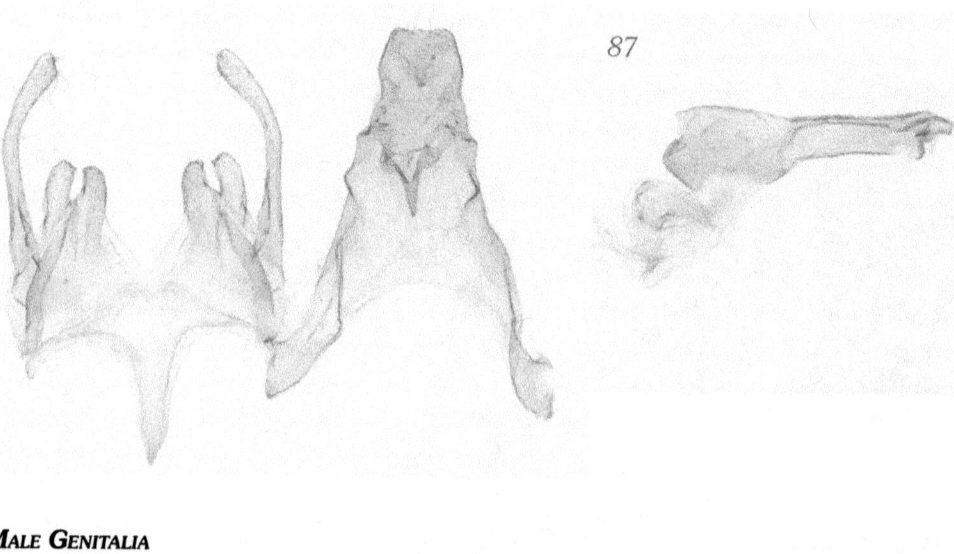

Male Genitalia

85. *Scrobipalpa deutschi* sp. n. – Holotype, Russia, gen. slide 5457 H.H. (NUPP).
86. *Scrobipalpa disjectella* (Staudinger) – Lectotype, Spain, GU 03/1185 P.H. (MNHU).
87. *Scrobipalpa mixta* sp. n. – Holotype, Macedonia, GU 04/1217 P.H. (ZSM).

88

89

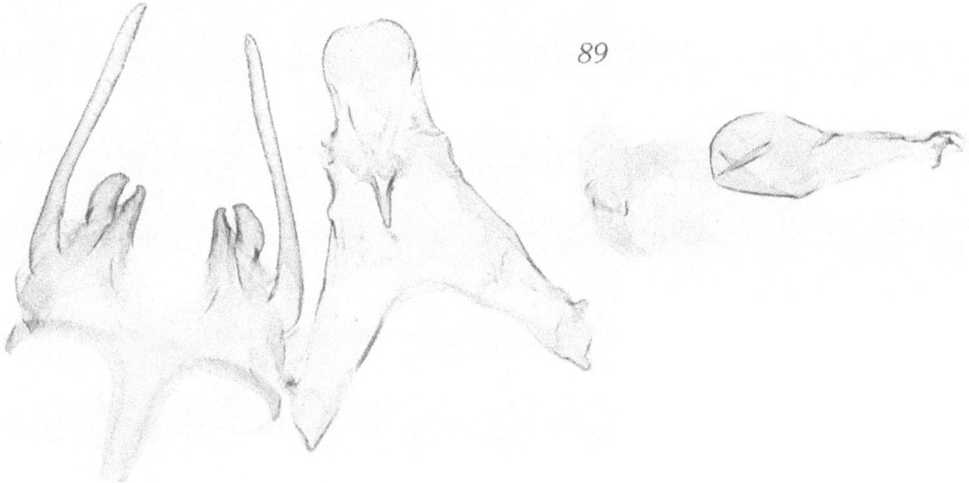

89a

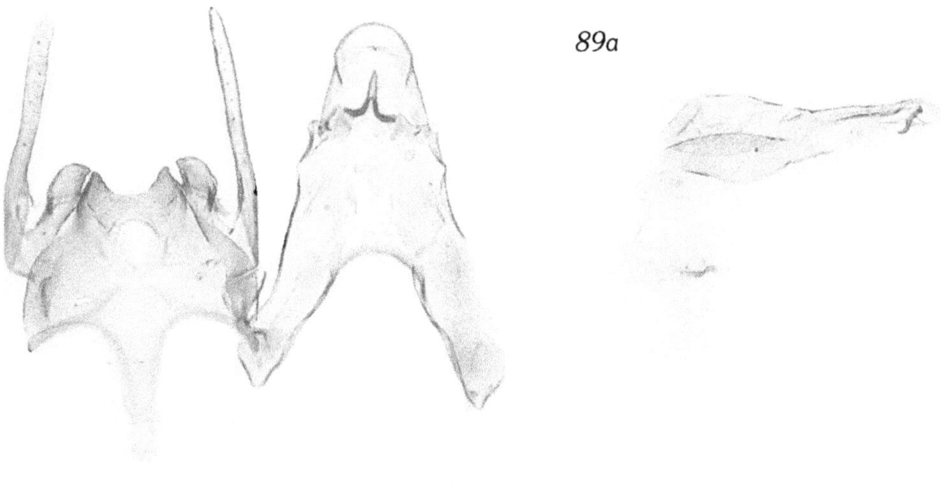

MALE GENITALIA

88. *Scrobipalpa rebeli* (Preissecker) – Austria, gen. slide 16539 (NHMW).
89. *Scrobipalpa gallicella* (Constant) – Italy, gen. slide 4805 O.K. (ZMUC).
89a. *Scrobipalpa gallicella* (Constant) – Russia, gen. slide 6427 H.H. (ZMUC).

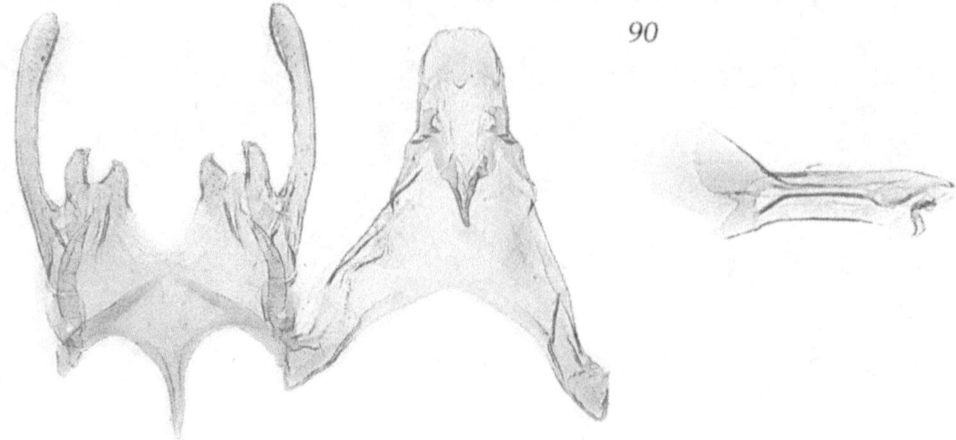

90

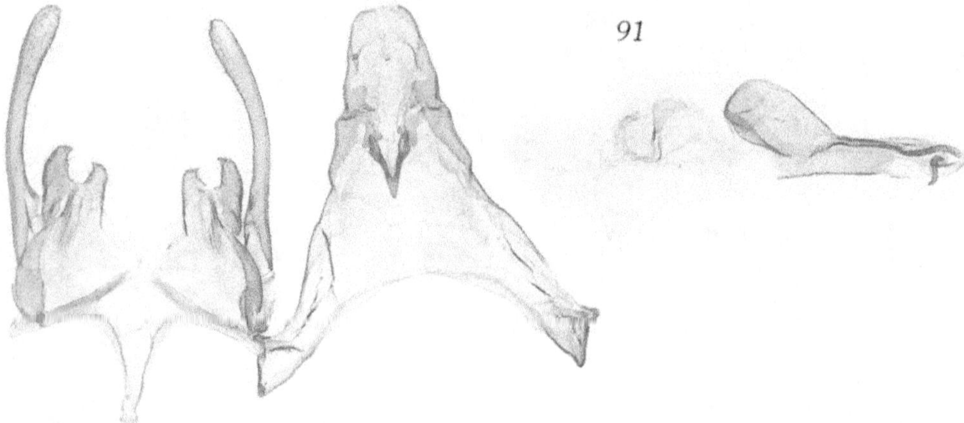

91

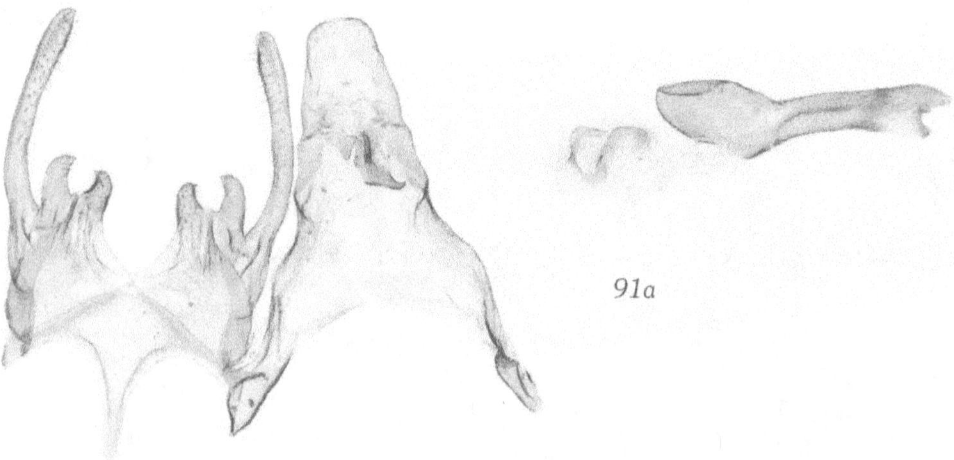

91a

Male Genitalia

90. *Scrobipalpa ustulatella* (Staudinger) – Lectotype, Russia, GU 09/1274 (MNHU).
91. *Scrobipalpa postulatella* sp. n. – Paratype, Spain, GU 01/1073 (ZMUC).
91a. *Scrobipalpa postulatella* sp. n. – Paratype, Spain, GU 09/1300 (ZMUC).

92

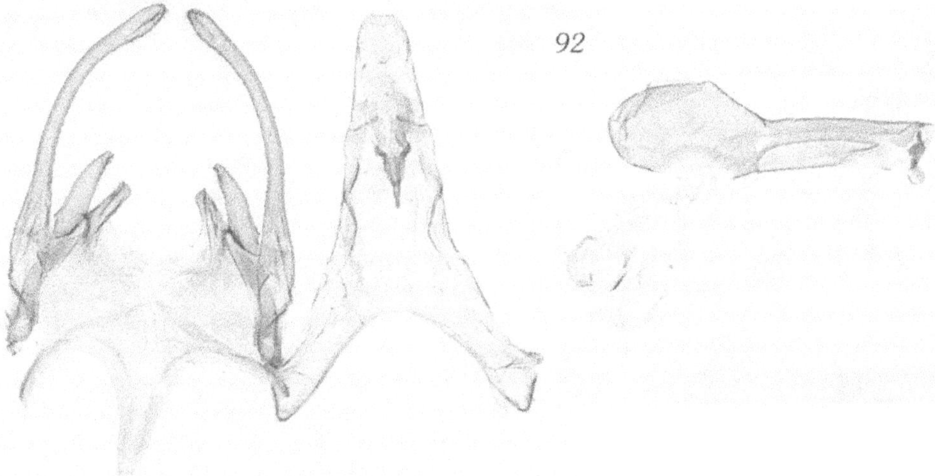

93

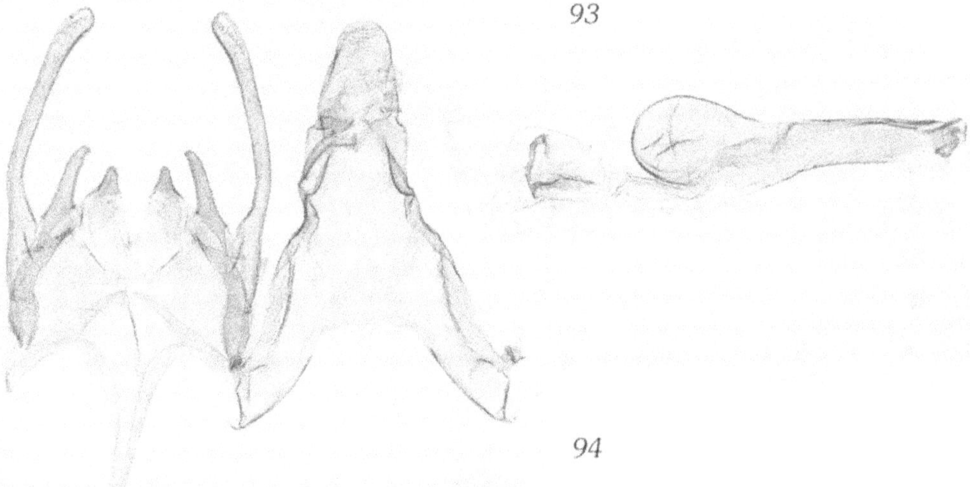

94

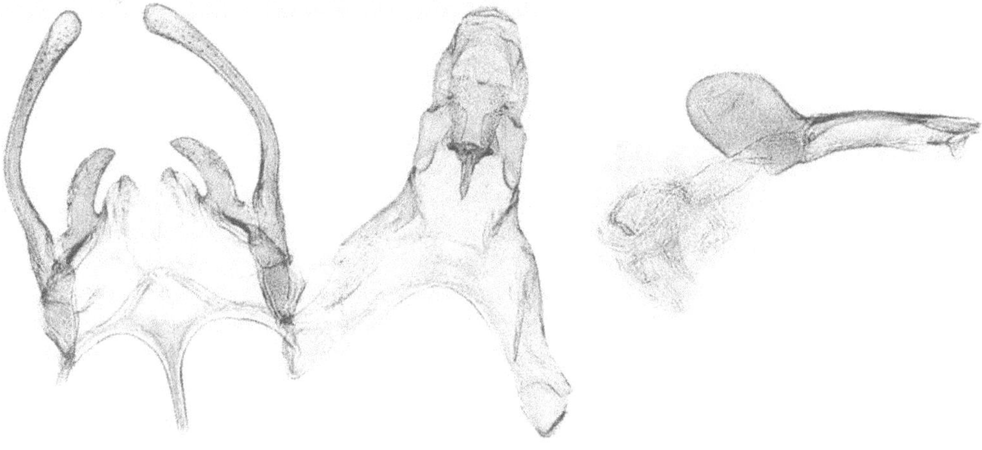

MALE GENITALIA

92. *Scrobipalpa nitentella* (Fuchs) – Denmark, GEL 944 P.H. (TLMF).
93. *Scrobipalpa costella* (Humphreys & Westwood) – Denmark, GEL 922 (TLMF).
94. *Scrobipalpa hyoscyamella* (Stainton) – Austria, GEL 1126 P.H. (TLMF).

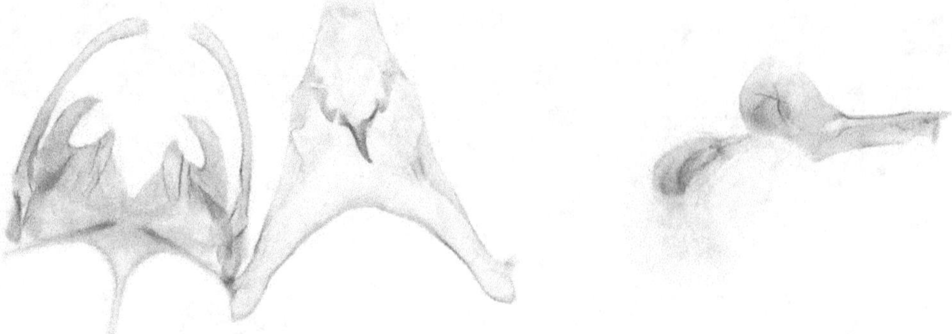

95

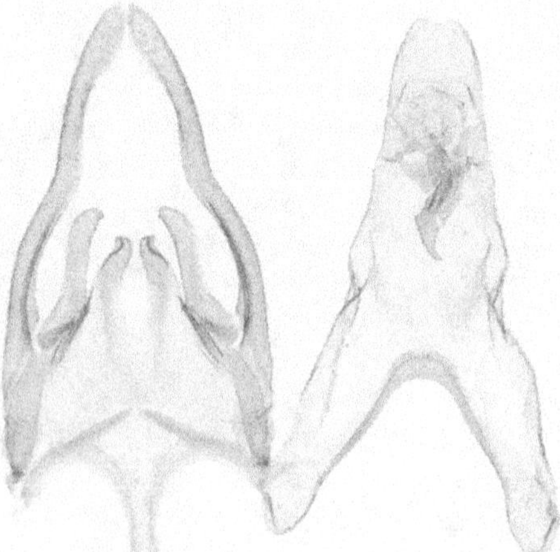

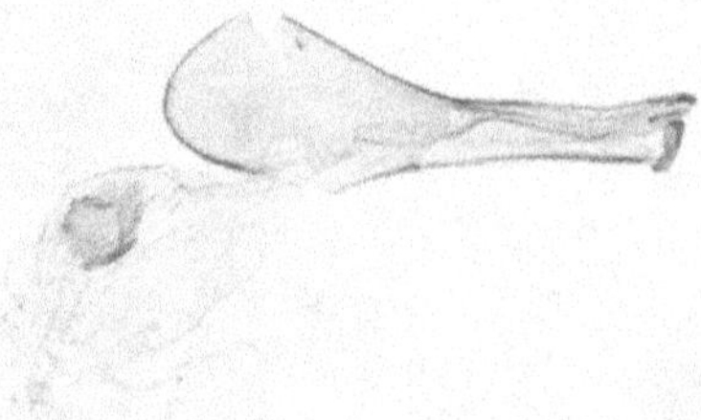

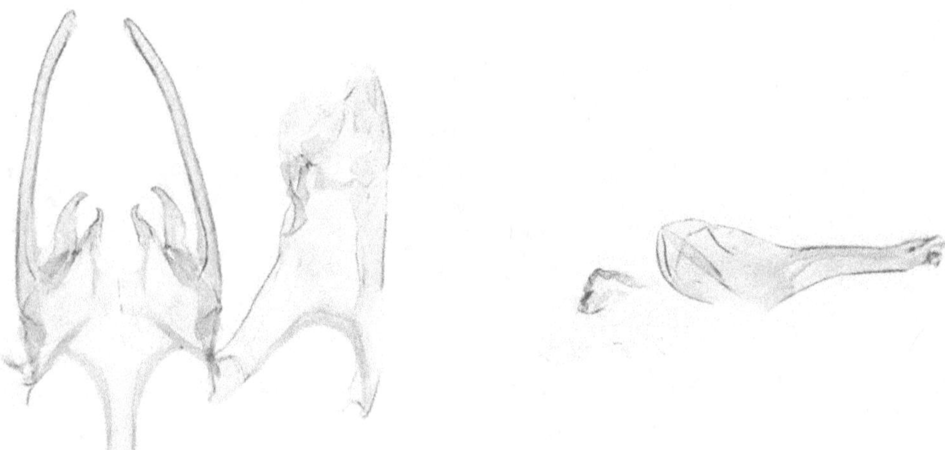

Male Genitalia

94a. *Scrobipalpa cf. hyoscyamella* (Stainton) – Ukraine, GU 01/1061 P.H. (ZMKU).
95. *Scrobipalpa portosanctana* (Stainton) – Spain, GU 01/1033 P.H. (BALD).
96. *Scrobipalpa vicaria* (Meyrick) – Spain, gen. slide 2910 H.H. (ZMUC).

97

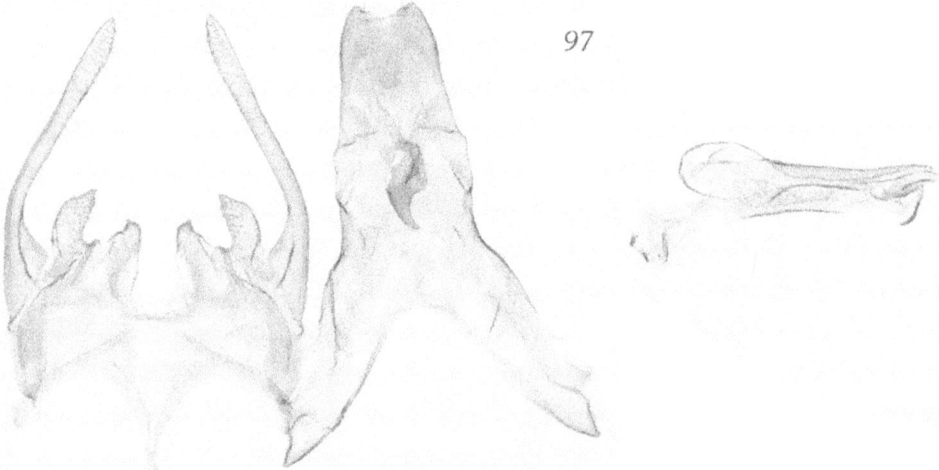

98

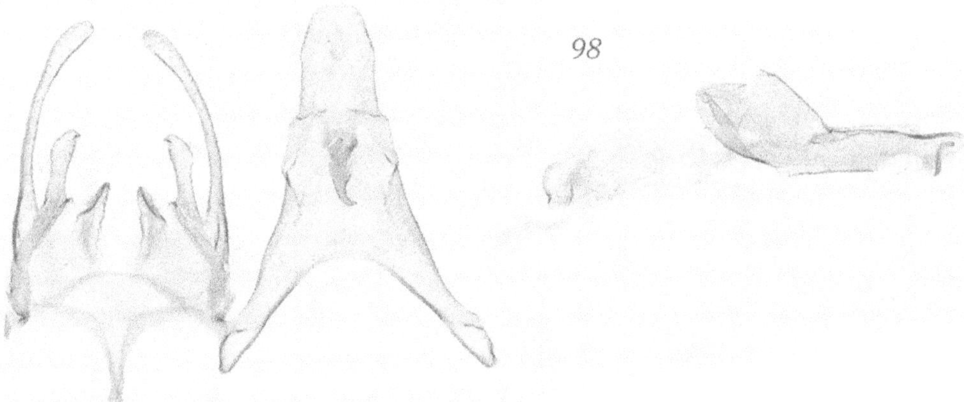

99

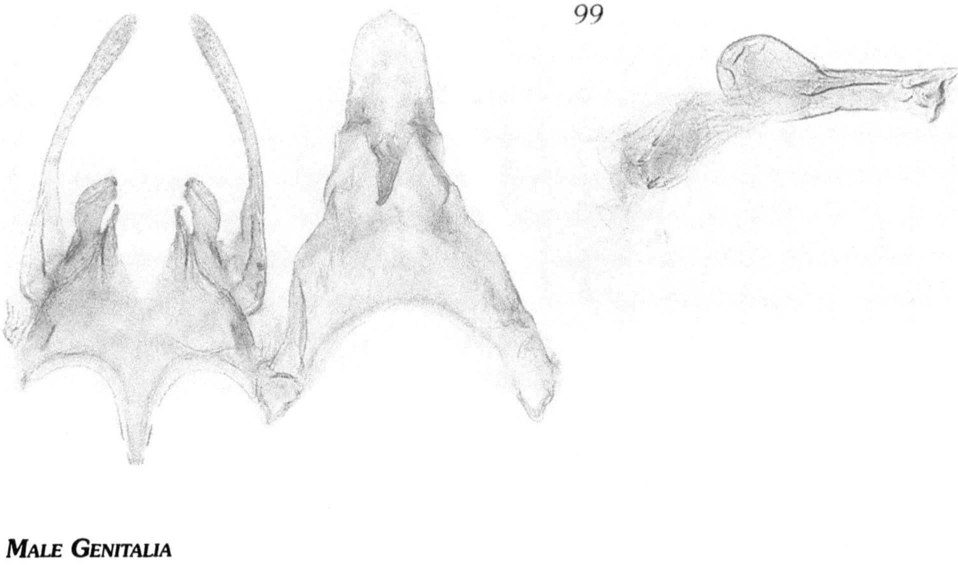

Male Genitalia

97. *Scrobipalpa ocellatella* (Boyd) – Macedonia, GU 00/899 P.H. (TLMF).
98. *Scrobipalpa pulchra* Povolný – Turkey, GU 01/1044 P.H. (ZMUC).
99. *Scrobipalpa gecko* (Walsingham) – Algeria, gen. slide 26586 (BMNH).

100

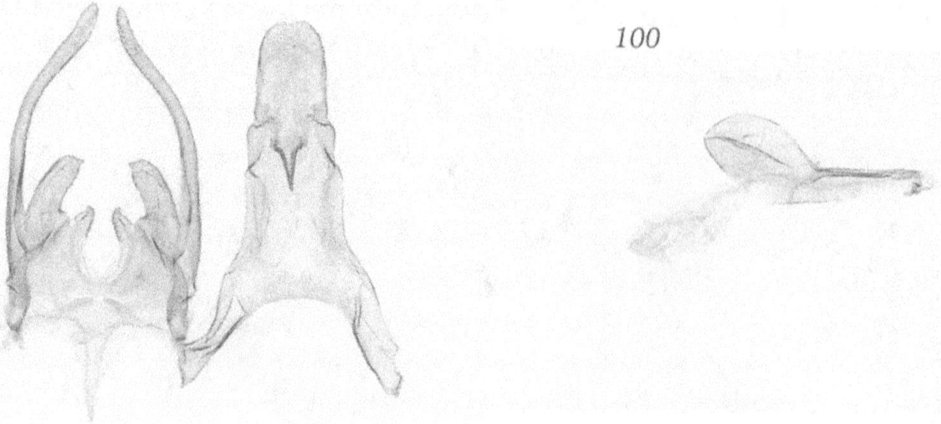

101

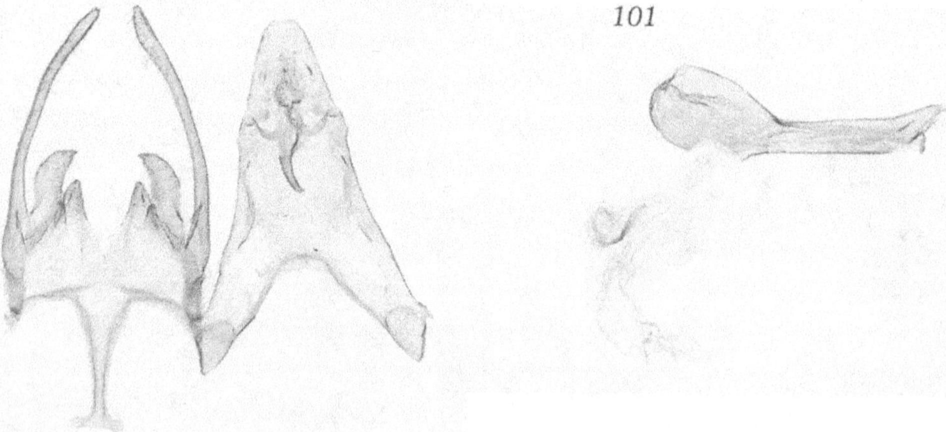

102

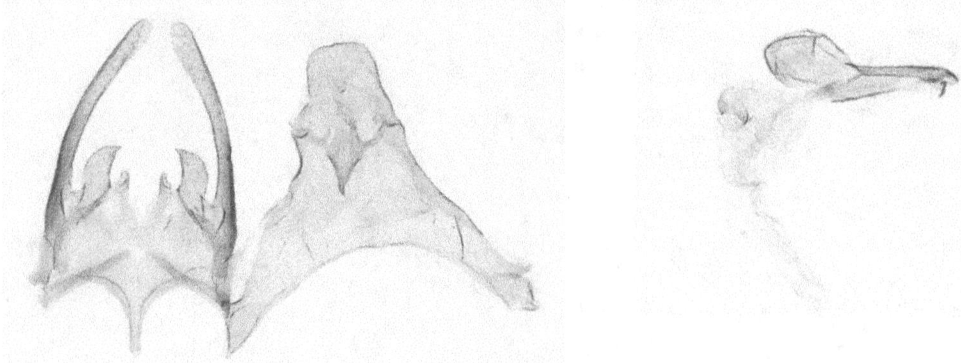

Male Genitalia

100. *Scrobipalpa hannemanni* Povolný – Russia, GU 01/1089 P.H. (JUNN).
101. *Scrobipalpa erichi* Povolný – Slovakia, GU 02/1056 P.H. (ZMUC).
102. *Scrobipalpa divisella* (Rebel) – Spain, GU 01/986 P.H. (GRÜN).

103

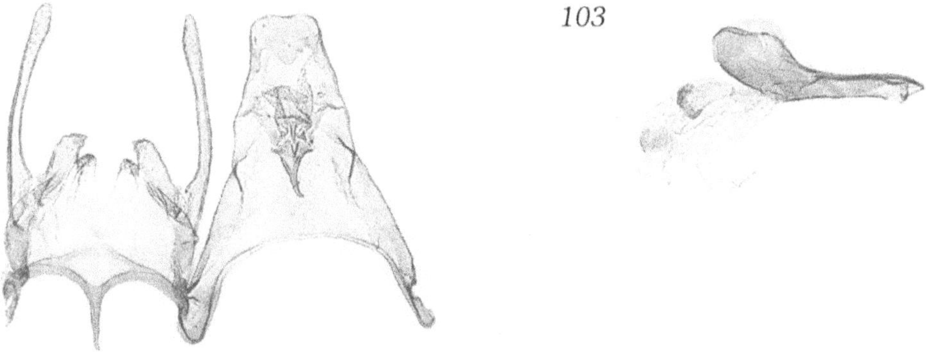

104

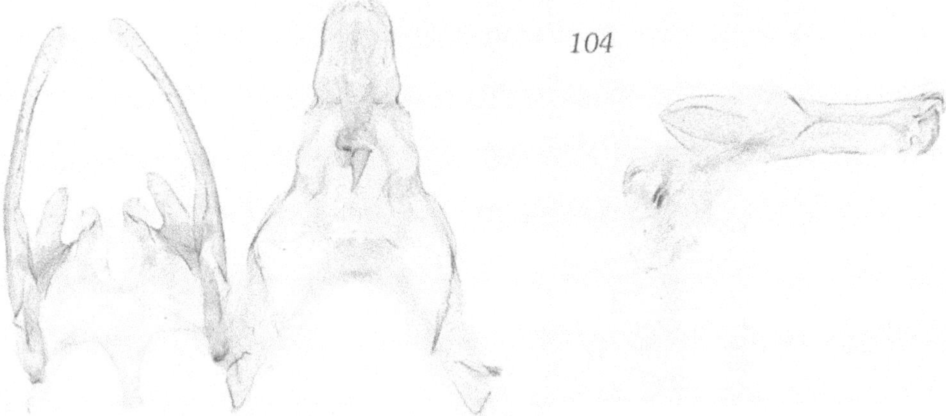

105

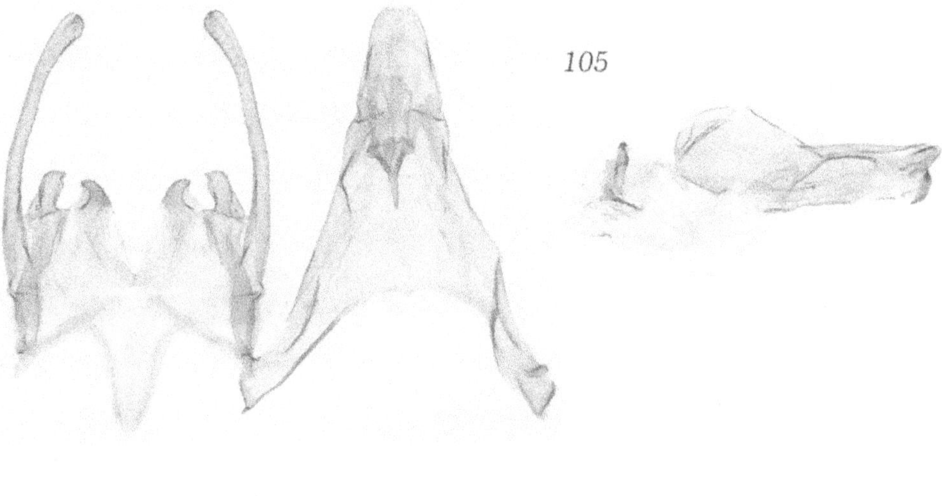

Male Genitalia

103. *Scrobipalpa voltinella* (Chrétien) – Paralectotype, France, GU 09/1312 P.H. (MZMB).
104. *Scrobipalpa suaedicola* (Mabille) – France, GU 01/1008 P.H. (ZMUC).
105. *Scrobipalpa monochromella* (Constant) – Spain, GU 99/880 P.H. (ZSM).

106

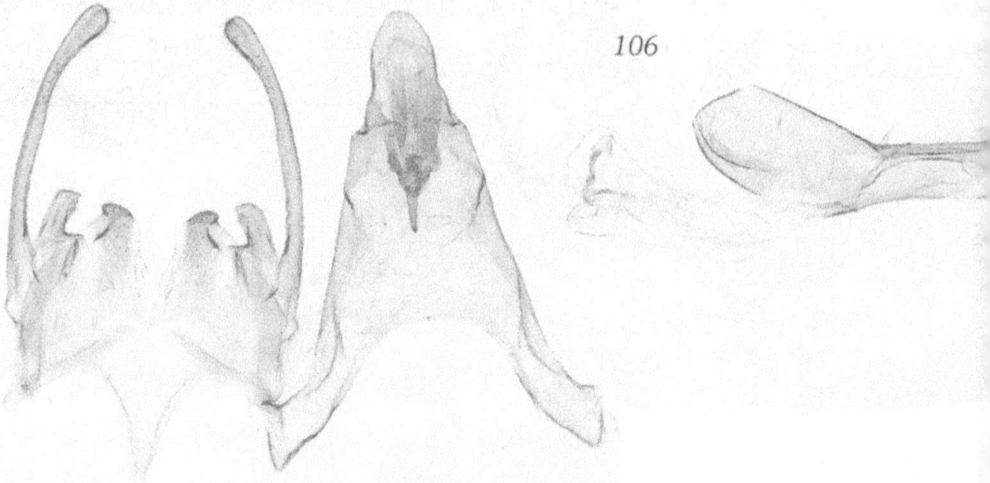

107

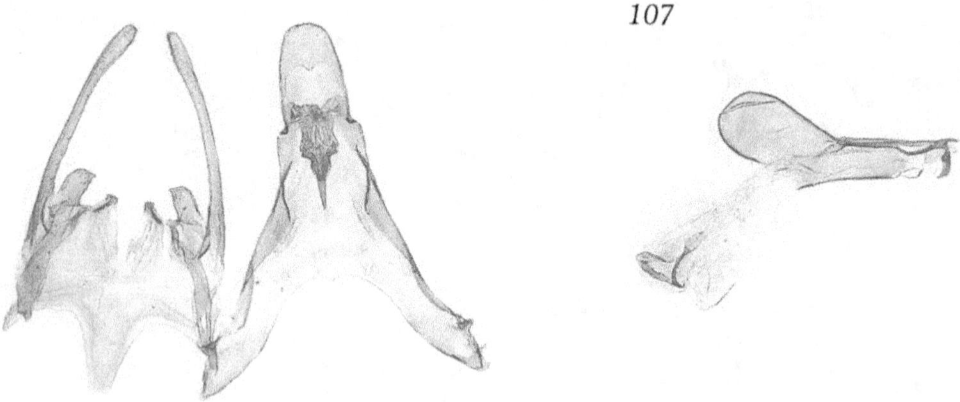

107a

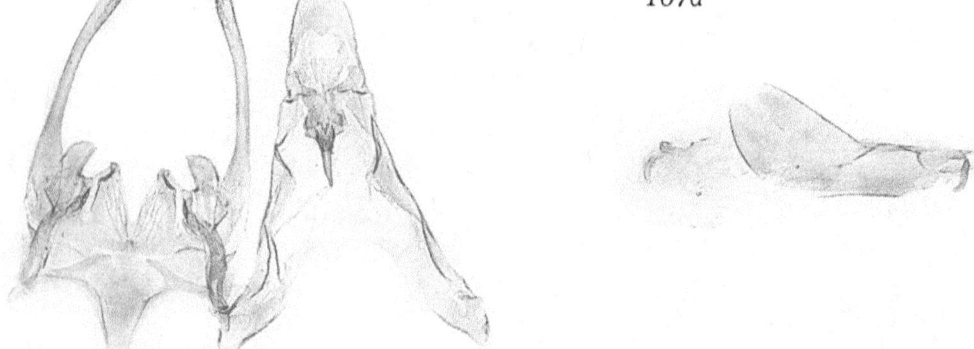

MALE GENITALIA

106. *Scrobipalpa samadensis* (Pfaffenzeller) – Denmark, GEL 411 (TLMF).
107. *Scrobipalpa salinella* (Zeller) – Spain, GU 09/1279 P.H. (ZMUC).
107a. *Scrobipalpa salinella* (Zeller) – France, GEL 1002 P.H. (TLMF).

108

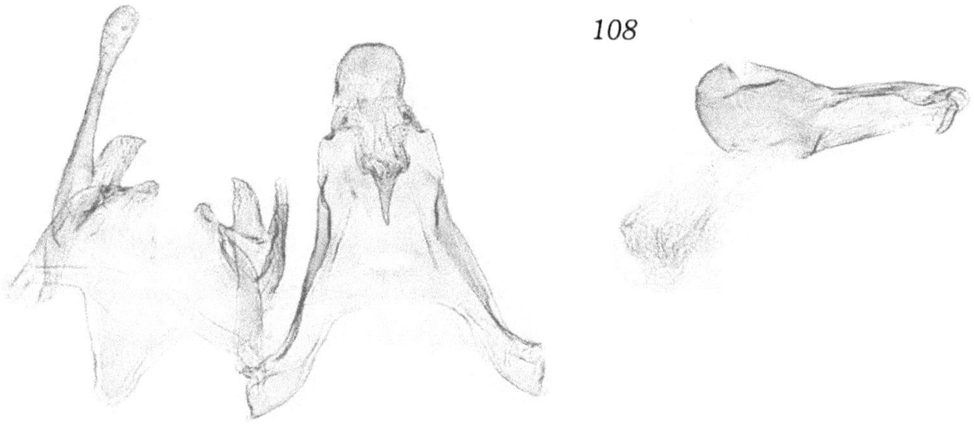

108a

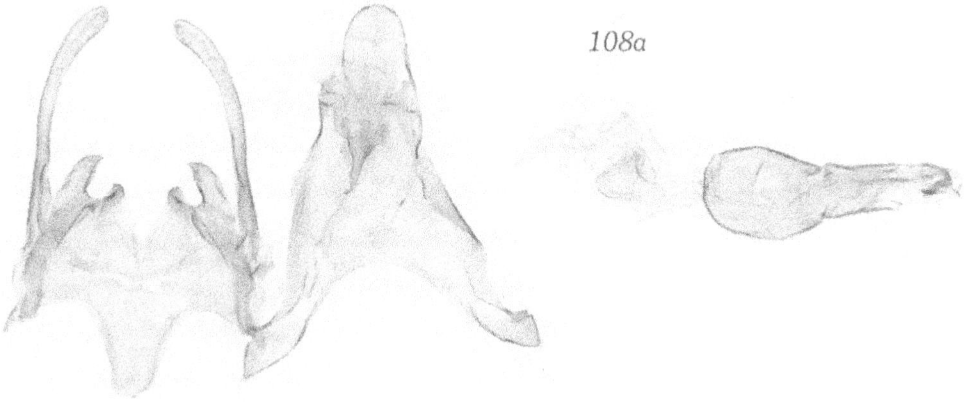

109

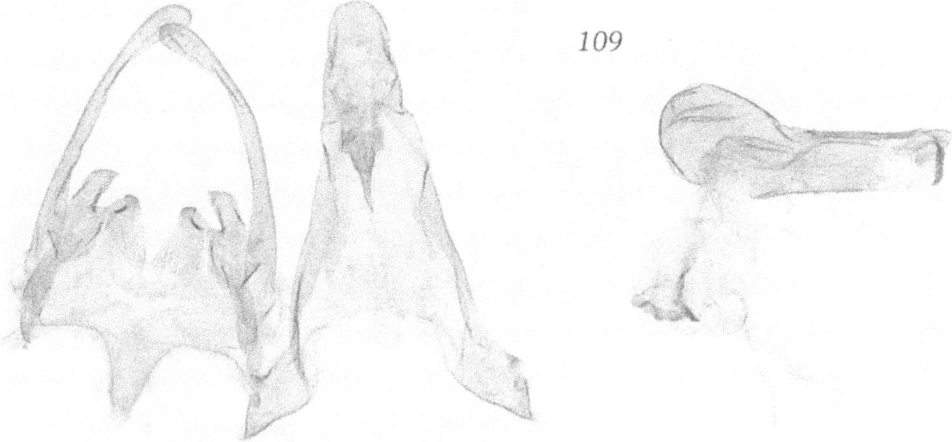

Male Genitalia

108. *Scrobipalpa spergulariella* (Chrétien) – France, GEL 1125 P.H. (TLMF).
108a. *Scrobipalpa spergulariella* (Chrétien) – France, GEL 984 P.H. (TLMF).
109. *Scrobipalpa salicorniae* (Hering) – Austria, GEL 940 P.H. (TLMF).

110

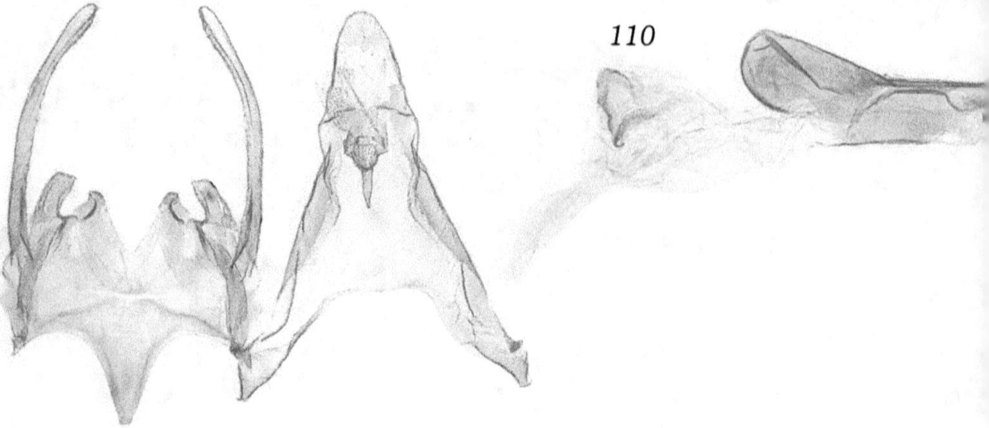

111

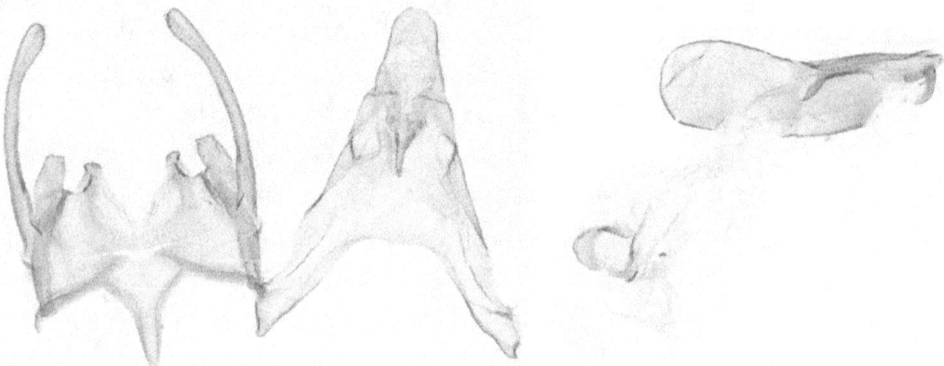

112

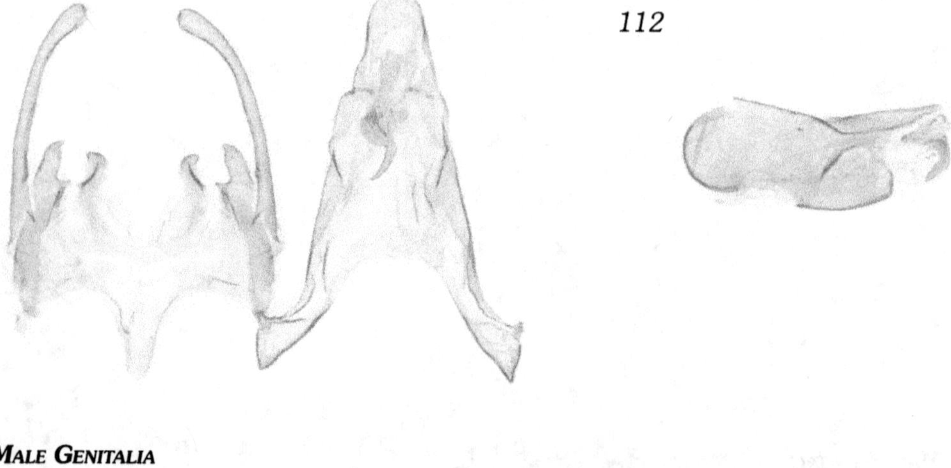

MALE GENITALIA

110. *Scrobipalpa halimioniella* sp. n. – Paratype, France, GU 09/1275 P.H. (ZMUC).
111. *Scrobipalpa thymelaeae* (Amsel) – Paratype, Italy, GU 01/998 P.H. (LNK).
112. *Scrobipalpa halymella* (Millière) – France, GEL 986 P.H. (TLMF).

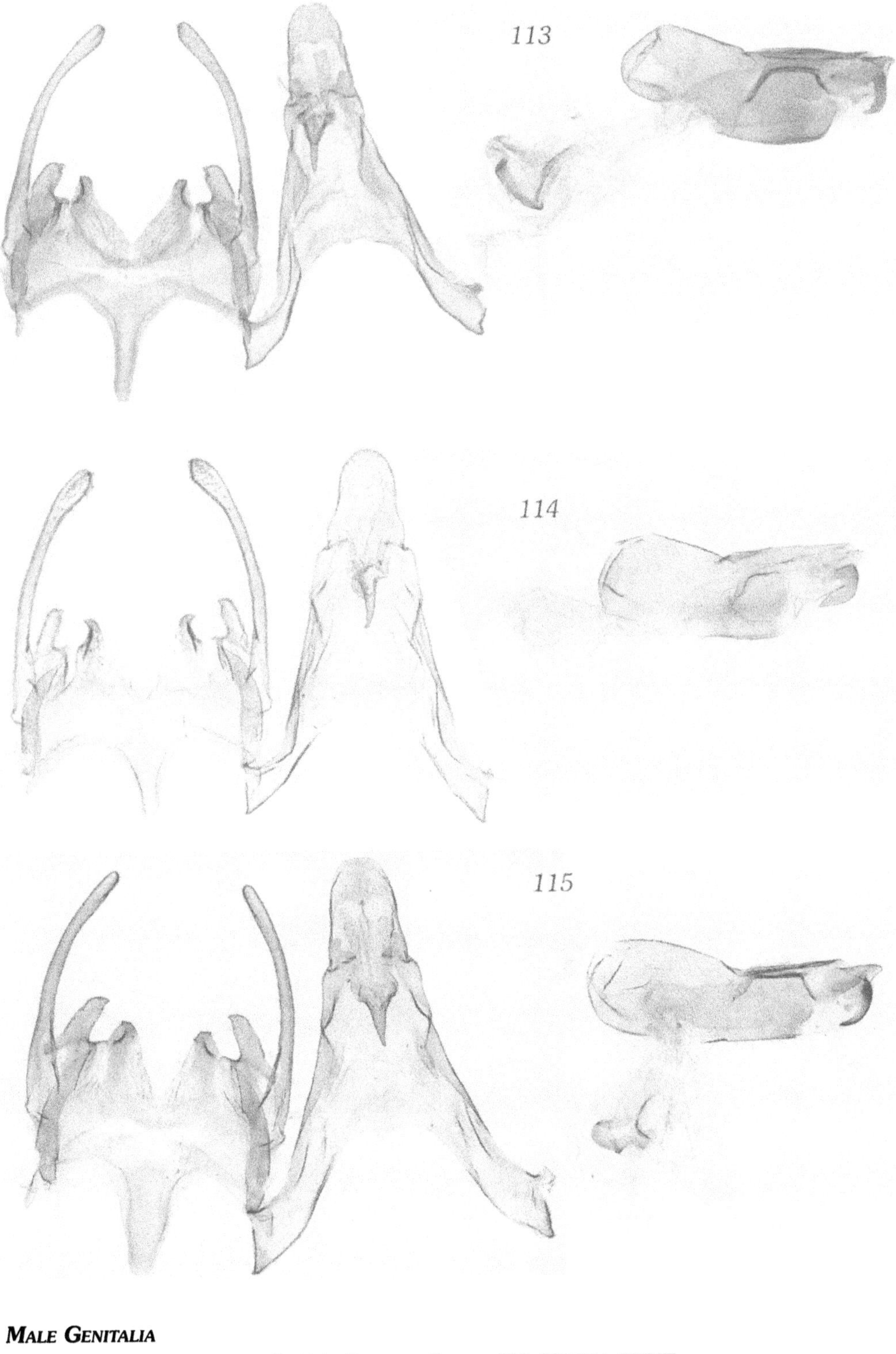

MALE GENITALIA

113. *Scrobipalpa camphorosmella* Nel – Paratype, France, GEL 982 P.H. (TLMF).
114. *Scrobipalpa stabilis* Povolný – Spain, GU 04/1211 P.H. (LNK).
115. *Scrobipalpa instabilella* (Douglas) – Greece, GEL 936 P.H. (TLMF).

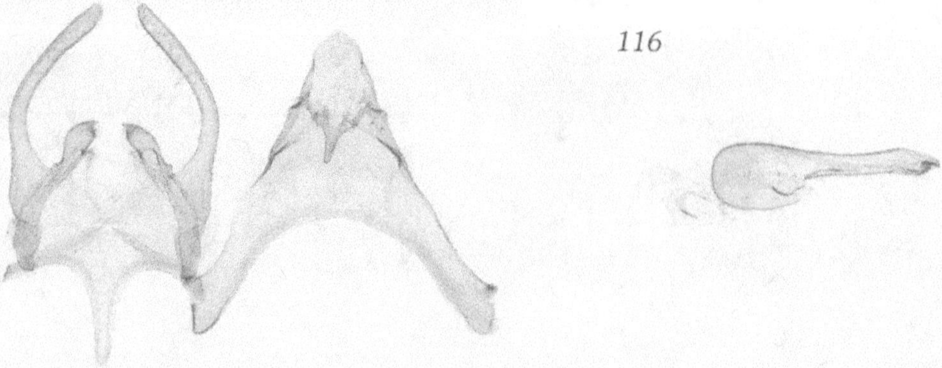

116

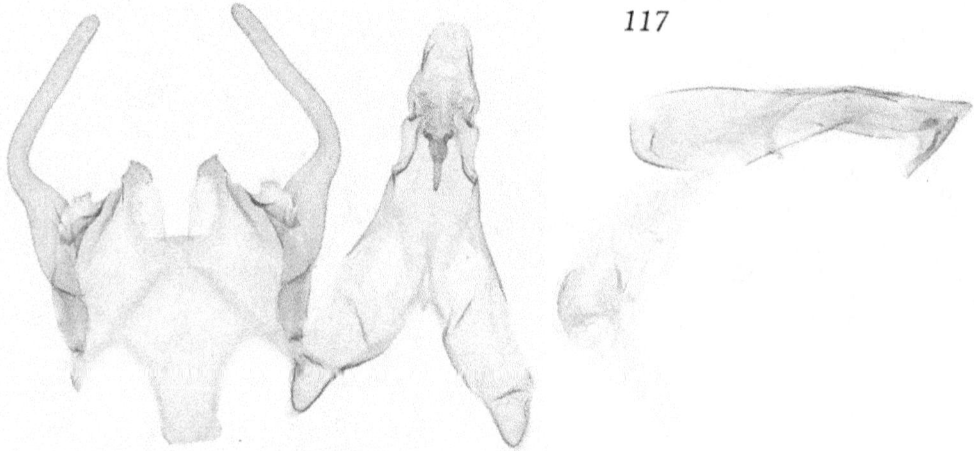

117

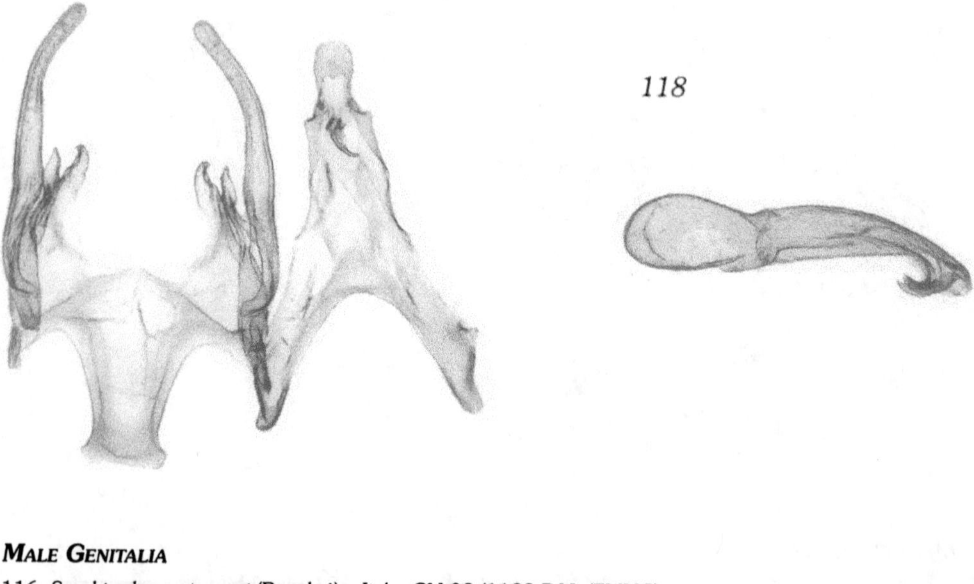

118

Male Genitalia

116. *Scrobipalpa peterseni* (Povolný) – Italy, GU 03/1183 P.H. (ZMUC).
117. *Scrobipalpa ergasima* (Meyrick) – Madeira, GU 01/1047 P.H. (ZMUC).
118. *Turcopalpa glaseri* Povolný – Paratype, Turkey, GU 09/1311 P.H. (remounted slide Gl. 4541 Povolný) (MZMB).

119

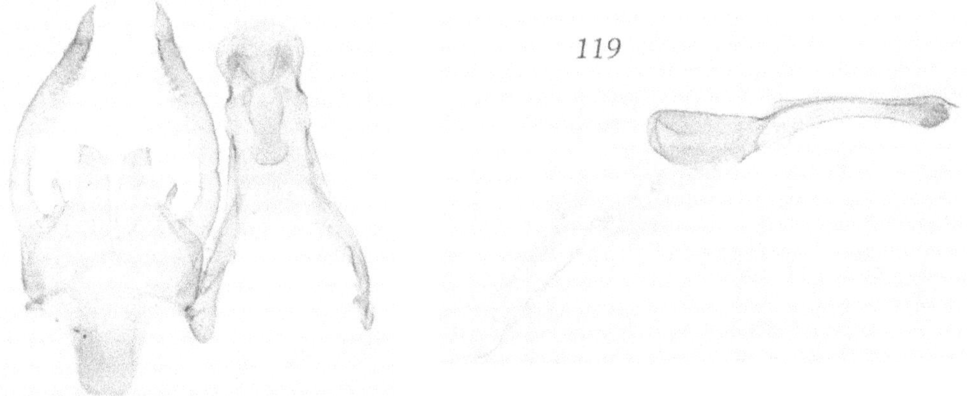

120

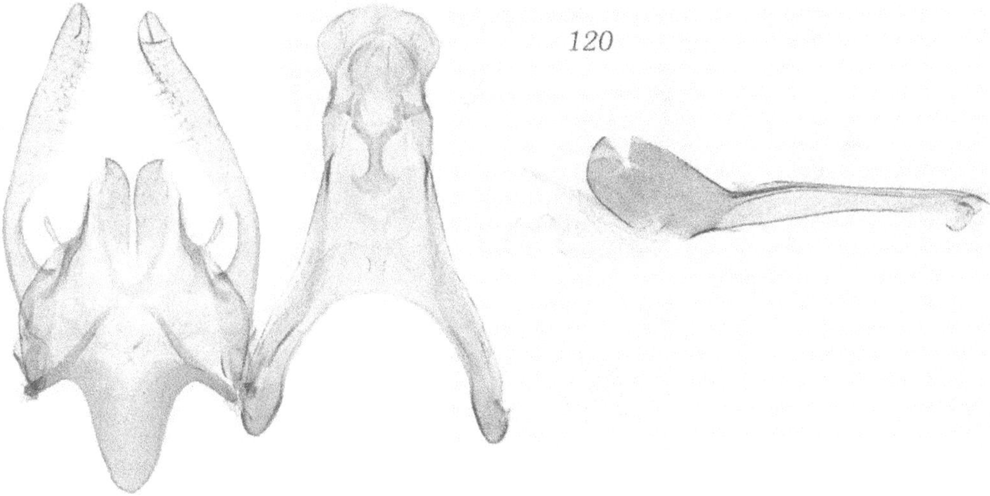

121

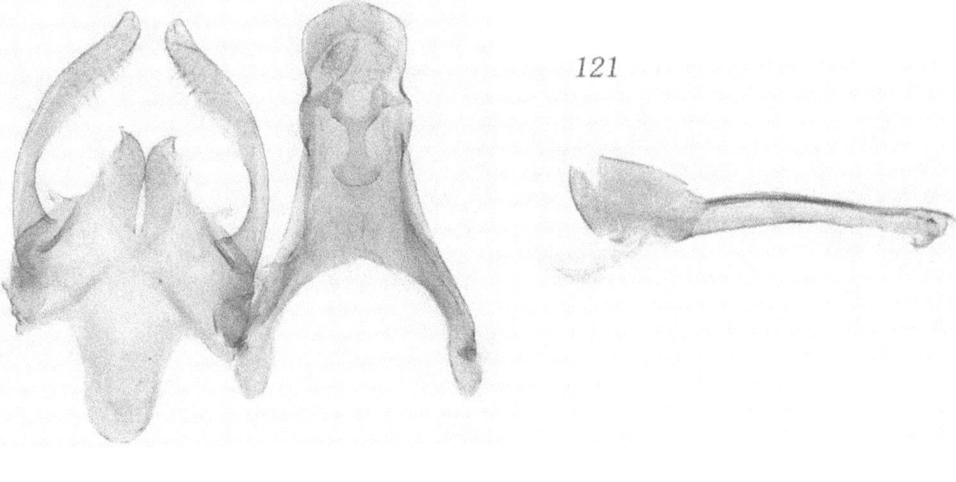

MALE GENITALIA

119. *Scrobipalpula psilella* (Herrich-Schäffer) – Austria, GEL 47 P.H. (TLMF).
120. *Scrobipalpula ramosella* (Müller-Rutz) – Switzerland, GU 95/533 P.H. (ETHZ).
121. *Scrobipalpula seniorum* Povolný – Macedonia, gen. slide 15304 (NHMW).

122

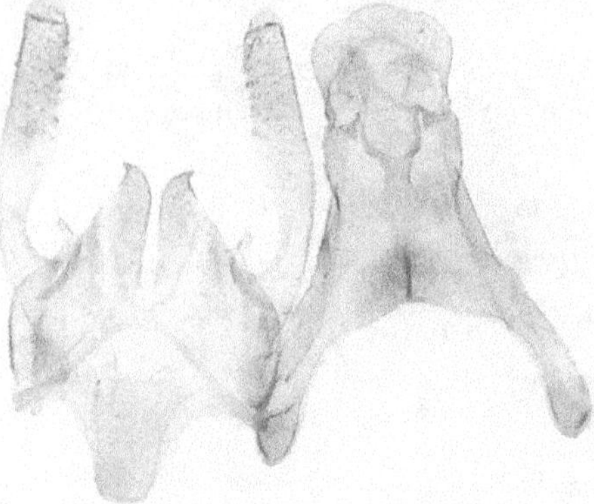

123

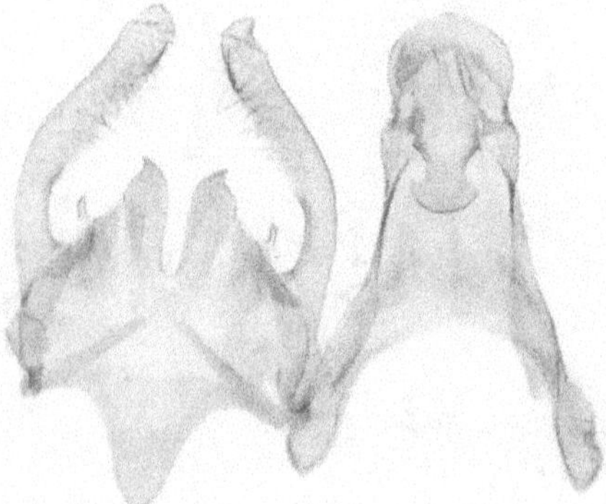

124

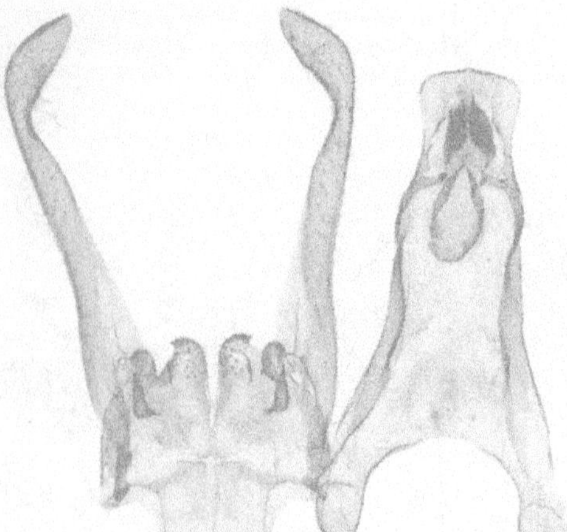

Male Genitalia

122. *Scrobipalpula diffluella* (Frey) – Austria, GEL 49 P.H. (TLMF).
123. *Scrobipalpula tussilaginis* (Stainton) – Austria, GEL 616 P.H. (TLMF).
124. *Phthorimaea operculella* (Zeller) – Spain, GU 01/988 P.H. (TLMF).

125

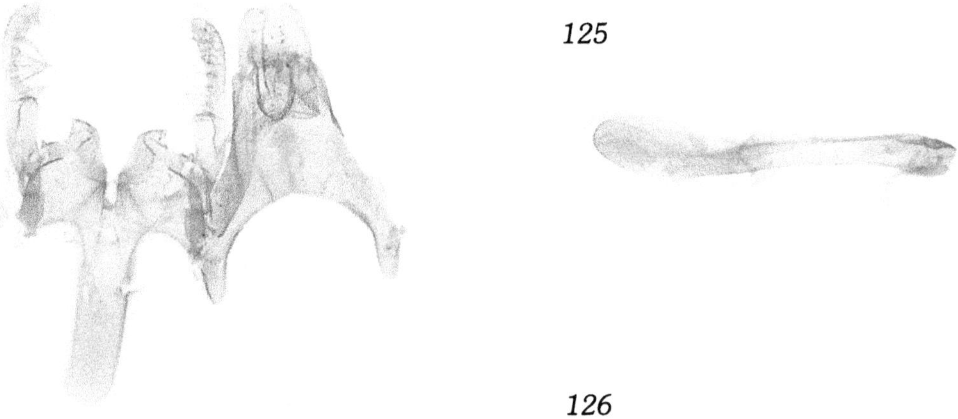

126

127

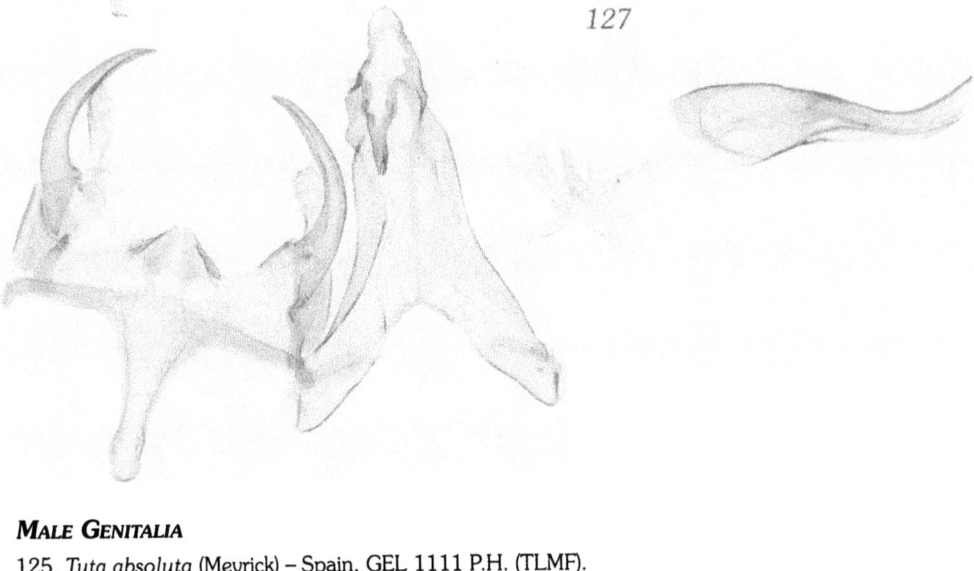

MALE GENITALIA

125. *Tuta absoluta* (Meyrick) – Spain, GEL 1111 P.H. (TLMF).
126. *Keiferia lycopsersicella* (Walsingham) – Dominicanic Republic, GU 09/1306 P.H. (ZMUC).
127. *Ephysteris promptella* (Staudinger) – Spain, GU 01/1031 P.H. (BALD).

128

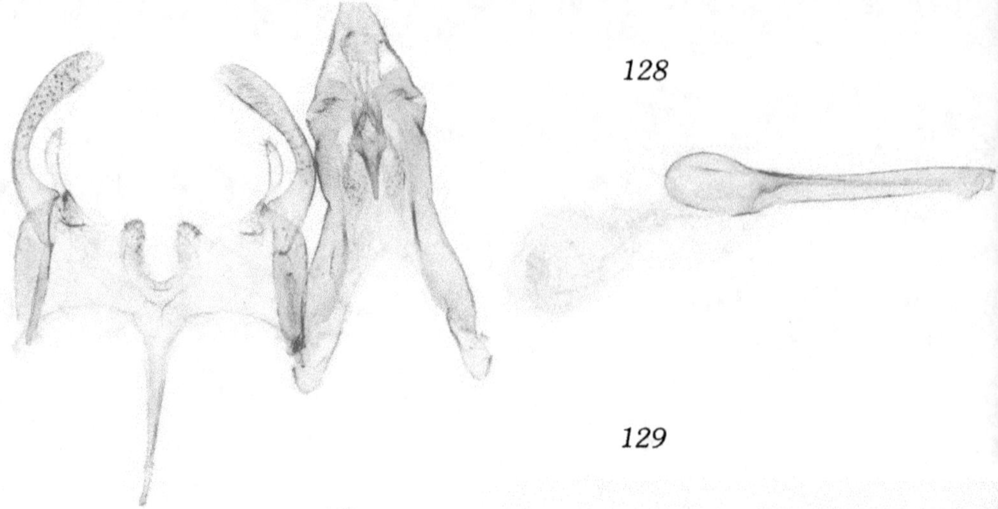

129

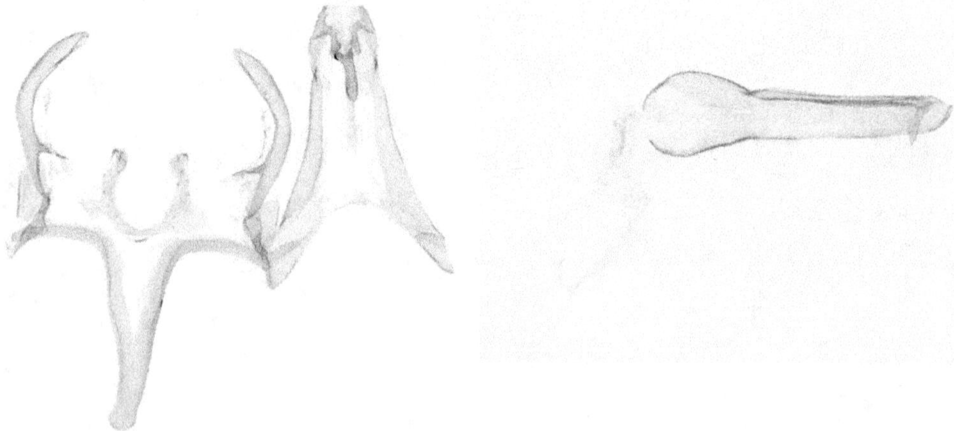

130

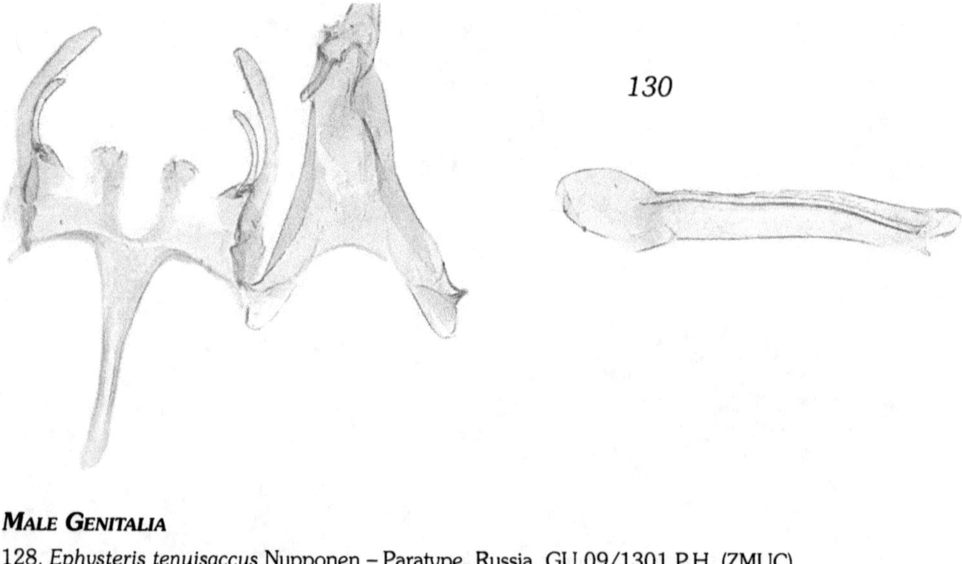

Male Genitalia

128. *Ephysteris tenuisaccus* Nupponen – Paratype, Russia, GU 09/1301 P.H. (ZMUC).
129. *Ephysteris deserticolella* (Staudinger) – Greece, GEL 633 P.H. (TLMF).
130. *Ephysteris insulella insulella* (Heinemann) – France, gen. slide 5007 H.H. (ZMUC).

130a

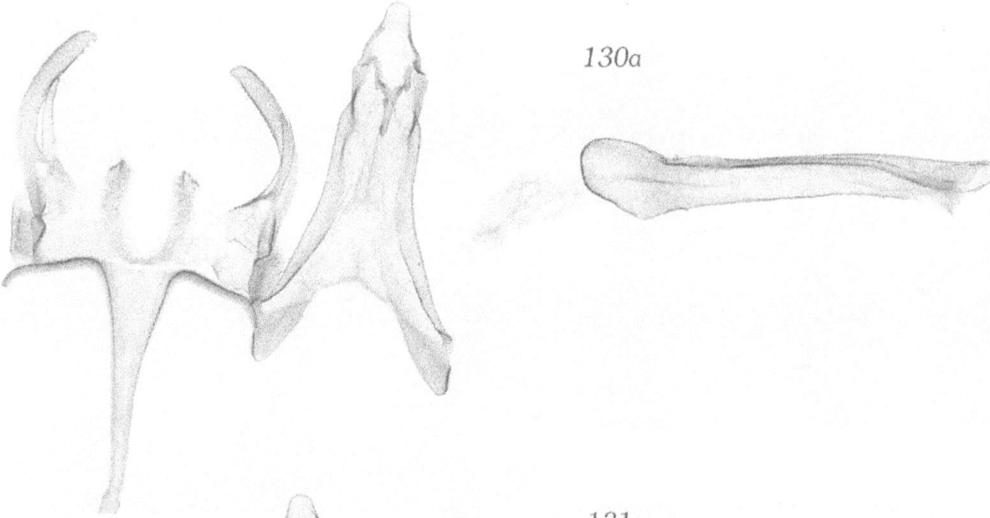

131

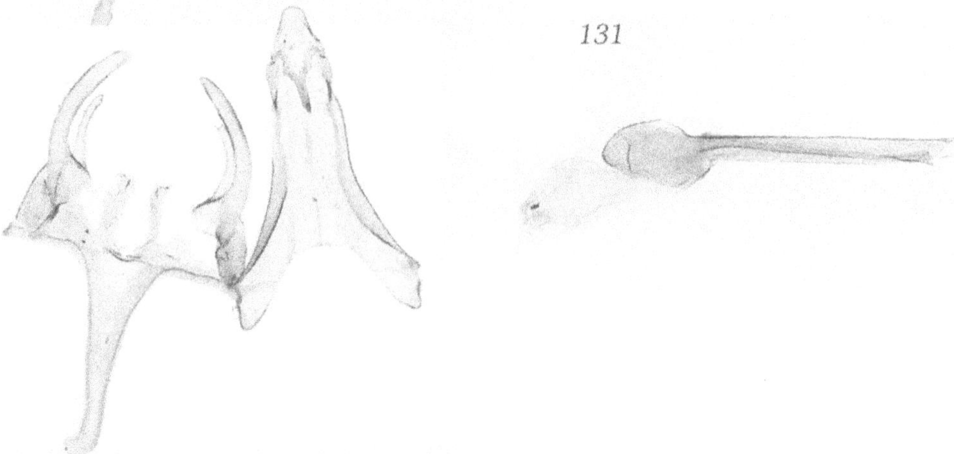

131a

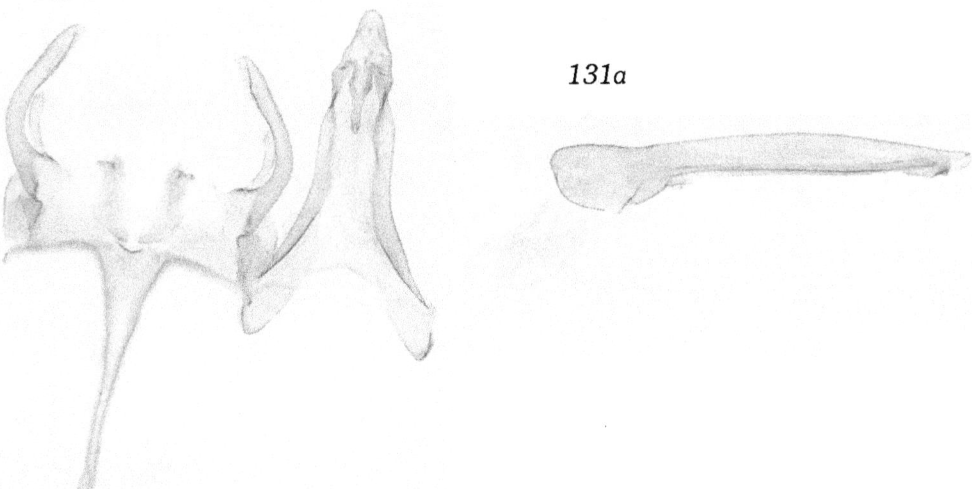

MALE GENITALIA

130a. *Ephysteris insulella praticolella* (Christoph) – Italy, GU 01/997 P.H. (ZSM).
131. *Ephysteris diminutella* (Zeller) – Italy, GU 06/1257 P.H. (ZSM).
131a. *Ephysteris diminutella* (Zeller) – Italy, GU 02/1131 P.H. (ZMUC).

131b

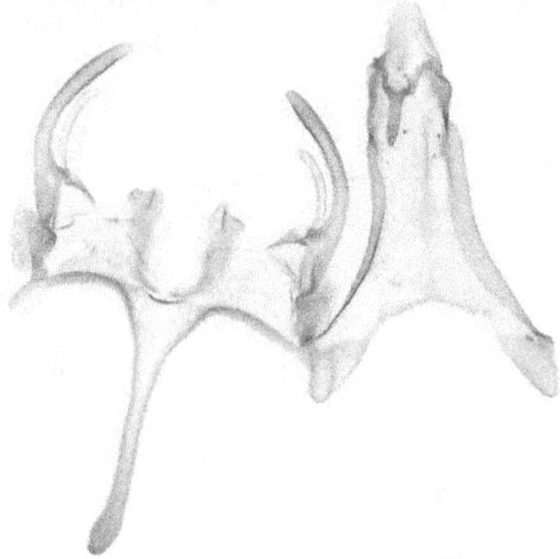

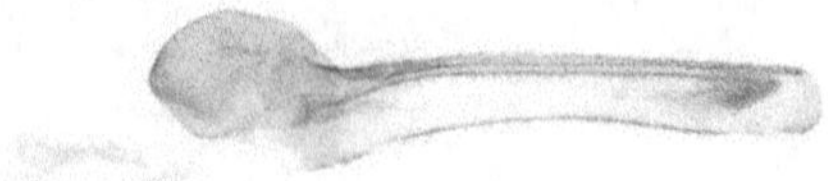

132

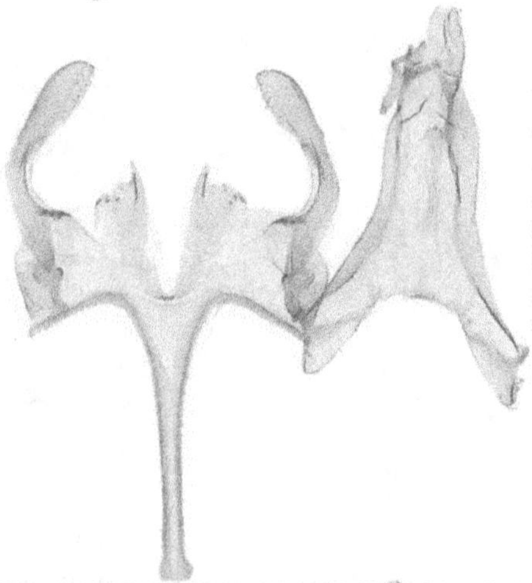

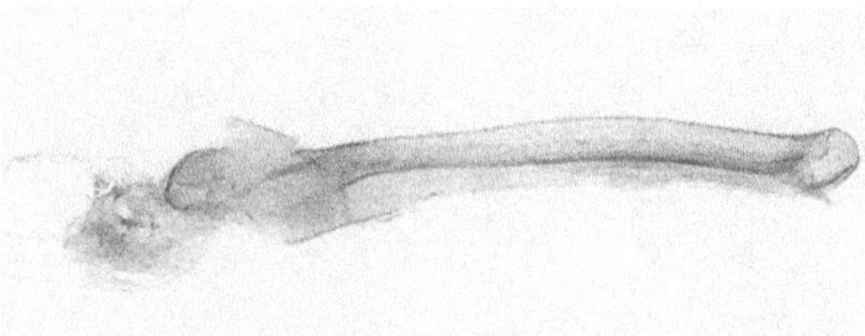

132a

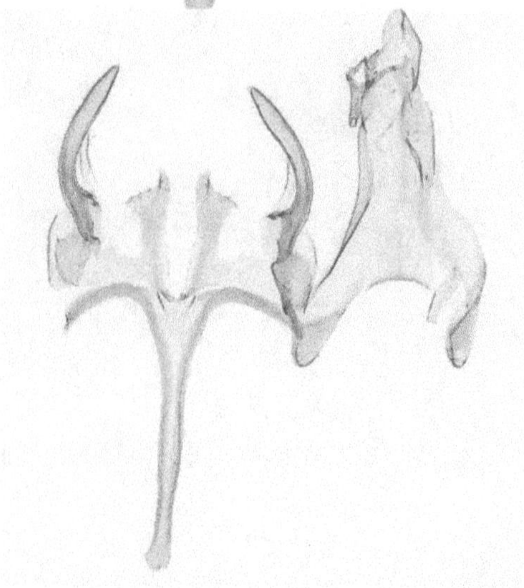

Male Genitalia

131b. *Ephysteris diminutella* (Zeller) – Spain, GU 08/1265 P.H. (ZMUC).
132. *Ephysteris inustella inustella* (Herrich-Schäffer) – Hungary, GU 01/1004 P.H. (ZSM).
132a. *Ephysteris inustella gredosensis* (Rebel) – Spain, gen. Slide 1421 H.H. (ZMUC).

133

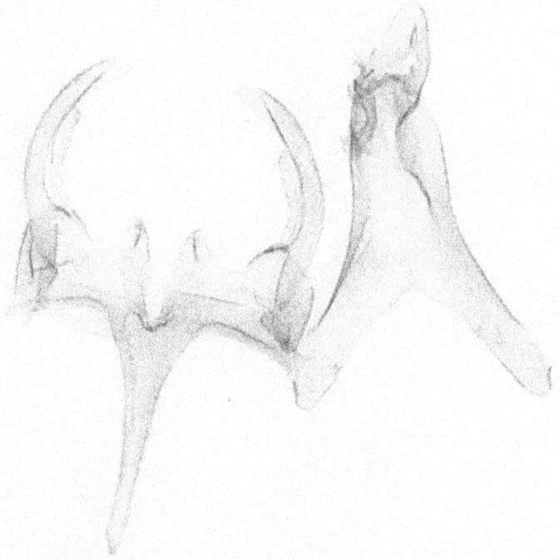

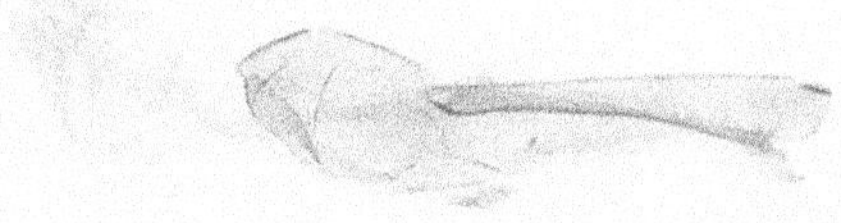

134

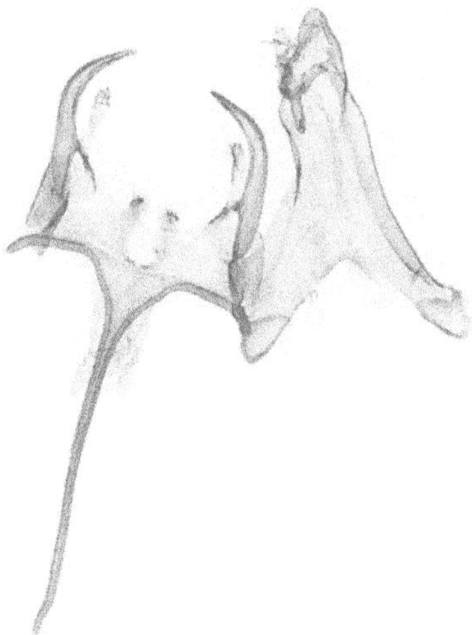

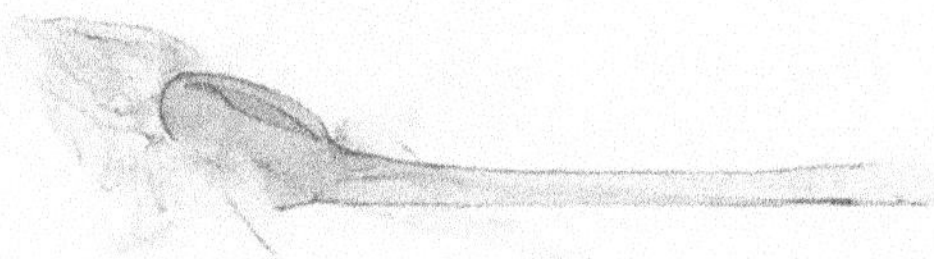

135

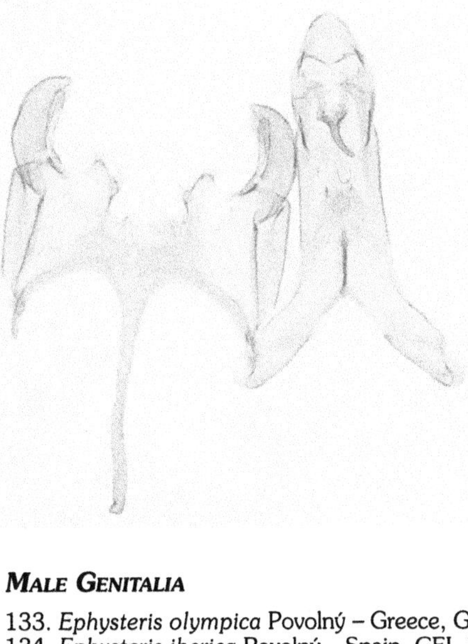

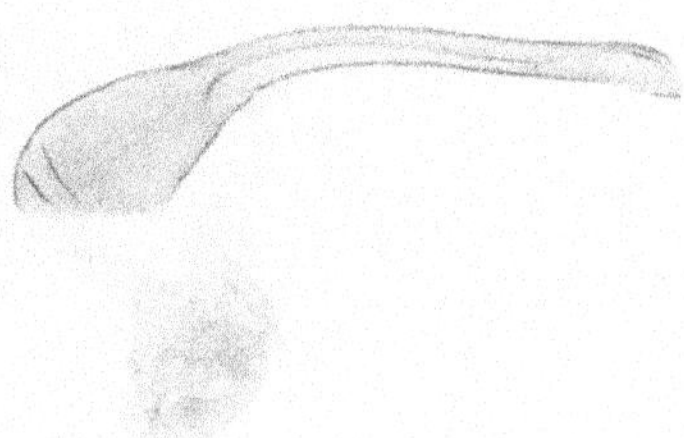

Male Genitalia

133. *Ephysteris olympica* Povolný – Greece, GU 01/1066 P.H. (ZSM).
134. *Ephysteris iberica* Povolný – Spain, GEL 1089 P.H. (TLMF).
135. *Ochrodia subdiminutella* (Stainton) – Spain (Canary Isl.), GU 01/1065 P.H. (ZSM).

135a

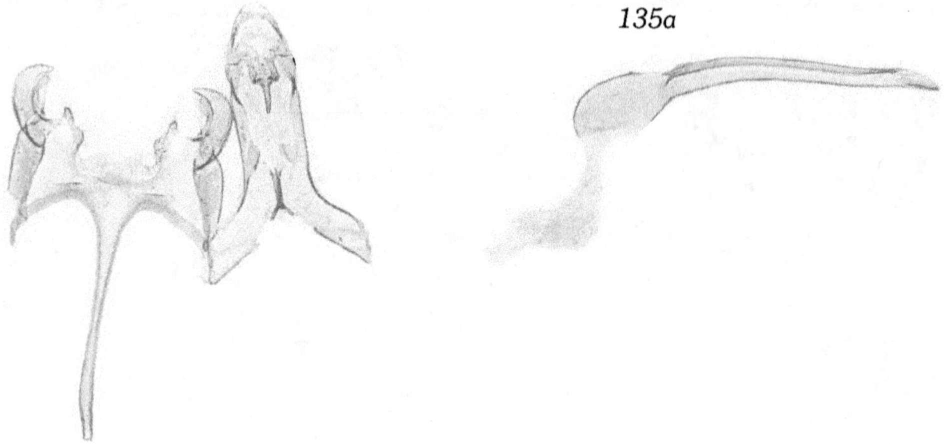

136

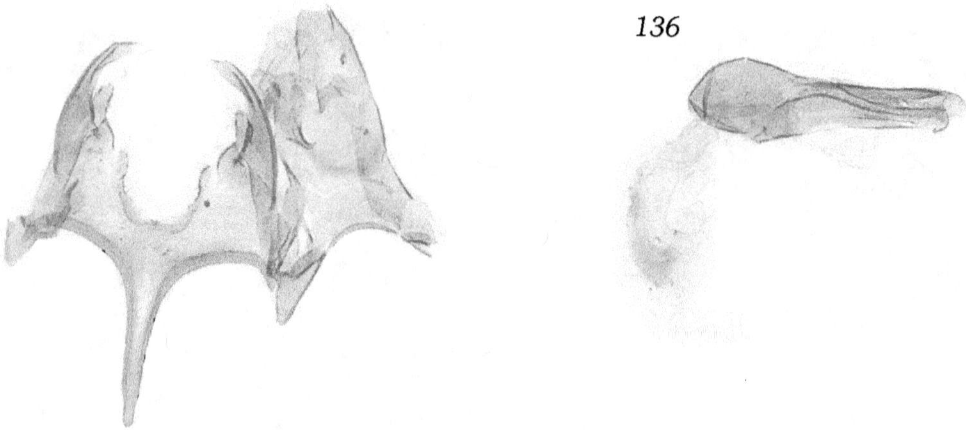

137

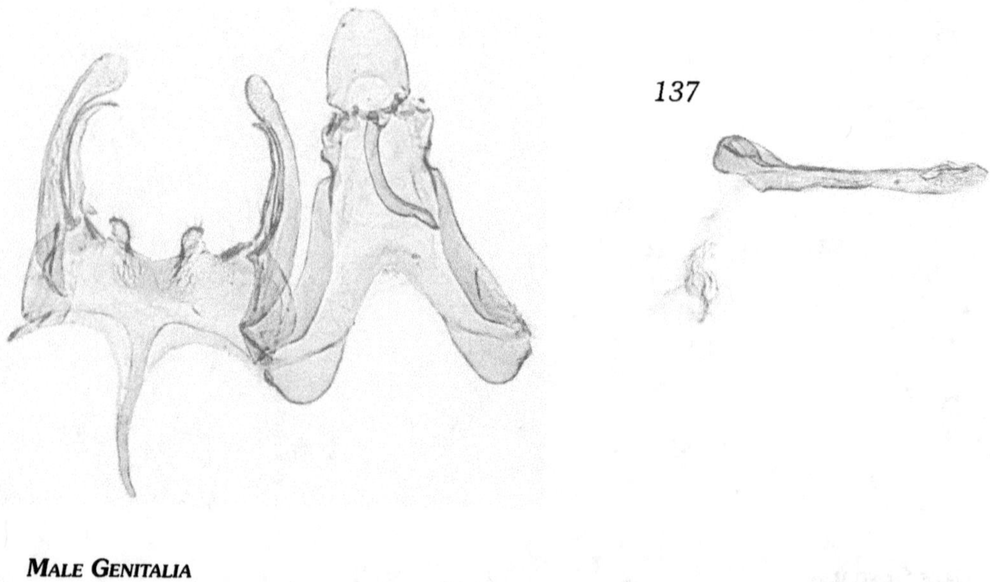

Male Genitalia

135a. *Ochrodia subdiminutella* (Stainton) – Spain, GEL 1127 P.H. (TLMF).
136. *Vladimirea glebicolorella* (Erschoff) – Ukraine (Crimea), gen. slide 527/07 Bidzilya (ZMKU).
137. *Microlechia rhamnifoliae* (Amsel & Hering) – Morocco, GU 09/1313 (ZMUC).

138

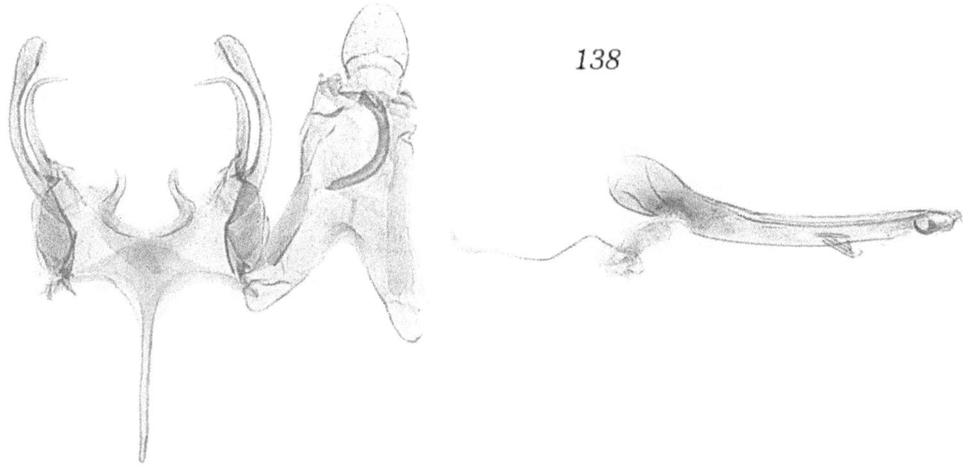

139

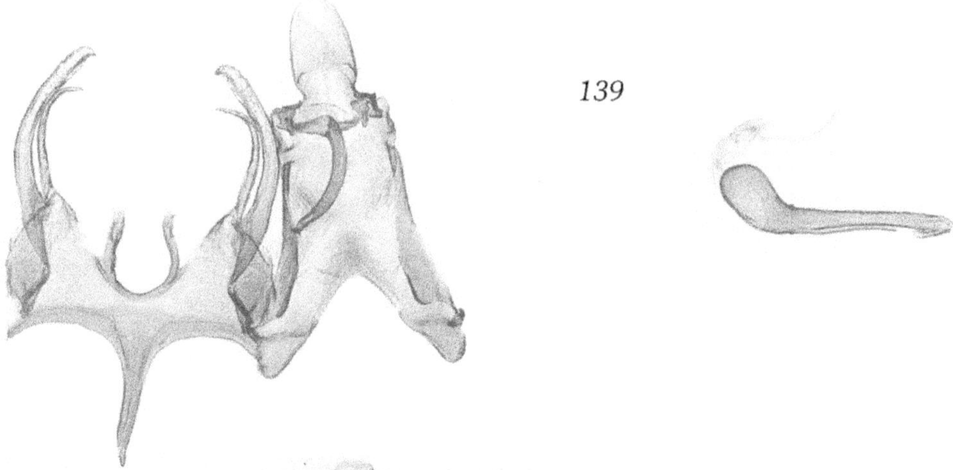

140

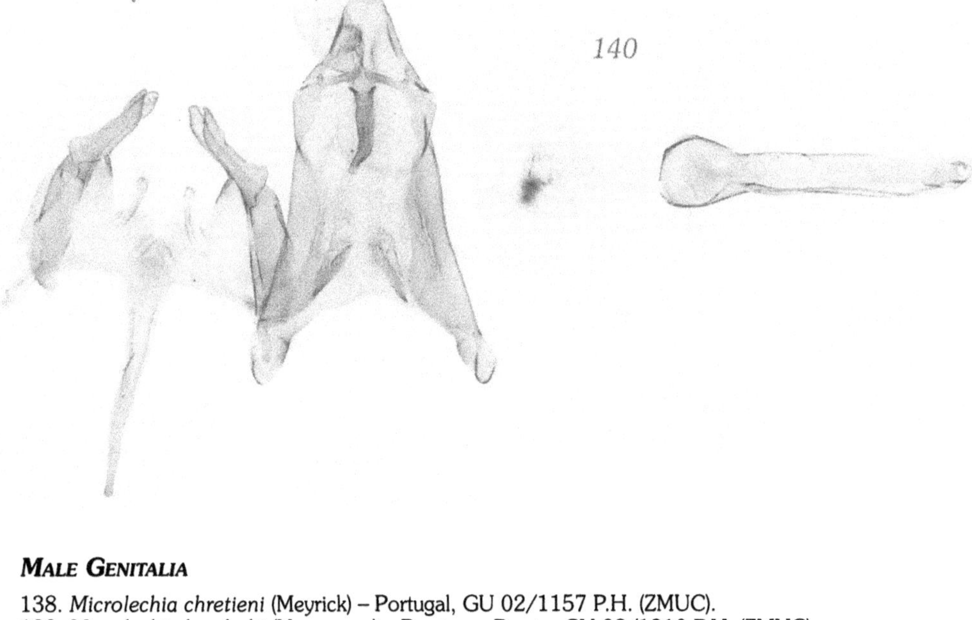

MALE GENITALIA

138. *Microlechia chretieni* (Meyrick) – Portugal, GU 02/1157 P.H. (ZMUC).
139. *Microlechia karsholti* (Nupponen) – Paratype, Russia, GU 09/1310 P.H. (ZMUC).
140. *Cosmardia moritzella* (Treitschke) – Switzerland, GEL 1031 P.H. (TLMF).

141

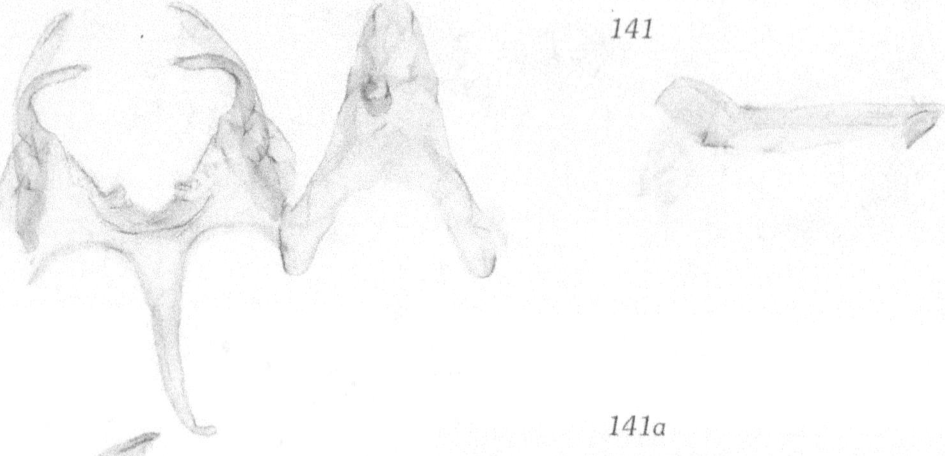

141a

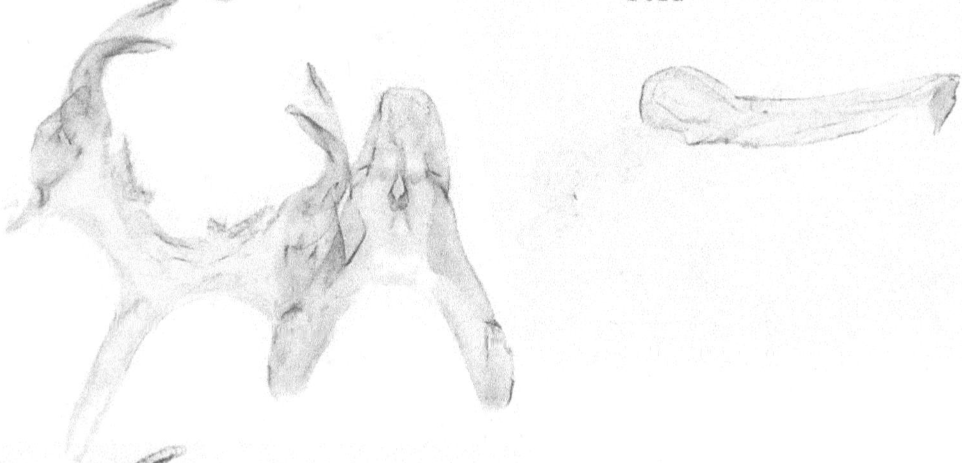

142

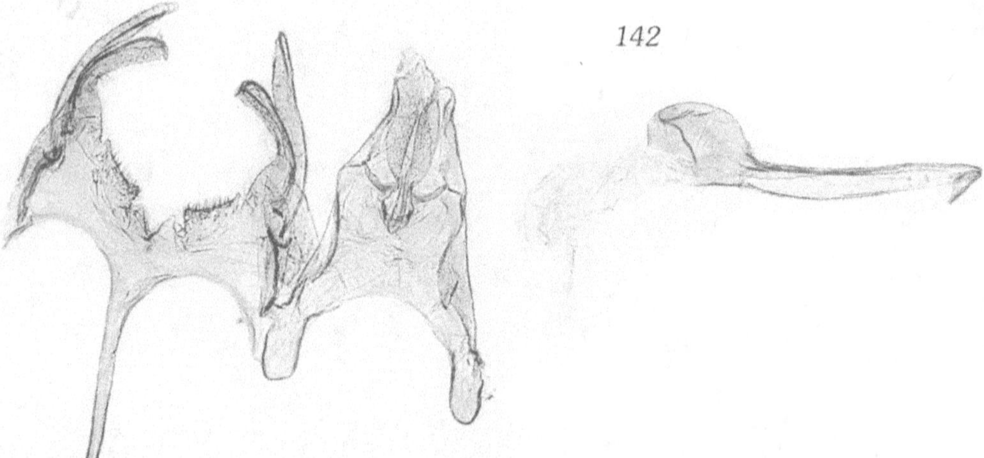

Male Genitalia

141. *Lutilabria lutilabrella lutilabrella* (Mann) – Croatia, GU 90/192 P.H. (ZIAP).
141a. *Lutilabria lutilabrella olympica* Huemer – Greece, GU 06/1248 P.H. (ZSM).
142. *Lutilabria volgensis* Anikin & Piskunov – Holotype, Russia, GU 03/1207 P.H. (ZIAP).

143

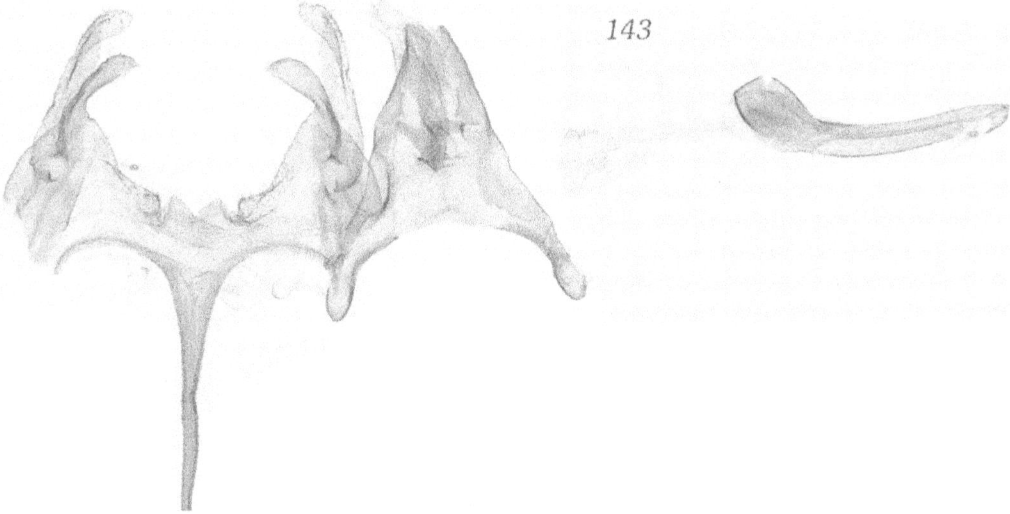

144

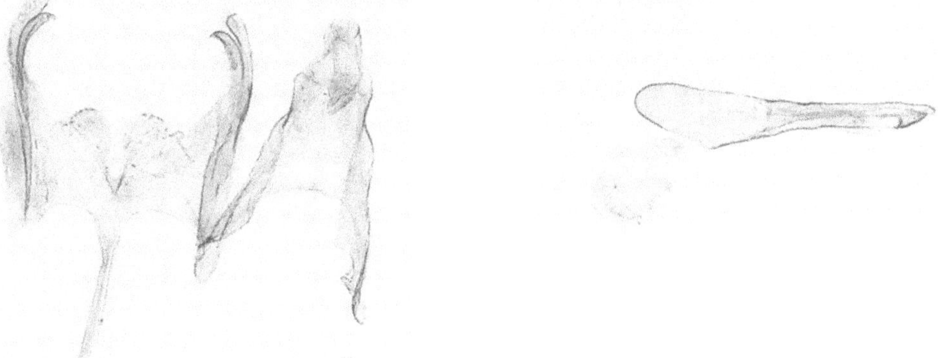

145

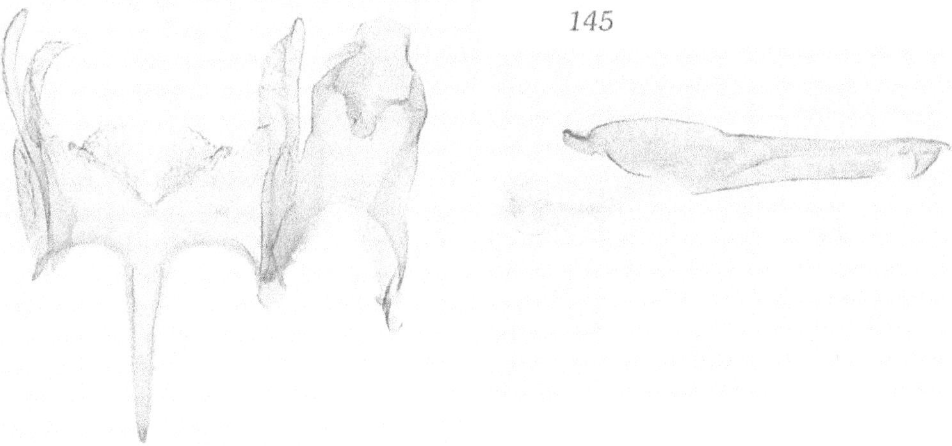

MALE GENITALIA

143. *Lutilabria prolata* Junnilainen & Nupponen – Romania, GU 99/866 P.H. (WIES).
144. *Klimeschiopsis kiningerella* (Duponchel) – Austria, gen. slide 16643 (NHMW).
145. *Klimeschiopsis discontinuella* (Rebel) – Italy, GEL 1040 P.H. (TLMF).

146

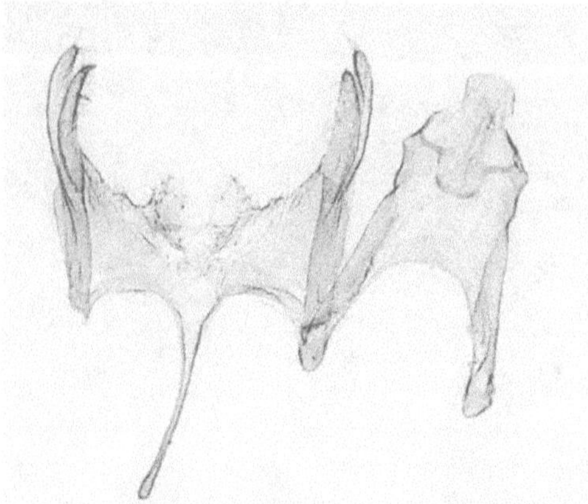

147

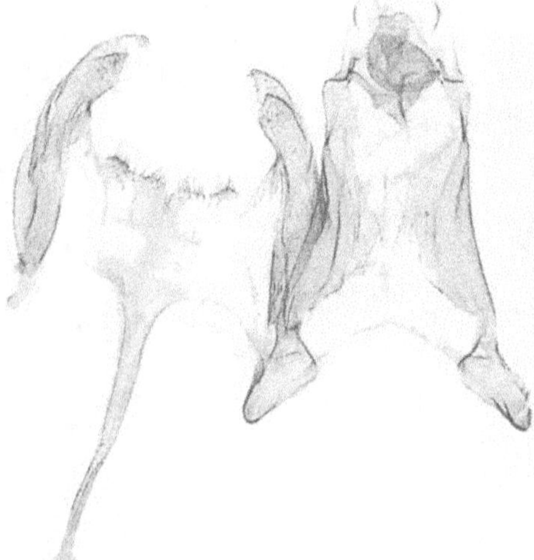

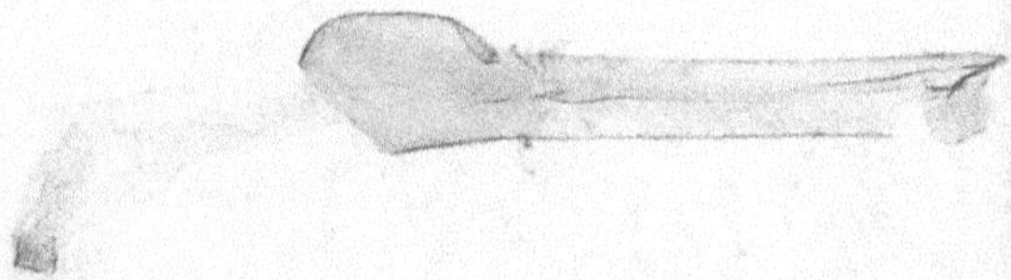

148

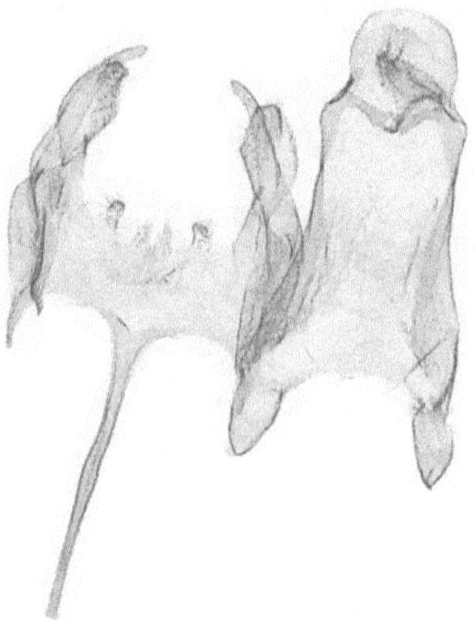

Male Genitalia

146. *Klimeschiopsis terroris* (Hartig) – Spain, gen. slide 3411 H.H. (ZMUC).
147. *Caryocolum fischerella* (Treitschke) – Germany, GEL 1082 P.H. (TLMF).
148. *Caryocolum tischeriella* (Zeller) – Austria, GEL 1083 P.H. (TLMF).

149

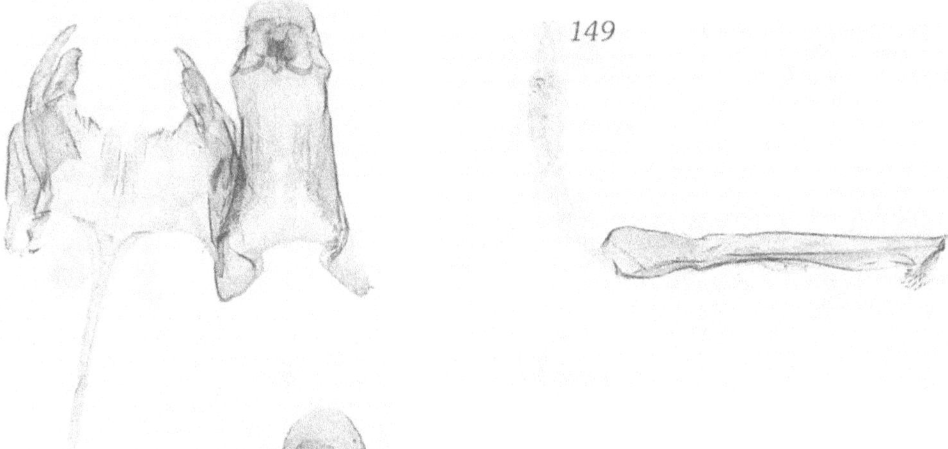

150

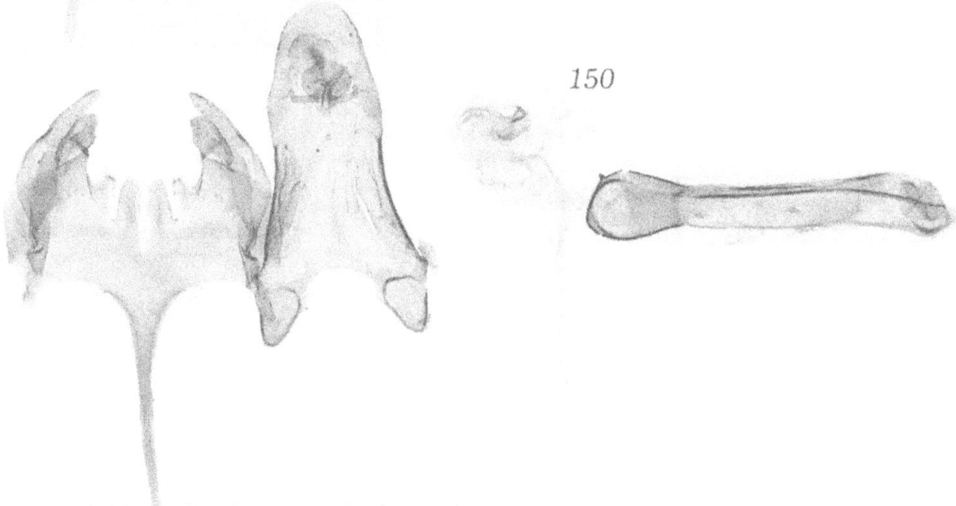

151

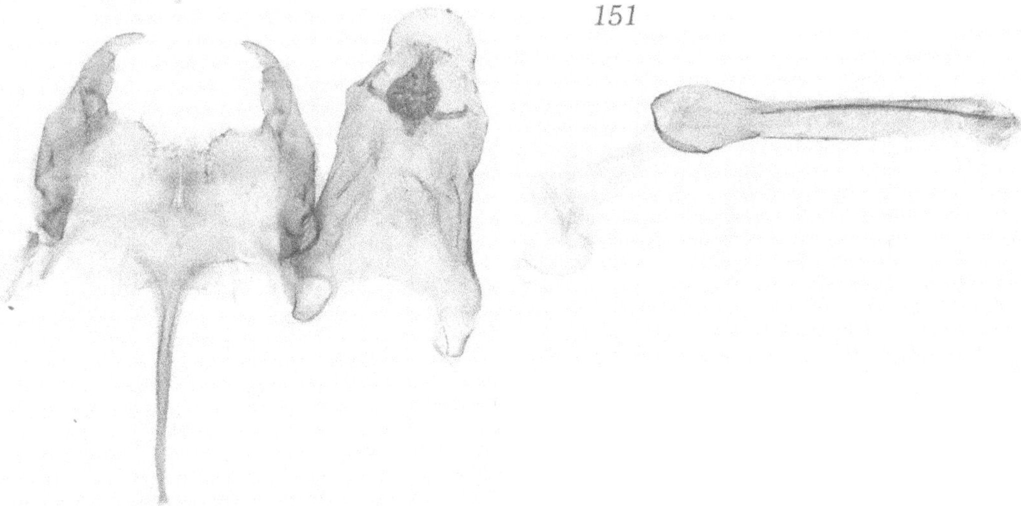

MALE GENITALIA

149. *Caryocolum alsinella* (Zeller) – Portugal, GU 05/1231 P.H. (CORL).
150. *Caryocolum viscariella* (Stainton) – Switzerland, GEL 423 P.H. (TLMF).
151. *Caryocolum albifaciella* (Heinemann) – Austria, GEL 419 P.H. (TLMF).

152

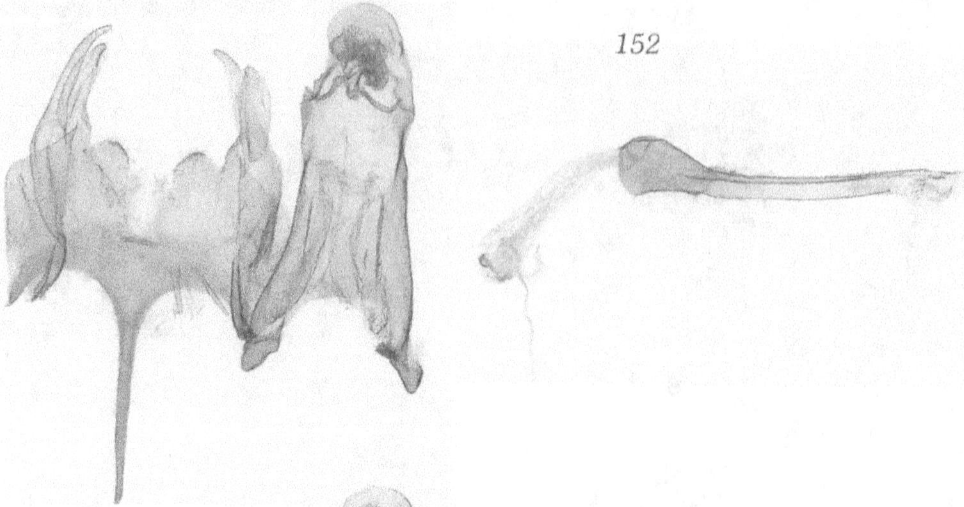

153

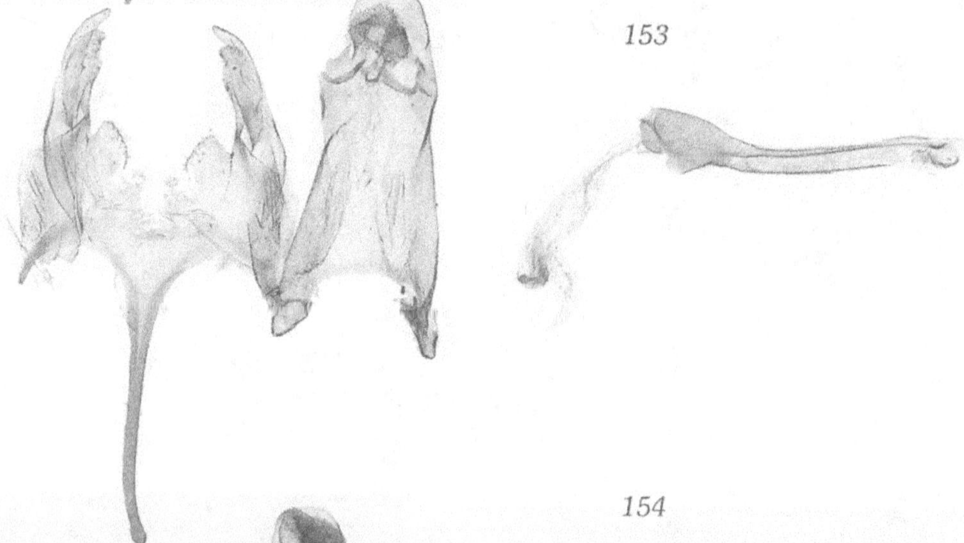

154

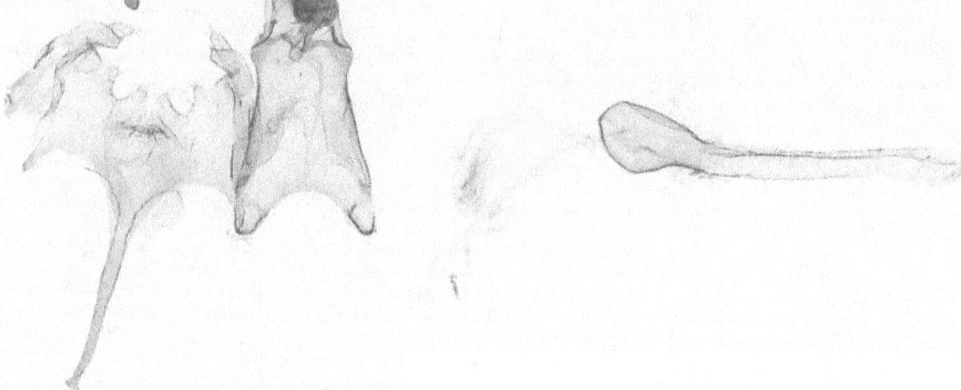

MALE GENITALIA

152. *Caryocolum vicinella* (Douglas) – England, gen. slide 24253 (BMNH).
153. *Caryocolum bosalella* (Rebel) – France (Corsica), gen. slide 13479 (NHMW).
154. *Caryocolum amaurella* (Hering) – Austria, GEL 1098 P.H. (TLMF).

155

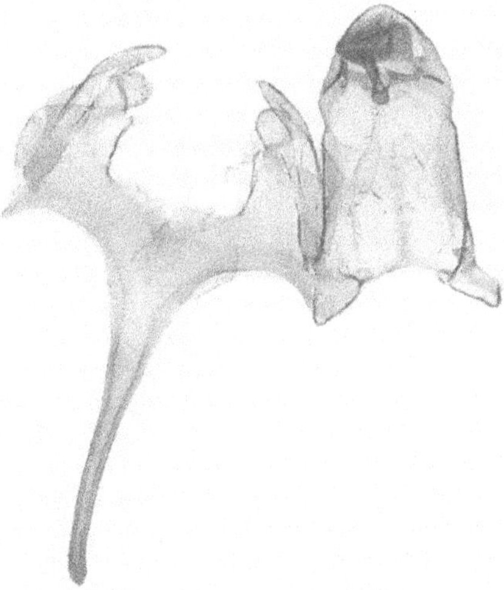

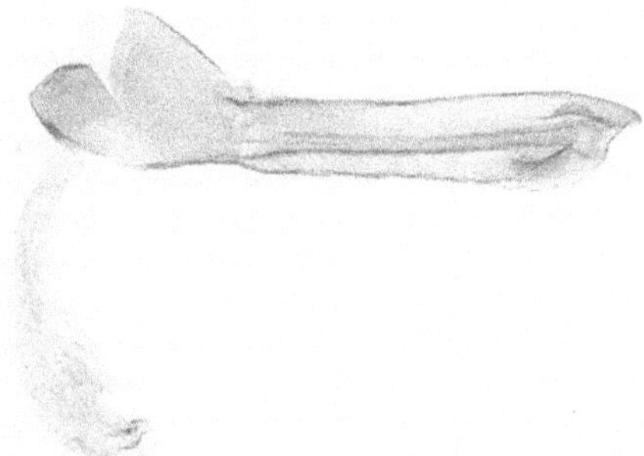

156

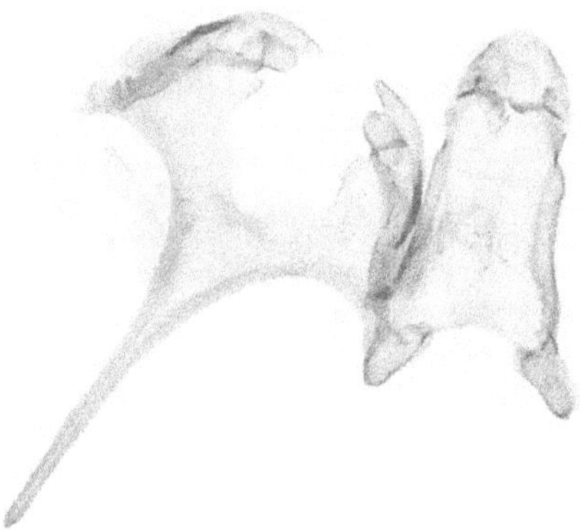

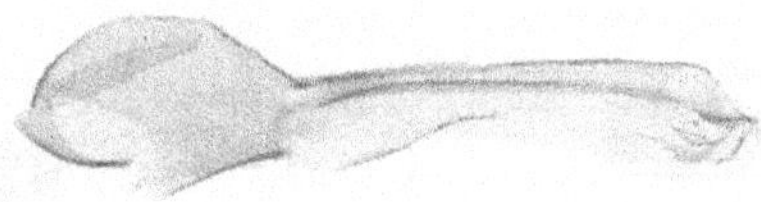

157

MALE GENITALIA

155. *Caryocolum oculatella* (Thomann) – Austria, GU 87/149 P.H. (TLMF).
156. *Caryocolum leucofasciatum* Huemer – Spain, GU 87/073 P.H. (TLMF).
157. *Caryocolum petryi* (Hofmann) – Switzerland, GU 86/089 P.H. (TLMF).

158

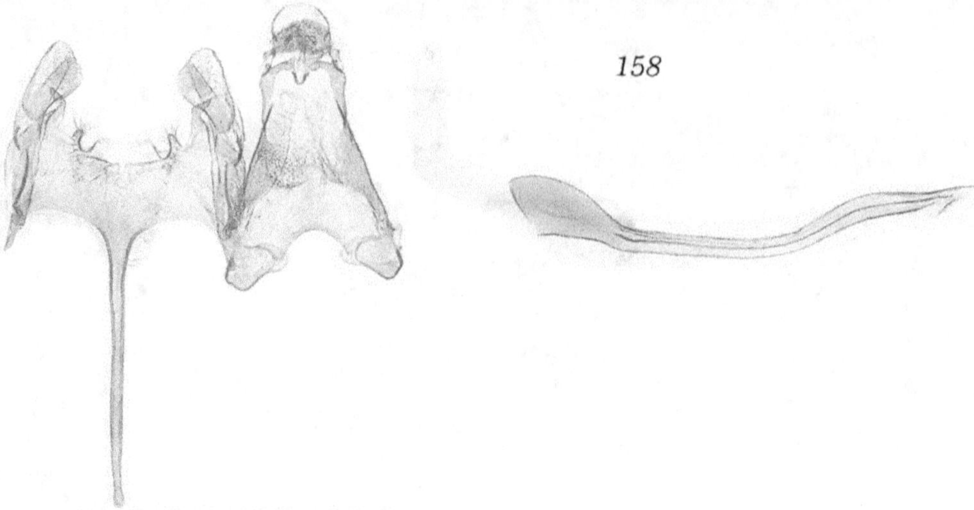

159

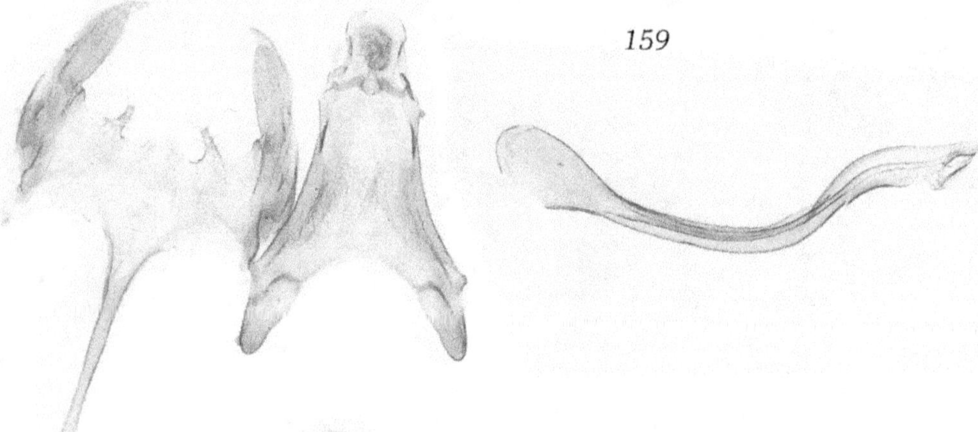

160

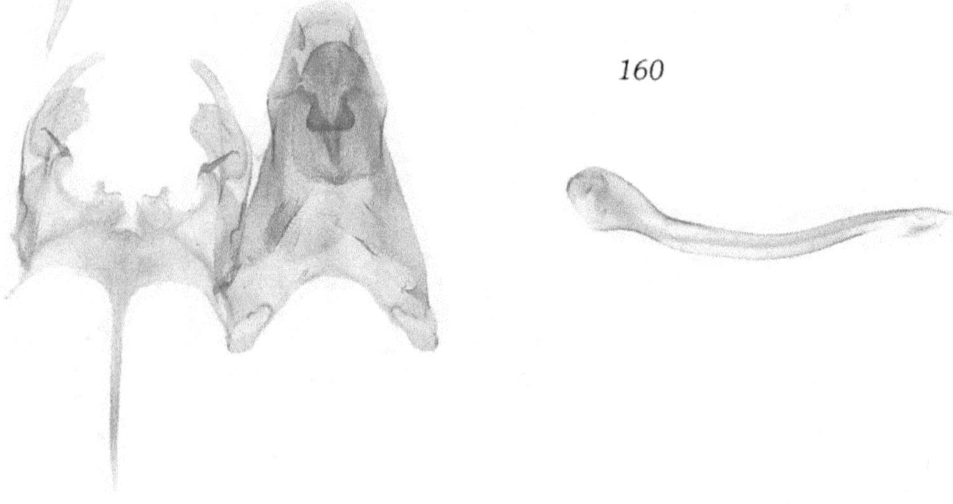

Male Genitalia

158. *Caryocolum baischi* sp. n. – Holotype, Greece, GEL 1121 P.H. (TLMF).
159. *Caryocolum repentis* Huemer & Luquet – Holotype, Austria, GEL 277 P.H. (TLMF).
160. *Caryocolum siculum* Bella – Paratype, Italy, GU 08/1263 (ZMUC).

161

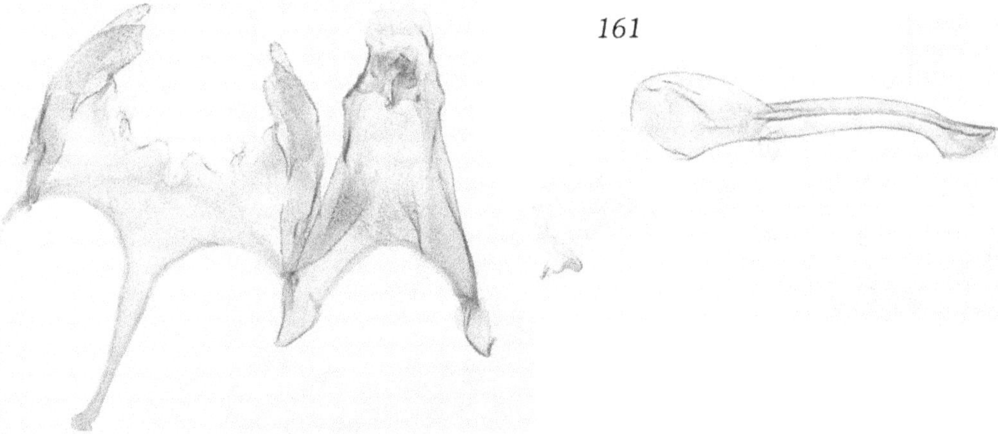

162

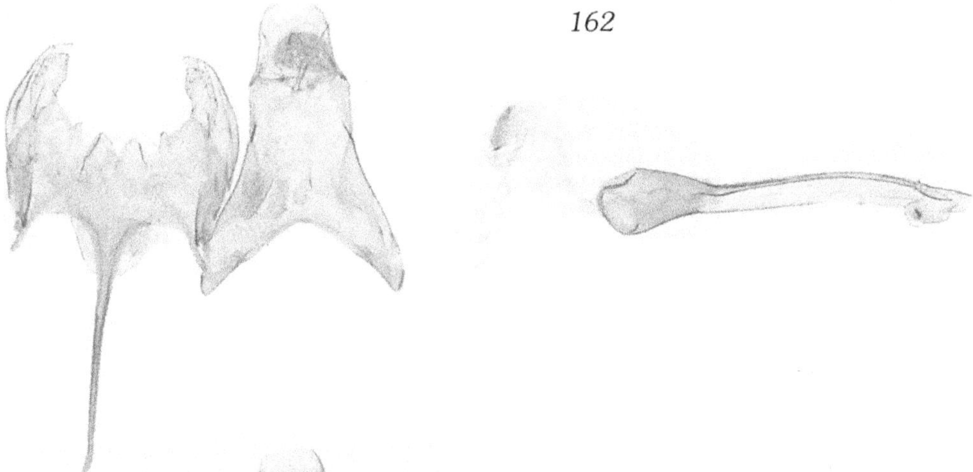

163

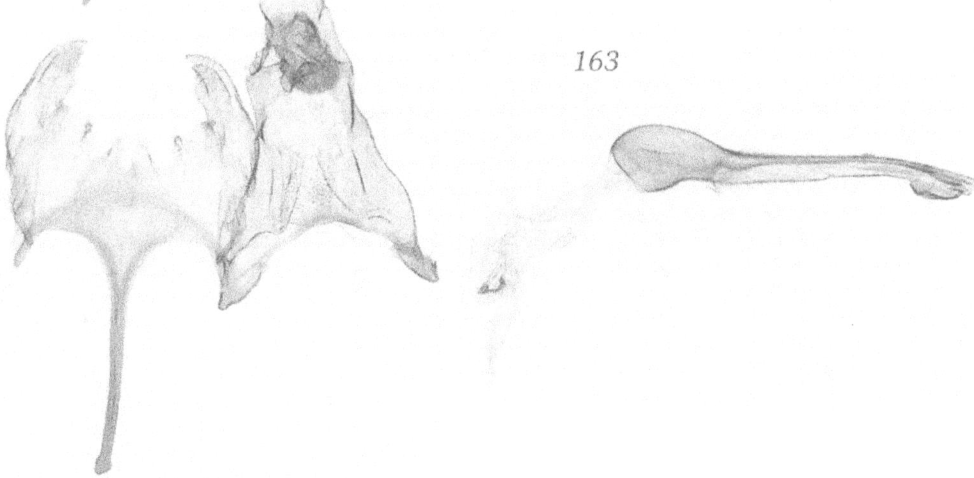

Male Genitalia

161. *Caryocolum inflativorella* (Klimesch) – Paralectotype (*Lita xuthella*), Macedonia, gen. slide 13075 (NHMW).
162. *Caryocolum saginella* (Zeller) – Italy, GEL 1090 P.H. (TLMF).
163. *Caryocolum cauligenella* (Schmid) – France, gen. slide 24407 (BMNH).

164

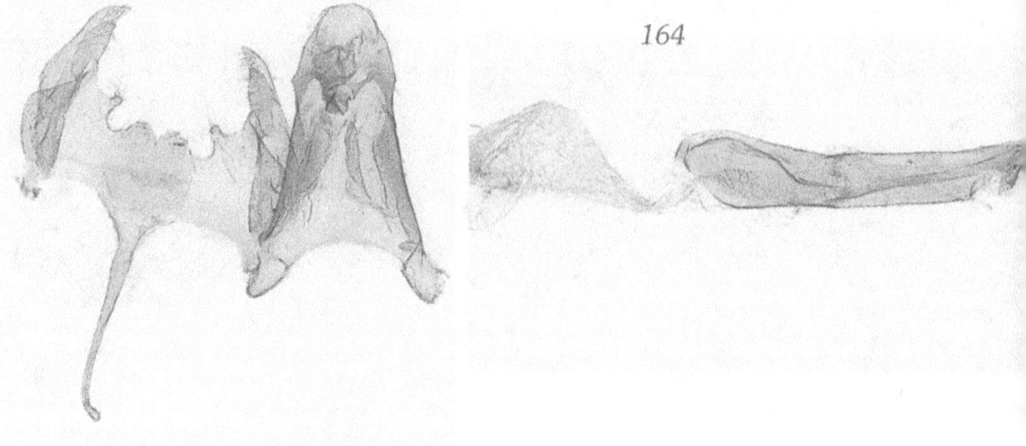

165

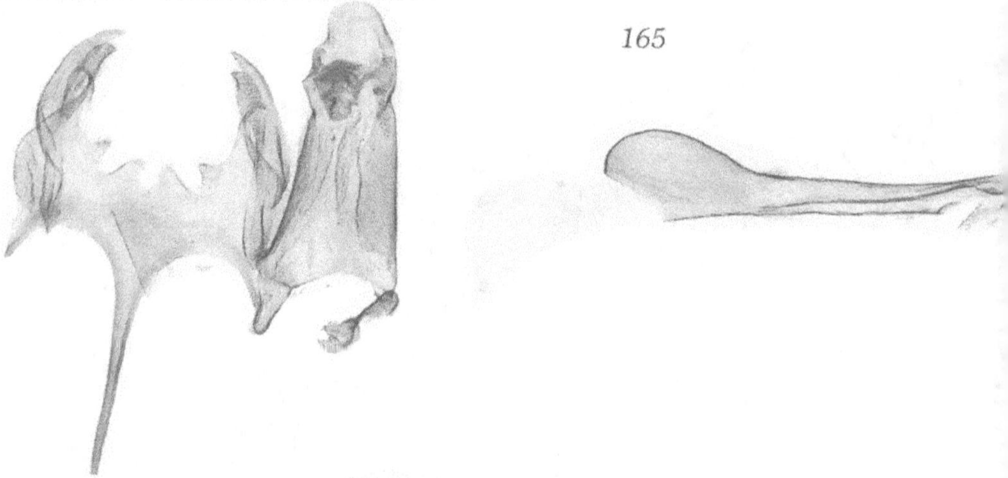

166

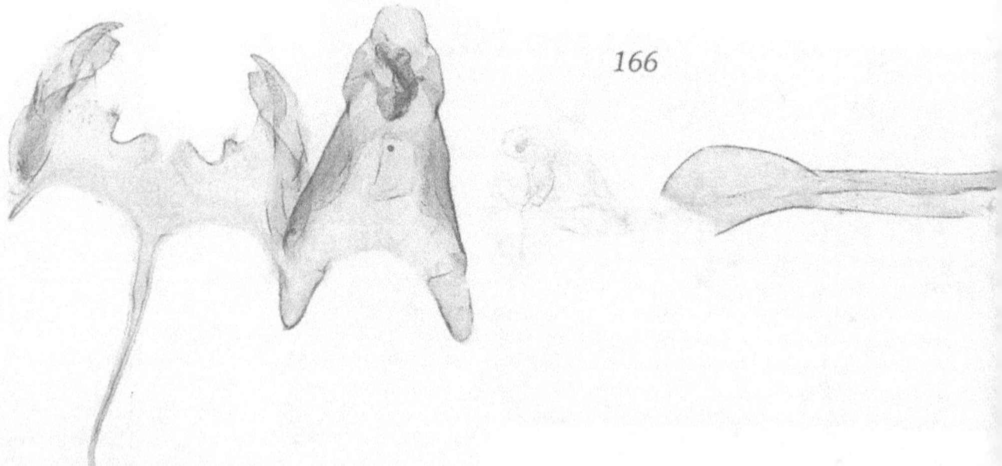

Male Genitalia

164. *Caryocolum trauniella* (Zeller) – Slovenia, GU 06/1256 P.H. (ZSM).
165. *Caryocolum peregrinella* (Herrich-Schäffer) – France, GU 86/080 P.H. (TLMF).
166. *Caryocolum delphinatella* (Constant) – Italy, GEL 1093 P.H. (TLMF).

167

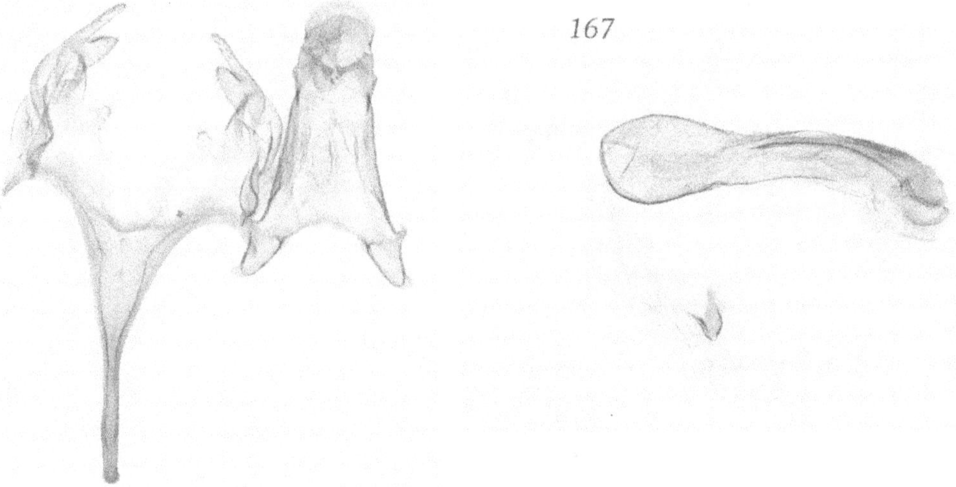

168

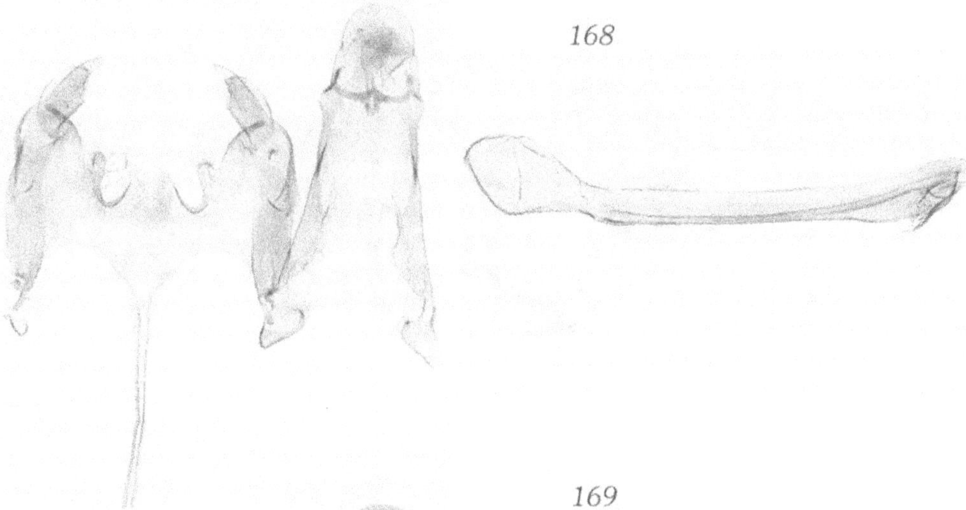

169

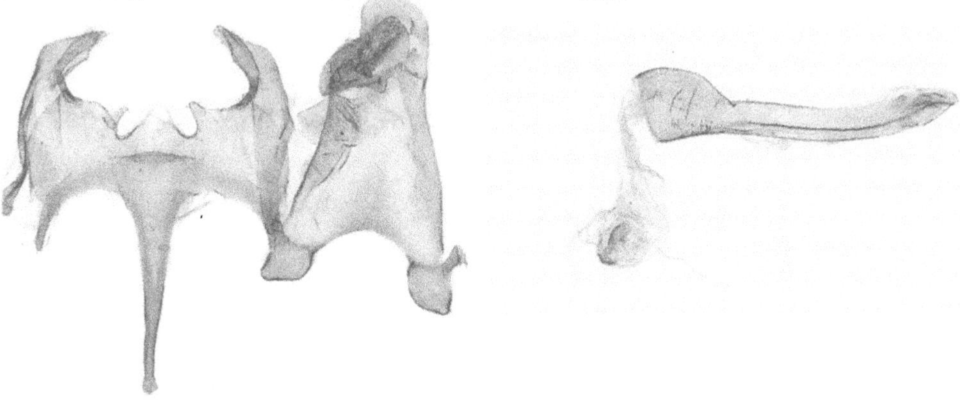

MALE GENITALIA

167. *Caryocolum provinciella* (Stainton) – Greece, GEL 980 P.H. (TLMF).
168. *Caryocolum mucronatella* (Chrétien) – Greece, GU 88/044 P.H. (ZMUC).
169. *Caryocolum leucomelanella* (Zeller) – Italy, GU 86/075 P.H. (TLMF).

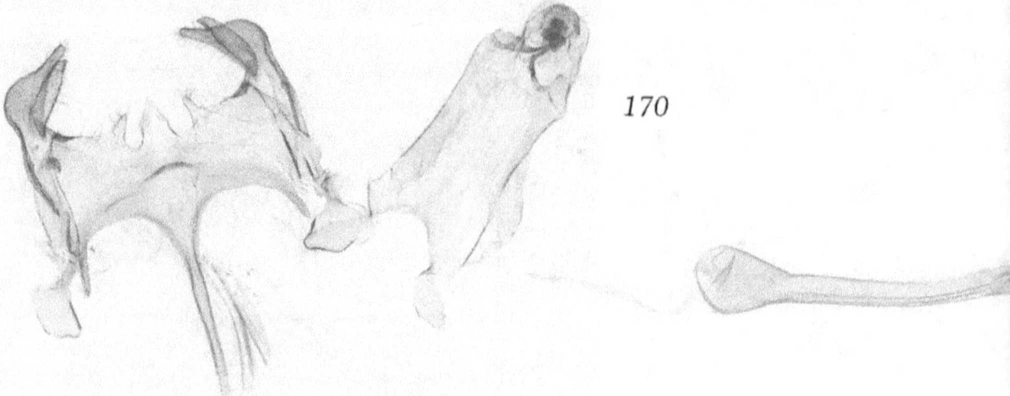

170

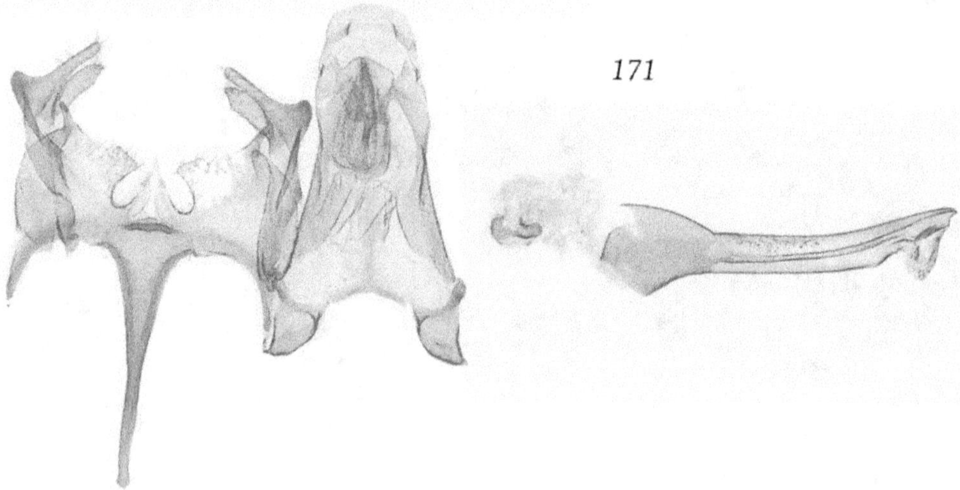

171

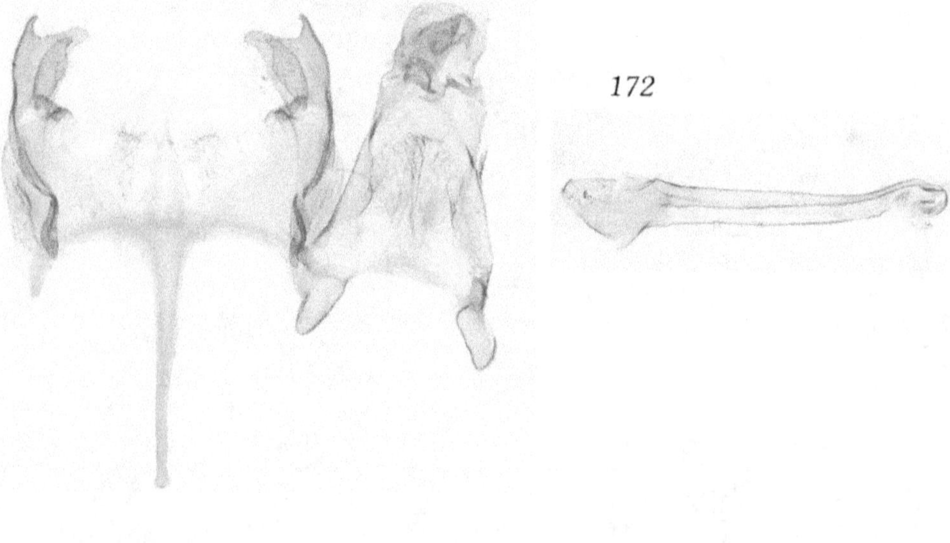

172

MALE GENITALIA

170. *Caryocolum mazeli* Huemer & Nel – Holotype, France, gen. slide 13520 J.N. (TLMF).
171. *Caryocolum leucothoracellum* (Klimesch) – Italy, GU 86/045 P.H. (TLMF).
172. *Caryocolum schleichi schleichi* (Christoph) – Turkey, GU 90/132 P.H. (ZMUC).

172a

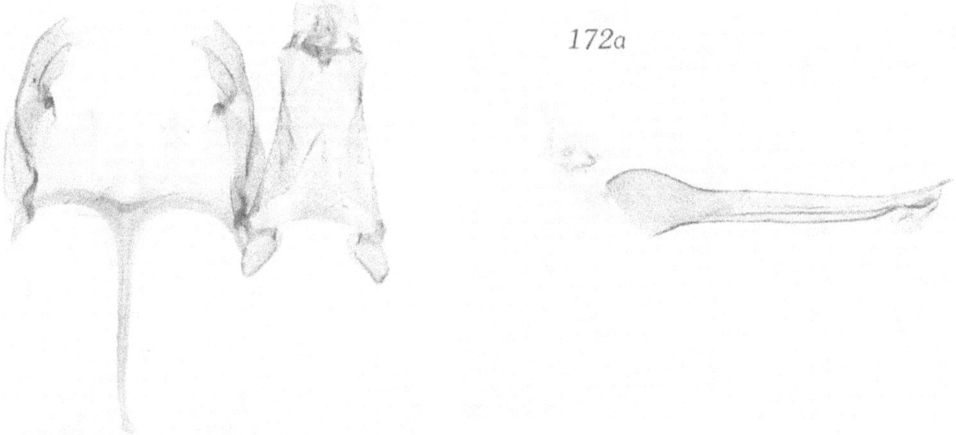

172b

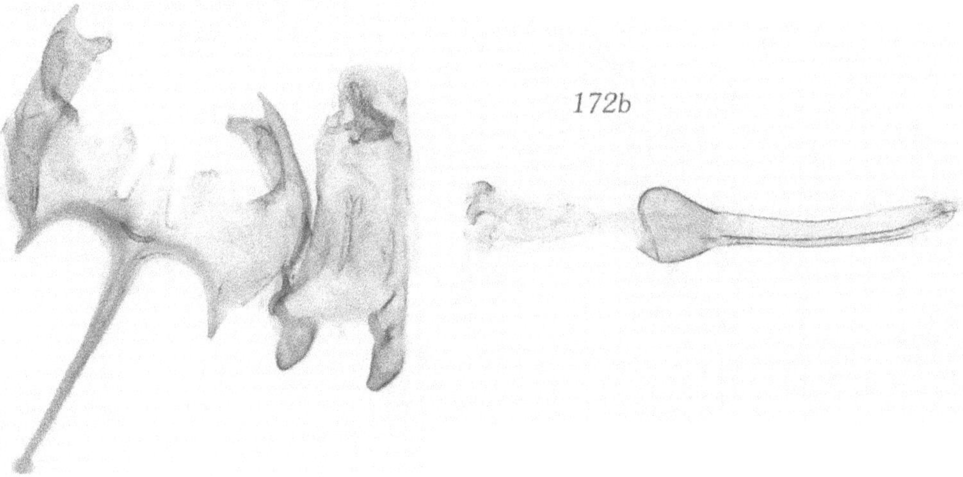

172c

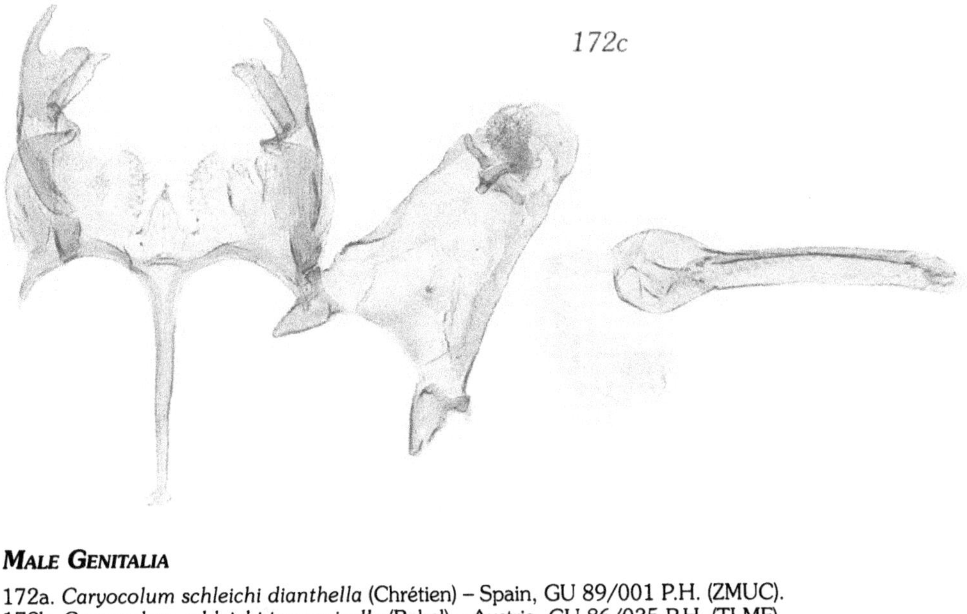

MALE GENITALIA

172a. *Caryocolum schleichi dianthella* (Chrétien) – Spain, GU 89/001 P.H. (ZMUC).
172b. *Caryocolum schleichi improvisella* (Rebel) – Austria, GU 86/035 P.H. (TLMF).
172c. *Caryocolum schleichi arenariella* (Benander) – Sweden, gen. slide 860 H.H. (ZMUC).

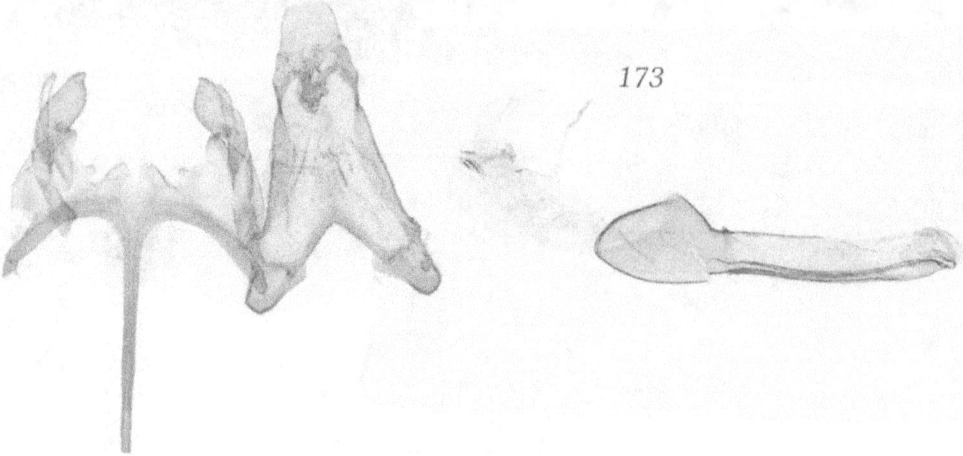

173

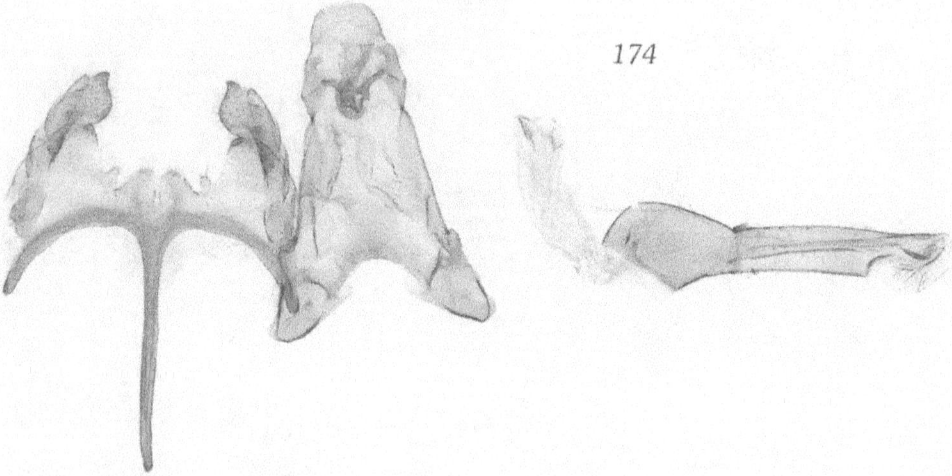

174

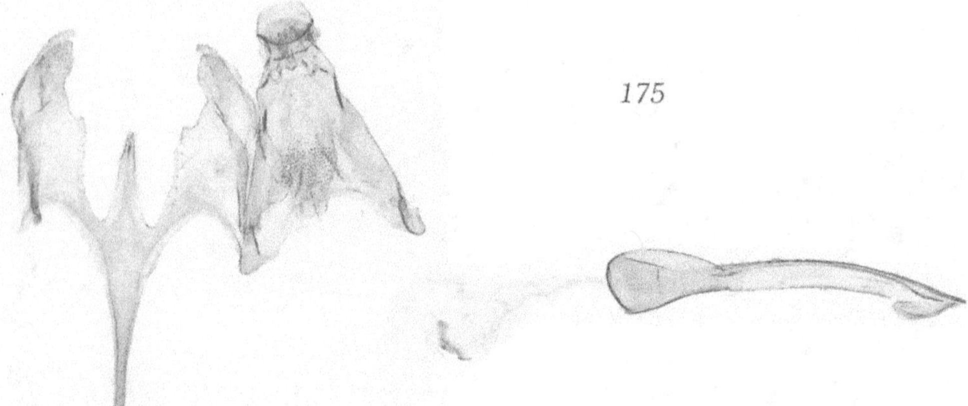

175

MALE GENITALIA

173. *Caryocolum marmorea* (Haworth) – Italy, GU 86/053 P.H. (TLMF).
174. *Caryocolum pullatella* (Tengström) – Norway, gen. slide 24237 (BMNH).
175. *Caryocolum stramentella* (Rebel) – Holotype (*Caryocolum emarginatum*), Italy, gen. slide 24414 (BMNH).

176

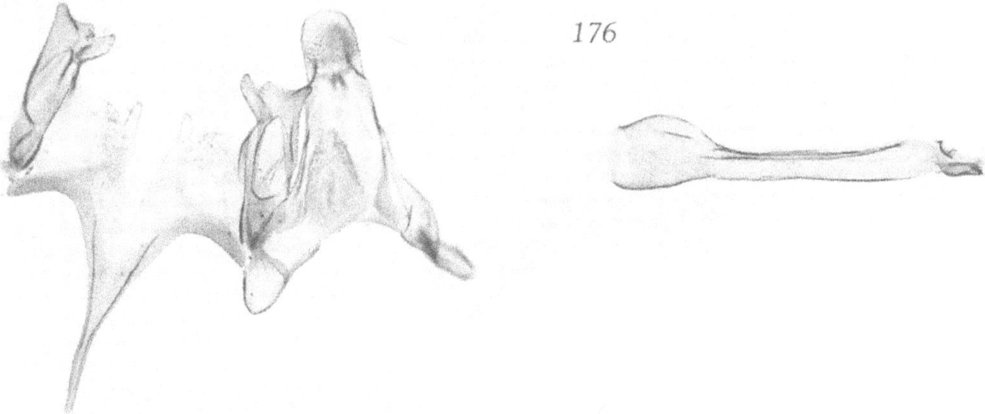

177

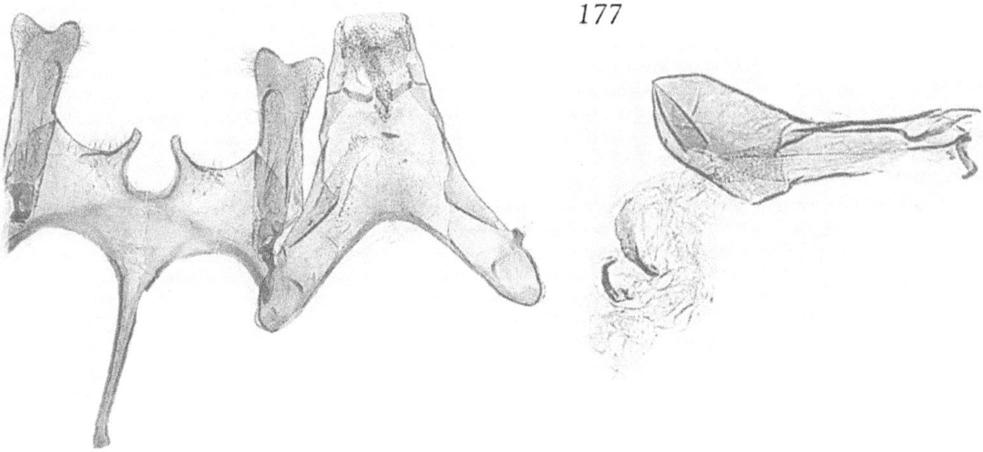

177a

MALE GENITALIA

176. *Caryocolum hispanicum* Huemer – Spain, GU 89/004 P.H. (TLMF).
177. *Caryocolum confluens* Huemer – Paratype, Greece, GU 86/380 P.H. (ZMUC).
177a. *Caryocolum cf. confluens* Huemer – Montenegro, GU 92/329 P.H. (TLMF).

178

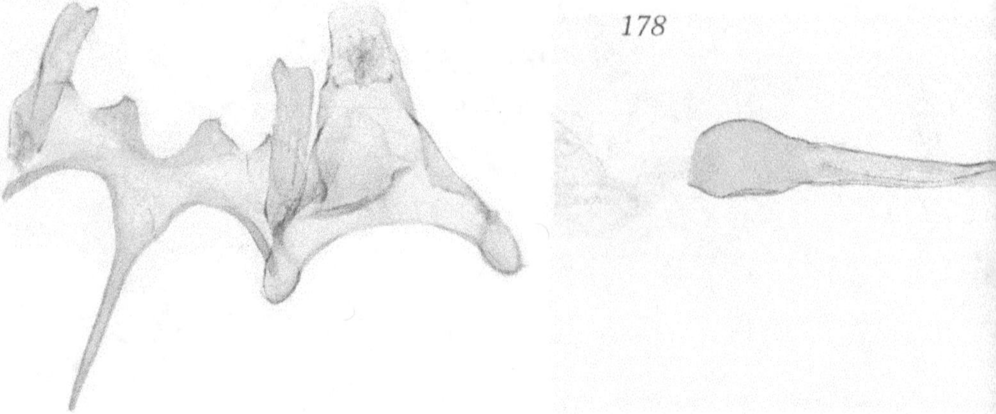

179

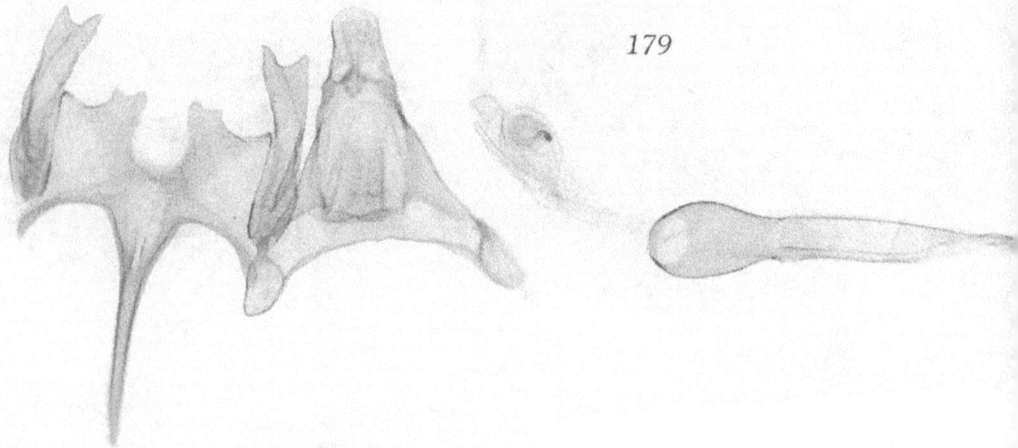

180

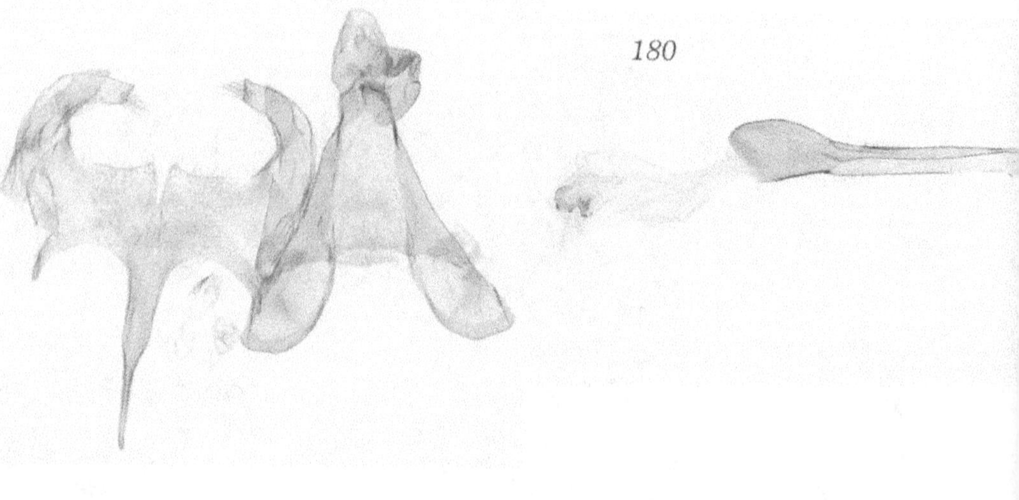

Male Genitalia

178. *Caryocolum gallagenellum* Huemer – Paratype, France, GU 88/056 P.H. (TLMF).
179. *Caryocolum fraternella* (Douglas) – Denmark, GU 88/059 P.H. (TLMF).
180. *Caryocolum klosi* (Rebel) – Germany, GU 86/241 P.H. (TLMF).

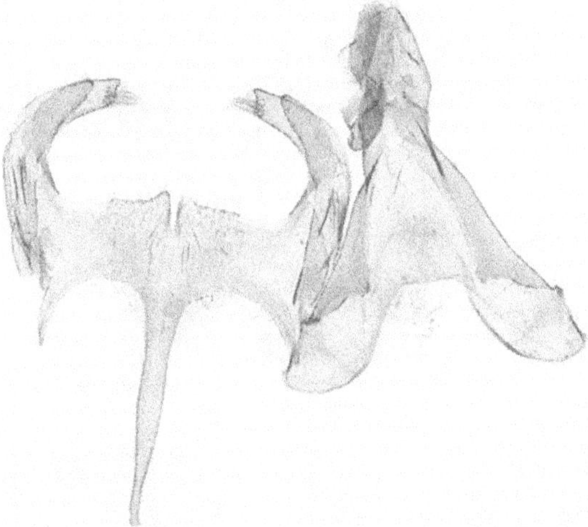

181

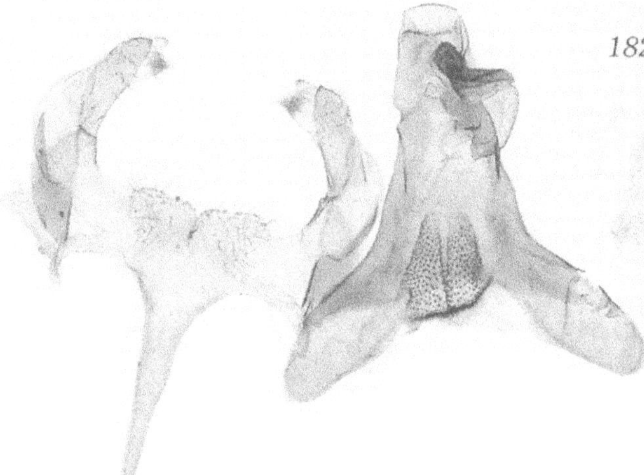

182

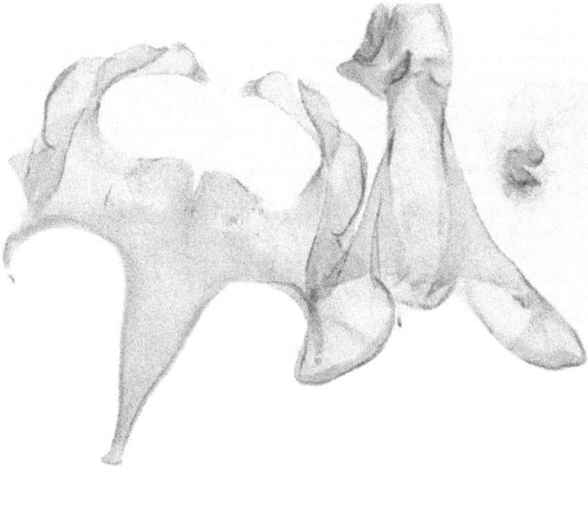

183

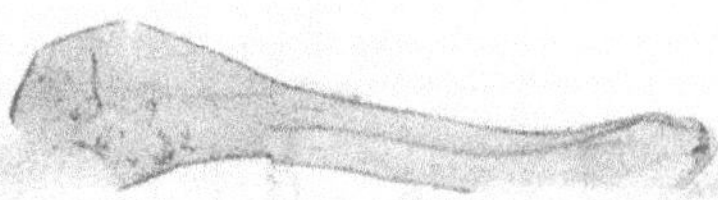

Male Genitalia

181. *Caryocolum interalbicella* (Herrich-Schäffer) – Austria, GEL 1088 P.H. (TLMF).
182. *Caryocolum laceratella* (Zeller) – Italy, GEL 1084 P.H. (TLMF).
183. *Caryocolum blandella blandella* (Douglas) – France, gen. slide 24558 (BMNH).

183a

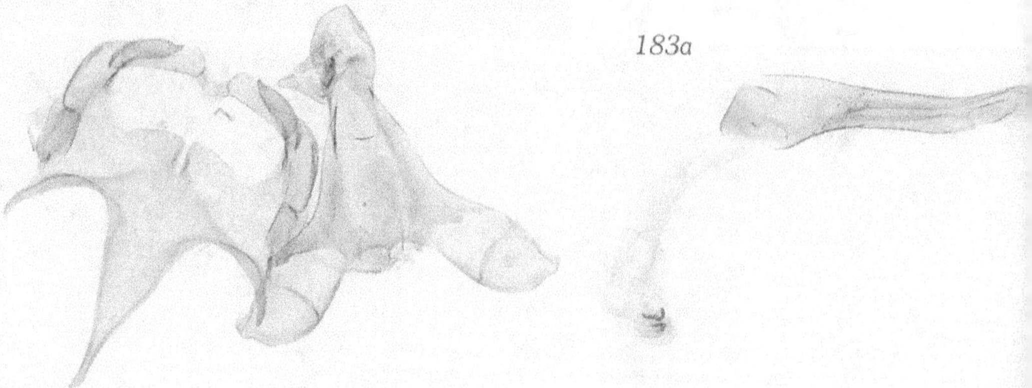

184

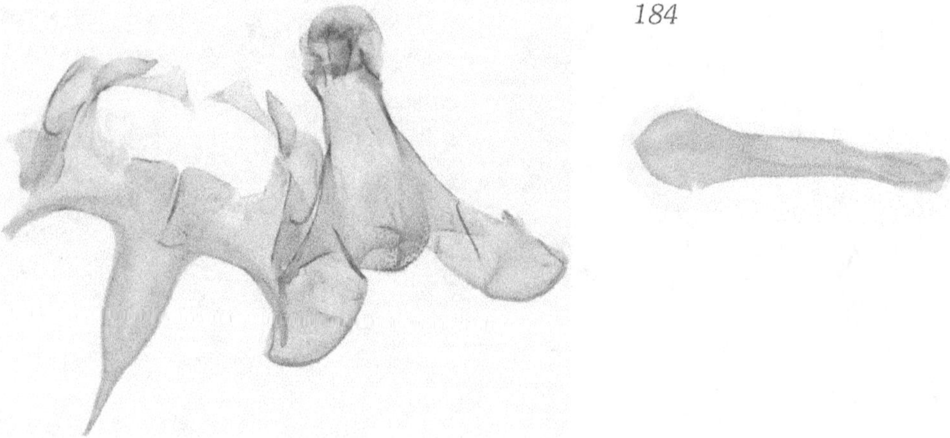

185

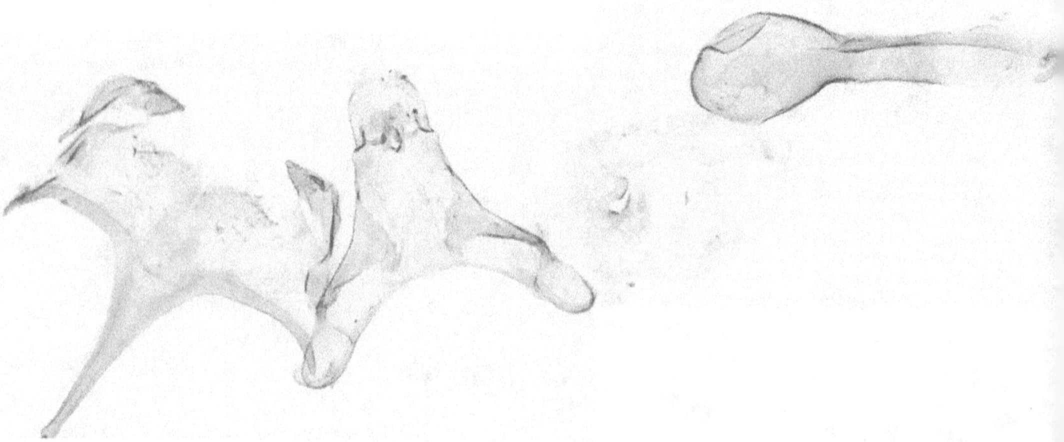

MALE GENITALIA

183a. *Caryocolum blandella horoscopa* (Meyrick) – Turkey, GU 87/155 P.H. (ZMUC).
184. *Caryocolum blandelloides* Karsholt – France (Corsica), GU 87/010 P.H. (TLMF).
185. *Caryocolum jaspidella* (Chrétien) – Morocco, GU 89/050 P.H. (TLMF).

186

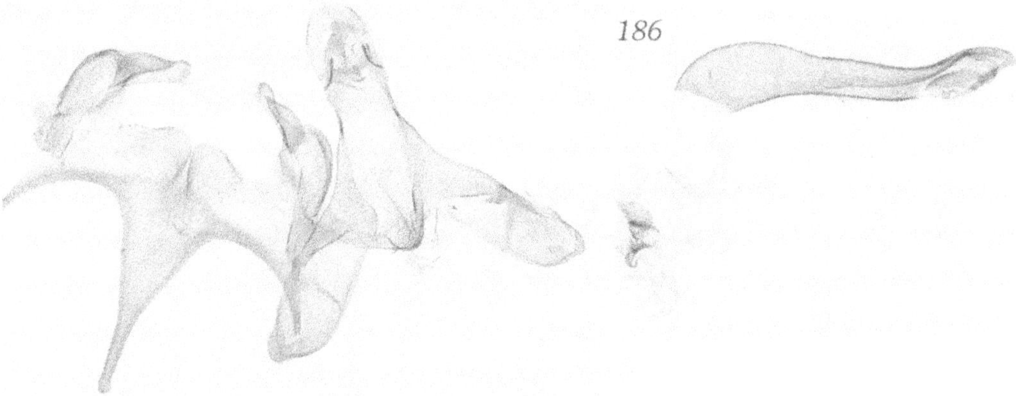

187

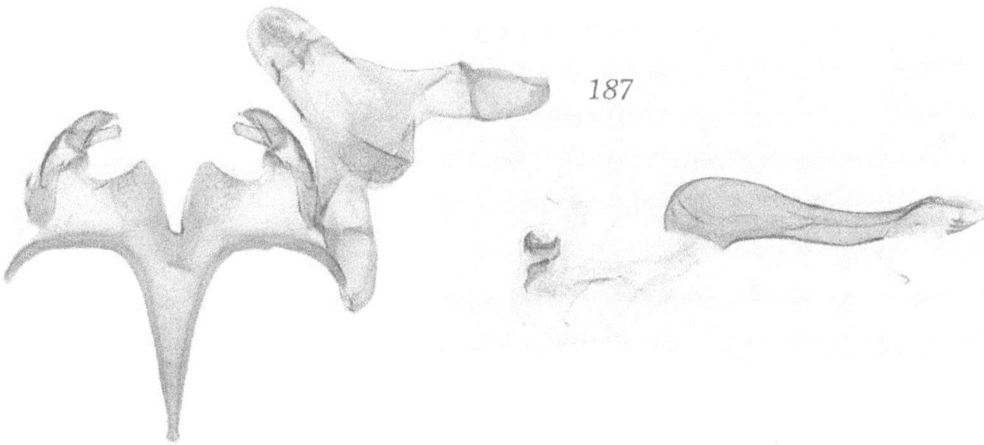

188

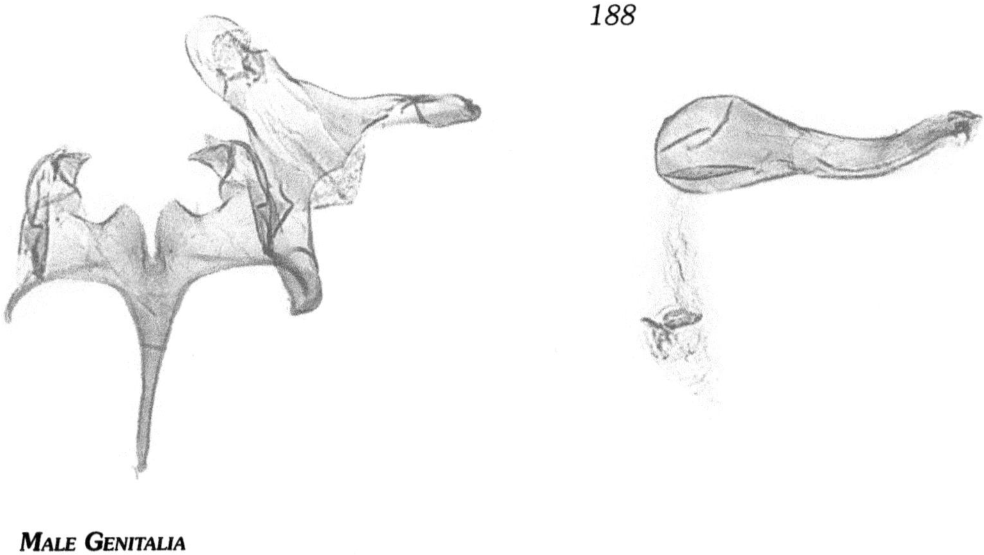

Male Genitalia

186. *Caryocolum proxima* (Haworth) – Austria, GU 88/214 P.H. (TLMF).
187. *Caryocolum blandulella* (Tutt) – Denmark, gen. slide 24529 (BMNH).
188. *Caryocolum arenbergeri* Huemer – Spain, GU 89/005 P.H. (ZMUC).

189

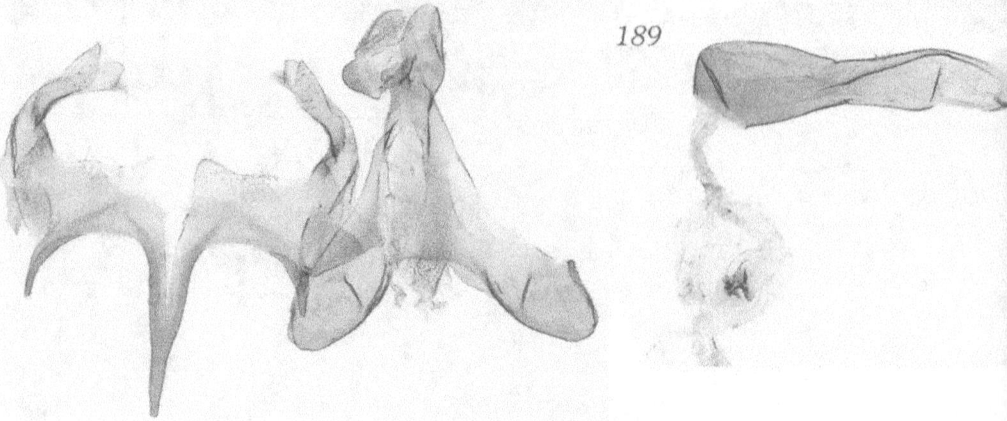

190

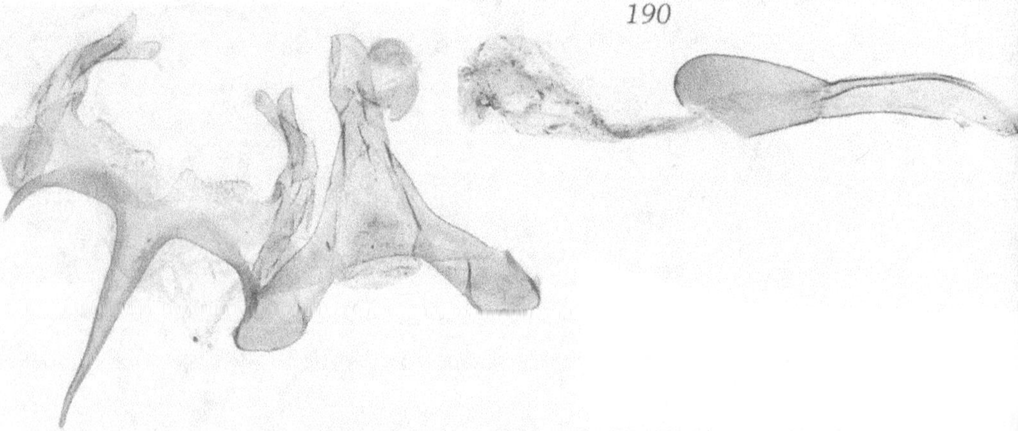

191

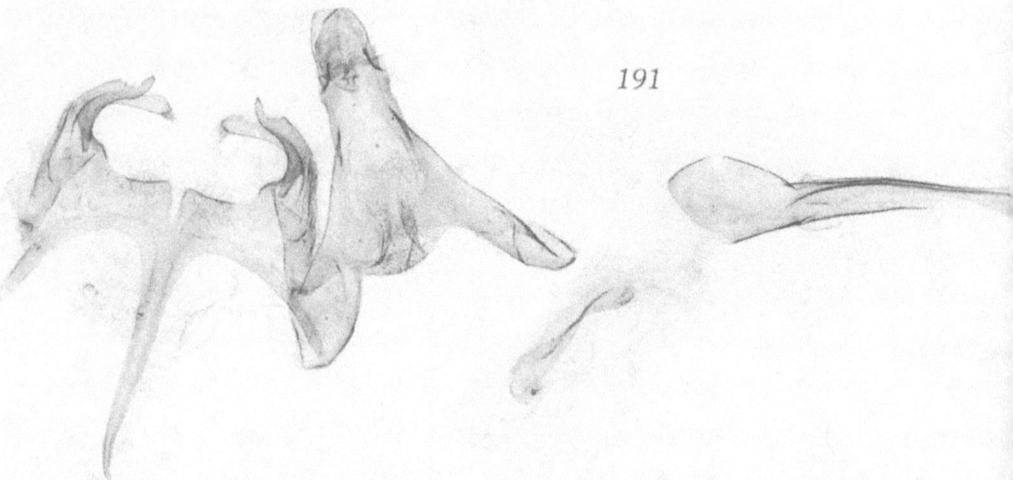

Male Genitalia

189. *Caryocolum tricolorella* (Haworth) – Germany, GU 86/032 P.H. (TLMF).
190. *Caryocolum fibigerium* Huemer – Paratype, Morocco, GU 89/276 P.H. (TLMF).
191. *Caryocolum junctella* (Douglas) – Austria, GEL 184 P.H. (TLMF).

192

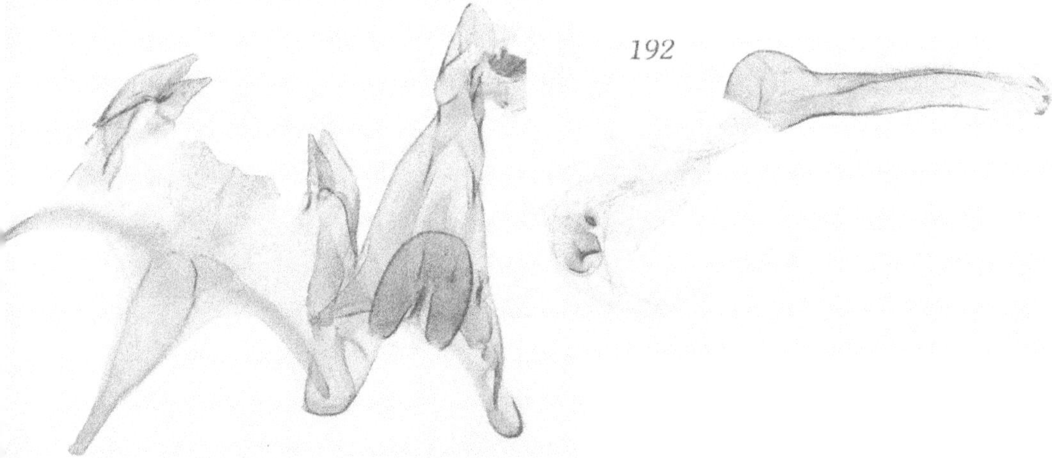

193

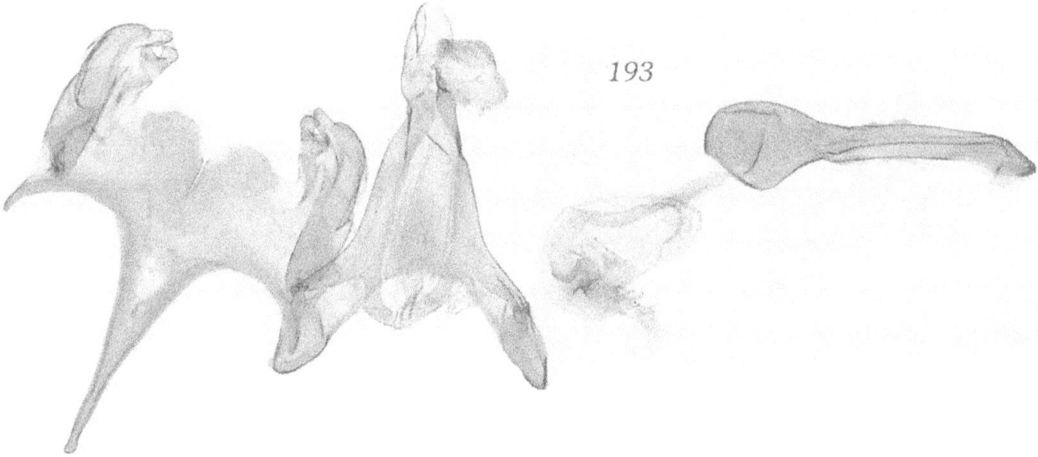

194

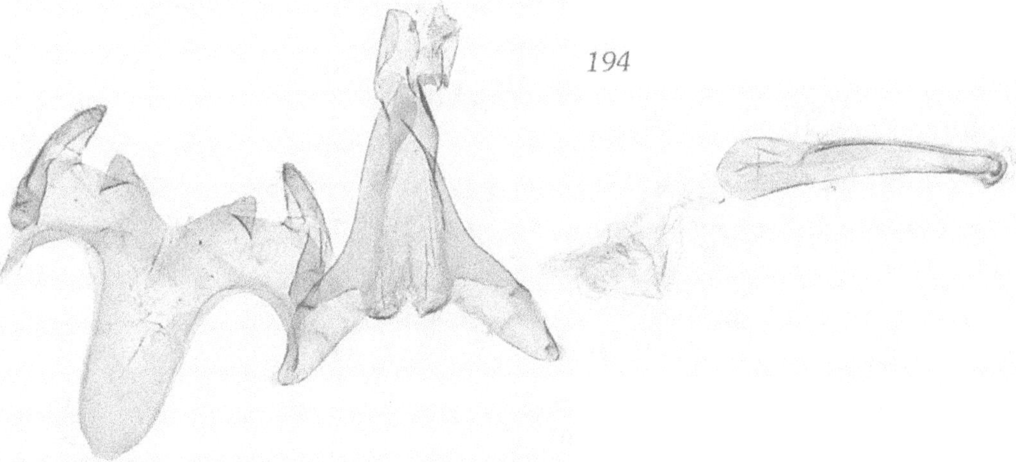

MALE GENITALIA

192. *Caryocolum cassella* (Walker) – Austria, GU 86/121 P.H. (TLMF).
193. *Caryocolum moehringiae* (Klimesch) – Austria, GU 86/135 P.H. (TLMF).
194. *Caryocolum petrophila* (Preissecker) – Austria, GEL 1094 P.H. (TLMF).

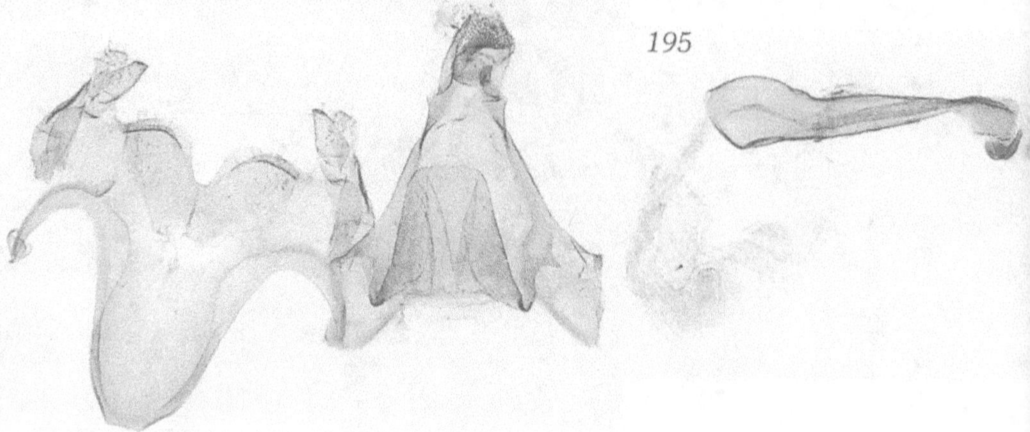

195

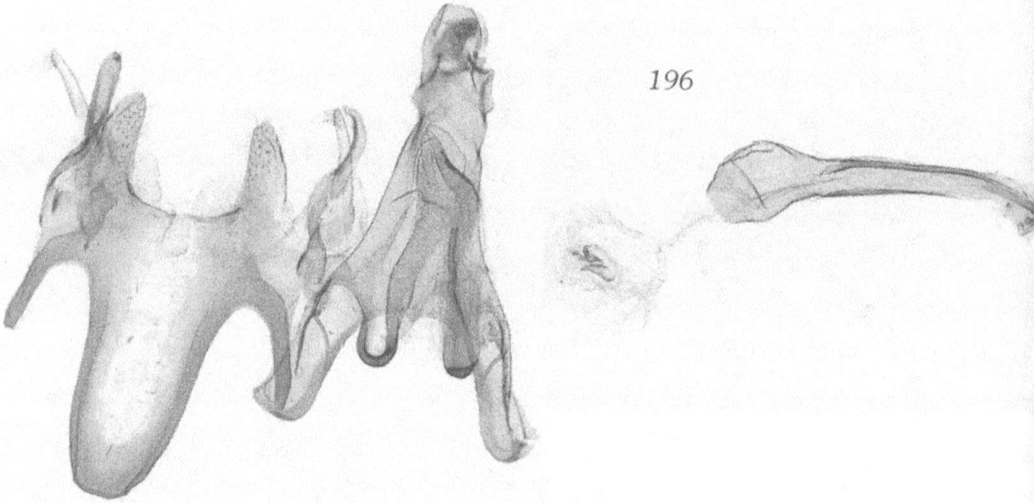

196

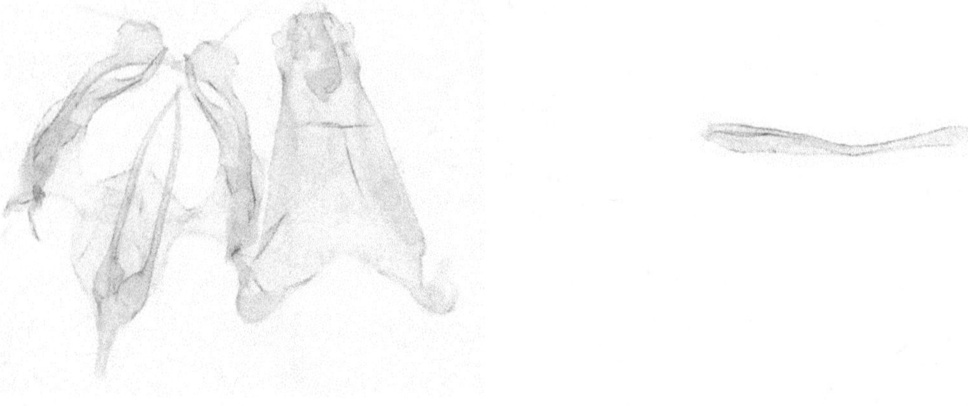

197

MALE GENITALIA

195. *Caryocolum huebneri* (Haworth) – Austria, GEL 1097 P.H. (TLMF).
196. *Caryocolum kroesmanniella* (Herrich-Schäffer) – locality unknown, GU 86/018 P.H. (TLMF).
197. *Tila capsophilella* (Chrétien) – Austria, GEL 1038 P.H. (TLMF).

198

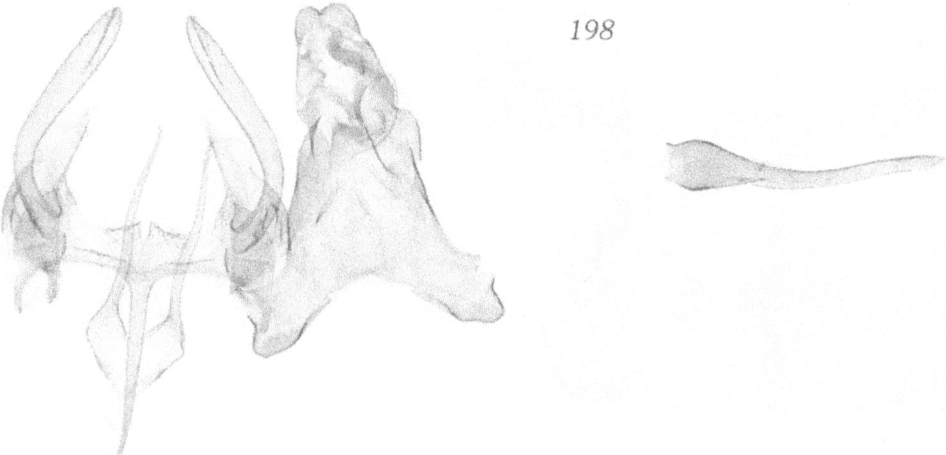

199

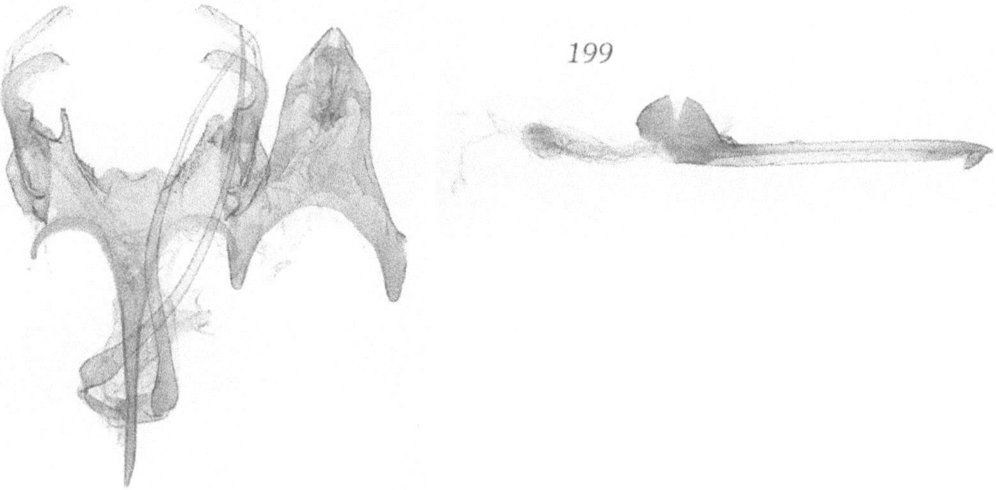

200

Male Genitalia

198. *Pogochaetia solitaria* Staudinger – Italy, GEL 1036 P.H. (TLMF).
199. *Agonochaetia terrestrella* (Zeller) – Switzerland, gen. slide 14978 (NHMW).
200. *Agonochaetia intermedia* Sattler – Austria, GEL 80 P.H. (TLMF).

202

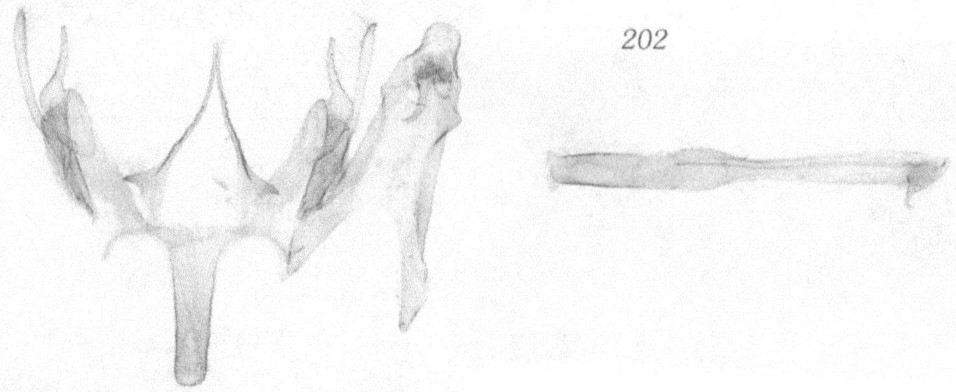

203

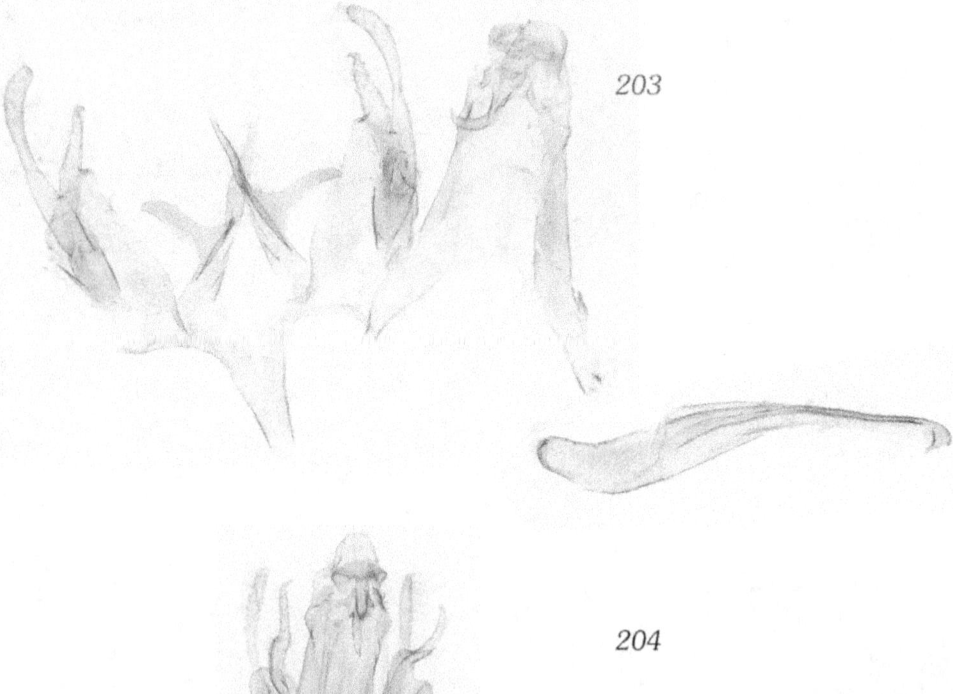

204

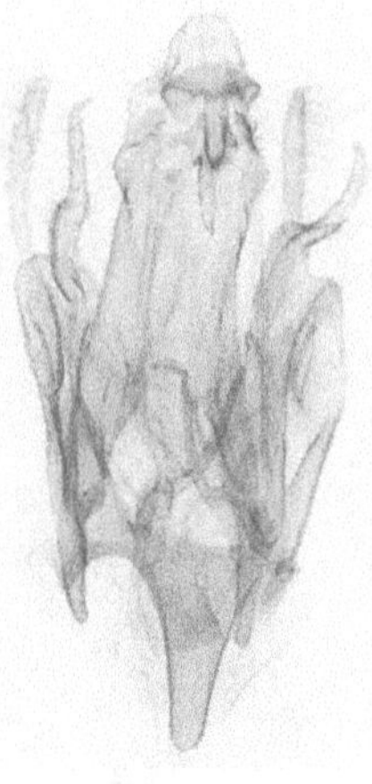

Male Genitalia

202. *Sattleria arcuata* Pitkin & Sattler – Spain, GEL 159 P.H. (TLMF).
203. *Sattleria pyrenaica* (Petry) – France, GEL 279 P.H. (TLMF).
204. *Sattleria angustispina* Pitkin & Sattler – Paratype, France, gen. slide 84 C. Gibeaux (without phallus) (MNHN).

205

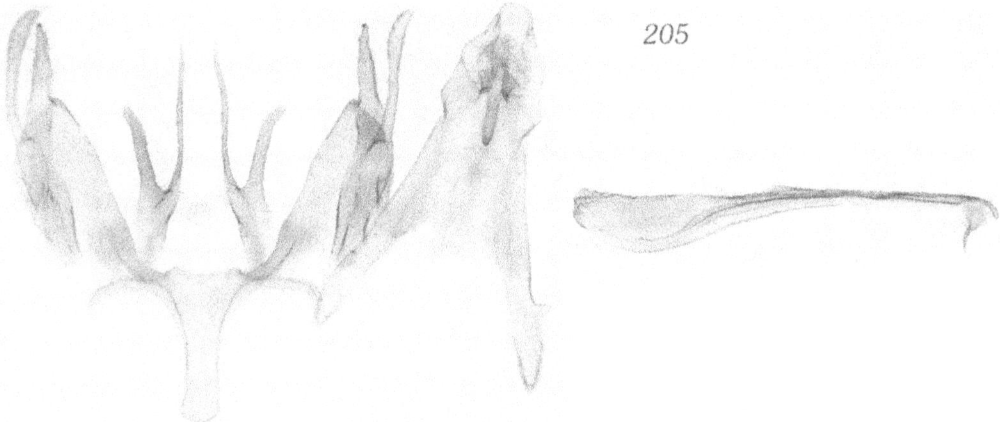

206

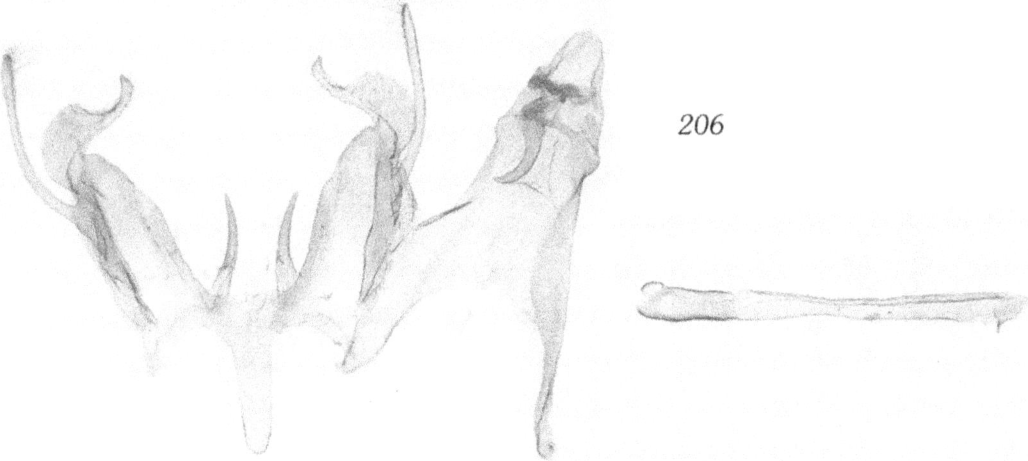

207

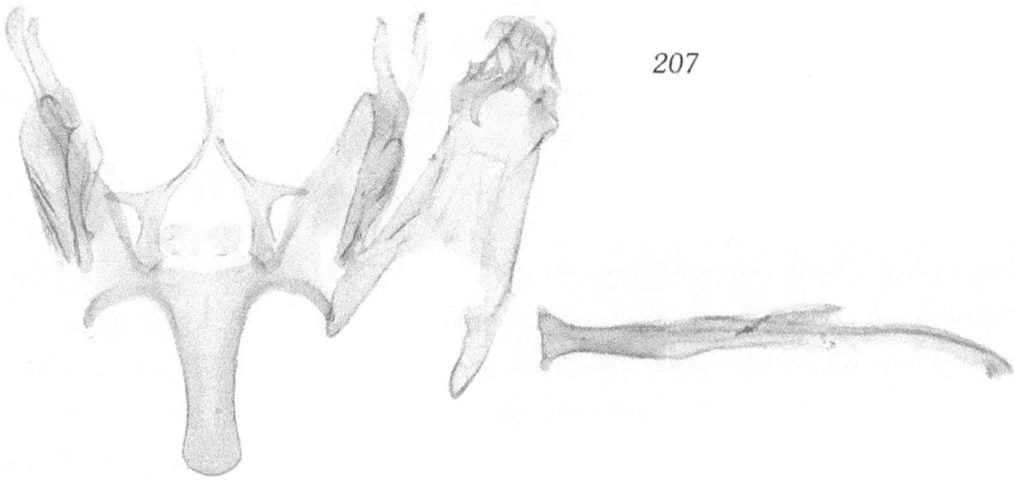

Male Genitalia

205. *Sattleria izoardi* Huemer & Sattler – Italy, GEL 1049 P.H. (TLMF).
206. *Sattleria breviramus* Pitkin & Sattler – France, GEL 154 P.H. (TLMF).
207. *Sattleria melaleucella* (Constant) – Austria, gen. slide 26 L.M.P. (TLMF).

208

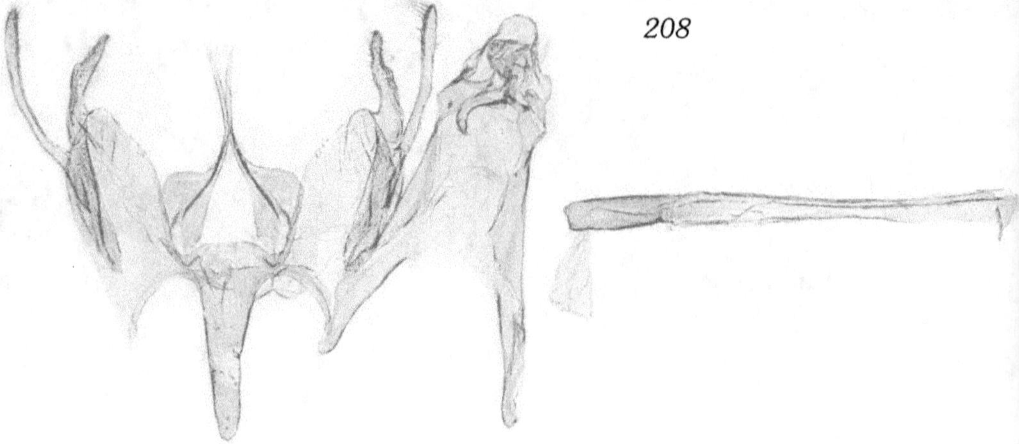

208a

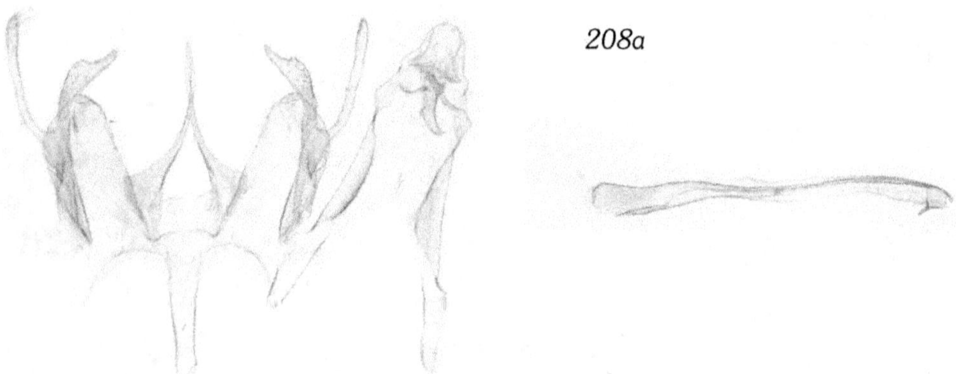

209

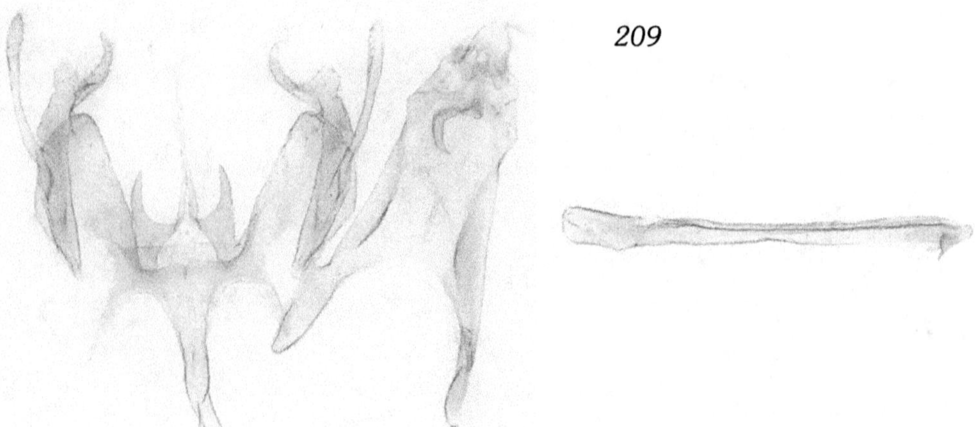

MALE GENITALIA

208. *Sattleria triglavica triglavica* Povolný – Montenegro, GEL 1129 P.H. (TLMF).
208a. *Sattleria triglavica basistrigella* Huemer – Switzerland, GU 91/228 P.H. (LNK).
209. *Sattleria styriaca* Pitkin & Sattler – Austria, GU 91/230 P.H. (LNK).

210

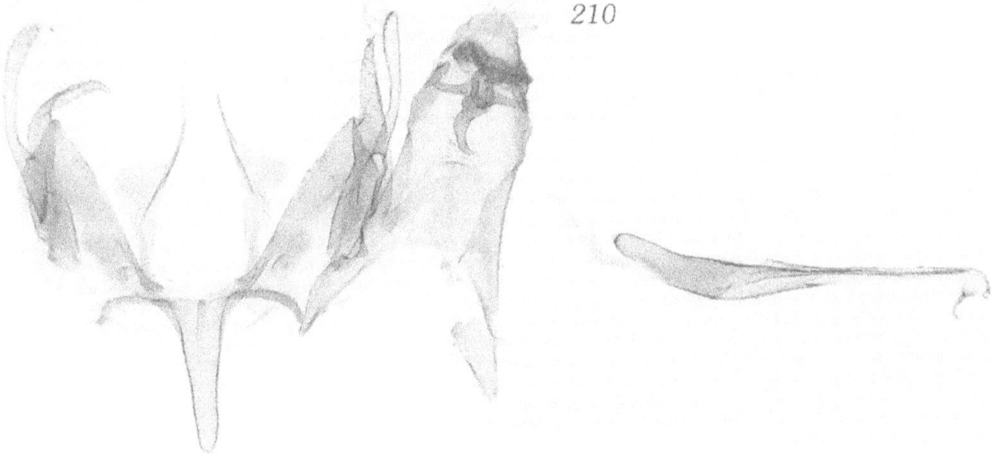

211

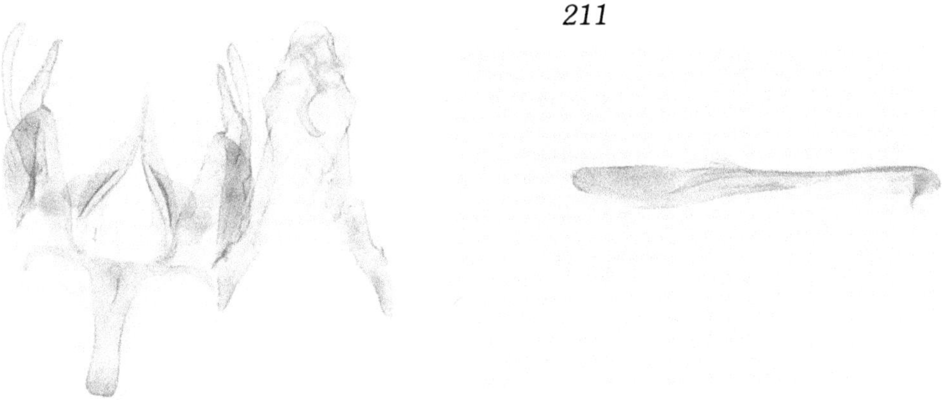

Male Genitalia

210. *Sattleria marguareisi* Huemer & Sattler – Paratype, France, GEL 150 P.H. (TLMF).
211. *Sattleria dzieduszyckii* (Nowicki) – Poland/Slovakia, GU 92/334 P.H. (LNK).

Female genitalia illustrations

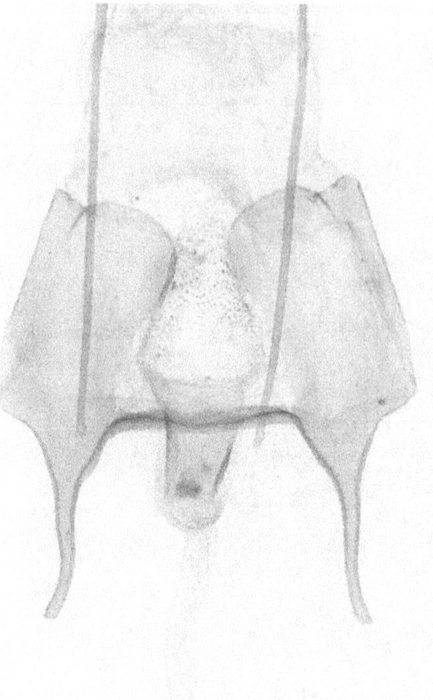

1

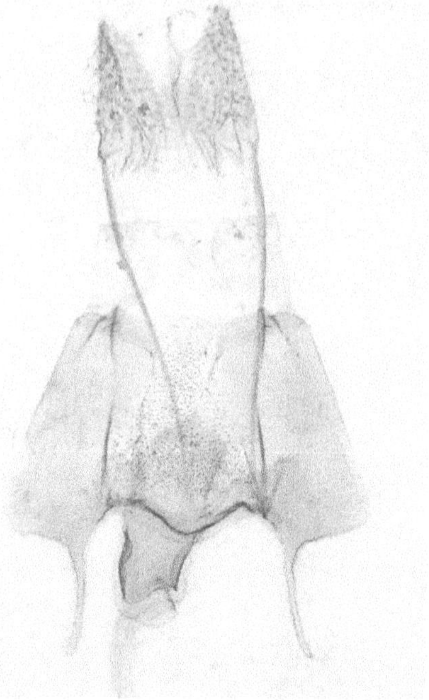

2

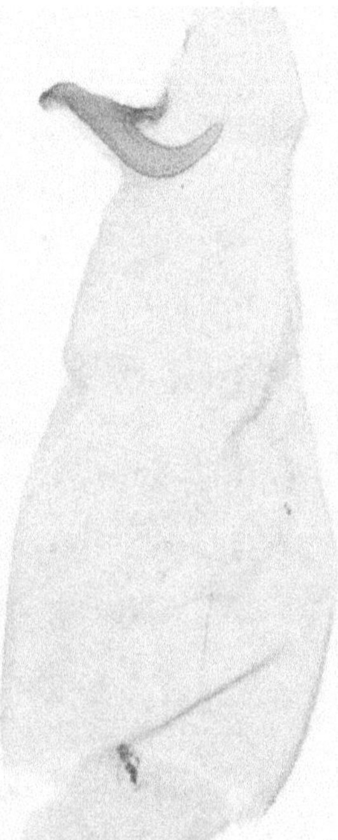

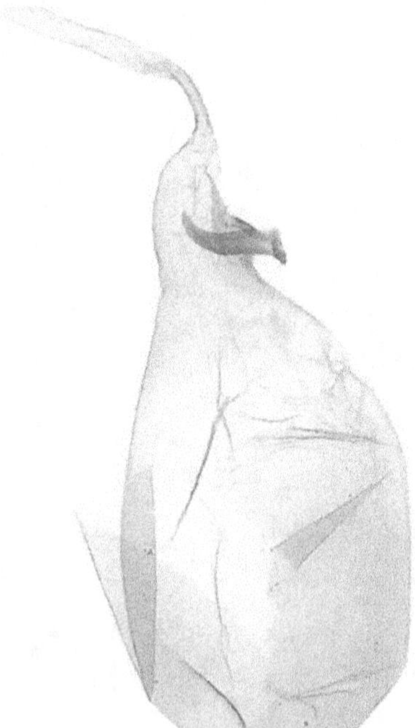

Female Genitalia

1. *Gnorimoschema soffneri* (Riedl) – Greece, GEL 637 P.H. (TLMF).
2. *Gnorimoschema herbichii* (Nowicki) – Finland, GU 08/1271 P.H. (ZMUC).

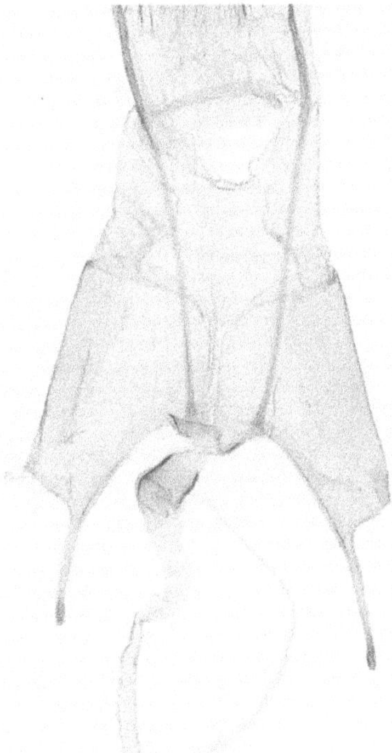

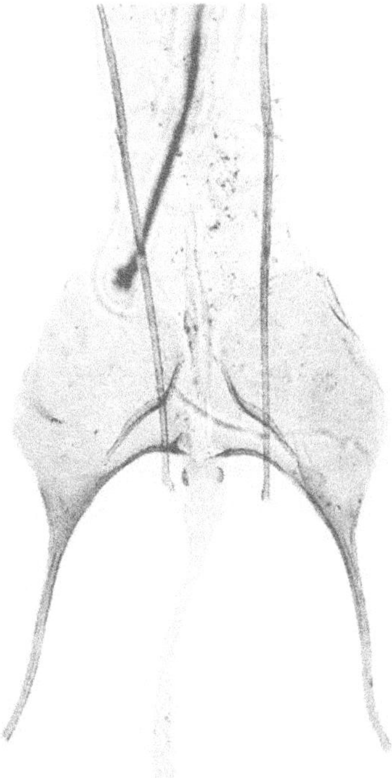

3 4

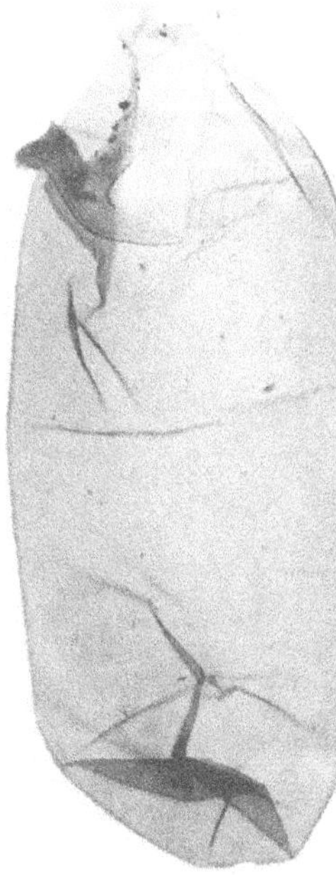

FEMALE GENITALIA

3. *Gnorimoschema bodillum* Karsholt & Nielsen – Paratype, Denmark, GEL 1080 P.H. (TLMF).
4. *Gnorimoschema nupponeni* sp. n. – Paratype, Russia, gen. slide 6426 H.H. (ZMUC).

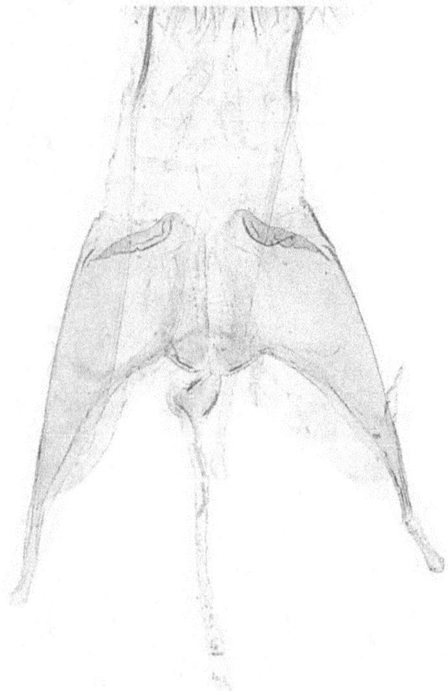

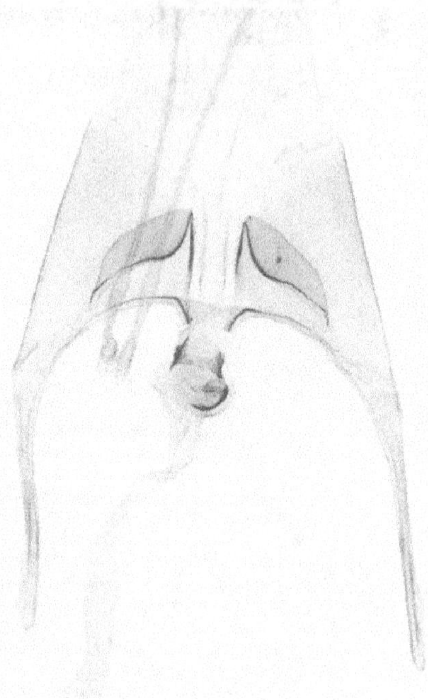

5

6

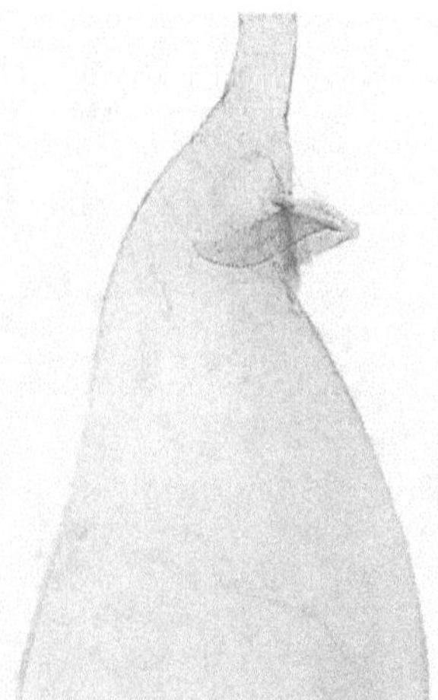

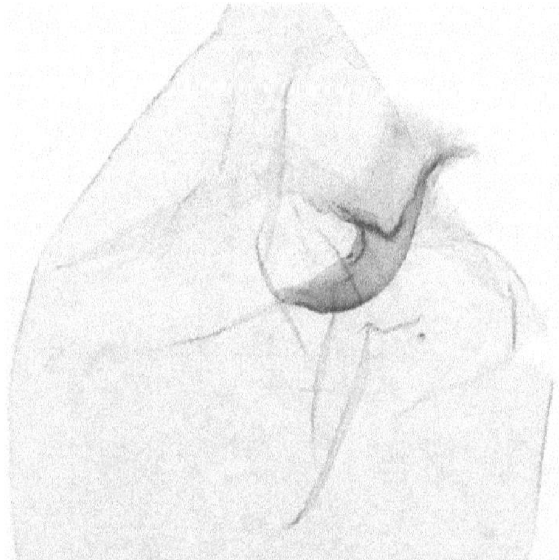

Female Genitalia

5. *Gnorimoschema robustella* (Staudinger) – Paralectotype, GU 06/1251 P.H. (ZMHU).
6. *Gnorimoschema steueri* Povolný – Germany, GEL 1078 P.H. (TLMF).

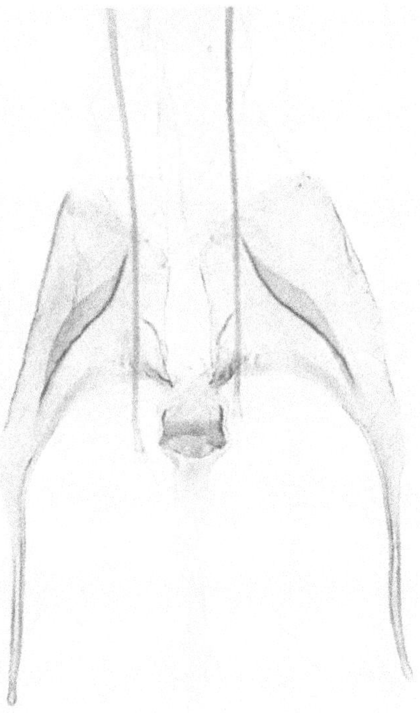

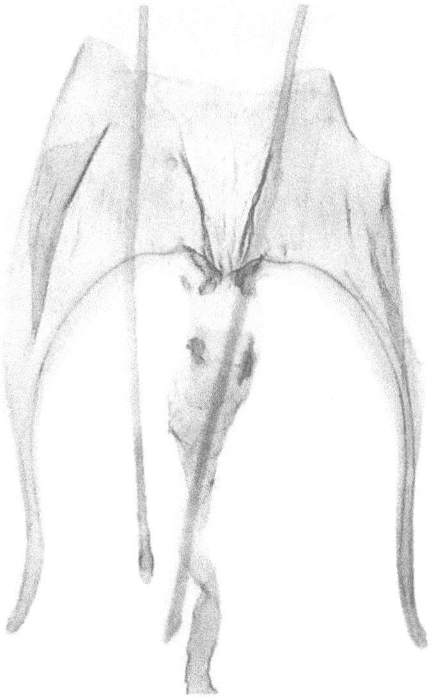

7

8

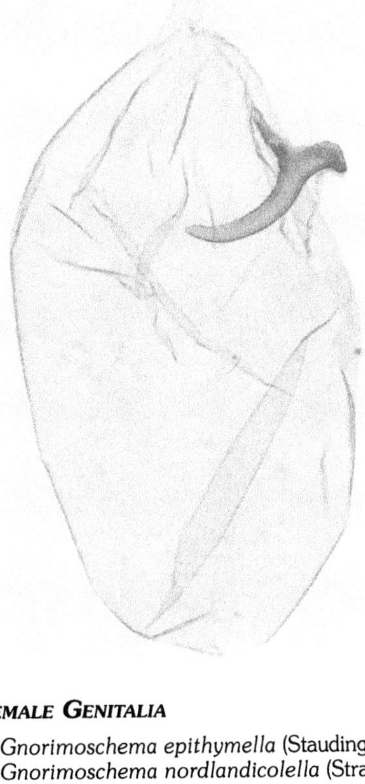

FEMALE GENITALIA

7. *Gnorimoschema epithymella* (Staudinger) – Austria, GEL 1079 P.H. (TLMF).
8. *Gnorimoschema nordlandicolella* (Strand) – Finland, gen. slide 2101 H.H. (ZMUC).

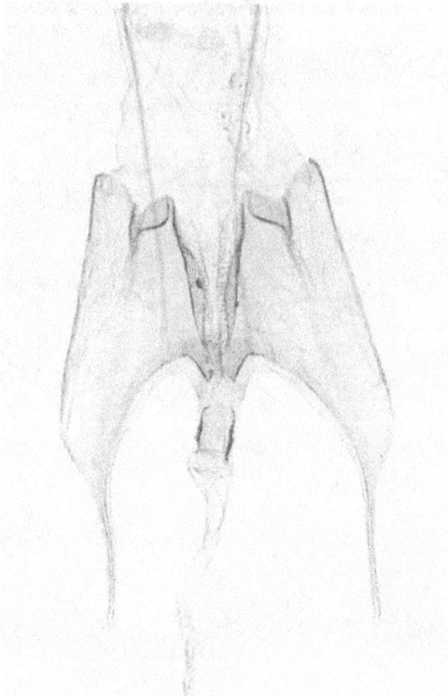

10

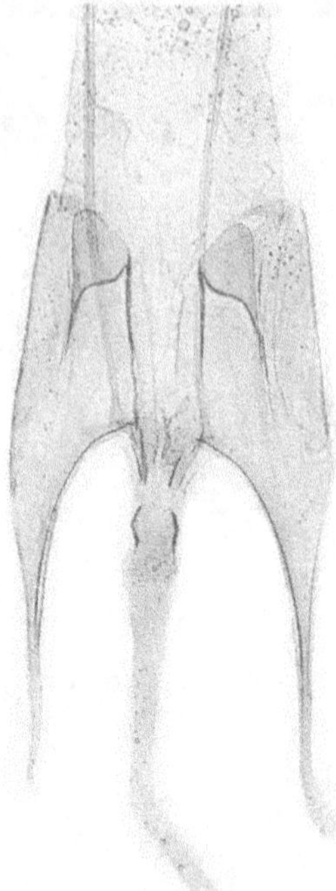

11

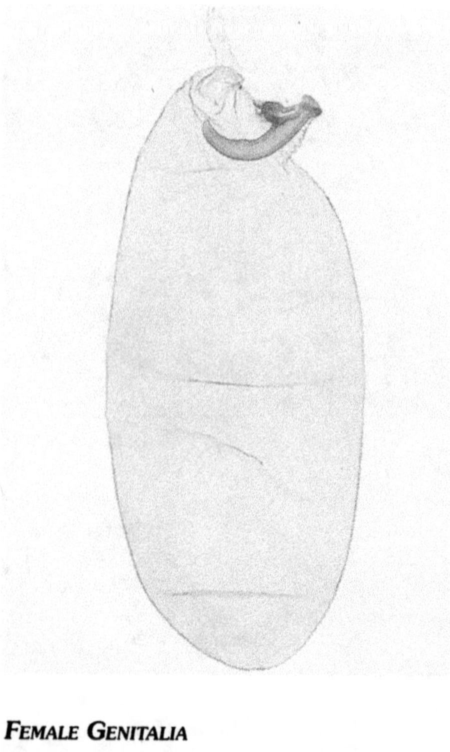

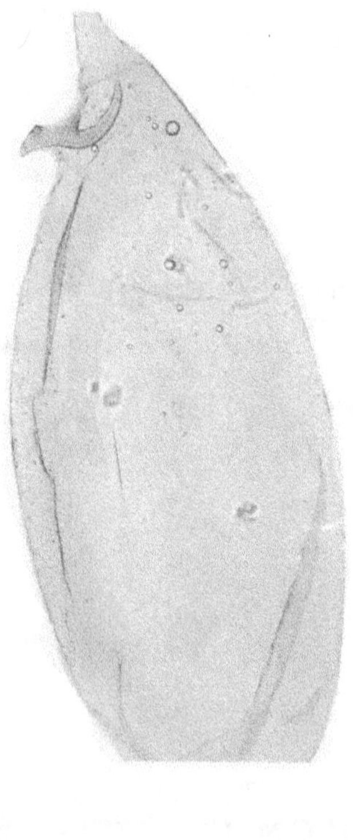

Female Genitalia

10. *Gnorimoschema valesiella* (Staudinger) – Switzerland, GU 06/1249 P.H. (ZSM).
11. *Gnorimoschema streliciella* (Herrich-Schäffer) – Poland, gen. slide 16641 (NHMW).

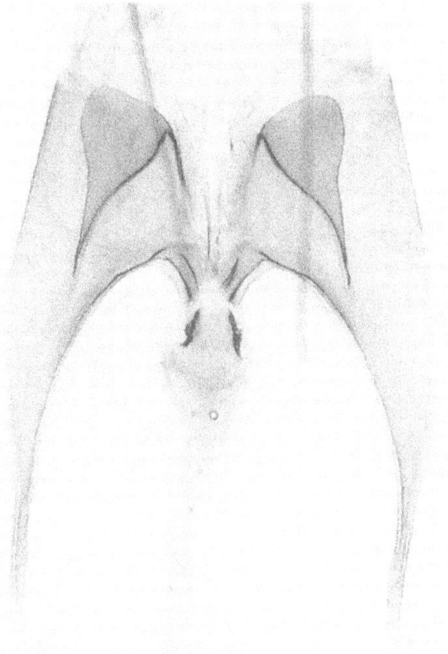

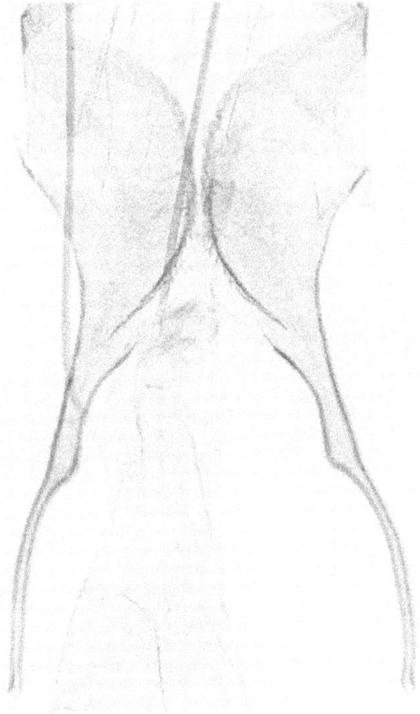

12

13

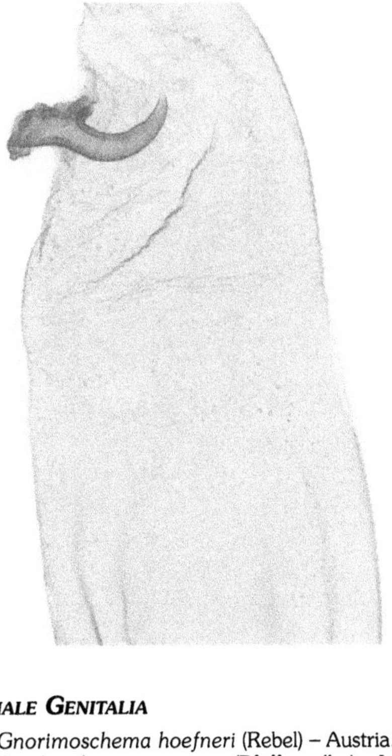

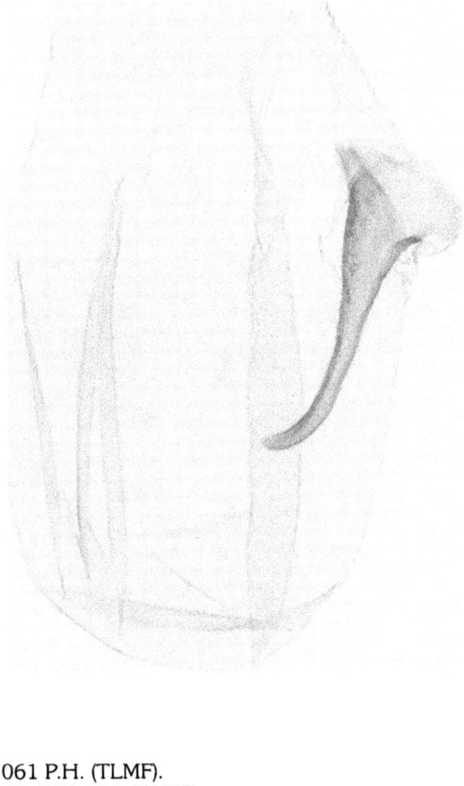

Female Genitalia

12. *Gnorimoschema hoefneri* (Rebel) – Austria, GEL 1061 P.H. (TLMF).
13. *Scrobipalpopsis petasitis* (Pfaffenzeller) – Italy, GU 01/983 P.H. (TLMF).

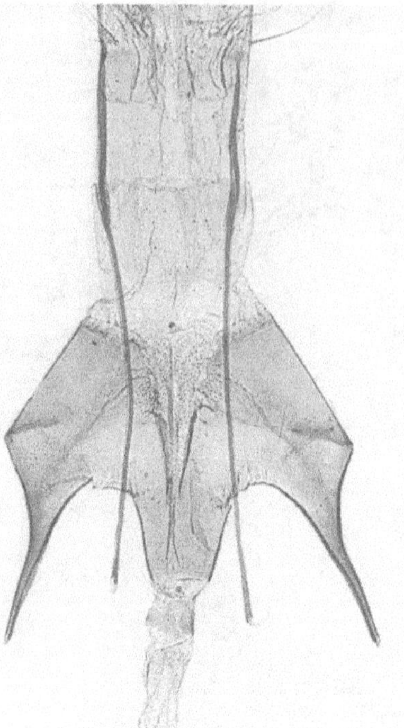

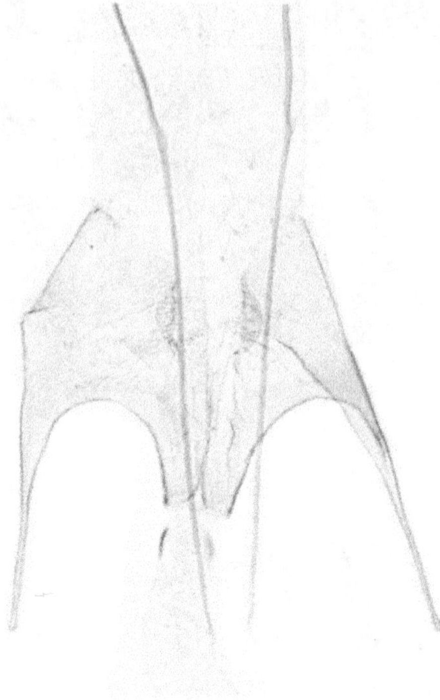

14

15

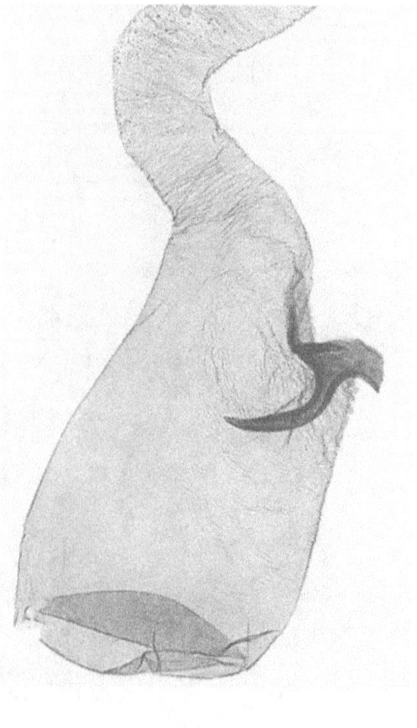

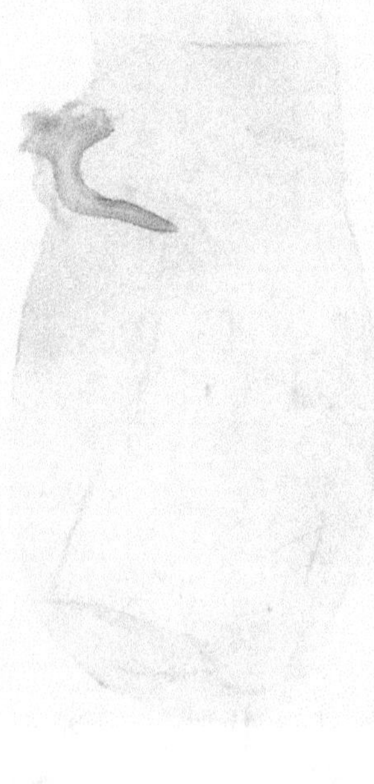

FEMALE GENITALIA

14. *Scrobipalpa aptatella* (Walker) – Egypt, gen. slide 26583 (BMNH).
15. *Scrobipalpa kasyi* Povolný – Greece, gen. slide 4999 O.K. (ZMUC).

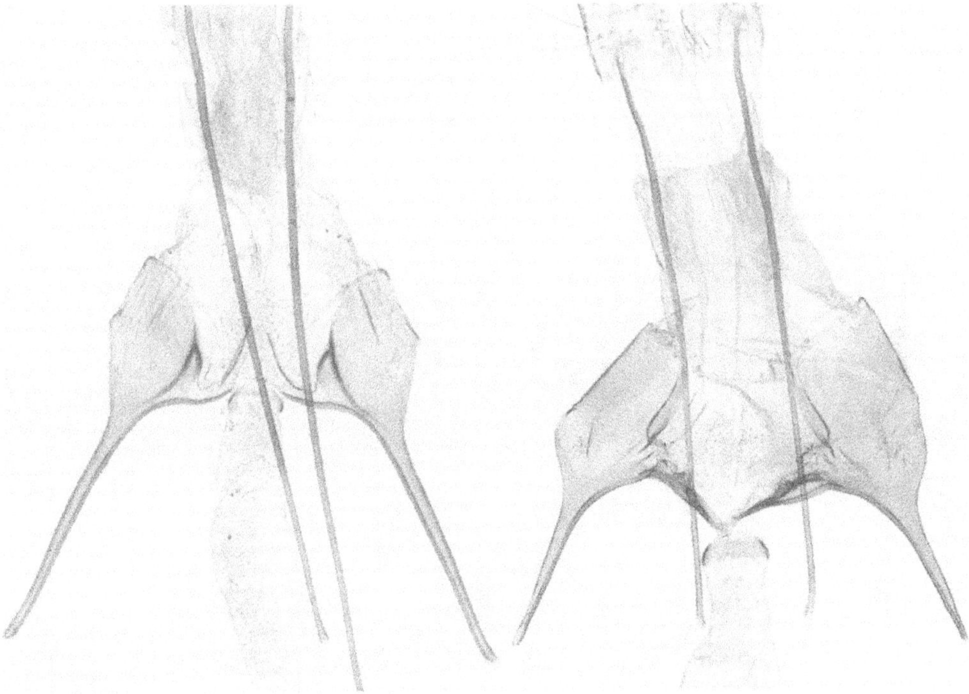

16 17

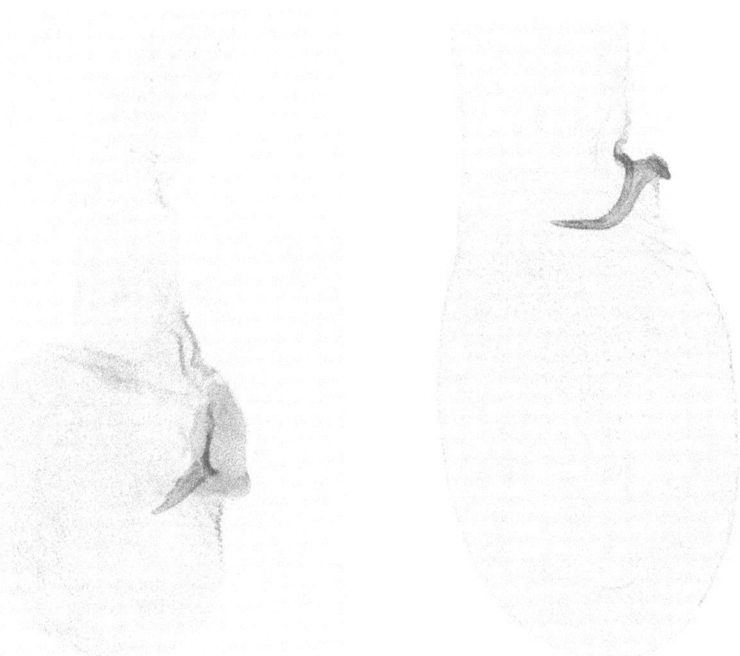

Female Genitalia

16. *Scrobipalpa notata* (Povolný) – Russia, GU 03/1196 P.H. (NUPP).
17. *Scrobipalpa acuminatella* (Sircom) – Germany, GEL 1100 P.H. (TLMF).

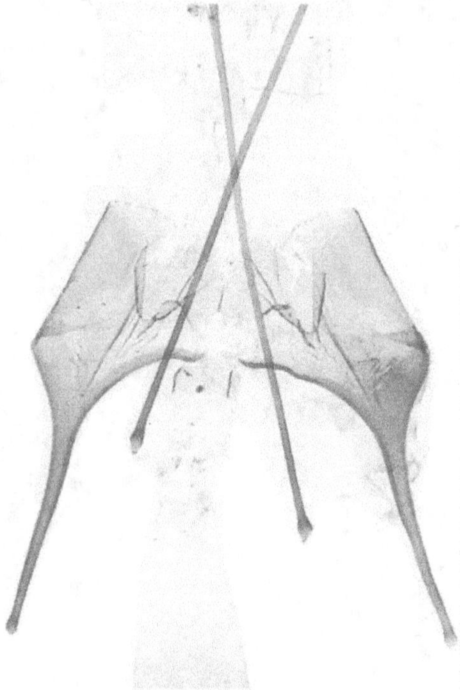

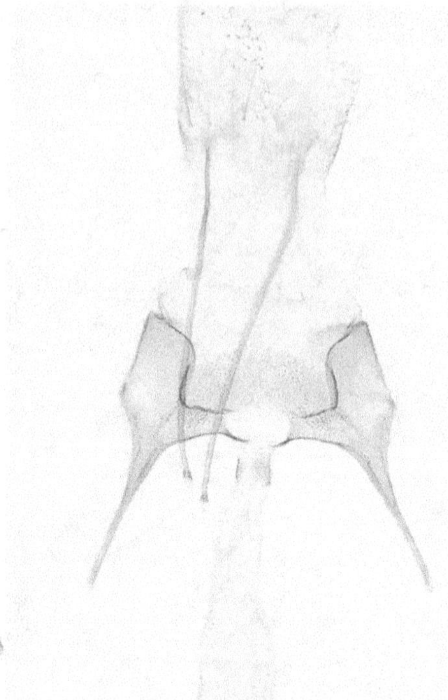

18

19

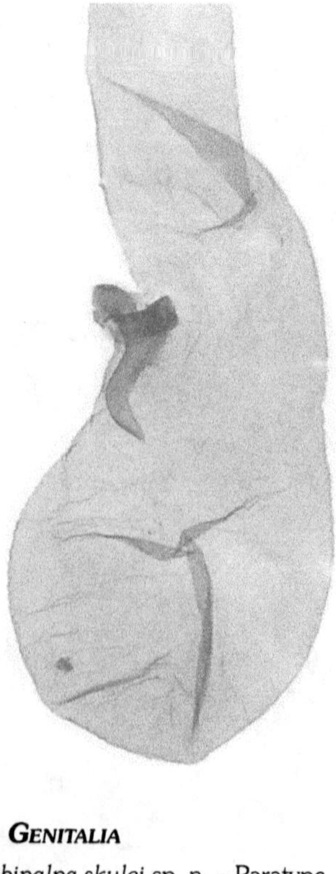

FEMALE GENITALIA

18. *Scrobipalpa skulei* sp. n. – Paratype, Slovenia, gen. slide 4764 H.H. (ZMUC).
19. *Scrobipalpa hungariae* (Staudinger) – Hungary, gen. slide 15426 (NHMW).

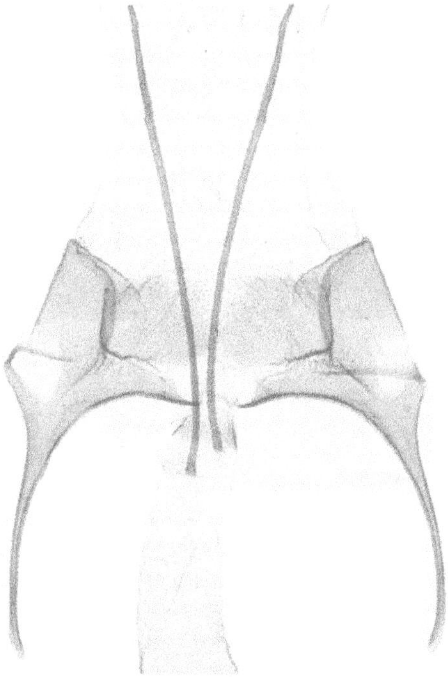

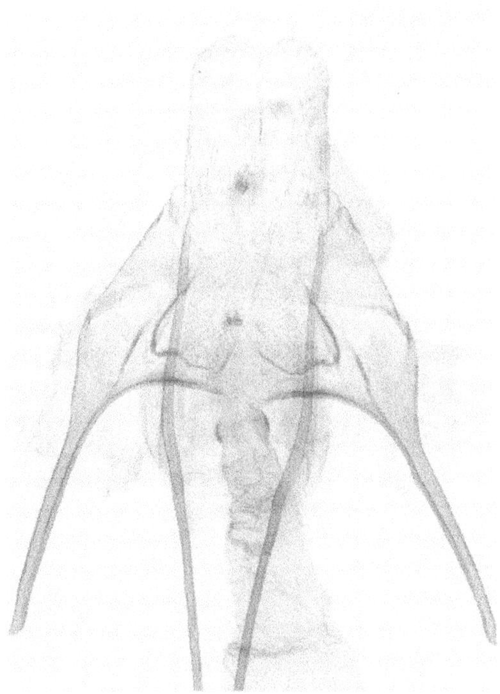

20

21

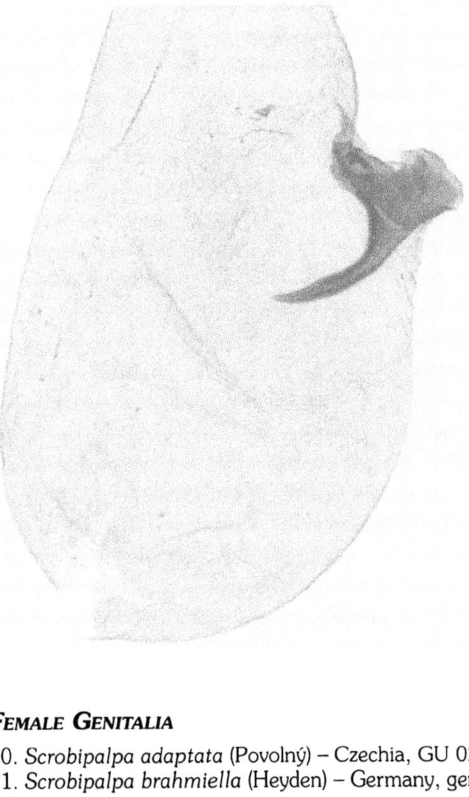

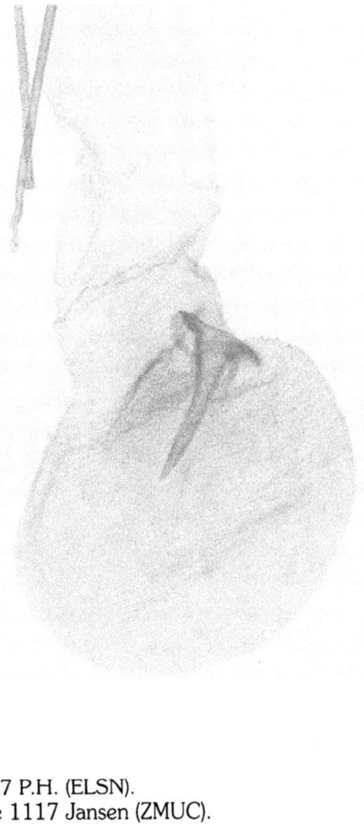

FEMALE GENITALIA

20. *Scrobipalpa adaptata* (Povolný) – Czechia, GU 02/1167 P.H. (ELSN).
21. *Scrobipalpa brahmiella* (Heyden) – Germany, gen. slide 1117 Jansen (ZMUC).

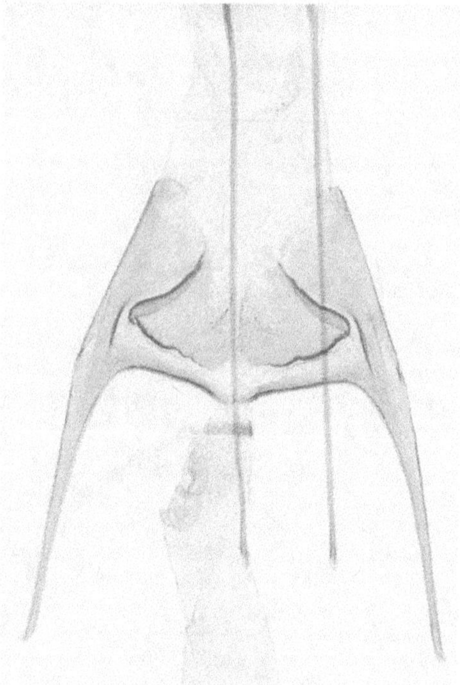

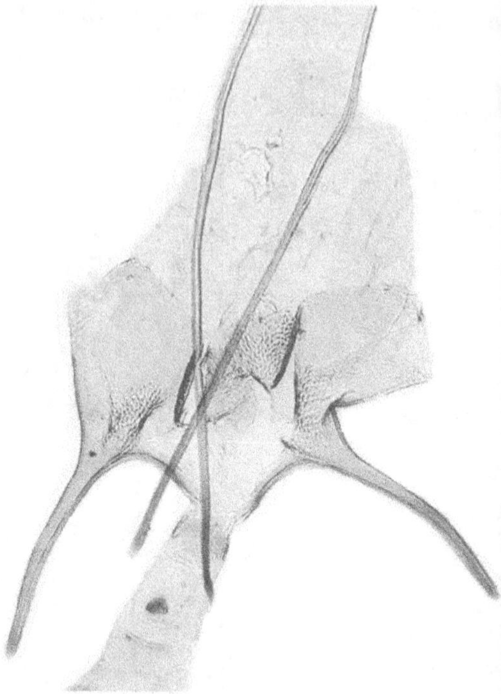

22

23

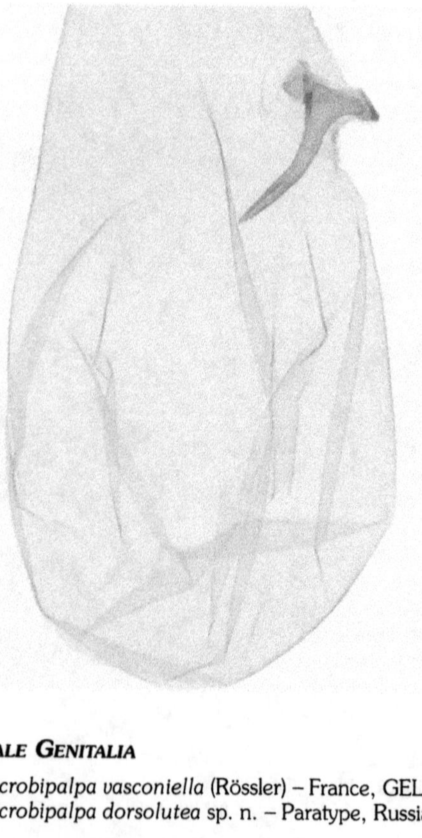

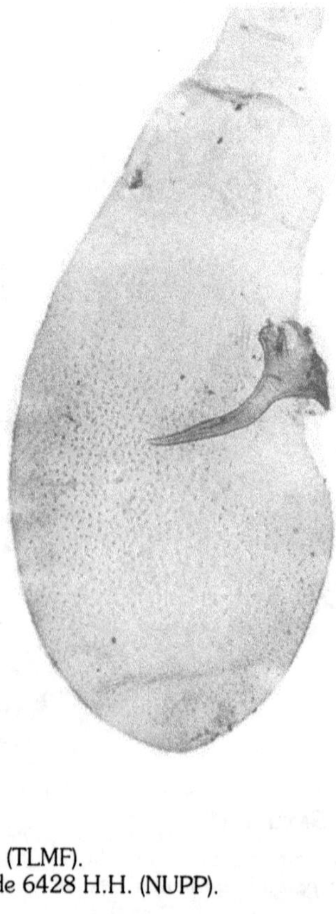

Female Genitalia

22. *Scrobipalpa vasconiella* (Rössler) – France, GEL 951 P.H. (TLMF).
23. *Scrobipalpa dorsolutea* sp. n. – Paratype, Russia, gen. slide 6428 H.H. (NUPP).

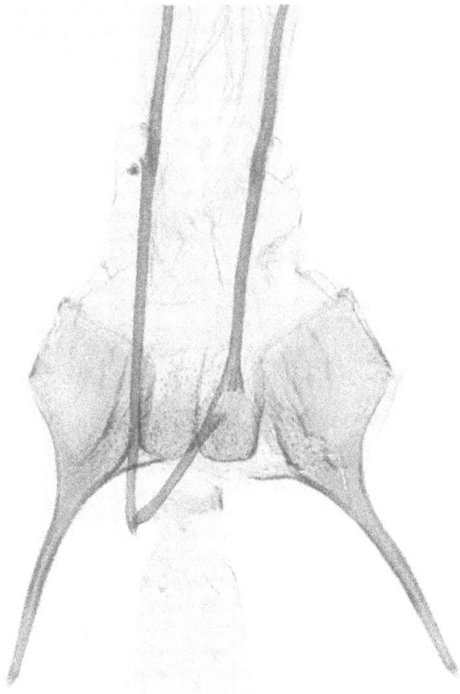

24

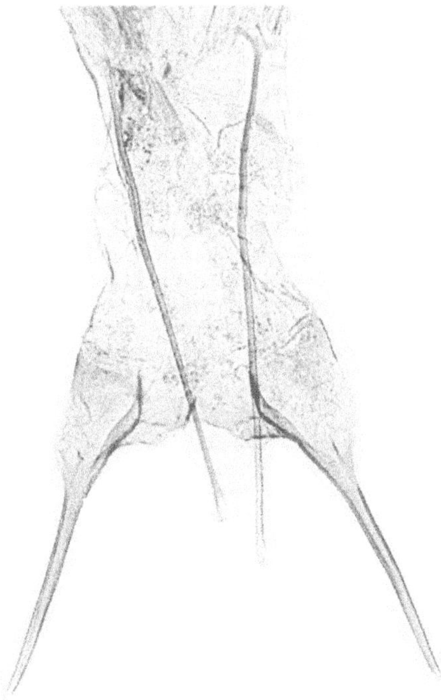

25

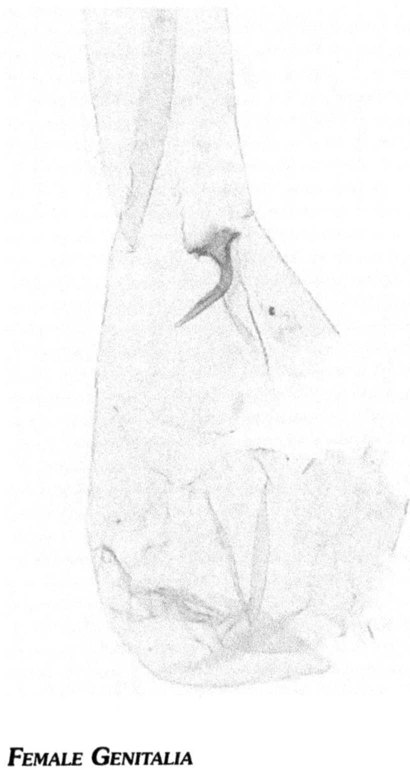

Female Genitalia

24. *Scrobipalpa amseli* Povolný – France, GU 01/1059 P.H. (LNK).
25. *Scrobipalpa hyssopi* Nel – France, gen. slide 19495 J.N. (TLMF).

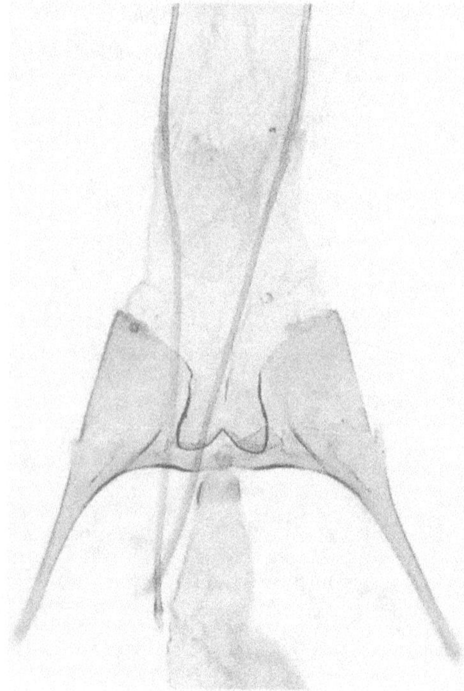

26

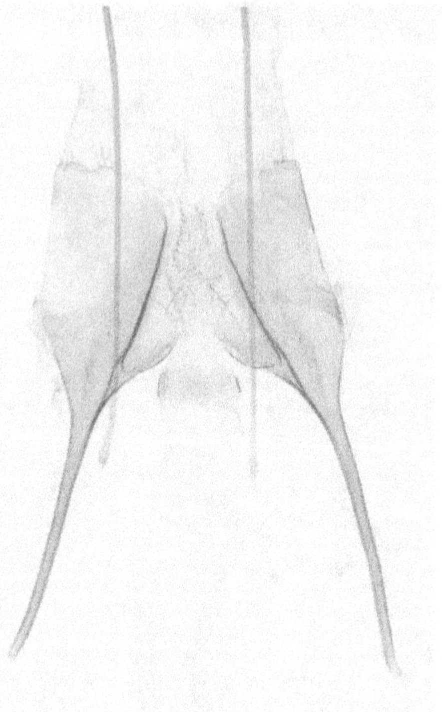

27

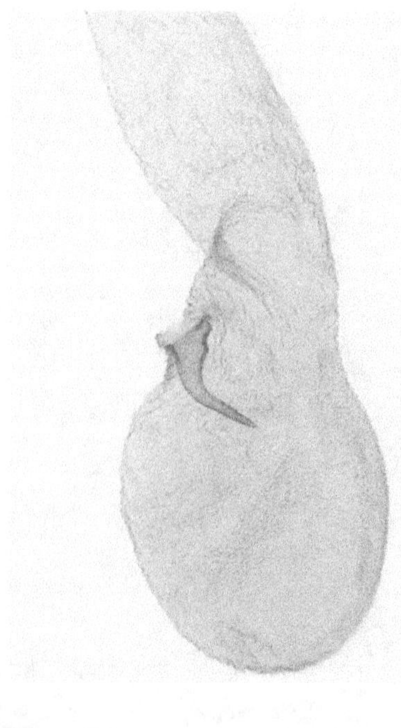

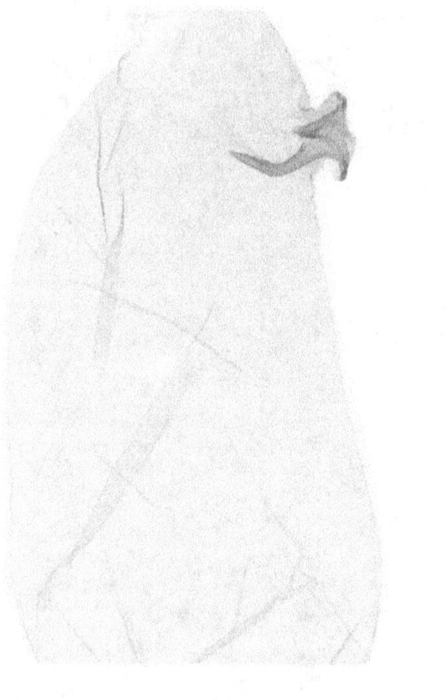

FEMALE GENITALIA

26. *Scrobipalpa montanella* (Chrétien) – Italy, GU 02/1146 P.H. (ZMUC).
27. *Scrobipalpa corleyi* sp. n. – Paratype, Portugal, GU 01/1094 P.H. (ZMUC).

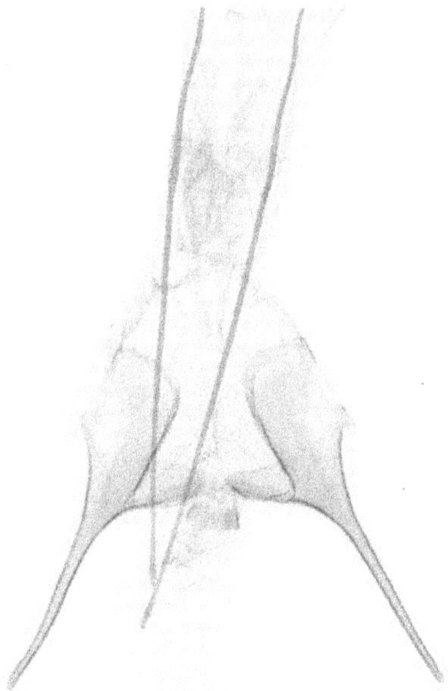

28

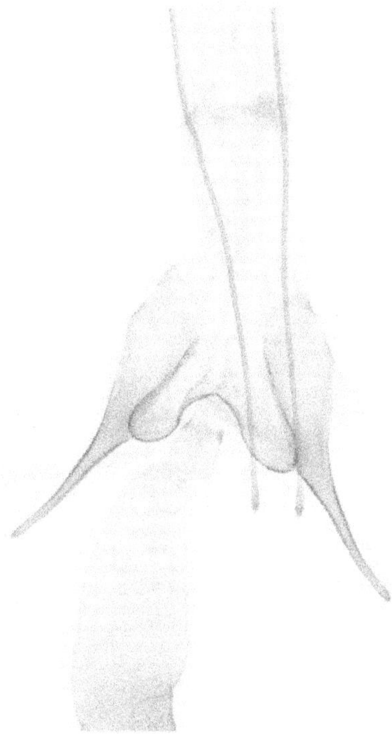

29

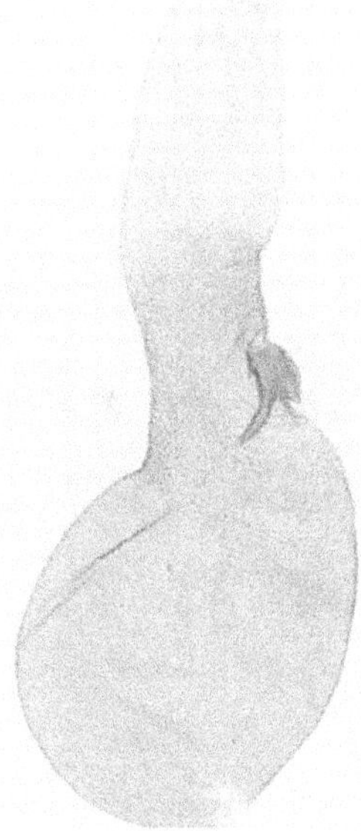

FEMALE GENITALIA

28. *Scrobipalpa chrysanthemella* (Hofmann) – Austria, GEL 994 P.H. (TLMF).
29. *Scrobipalpa proclivella* (Fuchs) – Austria, GEL 996 P.H. (TLMF).

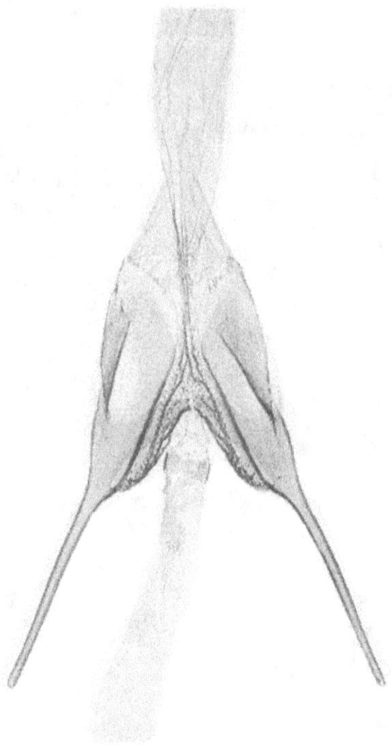

30

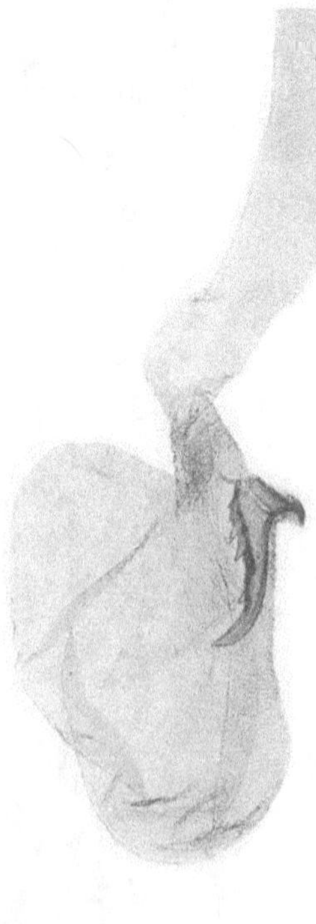

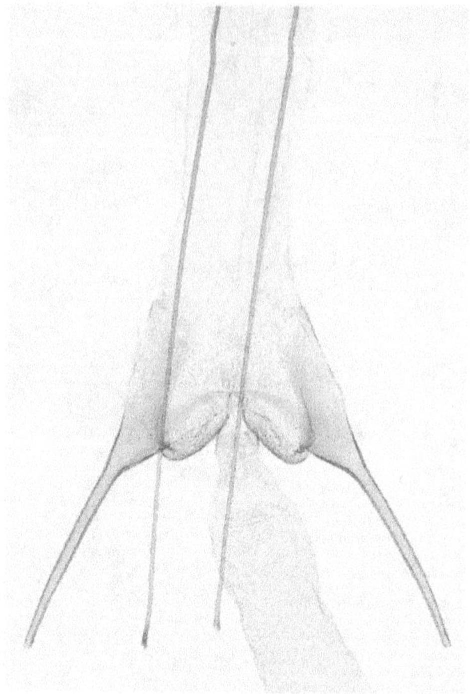

31

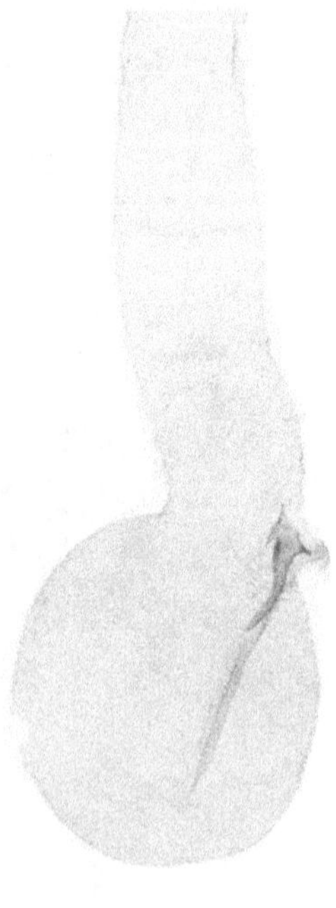

FEMALE GENITALIA

30. *Scrobipalpa frugifera* Povolný – Russia, GU 09/1320 P.H. (ANIK).
31. *Scrobipalpa oleksiyella* sp. n. – Paratype, Russia, GU 02/1122 P.H. (JUNN).

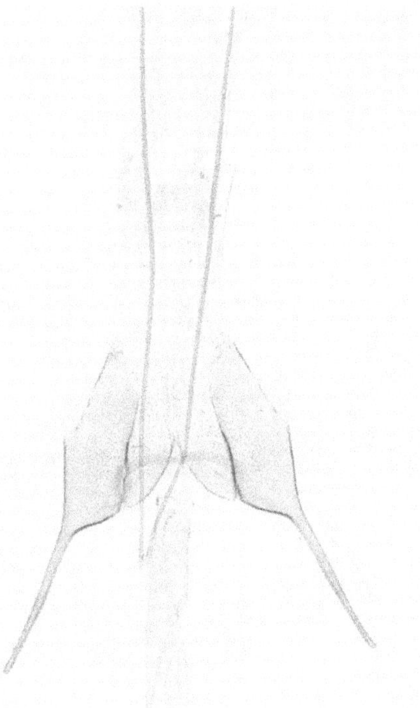

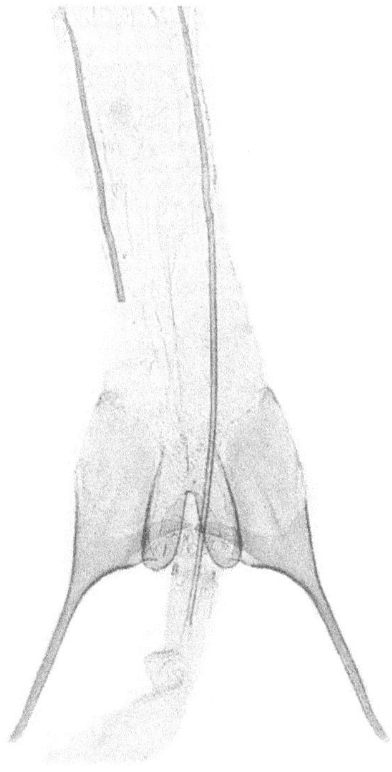

32 33

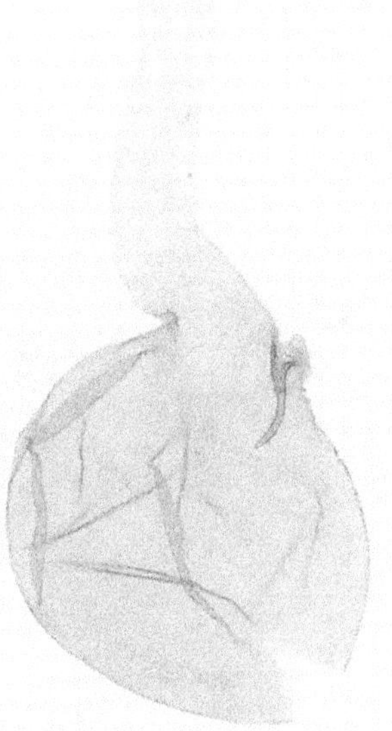

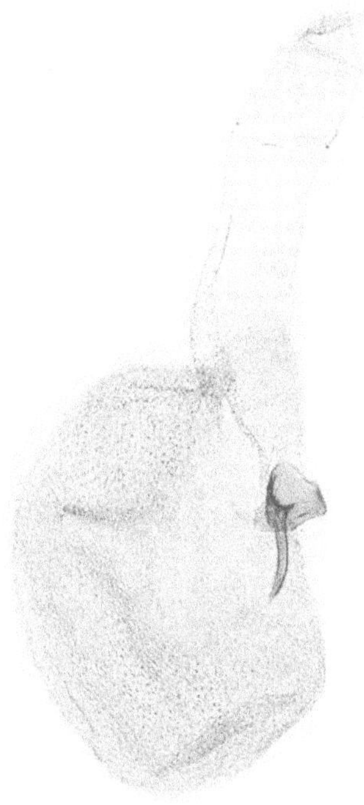

FEMALE GENITALIA

32. *Scrobipalpa smithi* Povolný & Bradley – Spain, GU 01/1039 P.H. (ZMUC).
33. *Scrobipalpa occulta* (Povolný) – Russia, GU 09/1314 P.H. (ZMUC).

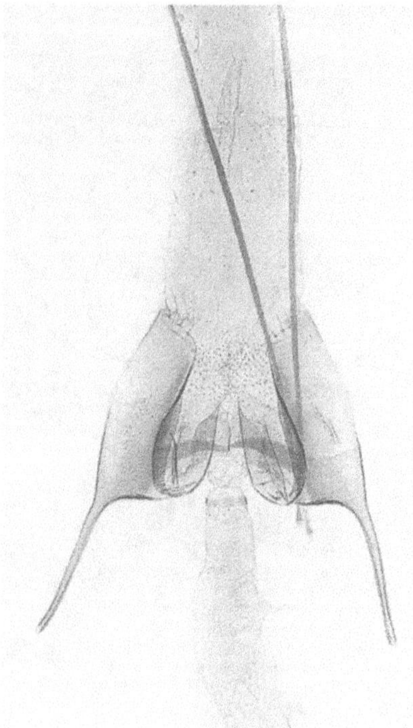

33a

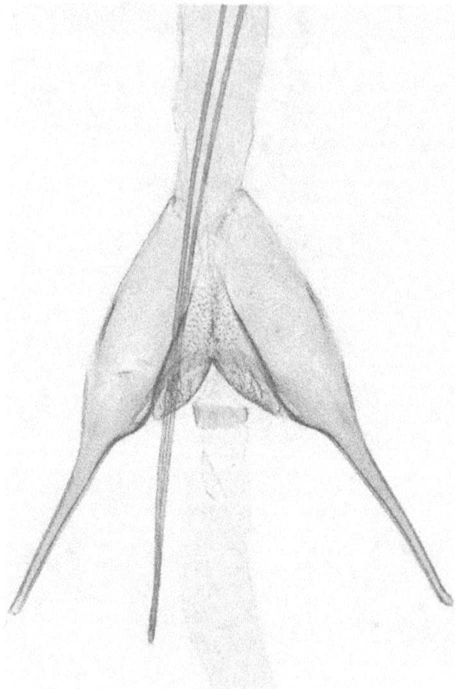

34

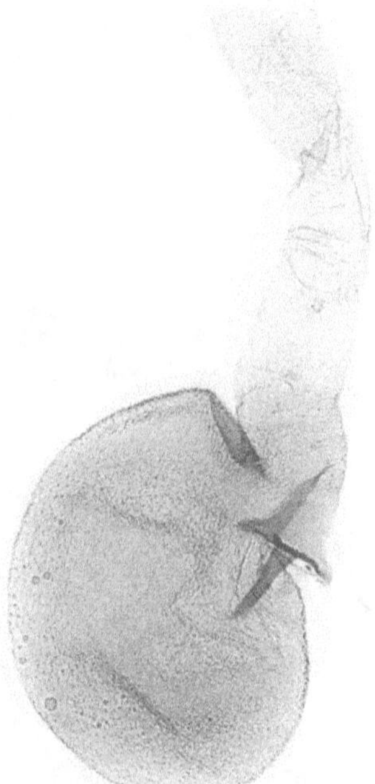

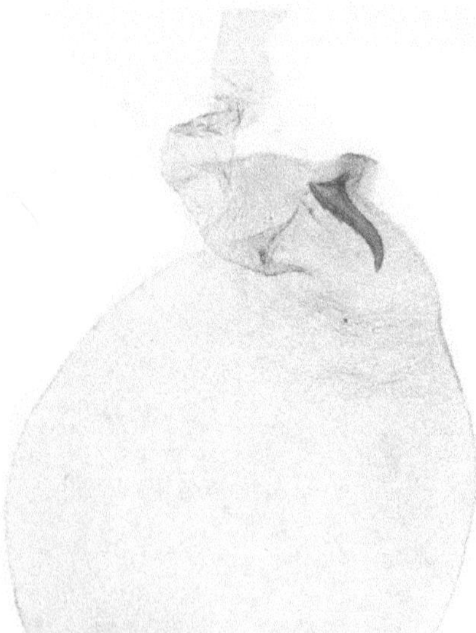

FEMALE GENITALIA

33a. *Scrobipalpa occulta* (Povolný) – Russia, gen. slide 26589 (BMNH).
34. *Scrobipalpa grisea* Povolný – Russia, GU 09/1305 P.H. (ZMUC).

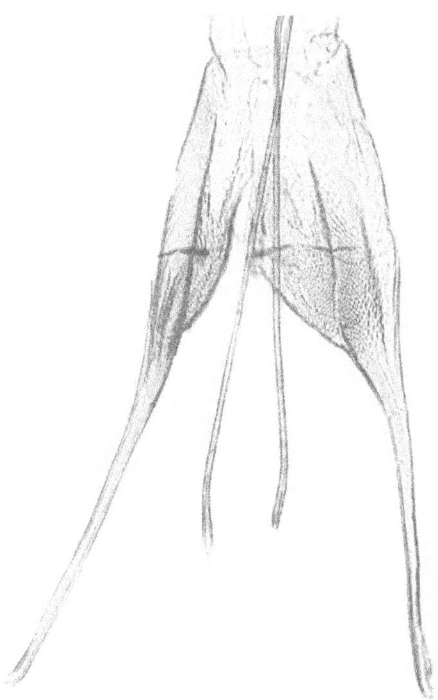

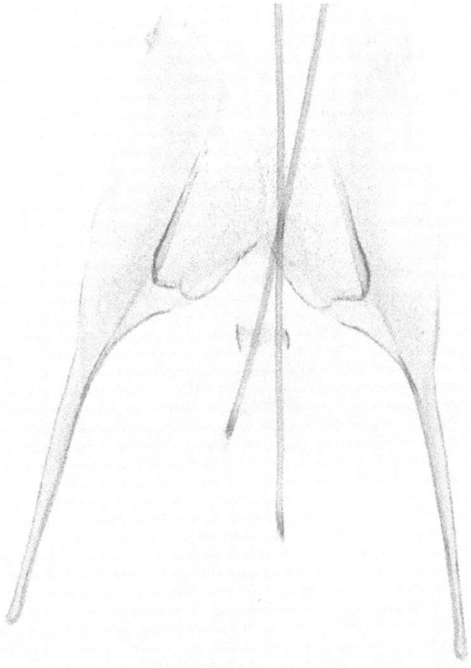

35

36

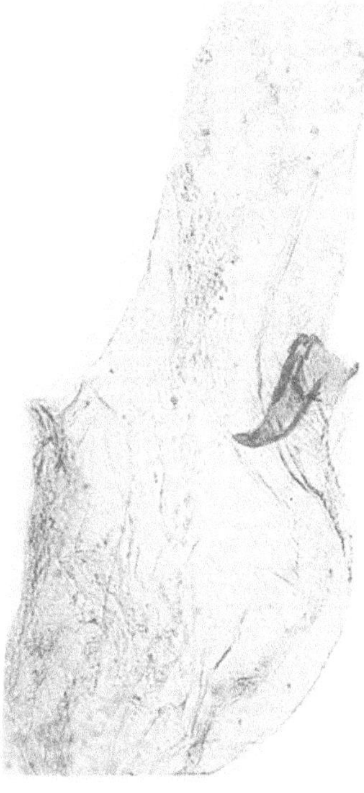

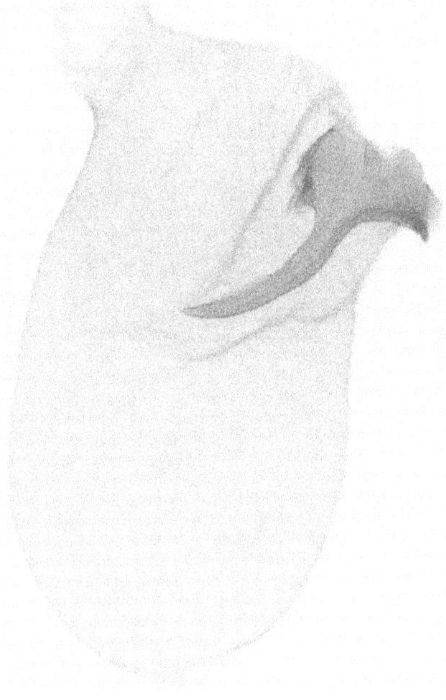

FEMALE GENITALIA

35. *Scrobipalpa usingeri* Povolný – Mongolia, gen. slide ZK. 5574 Povolný (MZMB).
36. *Scrobipalpa clintoni* Povolný – Denmark, GEL 992 P.H. (TLMF).

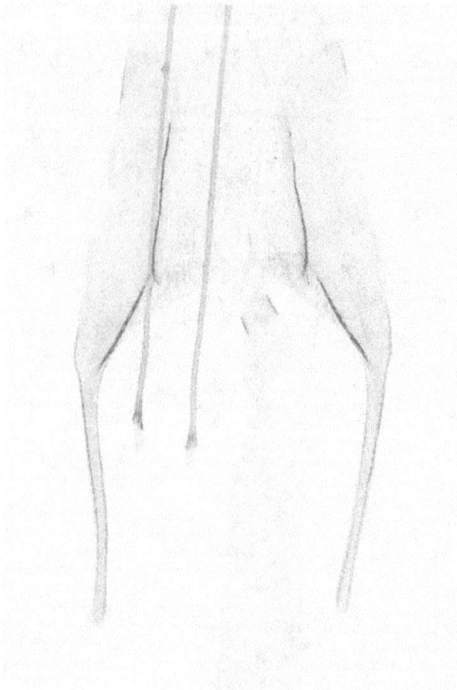

37

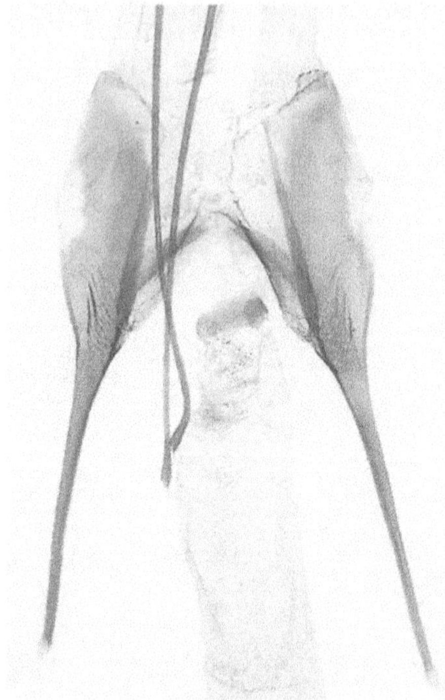

38

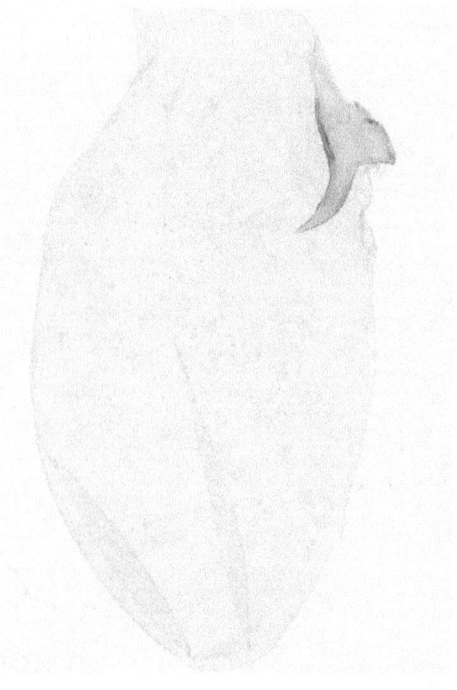

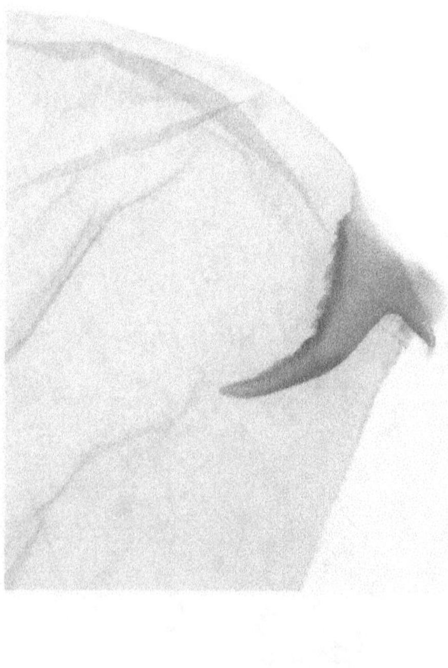

Female Genitalia

37. *Scrobipalpa reiprichi* Povolný – Norway, GU 01/1019 P.H. (ZMUC).
38. *Scrobipalpa obsoletella* (Fischer von Röslerstamm) – Italy, GU 01/1006 P.H. (TLMF).

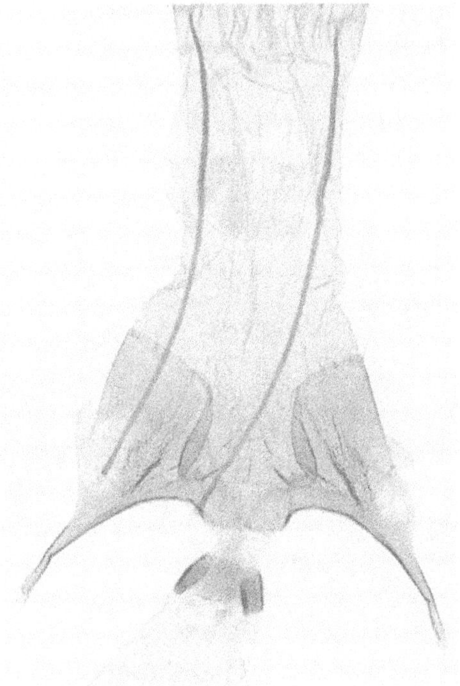

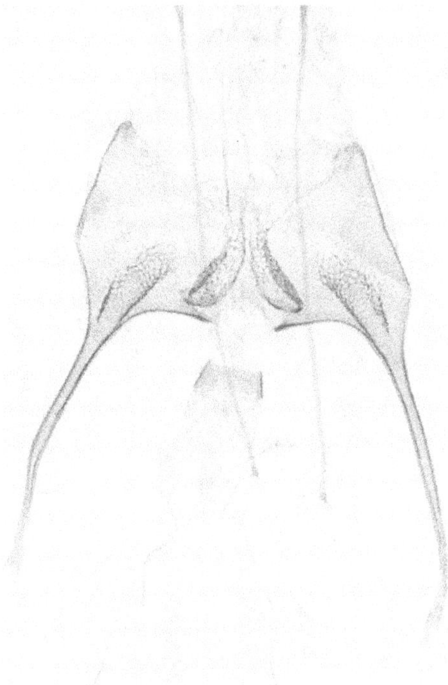

39

40

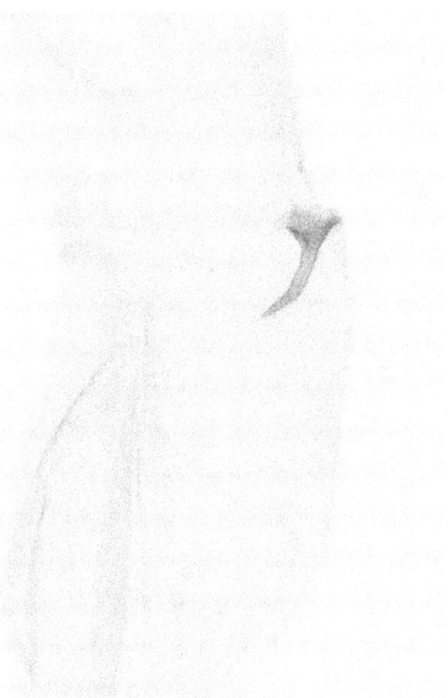

Female Genitalia

39. *Scrobipalpa feralella* (Zeller) – Austria, GEL 1017 P.H. (TLMF).
40. *Scrobipalpa halonella* (Herrich-Schäffer) – Austria, GU 04/1213 P.H. (ZSM).

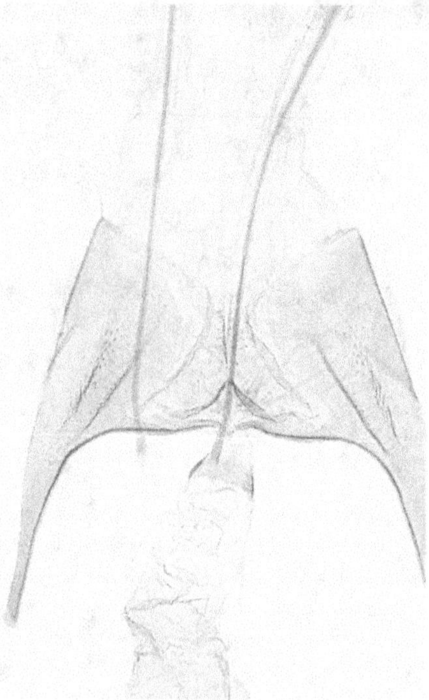

41

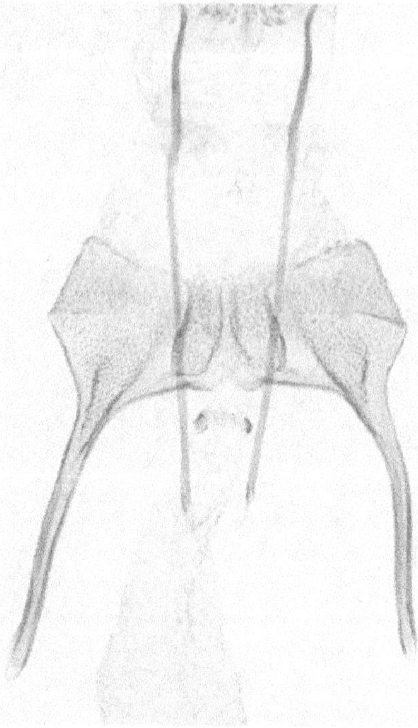

42

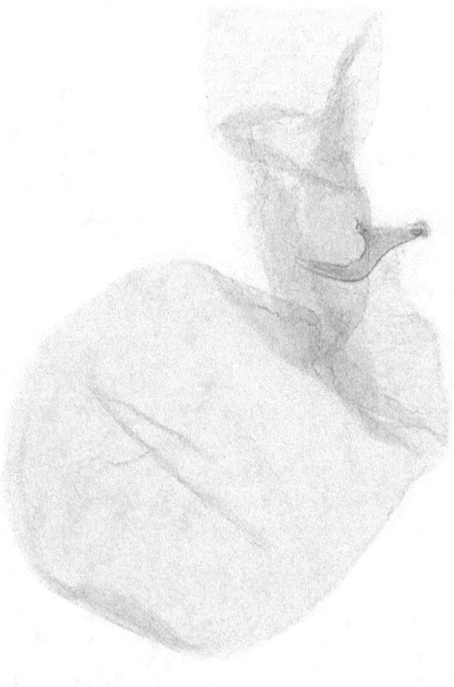

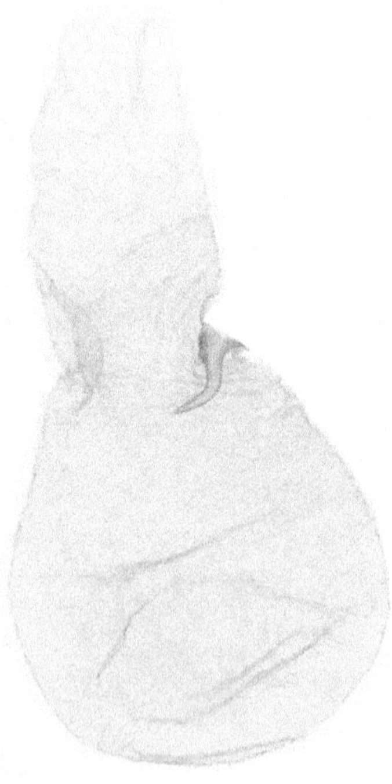

Female Genitalia

41. *Scrobipalpa perinii* (Klimesch) – Italy, GEL 947 P.H. (TLMF).
42. *Scrobipalpa phagnalella* (Constant) – Spain, GU 00/897 P.H. (ZSM).

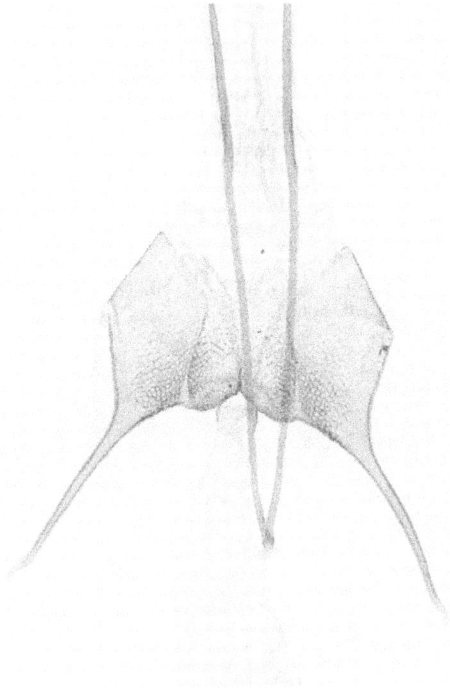

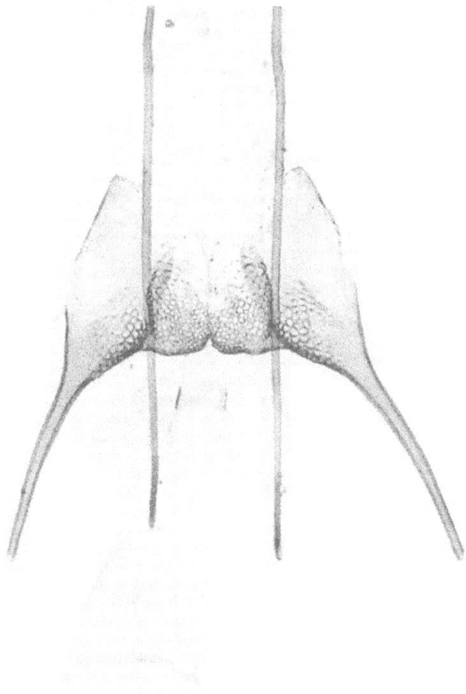

45 46

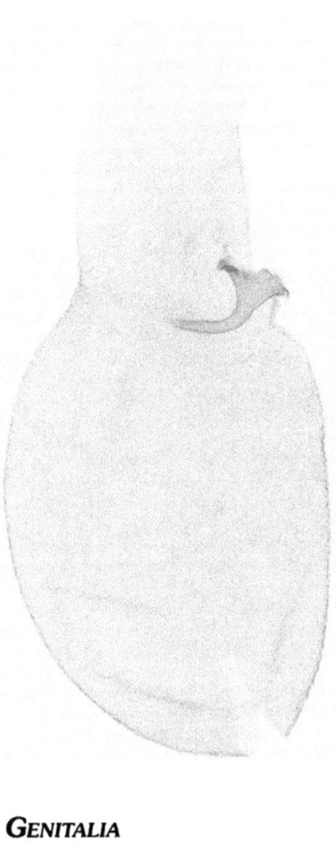

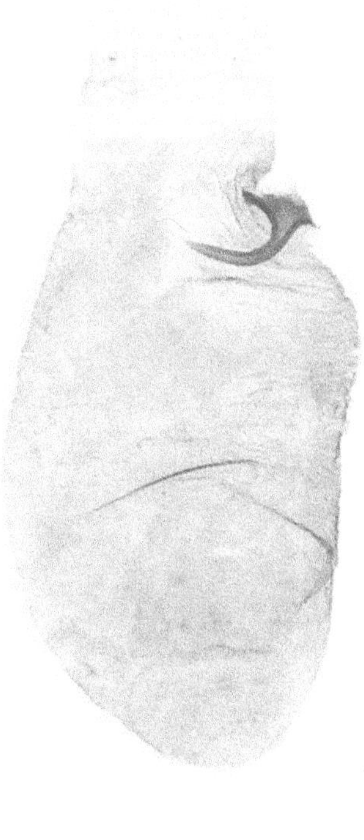

Female Genitalia

45. *Scrobipalpa heimi* sp. n. – Paratype, Russia, GU 01/1109 P.H. (JUNN).
46. *Scrobipalpa acuta* (Povolný) – Ukraine, gen. slide 2875 H.H. (ZMKU).

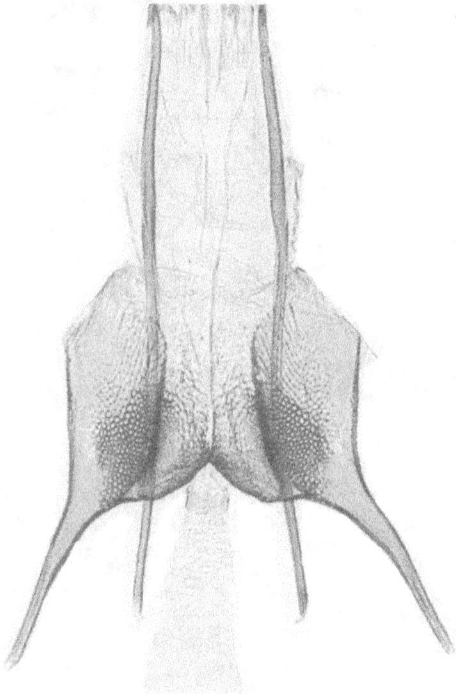

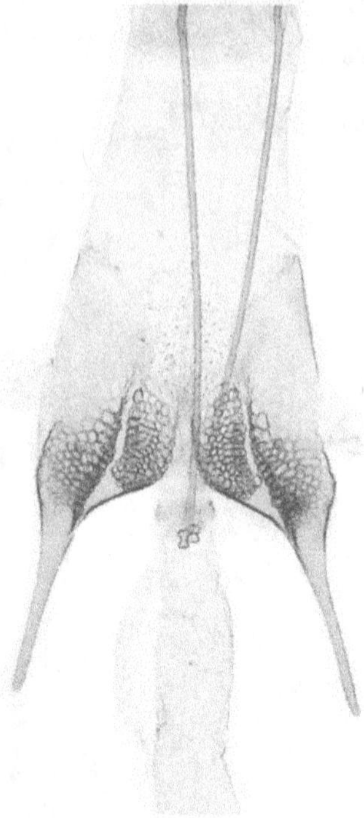

47

49

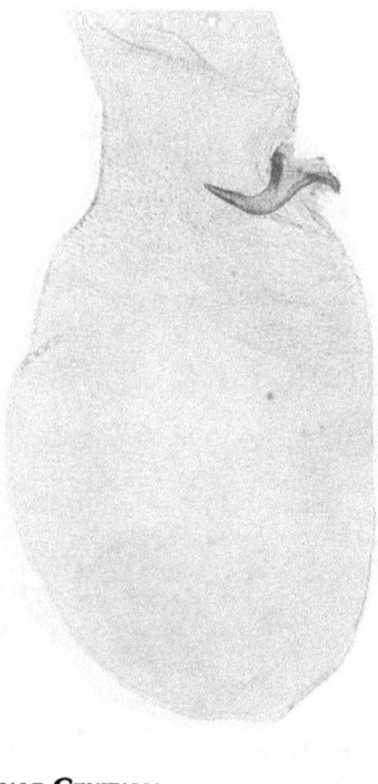

Female Genitalia

47. *Scrobipalpa soffneri* Povolný – Paratype, Bulgaria, GU 09/1315 P.H. (MZMB).
49. *Scrobipalpa murinella* (Duponchel) – Austria, GEL 1101 P.H. (TLMF).

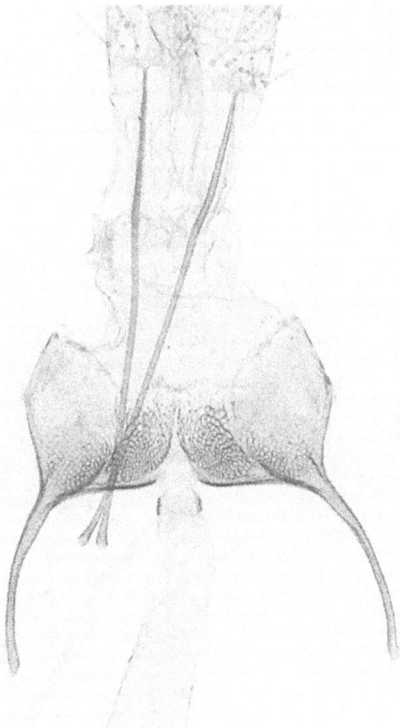

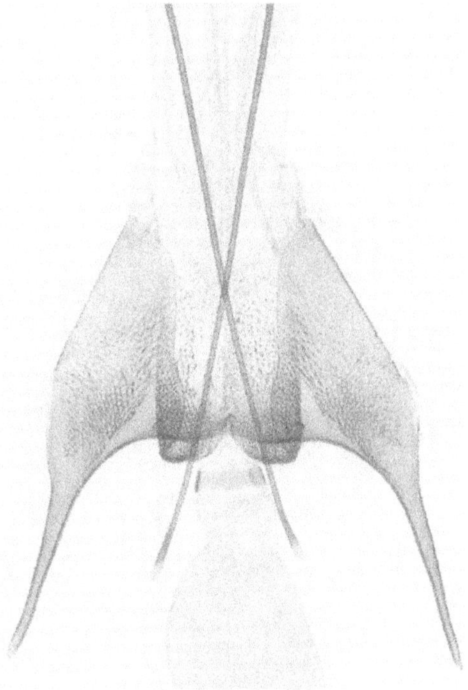

50 52

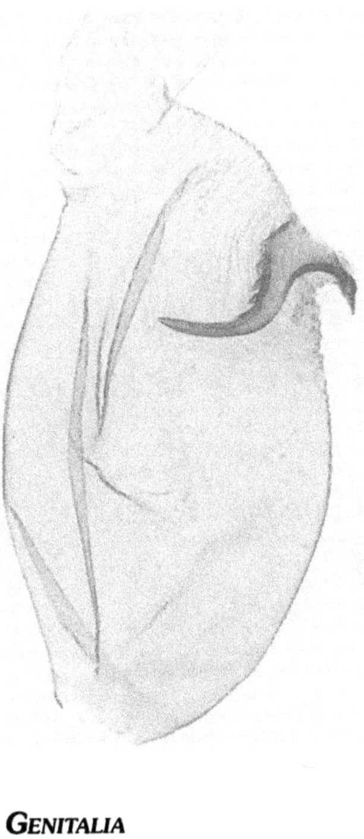

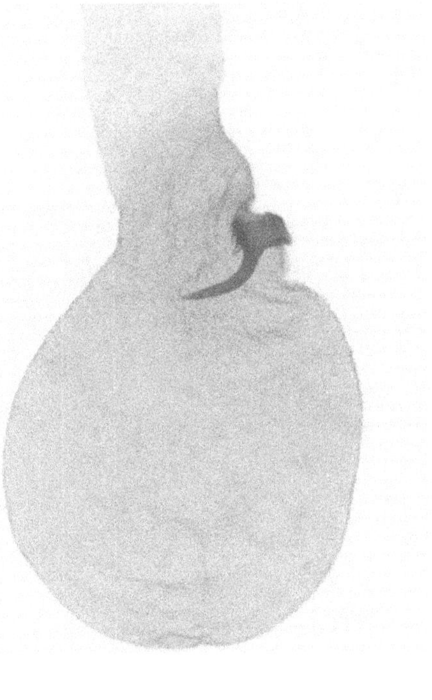

Female Genitalia

50. *Scrobipalpa wiltshirei* Povolný – Spain, GEL 1103 P.H. (TLMF).
52. *Scrobipalpa pauperella* (Heinemann) – Czechia, GEL 1065 (TLMF).

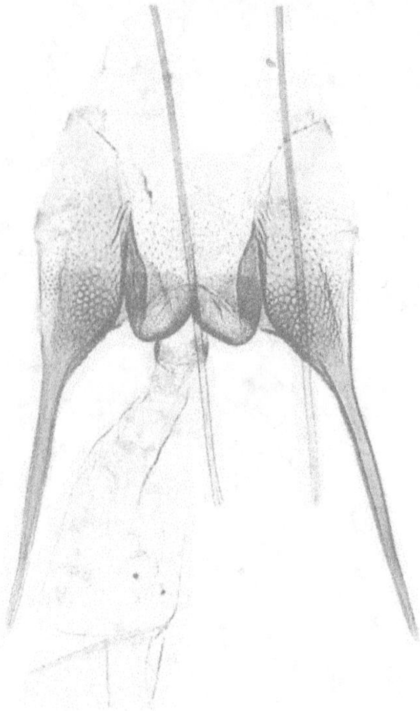

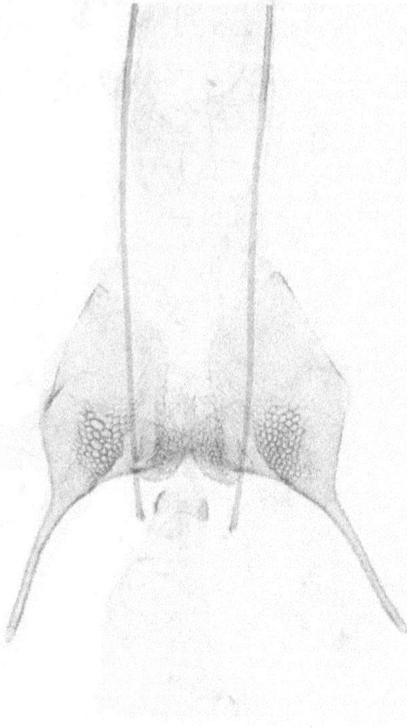

53

54

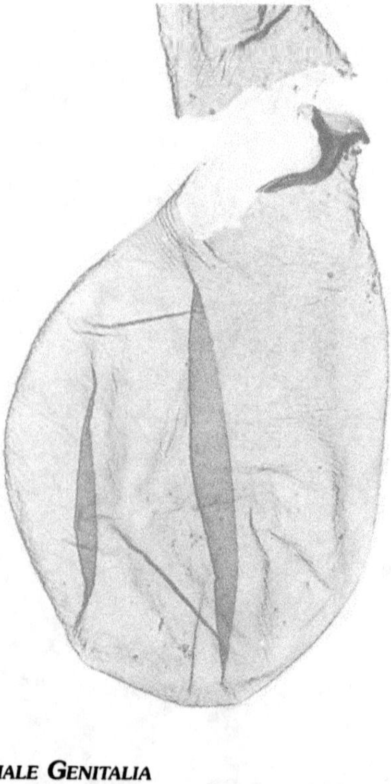

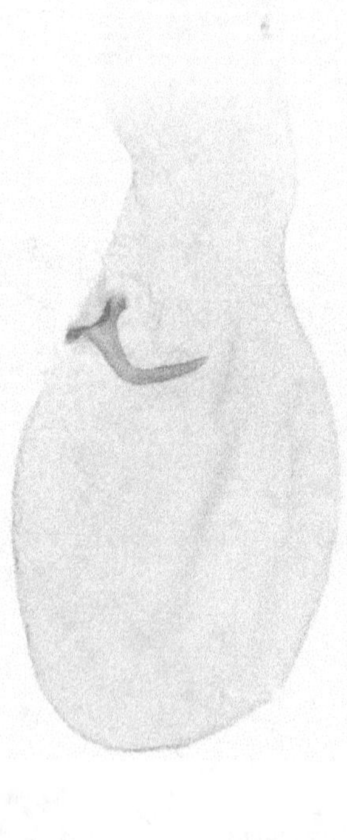

Female Genitalia

53. *Scrobipalpa spumata* (Povolný) – Russia, GU 01/1085 P.H. (JUNN).
54. *Scrobipalpa arenbergeri* Povolný – Croatia, GEL 974 P.H. (TLMF).

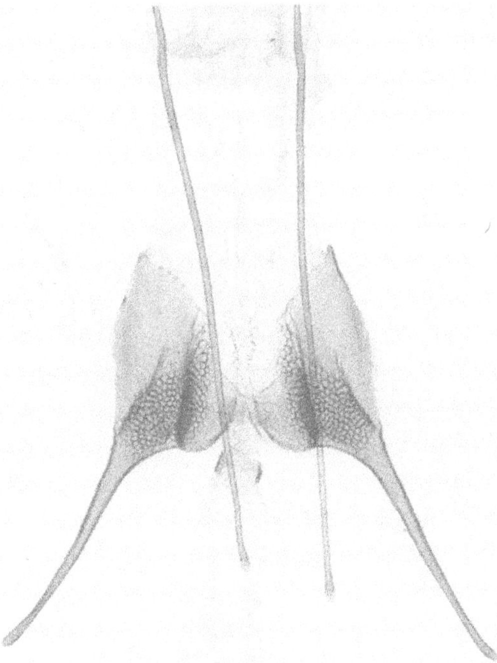

55

56

FEMALE GENITALIA

55. *Scrobipalpa nana* Povolný – Russia, GU 05/1235 P.H. (ZMKU).
56. *Scrobipalpa heretica* Povolný – Paratype, Turkey, gen. slide 4385 Povolný (LNK).

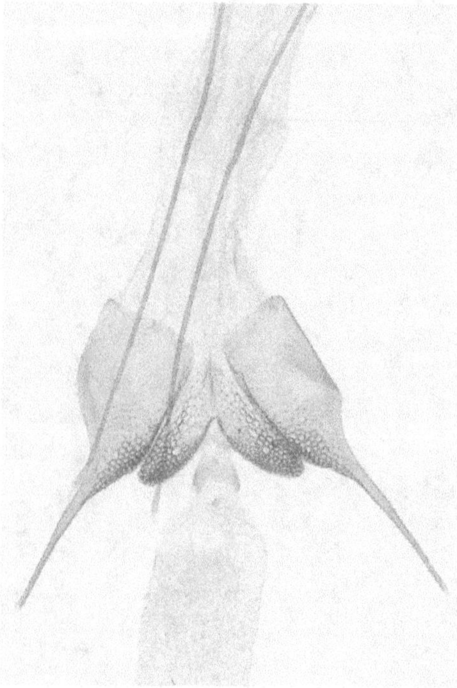

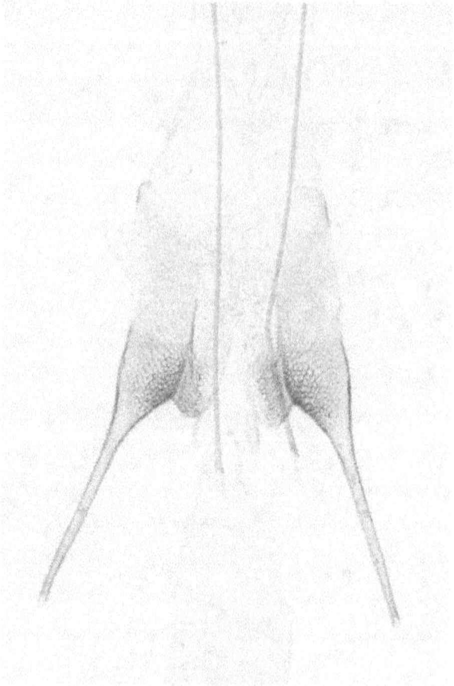

57

58

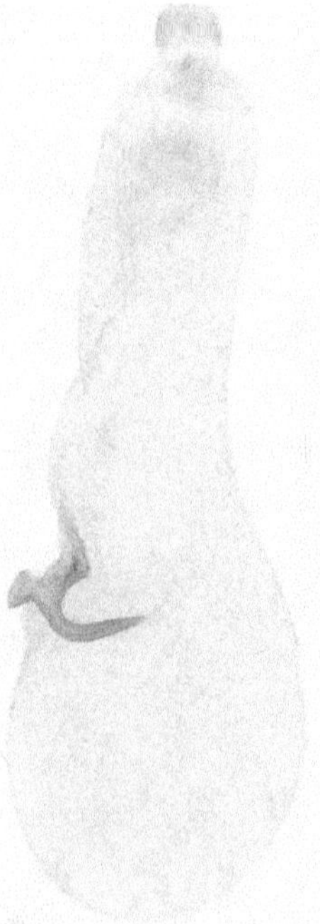

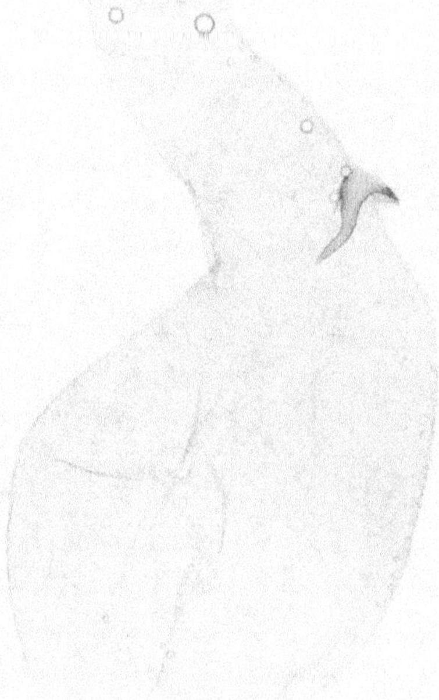

FEMALE GENITALIA

57. *Scrobipalpa bigoti* Povolný – Spain, GU 01/1018 P.H. (ZMUC).
58. *Scrobipalpa dorsoflava* (Povolný) – Russia, GU 02/1127 P.H. (JUNN).

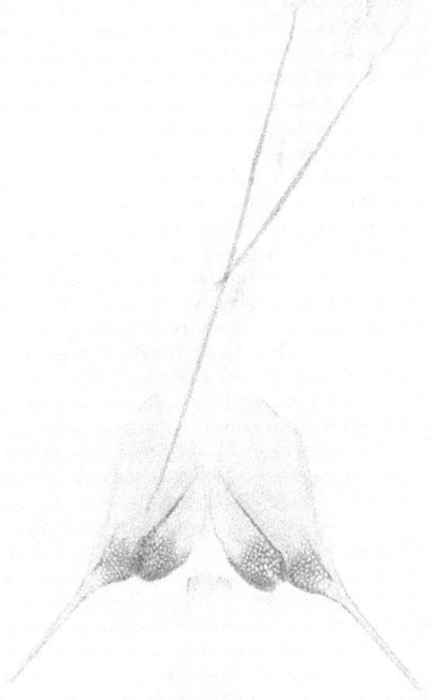

59

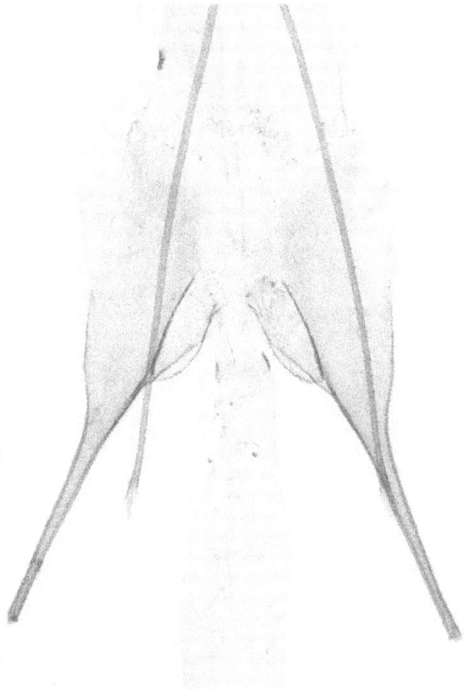

60

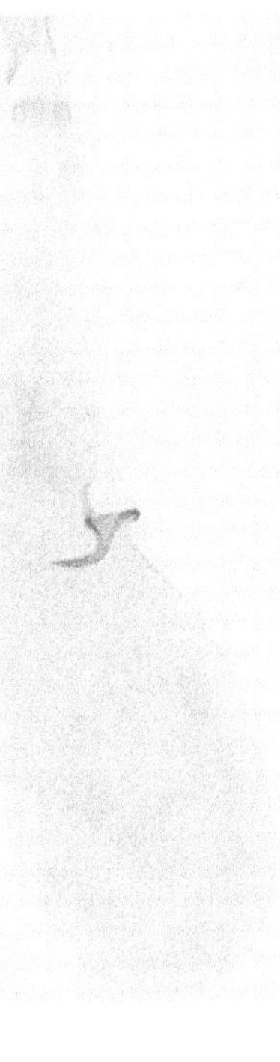

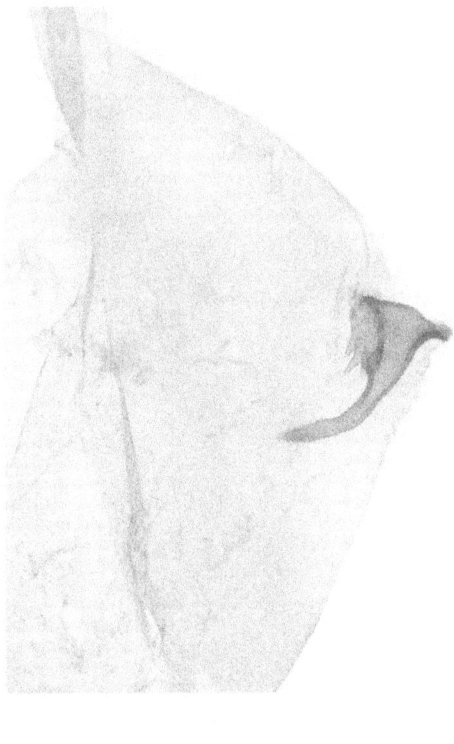

Female Genitalia

59. *Scrobipalpa magnificella* Povolný – Russia, GU 02/1117 P.H. (ZMUC).
60. *Scrobipalpa abstrusa* sp. n. – Paratype, Russia, gen. slide 2975 H.H. (ZMUC).

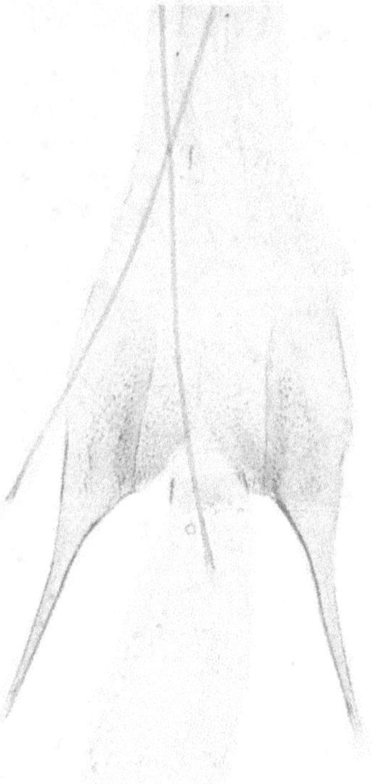

61

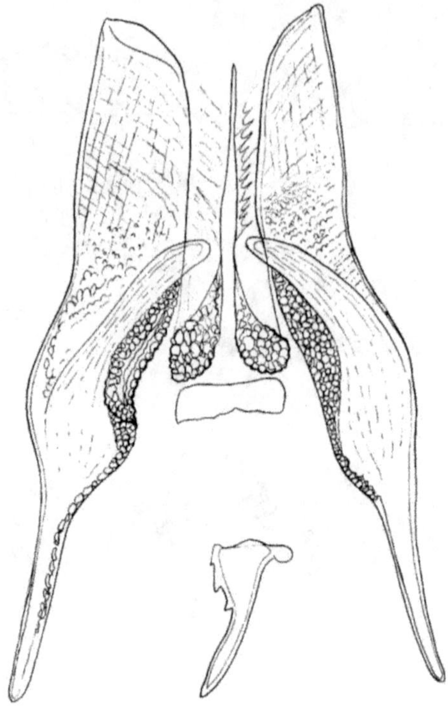

62

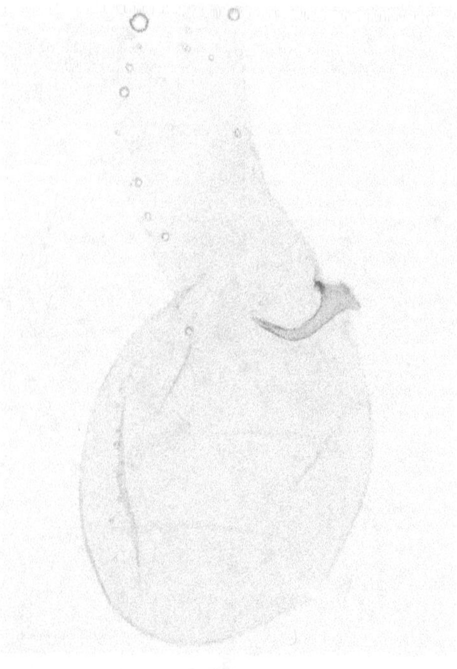

Female Genitalia

61. *Scrobipalpa superstes* Povolný – Portugal, GEL 1056 P.H. (TLMF).
62. *Scrobipalpa plesiopicta* Povolný – Russia, Povolný, 2002a: fig. 599.

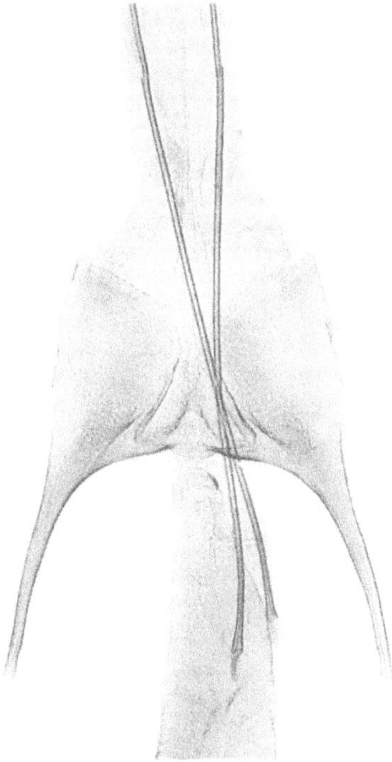

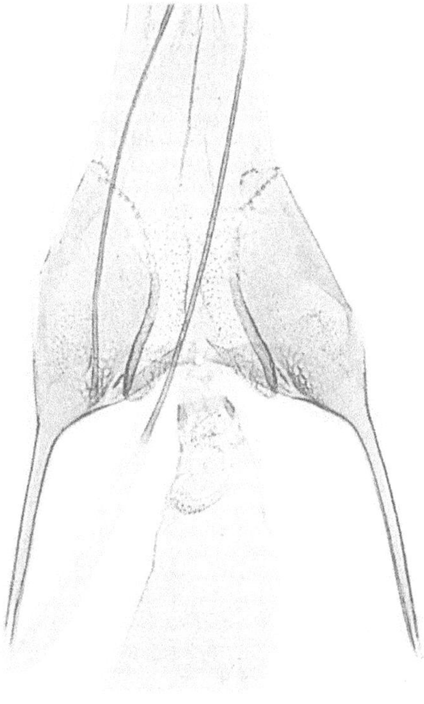

63

64

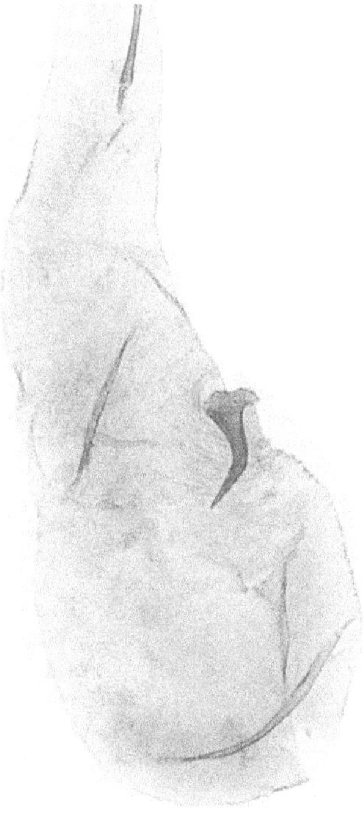

Female Genitalia

63. *Scrobipalpa bradleyi* Povolný – France, GU 01/1050 P.H. (GRÜN).
64. *Scrobipalpa selectella* (Caradja) – Greece, GEL 1114 P.H. (TLMF).

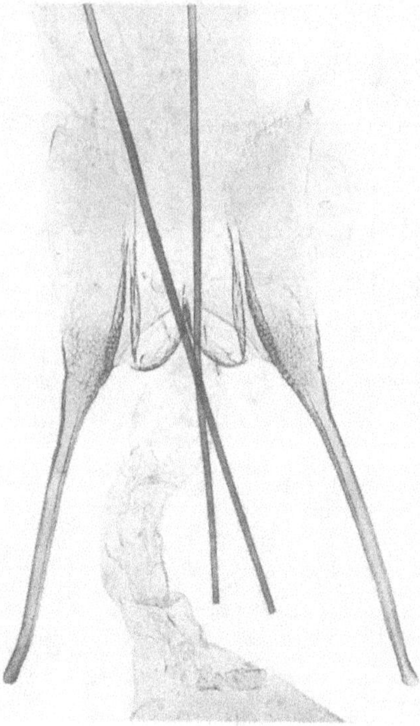

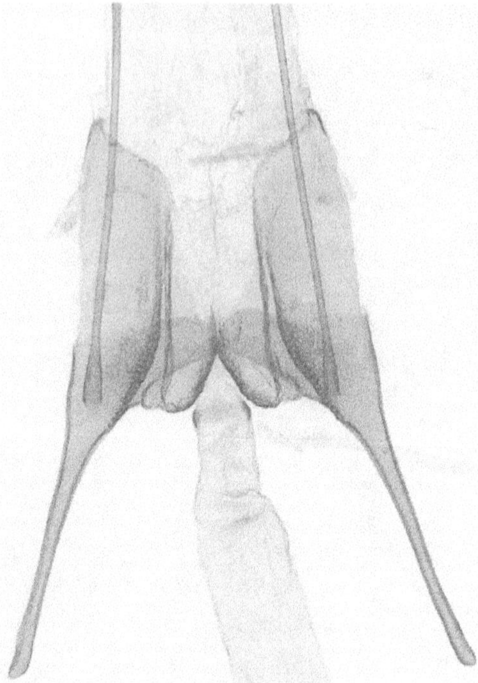

65 66

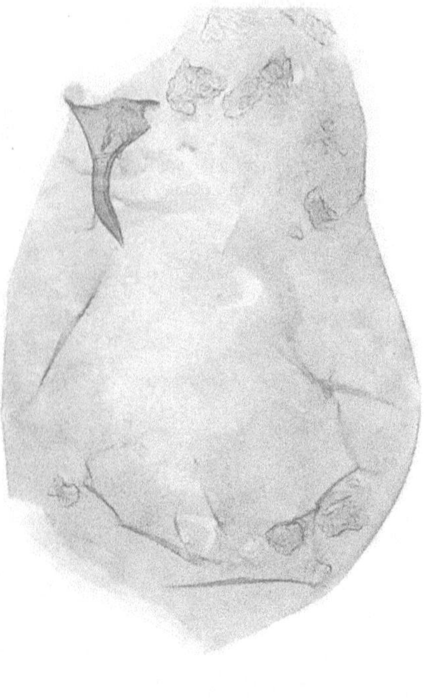

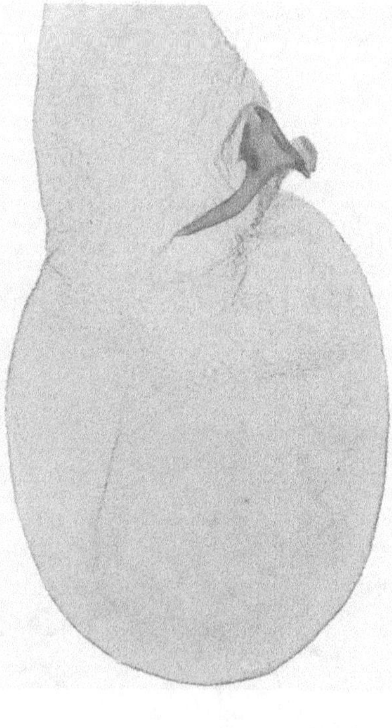

Female Genitalia

65. *Scrobipalpa alterna* (Falkovitsh & Bidzilya) – Russia, gen. slide 3175 H.H. (NUPP).
66. *Scrobipalpa lutea* Povolný – Turkey, GEL 1106 P.H. (TLMF).

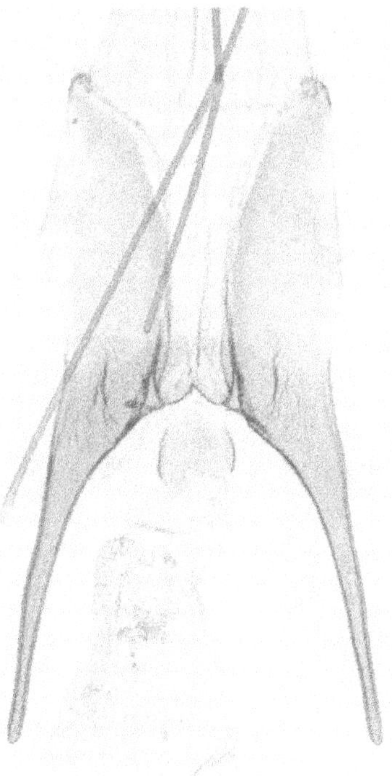

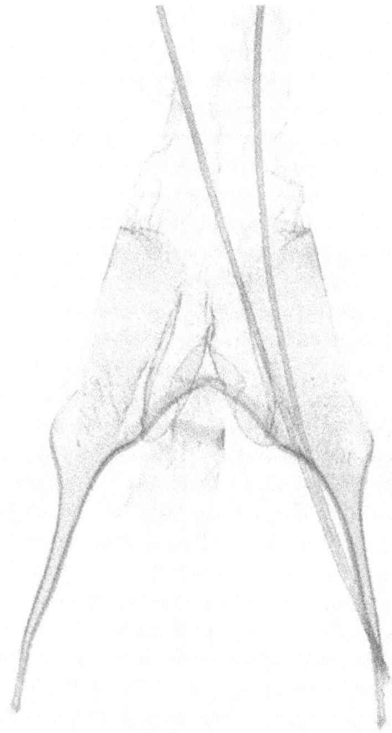

67 68

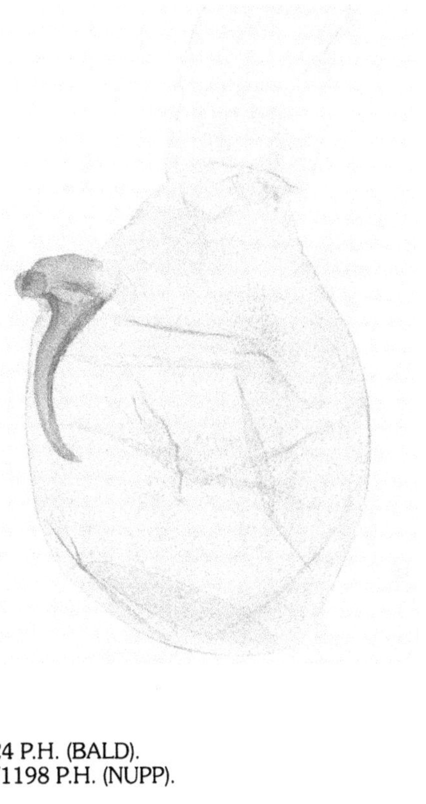

FEMALE GENITALIA

67. *Scrobipalpa niveifacies* Povolný – Spain, GU 01/1024 P.H. (BALD).
68. *Scrobipalpa indignella* (Staudinger) – Russia, GU 03/1198 P.H. (NUPP).

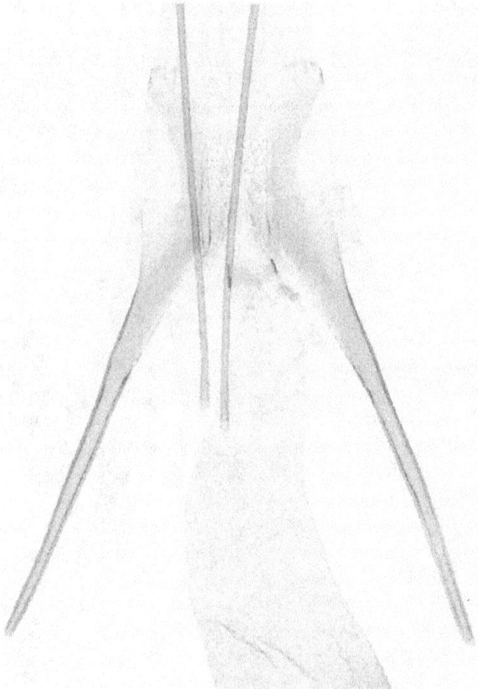

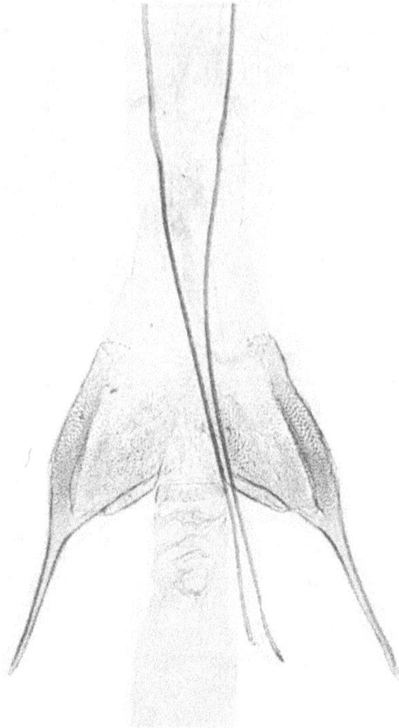

69

70

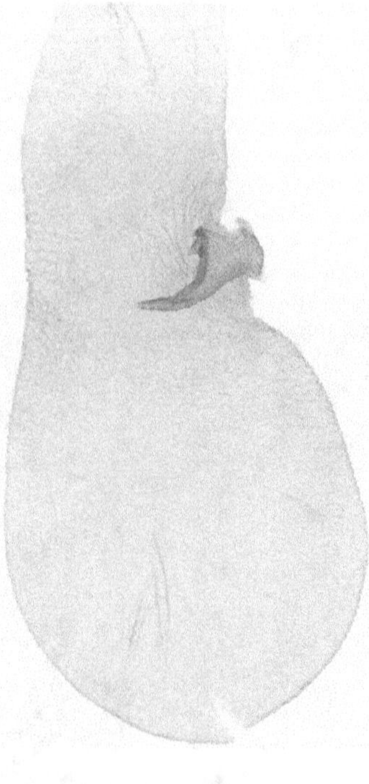

Female Genitalia

69. *Scrobipalpa punctata* (Povolný) – Russia, GU 01/1091 P.H. (JUNN).
70. *Scrobipalpa lagodes* (Meyrick) – Kazakhstan, gen. slide 40/05 Bidzilya (ZMKU).

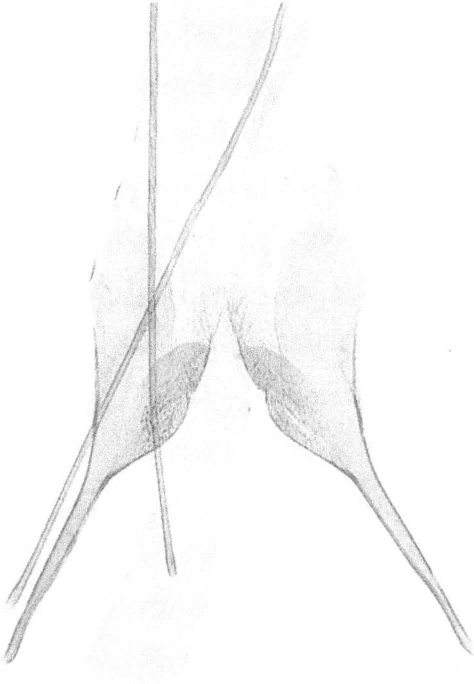

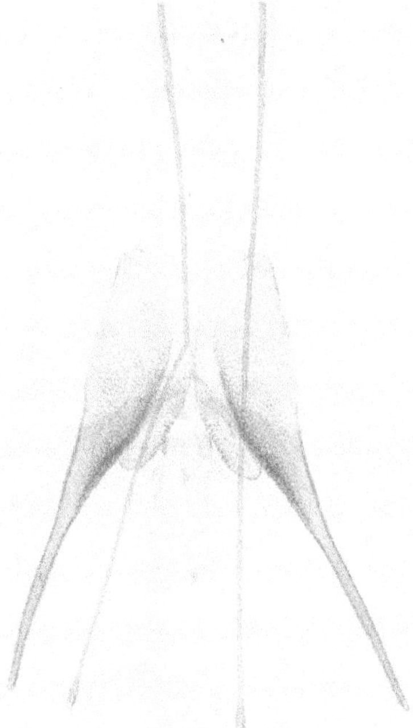

71 *72*

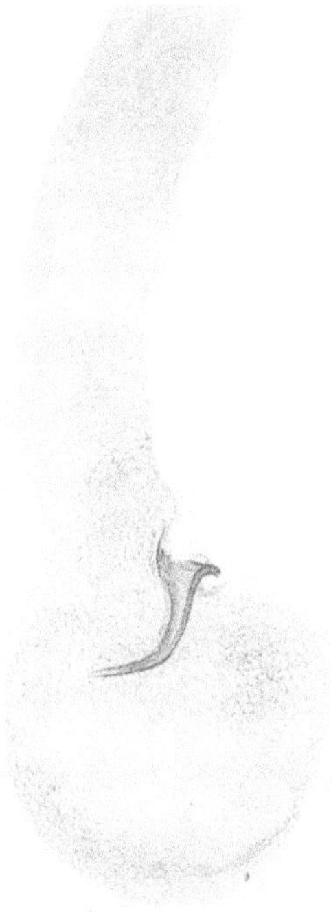

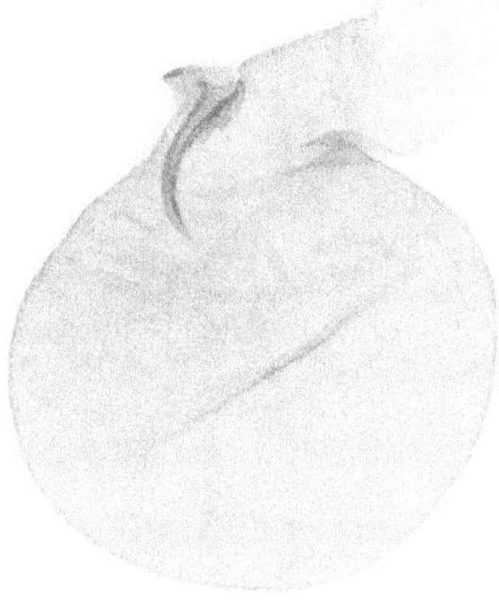

FEMALE GENITALIA

71. *Scrobipalpa deluccae* Povolný – Paratype, Lebanon, gen. slide 1368 (NHMW).
72. *Scrobipalpa atriplicella* (Fischer von Röslerstamm) – Germany, GEL 998 P.H. (TLMF).

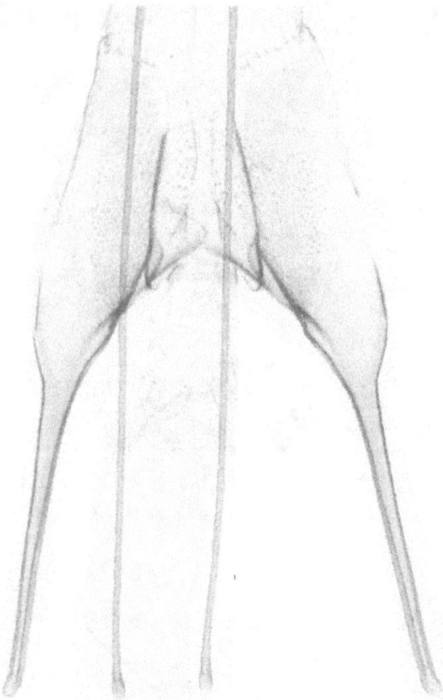

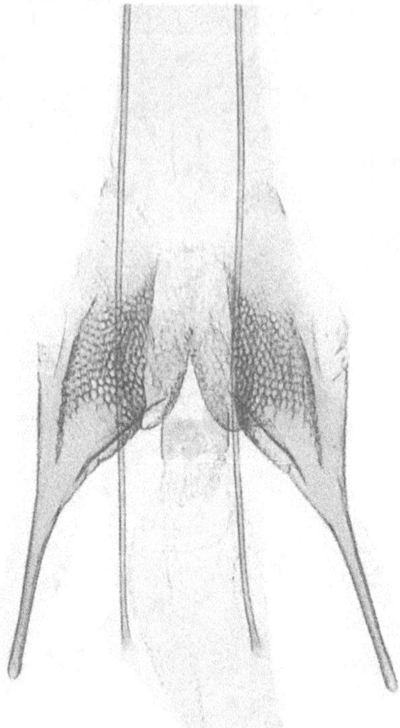

73

74

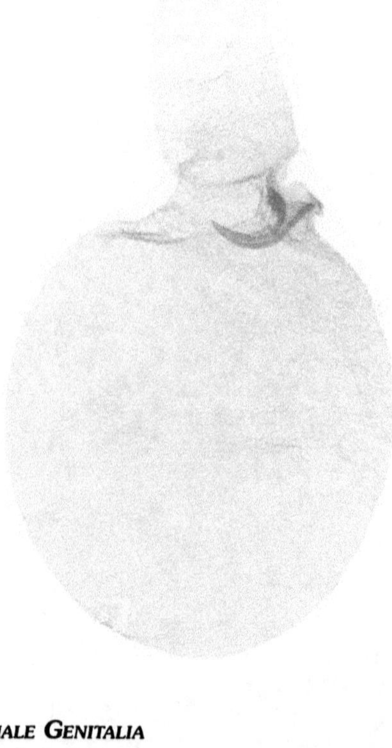

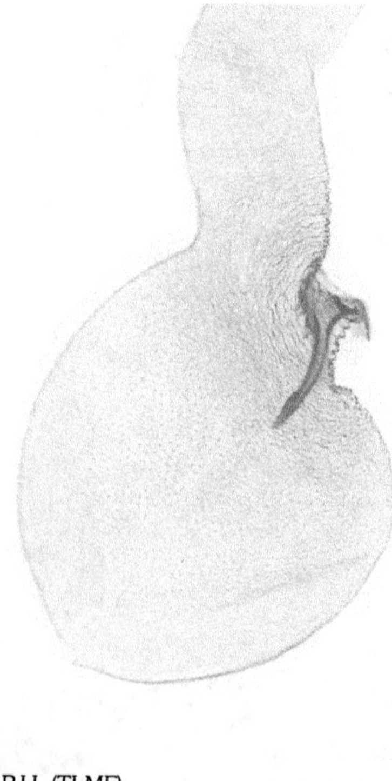

Female Genitalia

73. *Scrobipalpa suaedella* (Richardson) – Spain, GEL 977 P.H. (TLMF).
74. *Scrobipalpa solitaria* Povolný – Russia, GU 09/1282 P.H. (NUPP).

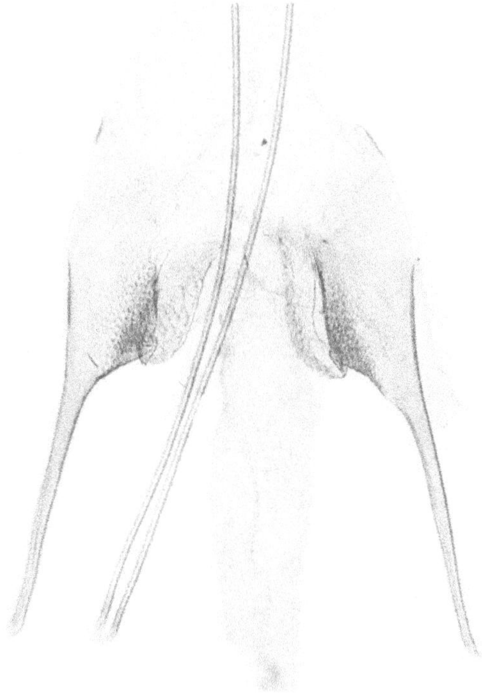

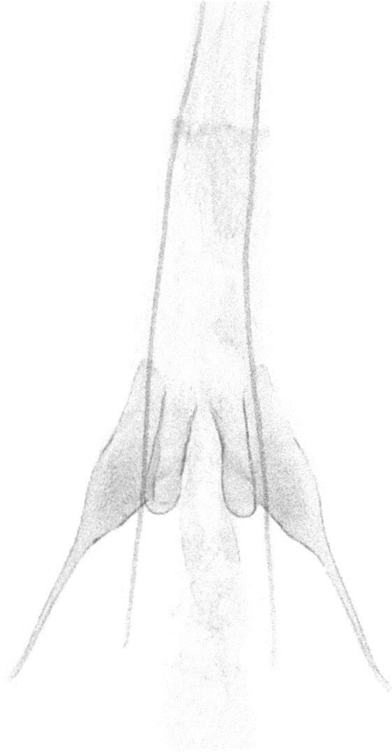

75 76

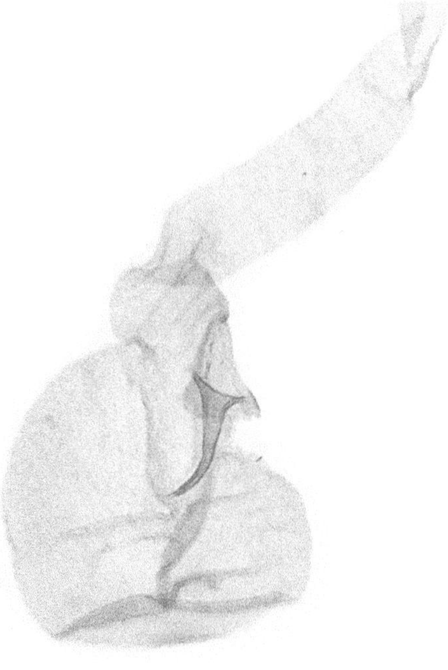

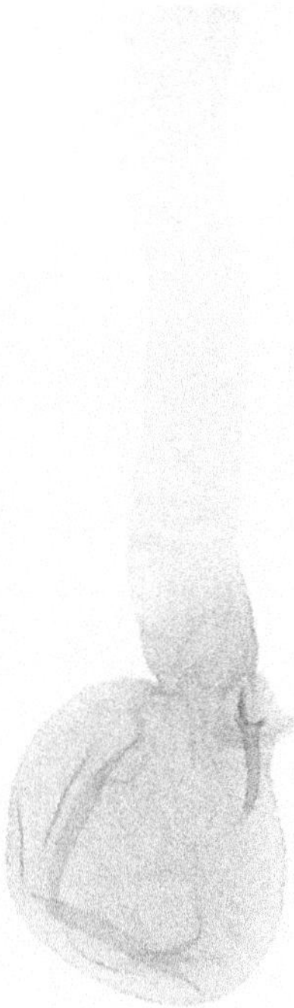

Female Genitalia

75. *Scrobipalpa dagmaris* Povolný – Uzbekistan, gen. slide 44/05 Bidzilya (ZMKU).
76. *Scrobipalpa suasella* (Constant) – France, GU 01/1054 P.H. (ZMUC).

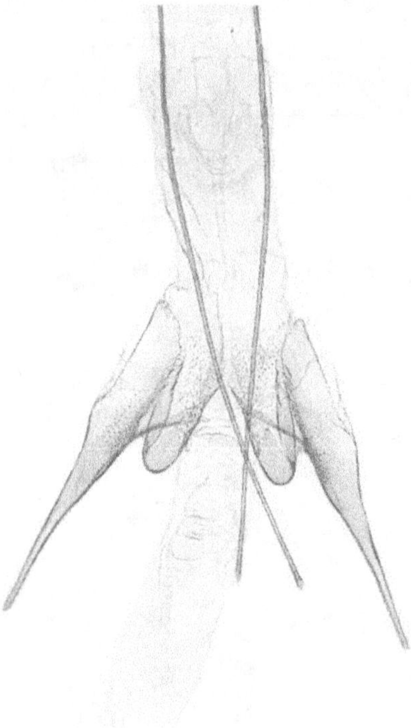

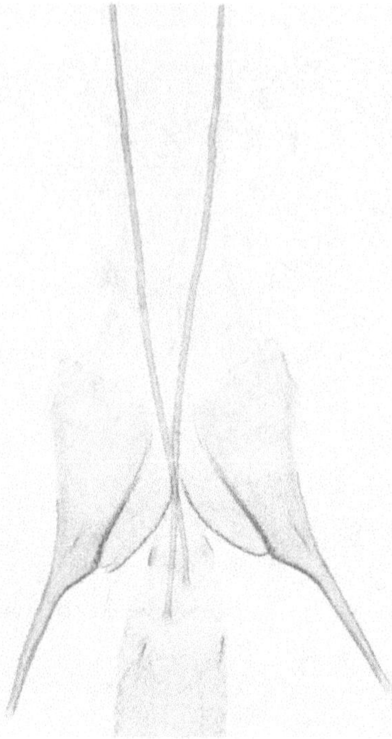

77

78

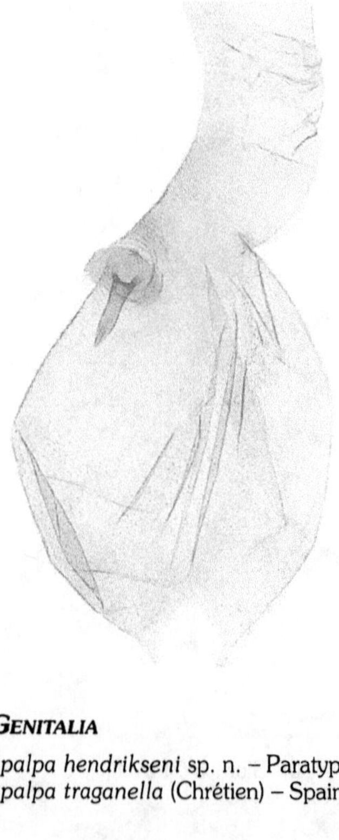

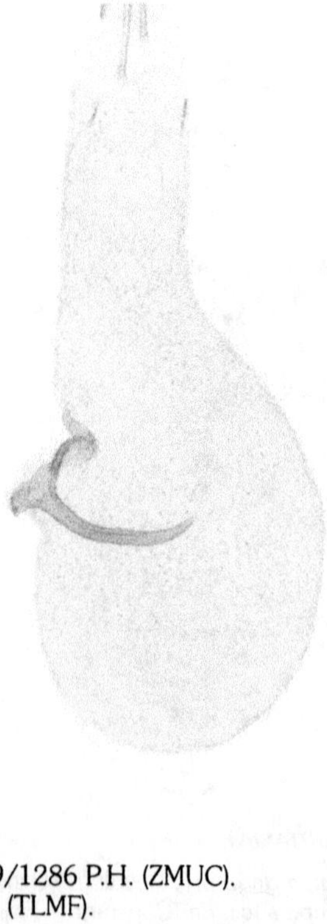

Female Genitalia

77. *Scrobipalpa hendrikseni* sp. n. – Paratype, Greece, GU 09/1286 P.H. (ZMUC).
78. *Scrobipalpa traganella* (Chrétien) – Spain, GEL 1012 P.H. (TLMF).

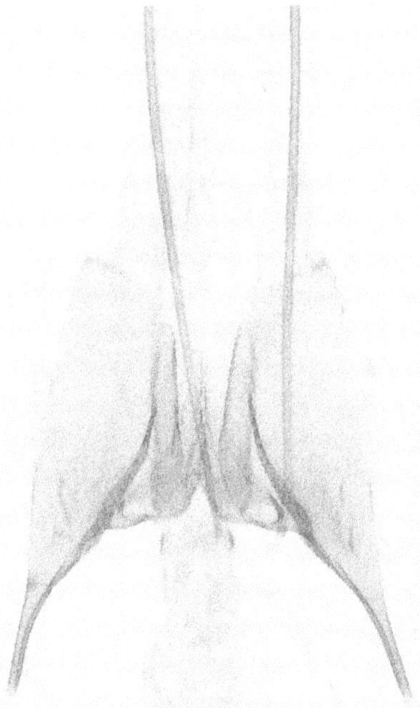

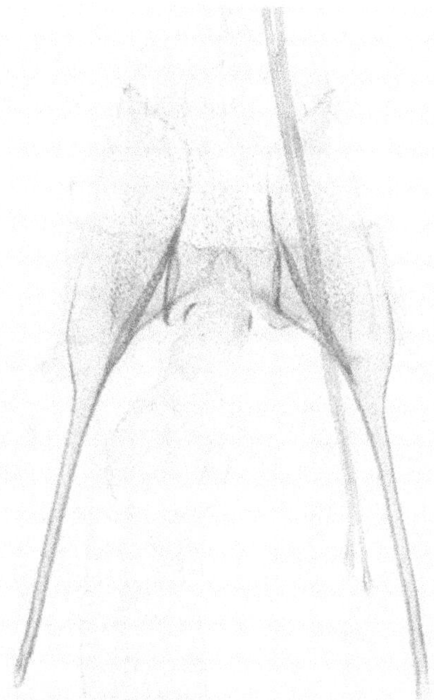

79

80

FEMALE GENITALIA

79. *Scrobipalpa bazae* Povolný – Spain, GU 01/1017 P.H. (ZMUC).
80. *Scrobipalpa artemisiella* (Treitschke) – Italy, GEL 58 P.H. (TLMF).

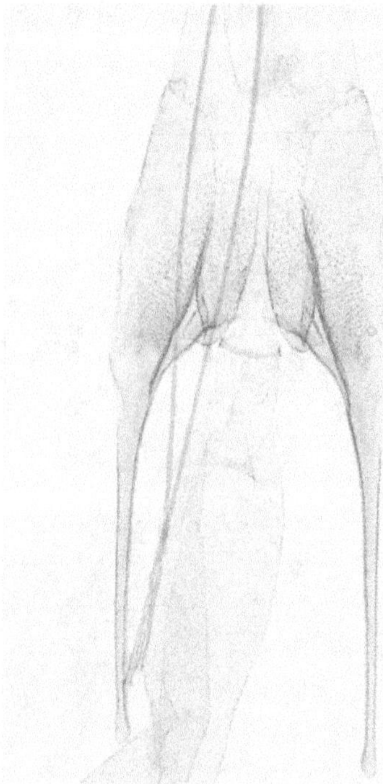

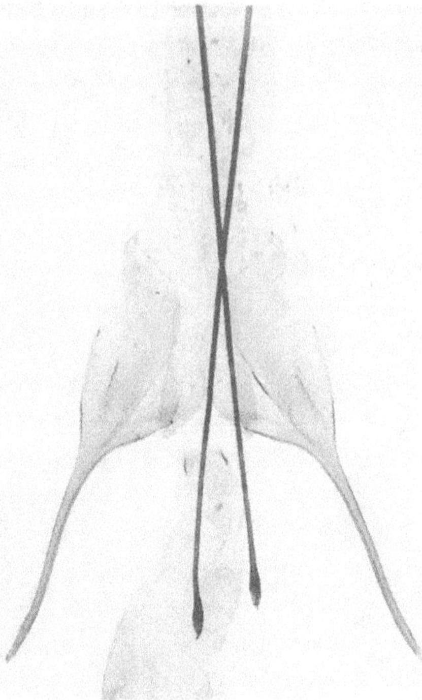

81

82

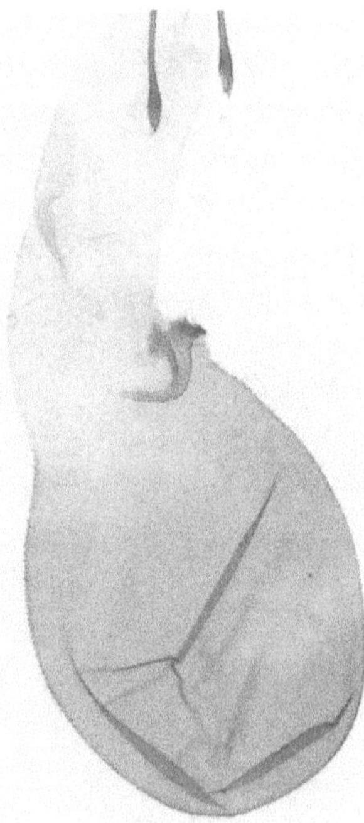

Female Genitalia

81. *Scrobipalpa stangei* (Hering) – Germany, gen. slide 16540 (NHMW).
82. *Scrobipalpa suaedivorella* (Chrétien) – Spain, gen. slide 4011 H.H. (ZMUC).

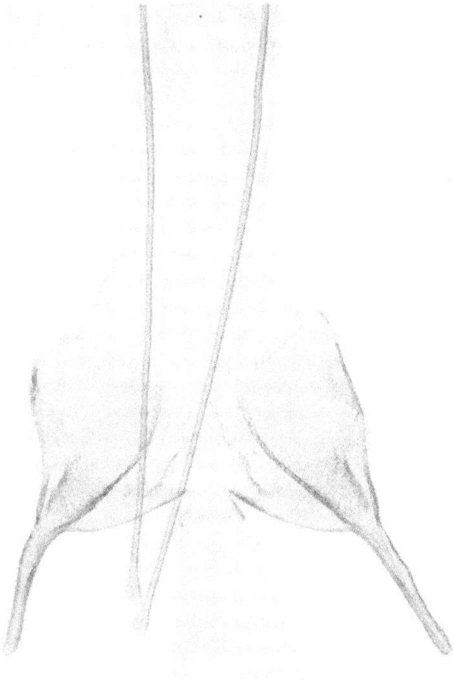

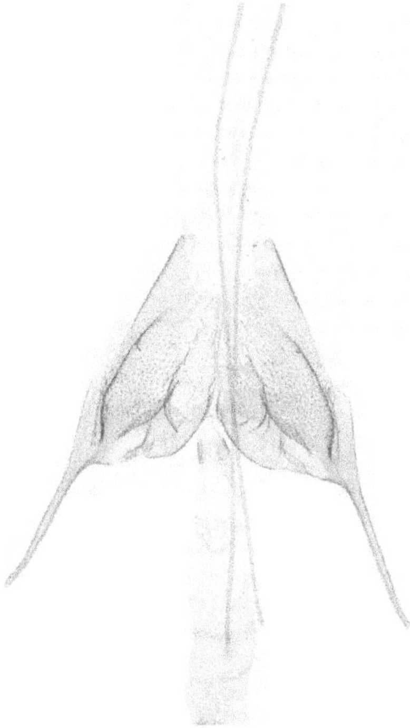

83

84

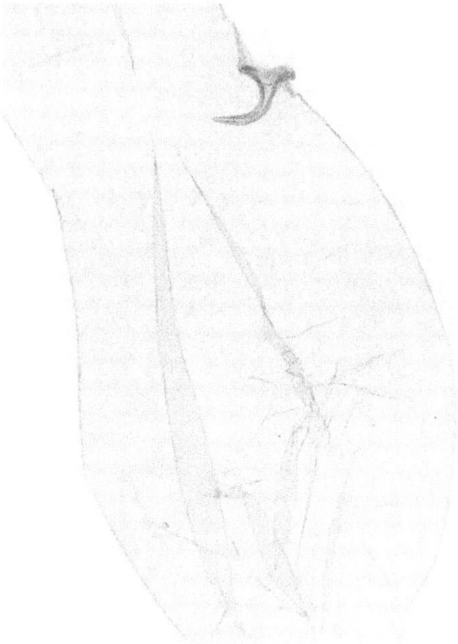

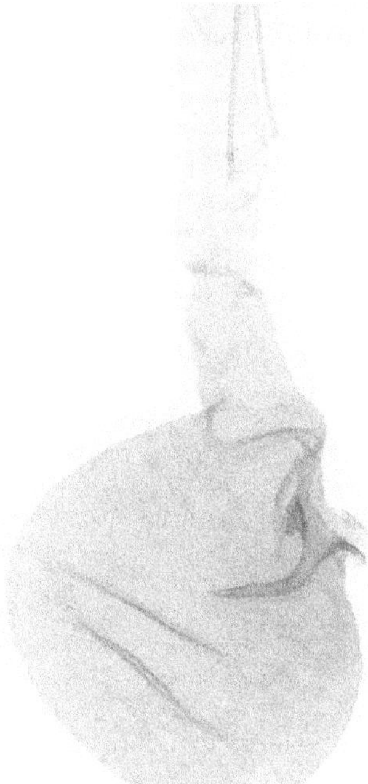

Female Genitalia

83. *Scrobipalpa bryophiloides* Povolný – Turkey, GEL 1013 P.H. (TLMF).
84. *Scrobipalpa algeriensis* Povolný & Bradley – Spain, GU 01/1048 P.H. (ZMUC).

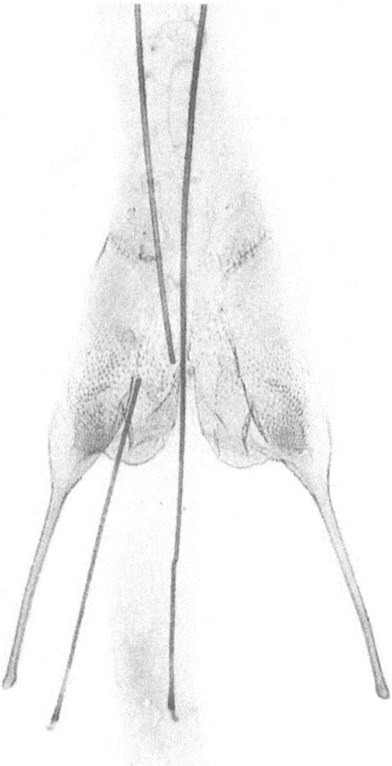

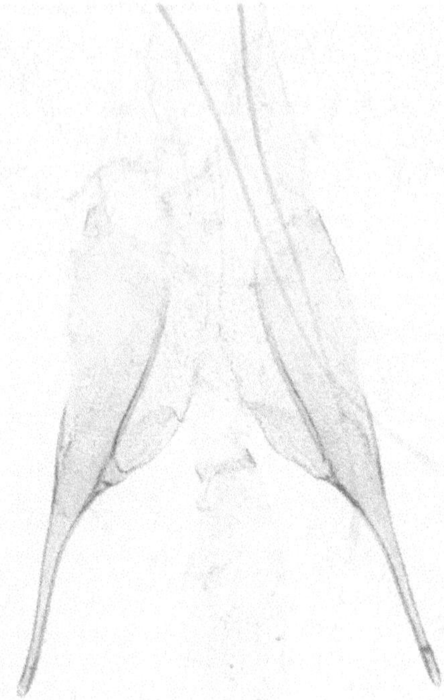

85

86

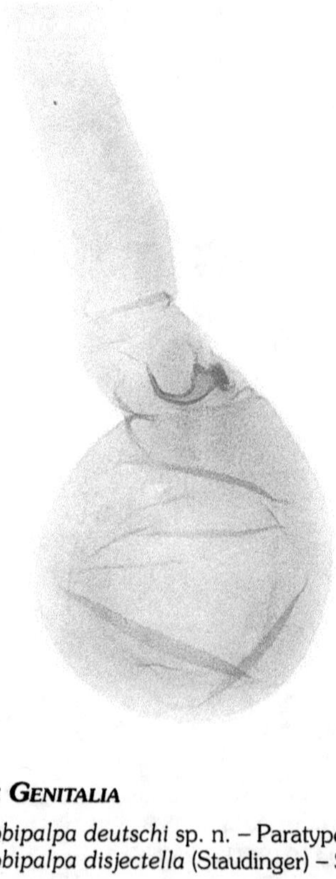

Female Genitalia

85. *Scrobipalpa deutschi* sp. n. – Paratype, Russia, gen. slide 3993 H.H. (NUPP).
86. *Scrobipalpa disjectella* (Staudinger) – Spain, GU 04/1212 P.H. (GRÜN).

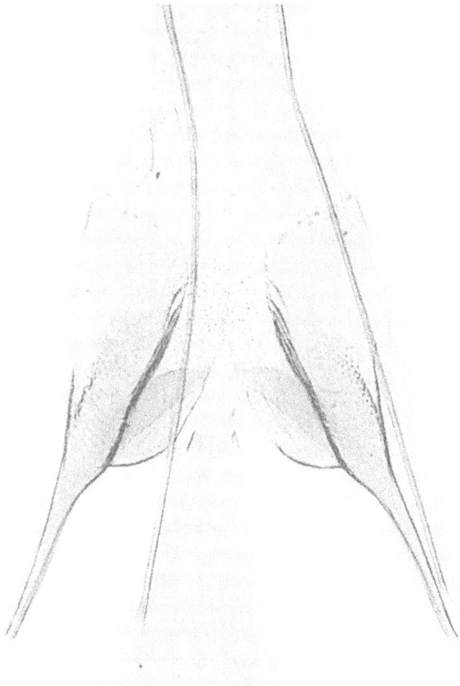

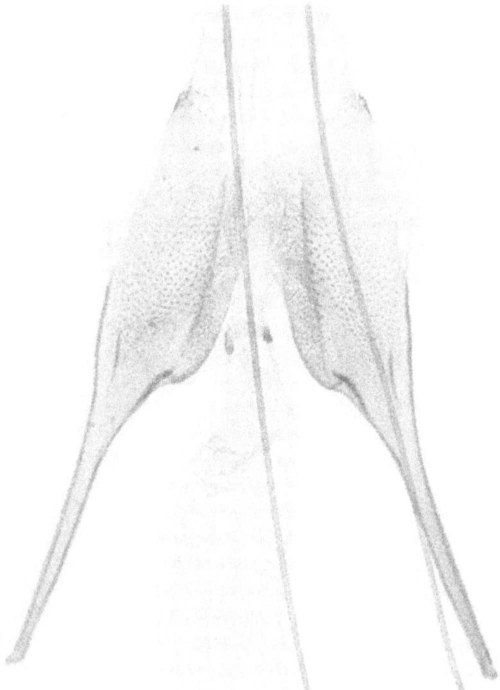

87

88

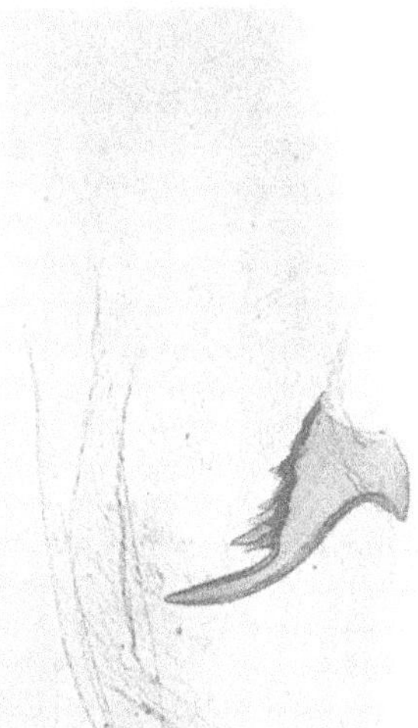

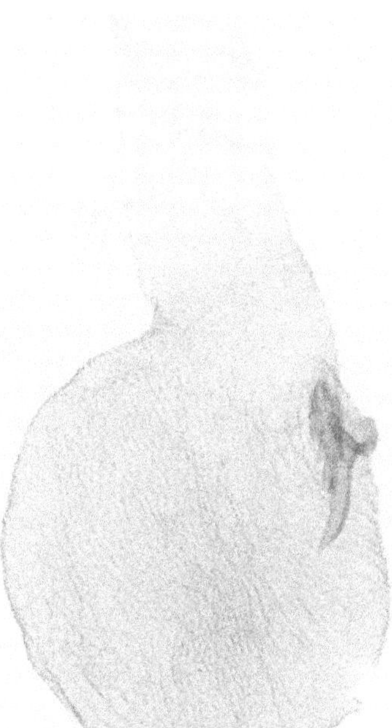

FEMALE GENITALIA

87. *Scrobipalpa mixta* sp. n. – Paratype, Macedonia, GU 03/1186 P.H. (ZSM).
88. *Scrobipalpa rebeli* (Preissecker) – Austria, GEL 978 P.H. (TLMF).

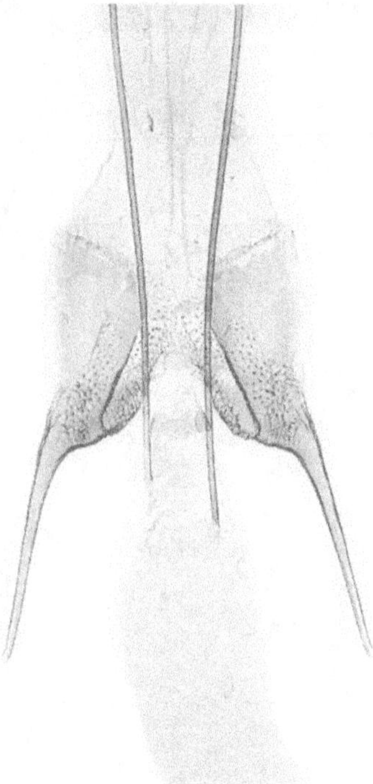

89

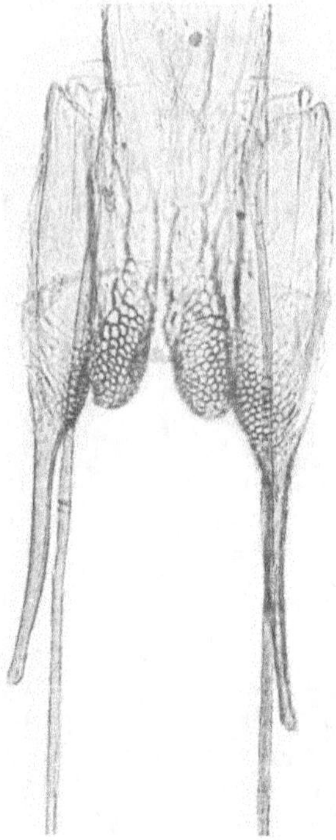

90

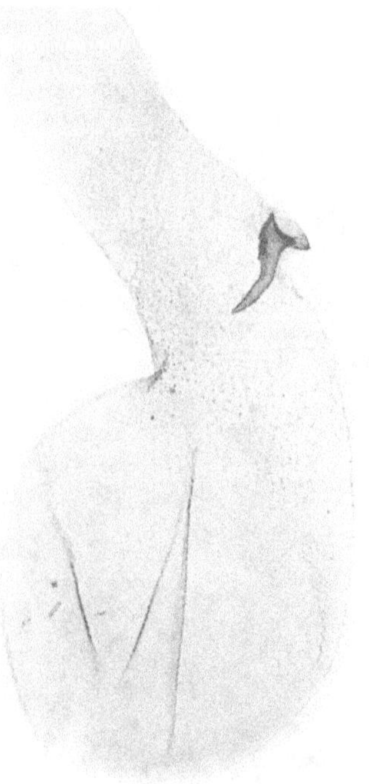

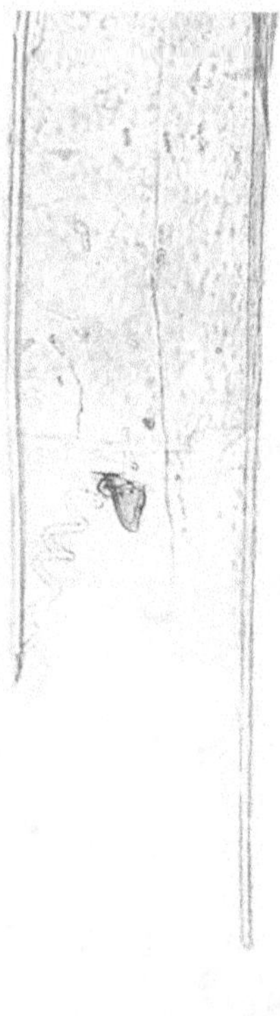

Female Genitalia

89. *Scrobipalpa gallicella* (Constant) – France, GEL 1116 P.H. (TLMF).
90. *Scrobipalpa ustulatella* (Staudinger) – Paralectotype, Russia, gen. slide St. 738 Povolný (MZMB).

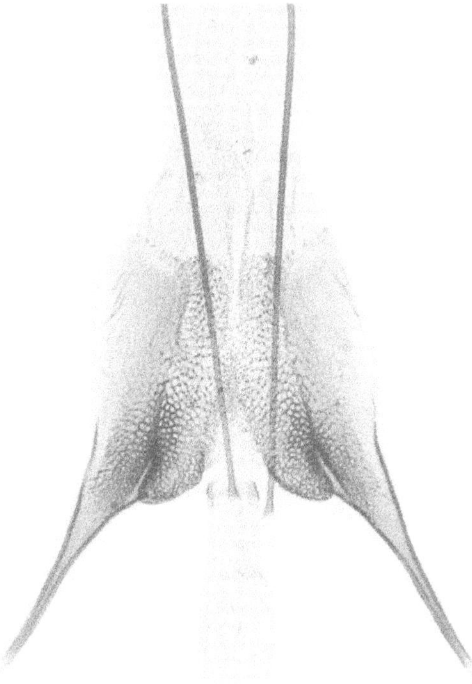

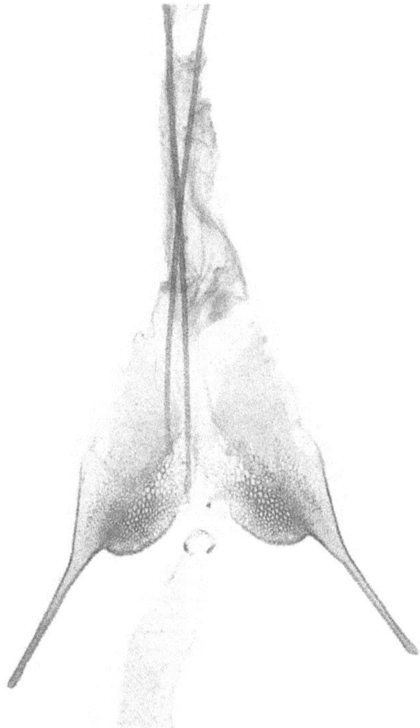

91 91a

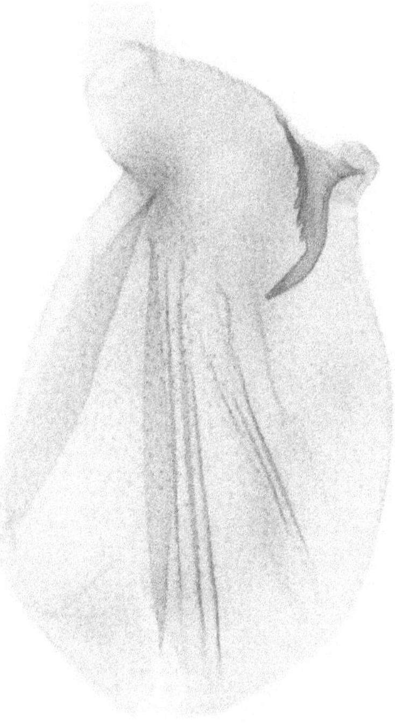

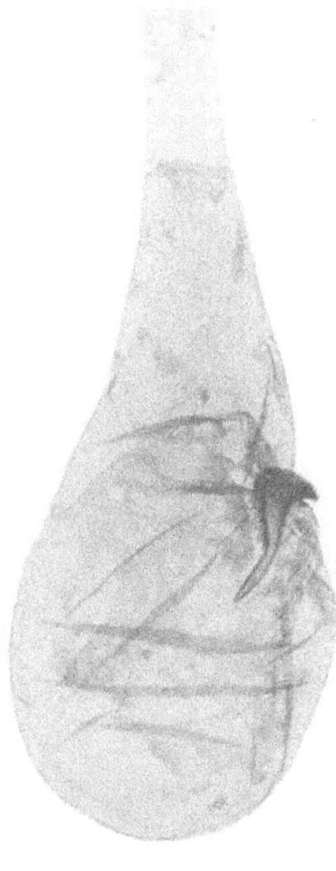

Female Genitalia

91. *Scrobipalpa postulatella* sp. n. – Holotype, Spain, GU 01/1074 (ZMUC)
91a. *Scrobipalpa postulatella* sp. n. – Spain, gen. slide 6343 H.H. (ZMUC)

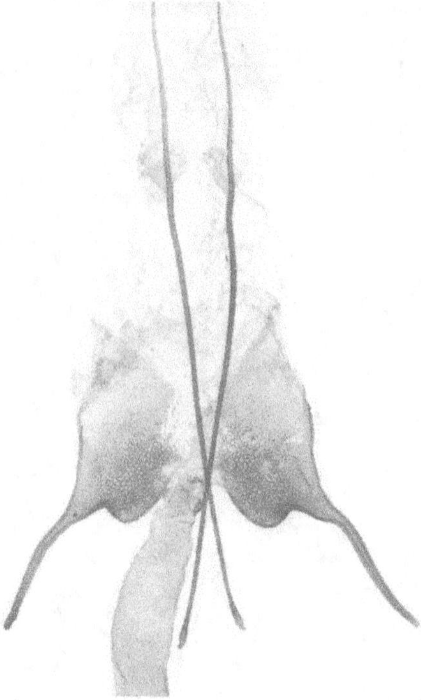

91b

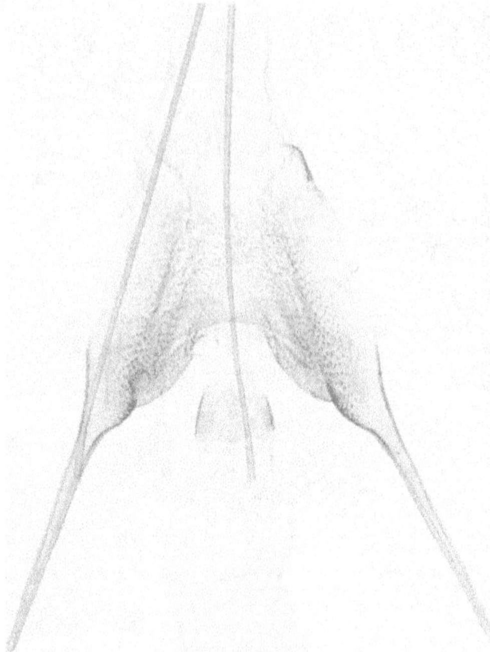

92

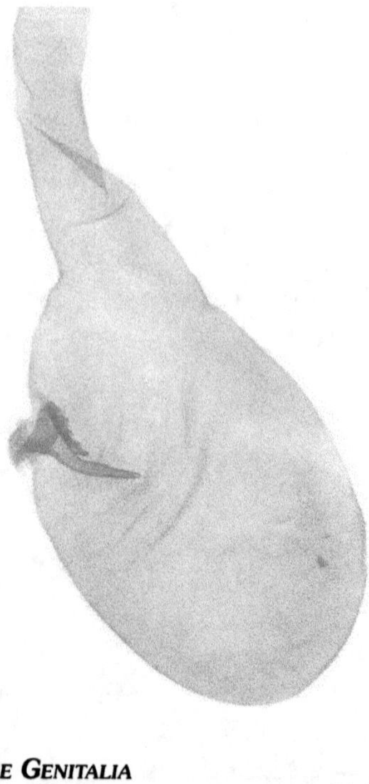

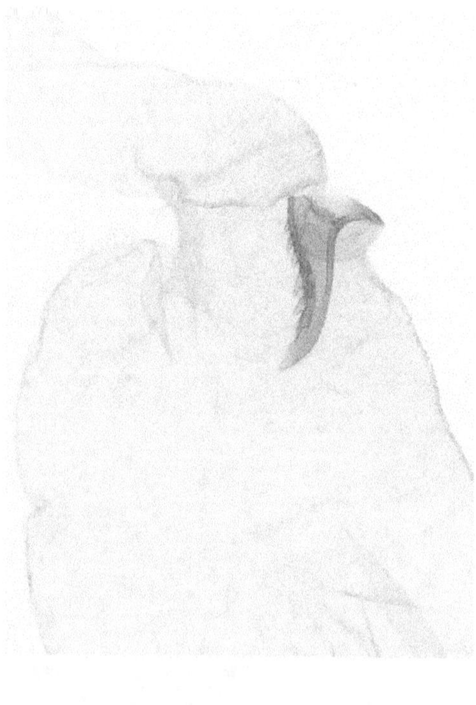

Female Genitalia

91b. *Scrobipalpa postulatella* sp. n. – Spain, gen. slide 6280 H.H. (ZMUC)
92. *Scrobipalpa nitentella* (Fuchs) – Denmark, GEL 945 P.H. (TLMF).

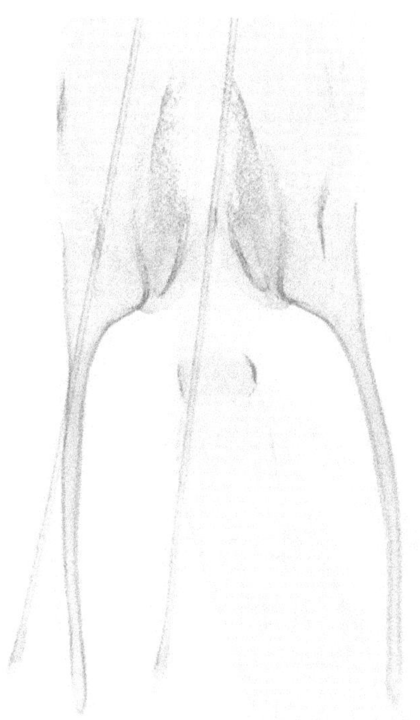

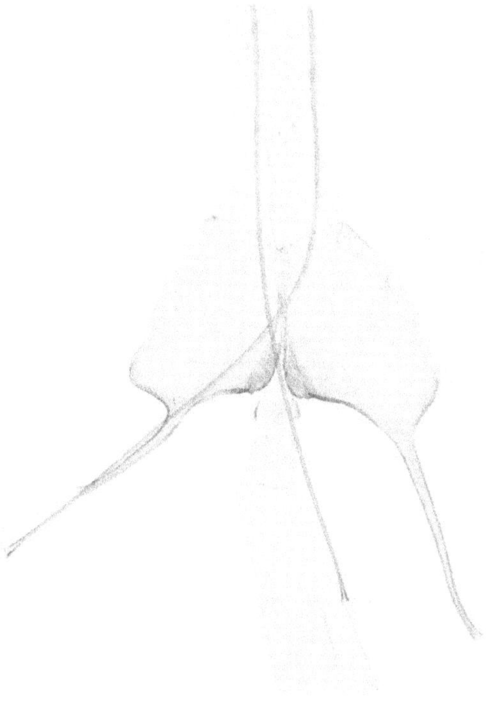

93

94

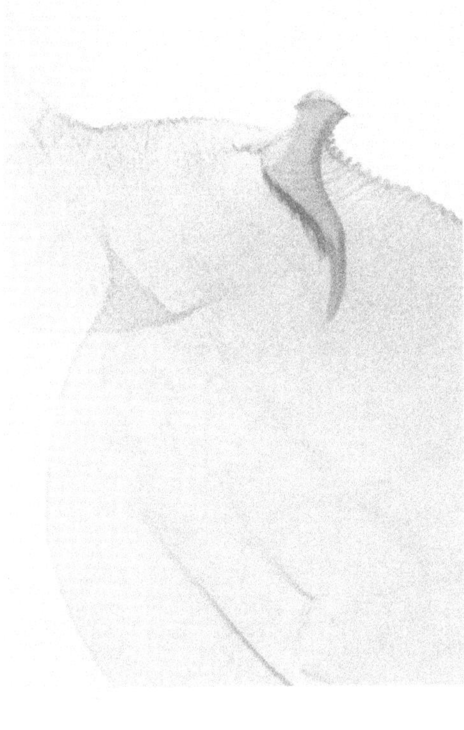

FEMALE GENITALIA

93. *Scrobipalpa costella* (Humphreys & Westwood) – Denmark, GEL 985 P.H. (TLMF).
94. *Scrobipalpa hyoscyamella* (Stainton) – Austria, GEL 1001 P.H. (TLMF).

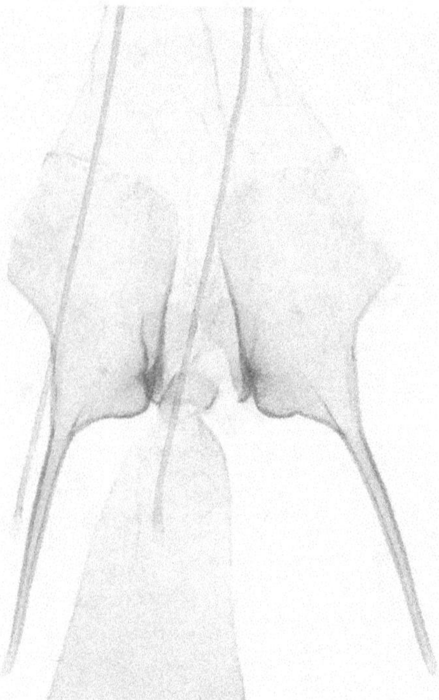

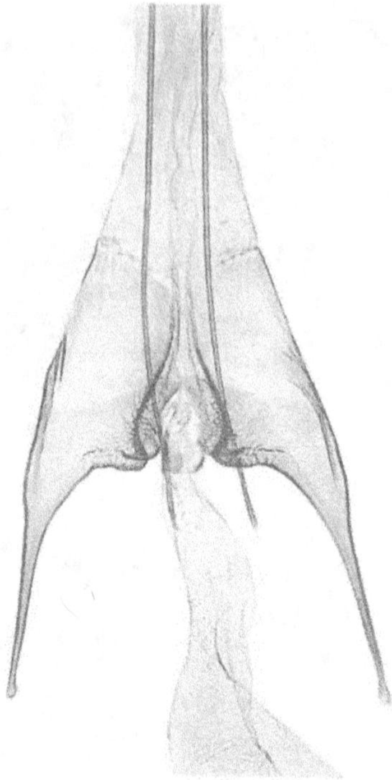

95

96

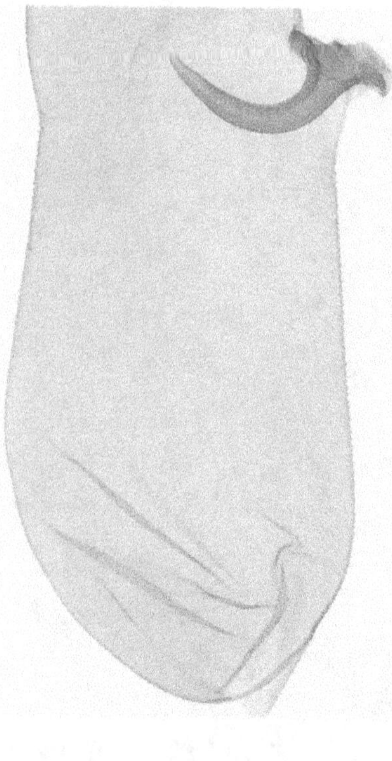

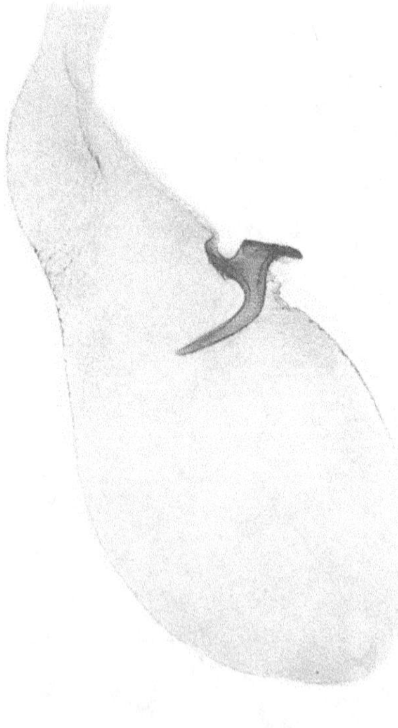

FEMALE GENITALIA

95. *Scrobipalpa portosanctana* (Stainton) – Madeira, GU 01/1057 P.H. (ZMUC).
96. *Scrobipalpa vicaria* (Meyrick) – Spain, GEL 1120 P.H. (TLMF).

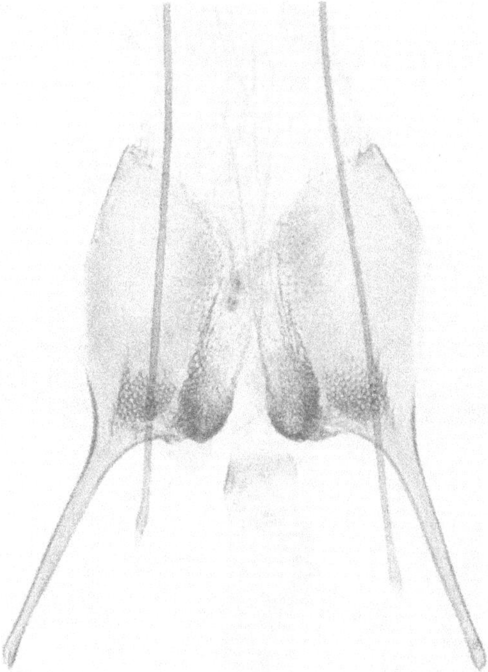

97

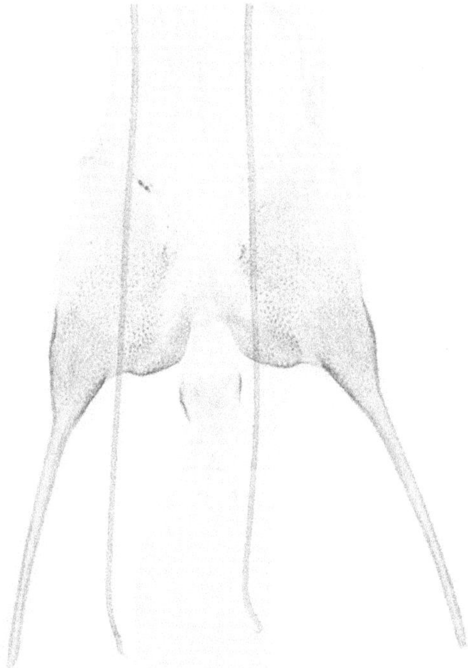

98

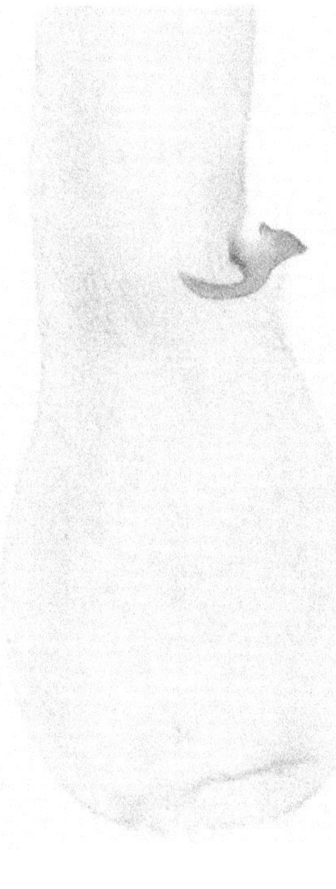

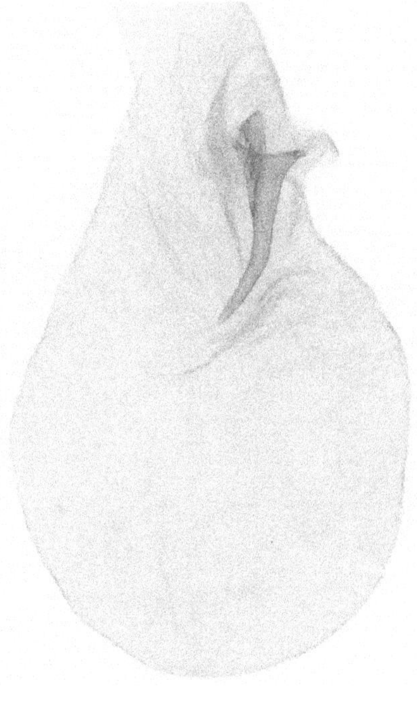

Female Genitalia

97. *Scrobipalpa ocellatella* (Boyd) – Italy, GEL 950 P.H. (TLMF).
98. *Scrobipalpa pulchra* Povolný – Turkey, GU 01/1043 P.H. (ZMUC).

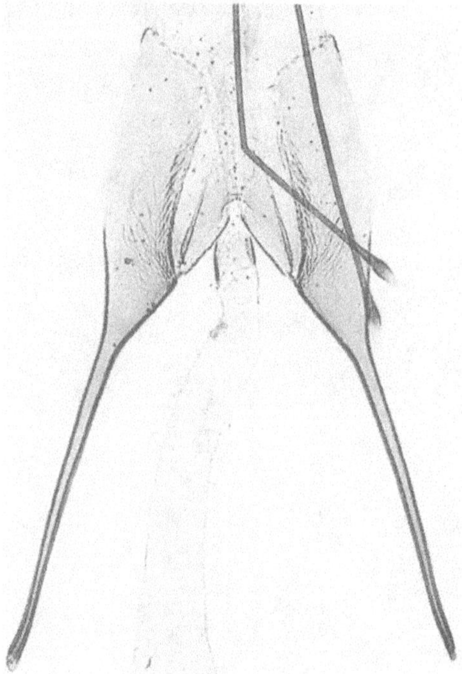

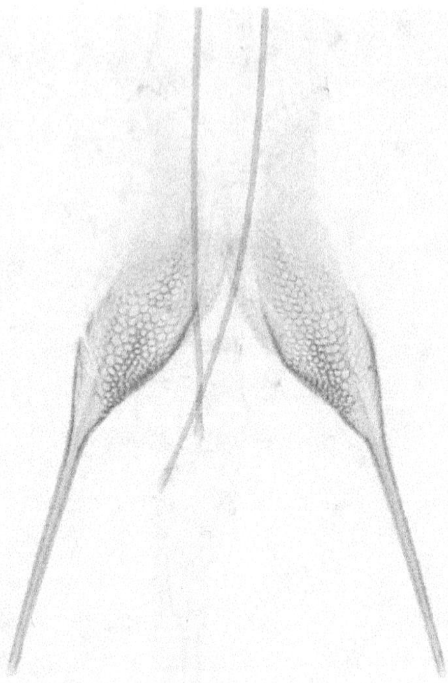

99

100

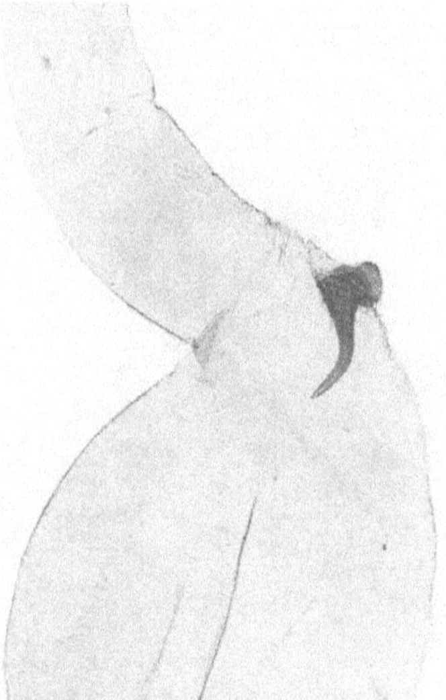

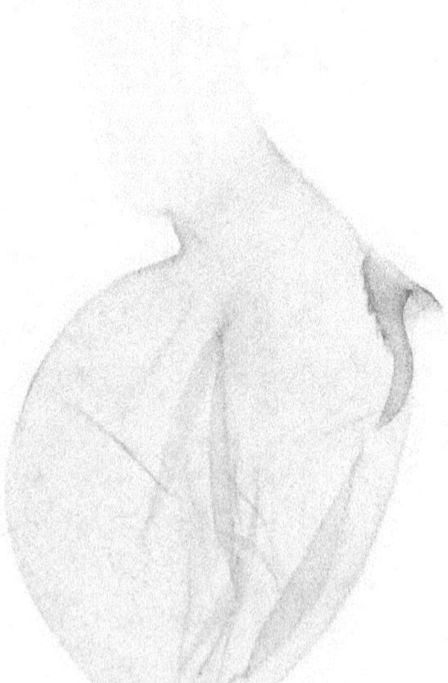

Female Genitalia

99. *Scrobipalpa gecko* (Walsingham) – Greece, GU 97/689 (ZMUC)
100. *Scrobipalpa hannemanni* Povolný – Russia, GU 01/1090 P.H. (JUNN).

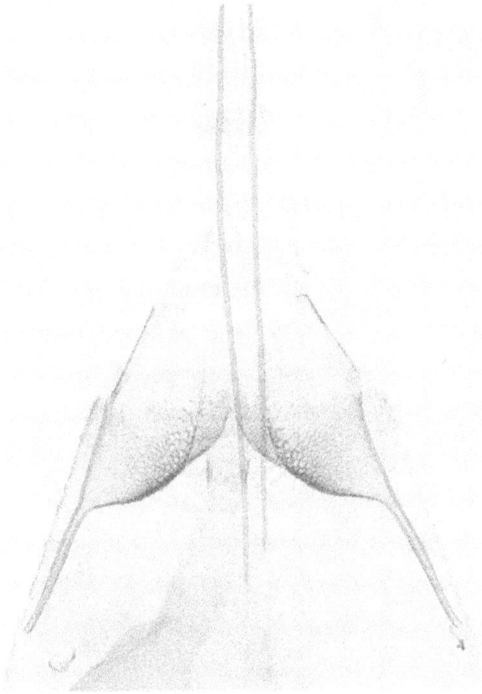

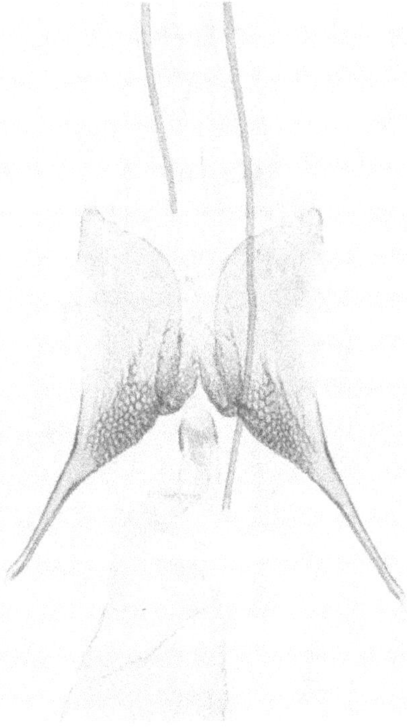

101 *102*

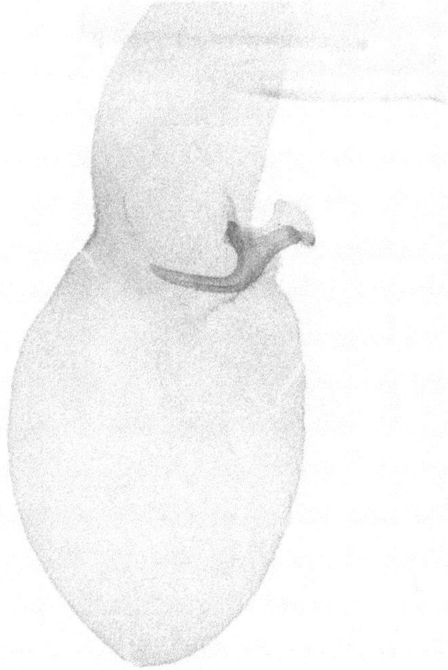

FEMALE GENITALIA

101. *Scrobipalpa erichi* Povolný – Ukraine, gen. slide 3668 O.K. (ZMUC).
102. *Scrobipalpa divisella* (Rebel) – Spain, GU 04/1208 P.H. (ZMUC).

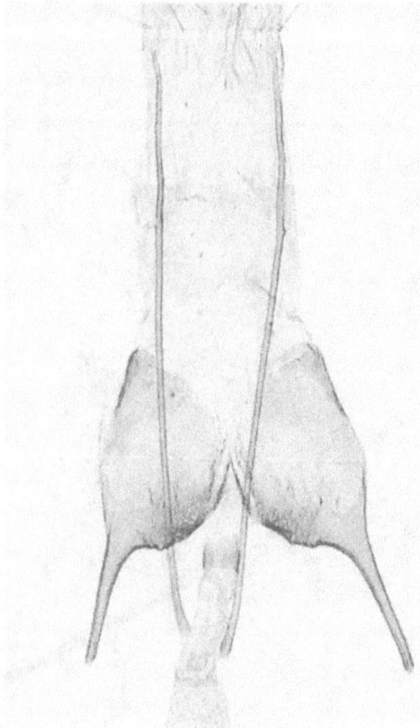

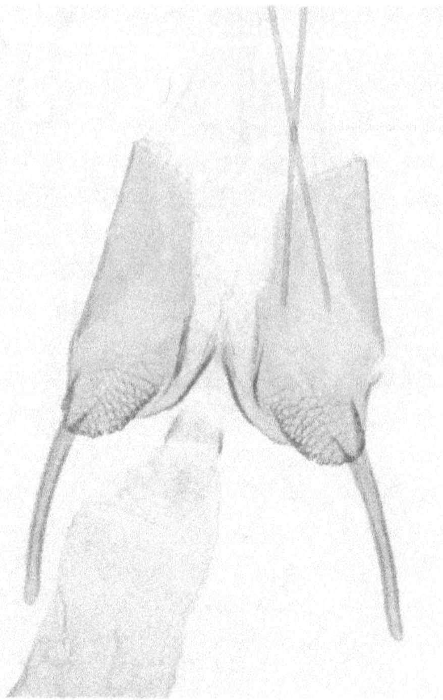

103

104

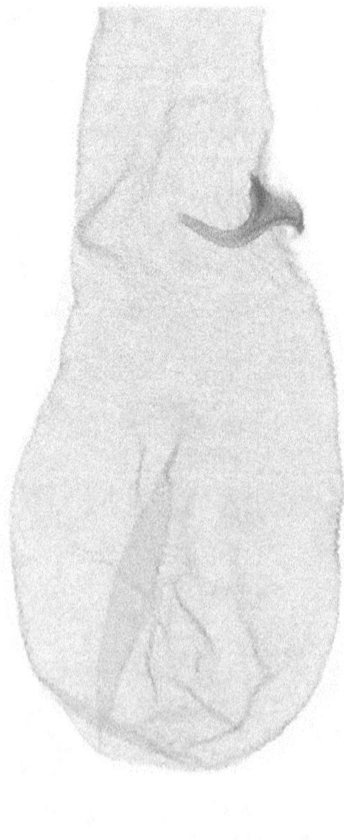

FEMALE GENITALIA

103. *Scrobipalpa voltinella* (Chrétien) – Paralectotype, France, GU 06/1246 (MNHN).
104. *Scrobipalpa suaedicola* (Mabille) – France, GU 01/1009 P.H. (ZMUC).

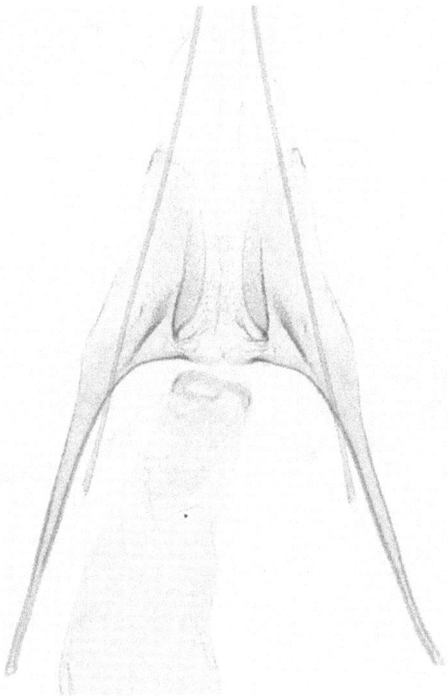
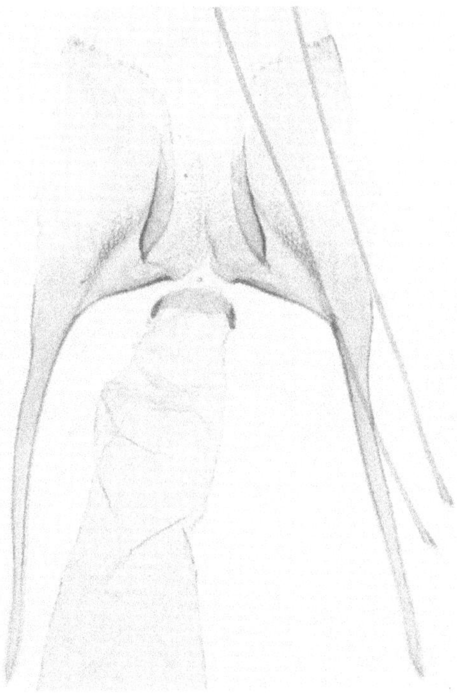

105 106

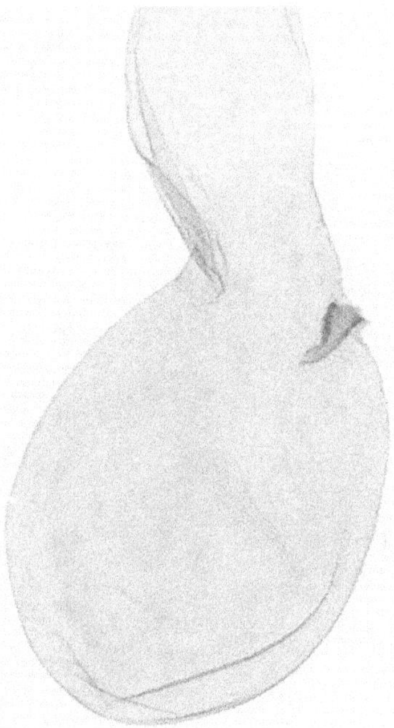
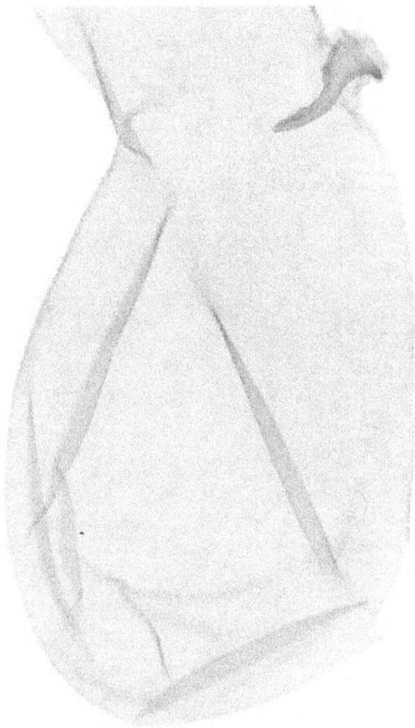

Female Genitalia

105. *Scrobipalpa monochromella* (Constant) – Spain, GU 99/881 P.H. (ZSM).
106. *Scrobipalpa samadensis* (Pfaffenzeller) – Austria, GEL 943 P.H. (TLMF).

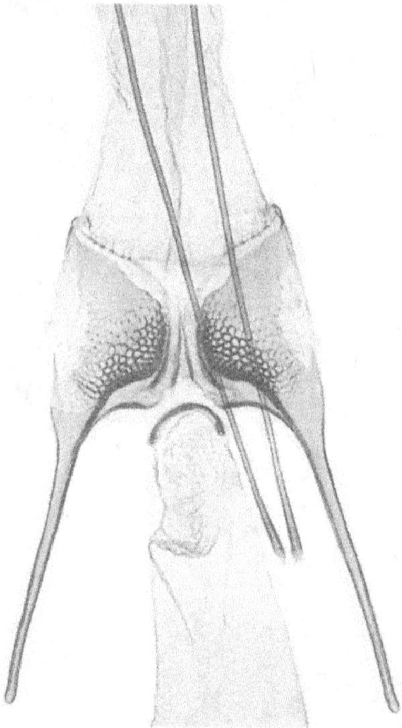

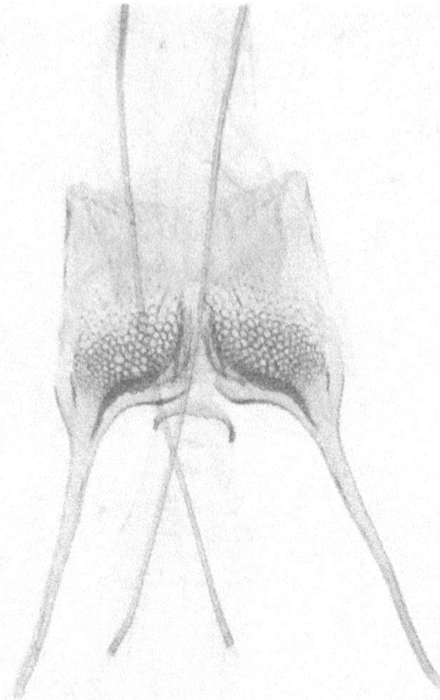

107

107a

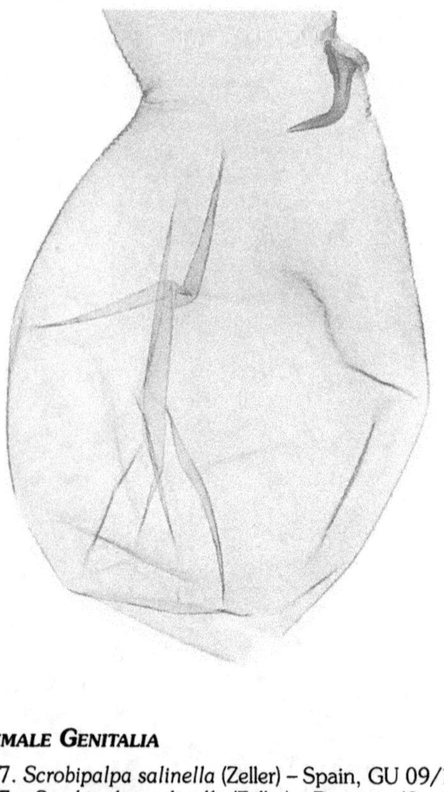

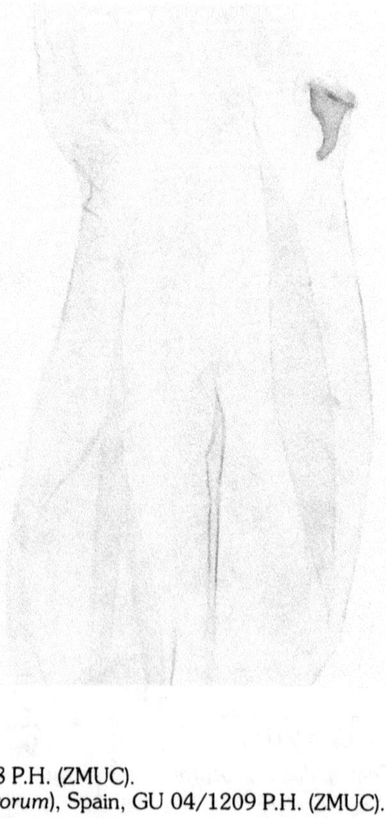

FEMALE GENITALIA

107. *Scrobipalpa salinella* (Zeller) – Spain, GU 09/1278 P.H. (ZMUC).
107a. *Scrobipalpa salinella* (Zeller) – Paratype (*S. glaserorum*), Spain, GU 04/1209 P.H. (ZMUC).

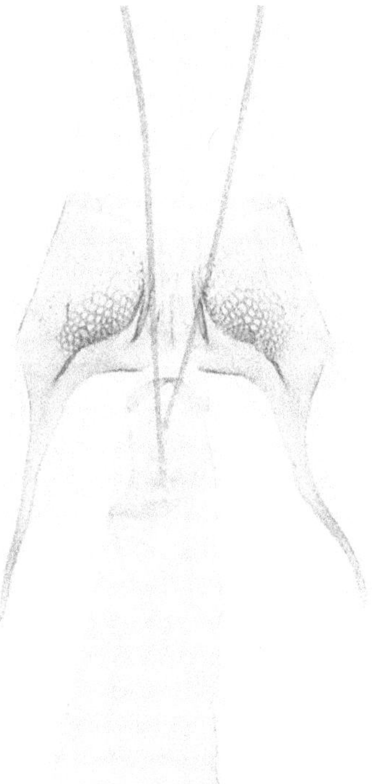

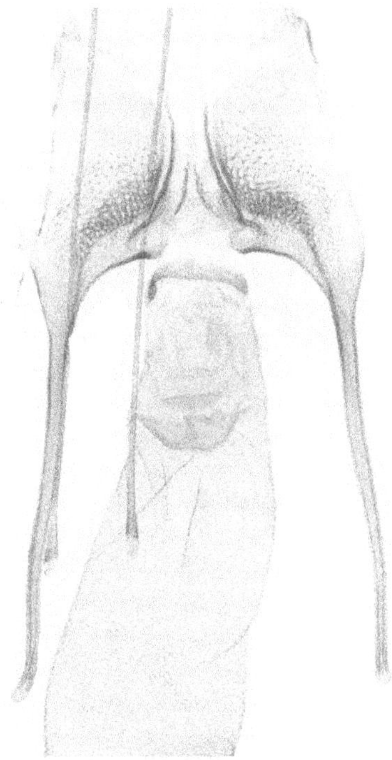

108

109

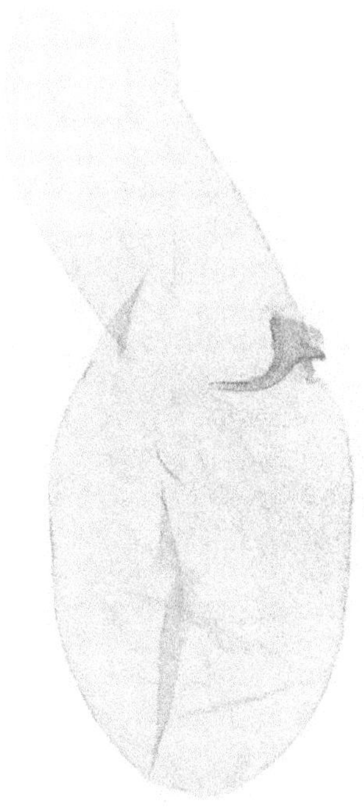

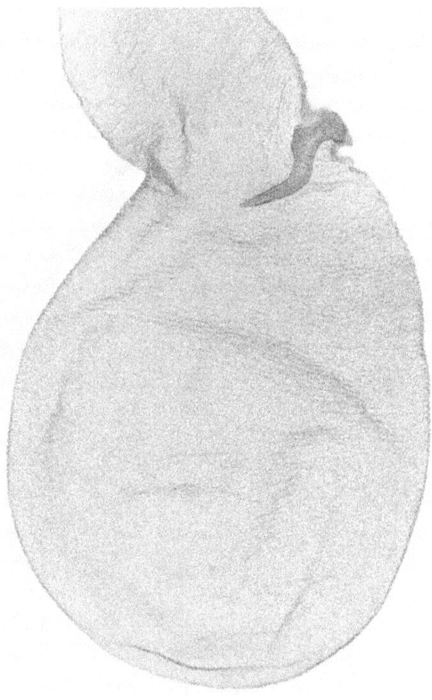

FEMALE GENITALIA

108. *Scrobipalpa spergulariella* (Chrétien) – France, GEL 983 P.H. (TLMF).
109. *Scrobipalpa salicorniae* (Hering) – Austria, GEL 941 P.H. (TLMF).

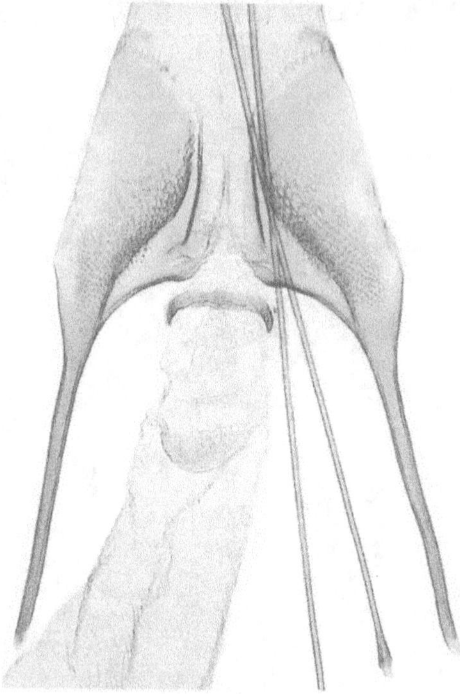

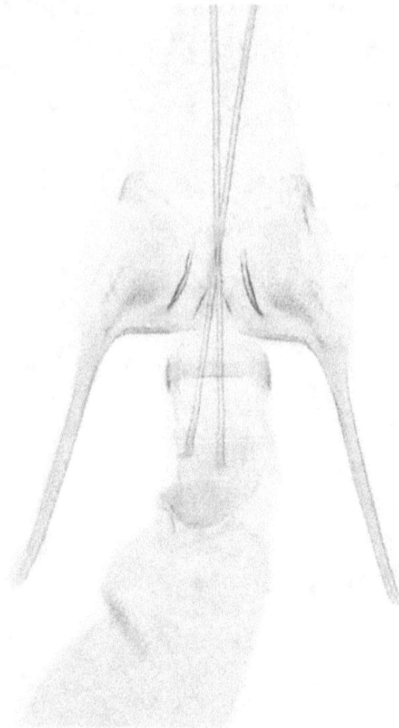

110

111

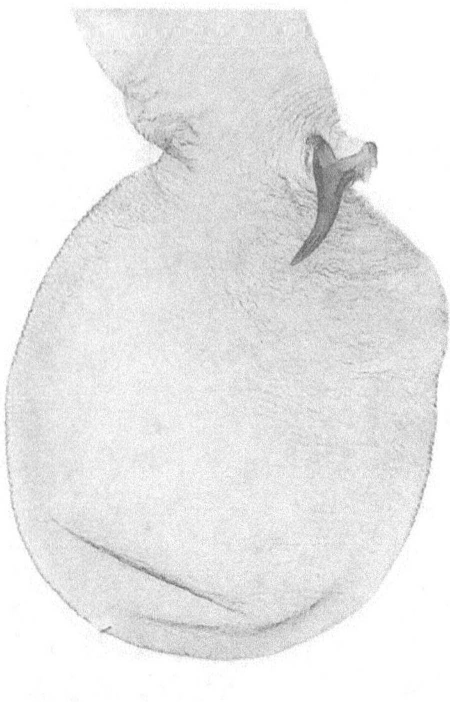

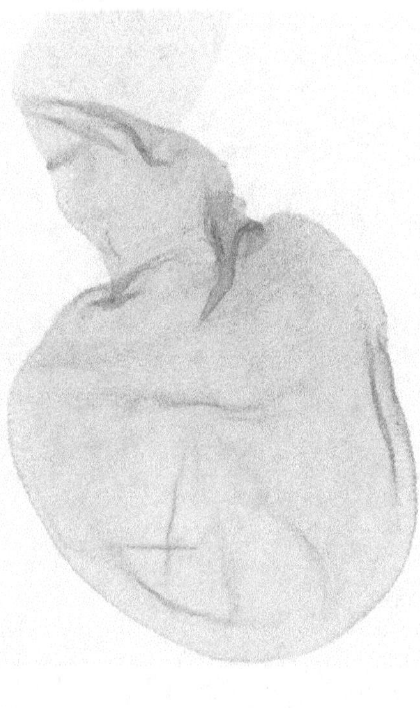

Female Genitalia

110. *Scrobipalpa halimioniella* sp. n. – Paratype, France, GU 09/1276 P.H. (ZMUC).
111. *Scrobipalpa thymelaeae* (Amsel) – Italy, gen. slide 16522 (NHMW).

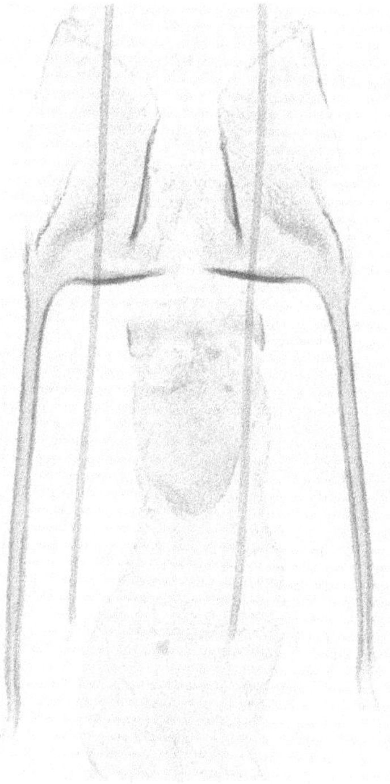

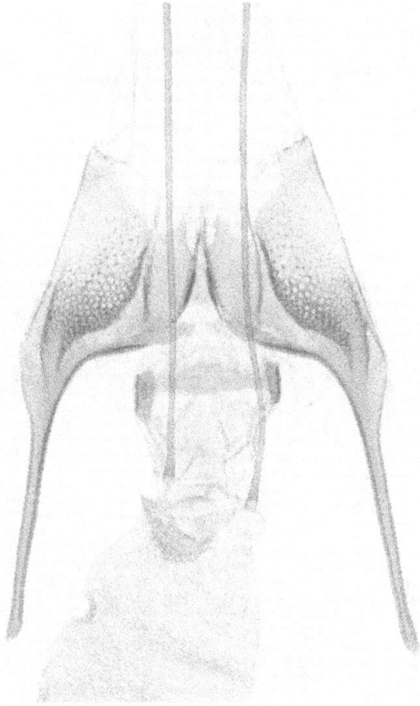

112 *113*

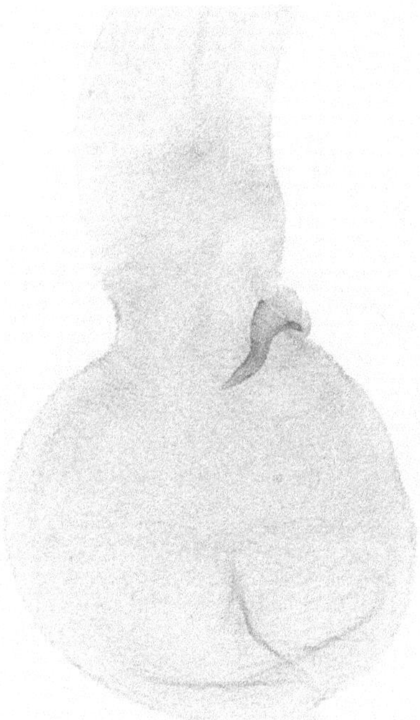

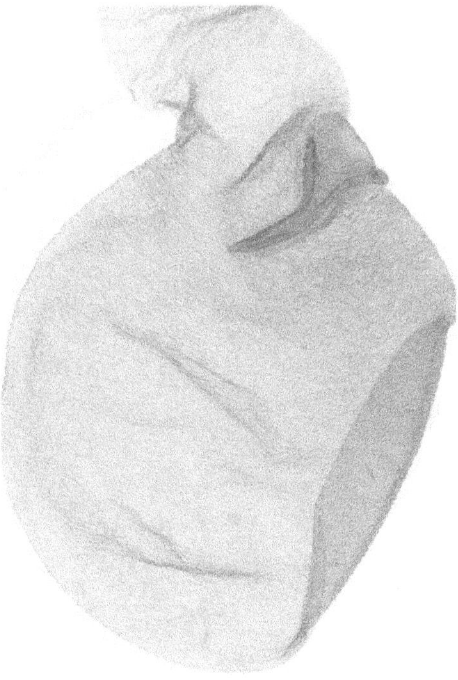

FEMALE GENITALIA

112. *Scrobipalpa halymella* (Millière) – France, GU 01/1012 P.H. (ZMUC).
113. *Scrobipalpa camphorosmella* Nel – Paratype, France, GEL 981 P.H. (TLMF).

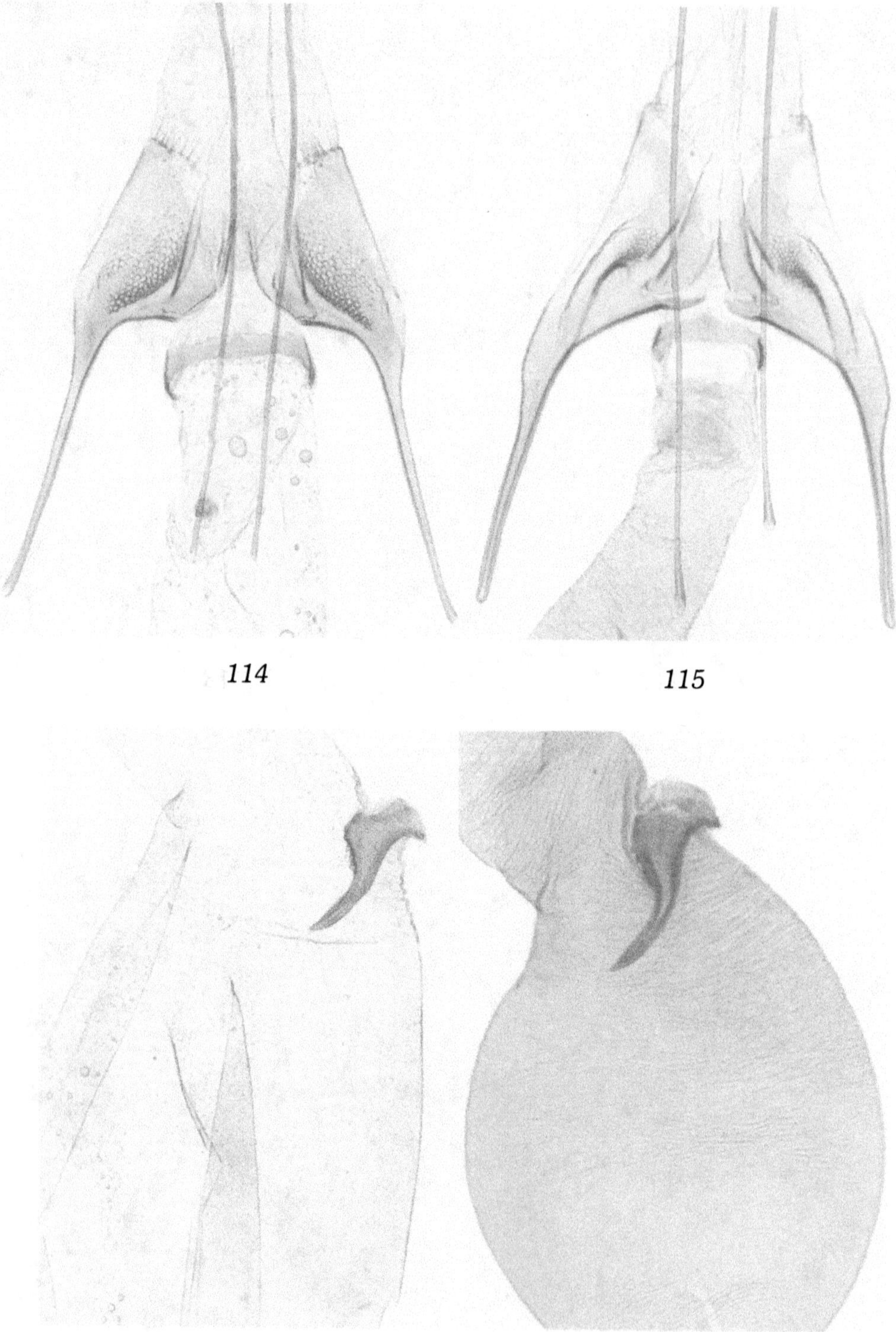

114 115

FEMALE GENITALIA

114. *Scrobipalpa stabilis* Povolný – Paratype, Spain, GU 04/1210 P.H. (LNK).
115. *Scrobipalpa instabilella* (Douglas) – Greece, GU 01/1034 P.H. (GRÜN).

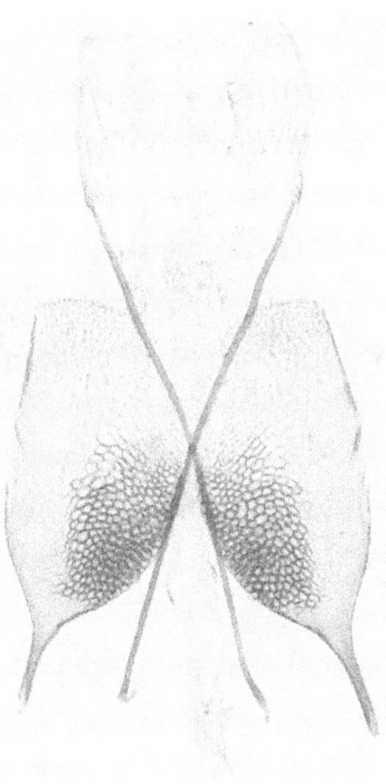

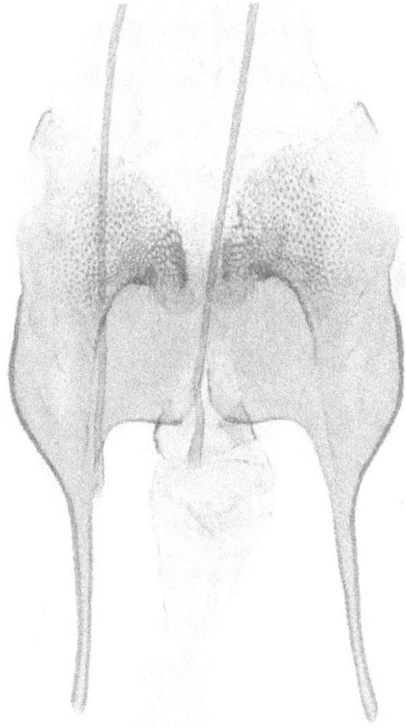

116

117

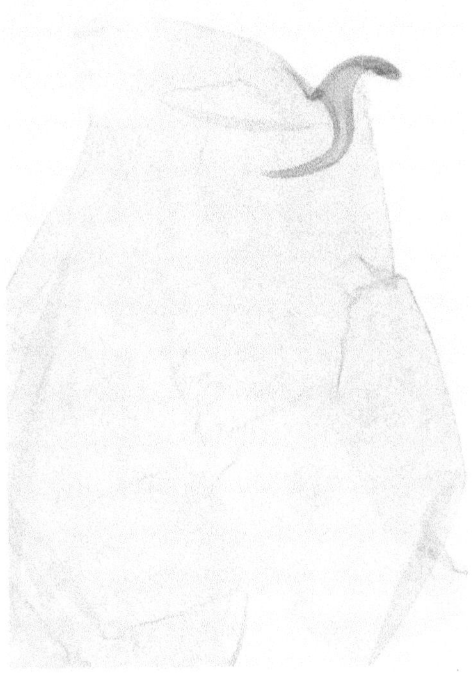

FEMALE GENITALIA

116. *Scrobipalpa peterseni* (Povolný) – Tunesia, gen. slide 2985 H.H. (ZMUC).
117. *Scrobipalpa ergasima* (Meyrick) – Greece, GEL 990 P.H. (TLMF).

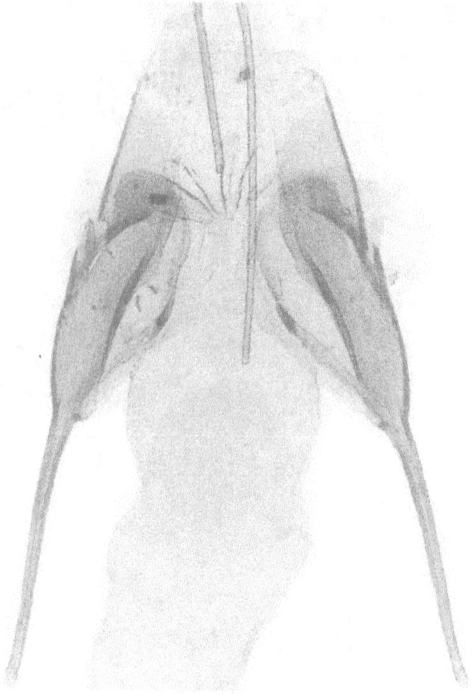

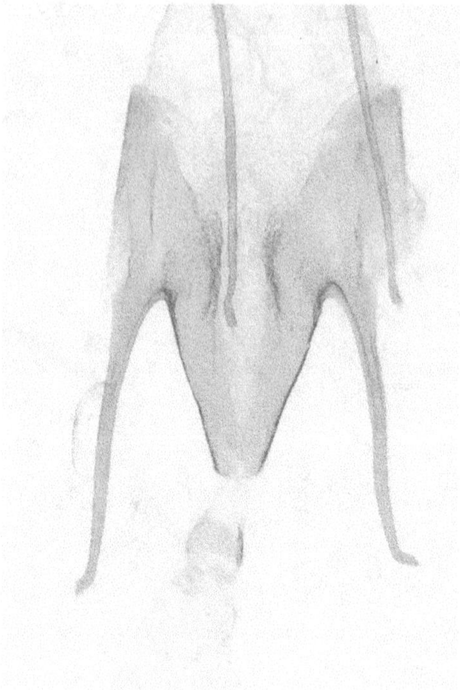

118 *119*

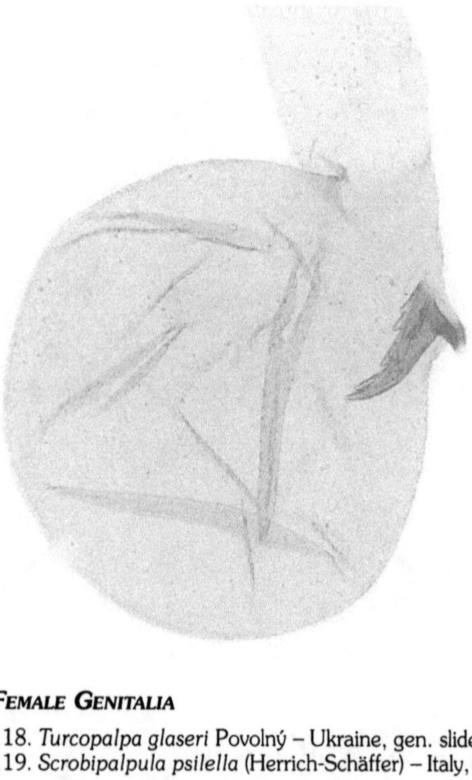

FEMALE GENITALIA

118. *Turcopalpa glaseri* Povolný – Ukraine, gen. slide 18/09 Bidzilya (ZMKU).
119. *Scrobipalpula psilella* (Herrich-Schäffer) – Italy, GEL 618 P.H. (TLMF).

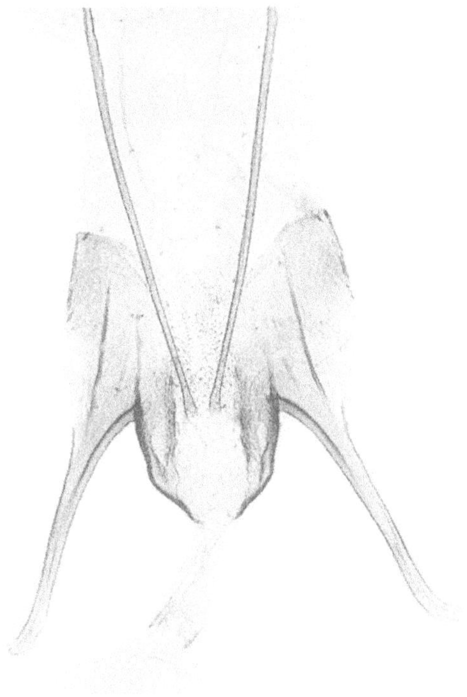
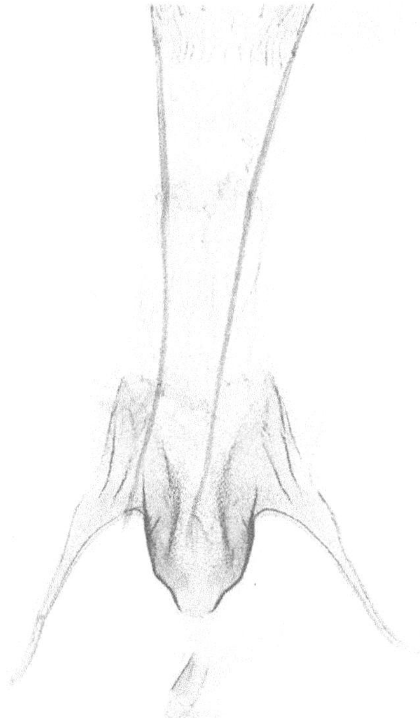

120 121

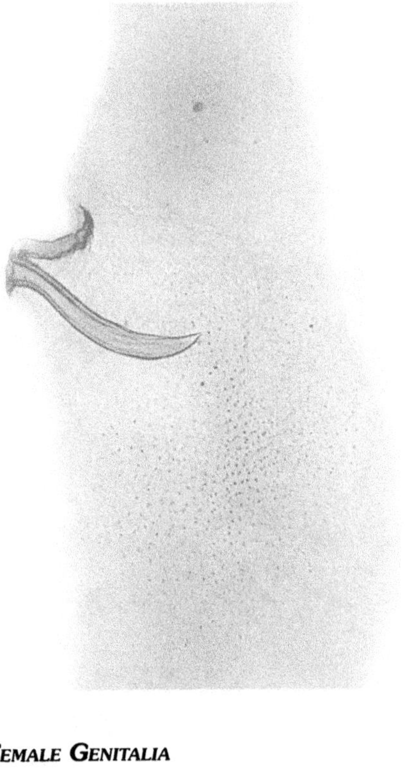
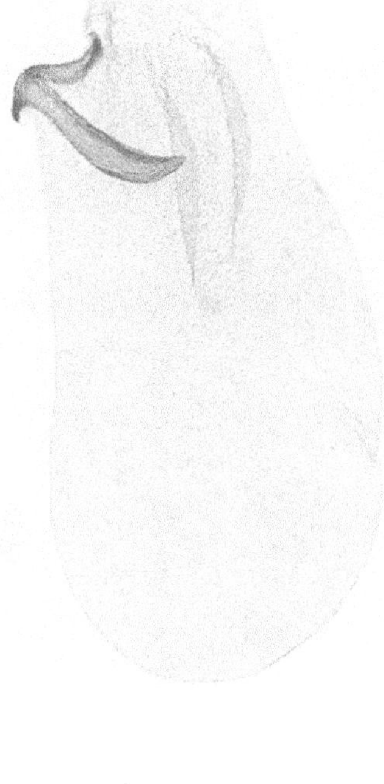

FEMALE GENITALIA

120. *Scrobipalpula ramosella* (Müller-Rutz) – Switzerland, GEL 627 P.H. (TLMF).
121. *Scrobipalpula seniorum* Povolný – Greece, gen. slide 16645 (NHMW).

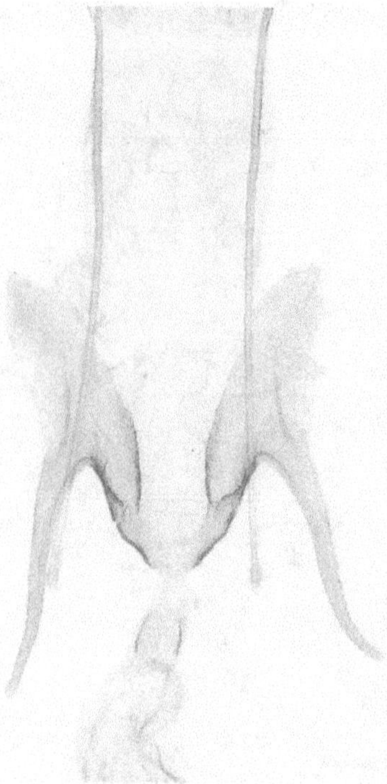

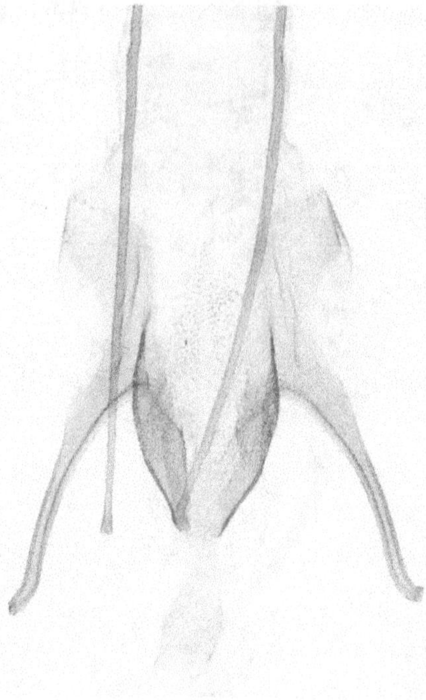

122

123

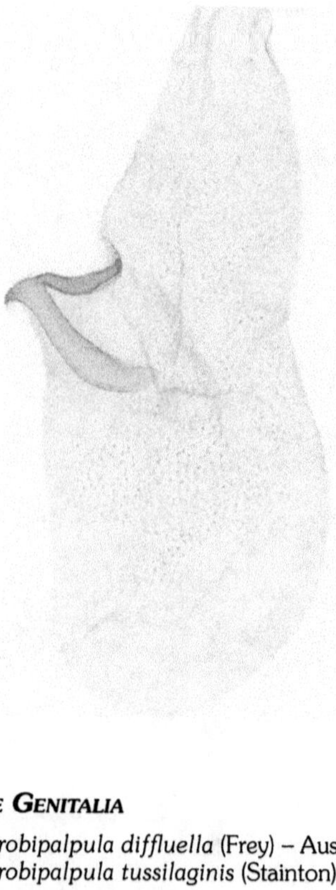

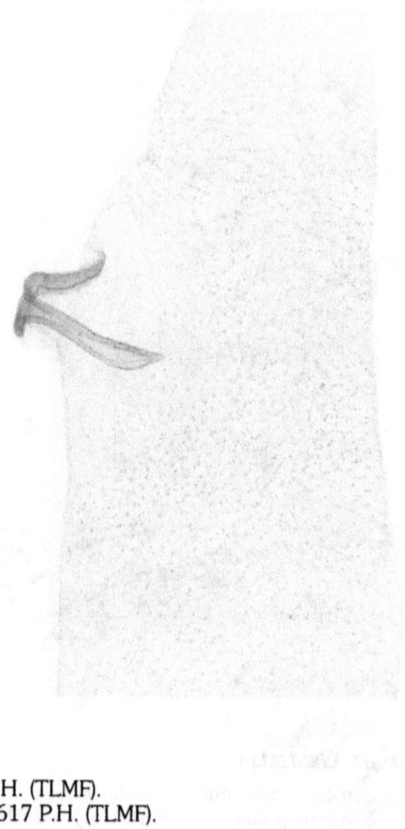

FEMALE GENITALIA

122. *Scrobipalpula diffluella* (Frey) – Austria, GEL 619 P.H. (TLMF).
123. *Scrobipalpula tussilaginis* (Stainton) – Austria, GEL 617 P.H. (TLMF).

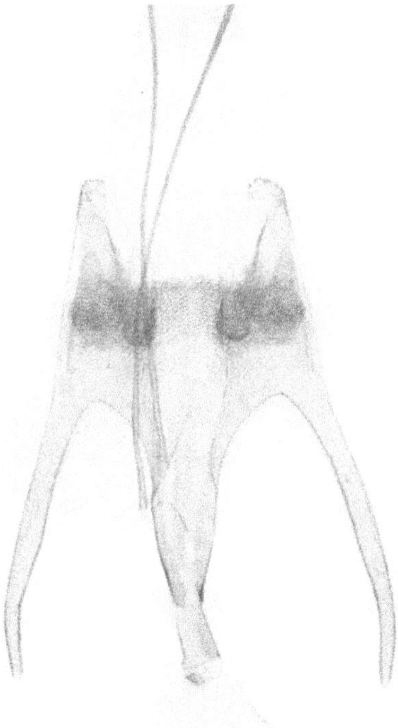

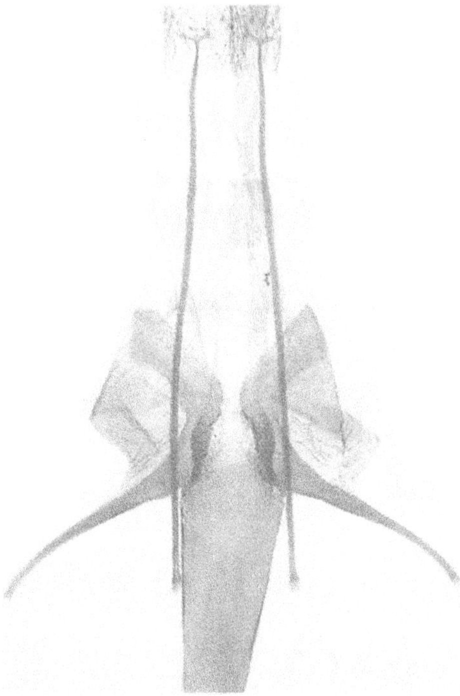

124

125

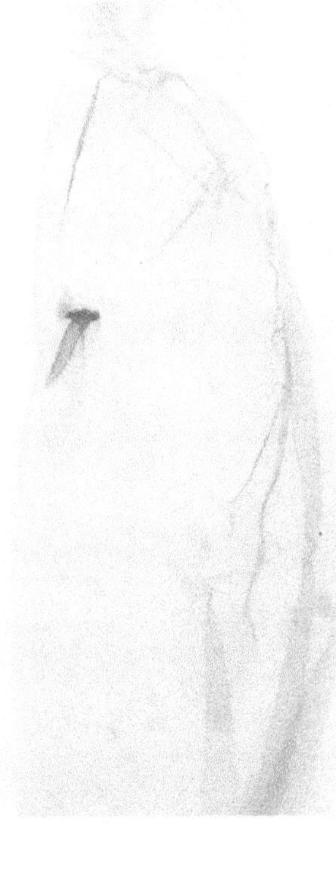

FEMALE GENITALIA

124. *Phthorimaea operculella* (Zeller) – Spain, GEL 1042 P.H. (TLMF).
125. *Tuta absoluta* (Meyrick) – Brasil, GU 08/1262 (ZMUC).

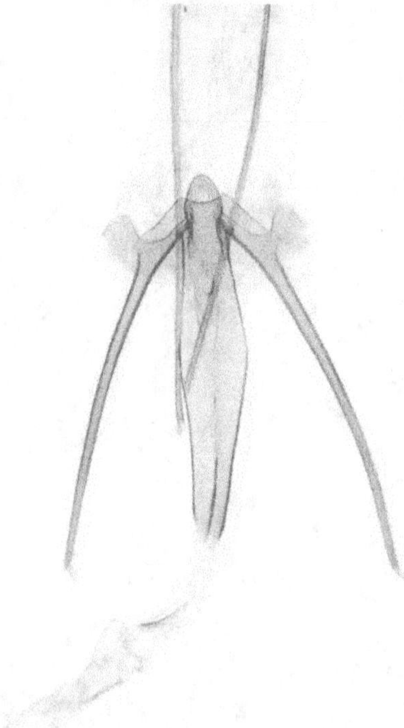

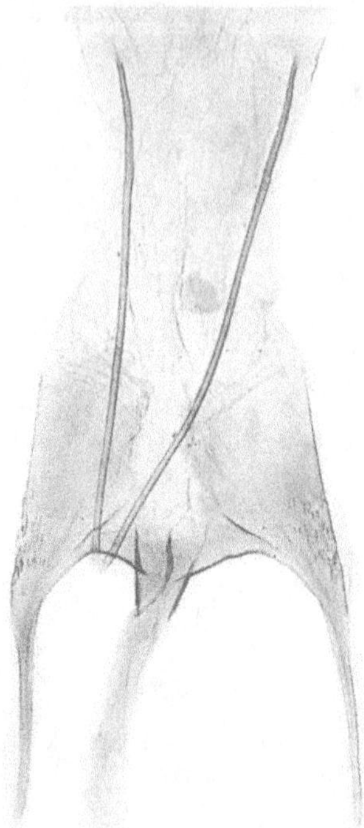

126

127

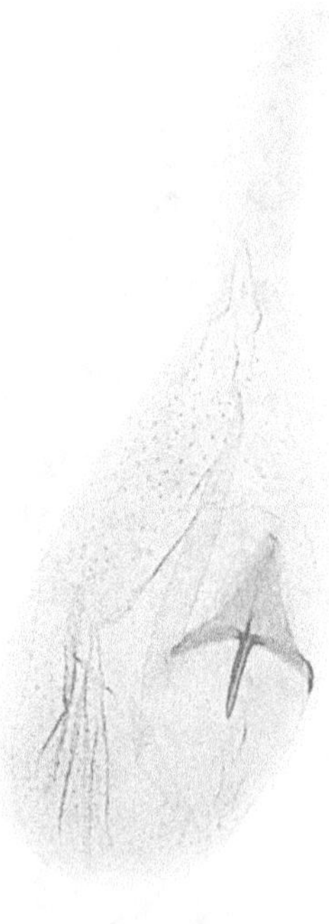

Female Genitalia

126. *Keiferia lycopsersicella* (Walsingham) – Dominicanic Republic, GU 09/1307 P.H. (ZMUC).
127. *Ephysteris promptella* (Staudinger) – Spain, GEL 1115 P.H. (TLMF).

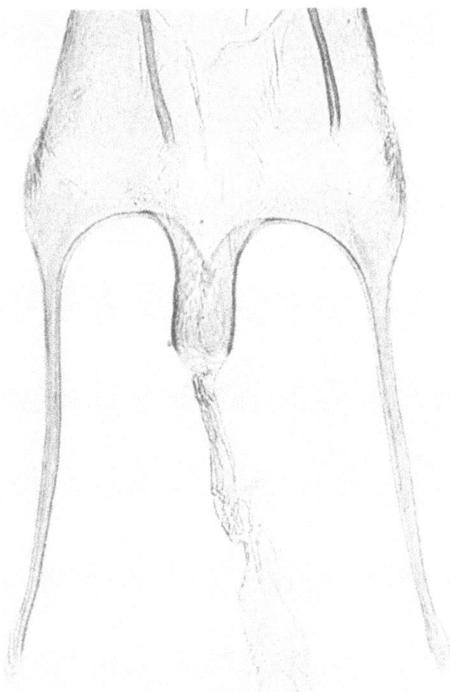

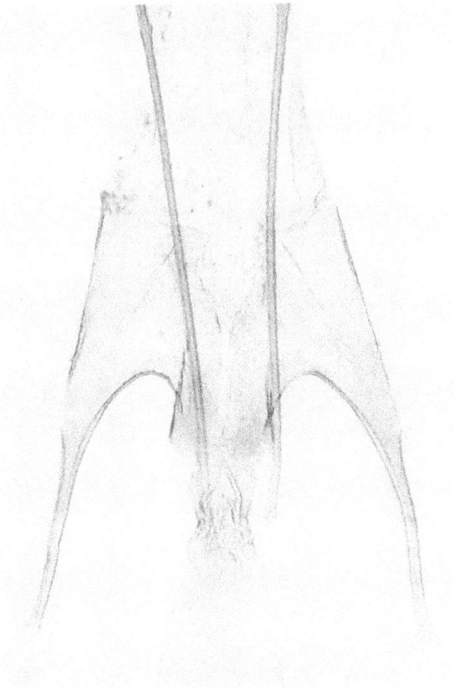

128 129

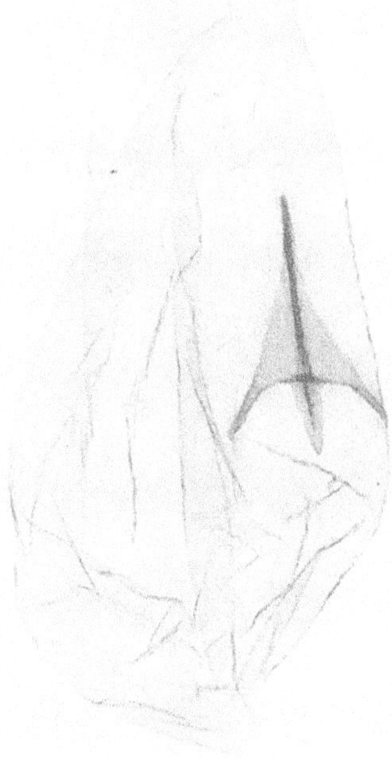

Female Genitalia

128. *Ephysteris tenuisaccus* Nupponen – Paratype, Russia, GU 09/1302 P.H. (ZMUC).
129. *Ephysteris deserticolella* (Staudinger) – Turkey, GU 06/1250 P.H. (ZSM).

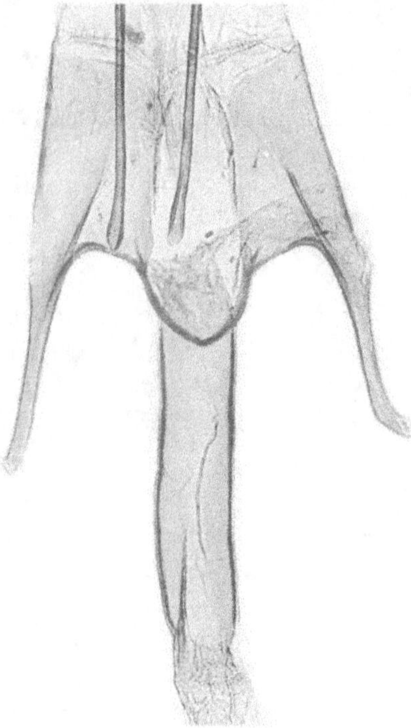

130

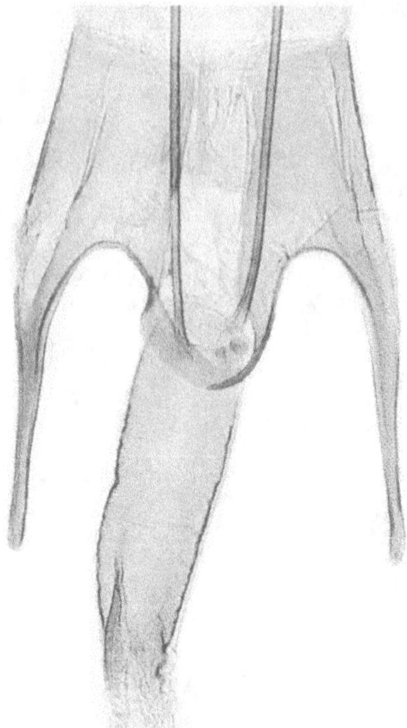

130a

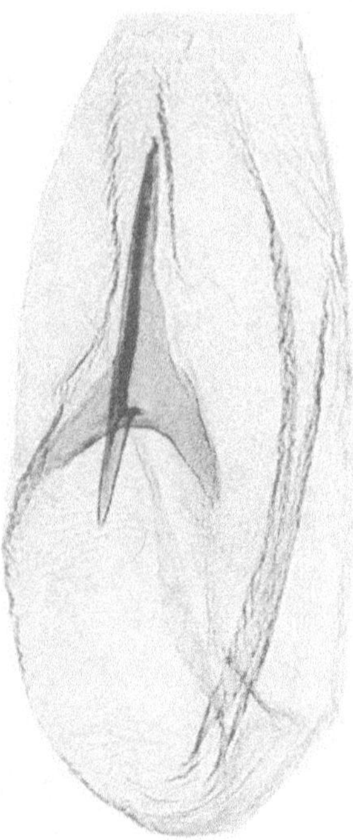

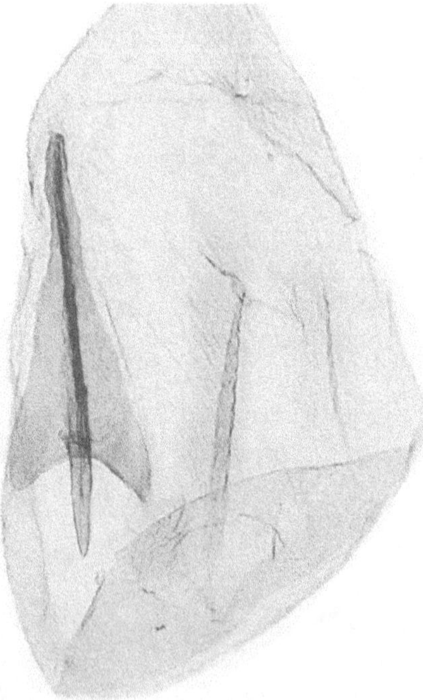

Female Genitalia

130. *Ephysteris insulella insulella* (Heinemann) – France, GEL 1124 P.H. (TLMF).
130a.*Ephysteris insulella praticolella* (Christoph) – Italy, GU 00/898 P.H. (ZSM).

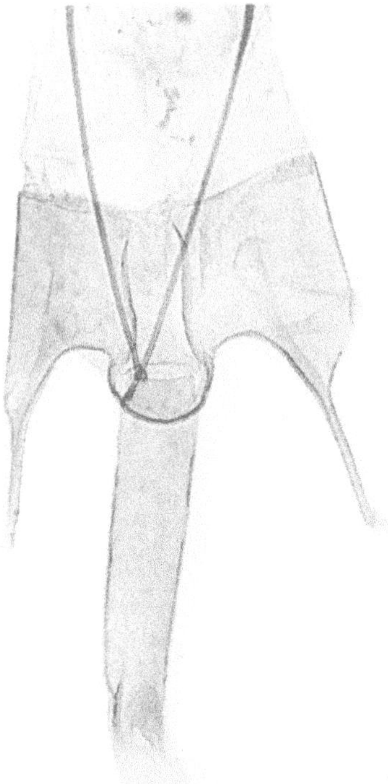

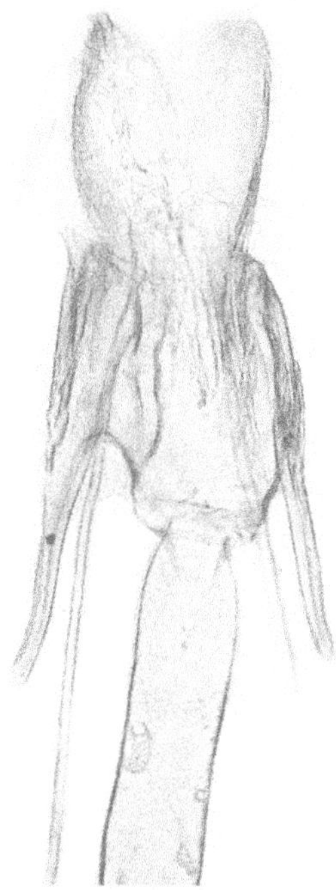

131

131a

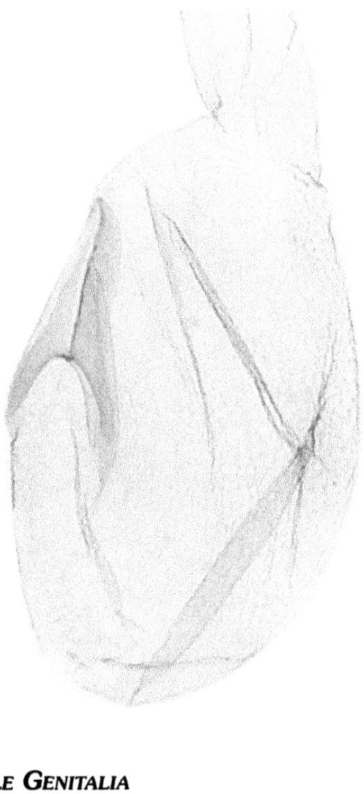

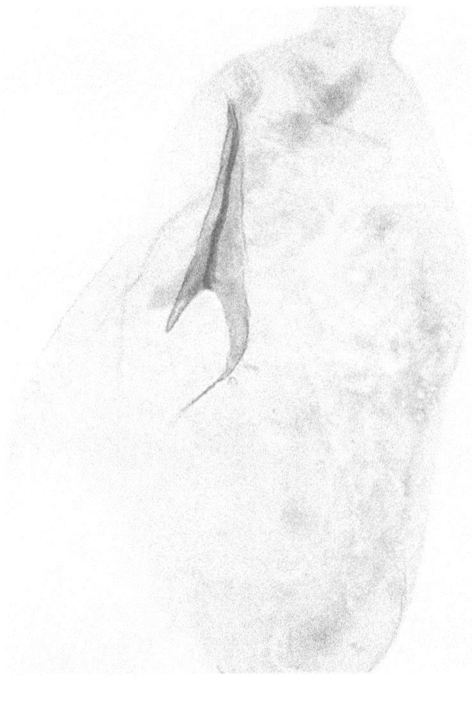

FEMALE GENITALIA

131. *Ephysteris diminutella* (Zeller) – Italy, GU 01/1062 P.H. (ZSM).
131a. *Ephysteris diminutella* (Zeller) – Paratype (*Ephysteris treskensis*), Mazedonia, 131 Ks. (MZMB).

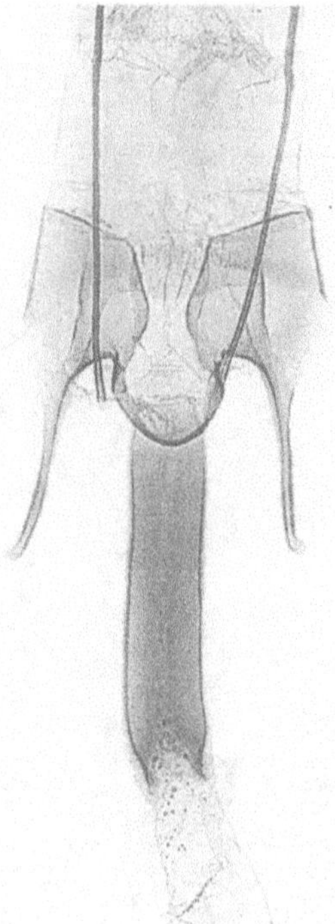

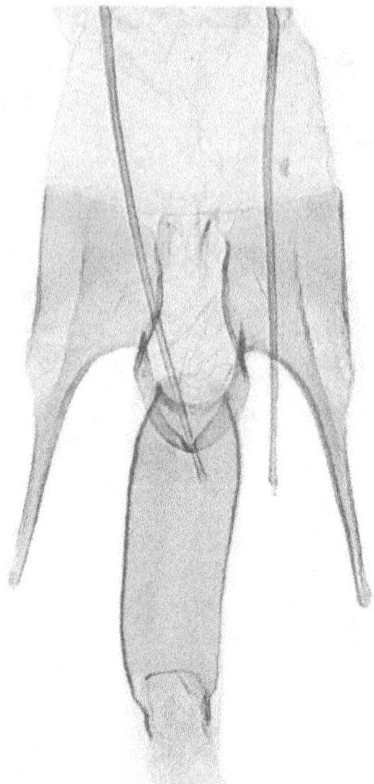

131b *131c*

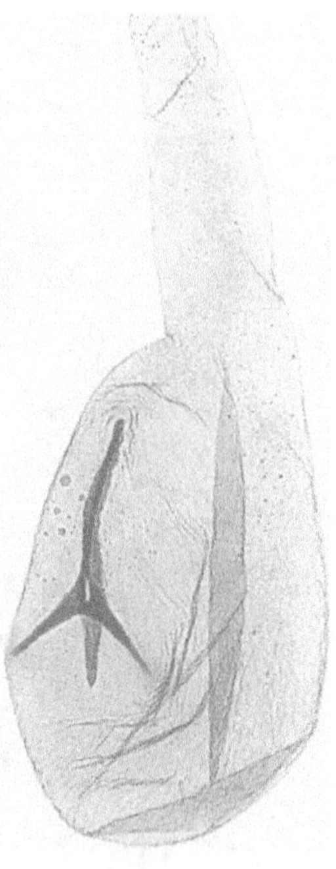

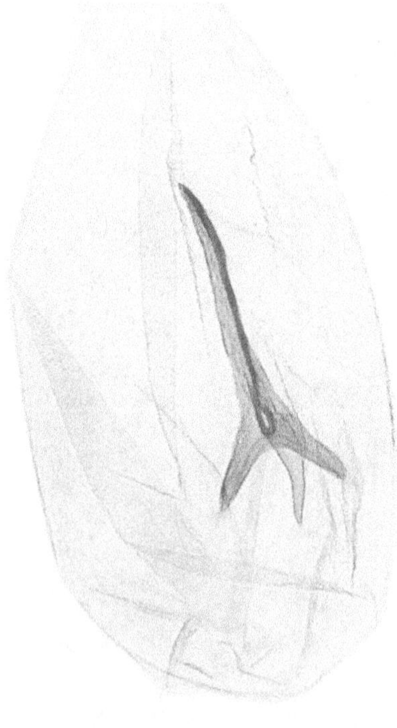

Female Genitalia

131b. *Ephysteris diminutella* (Zeller) – Italy, GU 02/1132 P.H. (ZMUC).
131c. *Ephysteris diminutella* (Zeller) – Spain, GU 08/1064 P.H. (ZMUC).

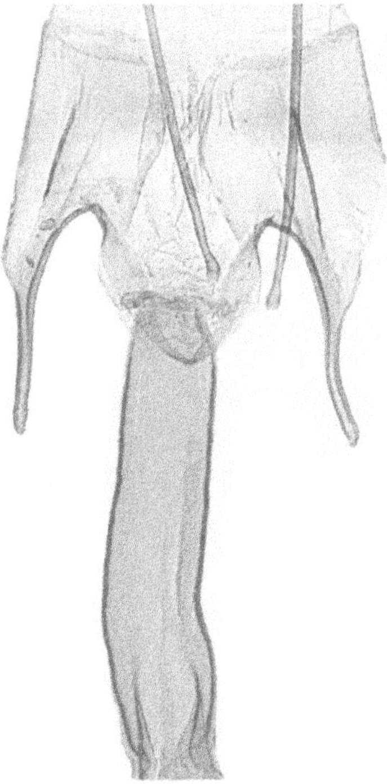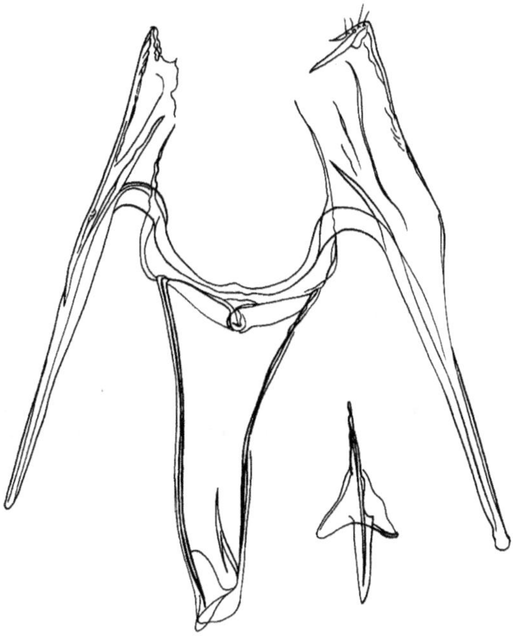

132 133

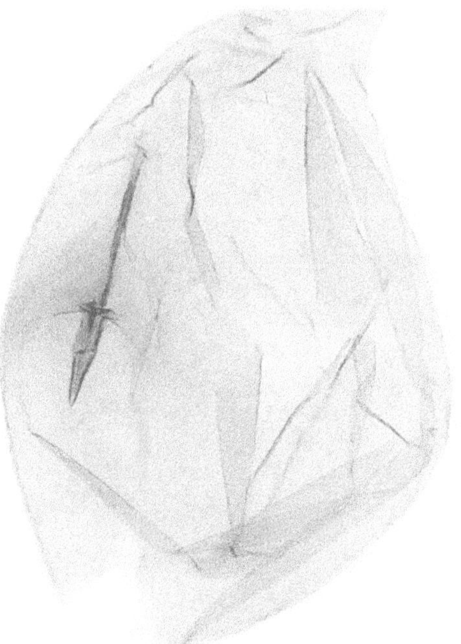

FEMALE GENITALIA

132. *Ephysteris inustella* (Herrich-Schäffer) – Hungary, GU 01/1005 P.H. (ZSM).
133. *Ephysteris olympica* Povolný – Paratype (*E. olympica monticola*), France, Povolný, 2002a: fig. 852.

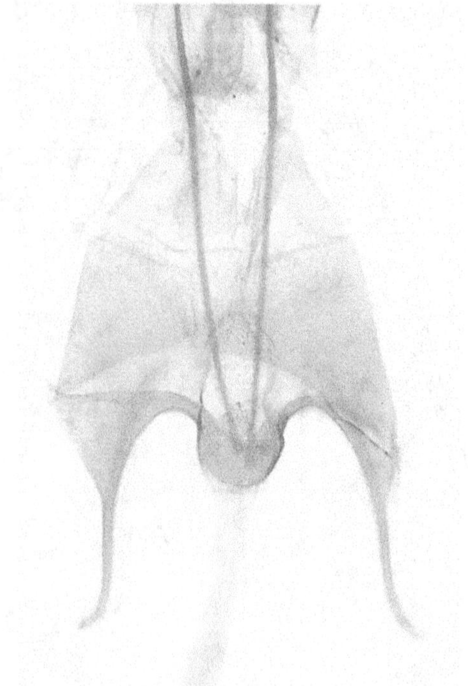
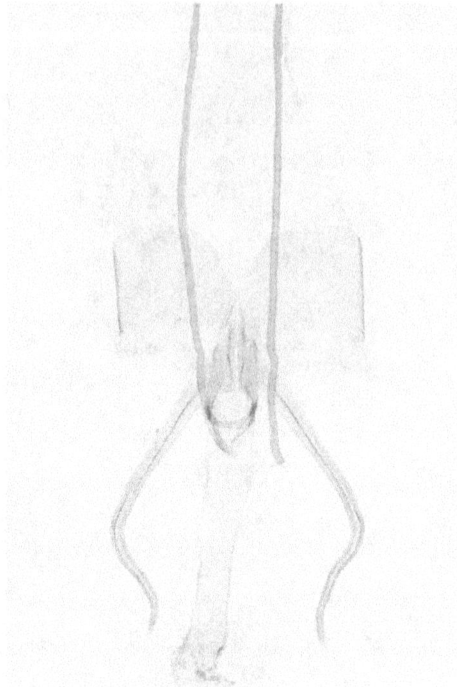

134 *135*

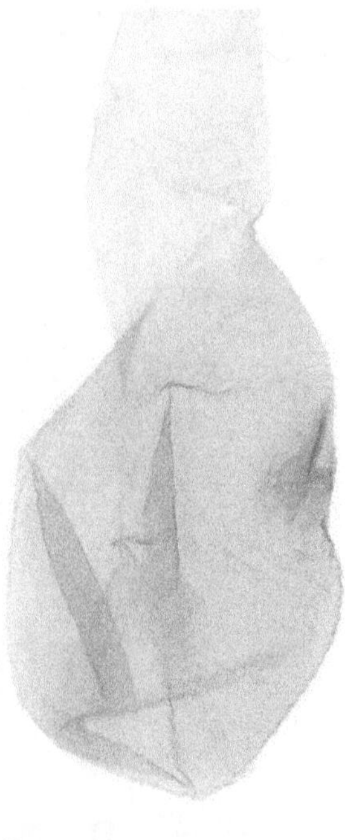

FEMALE GENITALIA

134. *Ephysteris iberica* Povolný – Spain, GU 08/1267 P.H. (ZMUC).
135. *Ochrodia subdiminutella* (Stainton) – Spain (Canary Isl.), GU 01/1064 P.H. (ZSM).

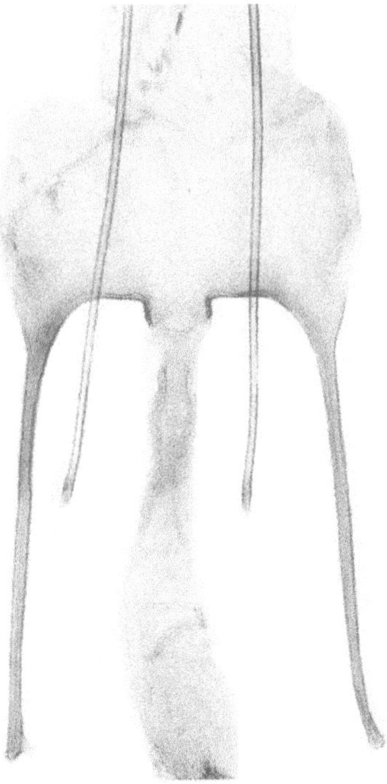

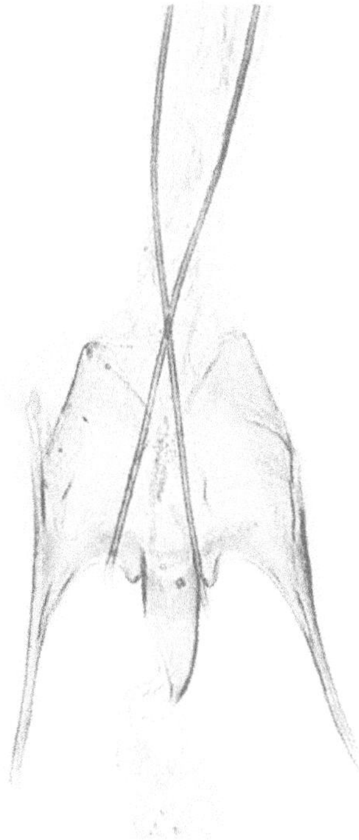

136 137

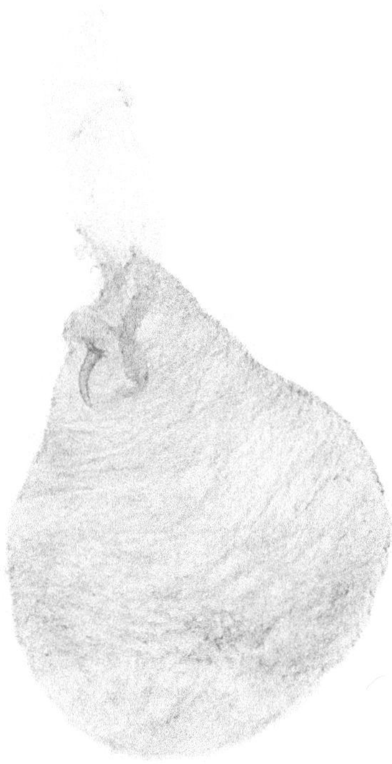

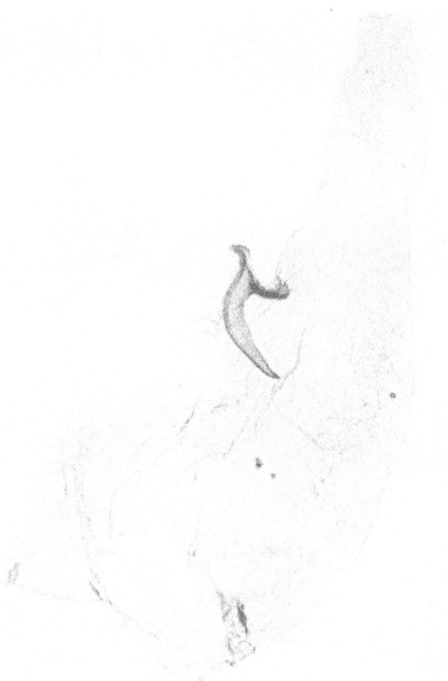

Female Genitalia

136. *Vladimirea glebicolorella* (Erschoff) – Ukraine (Crimea), gen. slide 521/07 Bidzilya (ZMKU).
137. *Microlechia rhamnifoliae* (Amsel & Hering) – Greece, GU 09/1308 P.H. (ZMUC).

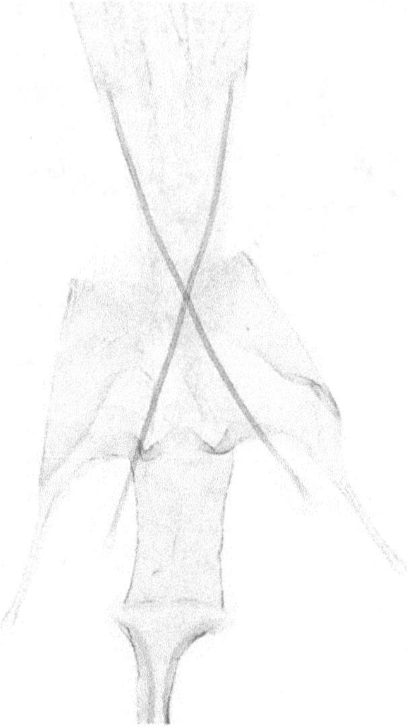

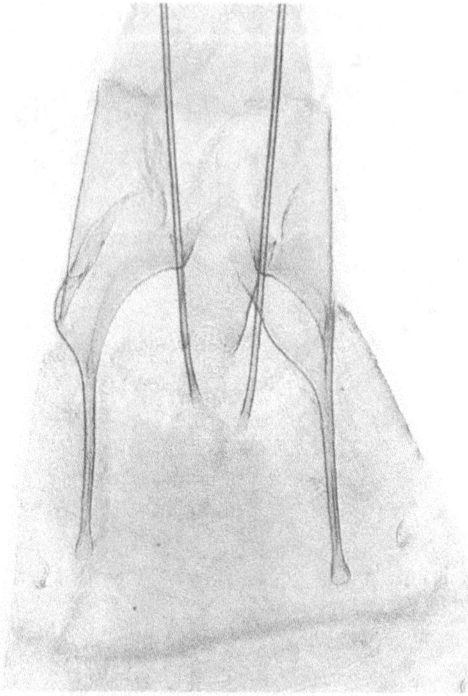

138

139

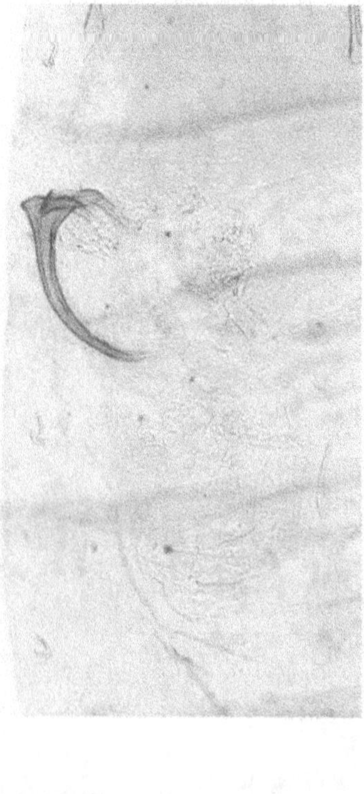

Female Genitalia

138. *Microlechia chretieni* Turati – Portugal, GU 02/1156 P.H. (ZMUC).
139. *Microlechia karsholti* (Nupponen) – Paratype, Russia, gen. slide 1/10.2009 Nupponen (NUPP).

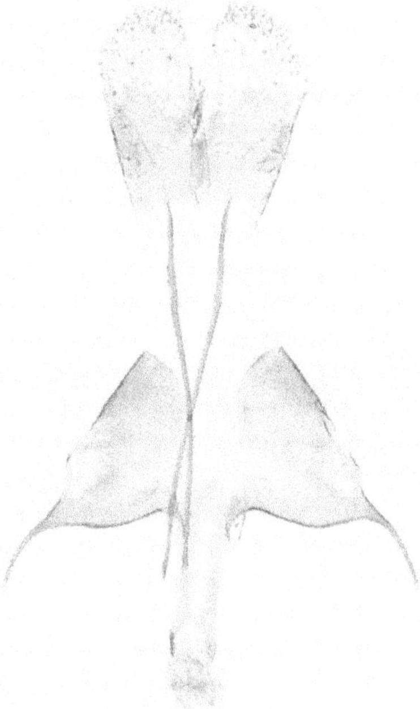

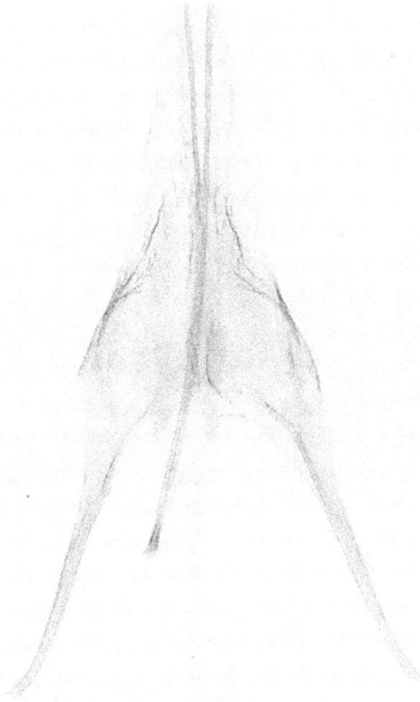

140

141

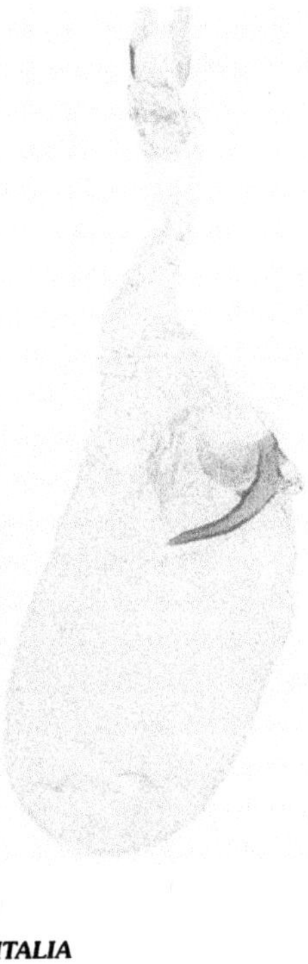

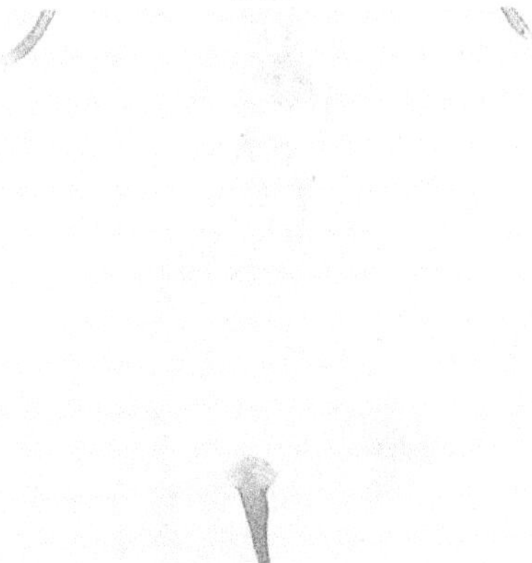

FEMALE GENITALIA

140. *Cosmardia moritzella* (Treitschke) – Switzerland, GEL 1032 P.H. (TLMF).
141. *Lutilabria lutilabrella lutilabrella* (Mann) – Croatia, GU 90/191 P.H. (ZIAN).

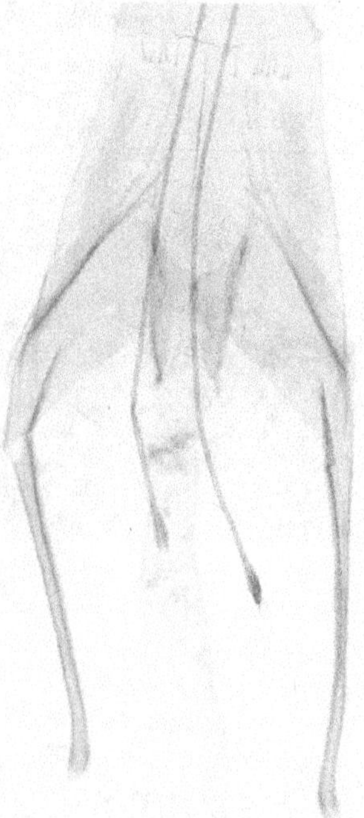

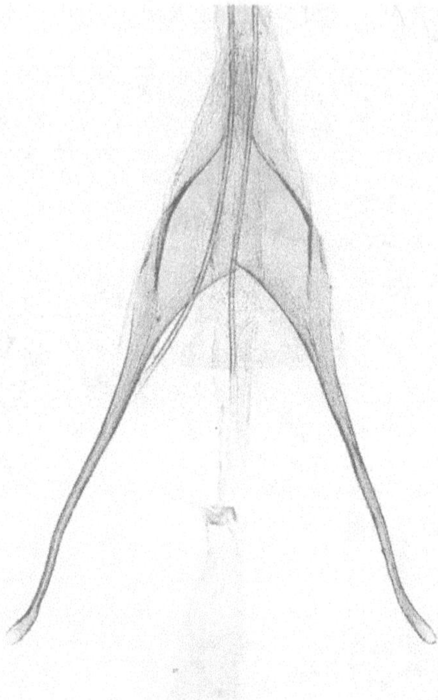

141a

142

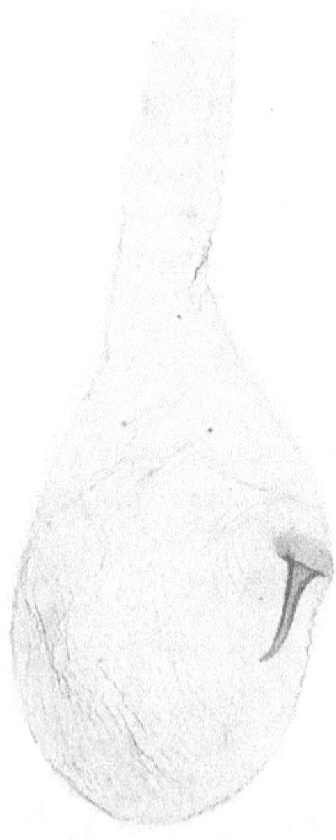

FEMALE GENITALIA

141a. *Lutilabria lutilabrella olympica* Huemer – Greece, GU 99/856 P.H. (ZSM).
142. *Lutilabria volgensis* Anikin & Piskunov – Paratype, Russia, GU 09/1304 P.H. (ZMUC).

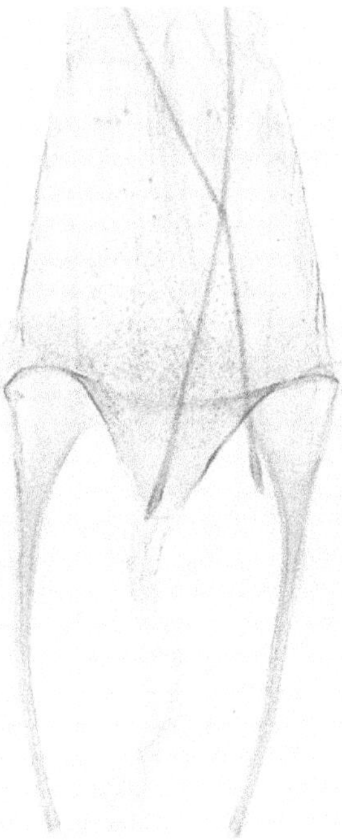
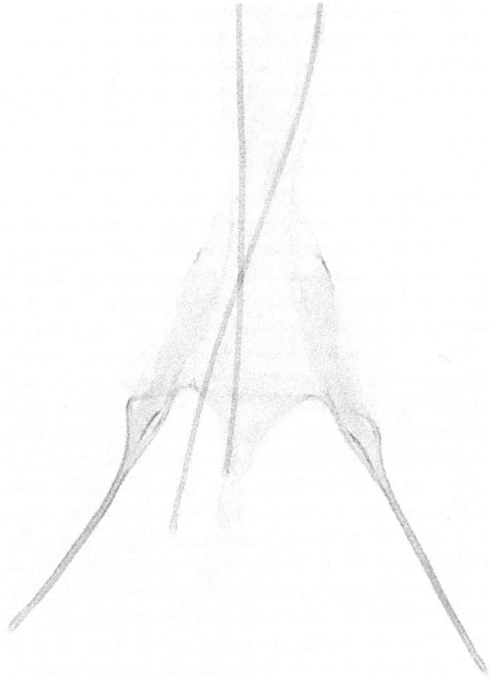

144 145

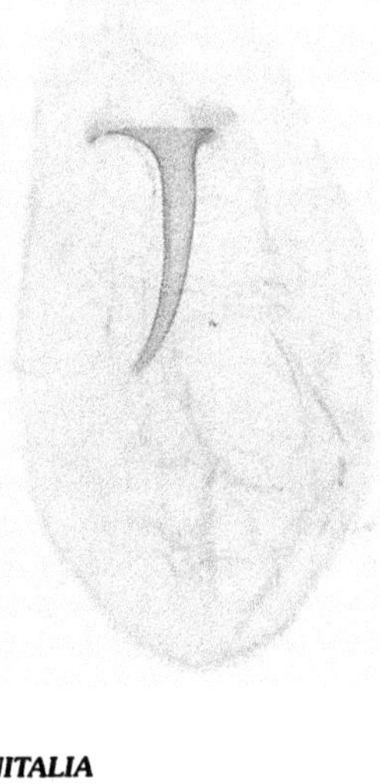
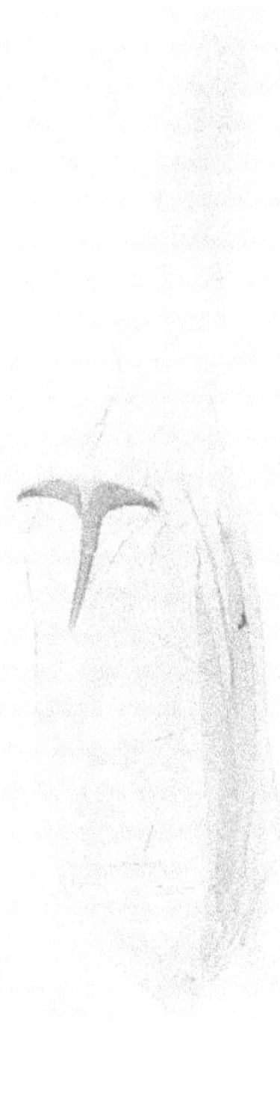

Female Genitalia

144. *Klimeschiopsis kiningerella* (Duponchel) – Austria, gen. slide 16644 (NHMW).
145. *Klimeschiopsis discontinuella* (Rebel) – Italy, GEL 1041 P.H. (TLMF).

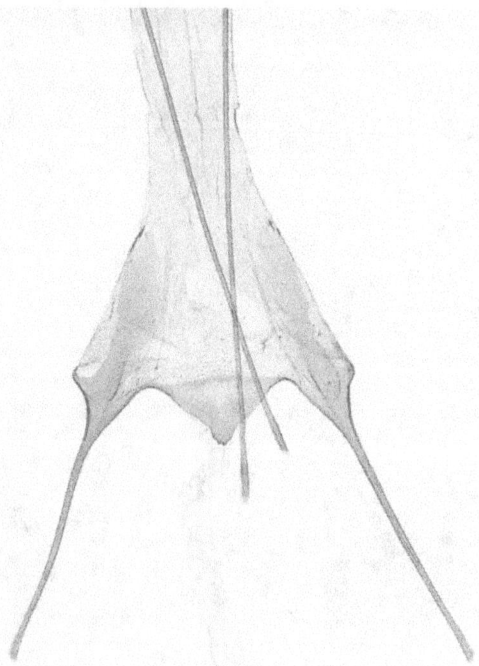

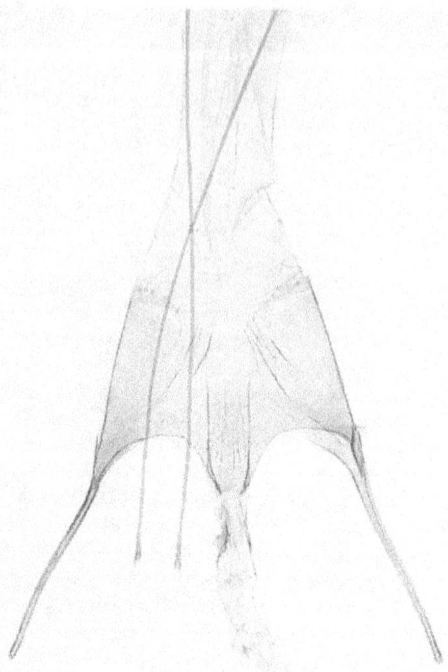

146

147

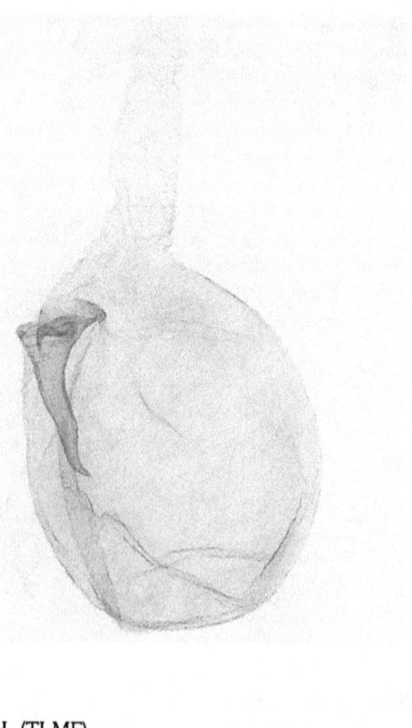

Female Genitalia

146. *Klimeschiopsis terroris* (Hartig) – Spain, GEL 1075 P.H. (TLMF).
147. *Caryocolum fischerella* (Treitschke) – Germany, GEL 1043 P.H. (TLMF).

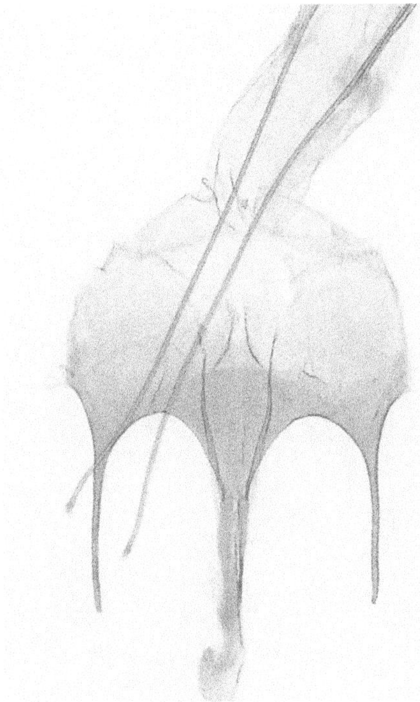

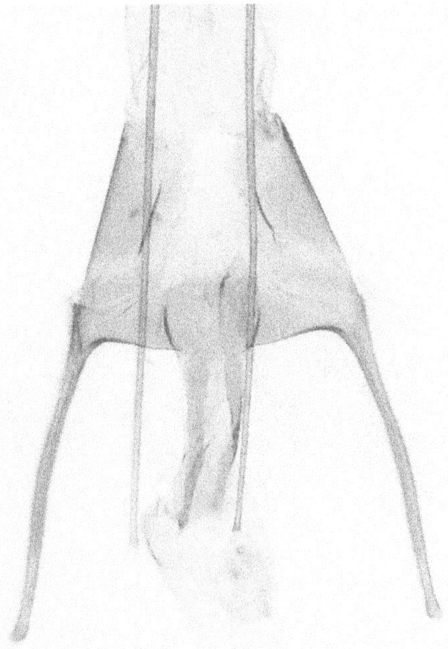

148 *149*

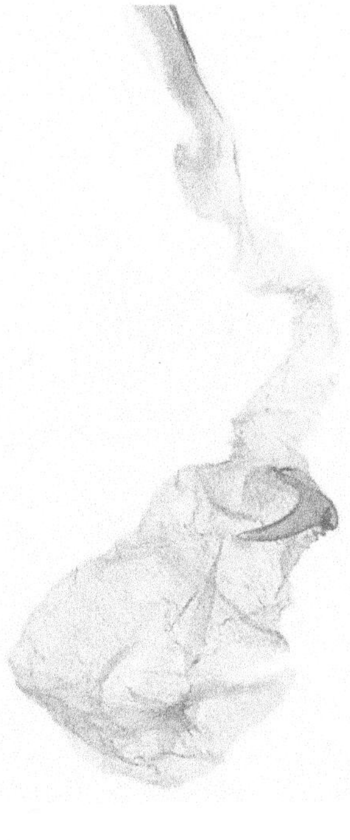

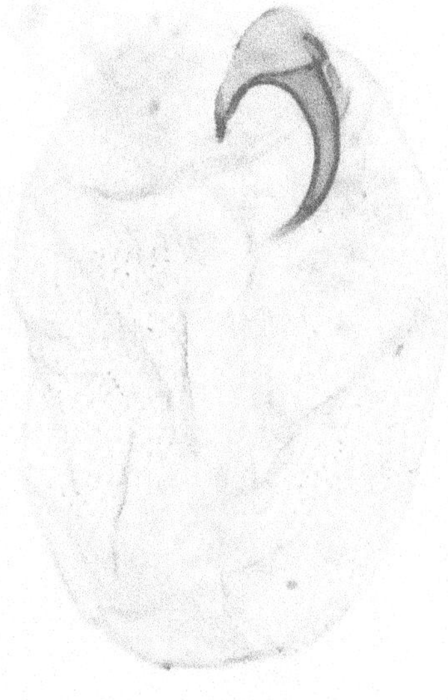

Female Genitalia

148. *Caryocolum tischeriella* (Zeller) – Germany, gen. slide 24595 (BMNH).
149. *Caryocolum alsinella* (Zeller) – Croatia, GU 86/343 P.H. (TLMF).

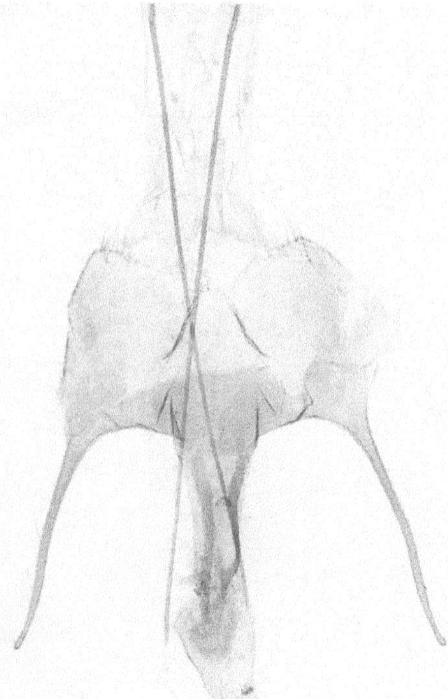

150

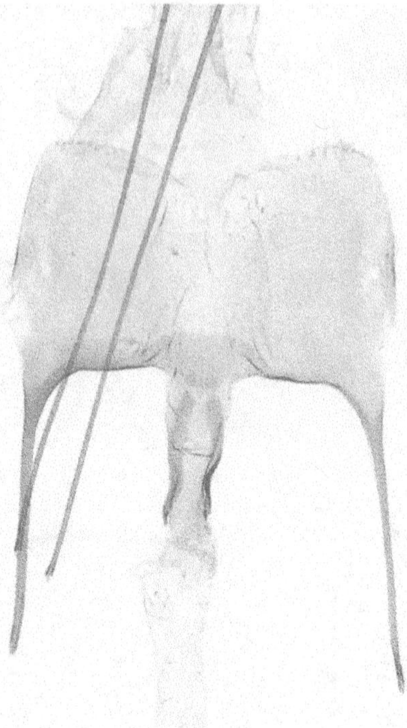

151

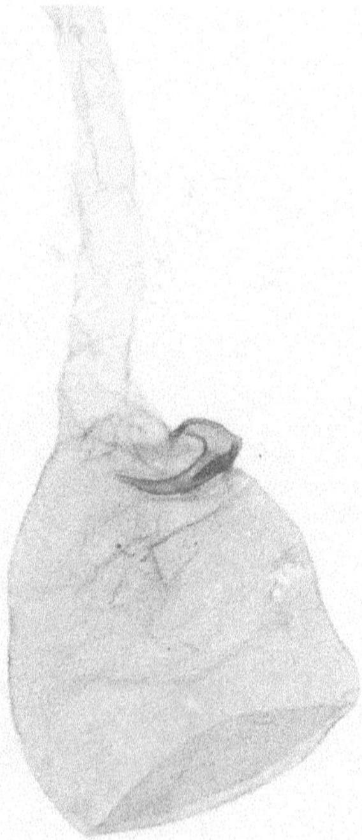

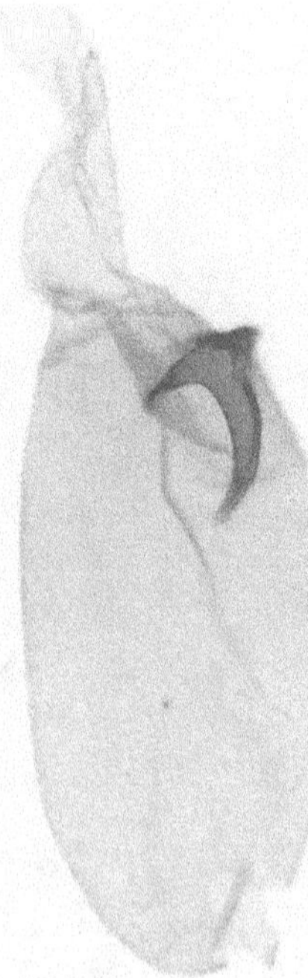

FEMALE GENITALIA

150. *Caryocolum viscariella* (Stainton) – Austria, GEL 425 P.H. (TLMF).
151. *Caryocolum albifaciella* (Heinemann) – Austria, GEL 429 P.H. (TLMF).

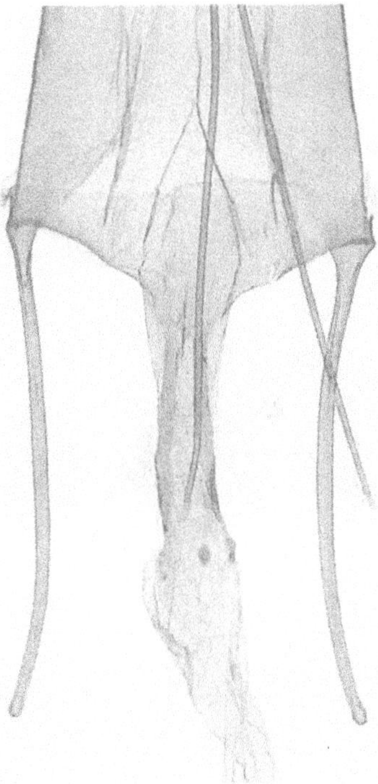

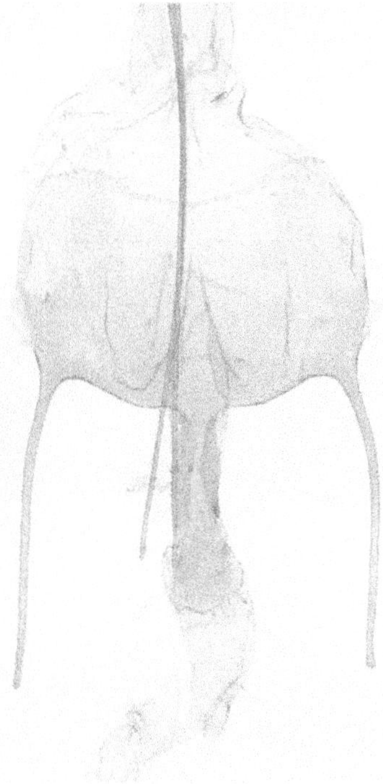

152

153

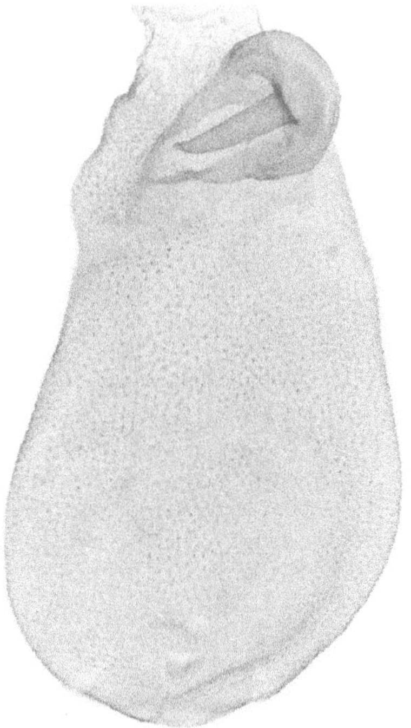

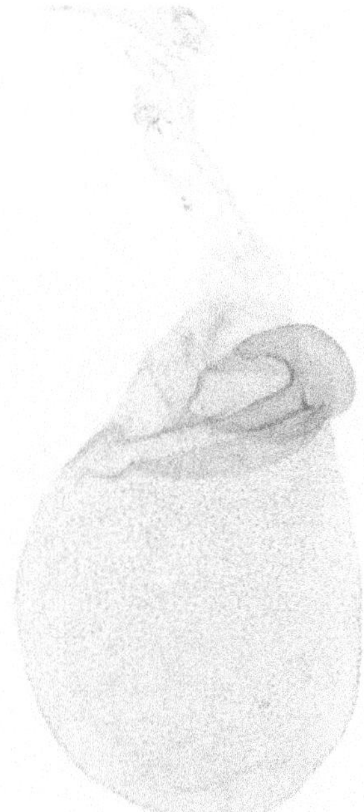

FEMALE GENITALIA

152. *Caryocolum vicinella* (Douglas) – Italy, GEL 1105 (TLMF).
153. *Caryocolum bosalella* (Rebel) – Italy (Sardinia), GU 86/418 P.H. (ZMUC).

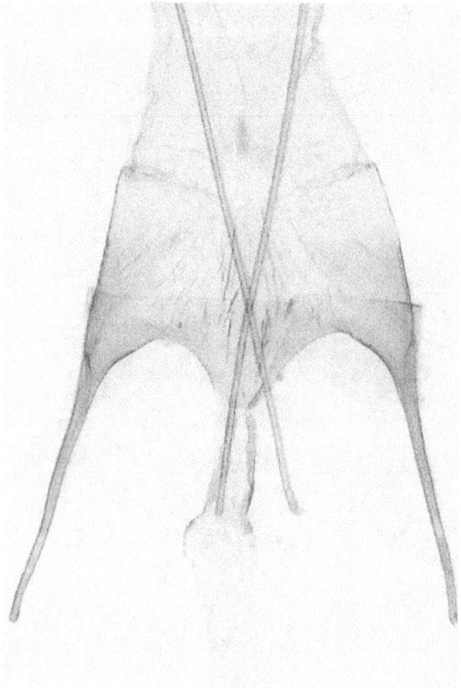

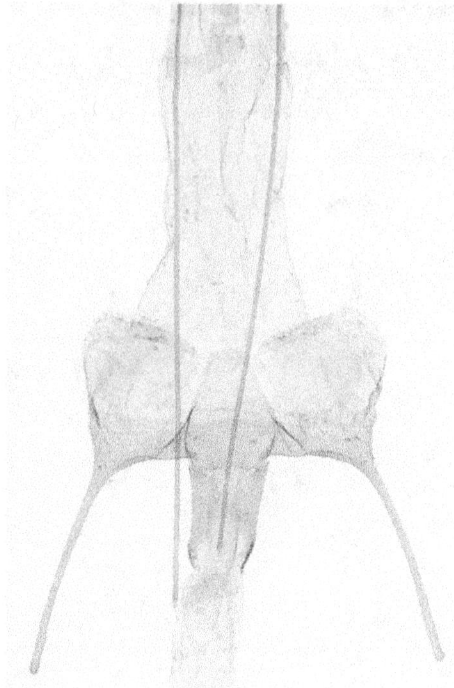

154

155

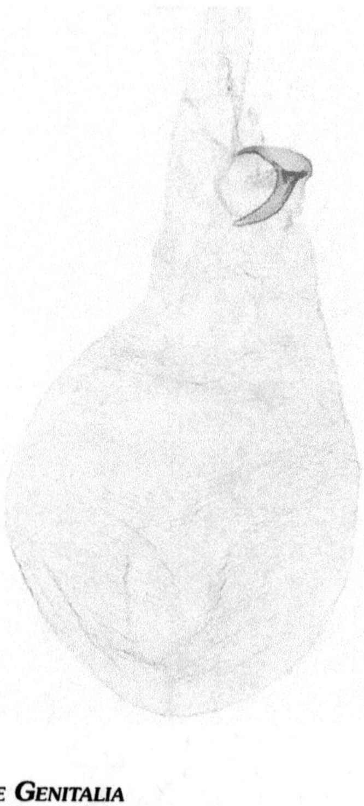

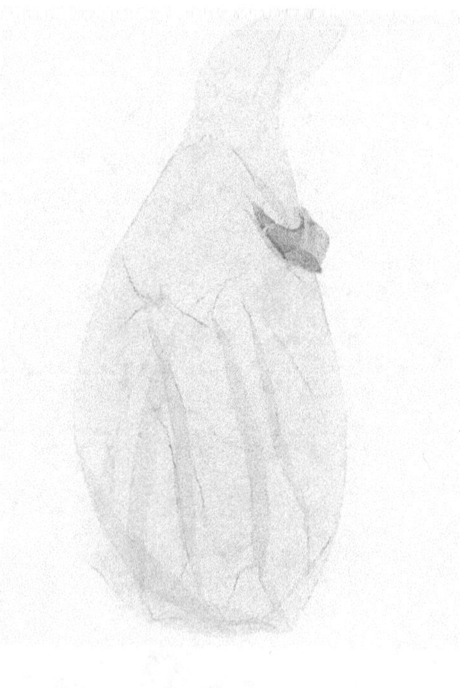

Female Genitalia

154. *Caryocolum amaurella* (Hering) – Switzerland, gen. slide 24602 (BMNH).
155. *Caryocolum oculatella* (Thomann) – Austria, GU 87/108 P.H. (TLMF).

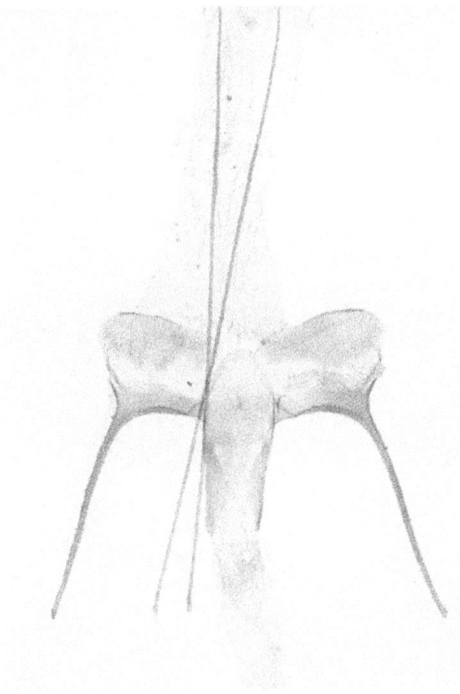

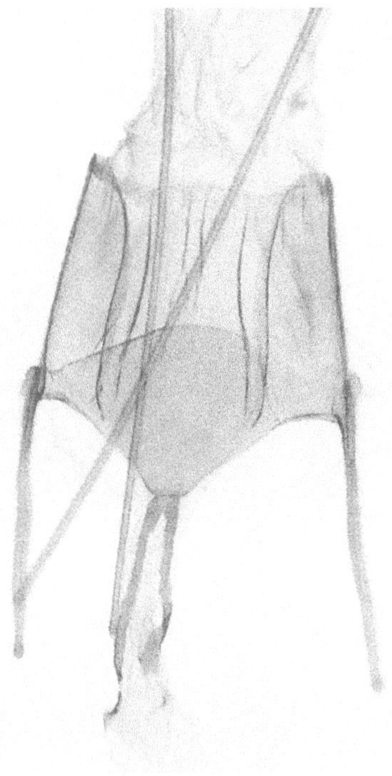

156

157

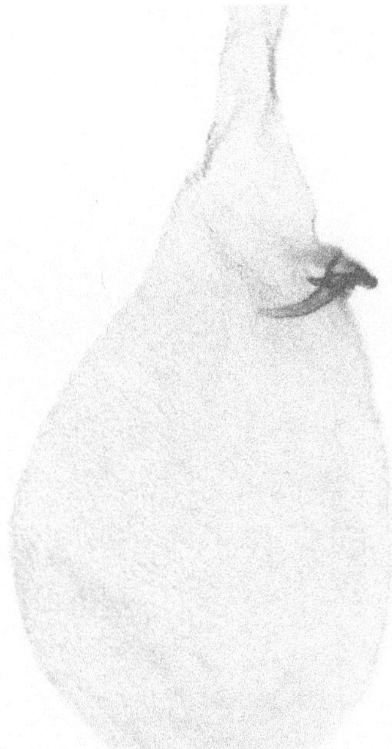

Female Genitalia

156. *Caryocolum leucofasciatum* Huemer – Spain, GU 87/074 P.H. (TLMF).
157. *Caryocolum petryi* (Hofmann) – France, GU 86/087 P.H. (TLMF).

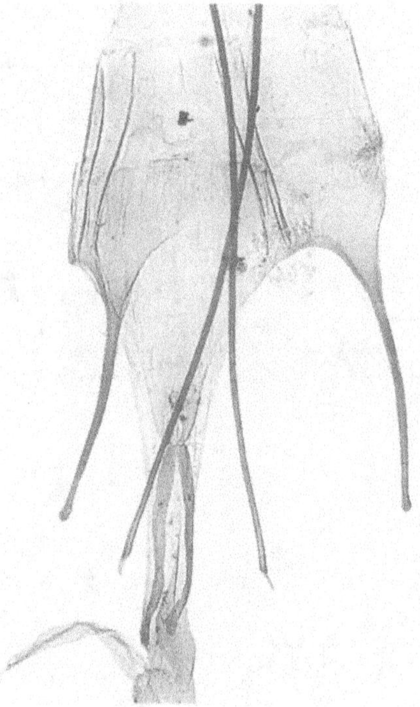 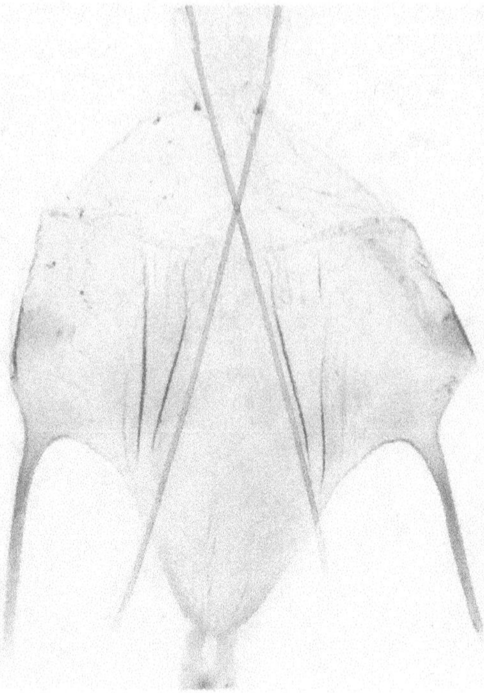

158 159

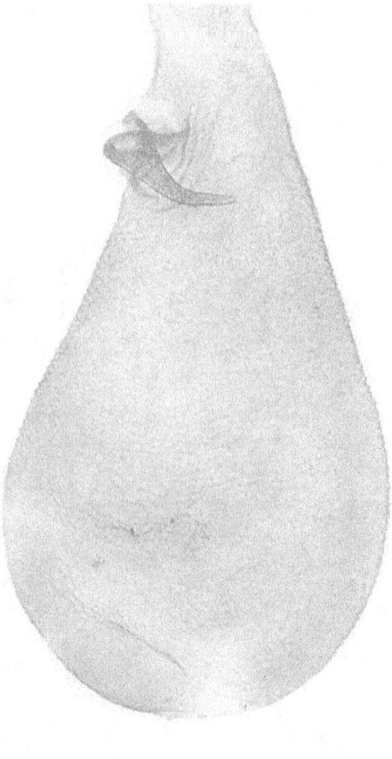 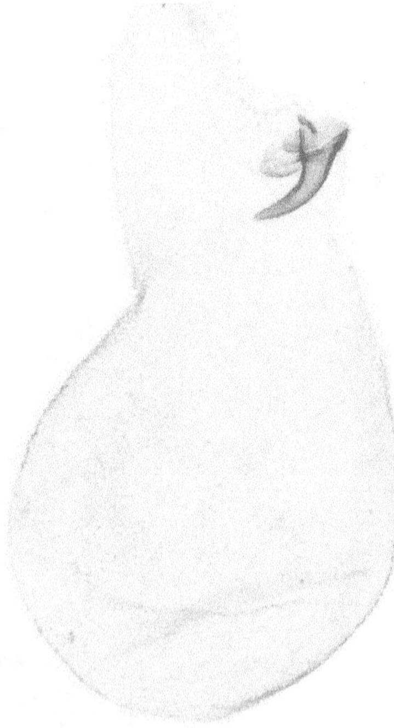

Female Genitalia

158. *Caryocolum baischi* sp. n. – Paratype, Greece, gen. slide 5919 H.H. (ZMUC).
159. *Caryocolum repentis* Huemer & Luquet – Paratype, Austria, GEL 278 (TLMF).

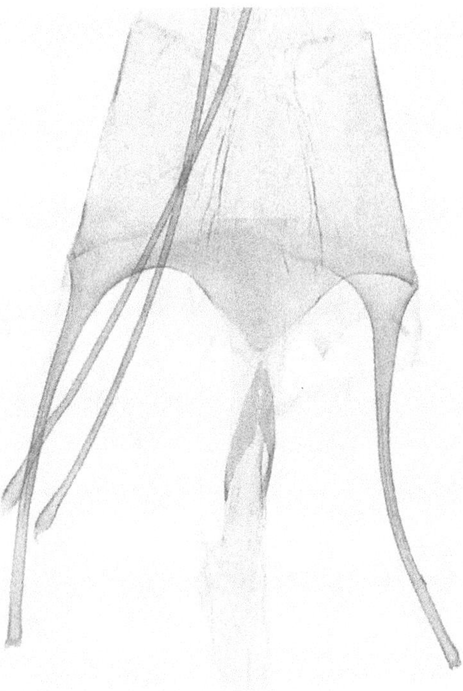

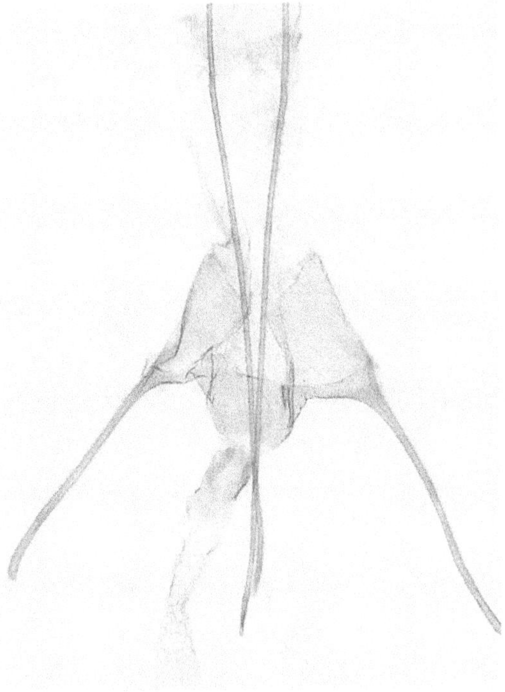

160

161

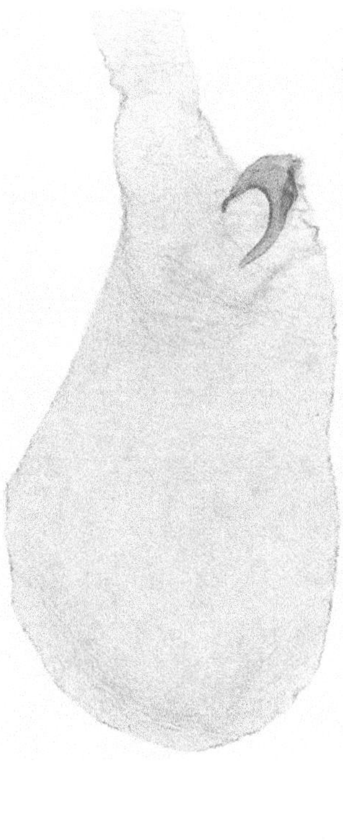

FEMALE GENITALIA

160. *Caryocolum siculum* Bella – Paratype, Italy (Sicily), GU 03/1202 P.H. (ZMUC).
161. *Caryocolum inflativorella* (Klimesch) – Hungary, gen. slide 24246 (BMNH).

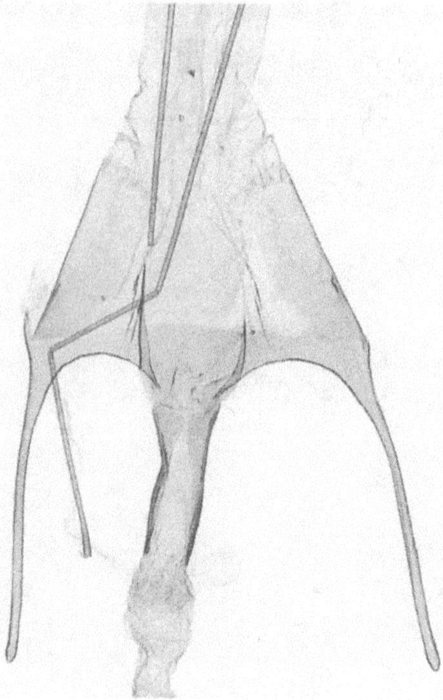

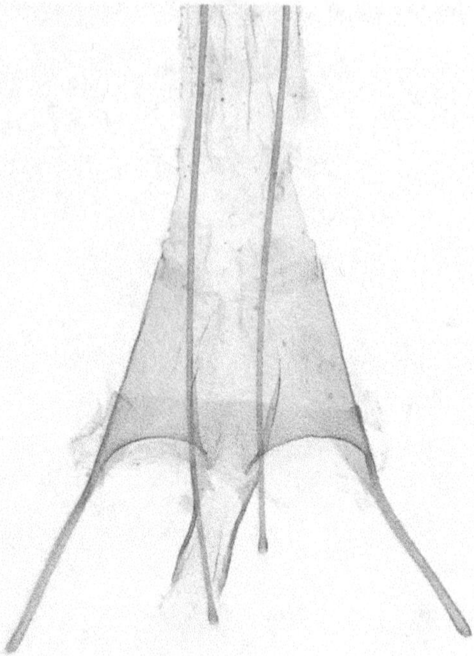

162 163

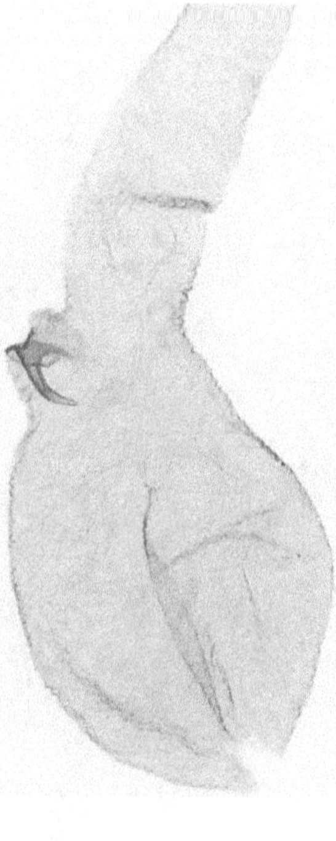

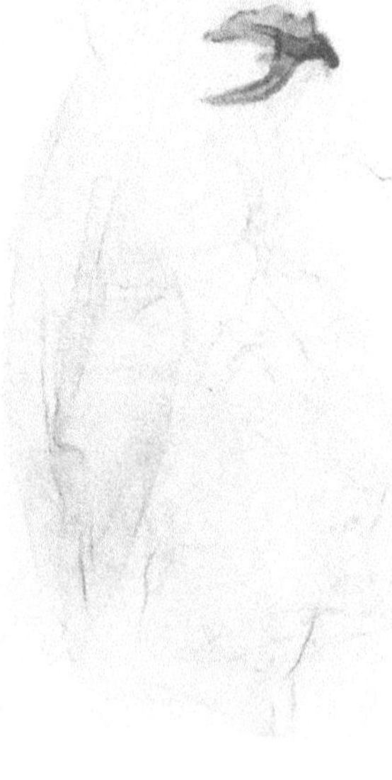

FEMALE GENITALIA

162. *Caryocolum saginella* (Zeller) – Italy, GEL 1091 P.H. (TLMF).
163. *Caryocolum cauligenella* (Schmid) – Austria, GU 88/213 P.H. (TLMF).

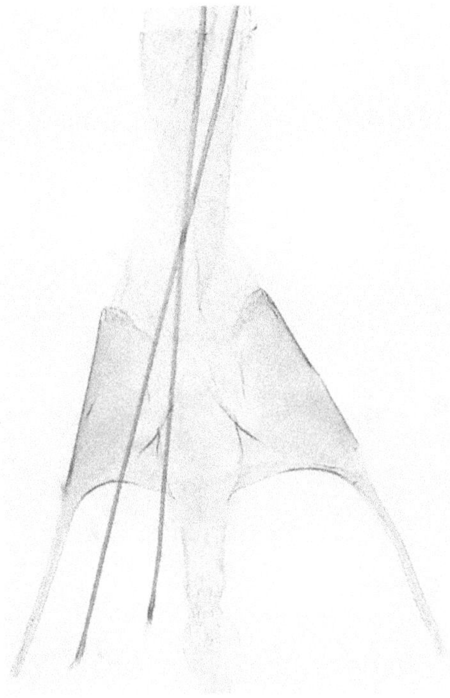

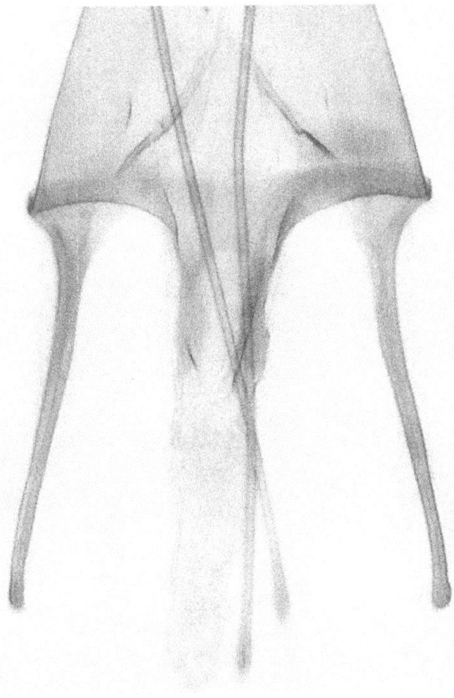

164

165

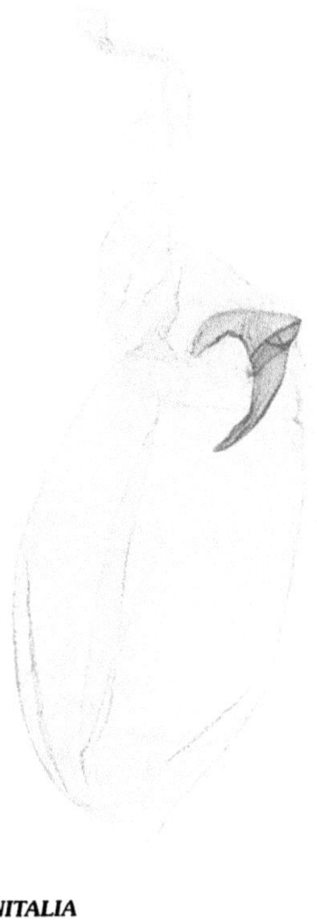

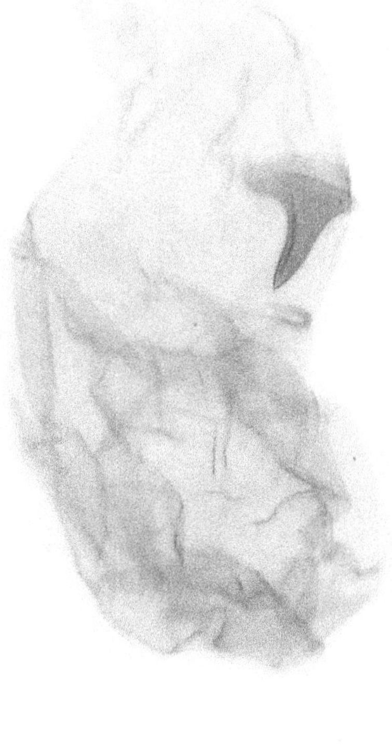

FEMALE GENITALIA

164. *Caryocolum trauniella* (Zeller) – Slovenia, GEL 1087 P.H. (TLMF).
165. *Caryocolum peregrinella* (Herrich-Schäffer) – Macedonia, GU 86/081 P.H. (TLMF).

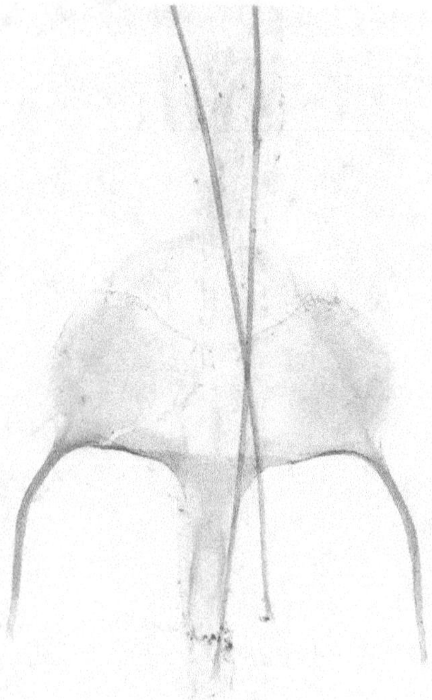

166

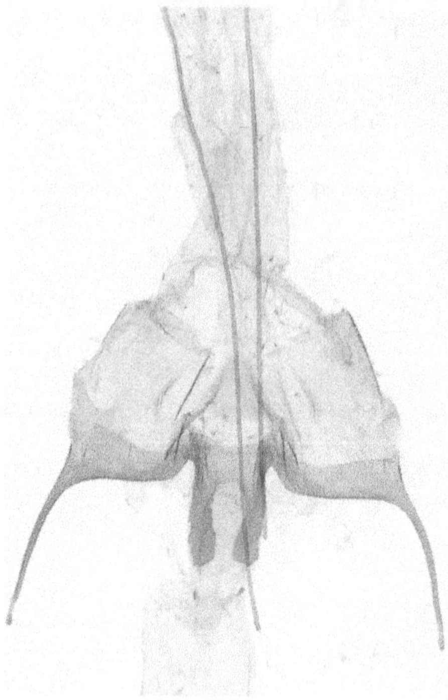

167

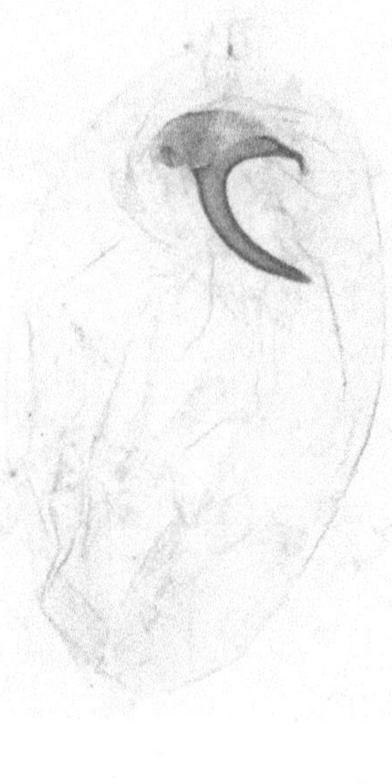

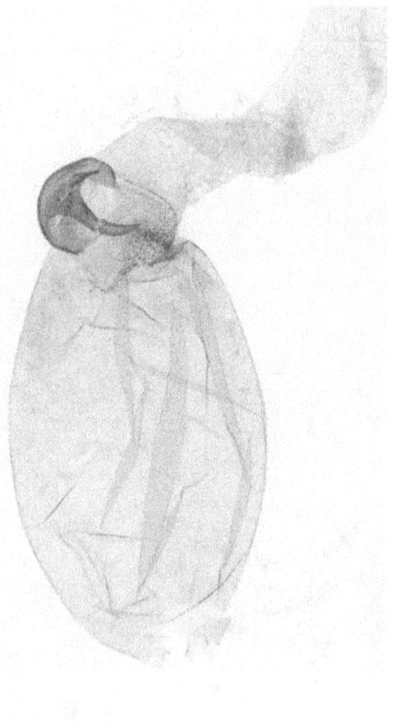

Female Genitalia

166. *Caryocolum delphinatella* (Constant) – France, GEL 158 P.H. (TLMF).
167. *Caryocolum provinciella* (Stainton) – Spain, GU 87/075 P.H. (TLMF).

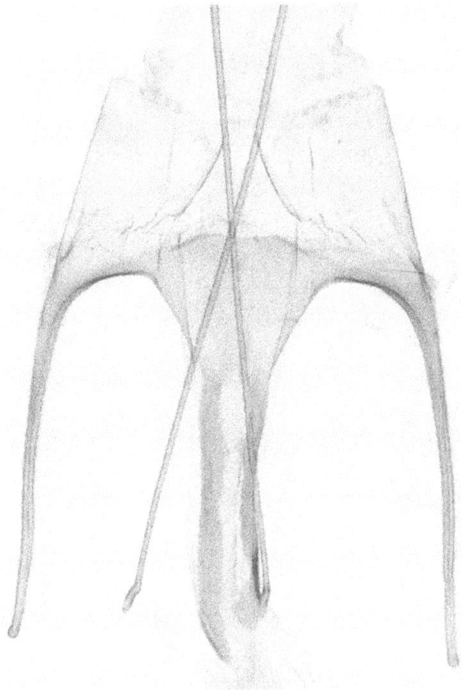

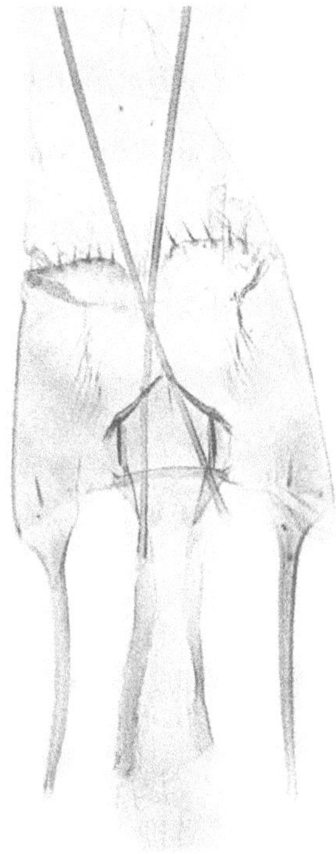

168

169

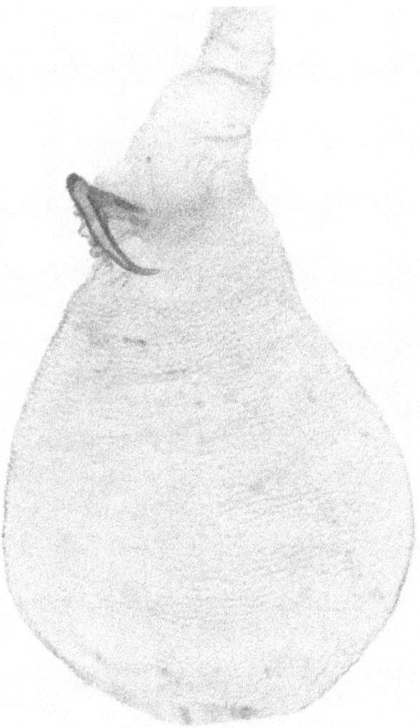

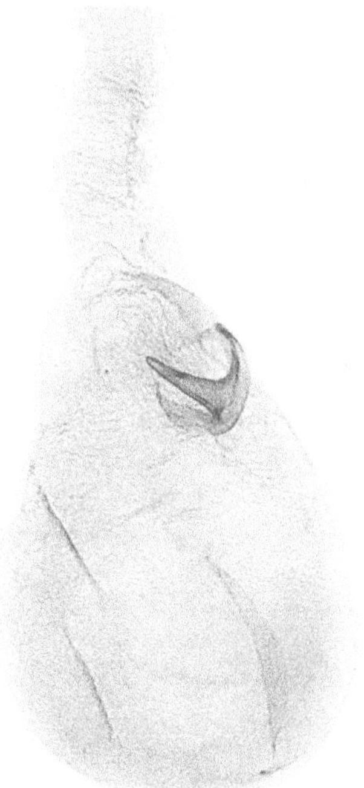

Female Genitalia

168. *Caryocolum mucronatella* (Chrétien) – Austria, GU 86/132 P.H. (TLMF).
169. *Caryocolum leucomelanella* (Zeller) – Italy, GEL 1073 P.H. (TLMF).

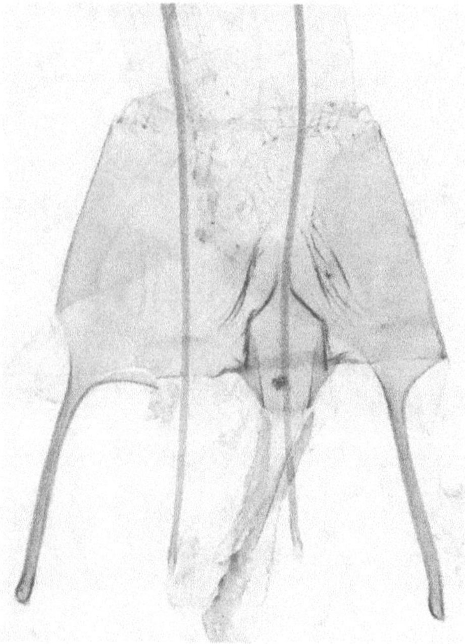

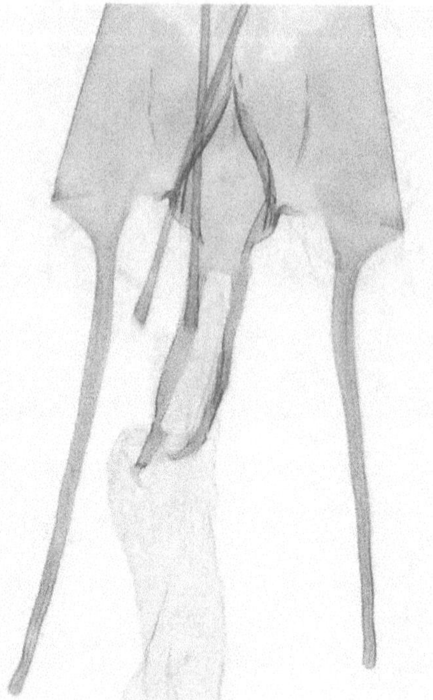

170

171

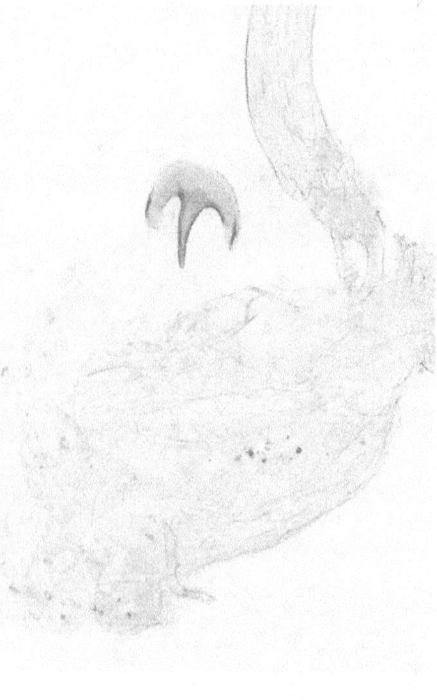

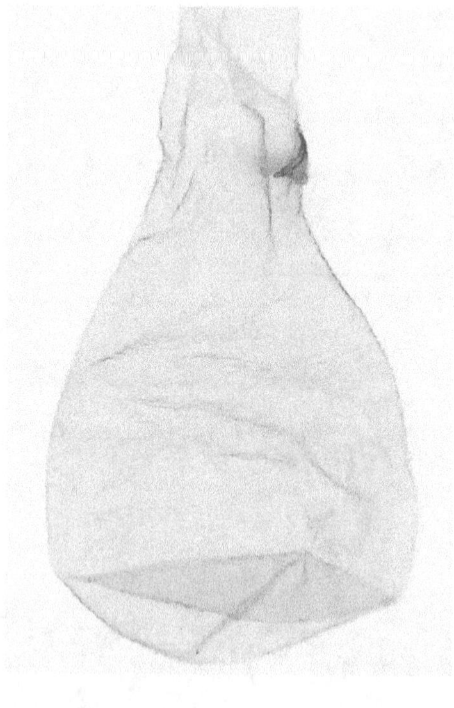

FEMALE GENITALIA

170. *Caryocolum mazeli* Huemer & Nel – Paratype, France, gen. slide 13493 J.N. (TLMF).
171. *Caryocolum leucothoracellum* (Klimesch) – France, gen. slide 24431 (BMNH).

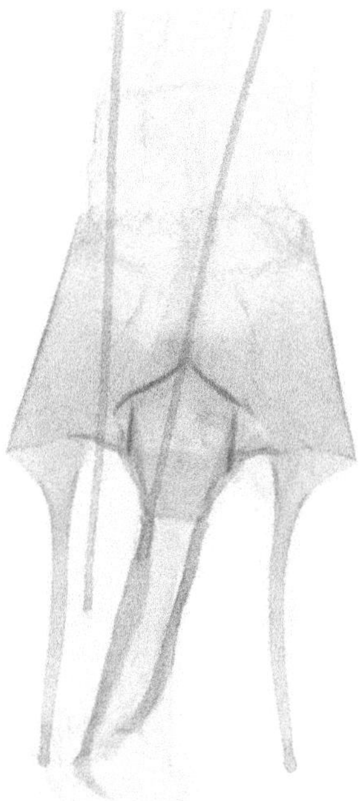

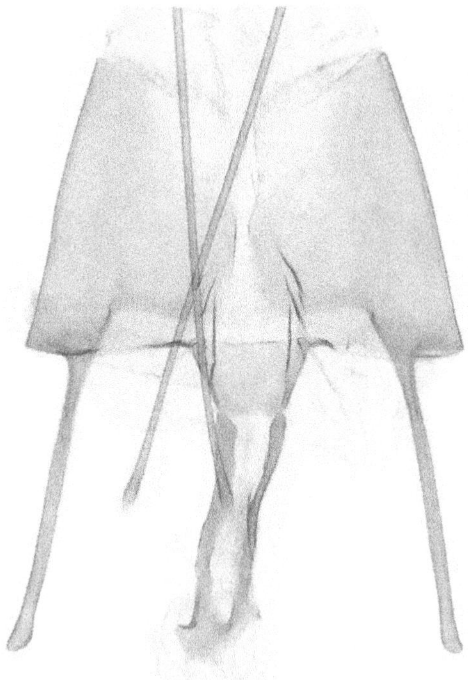

172

172a

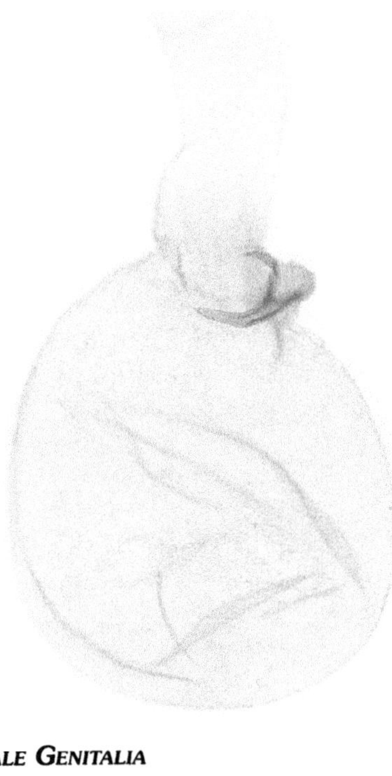

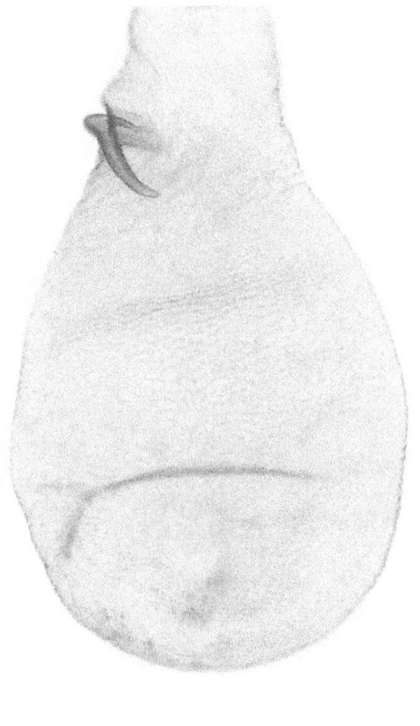

Female Genitalia

172. *Caryocolum schleichi schleichi* (Christoph) – Turkey, GU 88/040 P.H. (ZMUC).
172a. *Caryocolum schleichi improvisella* (Rebel) – Austria, GU 86/252 P.H. (TLMF).

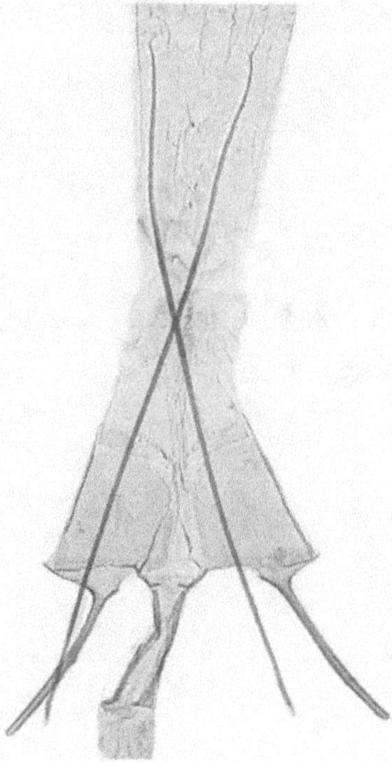

172b

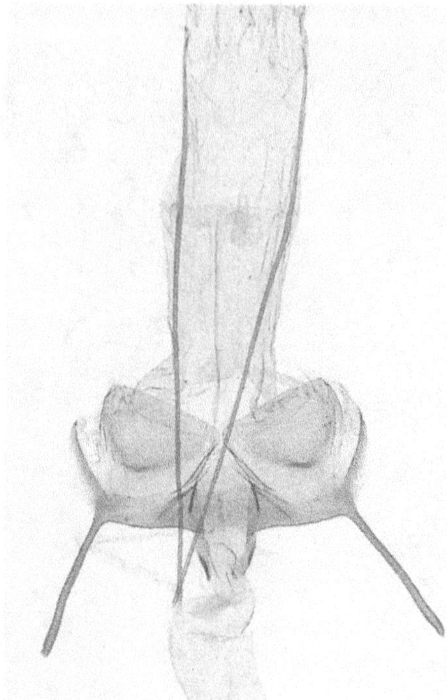

173

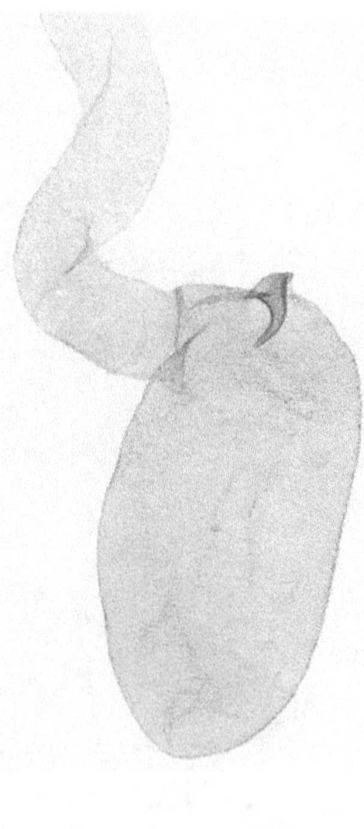

Female Genitalia

172b. *Caryocolum schleichi arenariella* (Benander) – Sweden, gen. slide 2055 H.H. (ZMUC).
173. *Caryocolum marmorea* (Haworth) – Italy, GEL 1099 P.H. (TLMF).

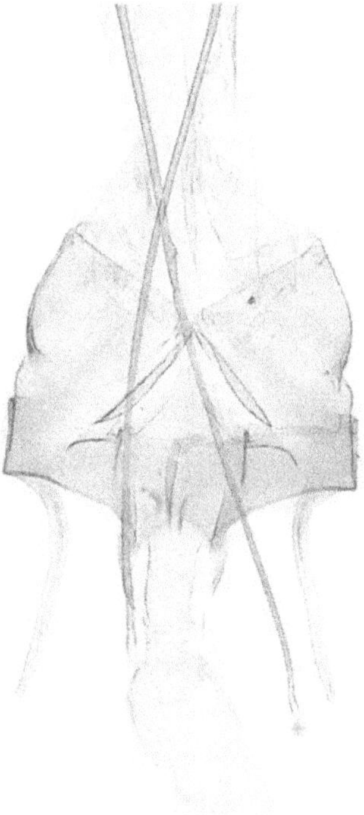

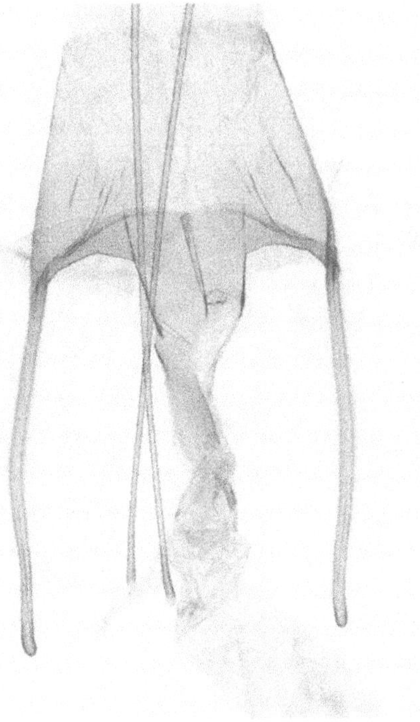

174

176

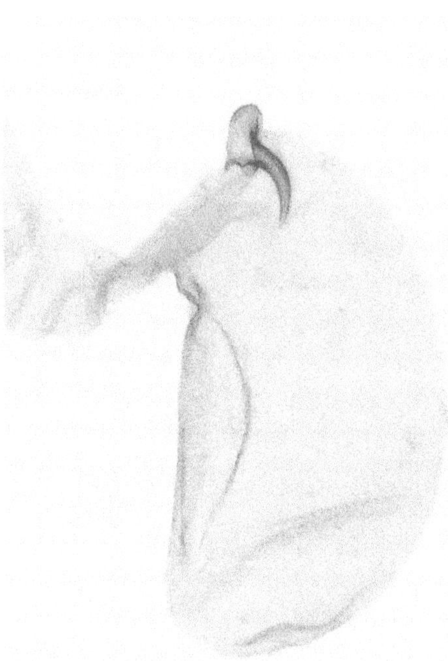

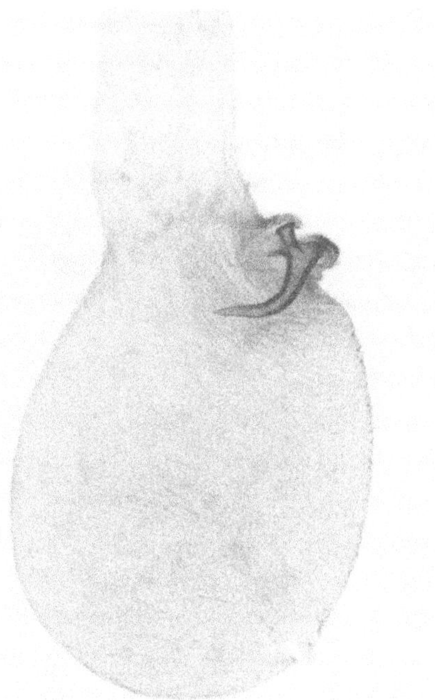

Female Genitalia

174. *Caryocolum pullatella* (Tengström) – Russia, gen. slide 24557 (BMNH).
176. *Caryocolum hispanicum* Huemer – Paratype, Spain, gen. slide 24219 (BMNH).

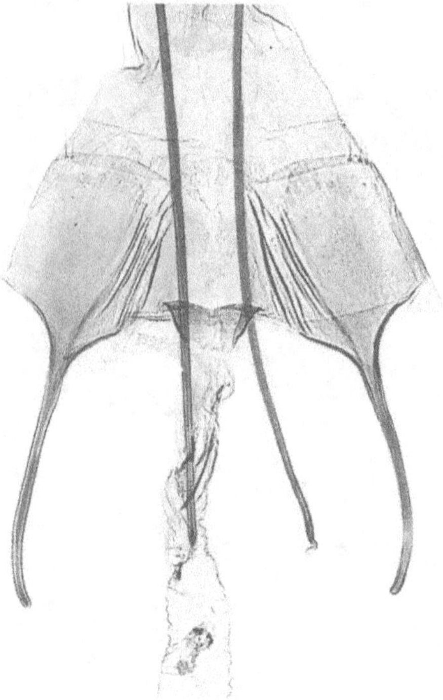

177

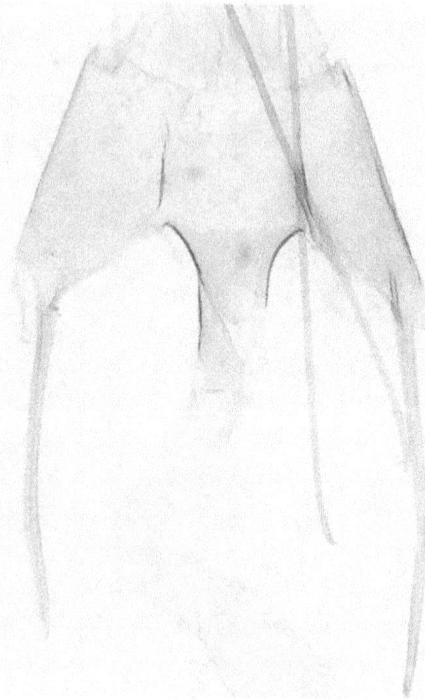

178

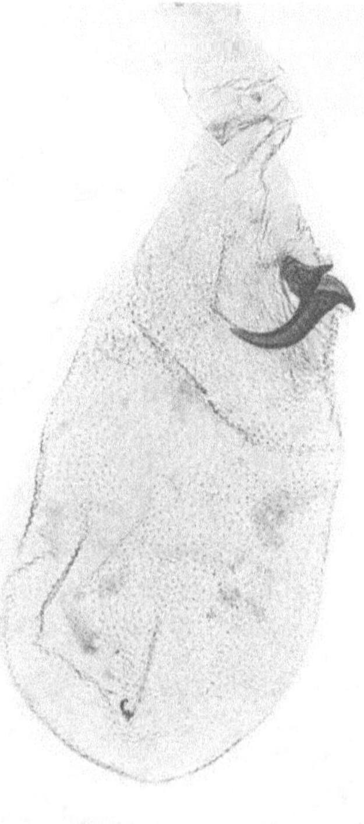

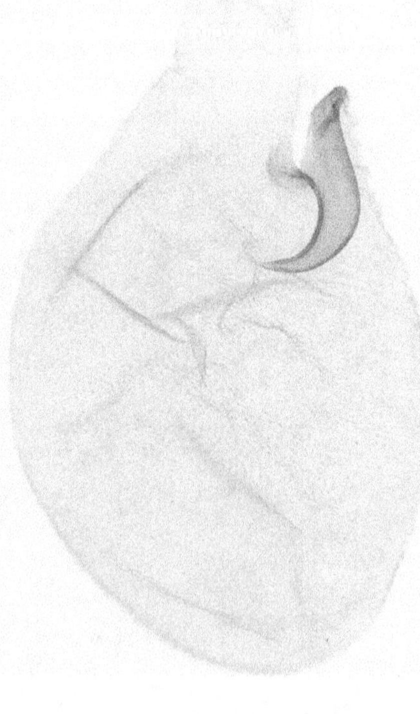

FEMALE GENITALIA

177. *Caryocolum confluens* Huemer – Holotype, Greece, GU 86/381 P.H. (ZMUC).
178. *Caryocolum gallagenellum* Huemer – Paratype, France, GU 88/084 P.H. (TLMF).

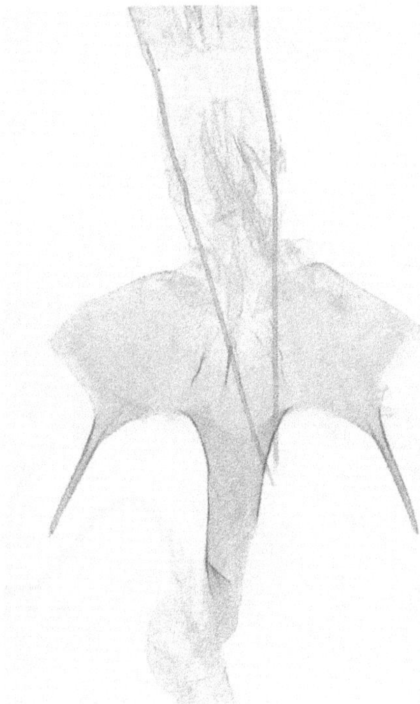

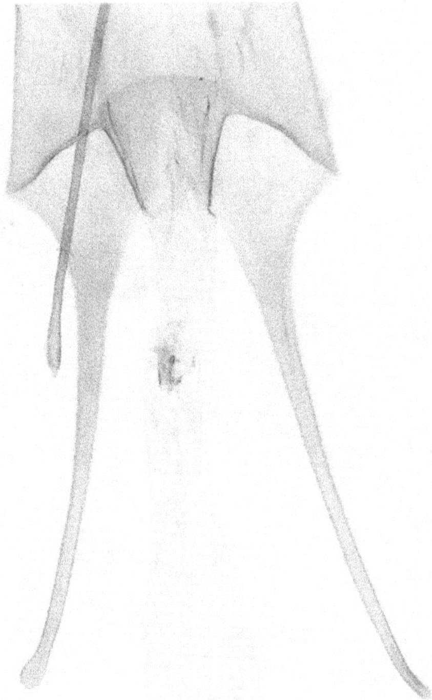

179

180

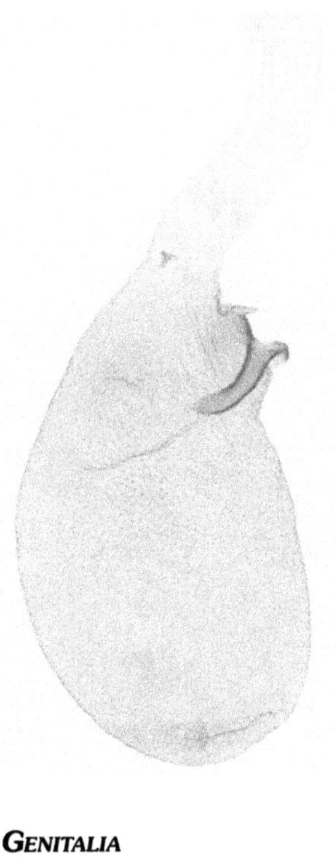

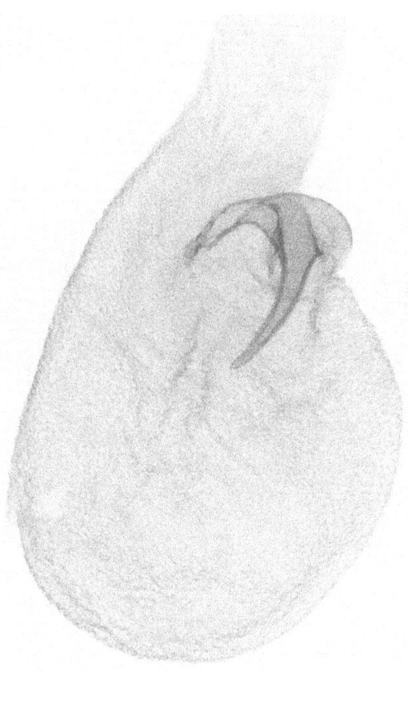

Female Genitalia

179. *Caryocolum fraternella* (Douglas) – Denmark, GU 88/087 P.H. (TLMF).
180. *Caryocolum klosi* (Rebel) – Germany, GU 86/259 P.H. (TLMF).

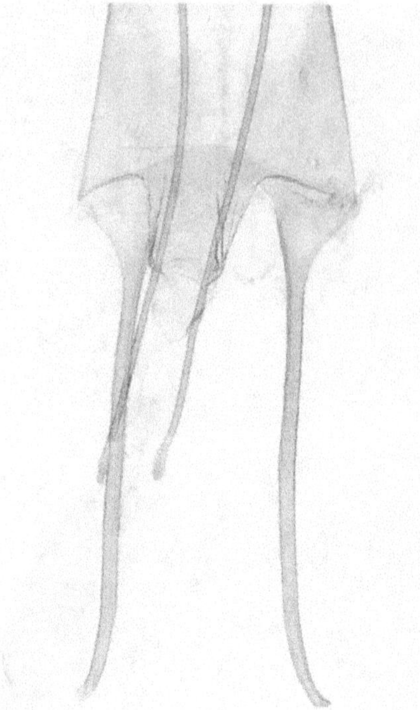

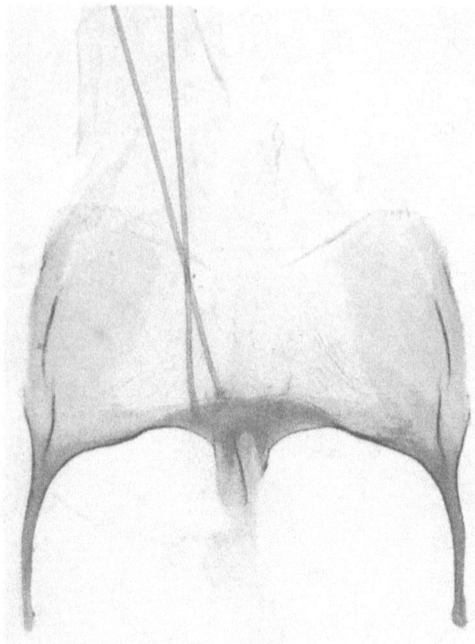

181 *182*

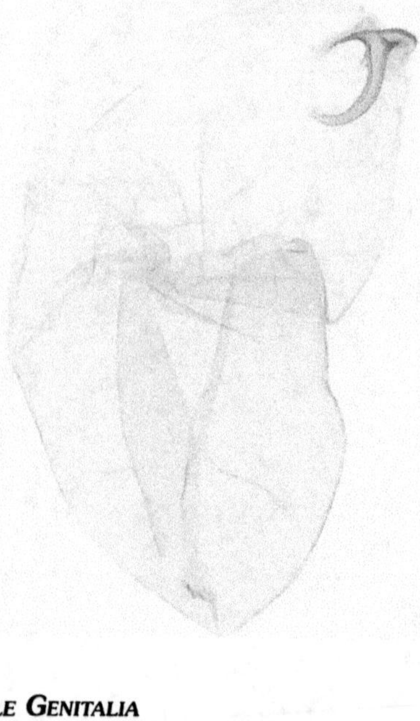

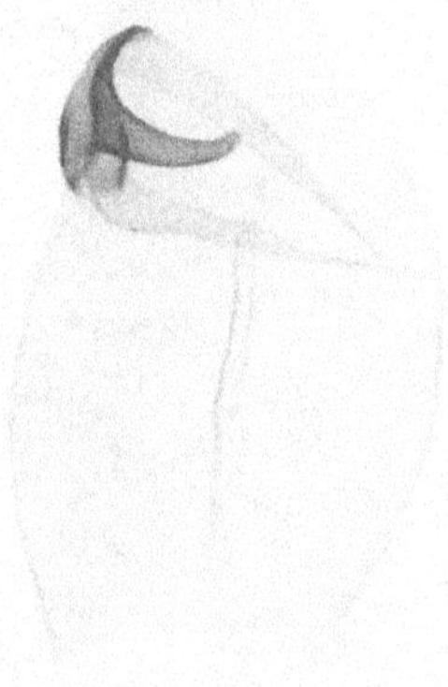

Female Genitalia

181. *Caryocolum interalbicella* (Herrich-Schäffer) – Italy, GU 86/280 P.H. (TLMF).
182. *Caryocolum laceratella* (Zeller) – Italy, GU 87/072 P.H. (TLMF).

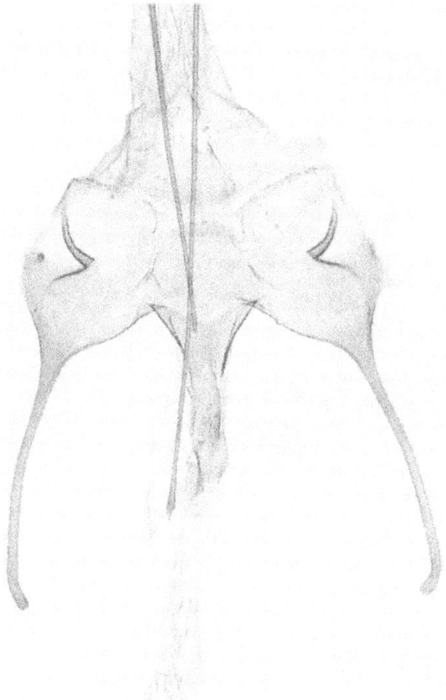

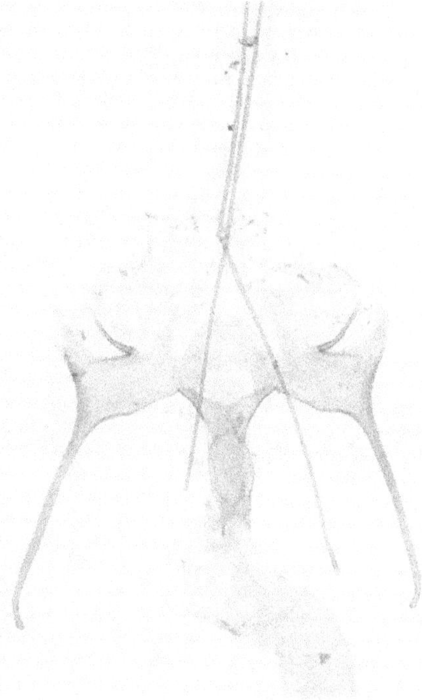

183

183a

FEMALE GENITALIA

183. *Caryocolum blandella blandella* (Douglas) – France, gen. slide 24587 (BMNH).
183. *Caryocolum blandella horoscopa* (Meyrick) – Greece, GU 87/093 P.H. (ZMUC).

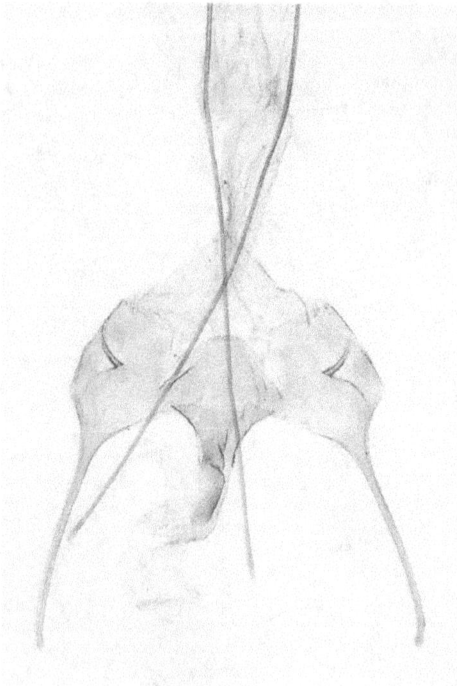

184

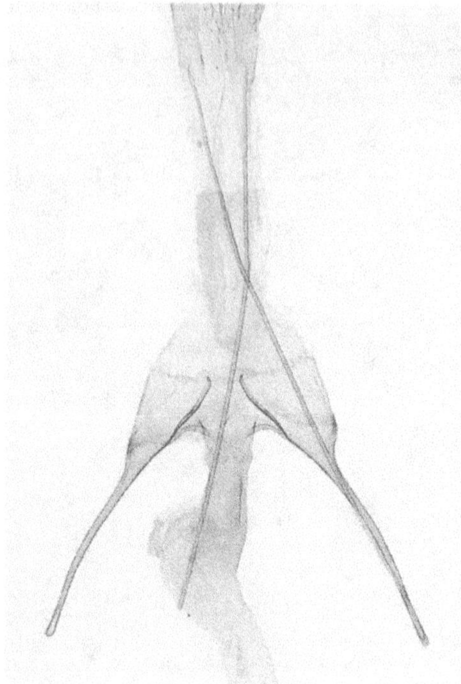

185

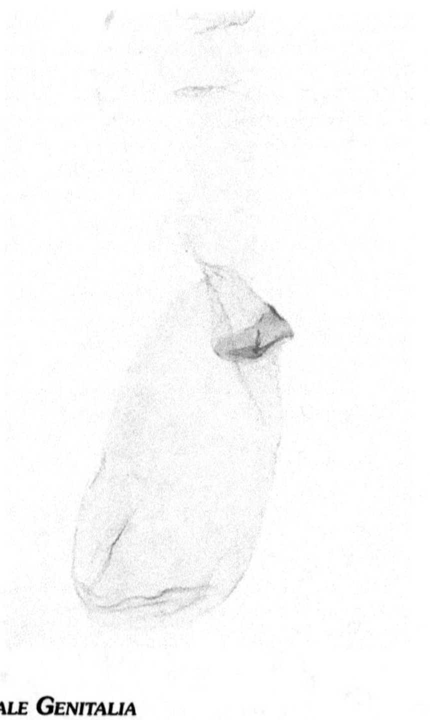

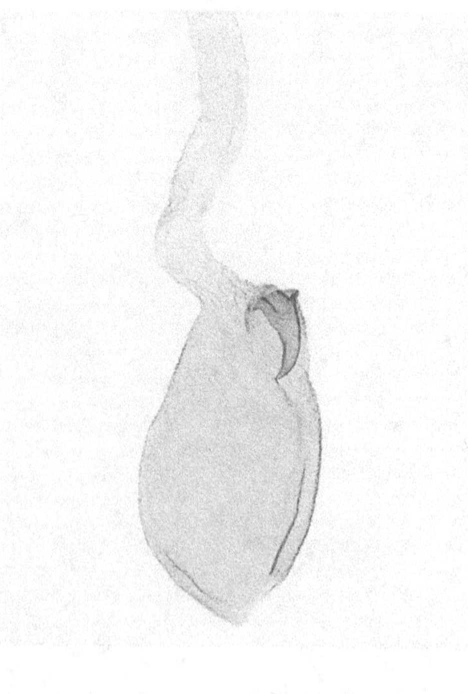

Female Genitalia

184. *Caryocolum blandelloides* Karsholt – Austria, GEL 1085 P.H. (TLMF).
185. *Caryocolum jaspidella* (Chrétien) – Morocco, GEL 1086 P.H. (TLMF).

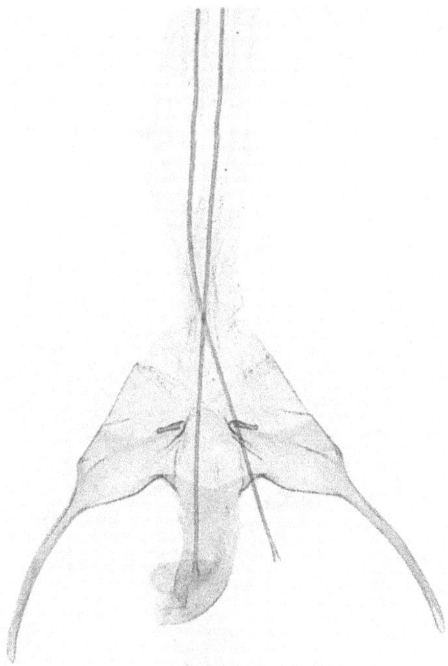

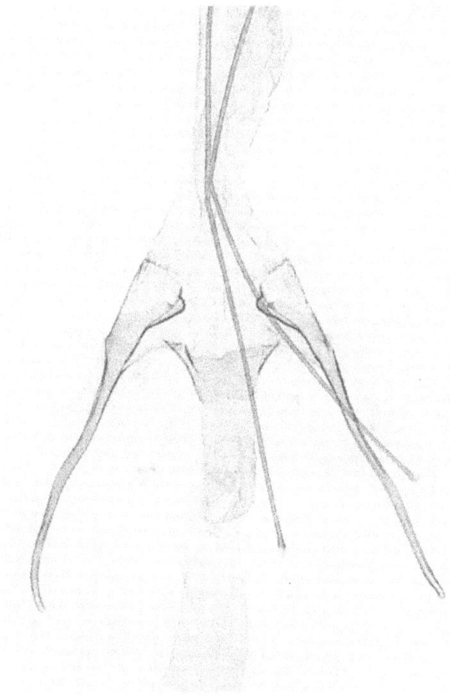

186

187

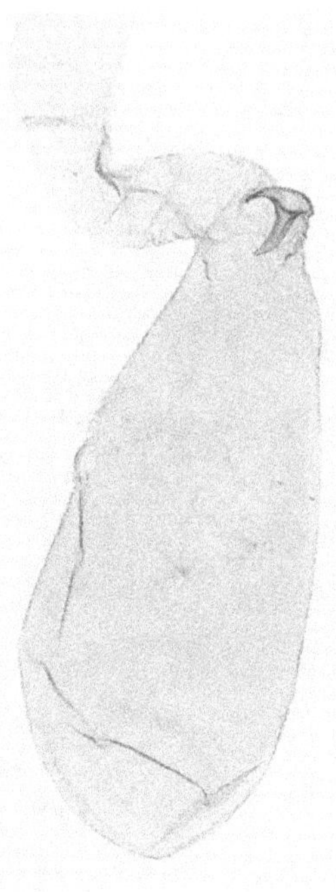

Female Genitalia

186. *Caryocolum proxima* (Haworth) – Portugal, GU 02/1128 (CORL).
187. *Caryocolum blandulella* (Tutt) – England, gen. slide 24528 (BMNH).

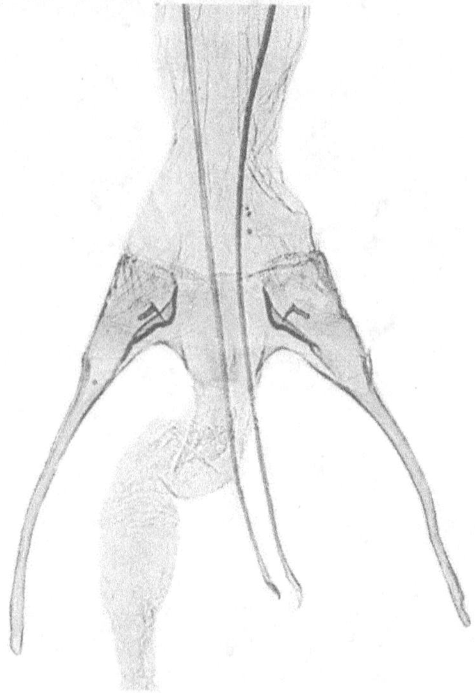

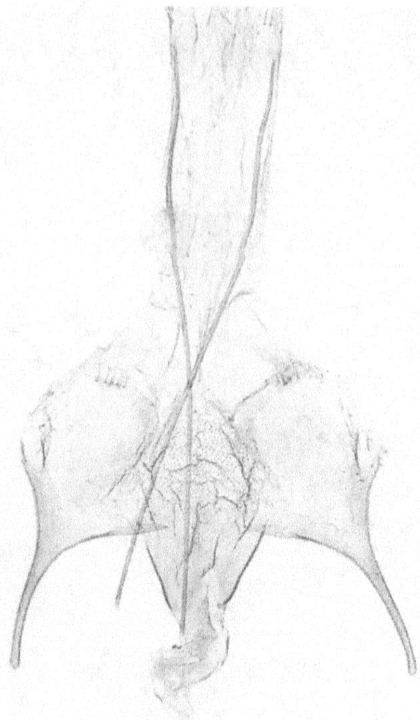

188

189

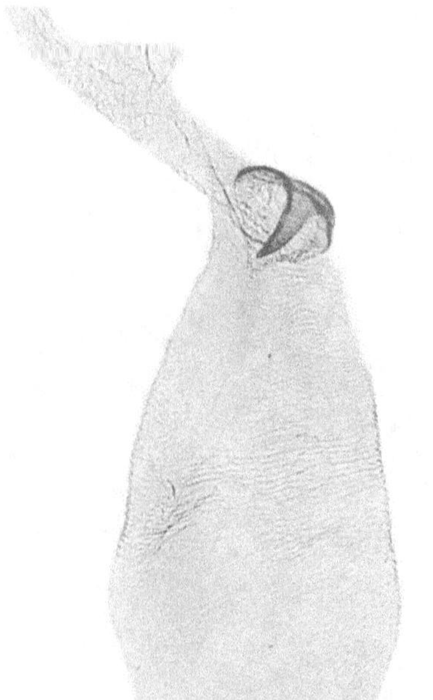

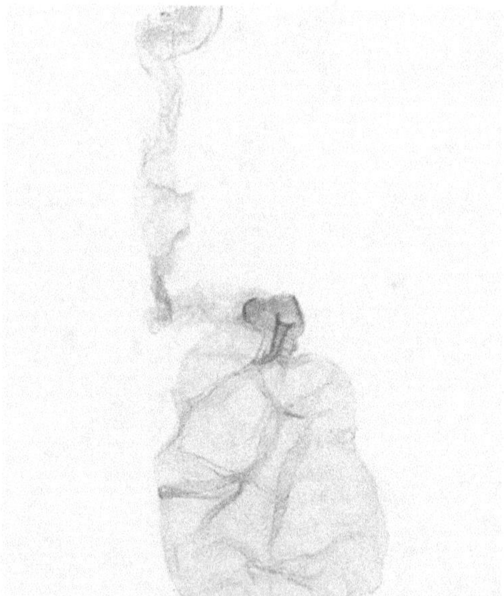

Female Genitalia

188. *Caryocolum arenbergeri* Huemer – Spain, GU 10/1323 P.H. (ZMUC).
189. *Caryocolum tricolorella* (Haworth) – Germany, GEL 1092 P.H. (TLMF).

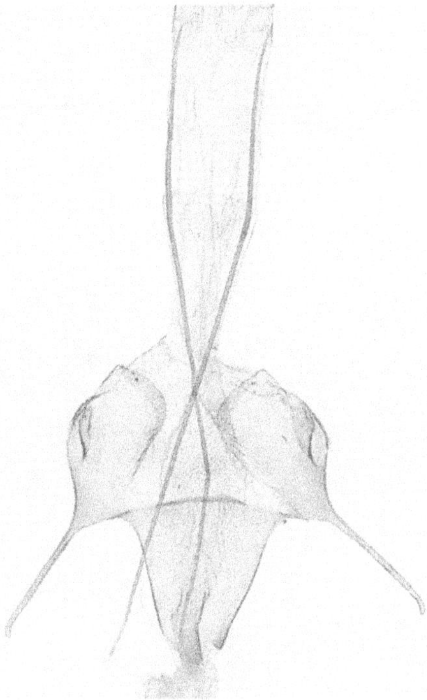

190

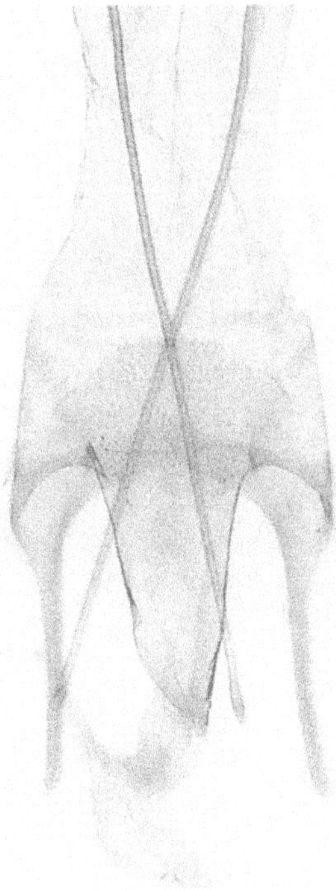

191

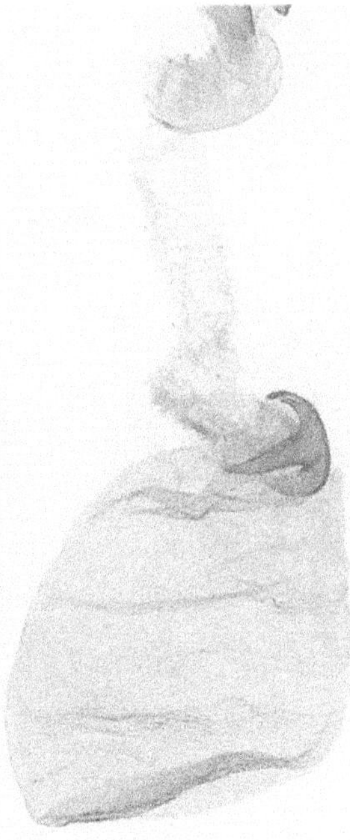

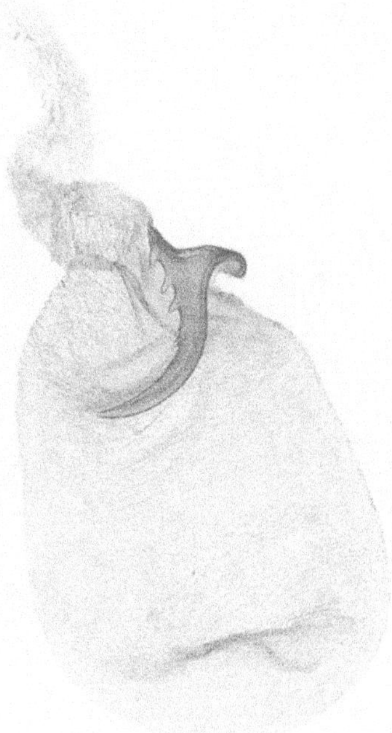

FEMALE GENITALIA

190. *Caryocolum fibigerium* Huemer – Spain, GEL 1095 (TLMF).
191. *Caryocolum junctella* (Douglas) – locality unknown, gen. slide 24260 (BMNH).

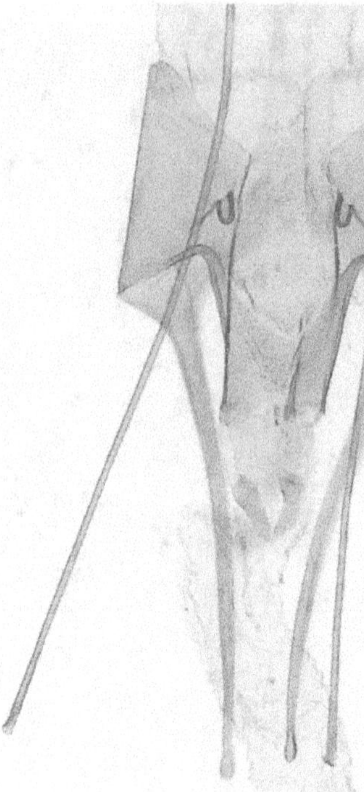

192

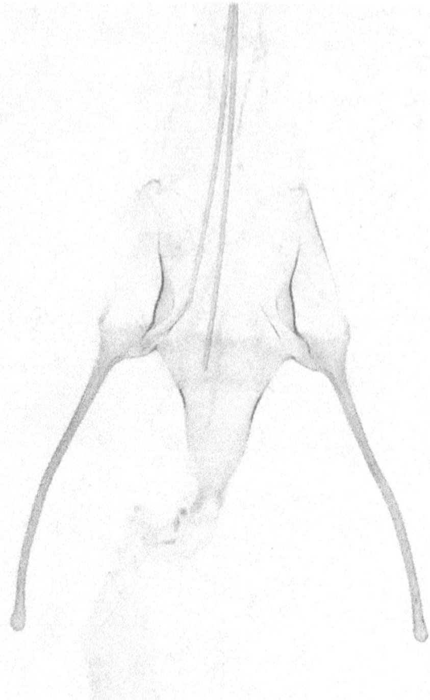

193

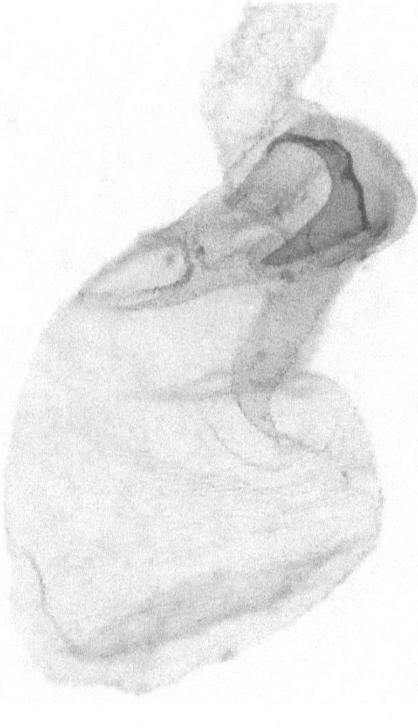

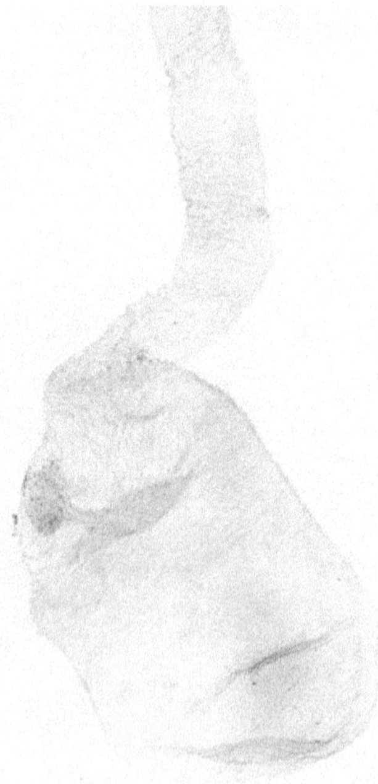

Female Genitalia

192. *Caryocolum cassella* (Walker) – Austria, GU 86/120 P.H. (TLMF).
193. *Caryocolum moehringiae* (Klimesch) – Greece, GU 88/170 P.H. (ZMUC).

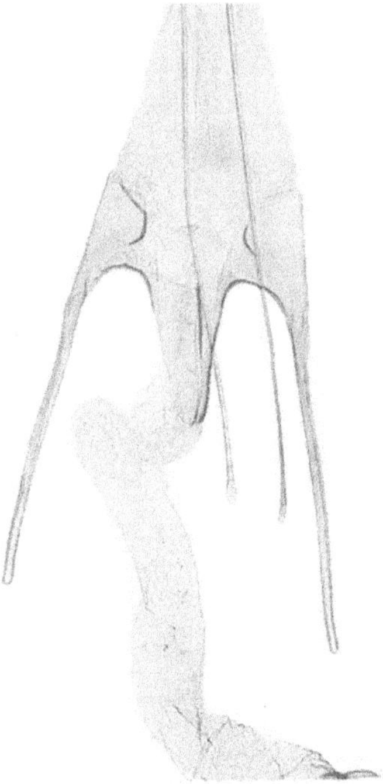

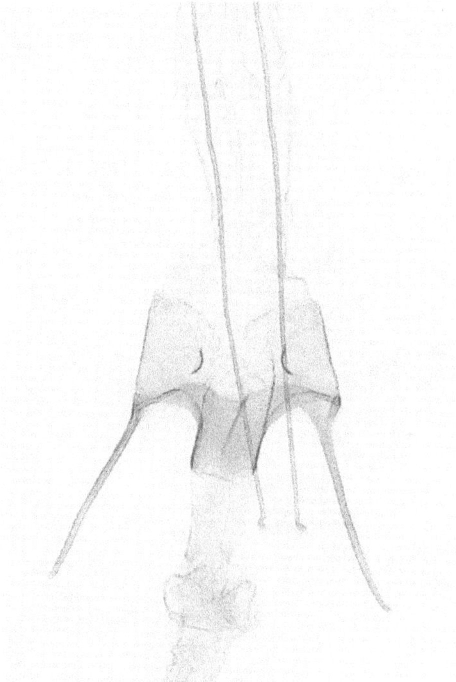

194

195

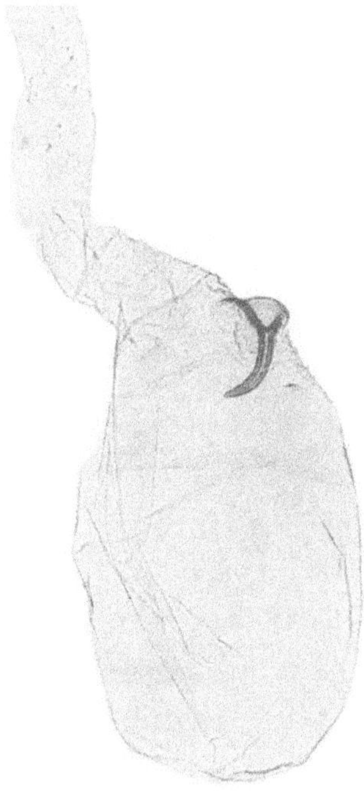

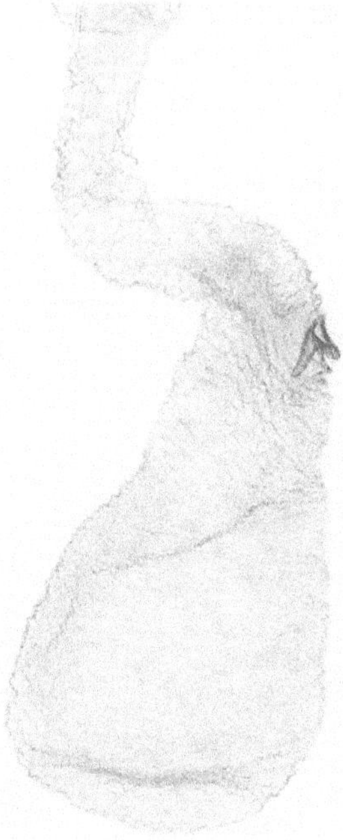

FEMALE GENITALIA

194. *Caryocolum petrophila* (Preissecker) – Austria, GEL 1130 P.H. (TLMF).
195. *Caryocolum huebneri* (Haworth) – Germany, gen. slide 24547 (BMNH).

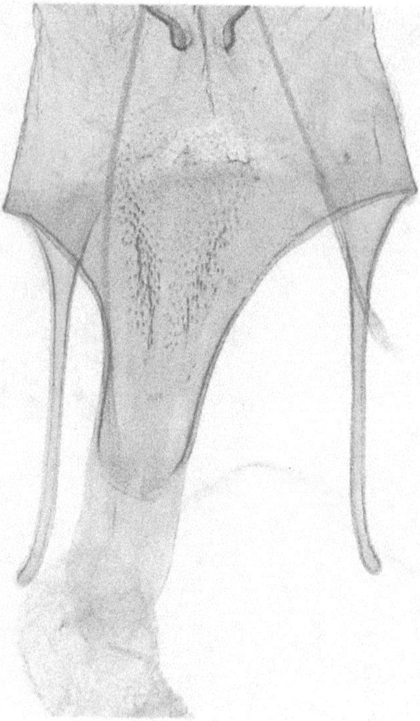

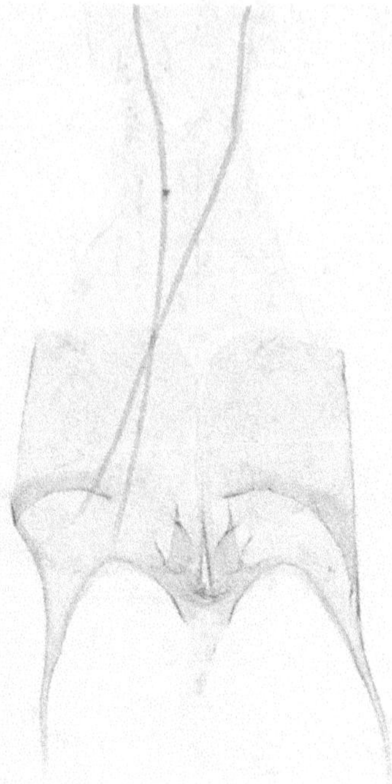

196 197

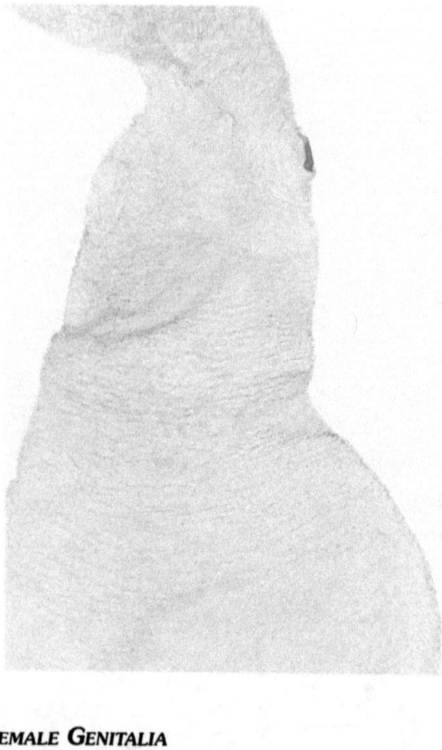

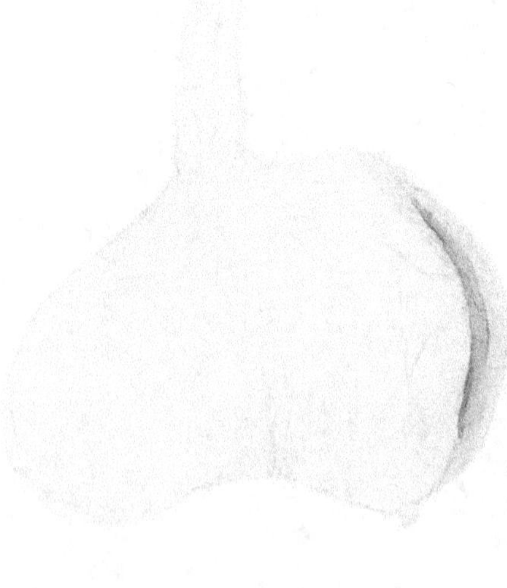

FEMALE GENITALIA

196. *Caryocolum kroesmanniella* (Herrich-Schäffer) – Germany, GEL 1096 P.H. (TLMF).
197. *Tila capsophilella* (Chrétien) – Austria, GEL 1039 P.H. (TLMF).

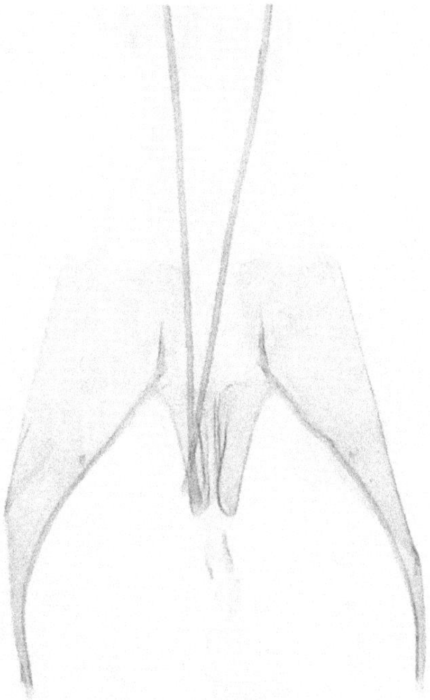

198

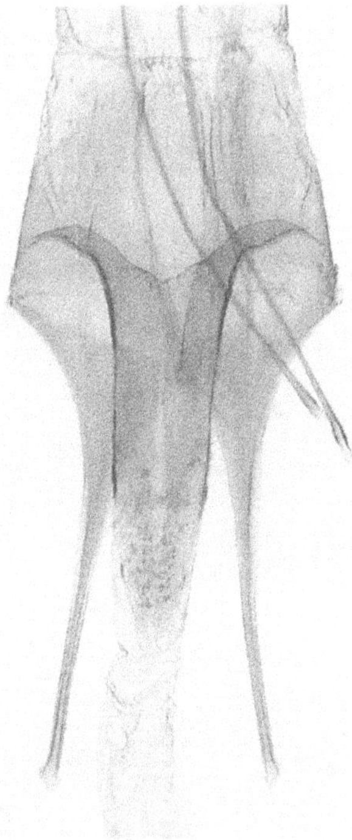

199

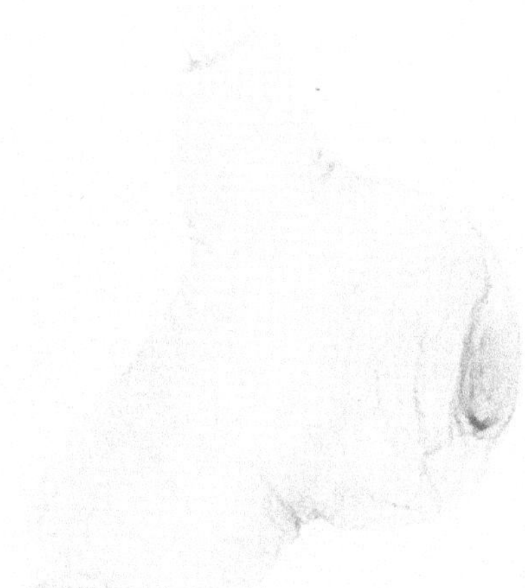

FEMALE GENITALIA

198. *Pogochaetia solitaria* Staudinger – Italy, GEL 1037 P.H. (TLMF).
199. *Agonochaetia terrestrella* (Zeller) – Switzerland, gen. slide 16367 (BMNH).

199a

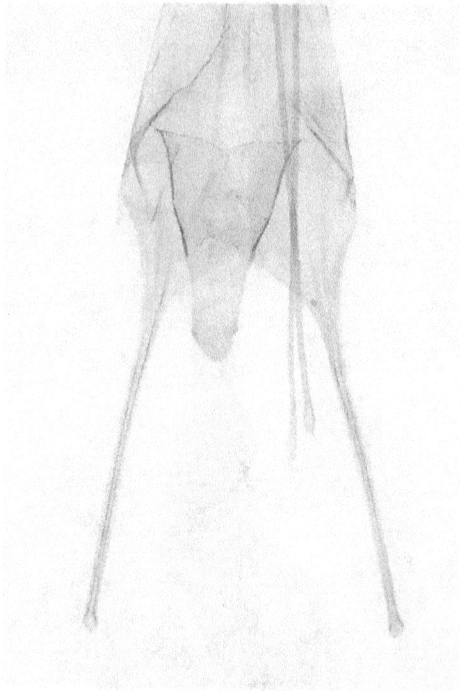

200

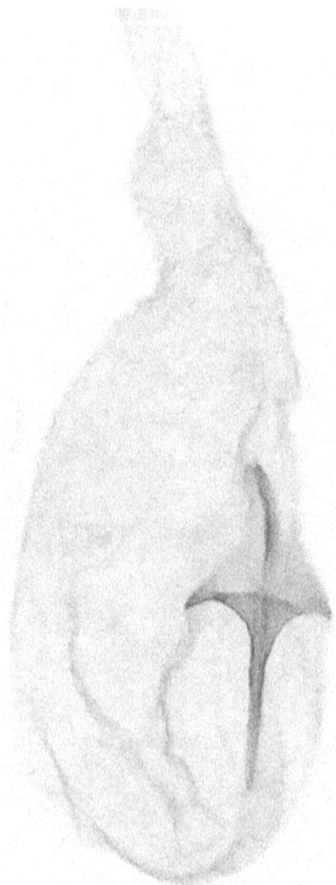

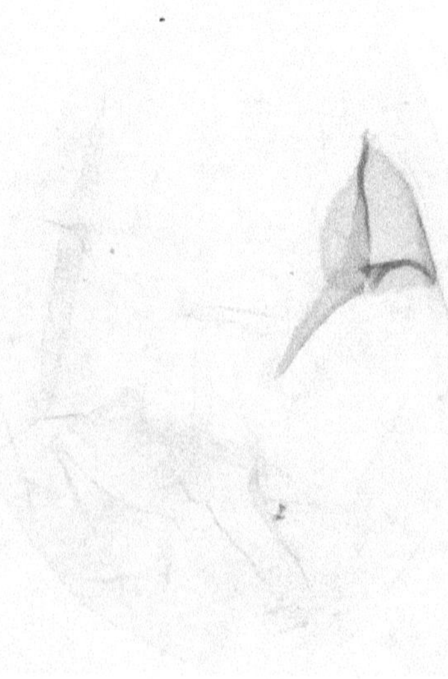

Female Genitalia

199a. *Agonochaetia terrestrella* (Zeller) – Switzerland, GU 06/1254 P.H. (BNMC).
200. *Agonochaetia intermedia* Sattler – Austria, GU 87/107 P.H. (TLMF).

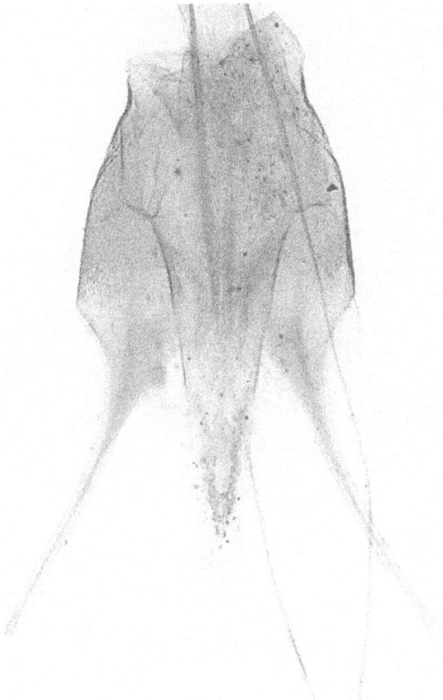

201

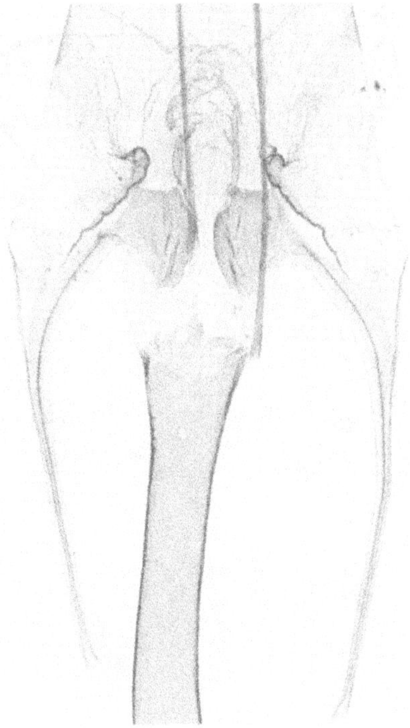

202

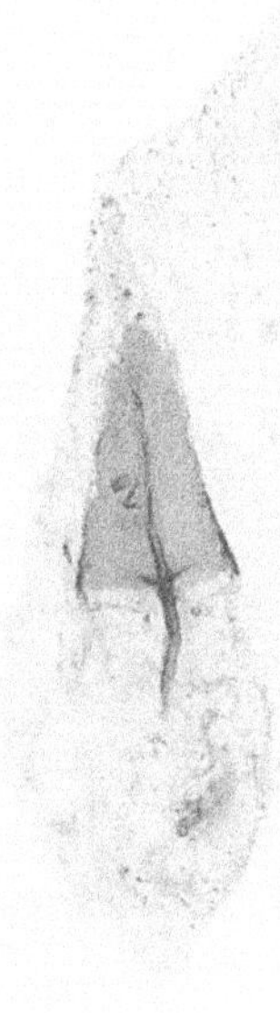

Female Genitalia

201. *Agonochaetia quartana* Povolný – Holotype, Bulgaria, 3222 Povolný (MZMB).
202. *Sattleria arcuata* Pitkin & Sattler – Spain, GEL 330 P.H. (TLMF).

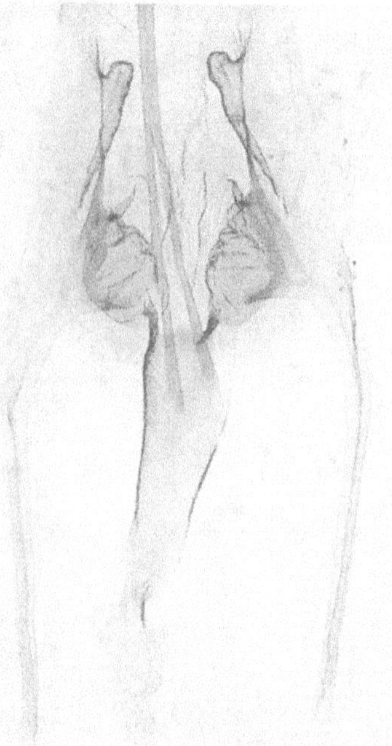

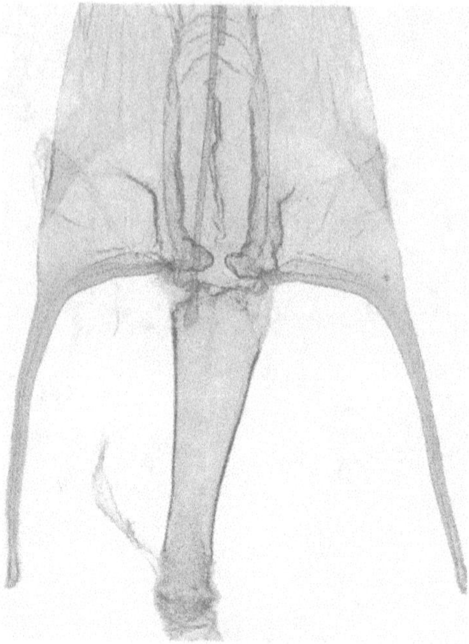

203

204

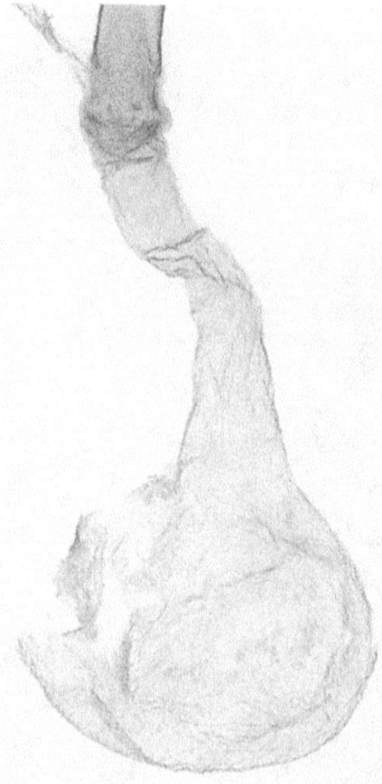

FEMALE GENITALIA

203. *Sattleria pyrenaica* (Petry) – ?Spain, GEL 331 P.H. (TLMF).
204. *Sattleria angustispina* Pitkin & Sattler – Paratype, France, gen. slide 53 L.M.P. (TLMF).

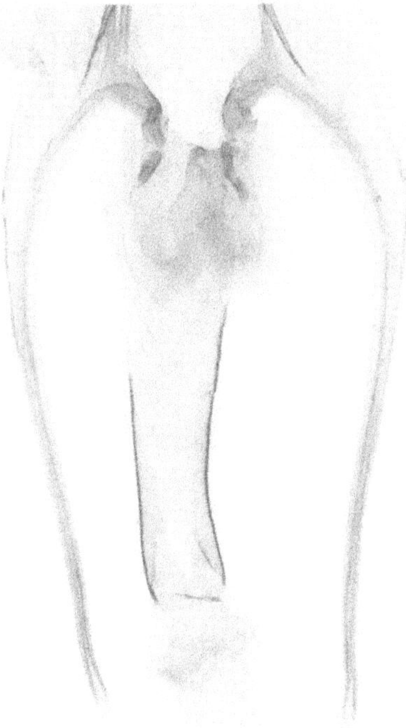

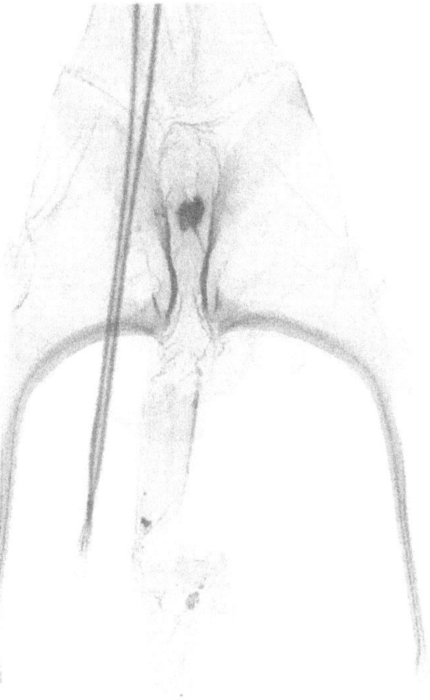

205

206

Female Genitalia

205. *Sattleria izoardi* Huemer & Sattler – Italy, GEL 1050 P.H. (TLMF).
206. *Sattleria breviramus* Pitkin & Sattler – France, gen. slide 13696 J.N. (TLMF).

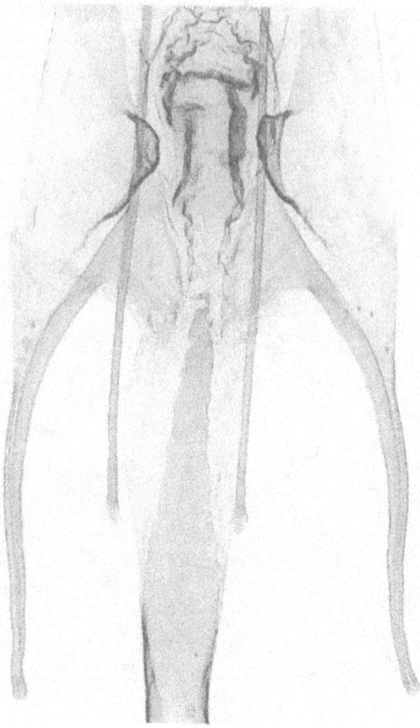

207

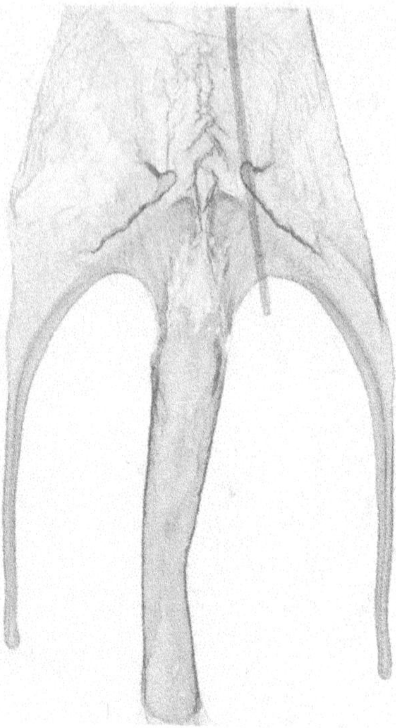

208

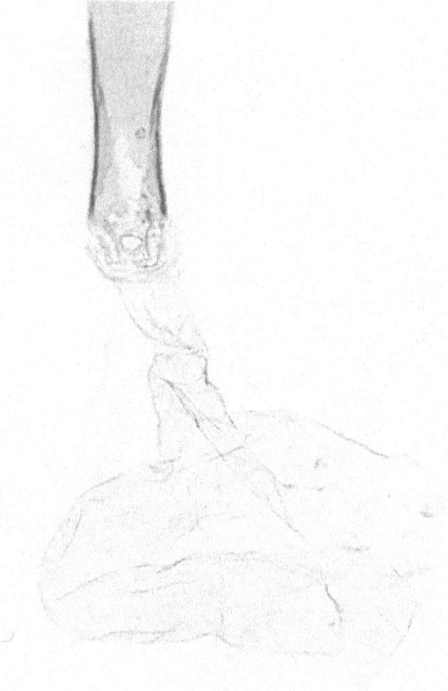

Female Genitalia

207. *Sattleria melaleucella* (Constant) – Austria, gen. slide 101 L.M.P. (TLMF).
208. *Sattleria triglavica basistrigella* Huemer – Switzerland, gen. slide 88 L.M.P. (TLMF).

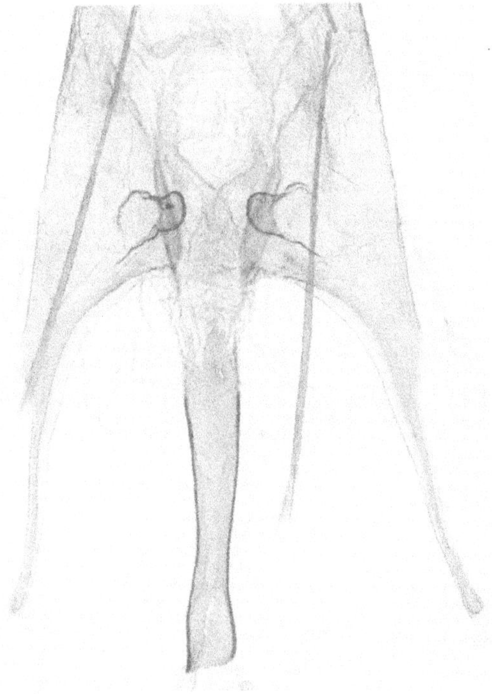

209

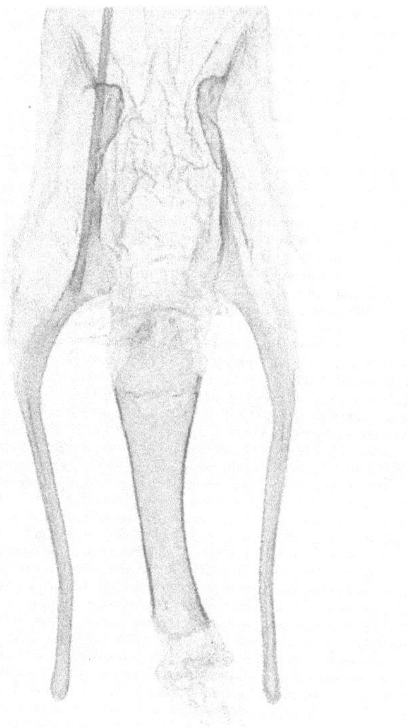

211

Female Genitalia

209. *Sattleria styriaca* Pitkin & Sattler – Austria, gen. slide 11240 (NHMW).
211. *Sattleria dzieduszyckii* (Nowicki) – Poland/Slovakia, gen. slide 98 L.M.P. (TLMF).

References

Agenjo, R., 1968. Catálogo Ordenador de los Lepidópteros de Espana, XIV y última entrega. Familias: Gelechiidae, Holcopogonidae, Carposinidae (nueva edición), Tortricidae, Cochylidae (nueva edición), Zygaenidae (nueva edición), Drepanidae (nueva edición), Thyatiridae (=Cymatophoridae) (nueva edición).- *Graellsia* **23** (1967) [apendix without pagination].

Agri Pest, 2010. *Tuta absoluta information network*.- http://www.tutaabsoluta.com/index.php (visited 7.5.2010).

Aguiar, A. M. F. & Karsholt, O., 2006. Lepidoptera. Systematic catalogue of the entomofauna of the Madeira Archipelago and Selvagens Islands 1.- *Bolm Mus. Munic. Funchal*, Supl. **9**: 5-139.

Amsel, H. G., 1935a. Weitere Mitteilungen über palästinensische Lepidopteren.- *Veröff. Dt. Kol. Übersee-Mus. Bremen* **1**: 223-277.

– 1935b. Neue palästinensische Lepidopteren.- *Mitt. Zool. Mus. Berlin* **20**: 271-319, pls 9-12.

– 1939. See Hartig, F. & Amsel, H. G., 1939.

– 1952. Lepidoptera Sardinica. Parte III. Descrizione di specie nuove ed osservazioni sistematiche di carattere generale.- *Fragm. ent.* **1** (1951): 101-144.

Amsel, H. G. & Hering, E. M., 1931. Beitrag zur Kenntnis der Minenfauna Palästinas.- *Dt. ent. Z.* **1931**: 113-152, pls 1-2.

Anikin, V. V. & Piskunov, V. I. 1996. New species of gelechiid moths from Saratov Province, Russia (Lepidoptera: Gelechiidae).- *Zoosyst. Rossica* **4**: 171-175.

– 1998. Two species of the genus *Scrobipalpa* Janse, 1951 (Lepidoptera: Gelechiidae: Gnorimoschemini) new for the fauna of Russia.- *Russian Entomol. J.* **7**(1-2): 69-70.

– 2001. Gelechiid moths (Lepidoptera, Gelechiidae) of lower Volga. [In Russian.]- *Izvestja saratovskogo gosudarstvenhogo Universiteta, novajya seria* **2001**: 259-264.

– 2003. Species of gelechiid moths (Lepidoptera, Gelechiidae), new to the European fauna. [In Russian.]- *Povolzhskii ekologicheskii zhurnal* **1**: 70-71.

Anikin, V. V., Sachkov, S. A. & Zolotuhin, V. V. 1999. "Fauna Lepidopterologica Volgo-Uralensis" 150 years later: changes and additions. Part 4. Coleophoridae, Gelechiidae, Symmocidae and Holcopogonidae (Insecta, Lepidoptera).- *Atalanta* **29** (1/4): 295-336.

Anonymous, 2006. *Tuta absoluta*. Data sheets of quarantine pests.- **35**: 434–435.

– 2009. *Tuta absoluta* (fam. Gelechiidae); de tomatenmineermot.- Plant Protection Service, National Reference Laboratory, Min.Agric., Nature & Food Qual. the Netherlands, 3 pp.

– 2010. Tomatenminiermotte.- http://www.fera.defra.gov.uk/showNews.cfm?id=402 (visited 11.3.2010).

Bankes, E. R., 1894. *Lita instabilella*, Dgl., and its nearest British Allies.- *Entomologist's mon. Mag.* **30**: 80-83, 125-128, 188-194.

– 1909. *Gelechia vicinella* Dgl., conspecific with *G. leucomelanella* Z.- *Entomologist's mon. Mag.* **20**: 263-267.

Bella, S., 2008. *Caryocolum siculum* sp. n. (Gelechiidae), feeding on *Gypsophila* (Caryophyllaceae) in Sicily.- *Nota lepid.* **31**: 69-75.

Benander, P., 1926. Über *Lita leucomelanella* Z. und eine dieser nahestehende Art.- *Ent. Tidskr.* **47**: 171-177, figs. 1-16.

– 1928. Fjärilar, Lepidoptera. II. Småfjärilar, Microlepidoptera. Tredje familjegruppen: Malfjärilar, Tineina. 1. Familjen Gelechiidae.- *Svensk Insektfauna* **10**: 1-97, pls 1-7.

– 1929. Zur Biologie einiger Kleinschmetterlinge III.- *Ent. Tidskr.* **50**: 123-145.

– 1937. Notiser om svenska Microlepidoptera, med beskrivning av *Lita arenariella* n. sp.- *Opusc. ent* **2**: 29-32.

– 1965. Notes on larvae of Swedish Microlepidoptera II.- *Opusc. ent.* **30**: 1-23.

Bidzilya, A. V., 1997. New records of gelechiid moths (Lepidoptera, Gelechiidae) from the Ukraine.- *Russian Entomol. J.* **6**: 123-125.

– 2000. New records of gelechiid moths from the Southern Siberia with description of three new species (Lepidoptera).- *Beitr. Ent.* **50**: 385-395.
– 2001. Two new species of Gelechiid-Moths from Central Asia (Lepidoptera: Gelechiidae).- *SHILAP Revta. lepid.* **29**: 161-163.
– 2005. On the distribution of Gelechiid Moths (Lepidoptera, Gelechiidae) in Siberia. Contribution 2.- *Proc. Zool. Mus. Kiev Taras Shevchenko Nat. Univ.* **3**: 7-19.
– 2009. On the distribution of gelechiid moths (Lepidoptera, Gelechiidae) in Siberia. Contribution 3.- *Proc. Zool. Mus. Kiev Taras Shevchenko Nat. Univ.* **5**: 3-13.
Bidzilya, A. V. & Budashkin, Y. I., 1998. New records of Lepidoptera from Ukraine. [In Russian]- *Proc. Zool. Mus. Kiev Taras Shevchnko Nat. Univ.* **2**: 59-68.
– 2004. New records of Microlepidoptera from the Ukraine. [In Russian]- *J. Ukr. ent. Soc.* **4** (3-4): 3-15.
– 2005. New records of Microlepidoptera from Ukraine. [In Russian]- *Proc. Zool. Mus. Kiev Taras Shevchenko Nat. Univ.* **3**: 20-30.
– 2009. New records of Microlepidoptera from Ukraine. [In Russian]- *Proc. Zool. Mus. Kiev Taras Shevchenko Nat. Univ.* **5**: 14-28.
Bidzilya, A. V., Budashkin, Y. I. & Kostjuk, I. Y., 1998. Additions to the fauna of Microlepidoptera of Transbaikalia.- *J. Ukr. ent. Soc.* **4** (1-2): 33-63.
Bidzilya, O. & Karsholt, O., 2008. New data on Anomologini from Palaearctic Asia (Gelechiidae).- *Nota lepid.* **31**: 199-213.
Bidzilya, O. & Li, H., 2010. Genus *Scrobipalpa* Janse (Lepidoptera, Gelechiidae) from China, with descriptions of 13 new species. *Zootaxa* **2513**: 1-26.
Billi, F., Tautel, C. & Varenne, T., 2010. Contribution à l'inventaire des Lépidoptères de Corse.- *Revue Assoc. roussill. Ent.* **19**: 85-88.
Bland, K. P., Corley, M. F. V., Emmet, A. M., Heckford, R. J., Huemer, P., Langmaid, J. R., Palmer, S. M., Parsons, M. S., Pitkin, L. M., Sattler, K. & Simpson, A. N. B., 2002. Gelechiinae. Pp. 118-204. *In* Emmet, A. M. & Langmaid, J. R. (eds): The moths and butterflies of Great Britain and Ireland. Vol. **4** (2) Gelechiidae. 277 pp. Colchester, Essex.
Boyd, T., 1858. Notes of an entomological tour in Cornwall.- *Entomologist's wkly Intell.* **4**: 142-144.
Boisduval, J. B. A. de, 1874. Note sur une nouvelle Maladie des Pommes de Terre.- *Journ. Soc. centr. Hort. Fr.* (2)**8**: 713-714.
Bradford, E. S., 1979. Gelechiidae. Pp. 115-132. *In* Emmet, A. M. (ed.): *A field guide to the smaller British Lepidoptera*. 271 pp. London.
Bruand d'Uzelle, C.T., 1851 ["1850"]. Catalogue systématique et synonymique des lépidopteres du Département du Doubs. Tinéides.- *Mém. Soc. Emul. Doubs* **3**(3): 23-58. [Published 11th March 1851 according to Viette (1977: 287)].
– 1858-59a. Classification des Tinéides et examen des caractères et de leur importance relative, d'après la mèthode naturelle.- *Annls Soc. ent. Fr.* (1857) (3) **5**: 807-826; (1858) **6**: 601-702 (1859).
Bryner, R., Grimm, K., Kopp, A. & Sonderegger, P., 2005. Für die Schweiz neue oder interessante Funde von Microlepidopteren.- *Mitt. ent. Ges. Basel* **54**: 123-132.
Bucheli, S. R., 2009. Annotated review and discussion of phylogenetically important characters for families and subfamilies of Gelechioidea (Insecta: Lepidoptera).- *Zootaxa* **2261**: 1-22.
Bucheli, S. R. & Wenzel, J. W., 2005. Gelechioidea (Insecta: Lepidoptera) systematics: A reexamination using combined morphology and mitochondrial DNA data.- *Molecular Phylogenetics and Evolution* **35**: 380-394.
Buhl, O., Falck, P., Jørgensen, B., Karsholt, O., Larsen, K. & Vilhelmsen, F., 1998. Fund af småsommerfugle fra Danmark i 1997 (Lepidoptera). - *Ent. Meddr.* 66: 105-115.
Buhl, O., Falck, P., Karsholt, O., Larsen, K. & Vilhelmsen, F., 2009. Fund af småsommerfugle fra Danmark i 2008 (Lepidoptera).- *Ent. Meddr.* **77**: 65-81.
– in press. Fund af småsommerfugle fra Danmark i 2009 (Lepidoptera).- *Ent. Meddr.* **78**.
Burmann, K., 1949. Drei für Österreich neue *Lita*-Arten aus Nordtirol.- *Z. wien ent. Ges.* **34**:

68-71.
– 1951. Etwas über *Lita cacuminum* Frey (Lepidoptera, Gelechiidae).- *VerZ. Ent. Ver. Innsbruck* **2**: 25-26.
– 1954a. *Gelechia dzieduszyckii* Now. nov. subspec. *fusca* (Lepidoptera, Gelechiidae).- *Z. wien. ent. Ges.* **65**: 345-352, pl. 18.
– 1954b. *Gelechia interalbicella* H. S. (Lepidoptera, Gelechidae).- *Ent. Z. Stuttgart* **64**: 202-205.
– 1990. Beiträge zur Microlepidopteren-Fauna Tirols. XIV. *Caryocolum* Gregor & Povolný, 1954 (Insecta: Lepidoptera, Gelechiidae).- *Ber. nat.-med. Ver. Innsbruck* **77**: 171-184.
Busck, A., 1900. New species of moths of the superfamily Tineina from Florida. - *Proc. U. S. natn. Mus.* **23**: 225-254, pl. 1.
– 1903. A revision of the American moths of the family Gelechiidae, with descriptions of new species.- *Proc. U. S. natn. Mus.* **25**: 767-930, pls 28-32.
– 1916. Descriptions of new North American Microlepidoptera.- *Proc. ent. Soc. Wash.* **18**: 147-154.
– 1928. *Phthorimaea lycopersicella*, n. sp., (Family Gelechiidae) a leaf feeder on tomato (Lep.).- *Proc. Hawaii. ent. Soc.* **7**: 171-176.
– 1939. Restriction of the genus *Gelechia* (Lepidoptera: Gelechiidae), with descriptions of new genera.- *Proc. U.S. natn. Mus.* **86**: 563-593, pls 58-71.
Buszko, J., Karsholt, O. & Rynarzewski, T., 2000. Gelechiidae.- Pp. 44-51. *In* Buszko, J. & Nowacki, J. (eds.): The Lepidoptera of Poland. A Distributional Checklist.- *Polish Entomological Monographs* **1**: 1-178.
Butler, A. G., 1880. On a collection of Lepidoptera Heterocera from Marlborough Province, New Zealand.- *Cist. Ent.* **2**: 541-562.
Caradja, A., 1920. Beitrag zur Kenntnis der geographischen Verbreitung der Microlepidopteren des palaearktischen Faunengebietes nebst Beschreibung neuer Formen. III. Teil.- *Dt. ent. Z. Iris* **34**: 75-179.
– 1932. Beiträge zur Lepidopteren-Fauna Grossrumäniens für das Jahr 1931.- *Bull. Sect. sc. Acad. Rom.* **15**: 35-46.
Chrétien, P., 1898. Nouvelles espèces de Microlépidoptères de France et d´Algèrie.- *Naturaliste* **20**: 177-178.
– 1900a. Description d´une nouvelle espèce de Microlépidoptère de France.- *Bull. Soc. ent. Fr.* **1900**: 138.
– 1900b. Description d´une nouvelle espèce de Microlépidoptère de France.- *Bull. Soc. ent. Fr.* **1900**: 223-224.
– 1901. Microlépidoptères du *Silene nutans* L.- *Naturaliste* **23**:17-18.
– 1908a. Microlépidoptères nouveaux pour la faune francaise.- *Naturaliste* **30**: 245-246.
– 1908b. Nouveaux Microlépidoptères de France et de Corse.- *Naturaliste* **30**: 258-261.
– 1908c. Description d´un Microlépidoptère nouveau d`Algérie.- *Bull. Soc. ent. Fr.* **1908**: 231-232.
– 1910. Lépidoptères nouveaux pour la faune francaise.- *Naturaliste* **32**: 260-262, 272-274.
– 1915. Contribution a la connaissance des Lépidoptères du Nord de l´Afrique.- *Annls Soc. ent. Fr.* **84**: 313-356.
– 1925. La legende de *Graellsia isabellae.* Appendice.- *Amateur Papillons* **2**: 241-247.
Christoph, H., 1872. Neue Lepidoptera des europaeischen Faunengebietes.- *Horae Soc. ent. ross.* **9**: 3-39, pl. 1-2.
Clarke, J. F., 1932. New Microlepidoptera from the Pacific Coast (Gelechiidae).- *Can. Ent.* **64**: 63-69.
– 1962. New Species of Microlepidoptera from Japan.- *Ent. News* **73**: 91-102.
– 1965. Microlepidoptera of Juan Fernandez Islands. - *Proc. U. S. natn. Mus.* **117**: 1-105, pl. 1.
– 1969. Catalogue of the Type specimens of Microlepidoptera in the British Museum (Natural History) described by Edward Meyrick. Gelechiidae (D-Z). **7**: 1-531, pls 1-265. British Museum (Nat. Hist.), London.

Clemens, B., 1860. Contributions to American Lepidopterology.- *Proc. Acad. nat. Sci. Philad.* **1860**: 156-174.

Constant, A., 1865. Description de quelques Lépidoptères nouveaux.- *Annls Soc. ent. Fr.* (4)**5**: 189-198, pl. 7.

– 1885. Notes sur quelques Lépidoptères nouveaux 2e Partie (1).- *Annls Soc. ent. Fr.* **1884**: 251-262, pl. 2.

– 1890a. Séance du 26 juin 1889 [Diagnoses d'espèces nouvelles de Lépidoptères de France].- *Bull. Soc. ent. Fr.* **1890**: 124-126.

– 1890b. Descriptions de Microlépidoptères nouveaux ou peu connus.- *Annls Soc. ent. Fr.* **69**: 5-16.

– 1895. Microlépidoptères nouveaux de la faune francaise.- *Bull. Soc. ent. Fr.* **1895**: 50-55.

Corley, M. F. V., 2002a. The larva of *Caryocolum fibigerium* Huemer, 1988 (Lepidoptera: Gelechiidae).- *Entomologist's Gaz.* **53**: 20.

– 2002b. A week in Serra da Estrela, Portugal in early September 2001, with additional Lepidoptera species for the Portugese fauna.- *Entomologist's Rec.* **114**: 97-104.

– 2005. Further additions to the Lepidoptera of Algarve, Portugal. II. (Insecta: Lepidoptera).- *SHILAP Revta. lepid.* **33**: 347-364.

– 2009. *In* Anonymous. 2008 Annual exhibition.- *Br. J. ent. nat. Hist.* **22**: 159-187, pls 1-3.

Corley, M. F. V., Gardiner, A. J., Cleere, N. & Wallis, P. D., 2000. Further additions to the Lepidoptera of Algarve, Portugal (Insecta: Lepidoptera).- *SHILAP Revta. lepid.* **28**: 245-319.

Corley, M. F. V., Marabuto, E. & Pires, P., 2007. New Lepidoptera for the fauna of Portugal (Insecta: Lepidoptera).- *SHILAP Revta. lepid.* **35**: 312-334.

Corley, M.F.V., Maravalhas, E. & Passos de Carvalho, J. 2006. Miscellaneous additions to the Lepidoptera of Portugal (Insecta: Lepidoptera).- *SHILAP Revta. lepid.* **34**: 407-427.

Courtois, J. H., 1988. *Scrobipalpa salinella* Zeller, espèce nouvelle pour la Lorraine (Lep., Gelechiidae).- *Alexanor* **14**: 234-238.

Dandria, D. & Catania, A. (2009): *Tuta absoluta* (Povolny, 1994), an important agricultural pest in Malta (Lepidoptera: Gelechiidae).- *Bull. ent. Soc. Malta* **2**: 57-60.

Derra, G., 1985. *Caryocolum hackeri* sp. n., eine neue Gelechiidae aus den Pyrenäen Frankreichs und Spaniens (Lepidoptera, Gelechiidae).- *Entomofauna, Linz* **6**: 373-379.

Diakonoff, A. N., 1954. Microlepidoptera of New Guinea. Results of the Third Archibold Expedition (American-Netherlands Indian Expedition 1938-1939). Part IV.- *Verh. K. ned. Akad. Wett., Afd. Natuurk., Reeks* 2 **50**(1): 1-191.

Dohrn, C. A., 1867. Einleitende Bemerkung der Redaction. *In* Staudinger, O.: *Gelechia petasitella* und *Phyllobrostis Hartmanni.- Stettin. ent. Ztg* **28**: 210-212.

Doubleday, H., 1859. *The Zoologist synonymic list of British butterflies and moths*, edn 2. 40 pp. London.

Douglas, J. W., 1846. Descriptions of ten new British moths.- *Zoologist, London* **4**: 1266-1270.

– 1850-1852a. On the British species of the genus *Gelechia* of Zeller. (Continued.)- *Trans. ent. Soc. Lond.* (N.S.) **1**: 14-21, 60-68 (1850); 101-108 (1851); 241-251 (1852).

– 1852b. Contributions towards the Natural History of British Microlepidoptera.- *Trans. ent. Soc. Lond.* (N.S.) **2**: 75-81, pls. 10-11.

Dumont, C., 1931. Contribution à l'étude des Lépidoptères du Nord de l'Afrique (suite). Description et éthologie de trois espèces nouvelles (Lep. Gelechiidae).- *Bull. Soc. ent. Fr.* **1931**: 148-153.

– 1932. Les Lépidoptères des gommiers du nord de l'Afrique.- *Livre du Centenarie. Soc. ent. Fr.*: 689-719.

Duponchel, P.-A.-J., 1842-1844. Nocturnes, supplement aux tomes quatrième et suivants. *In* Godart, J.-B.: *Histoire naturelle des Lépidoptères ou papillons de France.* **Suppl. 4**., 553 pp., pls 51-90. Paris.

Elmore J. C. & Howland A. F., 1943. Life history and control of the tomato pinworm.- *USDA Technical Bulletin* **841**: 30 pp.

Elsner, G., 1998. Gelechiidae.- Pp. 35-39. *In* Laštůvka, Z. (ed.): *Checklist of Lepidoptera of the Czech and Slovak Republics (Insecta, Lepidoptera)*. 117 pp. Brno.

Elsner, G., Huemer, P. & Tokár, Z., 1999. *Die Palpenmotten (Lepidoptera, Gelechiidae) Mitteleuropas. Bestimmung – Verbreitung – Flugstandort – Lebensweise der Raupen*.- 208 pp. Bratislava.

EPPO, 2009. *List of pests recommended for regulation as quarantine pests A2*. Version 2009-9. European and Mediterranean Plant Protection Organisation.- http://www.eppo.org/QUARANTINE/listA2.htm (visited 17.5.2010).

Erschoff, N. G., 1874. Lepidoptera. *In* Fedchenko, A. P.: Puteshestvie v Turkestan. [Journey in Turkestan.] [In Russian.]- *Izv. imp. Obshch. Lyub. Estest. Antrop. Etnogr.* **11**(2): i-vi, 1-127, pls 1-6.

– 1877. Diagnosen neuer Lepidopteren aus den verschiedenen Provinzen des Russischen Reiches.- *Horae Soc. ent. ross.* **12**: 336-348.

Espinosa, B. & Sannino, L. (2009): *Tuta, Keiferia e Phthorimaea*, tignole da tenere sotto controllo.- L'Informatore Agrario **29/2009**: 56-58.

Fabricius, J. C., 1794. *Entomologia systematica emendata et aucta* **3**(2): 349 pp. Hafniae.

Falkovitsh, M. I. & Bidzilya, O. V., 2003. Gelechiidae (Lepidoptera) reared from the larvae collected in Kyzylkum desert, with descriptions of new species. [In Russian]- *Proc. Zool. Mus. Kiev Taras Shevchenko Nat. Univ.* **1** (1): 113-147.

– 2006. The Turanian gelechiid moths of the tribe Gnorimoschemini (Lepidoptera, Gelechiidae) living on plants of the family Chenopodiaceae, with descriptions of new species. [In Russian]- *Proc. Zool. Mus. Kiev Taras Shevchenko Nat. Univ.* **4**: 62-104.

– 2009. A list of gelechiid moths (Lepidoptera, Gelechiidae) of the Southern Kyzylkum. [In Russian]- *Proc. Zool. Mus. Kiev Taras Shevchenko Nat. Univ.* **5**: 65-98.

Fazekas, I., 2002. Systematisches und synonymisches Verzeichnis der Microlepidopteren Ungarns (Lepidoptera: Microlepidoptera).- *Folia hist.-nat. Mus. Matr.* **26**: 289-327.

Fera, 2009. Fera confirms the first outbreak in the UK of *Tuta absoluta* - the South American tomato moth.- *The Food and Enviromental Research Agency*.- http://www.fera.defra.gov.uk/showNews.cfm?id=402 (visited 4.3.2010).

Fischer von Röslerstamm, J. E., 1834-1843. *Abbildungen zur Berichtigung und Ergänzung der Schmetterlingskunde, besonders der Microlepidopterologie als Supplement zu Treitschke's und Hübner's europäischen Schmetterlingen mit erläuterndem Text*.- 304 pp, 100 pls. Leipzig.

Frey, H., 1867. Die schweizerischen Microlepidopteren.- *Mitt. schweiz. ent. Ges.* **2**: 286-303.

– 1870. Ein Beitrag zur Kenntnis der Microlepidopteren.- *Mitt. schweiz. ent. Ges.* **3**: 244-256, 277-289.

– 1880. *Die Lepidopteren der Schweiz*. xxvi + 454 pp. Leipzig.

Fuchs, A., 1886. Microlepidopteren des unteren Rheingau's, nebst einer allgemeinen topographisch-lepidopterologischen Einleitung.- *Stettin. ent. Ztg* **47**: 39-83.

– 1901. Vier neue Kleinfalter der europäischen Fauna.- *Stettin. ent. Ztg* **62**: 382-387.

– 1902. Neue Geometriden und Kleinfalter des europäischen Faunengebietes.- *Stettin. ent. Ztg* **63**: 317-330.

Gaede, M., 1937. Familia: Gelechiidae. *In* F. Bryk (ed.): *Lepidopterorum Catalogus* **79**: 1-630. Berlin.

Gaedike, R. & Heinicke, W. (eds.), 1999. Verzeichnis der Schmetterlinge Deutschlands (Entomofauna Germanica 3).- *Ent. Nachr. Ber.*, Suppl. **5**: 1-126.

Gerstberger, M., 2003. Halotopbionte Schmetterlingsarten (Insecta: Lepidoptera) der Naturschutzgebiete „Artener Solgraben" und „Schlossberg-Solwiesen" bei Auleben in Thüringen.- *Rudolstädter nat.-hist. Schr.* **11**: 135-140.

Geyer, C., [1832-1836]. Plate 70-71. *In* Hübner, J. 1796-1836: *Sammlung europäischer Schmetterlinge* **8**. Tineae-Schaben. [78] pp., 71 pls. Augsburg.

Gozmány, L. A., 1951. Neue Kleinschmetterlinge I.- *Folia ent. hung.* (S. N.) **4**: 17-24.

– 1954. Studies on Microlepidoptera.- *Annls hist.-nat. Mus. natn. hung.* **5**: 273-285.

– 1960. The Results of the Zoological Collecting Trip to Egypt in 1957, of the Natural History Museum, Budapest. 8. Egyptian Microlepidoptera II.- *Annls hist.-nat. Mus. natn. hung., pars Zool.* **52**: 411-421.

Gregersen, K. & Kjeldgaard, S., 2009. Om forekomsten af *Gnorimoschema herbichii* (Nowicki, 1864) i Danmark (Gelechiidae).- *Lepidoptera* **9**: 235-242.

Gregor, F. & Povolný, D., 1954. Systematische und zoogeographische Studie über die Gruppe der Arten *Gnorimoschema* Busck mit Rücksicht auf die richtige Diagnostik des Schädlings *Gnorimoschema ocellatellum* Boyd.- *Zool. Ent. Listy* **3**: 83-96, pl 7, map.

– 1955. Nové a významné nálezi Lepidopter z Československa.- *Čas. Morav. mus. Brne* **40**: 114-*129, pls 1-7.*

Hackman, W., 1946. Neue *Phthorimaea*-Arten (Lepid. Gelechiidae) aus Ostfennoskandien.- *Notulae Entomol.* **26**: 59-64.

Hammad, S. M., El-Din Hussein Youssef, K. & Abdel-Halim Assem, M., 1968. The biology of the sugar beet mining moth.- *Bull. Soc. ent. Egypte* **12**: 49-51.

Hartig, F., 1938. Sechs neue Mikrolepidopteren.- *Z. öst. EntVer.* **23**: 82-83, 88-90, pl. 7.

– 1941. Neue Microlepidopteren.- *Mitt. münch. ent. Ges.* **31**: 154-163, pls 5-8.

Hartig, F. & Amsel, H. G., 1939. Contributo alla conoscenza della fauna entomologica della Sardegna. Nuove forme di Lepidotteri.- *Memorie Soc. ent. ital.* **17**: 63-84, pls 3-5.

– 1952. Lepidoptera Sardinica. Parte II. Microlepidoptera.- *Fragm. ent.* **1** (1951): 43-100.

Hauder, F., 1917. See Hauder, F., & Rebel, H., 1917.

Hauder, F. & Rebel, H., 1917. Versammlung der Sektion für Lepidopterologie.- *Verh. zool.-bot. Ges. Wien* **67**: (29)-(35).

Hausenblas, D., 2006. Korrekturen und Ergänzungen zur Mikrolepidopterenfauna Baden-Württembergs und angrenzender Gebiete.- *Mitt. ent. Ver. Stuttgart* **41**: 3-27.

– 2009. Korrekturen und Ergänzungen zur Mikrolepidopterenfauna Baden-Württembergs und angrenzender Gebiete – 2. Beitrag.- *Mitt. ent. Ver. Stuttgart* **44**: 81-106, 1 pl.

Haworth, A. H., 1812. A brief account of some rare insects announced at various times to the Society, as new to Britain.- *Trans. ent. Soc. Lond.* **1**: 332-340.

– 1828. *Lepidoptera Britannica.* Part **4** (1828): 512-609. Londini.

Heckford, R. J., 2000. *Caryocolum marmoreum* (Haworth) (Lepidoptera: Gelechiidae): some apparently unrecorded observations on the early larval stages.- *Entomologist's Gaz.* **51**: 194.

Heckford, R. J. & Sattler, K., 1999. *Scrobipalpula diffluella* (Frey, 1870) in Britain (Lepidoptera: Gelechiidae).- *Entomologist's Gaz.* **50**: 255-260.

Heikkilä, M. & Kaila, L., 2010. Reassessment of the enigmatic Lepidopteran family Lypusidae (Lepidoptera: Tineoidea; Gelechioidea).- *Systematic Entomology* **35**: 71-89.

Heinemann, H. v., 1870. *Die Schmetterlinge Deutschlands und der Schweiz.* 2. Abteilung Kleinschmetterlinge. **2**. Die Motten und Federmotten. Heft 1. 388 pp. Braunschweig.

Hering, E., 1889. Beiträge zur Mitteleuropäischen Micro-Lepidopterenfauna.- *Stettin. ent. Ztg* **50**: 290-320.

– 1891. Ergänzungen und Berichtigungen zu F. O. Büttner's Pommerschen Microlepidopteren.- *Stettin. ent. Ztg* **52**: 135-227.

Hering, E. M., 1924. Beitrag zur Kenntnis der Microlepidopteren-Fauna Finlands.- *Notulae Entomol.* **4**: 75-84.

– 1933. *Lita petryi* Hering in Schweden.- *Ent. Tidskr.* **54**: 86-88.

– 1957. *Bestimmungstabellen der Blattminen von Europa einschliesslich des Mittelmeerbeckens und der Kanarischen Inseln.* **1**: 1-648; **2**: 651-1185; **3**: 1-221. S'Gravenhage.

Herrich-Schäffer, G. A. W., 1847-1855. *Systematische Bearbeitung der Schmetterlinge von Europa.* **5**, 394 pp., 124 + 7 + 1 pls. Regensburg.

Heyden, C. v., 1862. Fragmente aus meinen entomologischen Tagebüchern.- *Stettin. ent. Ztg* **23**: 171-176.

Hoare, R. J. B., Langmaid, J. R., Simpson, A. N. B. & Young, M. R., 1999. *Caryocolum blandelloides* Karsholt, 1981 (Lepidoptera: Gelechiidae) in the British Isles.- *Entomologist's Gaz.* **50**: 149-154.

Hodges, R. W., 1978. Gelechioidea in part: Cosmopterigidae. *In* Dominick, R. B. *et al.*, *The Moths of America North of Mexico including Greenland.* ix + 166 pp., 6 pls. London.
– 1983. Gelechiidae. Pp. 19-25. *In* Hodges, R. W., (ed.): *Checklist of the Lepidoptera of America North of Mexico.* xxiv + 284 pp. London.
– 1986. Gelechioidea: Gelechiidae (in part). *In* Dominick, R. B. *et al.*, *The Moths of America North of Mexico including Greenland* fasc. **7**(1): 1-195, vii-xiii. Washington.
– 1998. The Gelechioidea. Pp. 131-158. *In* Kristensen, N. P., (ed.): *Lepidoptera, Moths and Butterflies. Volume 1: Evolution, Systematics, and Biogeography. Handbook of Zoology. Volume IV Arthropoda: Insecta Part* **35**. vi + 491 pp. Berlin, New York.
Hodges, R. W. & Becker, V. O., 1990: Nomenclature of some neotropical Gelechiidae (Lepidoptera).- *Proc. ent. Soc. Wash.* **92**: 76-85.
Hodgkinson, J. B., 1897. A new British species of the Gelechiidae: *Lita intermediella* Hodgkinson.- *Entomologist's Rec. J. Var.* **9**: 36.
Hofmann, E., 1867. Drei neue Gelechien und ein neuer *Chauliodus*.- *Stettin. ent. Ztg* **28**: 200-207.
Hofmann, O., 1899. *Lita petryi* mihi.- *Stettin. ent. Ztg* **60**: 139-144.
Homan, R., 2010. *Tuta absoluta* (Meyrick, 1917: Lep. Gelechiidae): an adventive species new to Britain.- *Entomologist's Rec. J. Var.* 122: 31-32.
Howard, L. O., 1897. A leaf-miner of tobacco.- *Bull. U. S. Bur. Ent.* (n. s) **7**: 76.
Huemer, P., 1986. Neufunde von Kleinschmetterlingen aus Vorarlberg (Österreich) (Insecta: Lepidoptera).- *Ber. nat.-med. Ver. Innsbruck* **73**: 147-154.
– 1987. *Caryocolum blandelloides* Karsholt, 1981, neu für Österreich und Mitteleuropa (Insecta: Lepidoptera: Gelechiidae).- *Ber. nat.-med. Verein Innsbruck* **74**: 207-209.
– 1988. A taxonomic revision of *Caryocolum* (Lepidoptera: Gelechiidae).- *Bull. Br. Mus. nat. Hist.* (Ent.) **57**: 439-571.
– 1989a. Eine neue, gallenerzeugende *Caryocolum*-Art aus Mitteleuropa (Lepidoptera: Gelechiidae).- *Nota lepid.* **12**: 21-28.
– 1989b. Two new species of the genus *Caryocolum* from Spain.- *SHILAP Revta. lepid.* **17**: 197-204.
– 1989c. Bemerkenswerte Funde von *Caryocolum*-Arten aus den Südalpen und dem Mediterraneum (Lepidoptera, Gelechiidae).- *NachrBl. bayer. Ent.* **38**: 37-40.
– 1989d. Neue und wenig bekannte Arten der Gattung *Caryocolum* Gregor & Povolný, 1954, aus Südwestasien (Lepidoptera, Gelechiidae).- *Mitt. münch. ent. Ges.* **79**: 127-142.
– 1989e. Das Weibchen von *Agonochaetia intermedia* Sattler, 1968 (Lepidoptera, Gelechiidae).- *Mitt. schweiz. ent. Ges.* **62**: 387-389.
– 1991. *Caryocolum delphinatella* (Constant) sp. rev., a senior synonym of *C. fiorii* (Klimesch) (Lepidoptera: Gelechiidae).- *Tijdschr. Ent.* **134**: 31-34.
– 1993a. Bemerkungen zu den paläarktischen Arten der Gattung *Lutilabria* Povolný (Lepidoptera, Gelechiidae).- *Dt. ent. Z.* (N. F.) **40**: 341-347.
– 1993b. *Caryocolum viscariella* (Stainton) und *C. albifaciella* (Heinemann) sp. rev., zwei distincte Arten (Lepidoptera: Gelechiidae).- *Z. ArbGem. öst. Ent.* **45**: 27-34.
– 1993c. The British species of *Caryocolum* Gregor & Povolný.- *Br. J. Ent. Nat. Hist.* **6**: 145-157.
– 1996. *Gnorimoschema nilsi* sp. n. - eine bemerkenswerte neue Schmetterlingsart aus dem Gebiet des Nationalparks Hohe Tauern (Osttirol, Österreich) (Lepidoptera: Gelechiidae).- *Wissenschaftliche Mitt. Nationalpark Hohe Tauern* **2**: 75-86.
– 1997. *Sattleria basistrigella* (Müller-Rutz, 1934), ein invalider Name (Lepidoptera: Gelechiidae).- *Mitt. schweiz. ent. Ges.* **70**: 285-288.
– 1998. Die monophage Schmetterlingsfauna (Lepidoptera) des Kriechenden Gipskrautes (*Gypsophila repens*) im Gebiet des Nationalparks Hohe Tauern.- *Wissenschaftliche Mitt. Nationalpark Hohe Tauern* **4**: 47-56.
Huemer, P. & Erlebach, S., 2003. *Cosmardia moritzella* (Treitschke, 1835), ein interessanter Kleinschmetterlings-Neufund für die Fauna Südtirols (Lepidoptera: Gelechiidae).- *Gredleriana*

3: 33-36.
Huemer, P. & Jakšic, P., 1996. Gelechiidae (Insecta, Lepidoptera).- *Fauna Durmitora* **5**: 81-89.
Huemer, P. & Karsholt, O., 1998. A review of the Old World *Scrobipalpula* (Gelechiidae), with special reference to central and northern Europe.- *Nota lepid.* **21**: 37-65.
– 1999. Gelechiidae I (Gelechiinae: Teleiodini, Gelechiini).- *In* Huemer, P., Karsholt, O. & Lyneborg, L. (eds): *Microlepidoptera of Europe* **3**: 1-356. Stenstrup.
– 2008. Bemerkungen zur Synonymie und Verbreitung des Kleinschmetterlings *Scrobipalpa feralella* (Zeller, 1872) = *Scrobipalpa rebeliella* (Hauder, 1917) syn. nov. (Lepidoptera).- *Wiss. Jb. tirol. Landesmuseen* **1**: 200-203.
Huemer, P. & Luquet, G., 1992a. De l'identité de *Lita repentella* Chrétien, 1908 (Lepidoptera, Gelechiidae).- *Nota lepid.* **15**: 29-40.
Huemer, P. & Mayr, T., 2000. Bemerkenswerte Erstnachweise von Schmetterlingen (Lepidoptera) für Vorarlberg.- *Vorarlberger Naturschau* **8**: 113-128.
Huemer, P. & Nel, J., 2005. *Caryocolum mazeli* sp. n., a new species from southern France (Lepidoptera, Gelechiidae).- *Bull. Soc. ent. Fr.* **110**: 125-127.
Huemer, P. & Sattler, K., 1989. Das brachyptere Weibchen von *Caryocolum laceratella* (Zeller, 1868) (Lepidoptera, Gelechiidae).- *Nota lepid.* **11**: 256-624.
– 1992. Ergänzende Bemerkungen zur Speziation alpiner *Sattleria* (Lepidoptera: Gelechiidae).- *Z. ArbGem. öst. Ent.* **44**: 59-73.
Huemer, P. & Tarmann, G., 1993. Die Schmetterlinge Österreichs (Lepidoptera). Systematisches Verzeichnis mit Verbreitungsangaben für die einzelnen Bundesländer. *Veröff. tirol. Landesmus. Ferdinandeum Innsb.*, Suppl. **5**: 1-224.
Huismann, K. J. & Koster, J. C., 1997. Nieuwe en interessante Microlepidoptera uit Nederland in het jaar 1994 (Lepidoptera).- Ent. Ber., Amst. 57: 45-65.
Humphreys, H. N. & Westwood, J. O., 1845. *British Moths and their Transformations* **2.** xix + 268 pp., pls 57-124. London.
ICZN, 1999. International code of zoological nomenclature. 4th ed. xxix + 306 pp. London.
Ivinskis, P., 2004. *Lepidoptera of Lithuania. Annotated catalogue*. 380 pp. Vilnius.
Janse, A. J. T., 1949-63. Gelechiadae. *Moths of South Africa* **5**: 1-464, 202 pls (1949-54); **6**: 1-284 pp., 138 pls (1958-63). Pretoria.
Jansen, M., 1999. The genus *Scrobipalpa* in the Netherlands (Lepidoptera: Gelechiidae).- *Nederlandse Faunistische Mededelingen* **9**: 29-78.
Joannis, J. de, 1910. Descriptions de deux Géléchides nouvelles de France (Lep.).- *Bull. Soc. ent. Fr.* **15**: 295-298.
Junnilainen, J., Karsholt, O., Nupponen, K., Kaitila, J.-P., Nupponen, T., & Olschwang, V., 2009. The gelechiid fauna of the Southern Ural Mountains, part II: list of recorded species with taxonomical notes (Lepidoptera: Gelechiidae).- *Zootaxa* **2367**: 1-68.
Junnilainen, J. & Nupponen, K., 2010. The gelechiid fauna of the Southern Ural Mountains, part I: descriptions of seventeen new species (Lepidoptera: Gelechiidae).- *Zootaxa* **2366**: 1-34.
Jürivete, U., 2007. Eesti faunale uutest pisiliblikatest (Microlepidoptera) aastal 2005.- *Lepinfo, Tartu*, **17**: 6-15.
Jürivete, U. & Õunap, E., 2008. *Eesti Liblikad Katalog. Estonian Lepidoptera Catalogue.* 175 pp. Talin.
Kaila, L., 2004. Phlogeny of the superfamily Gelechioidea (Lepidoptera: Ditrysia): an exemplar approach.- *Cladistics* **20**: 303-340.
Kaitila, J.-P., 1996. Suomen jäytäjäkoiden (Gelechiidae) elintavat.- *Baptria* **21**: 81-105.
Karsholt, O., 1981. Northern European species of the genus *Caryocolum* Gregor & Povolný, 1954, feeding on *Cerastium* and *Stellaria*, with the description of a new species (Lepidoptera: Gelechiidae).- *Ent. scand.* **12**: 251-270.
– 1992. *Caryocolum jaspidella* (Chrétien, 1908): Male genitalia, distribution and notes on its biology (Lepidoptera, Gelechiidae).- *Nota lepid.* **14**: 302-307.
– 1995. Kommentiertes Verzeichnis der Symmocidae, Blastobasidae und Gelechiidae Ost-

deutschlands (Lepidoptera).- *Beitr. Ent.* **45**: 137-154.
– 2004-2009. Gelechiidae. *In:* Karsholt, O. & Nieukerken, E. J. van (eds.). Lepidoptera, Moths.- *Fauna Europaea,* version 1.1 (2004), version 2.0 (2009).- http://www.faunaeur.org
– 2005. Gelechiidae. *In* Baldizzone, G.: I Microlepidotteri del Parco Naturale Alpi Marittime (Italia, Piemonte) (Lepidoptera).- *Boll. Mus. Reg. Sci. nat. Torino* **22**(1): 112-141.
Karsholt, O., Kristensen, N. P., Ahola, M. & Simonsen, T., in press. Lepidoptera. *In* Böcher, J. & Kristensen, N. P. (eds): *The Greenland Entomofauna. An Identification Manual of Insects and their Allies*.- Stenstrup.
Karsholt, O., Larsen, K. & Aarvik, L., 1986. A remarkable disjunction: *Scrobipalpa reiprichi* Povolný, 1984 discovered in Norway, with remarks on the characteristics of the species (Lepidoptera, Gelechiidae).- *Nota lepid.* **9**: 191-199.
Karsholt, O., & Nielsen, E. S., 1974. A new species of *Gnorimoschema* from Northern Europe (Lepidoptera, Gelechiidae).- *Notulae Entomol.* **54**: 91-96.
Karsholt, O. & Razowski, J. (eds.), 1996. *The Lepidoptera of Europe. A distributional checklist*.- 380 pp. Stenstrup.
Karsholt, O. & Riedl, T., 1996. Gelechiidae, excl. Gnorimoschmini. Pp. 103-113, 118-122, 310, 312. *In* Karsholt, O. & Razowski, J. (eds): *The Lepidoptera of Europe. A distributional checklist*.- 380 pp. Stenstrup.
Karsholt, O. & Sattler, K., 1998. A new brachypterous *Ephysteris* Meyrick from the island of Madeira (Lepidoptera: Gelechiidae).- *Entomologist's Gaz.* **49**: 35-47.
Kieffer, J. J. & Jørgensen, P., 1910. Gallen und Gallentiere aus Argentinien.- *Zentbl. Bakt. ParasitKde* (2 Abt.) **27**: 362-444.
Klimesch, J., 1938a. *Lita inflativorella* nov. spec. (Lep., Gelechiidae).- *Z. öst. EntVer.* **23**: 1-4.
– 1938b. Piccolo contributo alla conoscenza dei Microlepidottero della Valle Venosta.- *Studi trent. Sci. nat.* **19**: 38-51.
– 1943. Die Raupe der *Lita cacuminum* Frey (Lep., Gelechiidae).- *Z. wien. ent. Ges.* **28**: 153-156.
– 1951a. Contributo alla fauna Lepidotterologica del Trentino.- *Studi trent. Sci. nat.* **27**: 11-78, 2 pls.
– 1951b. Über Microlepidopteren des Traunsteingebietes in Oberösterreich.- *Z. wien. ent. Ges.* **36**: 101-117.
– 1951c. *Phthorimaea* (*Lita*) *saginella* Z. (= *coussonella* Chret.) (Lep., Gelechiidae).- *Entomol. NachrBl. öst. schweiz. Ent.* **3**: 191-192.
– 1953-1954. Die an Caryophyllaceen lebenden europäischen *Gnorimoschema* Busck (= *Phthorimaea* Meyr.)-Arten.- *Z. wien. ent. Ges.* **38** (1953): 225-239, 272-282, 311-319; **39** (1954): 273-288, 335-341, 357-362.
– 1958. Beiträge zur Kenntnis der blattminierenden Insektenlarven des Linzer Gebietes und Oberösterreichs. III. Gelechiidae, Acrolepiidae.- *Naturk. Jb. Stadt Linz* **1958**: 265-279.
– 1968. Die Lepidopterenfauna Mazedoniens, IV Microlepidopteren.- *Posebno Izd. prirod. Muz. Skopje* **5**: 1-203.
– 1984. Beiträge zur Kenntnis der Microlepidopteren-Fauna des Kanarischen Archipels. 6. Beitrag: Gelechiidae.- *Viraea* **13**: 145-182.
Kozlov, M. V. & Kullberg, J., 2006. Lepidoptera of Terskij seashore of the Kola Peninsula, Northwestern Russia. *Ent. fenn.* **17**: 123-129.
Kovács, Z. & Kovács, S., 2000. The commented checklist of Gelechiidae (Lepidoptera) of Romania.- *Entomol. rom.* **4**: 11-44.
Kristensen, N. P., (ed.) 2003. *Lepidoptera, Moths and Butterflies Volume 2: Morphology, Physiology, and Development. Handbook of Zoology. Volume IV Arthropoda: Insecta Part* **36**. I-xii + 564 pp. Berlin, New York.
Kullberg, J., Albrecht, A., Kaila, L. & Varis, V., 2002. Checklist of Finnish Lepidoptera – Suomen perhosten luettelo.- *Sahlbergia* **6**: 45-190.
Kuznetzov, V. I & Stekolnikov, A. A., 1984. System and phylogenetic associations of families

and superfamilies of gelechioid Lepidoptera of the infraorder Papilionomorpha (Lepidoptera: Copromorphoidea, Elachistoidea, Coleophoroidea, Gelechioidea) with use of functional morphology of genitalia of. [In Russian.]- *Trudy zool. Inst. Leningrad* **122**: 3-68.

– 2001. New approaches to the system of Lepidoptera of world fauna (on the basis of the functional morphology of abdomen). [In Russian.]- *Proc. zool. Inst. St. Petersburg* **282**: 1-462.

Langmaid, J. R., 1995. An alternative feeding habit by the larva of *Scrobipalpa obsoletella* (Fischer von Röslerstamm) (Lepidoptera: Gelechiidae).- *Entomologist´s Gaz.* **46**: 277.

Larsen, C. S., 1927. Tillæg til Fortegnelse over Danmark Microlepidoptera.- *Ent. Meddr.* **17**: 7-211.

Lee, S., Hodges, R. W. & Brown, R. L., 2009. Checklist of Gelechiidae (Lepidoptera) in America North of Mexico.- *Zootaxa* **2231**: 1-39.

Leraut, P., 1997. Liste systématique et synonymique des Lépidoptères de France, Belgique et Corse (deuxième edition).- *Alexanor* (Suppl.).- 526 pp. Paris.

Lhomme, L., 1946-1949. *Catalogue des Lépidoptères de France et de Belgique* **2**(4): 489-648 [1946-1948]; **2**(5): 649-808 [1948-1949]. Douelle (Lot).

Li, H. & Bidzilya, O., 2008. A review of the genus *Ephysteris* Meyrick, 1908 from China, with descriptions of two new species (Lepidoptera : Gelechiidae).- *Zootaxa* **1733**: 45-56.

Li, H. & Yang, C., 1996. Two species of gelechiid moths new to China (Lepidoptera: Gelechiidae). [In Chinese.]- *Acta agric. bor.-occid. Sinica* **5**: 27-30.

Li, H. & Zheng, Z., 1995. A new species of the genus *Caryocolum* (Lepidoptera: Gelechiidae) from China.- *Entomotaxonomia* **17**: 293-295.

Liu, Y. & Pai, K., 1979. A study on the Gelechiid moths in Tailing forest areas of Heilunkiang province.- *Scient. silv. sin.* **15**: 276-280, pls 4-6.

Lower, O. B. 1900. Descriptions of new Australian Lepidoptera.- *Proc. Linn. Soc. N. S. Wales* **25**: 403-423.

Lucas, D, 1940. Contribution à l'étude des Lépidoptères de l'Afrique du Nord.- *Bull. Soc. ent. Fr.* **44** (1939): 226-229.

– 1950. Contribution à la Faune des Lépidoptères de l'Afrique du Nord.- *Bull. Soc. ent. Fr.* **54** (1949): 143-144.

Lvovsky, A. L. & Piskunov, V. I., 1989. The gelechiid moths (Lepidoptera, Gelechiidae) of the Transaltai Gobi. [In Russian.]- *Nasek. Mongolii* **10**: 521-571.

Mabille, P., 1906. Essai sur la faune de l'Ile d'Oléron.- *Annls Soc. ent. Fr.* **75**: 37-56, pl. 3, fig. 5.

Mann, J., 1857. Verzeichnis der im Jahre 1853 in der Gegend von Fiume gesammelten Schmetterlinge.- *Wien. ent. Monatsschr.* **1**: 139-160, 161-189.

– 1867. Schmetterlinge gesammelt im Jahre 1867 in der Umgebung von Bozen und Trient.- *Verh. zool.-bot. Ges. Wien* **17**: 829-844.

– 1872. Beschreibung sieben neuer Arten Microlepidopteren.- *Verh. zool.-bot. Ges. Wien* **22**: 35-40.

– 1878. See Mann, J. & Rogenhofer, A., 1878.

Mann, J. & Rogenhofer, A., 1878. Zur Lepidopteren-Fauna des Dolomiten-Gebietes.- *Verh. zool.-bot. Ges. Wien* **27**: 491-500.

Mariani, M., 1941-1943. Fauna Lepidopterorum Italiae. Parte 1 - Catalogo ragionata dei Lepidotteri d'Italia.- *Giorn. Sc. Nat. Eco. Palermo* **42** (3): 1-79 (1941), 81-273 (1943), 1 pl.

Meess, A., 1910. XXXVIII. Fam. Gelechiidae. *In* Spuler, A., 1903-1910: *Die Schmetterlinge Europas* **2**, [vi]+523 pp., colour pls 56-91. Stuttgart.

Mendes, C. de Azevedo. 1904. Lepidopteros de Portugal. II. Lepidopteros da região de S. Fiel (Beira Baixa).- *Brotéria, Lisboa* **3**: 223-254.

Meyrick, E., 1891. A fortnight in Algeria, with descriptions of new Lepidoptera.-*Entomologist´s mon. Mag.* **27**: 9-13, 55-62.

– 1902. A new genus of Gelechiadae.- *Entomologist´s mon. Mag.* **38**: 103-104.

– 1904. Descriptions of Australian Microlepidoptera. XVIII. Gelechiadae.- *Proc. Linn. Soc. N.S.W.* **29**: 255-441.

– 1908. Descriptions of African Micro-Lepidoptera.- *Proc. zool. Soc. Lond.* **1908**: 716-756.
– 1909. New South African Micro-Lepidoptera.- *Ann. S. Afr. Mus.* **5**: 349-379.
– 1910. Notes and descriptions of Indian Micro-Lepidoptera.- *Rec. Indian Mus.* **5**: 217-232.
– 1912-1937. *Exotic Microlepidoptera* **1**: 1-640 (1912-1916), **2**: 1-640 (1916-1923), **3**: 1-640 (1923-1930), **4**: 1-642 (1930-1936), **5**: 1-160 (1936-1937). Marlborough.
– 1914. Description of South African Micro-Lepidoptera.- *Ann. S. Afr. Mus.* **10**: 243-257.
– 1917. Descriptions of South American Micro-Lepidoptera.- *Trans. Ent. Soc. Lond.* **1917**: 1-52.
– 1921. Descriptions of South African Micro-Lepidoptera.- *Ann. Transv. Mus.* **8**: 49-148.
– 1934. A new micro-lepidopteron (Lepidoptera Tineina) from Greenland.- *Entomologist* **67**: 59.
Michaelis, H. N., 1977. Records of Gelechiidae (Lep.) from North Wales.- *Entomologist's Gaz.* **28**: 217-222.
Millière, P., 1864. Iconographie et Description de Chenilles et Lépidoptères inédits.- *Annls Soc. Linn. Lyon* (N. S.) **10**: [187]-244, pls 37-44.
– 1876. *Catalogue raisonné des Lépidoptères des Alpes-Maritimes* **3**: 249-455.
Minet, J., 1990. Remaniement partiel de la classification des Gelechioidea, essentiellement en function de caractères primaginaux (Lepidoptera Ditrysia).- *Alexanor* **16**: 239-255.
Mutanen, M., Itämies, J., Junnilainen, J., Kaitila, J., Kullberg, J., Mutanen, T. & Välimäki, P., 2003. Huomionarvoiset pikkuperhoshavainnot 2000 [Noteworthy records of Finnish Microlepidoptera (Micropterygidae-Pyralidae) in 2000].- *Baptria* **2003**(1): 2-16.
Mutanen, M., Mutanen, T., Kullberg, J., Kaitila, J.-P., Laasonen, E. M. & Nupponen, K., 2003. Pikkuperhoshavainnot 2006-2007 [Noteworthy records of Finnish Microlepidoptera (Micropterygidae-Pyralidae) in 2006-2007].- *Baptria* **2008**(4): 117-130.
Müller-Rutz, J., 1914. See Vorbrodt, K. & Müller-Rutz, J., 1913-1914.
– 1922. Die Schmetterlinge der Schweiz (4. Nachtrag).- *Mitt. schweiz. ent. Ges.* **13**: 217-259.
– 1934. Ueber Microlepidopteren. Beschreibung neuer Arten und Formen aus der Schweiz, sowie ergänzende Angaben über ältere, weniger bekannte Arten.- *Mitt. schweiz. ent. Ges.* **16**: 118-128, pl. 1.
Nel, J., 1999a. Sur la biologie et le statut de quelques espèces francaises du genre *Scrobipalpa* Janse, 1951, appartenant au complexe *salinella-instabilella* (Lepidoptera, Gelechiidae).- *Alexanor* **20** (1998): 455-465.
– 1999b. Espèces nouvelles ou rarement signalées de microlépidoptères de France (Lepidoptera).- *Bull. Soc. ent. Fr.* **104**: 347-355.
– 2001. Espèces nouvelles ou rarement signalées de microlépidopères des Alpes Méridionales francaises (Lepidoptera, Alucitidae, Gelechiidae, Elachistidae, Ochsenheimeriidae).- *Bull. Soc. ent. Fr.* **106**: 101-104.
– 2003a. Microlépidoptères nouveaux ou rarement signalés de la fauna de France (Lepidoptera).- *Bull. Soc. ent. Fr.* **108**: 81-86.
– 2003b. Description de quatre nouvelles espèces de microlépidoptères découvertes dans le Midi de la France (Lepidoptera: Tineidae, Gelechiidae, Epermeniidae).- *Revue Assoc. roussill. Ent.* **12** (2): 46-53.
– 2003c. Entomologie prospectrice: contribution à la connaisance d'espèces méconnues ou nouvelles pour la France (Lepidoptera).- *Bull. Soc. ent. Fr.* **108**: 393-400.
– 2006. Microlépidoptères nouveaux ou peu connus de la faune de France (Lepidoptera, Adelidae, Tineidae, Cosmopterigidae, Elachistidae et Gelechiidae).- *Revue de l'Association Roussillonnaise d'Entomologie* **15**: 47-53.
– 2009. Confirmation de la présence de *Tuta absoluta* (Meyrick, 1917), ravageur de la Tomate, dans le sud de la France (Lep., Gelechiidae).- *Bull. Soc. ent. Fr.* **114**: 427-428.
Nel, J. & Varenne, T., 2003. Deux microlépidoptères nouveaux ou méconnus de France, observés dans le Parc national Mercantour (Alpes-Maritimes) (Lep., Gelechiidae).- *Bull. Soc. ent. Fr.* **108**: 87-88.
Novickij, S., 1924. *Lita paniculatella.* (Rössler i. l.) n. sp.- *Sbornik Ent. odd. Narad. Mus.*

Prag (2) **19**: 141-146.
Nowicki, M., 1864. *Microlepidopterorum Species novae*. 32 pp., 1 pl. Cracoviae.
Nye, I. W. B. & Fletcher, D. S., 1991. *The Generic Names of Moths of the World*. Volume **6** Microlepidoptera. 368 pp. London.
Oberthür, C., 1888. Faunes entomologiques. Descriptions d'insects nouveaux ou peu connus.- *Études d'Entomologie* **12**: i-ix, 1-46, pls 1-7. Rennes.
Ostrauskas, H. & Ivinskis, P., 2010. Records of *Tuta absoluta* (Meyrick, 1917) (Lepidoptera: Gelechiidae) in Lithuania.- *Acta Zoologica Lituanica* **20**: 151-155.
Palm, N.-B., 1947. Microlepidoptera, Neuroptera, and Trichoptera from Medelpad and Norbotten, Sweden.- *Opusc. Ent.* **12**: 35-49.
Park, K.-T., 1993. Two species of the genus *Caryocolum* (Gelechiidae, Lepidoptera) new to Korea.- *Korean J. Ent.* **23**: 17-21.
Parsons, M. S., 1995. A review of the scarce and threatened ethmiine, stathmopodine and gelechiid moths of Great Britain.- *UK Nature Conservation* **16**: 1-130.
Passos de Carvalho, J., 1995. Microlepidoptera of the Madeira and Selvagem Archipelagoes (Insecta, Lepidoptera).- *Bolm Mus. munic. Funchal* Suppl. **4**: 559-588.
Pastorális, G., 2008. A checklist of Microlepidoptera occured in the territory of Hungary, no. 2 (Lepidoptera: Microlepidoptera).- *Acta nat. pannon.* **3**, Suppl. **2**: 82-160.
Patočka, J., 1989. Über die Puppen der Mitteleuropäischen Gelechiidae (Lepidoptera). 5. Teil, Tribus Gnorimoschemini.- *Věst. Čs. Společ. Zool.* **53**: 123-140.
Patočka, J. & Turčáni, M., 2005. *Lepidoptera pupae. Central European species*. **1** (text volume): 1-542; **2** (plate volume): 1-321. Stenstrup.
Pelham-Clinton, E. C., 1989. *Scrobipalpula tussilaginis* (Frey) (Lepidoptera: Gelechiidae) new to the British Isles.- *Entomologist's Gaz.* **40**: 103-108.
Petry, A., 1904a. Beschreibung neuer Microlepidopteren aus Korsika.- *Stettin. ent. Ztg* **65**: 242-254.
– 1904b. Zwei neue Gelechiiden aus den Central-Pyrenäen.- *Dt. ent. Z. Iris* **17**: 1-6.
– 1912. Zur Biologie der *Lita vicinella* Frey non Douglas.- *Dt. ent. Z. Iris* **26**: 115-118.
Peyerimhoff, H. de, 1871. Monsieur le Rédacteur!- *Mitt. schweiz. ent. Ges.* **3**: 409-415.
Pfaffenzeller, F., 1867. *Gelechia Petasitis*, n. sp.- *Stettin. ent. Ztg* **28**: 79.
– 1870. Neue Tineinen.- *Stettin. ent. Ztg* **31**: 320-324.
Pierce, F. N. & Metcalfe, J. W., 1935. *Phthorimaea obsoletella* F. v. R. and an allied species.- *Entomologist* **68**: 97-99.
Pinker, R., 1953. Beobachtungen in den Julischen Alpen.- *Z. wien. ent. Ges.* **38**: 177-181.
Piskunov, V. I., 1975. New data on the fauna of Gelechiidae (Lepidoptera) from the European part of the U.S.S.R. [In Russian.]- *Ent. Obozr.* **54**: 857-871.
– 1988. Results of study of the type specimens of gelechiid moths (Lepidoptera, Gelechiidae) from the collection of the Zoological Museum, Moscow State University, and the Zoological Institute, Academy of Sciences of the USSR. [In Russian.]- *Ent. Obozr.* **67**: 360-368.
– 1990. Second addition to the fauna of the gelechiid moths (Lepidoptera, Gelechiidae) of Mongolia. [In Russian.]- *Nasek. Mongolii* **11**: 286-307.
– 1995. New data on systematics of Xyloryctidae and Gelechiidae (Lepidoptera).- *Fauna y Sistematica, Minsk* **1**: 248-255.
– 2000. Three rare species of gelechiid moths (Lepidoptera, Gelechiidae) for Palaearctic fauna. [In Russian.]- *Vesnik Vitebskaya dzjiurzhannaga Universiteta imija P. M. Masherava Naukovy chabopiz* **1(15)**: 117-118.
Piskunov, V. I. & Anikin, V. V., 2001. New species of Gelechiidae for Russia fauna (Lepidoptera, Gelechiidae). [In Russian.]- *Entomological and Parasitological investigations in Povoltzh'e region. Collected proceedings. Saratov State University Press* **1**: 154-155.
Piskunov, V. I. & Uskov, M. V., 2006. Two new species of gelechiid moths (Lepidoptera: Gelechiidae) from Center of European Russia. [In Russian.]- *Eversmannia* **5**: 3-5.
Pitkin, L. M. (1984). Gelechiid moths of the genus *Mirificarma*.- *Bull. Br. Mus. nat. Hist.* (Ent.) **48**: 1-70.

Pitkin, L. M. & Sattler, K., 1991. *Sattleria*: a European genus of brachypterous alpine moths (Lepidoptera: Gelechiidae).- *Bull. Br. Mus. nat. Hist.* (Ent.) **60**: 205-241.

Ponomarenko, M. G., 2005. Gelechiid moths of the Palaearctic: functional morphology of the male genitalia, phylogeny and taxonomy (Lepidoptera, Gelechiidae). [In Russian.]- *Meetings in memory of N. A. Cholodovsky* **58**(1): 139 pp. St. Petersburg.

– 2006. The subfamily Dichomeridinae (Lepidoptera, Gelechiidae): the phylogeny, classification and position in the system of the Gelechiidae. [In Russian.]- *Ent. Obozr.* **85**: 375-384.

– 2008. Gelechiidae. Pp. 87-106, 327-329. *In* Sinev, S. Yu., (ed.): *Katalog Cheshuekrylykh (Lepidoptera) Rossii (Catalogue of the Lepidoptera of Russia)*. [In Russian.]- 424 pp. St. Petersburg - Moscow.

– 2009. *Gelechiid moths of the subfamily Dichomeridinae (Lepidoptera: Gelechiidae) of the world fauna*. [In Russian.]- 389 pp. Vladivostok.

Ponomarenko, M. G. & Park, K.-T., 2007. Two new species and a new record of Gelechiidae (Lepidoptera) from Korea.- *Zootaxa* **1437**: 55-60.

Popescu-Gorj, A., 1992. Le catalogue des types de Lépidoptères gardés dans les collections du Muséum d'Histore naturelle "Grigore Antipa" (Bucarest) (Fam. Micropterigidae - Pterophoridae).- *Trav. Mus. Hist. nat. "Gr. Antipa"* **32**: 131-184.

Povolný, D., 1955. Zur Bionomie der Rübenmotte *Gnorimoschema ocellatellum* Boyd in der ČSR. [in Czech.]- *Zool. Ent. Listy* **4**: 281-288.

– 1964. Gnorimoschemini trib. nov. - eine neue Tribus der Familie Gelechiidae nebst Bemerkungen zu ihrer Taxonomie (Lepidoptera).- *Acta ent. bohemoslovaca* **61**: 330-359, 3 pls.

– 1965. Neue und wenig bekannte palaearktische Arten und Gattungen der Tribus Gnorimoschemini nebst Bemerkungen zu ihrer Taxonomie (Lepidoptera, Gelechiidae).- *Acta ent. bohemoslovaca* **62**: 480-495.

– 1966a. A type revision of some old-world species of the tribe Gnorimoschemini with a special regard to pests (Lepidoptera).- *Acta ent. bohemoslovaca* **63**: 128-148.

– 1966b. Neubeschreibungen von palaearktischen *Scrobipalpa*-Arten (Lepidoptera, Gelechiidae).- *Acta ent. bohemoslovaca* **63**: 394-408.

– 1966c. Die taxonomisch geklärten palaearktischen *Gnorimoschema*-Arten (Lep. Gel.) nebst Beschreibung einer neuen Art.- *Ark. Zool.* **18**: 393-404, pls 1-11.

– 1967a. Ein kritischer Beitrag zur taxonomischen Klärung einiger palaearktischer Arten der Gattung *Scrobipalpa* (Lepidoptera, Gelechiidae).- *Acta sc. nat. Brno* **1**: 209-250.

– 1967b. Genitalia of some nearctic and neotropic members of the tribe Gnorimoschemini (Lepidoptera, Gelechiidae).- *Acta ent. Mus. natn. Pragae* **37**: 51-127.

– 1967c. *Vladimirea*, gen. nov., eine neue Gattung der palaearktischen Gnorimoschemini und ihre Angehörigen (Lepidoptera, Gelechiidae).- *Acta ent. Mus. natn. Pragae* **37**: 147-159.

– 1967d. Die stammesgeschichtlichen Beziehungen der Tribus Gnorimoschemini im Weltrahmen (Lepidoptera, Gelechiidae).- *Acta ent. Mus. natn. Pragae* **37**: 161-232.

– 1968a. *Scrobipalpa clintoni* sp. nov. (Lepidoptera, Gelechiidae) a surprising discovery from Scotland.- *Entomologist's Gaz.* **19**: 113-118.

– 1968b. Neue und wenig bekannte Taxone aus der Tribus Gnorimoschemini Povolný, 1964 (Lepidoptera, Gelechiidae).- *Acta sc. nat. Brno* **2**(3):1-23, pls 1-21.

– 1969a. Ergebnisse der zoologischen Forschungen von Dr. Z. Kaszab in der Mongolei (Nr. 218), (Tribus Gnorimoschemini, Lep., Gelechiidae).- *Acta sc. nat. Brno* **3**(12): 1-28, 32 pls.

– 1969b. Bedeutsame faunistische und taxonomische Entdeckungen bei den *Scrobipalpa*-Arten (Lepidoptera, Gelechiidae) in Mitteleuropa.- *Z. wien. ent. Ges.* **54**: 71-78.

– 1969c. Einige wichtige Befunde palaearktischer Gnorimoschemini aus Afghanistan und China (Lepidoptera, Gelechiidae).- *Acta ent. bohemoslovaca* **66**: 373-380.

– 1969d. Gnorimoschemini (Lep. Gel.) der Ausbeute von F. Brandt aus Iran, 1936–1938.- *Ent. Tidskr.* **90**: 1-13.

– 1971a. Zur Fauna der Tribus Gnorimoschemini (Lepidoptera, Gelechiidae) in Nordwestafrika.- *Acta ent. bohemoslovaca* **68**: 23-44.

– 1971b. Bedeutsame faunistische und taxonomische Entdeckungen bei den *Scrobipalpa*-Arten

(Lep., Gelechiidae) in Mitteleuropa.- *Z. wien. ent. Ges.* **54**: 71-78.
- 1971c. Eine notwendige nomenklatorische Änderung im Rahmen der Gattung *Scrobipalpa* Janse (Lepidoptera, Gelechiidae).- *Acta ent. bohemoslovaca* **68**: 207.
- 1972. Zur Fauna der Tribus Gnorimoschemini (Lepidoptera, Gelechiidae) im Iran.- *Acta ent. bohemoslovaca* **69**: 186-206.
- 1973a. Ergebnisse der Zoologischen Forschungen von Dr. Z. Kaszab in der Mongolei (Nr. 267) (Tribus Gnorimoschemini, Lepidoptera, Gelechiidae).- *Acta sc. nat. Brno* **7**: 1-42.
- 1973b. *Turcopalpa glaseri* gen. n., sp. n. und Taxonomie einiger Arten der Gattung *Scrobipalpa* Janse (Lepidoptera, Gelechiidae) aus der Türkei.- *Acta ent. bohemoslovaca* **70**: 98-116.
- 1973c. *Scrobipalpa arenbergeri* sp. n. und *Scrobipalpa bigoti* sp. n. – zwei bisher unbekannte europäische Arten der Tribus Gnorimoschemini (Lepidoptera, Gelechiidae).- *Acta ent. bohemoslovaca* **70**: 282-287.
- 1975. *Gnorimoschema steueri* sp. n. – eine neue Gelechiiden-Art aus Mitteleuropa (Lepidoptera, Gelechiidae).- *Acta ent. bohemoslovaca* **72**: 190-195, pl. 1.
- 1976. Neue und wenig bekannte Arten der Tribus Gnorimoschemini (Lepidoptera, Gelechiidae) aus Iran.- *Acta ent. bohemoslovaca* **73**: 36-51.
- 1977a. Drei neue Arten der Tribus Gnorimoschemini (Lepidoptera, Gelechiidae) aus dem westpaläarktischen Eremialgürtel.- *Acta Univ. Agric. Brno* **25**: 169-175.
- 1977b. Zur Fauna der Tribus Gnorimoschemini (Lepidoptera, Gelechiidae) der Iberischen Halbinsel.- *Acta ent. bohemoslovaca* **74**: 184-204.
- 1977c. Neue Arten der Gattungen *Scrobipalpula* Povolný, 1964 und *Scrobipalpa* Janse, 1951.- *Dt. ent. Z.* (N.F.) **24**: 141-149.
- 1977d. A contribution to the knowledge of the tribe Gnorimoschemini of Far East (Lepidoptera, Gelechiidae).- *Acta ent. Mus. natn. Pragae* **39**: 217-229.
- 1977e. Neue Funde von Gnorimoschemini (Lepidoptera, Gelechiidae) aus Iran und Afghanistan.- *Acta ent. bohemoslovaca* **74**: 322-338.
- 1978a. A revision of the genus *Megalocypha* Janse (Lepidoptera, Gelechiidae).- *Acta ent. bohemoslovaca* **75**: 42-57.
- 1978b. Gnorimoschemini aus der Mongolei (Lepidoptera: Gelechiidae).- *Acta zool. Acad. Sci. hung.* **24**: 177-186.
- 1979a. Eine Ausbeute der Tribus Gnorimoschemini aus Tunis (Lepidoptera: Gelechiidae).- *Folia ent. hung.* **32**: 111-119.
- 1979b. On some little known moths of the family Gelechiidae (Lepidoptera) as pests of crops.- *Acta Univ. Agric. Brno* **27**: 139-165.
- 1980. Insects of Saudi Arabia Lepidoptera: Fam. Gelechiidae, Tribus Gnorimoschemini.- *Fauna of Saudi Arabia* **2**: 241-251.
- 1981. Über neue und wenig bekannte Arten der Tribus Gnorimoschemini (Lep., Gelechiidae) aus dem Mediterraneum.- *Acta Univ. Agric. Brno* **29**: 365-397.
- 1982. Zur artspezifischen Identität mancher westpaläarktischer Gnorimoschemini.- *Nota lepid.* **5**: 121-132.
- 1983a. Eine Typenrevision der von französischen Autoren beschriebenen Gnorimoschemini (Lepidoptera, Gelechiidae).- *Acta ent. Mus. natn. Pragae* **41**: 159-187.
- 1983b. *Scrobipalpa arenbergeri* Povolný, 1973 – ein Synonym von *Scrobipalpa klimeschi* Povolný, 1967 (Lepidoptera, Gelechiidae).- *Z. ArbGem. öst. Ent.* **35**:14-16.
- 1984. *Scrobipalpa* (*Euscrobipalpa*) *reiprichi* sp. n. (Lepidoptera, Gelechiidae) a surprising discovery from Eastern Slovakia.- *Acta ent. bohemoslovaca* **81**: 453-457.
- 1987a. Kritische Bemerkungen zum Differenziationsprozeß von *Sattleria dzieduszyckii* (Nowicki, 1864) in den europäischen Hochgebirgen.- *Atalanta* **17**: 85-104, 1 pl.
- 1987b. *Scrobipalpa* (*Euscrobipalpa*) *dagmaris* sp. n. und andere interessante Entdeckungen bei den europäischen Gnorimoschemini (Lepidoptera, Gelechiidae).- *Nota lepid.* **10**: 79-86.
- 1987c. Gnorimoschemini of southern South America III: the scrobipalpuloid genera (Insecta, Lepidoptera, Gelechiidae).- *Steenstrupia* **13**: 1-91.

– 1988. *Scrobipalpa nitentella* (Fuchs, 1902) species revocata und *S. indignella* (Staudinger, 1879) species bona sowie sonstige irrtümliche Urteile über die Gattung *Scrobipalpa* (Lepidoptera, Gelechiidae).- *Dt. ent. Z.* (N. F.) **35**: 369-378.
– 1989a. Bemerkungen zur generischen und subgenerischen Bewertung einiger Gnorimoschemini (Insecta, Lepidoptera: Gelechiidae).- *Reichenbachia* **26**: 157-167.
– 1989b. Gnorimoschemini of southern South America IV: the genera *Symmetrischema* and *Phthorimaea* (Lepidoptera, Gelechiidae).- *Steenstrupia* **15**: 57-104.
– 1990a. On *Agonochaetia quartana* sp. n. and its allies (Lepidoptera: Gelechiidae).- *SHILAP Revta. lepid.* **18**: 149-157.
– 1990b. Zur heutigen Kenntnis von Nahrungspflanzen der Tribus Gnorimoschemini (Lepidoptera, Gelechiidae).- *Acta Univ. Agric. Brno* **38**: 191-204.
– 1991a. Morphologie, Systematik und Phylogenie der Tribus Gnorimoschemini (Lepidoptera, Gelechiidae).- *Acta sc. nat. Brno* **25**(9-10): 1-103.
– 1991b. Zur Bionomie und ökonomischen Bedeutung der Tribus Gnorimschemini (Lepidoptera, Gelechiidae).- *Acta Univ. Agric. Brno* **39**: 275-291.
– 1992a. A critical review of the Palaearctic taxa of *Gnorimoschema* Busck (Lepidoptera, Gelechiidae).- *Acta ent. bohemoslovaca* **89**: 217-233.
– 1992b. Description de deux *Opacopsis* nouveaux du sud de la France (Lep. Gelechiidae Gnorimoschemini).- *Alexanor* **17**: 329-334.
– 1993. Zur Taxonomie und Nomenklatur der amerikanischen gnorimoschemoiden Gattungen *Tuta* Strand, *Tecia* Strand, *Scrobipalpopsis* Povolný und *Keiferia* Busck (Insecta: Lepidoptera: Gelechiidae).- *Reichenbachia* **30**: 85-98.
– 1994a. New taxa and records of *Gnorimoschema* Busck and *Gobipalpa* Povolný from Palaearctic Asia (Lepidoptera, Gelechiidae).- *Ent. fenn.* **5**: 57-64.
– 1994b. Gnorimoschemini of southern South America VI: identification keys, checklist of Neotropical taxa and general considerations (Insecta, Lepidoptera, Gelechiidae).- *Steenstrupia* **20**: 1-42.
– 1996a. Gnorimoschemini (except *Caryocolum*). Pp. 113-118, 311. *In* Karsholt, O. & Razowski, J. (eds): *The Lepidoptera of Europe. A distributional checklist.*- 380 pp. Stenstrup.
– 1996b. New taxa and faunistic records of *Ilseopsis* (*Euscrobipalpa*) Povolný and *Hedma* Dumont from Palaearctic Asia (Lepidoptera, Gelechiidae).- *Acta zool. fenn.* **200**: 19-40.
– 1998. New taxa and faunistic records of the tribe Gnorimoschemini from the Nearctic Region (Lepidoptera, Gelechiidae).- *Stapfia* **55**: 327-347.
– 2000. Towards the interpretation of the Palaearctic taxa of the *Scrobipalpula psilella* (Herrich-Schäffer, 1854) – complex (Lepidoptera, Gelechiidae, Gnorimoschemini).- *Acta Univ. Agric. Brno* **48**(4): 71-89.
– 2001a. Eighteen new species and additional records of the Palaearctic Gnorimoschemini (Insecta: Lepidoptera: Gelechiidae).- *Reichenbachia* **34**: 181-206.
– 2001b. A review of the taxa hierarchy in the euro-alpine genus *Sattleria* Povolný, 1965 (Lepidoptera, Gelechiidae, Gnorimoschemini).- *Acta Univ. Agric. Brno* **49**(3): 39-58.
– 2002a. *Iconographia tribus Gnorimoschemini (Lepidoptera, Gelechiidae) Regionis Palaearcticae.* 110 pp., 16 colour pls., 87 pls. Bratislava.
– 2002b. Synopsis of the genera of the tribe Gnorimoschemini (Lepidoptera: Gelechiidae).- *Lepidoptera News* **2002**(1-2): 37-48.
– 2004. The Guatemalan Potato Tuber Moth (*Scrobipalpopsis solanivora* Povolný, 1973) before the gateways of Europe (Lepidoptera, Gelechiidae).- *Acta Univ. Agric. Brno* **52**(1): 183-196.
Povolný, D. & Bradley, J. D., 1964. Two new species of *Scrobipalpa* Janse (Lep., Gelechiidae) from Algeria.- *Annls Mag. nat. Hist.* **7**: 553-557.
– 1965. *Scrobipalpa psilella* (Herrich-Schäffer) (Lep., Gelechiidae) – a species hitherto overlooked in the British Isles.- *Entomologist's Gaz.* **16**: 9-12, 3 figs.
Povolný, D. & Gregor, F., 1955. Einige neue Microlepidopteren aus dem Mediterrangebiet.- *Z. wien. ent. Ges.* **66**: 81-87, pl. 4.

Povolný, D. & Luquet, G., 1983. Commentaires sur quelques Gnorimoschemini nouveaux ou peu connus de France [Lepidoptera Gelechiidae].- *Alexanor* **13**: 63-74.

Povolný, D. & Šustek, Z., 1988. Versuch einer numerisch-taxonomischen Lösung der phylogenetischen Beziehungen im Rahmen der gelechioiden Tribus Gnorimoschemini (Lepidoptera).- *Stapfia* **16**: 209-247.

Powell, J. A. & Povolný, D., 2001. Gnorimoschemini moths of coastal dune and scrub habitats in California (Lepidoptera: Gelechiidae).- *Holarctic Lepidoptera* **8**, Suppl. 1: 1-53.

Preissecker, F., 1914. *In* Galvagni, E. & Preissecker, F.: Die lepidopterologischen Verhältnisse des niederösterreichischen Waldviertels. IV. Teil.- *Jahresb. wien. ent. Ver.* **25**: 1-68.

Pröse, H., 1979. Die Kleinschmetterlinge der Umgebung von Hof mit einem Überblick über die oberfränkische Fauna (Lepidoptera).- *Ber. nordoberfr. Ver. Natur, Geschichts-, Landeskunde (und Familienkunde)* **27**: 1-134.

Ragonot, M., 1879. Note sur la *Gelechia tabacella* Rag.- *Bull. Soc. Ent. Fr.* **1879**: cxlvi-cxlvii.

Rebel, H., 1891. Beitrag zur Microlepidopteren-Fauna Dalmatiens.- *Verh. zool.-bot. Ges. Wien* **41**: 610-639.

– 1893. Neue oder wenig gekannte Microlepidopteren des palaearktischen Faunengebietes.- *Stettin. ent. Ztg* **54**: 37-59.

– 1899. Zweiter Beitrag zur Lepidopteren-Fauna Südtirols.- *Verh. zool.-bot. Ges. Wien* **49**: 158-185, pl. 4.

– 1901. Neue palaearctische Tineen.- *Dt. ent. Z. Iris* **13**: 161-188.

– 1903. Studien über die Lepidopterenfauna der Balkanländer. I. Teil. Bulgarien und Ostrumelien.- *Annln naturh. Mus. Wien* **18**: 123-347, pl. 3.

– 1904. Studien über die Lepidopterenfauna der Balkanländer. II. Teil Bosnien und Herzegowina.- *Annln naturh. Mus. Wien* **19**: 97-377, pls 4, 5.

– 1907. Neue palaearctische Microheteroceren.- *Dt. ent. Z. Iris* **19**: 227-242.

– 1909. Versammlung der Sektion für Lepidopterologie.- *Verh. zool.-bot. Ges. Wien* **59**: (330) - (333).

– 1912. Beitrag zur Lepidopterenfauna Unter-Aegyptens.- *Dt. ent. Z. Iris* **26**: 65-92.

– 1913. Ùj Gelechiida Magyarországból. (Eine neue Gelechiidae aus Ungarn).- *Rovart. Lap.* **20**: 173-175.

– 1917. See Hauder, F., & Rebel, H., 1917.

– 1918. Lepidopteren aus Mittelalbanien.- *Z. öst. EntVer.* **3**: 75-77, 85-88.

– 1927. Beiträge zur Lepidopterenfauna Aegyptens.- *Bull. Soc. ent. Egypte* **10** (1926): 179-191.

– 1928. Versammlung der Sektion für Lepidopterologie. Zur Lepidopterenfauna Cyperns. V. *Verh. zool.-bot. Ges. Wien* **78**: (29)-(34).

– 1935a. Neue Pterophoriden und Tineen aus der Sierra de Gredos (Kastilien).- *Z. öst. EntVer.* **20**: 9-14.

– 1935b. Neue Microlepidopteren aus Kleinasien.- *Mitt. münch. ent. Ges.* **25**: 39-41.

– 1936a. Neue Mikrolepidopteren von Sardinien.- *Dt. Ent. Z. Iris* **50**: 92-100.

– 1936b. Drei neue Mikrolepidopterenarten aus der Schweiz.- *Z. öst. EntVer.* **21**: 2-4.

– 1936c. *In* Osthelder, O. & Rebel, H.: Lepidopteren-Fauna von Marasch in türkisch Nordsyrien.- *Mitt. münch. ent. Ges.* **25**: 67-90.

– 1941. Über einige neue Mikrolepidopteren von der Balkanhalbinsel und besonders aus der Gegend des Ochrida-Sees in Mazedonien.- *Izv. tsarsk. prirodonauch. Inst. Sofia* **14**: 1-8.

Requena, E., 2009. Aproximació a la fauna dels gelèquids de Catalunya i Balears (Lepidoptera: Gelechiidae).- *Treballs de la Societat Catalana de Lepidopterologia* **16**: 5-77.

Reutti, C., 1898. Übersicht der Lepidopteren-Fauna des Grossherzogtum's Baden und der anstossenden Länder. 2. Ausg. Herausgegeben von A. Meess & A. Spuler.- *Beitr. Rhein. Naturg.* xii + 361 pp. Berlin.

Richardson, N. M., 1893. On some members of the *instabilella* group of the genus *Lita* (*Gelechia*, partim), with descriptions of *L. suaedella* n. sp., and *L. instabilella*, Douglas.- *Entomologist's mon. Mag.* **29**: 241-248.

Ridout, B. V., 1977. Two new synonymies of Microlepidoptera (Gelechiidae: Gnorimoschemini and Oecophoridae: Depressariinae).- *Entomologist´s Gaz.* **28**: 38-42.
Riedl, T., 1965. Matériaux pour la connaissance des Momphides paléarctiques (Lepidoptera). Partie III. Étude sur quelques Momphides européens.- *Polskie Pismo ent.* **35**: 419-468, 4 pls.
Robinson, G. S., Ackery, P. R., Kitching, I. J. Beccaloni, G. W. & Hernández, L. M., 2007. Hosts – a database of the world´s Lepidoptera hostplants.- http://www.nhm.ac.uk/research-curation/research/projects/hostplants/
Roditakis, E., Papachristos, D. & Roditakis, N. E., 2010. Current status of the tomato leafminer *Tuta absoluta* in Greece.- *Bulletin OEPP/EPPO Bulletin* **40**: 163–166.
Rössler, A., 1866. Verzeichniss der Schmetterlinge des Herzogthums Nassau, mit besonderer Berücksichtigung der biologischen Verhaltnisse und der Entwicklungsgeschichte.- *Jahrbuch. Nassau Ver. Naturkunde* **19-20**: 99-442.
– 1877. Verzeichnis um Bilbao gefundener Schmetterlinge.- *Stettin. ent. Ztg* **38**: 359-380.
Rye, E. C., 1881. Index to genera and subgenera recorded as new in this volume.- *Zool. Rec.* **16** (1879): 1-11.
Salin, T., 2004. Bulgaria.- *Baptria* **29**: 78-81.
Sannino, L. & Espinosa, B., 2009. *Keiferia lycopersicella*, una nuova tignola su pomodoro.- *L'Informatore Agrario* 4/2009: 69-70.
Sattler, K., 1968. Die systematische Stellung einiger Gelechiidae (Lep.).- *Dt. ent. Z.* (N. F.) **15**: 111-131.
– 1969. Die systematische Stellung von *Scrobipalpa hungariae* (Staudinger, 1871) (Lepidoptera: Gelechiidae).- *Acta zool. Acad. Sci. hung.* **15**: 391-396.
– 1973. A catalogue of the family-group and genus-group names of the Gelechiidae, Holcopogonidae, Lecithoceridae and Symmocidae (Lepidoptera).- *Bull. Br. Mus. nat. Hist.* (Ent.) **28**: 155-182.
– 1978. The identity of the genus *Athrips* Billberg, 1820 (Lep., Gelechiidae.).- *Dt. ent. Z.* (N. F.) **25**: 57-61.
– 1985. The systematic position of three Gelechiidae described by Constant and Lucas (Lepidoptera).- *Ent. gall.* **1**: 227-230.
– 1986. The systematic position of *Bryotropha indignella* Staudinger, 1879 (Lepidoptera, Gelechiidae).- *Dt. ent. Z.* (N. F.) **33**: 233-234, pl. 15.
– 1987a. The type-material of *Ephysteris olympica* Povolný (Lepidoptera: Gelechiidae).- *Entomologist´s Gaz.* **38**: 199-201, 1 pl.
– 1987b. Exhibit, Annual Exhibition, 1986.- *Proc. Trans. Br. ent. nat. Hist. Soc.* **20**: 53.
– 1987c. Die an Compositen gebundenen *Scrobipalpa*-Arten des östlichen Österreichs (Lepidoptera, Gelechiidae).- *Annln naturh. Mus. Wien* **88/89B**: 435-456.
– 1988. The systematic status of the genera *Ilseopsis* Povolný, 1965, and *Empista* Povolný, 1968 (Lepidoptera: Gelechiidae: Gnorimoschemini).- *Nota lepid.* **10**: 224-235.
– 1989. The taxonomic status of *Scrobipalpa klimeschi* Povolný, 1967, and *Lita pauperella* Heinemann, 1870 (Lepidoptera: Gelechiidae).- *Entomologist´s Gaz.* **40**: 7-12.
– 1992. New Syonymyms of European Gelechiidae (Lepidoptera).- *Ent. gall.* **3**: 107-112.
Sauter, W., 1961. Über einige von J. C. de la Harpe, J. Müller-Rutz und P. Weber aus der Schweiz beschriebene Kleinschmetterlinge (Lep.).- *Mitt. schweiz. ent. Ges.* **33**: 264-274.
Sauter, W. & Whitebread, S., 2005. Die Schmetterlinge der Schweiz (Lepidoptera). 9. Nachtrag.- *Mitt. schweiz. ent. Ges.* **78**: 59-115.
Savenkov, N. & Šulcs, I., 2004. New and rare lepidoptera for the Latvian fauna.- *Baptria* **2004** (2): 52-58.
– in press. *Latvijas Taruņi Katalogs. Latvian Lepidoptera Catalogue*. Talin.
Schantz, M. v., 1952. Beitrag zur Kenntnis dreier *Phthorimaea*-Arten (Lep., Gelechidae). *Phthorimaea hackmani* n. sp. - *Notulae Entomol.* **32**: 15-22.
Schawerda, K., 1921. Beiträge zur Lepidopterenfauna der kroatischen Küste und Neubeschreibungen.- *Dt. ent. Z. Iris* **35**:111-138.

Schmid, A., 1863. Beiträge zur Naturgeschichte der Schmetterlinge.- *Berl. ent. Z.* **7**: 57-66.

– 1885-1887. Die Lepidopteren-Fauna der Regensburger Umgegend mit Kehlheim und Wörth. III. Microlepidoptera.- *KorrespBl. naturw. Ver. Regensburg* **39**: 21-46, 75-95, 97-135, 151-201 (1885); **40**: 19-58, 83-98, 101-224 (1887 [„1886"]).

Schmid, J., 2006. Für die Schweiz neue Kleinschmetterlinge aus Graubünden (Lepidoptera, Tineidae, Gelechiidae, Oecophoridae, Tortricidae).- *Mitt. ent. Ges. Basel* **56**: 95-98.

– 2007. Schweizer Erstfunde von Kleinschmetterlinge aus Graubünden (Lepidoptera: Nepticulidae, Tineidae, Bucculatricidae, Agonoxenidae, Blastobasidae, Gelechiidae, Tortricidae, Pyralidae).- *Mitt. ent. Ges. Basel* **57**: 124-131.

Schütze, K.T., 1926. *Lita viscariae* n. sp.- *Dt. ent. Z. Iris* **40**: 171-175.

– 1931. *Die Biologie der Kleinschmetterlinge unter besonderer Berücksichtigung ihrer Nährpflanzen und Erscheinungszeiten.* 235 pp. Frankfurt/Main.

Seliger, R., 2008. *Scrobipalpa costella* (Humphreys & Westwood, 1845) – eine neue Palpenmotte für die Fauna Deutschlands (Lep., Gelechiidae).- *Melanargia* **20**: 45-47.

Sigsgaard, L., Jacobsen, S. E. & Christiansen, J. L., 2008. Quinoa, *Chenopodium quinoa*, provides a new host for native herbivores in northern Europe: Case studies of the moth, *Scrobipalpa atriplicella*, and the tortoise beetle, *Cassida nebulosa*.- *J. Insect Science* **8**(49): 1-3.

Sinev, S. Yu., 1992. On the systematics and phylogeny of the Gelechioidea s.l. (Lepidoptera). [In Russian.]- *Ent. Obozr.* **71**: 143-159.

Sircom, J., 1850. Descriptions of Two New Species of Tineidae.- *Zoologist* **8**: lxxii.

Skala, H., 1950. Über minierende Gelechiidae.- *Z. wien. ent. Ges.* **35**: 115-116.

Snellen, P. C. T., 1882. *De vlinders van Nederland. Microlepidoptera.* 1196 pp. Leiden.

Sommerer, M., 2002. To agree or not to agree – the question of gender agreement in the International Code of Zoological Nomenclature.- *Nota lepid.* **25**: 191-204.

Sorhagen, L., 1886. *Die Kleinschmetterlinge der Mark Brandenburg und einiger angrenzenden Landschaften mit besonderer Berücksichtigung der Berliner Arten*, x + 368 pp. Berlin.

Stainton, H. T., 1851. *A Supplementary Catalogue of the British Tineidae & Pterophoridae.* iv + 13 pp. London.

– 1855. The new British species in 1854.- *Entomologist's Annual* **1855**: 40-50.

– 1859. Notes on Lepidoptera collected in Madeira by T. V. Wollaston, Esq.; with descriptions of some new species.- *Ann. Mag. nat. Hist.* (Ser. 3) **3**: 209-214.

– 1866. New British Tineina.- *Entomologist's Annual* **1866**: 167-171.

– 1867a. *The natural history of the Tineina* **10**, [xii] + 304 pp., pls 10-16. London, Paris, Berlin.

– 1867b. *The Tineina of Syria and Asia Minor.* vi + 84 pp. London.

– 1869. *The Tineina of Southern Europe.* 370 pp. London.

– 1871. New British Tineina in 1870.- *Entomologist's Annual* **1871**: 96-100.

– 1883. On two of the species of *Gelechia* which frequent our salt-marshes.- *Entomologist´s mon. Mag.* **19**: 251-253.

Staudinger, O., 1859. Diagnosen nebst kurzen Beschreibungen neuer andalusischer Lepidopteren.- *Stettin. ent. Ztg* **20**: 211-259.

– 1867. *Gelechia petasitella* und *Phyllobrostis Hartmanni*.- *Stettin. ent. Ztg* **28**: 210-212.

– 1871. Beschreibung neuer Lepidopteren des europäischen Faunengebiets.- *Berl. ent. Z.* **14**: (1870) 273-330.

– 1877. Neue Lepidopteren des europäischen Faunengebiets aus meiner Sammlung.- *Stettin. ent. Ztg* **38**: 175-208.

– 1879. Lepidopterenfauna Kleinasiens.- *Horae Soc. ent. ross.* **15**: 159-435.

Stehr, F. (ed.), 1987. *Immature Insects*.- xiv + 754 pp. Dubuque, Iowa.

Steuer, H., 2002. 2. Nachtrag zur Schmetterlingsfauna um Bad Blankenburg (Thüringen) (Lep.).- *Ent. Nachr. Ber.* **46**: 167-172.

Strand, E., 1902. Eine neue norwegische Gelechide (*Gelechia* (*Lita*) *nordlandicolella* Strand n. sp.).- *Ent. Z. Guben* **16** (6): 21.

– 1911. Sechs neue Gelechiidae aus Argentina.- *Berl. ent. Z.* **55**: 165-173.

Stüning, D., 1988. Biologisch-ökologische Untersuchungen an Lepidopteren des Supralitorals der Nordseeküste.- *Faun.-ökol. Mitt.*, Suppl. **7**: 1-118.
Süssner, L., 1966: Bemerkenswerte Funde von Microlepidopteren in Württemberg.- *Mitt. ent. Ver. Stuttgart* **1**: 91-103.
Svensson, I., 1991. Anmärkningsvärda fynd af Microlepidoptera i Sverige 1990.- *Ent. Tidskr.* **112**: 65-72.
– 2010. Anmärkningsvärda fynd af Microlepidoptera i Sverige 2009.- *Ent. Tidskr.* **131**: 17-27.
Szabóky, C., Kun, A. & Buschmann, F., 2006. Addenda and corrigenda to the Checklist of the Fauna of Hungary, Microlepidoptera.- *Folia ent. hung.* **67**: 69-83.
Szabóky, C., Tokár, Z., Liška, J. & Pastorális, G., 2009. New data on the Microlepidoptera fauna of Hungary, part XII (Lepidoptera: Lypusidae, Bucculatricidae, Yponomeutidae, Depressariidae, Coleophoridae, Blastobasidae, Autostichidae, Gelechiidae, Tortricidae).- *Folia ent. hung.* **70**: 139-146.
Tabell, J., 2008. Havaintoja *Monochroa arundinetellan* (Stainton, 1858) ja *Scrobipalpula diffluellan* (Frey, 1870) (Gelechiidae) elintavoista.- *Baptria* **33**: 136-138.
Teich, C. A., 1889. Baltische Lepidopteren-Fauna.- *Arb. Naturf.-Ver. Riga* (N.F.) **6**: i-ix, 1-152.
Tengström, J. M. J., 1848. Bidrag till Finlands fjäril-fauna.- *Notis. Sällsk. Fauna Flora fenn. Förh.* **1**: 69-164.
Thomann, H., 1929. *Lita ochraceella* n. sp. (Mikrolep.), eine neue Gelechiide aus dem Unterengadin.- *Mitt. schweiz. ent. Ges.* **14**: 191-192.
– 1930. Berichtigung.- *Mitt. schweiz. ent. Ges.* **15**: 24.
– 1956. Die Psychiden und Mikrolepidopteren des Schweizerischen Nationalparkes und der angrenzenden Gebiete.- *Ergeb. wiss. Unters. schweiz. Nationalparks* **5** (35): 379-446.
Threlfall, J. H. & Stainton, H. T., 1887. *Gelechia semidecandrella* (n.sp.).- *Entomologist's mon. Mag.* **23**: 233-234.
Thunberg, C. P., 1788. *D. D. Museum Naturalium Academiae Upsaliensis* **6**: 69-84, 1 pl.
Thurner, J., 1941. Die Schmetterlinge der Ochrid-Gegend in Mazedonien II. Teil: Microlepidoptera.- *Izv. tsarsk. prirodonauch. Inst. Sofia* **14**: 9-35.
Tokár, Z., Richter, I., Richter, I., Liska, J., Pastorális, G., Kosorin, F., Elsner, G., Bohm, S. & Nemy, J., in press. Faunistic records from Slovakia. Lepidoptera: Micropterigidae, Eriocraniidae, Nepticulidae, Prodoxidae, Incurvariidae, Tineidae, Psychidae, Bucculatricidae, Gracillariidae, Yponomeutidae, Acrolepiidae, Elachistidae, Scythrididae, Oecophoridae, Coleophoridae, Blastobasidae, Autostichidae, Gelechiidae, Tortricidae, Pyralidae, Crambidae.- *Entomofauna Carpathica.*
Toll, S., 1936. Untersuchung der Genitalien bei *Pyrausta purpuralis* L. und *ostrinalis* Hb., nebst Beschreibung 11 neuer Microlepidopteren-Arten.- *Annales Mus. zool. Pol.* **11**: 403-413, pls 47-49.
– 1947. Beitrag zur Mikrolepidopterenfauna von Nordost-Persien.- *Z. wien. ent. Ges.* **58**: 107-116, pls 4-6.
Treitschke, F., 1833. *Die Schmetterlinge von Europa.* **9**(2): 1-294, **10**(3): 1-303. Leipzig.
Turati, E., 1924. Spedizione Lepidotterologica in Cirenaica 1921 - 1922.- *Atti Soc. ital. Sci. nat.* **68**: 21-191, pls 1-6.
– 1930. Novità di Lepidotterologia in Cirenaica. III.- *Atti Soc. ital. Sci. nat.* **69**: 46-92, pl. 2.
Turner, A. J., 1919. The Australian Gelechianae (Lepidoptera).-*Proc. R. Soc. Qd* **31**: 108-172.
Tutt, J. W., 1887a. On the '*Lita*' group of the Gelechiidae.- *Entomologist* **20**: 28-31.
– 1887b. Description of a new *Gelechia* of the '*Lita*' group, closely allied to *G. maculea*.- *Entomologist's mon. Mag.* **24**: 105-106.
Vávra, J., 1999. Notes on bionomics of *Gnorimoschema steueri* (Lepidoptera: Gelechiidae).- *Klapalekiana* **35**: 157-159.
Vives Moreno, A., 1991. *Catálogo Sistemático y Sinonimico de los Lepidopteros de la Peninsula Ibérica y Baleares (Insecta: Lepidoptera).* 378 pp. Ministerio de Agricultura, Pesca y Alimentación. Paracuellos del Jarama (Madrid).
– 2003. Una tribu, ocho géneros y dieciocho especies nuevos para la fauna espanola. *Yp-*

solopha cajaliella Vives, sp. n., para la Ciencias en España (Insecta : Lepidoptera).- *SHILAP Revta. lepid.* **29**: 93-110.

Vorbrodt, K. & Müller-Rutz, J., 1913-1914. *Die Schmetterlinge der Schweiz.* **2**, 726 pp. Bern.

Walker, F., 1863. *List of the Specimens of Lepidopterous Insects in the Collection of the British Museum.* **28**: 229-561. London.

– 1864a. *List of the Specimens of Lepidopterous Insects in the Collection of the British Museum.* **29**: 563-835. London.

– 1864b. *List of the Specimens of Lepidopterous Insects in the Collection of the British Museum.* **30**: 837-1096. London.

Walsingham, T. de Grey, 1897. Revision of the West-Indian Micro-Lepidoptera, with Descriptions of new Species.- *Proc. zool. Soc. Lond.* ***1897***: 54-183.

– 1898-1920. New Corsican and French Micro-Lepidoptera.- *Entomologist's mon. Mag.* **34**: 131-134, 166-172 (1898); **36**: 152-153, 216-220 (1900); **37**: 177-184 (1901); **38**: 81 (1902); **46**: 231-233 (1910); **56**: 9-10 (1920).

– 1904-1911. Algerian Microlepidoptera.- *Entomologist's mon. Mag.* **40**: 214-223, 265-273 (1904); **41**: 37-41, 124-128 (1905); **43**: 6-10, 55-58, 125-129, 147-154, 187-195 (1907); **47**: 12-15, 187-191 (1911).

– 1908. Microlepidoptera of Tenerife.- *Proc. zool. Soc. Lond.* **1907**: 931-1034, pls 52-53.

Wocke, M. F., 1862. Microlepidoptera. *In* Wocke, M. F. & Staudinger, O. 1861-1862: Reise nach Finmarken.- *Stettin. ent. Ztg* **22**: 325-402 (1861); **23**: 30-78, 233-257 (1862).

Wollaston, T. V., 1858. Brief diagnostic characters of undescribed Madeiran Insecta.- *Annls Mag. nat. Hist.* (3) **1**: 117-123.

Wörz, A., 1954. Die Lepidopterenfauna von Württemberg II. Microlepidopteren. Kleinschmetterlinge.- *Jahresh. Ver. vaterl. Naturk. Württemberg* **109**: 83-130.

Zeller, P.C., 1839. Versuch einer naturgemässen Eintheilung der Schaben.- *Isis, Leipzig* **1839**: 167-220.

– 1847. Bemerkungen über die aus einer Reise nach Italien und Sicilien beobachteten Schmetterlingsarten.- *Isis, Leipzig* **1847**: 121-159, 213-233, 284-308, 401-457, 481-522, 561-594, 641-673, 721-771, 801-859, 881-914.

– 1868. Lepidopterologische Ergebnisse einer Reise in Oberkärnthen.- *Stettin. ent. Ztg* **29**: 121-149.

– 1872. Bemerkungen über einige Graubündner Lepidoptern.- *Stettin. ent. Ztg* **33**: 97-210.

– 1873. Beiträge zur Kenntniss der nordamericanischen Nachtfalter, besonders der Microlepidopteren.- *Verh. zool.-bot. Ges. Wien* **23**: 201-334.

Zerkowitz, A., 1946. The Lepidoptera of Portugal.- *Jl N. Y. ent. Soc.* **54**: 51-87, 115-165, 211-261.

Zetterstedt, J. W., 1838-1840. *Insecta lapponica*: 1-867 (1838); 874-1013 (1839); 1022-1139 (1840).

Zimmermann, F. 1923. Über die Fauna der Halophytenstandorte Südmährens.- *Verh. zool.-bot. Ges. Wien* **72**(1922): (15)-(18).

– 1926. III. Nachtrag zur Lepidopterenfauna von Mähren.- *Lotos* **74**: 19-28.

Index to entomological genus-group names

The numbers refer to page numbers. Numbers in bold refer to the main treatment. Synonyms are in italics.

Index to entomological species-group names

The numbers refer to species numbers. Numbers in bold refer to the main treatment, other numbers are secondary entries. Synonyms are in italics. Capital letters refer to incorrect and/or uncertain records following the major treatment of species

Index to plant names

The numbers refer to the species numbers. Synonyms are not listed.

www.ingramcontent.com/pod-product-compliance
Lightning Source LLC
LaVergne TN
LVHW021041011026
849160LV00007BA/230